TABLE OF CONTENTS

Preface

Acknowledgements

Chapter 1 Introduction

1.1 Historical background

    1.1.1 Discovery of "Ice Ages"

    1.1.2 Development of understanding of the dynamics of glaciers

1.2 Terminology and classification of glacigenic sediments

    1.2.1 Terminology and non-genetic classification of poorly sorted sediments

    1.2.2 Terminology of glacigenic sediments

    1.2.3 Genetic classification of tills

    1.2.4 Genetic classification of glaciomarine sediments

1.3 Facies analysis of glacigenic sediments

    1.3.1 The facies concept

    1.3.2 Glacial sedimentary facies

    1.3.3 Interpretation of facies

    1.3.4 Lithofacies coding

    1.3.5 Facies associations and facies models

    1.3.6 Stratigraphic (or facies) architecture

    1.3.7 Sequence stratigraphy

1.4 Other methods of analysis of glacigenic sediments

    1.4.1 Grain-size distribution

    1.4.2 Clast shape

    1.4.3 Clast fabric

    1.4.4 Surface features on clasts

    1.4.5 Surface features on sand- and silt-sized grains

    1.4.6 Mineralogy

    1.4.7 Geochemistry

    1.4.8 Palaeomagnetism

    1.4.9 Geotechnical properties

1.5 Evidence of glaciation in the geological record

1.6 Extent of glacier ice today

Chapter 2 Glacier dynamics

2.1 Inroduction

2.2 Formation and properties of glacier ice

    2.2.1 Derivation of glacier ice

2.2.2 Density of glacier ice

2.2.3 Crystallography of glacier ice

2.2.4 Deformation of a single ice crystal

2.2.5 Deformation of polycrystalline ice

2.2.6 Flow law for polycrystalline ice

2.3 Morphology of glaciers

2.4 Cold and warm glaciers

2.5 Glacier hydrology

2.5.1 Sources of meltwater

2.5.2 Water flow through a glacier

2.5.3 Ice-constrained lakes

2.5.4 Water discharge from glaciers

2.5.5 Meltwater as a geological agent

2.6 Glacier fluctuations in response to climate

2.6.1 Mass balance

2.6.2 Effect of mass balance changes on flow

2.7 Glacier flow

2.7.1 Internal deformation

2.7.2 Basal sliding

2.7.3 Deformable beds

2.7.4 Long- and short-term variations of glacier flow

2.8 Structure of glaciers and ice sheets

2.8.1 Primary structures

2.8.2 Secondary structures resulting from ductile deformation

2.8.3 Secondary structures resulting from brittle deformation

2.8.4 Secondary structures of composite origin

2.8.5 Structural patterns in glaciers and ice sheets

2.9 Glacier surges

2.9.1 Characteristics of surge-type glaciers

2.9.2 Mechanisms of glacier surging: rigid bed hypotheses

2.9.3 Mechanisms of surging: deformable bed hypothesis

2.9.4 Do ice sheets surge?

2.9.5 Evidence of surging in the geological record

2.10 Debris in transport

2.10.1 High level transport

2.10.2 Basal transport

2.10.3 Debris incorporated into an englacial position

Chapter 3. Glacial erosional processes and landforms

3.1 Introduction

iv

3.2 Processes of glacial erosion

3.3 Landforms resulting from erosion by glaciers

   3.3.1 Modes of erosion

   3.3.2 Glacial abrasion

   3.3.3 Abrasion and rock fracture combined

   3.3.4 Rock crushing

   3.3.5 Erosion by ice combined with frost action

3.4 Structural geological controls on glacial erosion

3.5 Preservation potential of erosional forms

Chapter 4. Environments of terrestrial glacial deposition

4.1 Introduction

4.2 Processes of terrestrial glacial deposition

   4.2.1 Subglacial lodgement

   4.2.2 Melt-out and flowage

   4.2.3 Sublimation of debris-rich ice

   4.2.4 Ice push

   4.2.5 Subglacial chemical processes

4.3 Terrestrial glacigenic facies and facies associations

   4.3.1 Boundary relations and geometry of glacigenic units

   4.3.2 Sedimentary structures

   4.3.3 Glaciotectonic structures

   4.3.4 Facies

   4.3.5 Facies associations

   4.3.6 Facies architecture

4.4 Physical, chemical and mineralogical characteristics

   4.4.1 Grain size distribution

   4.4.2 Shape an surface features of clasts

   4.4.3 Clast fabric

   4.4.4 Mineralogy of the matrix

   4.4.5 Geochemistry

   4.4.6 Characteristics of quartz and heavy mineral grains

4.5 Landforms in terrestrial glacial depositional environments

   4.5.1 Debris forms on the glacier surface

   4.5.2 Landforms formed subglacially, parallel to ice flow

   4.5.3 Landforms formed subglacially, transverse to flow

   4.5.4 Non-orientated landforms, formed subglacially

   4.5.5 Supraglacial landforms, parallel to ice flow

   4.5.6 Non-orientated landforms, formed supraglacially

   4.5.7 Ice-marginal landforms, transverse to flow

    4.5.8 Non-orientated ice-marginal and proglacial landforms
4.6 Preservation potential of terrestrial glacigenic sediments and landforms

Chapter 5. Glaciofluvial processes and landforms

5.1 Introduction
5.2 Glaciofluvial processes
    5.2.1 Meltwater erosion
    5.2.2 Transport of sediments and solutes
    5.2.3 Deposition of glaciofluvial sediments
5.3 Glaciofluvial landforms
    5.3.1 Landforms resulting from subglacial meltwater erosion
    5.3.2 Landforms resulting from subglacial meltwater deposition
    5.3.3 Ice-marginal landforms resulting from stream erosion
    5.3.4 Depositional features derived from meltwater in contact with the ice margin
    5.3.5 Landforms resulting from meltwater erosion in a proglacial setting
    5.3.6 Depositional landforms derived from meltwater in a proglacial setting
5.4 Glaciofluvial facies and facies associations
    5.4.1 Lithofacies and their interpretation
    5.4.2 Braided river facies associations
5.5 Preservation potential of fluvioglacial sediments

Chapter 6. Glaciolacustrine processes and sediments

6.1 Introduction
6.2 Pleistocene-Early Holocene glacial lakes
6.3 Physical character of glacial lakes
6.4 Glaciolacustrine sedimentary processes
6.5 Landforms resulting from glaciolacustrine deposition
    6.5.1 Deltas
    6.5.2 Delta moraines
    6.5.3 De Geer moraines
    6.5.4 Shorelines or strandlines
6.6 Glaciolacustrine facies
    6.6.1 Deltaic sediments
    6.6.2 Lake bottom sediments
    6.6.3 Rain-out diamict and subaquatic gravity flows
6.7 Glaciolacustrine facies associations

Chapter 7. Glaciomarine processes and sediments: Part I, Fjords

7.1 Introduction to the glaciomarine environment

7.2 Fjord environments with active glaciers

7.3 Processes controlling fjord sedimentation

    7.3.1 Deposition from glaciers

    7.3.2 marine processes

    7.3.3 Influences on fjord sedimentation from land sources

7.4 Patterns and rates of sedimentation in fjords

7.5 Textural characteristics of fjord sediments

7.6 Facies analysis of fjord sediments

    7.6.1 Facies associations forming today in Alaskan fjords

    7.6.2 Plio-Pleistocene facies association from an Antarctic fjord

    7.6.3 Submarine debris-flow facies association from a late Palaeozoic sequence in South Africa

7.7 Depositional and soft-sediment erosion features in fjords

    7.7.1 Morainal banks

    7.7.2 Grounding-line fans

    7.7.3 Ice-contact deltas

    7.7.4 Fluvio-deltaic complexes

    7.7.5 Proglacial laminites

    7.7.6 Fjord-bottom sediment complexes

    7.7.7 Beach and tidal-flat features

    7.7.8 Iceberg turbate deposits

7.8 Stratigraphic architecture in fjords

Chapter 8. Glaciomarine processes and sediments: Part II, Continental shelf and deep sea

8.1 Introduction

8.2 Transport of glacigenic sediment out of ice sheet system

    8.2.1 Nature of sediments arriving at the grounding zone

    8.2.2 Ice-ocean sediment transfer routes

8.3 Processes controlling sedimentation on glacier-influenced continental shelves

    8.3.1 Delivery of sediment to the glaciomarine environment

    8.3.2 Ice-ocean interactions

    8.3.3 Oceanographic processes

    8.3.4 Processes of reworking

    8.3.5 Diagenesis

8.4. Physical, chemical and mineralogical characteristics

    8.4.1 Sediment texture

    8.4.2 Clast shape

8.4.3 Surface features on gravel clasts

8.4.4 Clast fabric

8.4.5 Geochemistry

8.4.6 Mineralogy

8.5 Erosional features on the continental shelf and slope

8.5.1 Submarine troughs

8.5.2 Tunnel valleys

8.5.3 Iceberg and sea ice scours

8.5.4 Slope valleys

8.5.5 Boulder pavements

8.6 Depositional features on the continental shelf and slope

8.6.1 Fluviodeltaic complexes on the continental margin

8.6.2 Delta-fan complex

8.6.3 Subglacial deltas

8.6.4 Till tongues

8.6.5 Trough-mouth fans

8.6.6 Diamict(on/ite) aprons

8.6.7 Shelf moraines

8.6.8 Flutes and transverse ridges

8.6.9 Mass movement features

8.6.10 Raised beaches

8.7 Sediment distribution patterns

8.7.1 Seawards of Antarctic ice shelves

8.7.2 Continental shelf bordering mountain terrain with temperate glaciers

8.7.3 Continental shelf partly influenced by cold glaciers

8.8 Sedimentation rates

8.9 Sedimentary facies in glaciomarine environments

8.9.1 Sedimentary structures in glaciomarine sediments

8.9.2 Lithofacies in the glaciomarine environment

8.9.3 Glaciomarine facies associations

8.10 Stratigraphic architecture and depositional models of glacially influenced continental shelves

8.10.1 Antarctica

8.10.2 Arctic

8.11 Preservation potential of glaciomarine sediments

PREFACE

Glacial environments are scenically and scientifically among the most exciting and complex on Earth. Apart from phenomena associated directly with moving or stagnating glaciers, fluvial, aeolian, lacustrine and marine processes frequently interact with ice and its deposits. Glacial environments therefore possess a wide variety of landforms and sediment associations.

Some 10% of the Earth's land surface today is covered by ice, a figure that exceeded 30% during the Quaternary glaciations of the last two million years. In earlier geological history, the earth underwent glaciations of continent-wide extent on several occasions, some of them even more intense than those of the Quaternary Period. Yet these earlier glaciations have received little attention compared with other global climatic events. Furthermore, the study of ancient glacial sequences has, until recently, been relatively simplistic, and frequently revealed an unawareness of the complexities of glacial processes.

The prinipal aims of this book are as follows: (i) to examine the processes associated with contemporary ice-masses, especially those which can be observed or monitored in the field; (ii) to emphases the range and character of erosional and depositional landforms present at contemporary glacier margins, and those left by the Pleistocene ice sheets; (iii) to draw examples from both Quaternary and ancient glacial periods, thereby providing a link between what we see at the surface and the rock record; and (iv) to give equal emphasis to the glaciomarine environment which has been neglected in most textbooks.

The book deliberately avoids the use of most mathematical formulae since the emphasis is on what ccan be seen and interpreted on the ground, or otherwise sampled. There are other excellent texts, noted throughout the text, which provide rigorous mathematical treatment for those who desire it, but here it is intended that the subject matter be accessible to those without advanced mathematical training. It is intended that the book will be of value to undergraduates studying physical geography, geology, environmental science and countryside management, as well as to secondary/high school teachers and those who wish to understand better the processes that have shaped our lands. No previous knowledge of the subject is assumed, but an elementary knowledge of geological concepts.

By historical accident, Quaternary studies and glacial processes have been largely neglected in the traditional British geology departments. Rather, they commonly have formed a core element in physical geography courses, but here there has been a tendency top avoid rigorous application of sedimentological principles. In contrast, in North America, there has always been a much closer affinity between geomorphology and geology. The approach taken in this book is to integrate the approaches of the two disciplines as far as possible.

With the "greening" of earth sciences, and the development of employment opportunities in such areas as waste disposal, water management and extraction of aggregates, earth scientists need to know more about the superficial glacial deposits that cover so much of the populated areas of the Northern Hemisphere. Ancient glacial deposits are also of economic importance since, in some regions, they provide petroleum plays, as in the Permo-Carboniferous and late Precambrian of the Middle East and South America. Furthermore, recent research, especially deep-drilling on the continental shelves, has demonstrated the volumetric importance of glacial sediments.

In a relatively small book such as this, it has not been possible to reference relevant literature thoroughly. Instead, I have cited a few key references together with those that provide particularly good examples of the phenomena under review. Nor has it been possible to describe all components of the glacier system as one might wish; for example glaciofluvial and glaciolacustrine environments are treated only lightly to allow for more thorough treatment of less-readily obtainable information on the glaciomarine environment. However, I hope that, in the end, the variety and fascination of glacial environments will have come across.

Michael J. Hambrey
Liverpool, 1993

ACKNOWLEDGEMENTS

This book would not have been possible without the efforts of many colleagues throughout the world in promoting the subject. Foremost among them are W. H. Theakstone who was my mentor when I first began working in glacial environments; A. G. Milnes who convinced me of the strong linkage between glaciology and structural geology; and W. B. Harland who led me into investigations of ancient glacial sediments in areas where modern ones were being formed.

I am indepted to leaders of various expeditions for opportunities to be in glacial environments: P. Worsley in northern Norway; the late F. Müller in the Canadian Arctic; W. B. Harland in Svalbard; N. Henriksen in East Greenland; P. J. Barrett, W. U. Erhmann, D. K. Fütterer and G. Kuhn in Antarctica. The leaders of various conference- or workshop-based field excursions have also played an important rôle in providing a forum for stimulating discussion and opportunities to see glacial phenomena in many different settings.

Several colleagues, with whom I have undertaken joint research, have, over the years, provided a sounding board for ideas about glacial environments, notably J. A. Dowdeswell, I. J. Fairchild, A. C. M. Moncrieff and M. J. Sharp. In particular, I am grateful to the following for reviewing the whole of this volume: I. J. Fairchild, P. J. Gibbard and D. Huddart.

Lastly, I acknowledge the help given by several funding agencies and institutions who have supported my work in regions influenced by glaciers today and in the past: University of Manchester, Swiss Federal Institute of Technology (Zürich), University of Cambridge (especially the Scott Polar Research Institute, where the bulk of this book was written), Liverpool John Moores University, Alfred Wegener Institute for Polar and Marine Research (Bremerhaven), Victoria University of Wellington, Natural Environment Research Council, The Royal Society, The Geological Survey of Greenland and the Cambridge Arctic Shelf Programme.

For permission to reproduce figures for this volume, the following publishers are thanked. All photographs were taken by the author, except where otherwise stated.

CHAPTER 1

INTRODUCTION

1.1 Historical background

1.1.1 Discovery of "Ice Ages"

Today it is common knowledge that the Earth once experienced an ice age, during which ice sheets spread over large parts of Eurasia and North America. We are all familiar with the sensationalist press reports that have inevitably arisen after exceptionally severe blizzards, debating whether a new ice age has come upon us, even if these have now been replaced by equally alarmist statements that human-induced global warming will cause catastrophic melting of the remaining polar ice sheets. However, when the concept of widespread glaciation in the past was first mooted in the early 19th century, it met with much opposition. The battles and personality conflicts that arose as a result throughout the first half of the 19th century, provide a fascinating insight into how geology evolved into a science and overcame rigid prejudices that were based on a misunderstanding of Old Testament accounts of natural disasters, such as Noah's flood (see more detailed accounts in Garwood 1932, Imbrie & Imbrie 1979 and Mills 1983).

The chief protagonist or the ice age theory in the early 1800s was the influential president of the Swiss Society of Natural Sciences, Agassiz, and it is he who came to be regarded as the "Father of Ice Ages". However, the idea that ice had once been much more extensive was not new. Agassiz, was not the first to believe that glaciers had been more extensive; indeed, for many years he was sceptical. It was others, not all of them scientists, who documented the evidence for more extensive ice. Probably the first to do this was a Swiss minister, Kuhn who, in 1787, interpreted local erratic boulders below the glaciers near Grindelwald as evidence of ice having been more extensive. In 1795, the leading geologist of the dat, Scot James Hutton published his *Theory of the Earth* in which he described how ice had transported great boulders of granite into the Jura Mountains. In 1815 a Swiss mountaineer and hunter, Perraudin, argued that glaciers extended much further down, and filled, the Val de Bagnes in the Alps. He expressed his views to a sceptical Charpentier who later became an ardent advocate of the glacial theory. Three years later, Perraudin tried to persuade the Swiss engineer, Venetz, but he too had doubts about the validity of the idea. However, slowly Venetz began to accept the hypothesis and by 1829 was able to argue from the distribution of moraine and erratics that glaciers once covered the Swiss plain, the Jura and other parts of Europe. Already, in 1824, Esmark had argued that the glaciers of Norway had once been more extensive.

None of these men made the intellectual leap to conceive of a period of widespread global cooling and ice sheet development. It was left to the famous German poet, Goethe to promote the idea of an ice age ("Eiszeit"). Taking note of the findings by scientists of erratics on the North German Plain, Goethe developed his ice age concept around 1823 in a novel, *Wilhelm Meister* (Cameron 1965).

Meanwhile, Charpentier, at last converted to the idea of more extensive ice, accepted Venetz's interpretation, and thereafter began to assemble a mass of evidence in its favour. Resistance was strong since it was generally held at the time that the large erratics were deposited by Noah's flood. By 1833, many scientists had come to accept the view of the leading British geologist of the day, Lyell, that the boulders had been rafted by icebergs. This theory had originated with the German mathametician, Wrede in 1804. The iceberg-rafting or "Drift" theory, which apparently explained far-travelled erratic boulders, was expounded by Lyell in perhaps the most influential geology textbook ever written, the *Principles of Geology*, published in 1833. This explanation embraced conveniently the Flood theory, by providing a mechanism by which sediment, and especially large boulders, could have been transported - hence the term "drift" for these deposits. Lyell was strongly supported by Darwin who, in a series of papers, became a prominent advocate of the theory until his death in 1882.

Charpentier had as a friend Agassiz who by now was one of Europe's leading scientists but, despite Charpentier's powers of persuasion, the young man was first unable to accept the glacial theory. Eventually, during a field trip to Bex, Agassiz was won over, and for the first time the glacial theory had a strong, forceful and influential character to promote it. Both Agassiz and Charpentier had by now acknowledged Goethe's theory of a great ice age, though most scientists of the day ignored it, perhaps because of the poet's lack of scientific credentials. Unfortunately, Agassiz developed the glacial theory further, taking liberties with Charpentier's work, and promoting it beyond the available evidence. Thus, when Agassiz presented his work to the Swiss Society of Natural Sciences at Neuchatel in 1837, he met with almost universal opposition. Nevertheless, he had a prominent ally in Germany at this time, Schimper, who provided him with much information about the former extent of glaciers in the Isar and Würm valley. Undaunted by the general opposition, Agassiz wrote up his work in the book *Etudes sur les Glaciers* (*Studies on Glaciers*) which was published in 1840, acknowledging the important work of his predecessors Venetz and Charpentier, but curiously not Schimper, and set about trying to convince other scientists and the public at large. His belief in an ice age which caused great devastation and extinguished many animal species, fitted in well with the philosophy of catastrophism that was prevalent in geology throughout the 18th and 19th centuries. The ice age concept merely substituted one catastrophe for another, the Great Flood. One of the major proponents of the Flood theory was Buckland who, in addition to being a clergyman, was a Professor of Mineralogy and Geology at Oxford University, and so was well placed to explore the links between geology and religion. Buckland was also one of the first geologists to focus specifically on the accumulations of unconsolidated mud, sand and gravel, that covered much of the British Isles, and which were referred to as "Diluvium" by those who believed in The Flood. Buckland's impressive account of these deposits was published in 1823 and gained him immense respect.

Buckland meanwhile found it impossible to explain all the evidence in terms of a great flood with icebergs floating around. In particular, he wanted to know where all the water had come from and where it had gone. After hearing Agassiz promote his ideas in another meeting in 1838, in Germany, Buckland joined Agassiz on a trip to the Alps, but remained unconvinced that glaciers were responsible for drift deposits elsewhere. In 1840, Agassiz took on the British, reading a paper to the Geological Society of London "On the evidences of the glaciation of Great Britain and Ireland". This visit was portrayed in the satirical magazine *Punch* as "a sporting tour in the search of moorhens (moraines)". By now, Buckland had changed his mind and, after having shown Agassiz drift deposits around Scotland and northern England, became a strong advocate of the glacial theory himself. Within just a few months Buckland had converted Lyell to the theory, but even with a trio of internationally renowned scientific heavyweights embracing it, wider opposition was not overcome immediately.

Indeed, Lyell, perhaps under the influence of Darwin, lapsed back into renewed support for the Drift theory.

Although by 1841, Forbes, a Professor of Natural History at Edinburgh University, and himself a key figure in the development of glaciological concepts, was able to write to Agassiz "You have made all the geologists glacier-mad here..", it was not for another twenty years that the majority of British geologists accepted the ice age theory, following publication of classic papers by Jameson in 1862 and Geikie in 1863.

Various reasons have been given as to why Agassiz had so much difficulty in overcoming entrenched beliefs. Apart from the religious views, it proved difficult to explain the widespread "shelly drifts" around the coasts of NW Europe. Furthermore, there was ignorance amongst geologists about glaciers themselves; it was not until 1852 that Greenland was found to have an ice sheet, and only towards the end of the century that Antarctic too was covered by one. Agassiz himself did not help his cause, because in his enthusiasm he envisaged glaciers in places where the evidence was non-existent, such as the Mediterranean or the Amazon basin.

In 1847 Agassiz moved to the USA, as Professor at Harvard University, and found that many American scientists had already accepted his theory. His arrival, however, did speed up its acceptance, and when Agassiz finally died in 1873, few scientists held out against it.

Following the establishment of an ice age explanation for the widespread unconsolidated sediments, that ultimately were equated with the Quaternary Period, it was only natural that geologists should seek for evidence of ice ages in the older rock record (Harland & Herod 1975). In 1855 Ramsey suggested that English Permian breccias were of glacial origin. Although he was incorrect in this, others began looking for Permian glacial sediments elsewhere, and by 1859 unequivocal deposits had been reported from India and Australia, and by 1870 from South Africa. In 1871 Precambrian glacigenic sediments ("tillites") were described from Scotland, and in 1891 from northernmost Norway where a striated pavement was also discovered. Evidence of older Precambrian and early Palaeozoic glaciations were found in the early 20th century.

Many discoveries of tillites were reported from around the world subsequently, but a phase of doubt, particularly with regard to the late Precambrian glaciation, entered the mind of many geologists as recently as the 1960's, perhaps because of the uncritical acceptance of many deposits as glacigenic at a time when the subdiscipline of sedimentology was revolutionizing the interpretation of sedimentary sequences. The few who maintained a pro-glacial stance for the Precambrian deposits (e.g. Harland 1964) have since been fully vindicated. A benchmark contribution that provided a thorough, objective account of the Precambrian glacial deposits of Scotland(Spencer 1971), set a standard which others have emulated with considerable success.

While the evidence for ancient ice ages was gradually being built up, the extent and number of glaciations in the Quaternary Period were being documented. Two schemes in particular became widely accepted round the turn of the century. In the Mid-West of North America Chamberlain and Leverett mapped four sheets of glacial drift, each representing a distinct "ice age". In the region to the north of the European Alps Penk and Brückner similarly derived four ice ages, but by associating gravel terraces at progressive lower levels with cold periods when deposition was rapid. Similar successions were subsequently derived for the Scandinavian and British ice sheets, and even New Zealand.

Despite the realization that successive glaciations tend to destroy the evidence of earlier ones, at least on land, the chronologies derived for the Mid-West and Europe were accepted uncritically and the two were inevitably correlated, despite the absence of dating evidence. These schemes survived intact until the deep sea record began to yield a different story.

When the first deep-sea sediment cores, going well back into the Quaternary Period, were obtained during the 1950's and onwards, oxygen-isotopic studies of planktonic foraminifera enabled palaeotemperatures and ice volume changes to be determined. The cores revealed rather more glacial periods than the commonly accepted terrestrial record, and land geologists for a long while ignored the evidence. In the 1960's, with the development of magnetostratigraphy and establishment of magnetic reversals, the much-needed method of dating Quaternary events had finally arrived. Sediment cores, analysed by an international team of scientists on a project called CLIMAP, gradually began to yield a climatic record that matched remarkably well the temporal changes in solar radiation derived by Milankovich from astronomical variables. The number of ice ages during the Quaternary Period thus proliferated, and the four-fold ice age chronology from the land areas finally was shown to be more incomplete than complete.

Until the early 1970's, it was assumed by many earth scientists that the period represented by ice ages was essentially equivalent to the Quaternary Period. In 1972, the Deep Sea Drilling Project extracted long cores from the Antarctic continental shelf in the Ross Sea (Hayes & Frakes 1975). To many people's surprise, there was evidence of glaciation going back 25 million years (m.y), to the late Oligocene Epoch. Antarctica has yielded further evidence of the antiquity of Cenozoic glaciation in the last 20 years. A succession of New Zealand drilling operations in the Ross Sea culminated in a drillhole in 1986 that showed that glacier ice was present at least as far back as earliest Oligocene time (36 m.y.). Most recently (1987/88), the Ocean Drilling Program in Prydz Bay has confirmed the existence of a large ice sheet over East Antarctica dating back to at least this far back. However, none of these drillholes penetrated the glacial/preglacial boundary.

Back in the northern hemisphere, the onset of ice-rafting, indicating the development of an ice sheet over Greenland reaching the coast, has been dated at 2.4 m.y. ago, with indications of ice-rafting in Baffin Bay from a Canadian Arctic source another million years earlier. Furthermore, glaciers in Alaska have been active since about 10 m.y. ago on the high mountains there. Thus we can no longer equate the Quaternary Period with the development of ice ages, only with intensification of the ice-cover.

1.1.2 Development of understanding of the dynamics of glaciers

To some extent, the development of ideas concerning the manner in which ice-masses themselves behave has taken place independently of the investigations of the products of glaciation, although there have always been some scientists who have taken an interest in both fields. The history of the development of glaciological ideas is just as interesting and full of conflicting views as the establishment of the ice age theory, and the reader is referred to Paterson (1981) and Clarke (1987) for fuller accounts.

The earliest descriptions of glaciers, in Icelandic literature, date from the 11th century. However, it was not until several centuries later that glaciers were recorded as being able to flow. By then the Earth was experiencing what became known as the Little Ice Age which peaked about 1750, and resulted in strong advances of glaciers in many parts of the world. These events are particularly well-documented where glaciers in the Alps and Norway destroyed pastures and even property, as well as being responsible for several disasters. Hence, there developed a strong scientific interest in glaciers, notably in Switzerland.

Fig. 1.1. The Mer de Glace in the French Alps, one of the earliest sites at which studies of glacier flow were first undertaken. The curving light and dark arcs are "ogives" or Forbes bands, each pair representing a year's movement through the ice-fall in the background. They were first described by James Forbes in the mid-19th C.

Prominent among the early pioneers was Scheuzer who, between 1706 and 1723, published several works of a geographical and scientific character. He took a special interest in legends of Alpine dragons, but also studied glaciers, proposing that water entered fissures in the ice, and on freezing expanded, causing the glacier to thrust forward - his so-called "dilation" theory. Altmann in 1751 and Grüner in 1760 explained that gravity was the cause of glacier motion, but assumed that this was accomplished entirely by ice sliding over its bed. In the late 18th century, Bordier suggested that ice can flow by internal deformation, somewhat like a viscous fluid.

Although Grüner had noted that stones on the surface of one of the Grindelwald glaciers had advanced fifty paces in six years, the first systematic measurements of glacier flow were not undertaken in the Alps until the 1830s.

Fig. 1.2. The Unteraargletscher in the Bernese Oberland of Switzerland. The first documented case of ice movement was recorded on this glacier when, over a period of several years, Hugi followed the displacement downglacier of a large block on the medial moraine on the right.

Foremost among these experimentalists, once gain, was Agassiz who showed that ice moves faster in the middle than at the sides. The British scientists Tyndall and Forbes, and the Swiss Hugi and De Saussure also became heavily involved in establishing the dynamics of glaciers in France and Switzerland by measuring ice movement and documenting surface structures. The Mer de Glace in France (Fig. 1.1) and the Unteraargletscher in Switzerland (Fig. 1.2) became favourite haunts. Forbes, having already fallen out with Agassiz after undertaking joint work on the Unteraargletscher, became involved in a heated dispute with Tyndall about the nature of glacier flow. Forbes believed that flow was of a viscous nature and considered ice to have many similarities to a metamorphic rock. Tyndall thought that motion resulted from the formation of small fractures that were subsequently healed by pressure melting and refreezing,

7

an idea that became known as the "regelation theory". Tyndall died in 1893 after his wife had unwittingly administered a lethal dose of choral, and with him the regelation theory also died. Forbes, on the other hand was essentially correct in linking glacier flow to fluid mechanics, although he did underestimate the rôle of basal sliding. His viscous flow theory motivated much laboratory experimental work and field measurements. Thus by the end of the century, the manner in which ice at the surface of a valley glacier flowedwas well known. In 1897, Reid in North America recorded the character of velocity vectors: inclined slightly downwards in the snow accumulation area and upwards in the ice ablation zone, elaborating this in a classic paper "The mechanics of glaciers". However, ice movement at depth posed a different problem. Even though Blümcke and Hess around the turn of the century, using stakes set into a Tyrôlean glacier, found that ice at depth moved faster than at the surface, many scientists for decades afterwards believed that the reverse was true, and invoked a mechanism called "extrusion flow". Demorest and Strieff-Becker were particularly forceful proponents of this idea. After the considerable progress achieved up to 1900 it was strange that such an idea should have taken such a strong hold.

The extrusion flow theory was not laid to rest until the 1950's following deep borehole measurements on Jungfraufirn in Switzerland and later on various Alaskan and Canadian glaciers, together with laboratory experiments on ice, and the application of modern ideas of solid-state physics, showed that ice deformed in a manner similar to other crystalline solids such as metals and rocks.

The foundations for our present understanding glacier deformation were laid by British physicists Glen and Nye in 1950's and 1960's, together with Lliboutry of France and Weertman of the USA, with their work on glacier sliding. Now we know that glacier ice deforms in a manner similar to plastic substances, and also slides on its bed. The rôle of the physicists in glaciology was expressed by Paterson (1981) in these terms: "a mere handful of mathematical physicists, who may seldom set foot on a glacier, have contributed far more to the understanding of the subject than have a hundred measurers of ablation stakes or recorders of advances and recessions of glacier termini". Unfortunately, since then, some mathematicians and physicists have tended to shift away from reality, and have derived equations that are virtually untestable in the field. However, the increasingly important rôle that glacier modellers are playing is bringing the observers and theoreticians back together again.

Major advances in other aspects of glaciology have taken place through the 20th century, such as the measurement of snow/ice density, accumulation and heat balance, the examination of glacier hydrology; the palaeoclimatic record in ice sheets from drill cores, not only in valley glaciers but in the polar ice sheets (eg. Koch & Wegener in Greenland in 1913, and Ahlmann in Svalbard and elsewhere between 1920 and 1940. From 1957, the International Geophysical Year, the Antarctic ice sheet has been investigated from all angles. Remote sensing techniques, such as radio-echo sounding and satellite imagery have revolutionised our understanding of the extent, thickness and character of the ice on that continent, as indeed elsewhere.

Developments in linking the sedimentary record to glaciological principles have been slow in coming. Geologists have been slow to understand the complexities of glacier dynamics, while many physicists have tended to assume that mathematics can provide all the answers and disregarded the evidence offered visually be glaciers. There are of course, exceptions and of these, Boulton has perhaps done more than most in explaining the development of glacial sedimentary sequences in glaciological terms, starting with a series of investigations at the margins of glaciers in Svalbard in the 1960's, and extending to the large northern hemisphere ice sheets in the last decade.

In any account of the development of the science of glaciology in the last fifty years one cannot ignore the rôle played by the International (formerly British) Glaciological Society, which through its *Journal of Glaciology* and more recently the *Annals of Glaciology*, has provided the principal focus for glacier research in all its aspects, and encouraging interdisciplinary approaches to the subject. A history of the Society and its rôle in the development of glaciology has been provided by Weertman (1987).

1.2 Terminology and classification of glacigenic sediments

Before beginning this discussion, it is necessary to define a few basic terms, and here the definitions of Dreimanis (1989) are broadly followed, although their use in the literature varies:

Glacigenic sediment (also glacigene, glaciogenic): "of glacial origin"; the term is used in a broad sense to embrace sediments with a greater or lesser component derived from glacier ice.

Glacial debris: material being transported by a glacier in contact with glacier ice.

Glacial drift: all rock material in transport by glacier ice, all deposits made by glacier ice, and all deposits predominantly of glacial origin deposited in the sea from icebergs, or from glacial meltwater.

Diamicton: a non-sorted or poorly sorted unconsolidated terrigenous sediment that contains a wide range of particle sizes (modified from Flint 1960).

Diamictite: the lithified equivalent of diamicton, and *diamict* embraces both (Harland et al. 1966). These terms, together with diamicton, have no genetic connotations.

Most investigations of glacigenic sediments, whether contemporary, Quaternary or ancient, have tended to use genetic terms for which no universal agreement has been reached. As a result, much confusion has ensued concerning the origin of a particular sediment. In the last decade it has been increasingly recognized that a study of a glacigenic sequence should begin with an objective description, before attempting to classify the sediments genetically.

1.2.1 Terminology and non-genetic classification of poorly sorted sediments

A variety of terms has been used in the past, mainly with reference to lithified deposits, to describe sediments without assuming a glacial origin. "Diamictite" has gradually found greater favour for lithified sediments than the synonymous "mixtite", whilst "Tilloid" has been used for "till-like rocks" in a variety of conflicting ways, and is also falling out of favour.

For the purposes of field investigation, a textural classification of diamictite has been devised by Moncrieff (1988) and a modified version is used here (Table 1.1). It is based on the proportions of sand and mud (as matrix) discernable using a hand lens or with the naked eye, against the proportion of gravel clasts. In Cenozoic glacigenic sediments, the biogenic component may make up a considerable proportion of the sediment. Following Ocean Drilling Program procedures (Barron, Larsen & Shipboard Scientific Party 1989), prefixes such as shelly, and diatomaceous, may be used where such components exceed 30%.

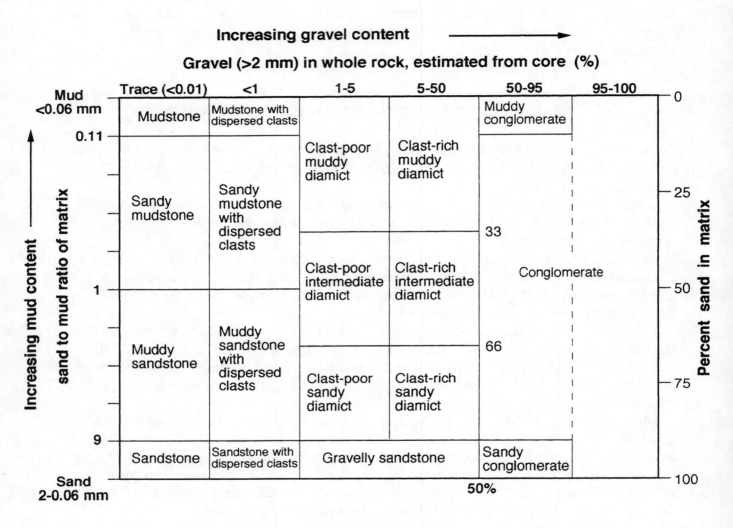

Table 1.1. *Non-genetic classification of poorly sorted sediments, based on Moncrieff (1989), but with maximum proportion of gravel in diamict reduced from 80 to 50% for compatibility with the Ocean Drilling Program's definition of diamict and conglomerate/breccia (Barron, Larsen & Shipboard Scientific Party 1989). The term "diamict" embraces both diamicton and diamictite. "Mud", as used in this context, covers all fine sediment, i.e. mixtures of clay and silt.*

1.2.2 Classification of sorted sediments

Various other sedimentary types, not restricted solely to glacial environments, should also be mentioned: these include gravel, sand, silt and clay with varying degrees of sorting; rhythmically bedded/laminated sediments of sand/silt/clay and non-cyclically deposited graded beds such as turbidites. The Wentworth (1922) scale of grain sizes is the most widely used, summarised as follows:

> boulder: >256 mm (< -8 phi units (ø))
> cobble: 64 to 256 mm (-6 to -8 ø)
> pebble: 4 to 64 mm (-2 to -6 ø)
> granule: 2 to 4 mm (-1 to -2 ø)
> sand: 0.0625 to 2 mm (4 to -1 ø)
> silt: 0.0039 to 0.0625 mm (8 to 4 ø)
> clay: < 0.0039 mm (< 8 ø)

"Gravel" embraces all classes > 2 mm. The term "mud" is used in various ways; in this book, usage is consistent with many sedimentological studies in meaning a mixture of silt + clay.

1.2.3 Terminology of glacigenic sediments

The question of terminology of till and till-like deposits has received a thorough airing repeatedly over the past three decades (e.g. Hambrey & Harland 1981, with reference to pre-Pleistocene sediments; Dreimanis 1989, for Quaternary sediments). Here, the most important terms used in this and subsequent chapters are summarized.

For an unsorted deposit with a wide range of grain sizes deposited directly from glacier ice, whether on land or beneath a floating glacier, and not subsequently modified the term *till* is applied. This term is an old Scottish word originally used by countryfolk to describe "a kind of coarse obdurate land", the soil developed on the stony clay that covers much of northern Britain" (Flint 1971: 148). It was adopted as a genetic term by Scottish geologists in the mid 19th century, and its use has spread across the English speaking world and into other languages. Some authors restrict the term to material deposited on land, but as it is often difficult to distinguish the environment of deposition, the deposit could be labelled as till only if one were sure of the nature of the environment, thus restricting its usefulness. The term *boulder clay*, which has been used in the British Isles as a synonym of till, is no longer favoured by most glacial geologists. *Moraine* has also often been used as a synonym for till, but it is best to restrict this term to the landform.

The term for a lithified glacial deposit, *tillite*, historically has evolved separately and is not strictly equivalent. Many authors have used the term to embrace sediments containing a signiciant proportion of iceberg-rafted material. Other authors have been more restrictive, although few would restrict it solely to material known to have been deposited on land, but would include lithified till-like material deposited beneath a floating glacier. Here, the terms *till* and *tillite* are used to include sediments released directly from a glacier, whether on land or through a water column, that have not been subject to reworking, such as by currents or gravity flowage resulting in disaggregation. Marine sediments, which include an ice-rafted component, are referred to as *glaciomarine sediments*.

Material which is released by ice into the sea, whether by continuous rain-out beneath a floating mass of glacier ice, or sporadically from icebergs, even if the proportion is small, are collectively referred to as *glaciomarine sediments* (also referred to in the literature as glacial-marine, glaci(-)marine). Thus, the broad inclusive definition proposed by Andrews & Matsch (1983: 2) and Borns & Matsch (1989: 263)is adopted (but anglicized) here:

"Glaciomarine sediment includes a mixture of glacial detritus and marine sediment deposited more or less contemporaneously. The glacier component may be released directly from glaciers and ice shelves or delivered to the marine depositional site from those sources by gravity, moving fluids, or iceberg rafting. The marine component comprises mainly terrigenous and biogenic ("biogenous" in North America) sediments. Glaciomarine sediments vary laterally from ice-proximal diamicton, gravel and sand facies, to an intermediate pebbly silt and mud facies, to distal marine environments where the glacial imprint is seen in ice-rafted debris particles usually in the -1 to 4 ø fraction".

Release of glacial debris and its deposition or redeposition			Depositional genetic varieties of till		
I. Environment	II. Position	III. Process	IV.By environment	V. By position	VI. By process
Glacio-terrestrial	Ice-marginal Frontal Lateral Supraglacial Subglacial Substratum	A. *Primary* Melting out Lodgement Sublimation Squeeze flow Subsole drag B. *Secondary* Gravity flow Slumping Sliding and rolling	Terrestrial non-aquatic till	Ice-marginal till Supraglacial till Subglacial till	A. *Primary till* Melt-out till Lodgement till Sublimation till Deformation till Squeeze flow till B. *Secondary till* Flow till

Table 1.2. Genetic classification of till in terrestrial settings (adapted from the INQUA classification, Dreimanis 1989: Table 11). The vertical columns are independent of each other and no correlation horizontally is implied. Not all combinations are feasible.

1.2.4 Genetic classification of tills

Tills (and tillites) are more variable than any other sediment known by a single name (Flint 1971: 154, Goldthwait 1971: 5). Not surprisingly, therefore, the meaning of till varies from one investigator to another. Two extreme views have been published. The first, by Harland et al. (1966) used the term in a very broad sense to include any poorly sorted sediment that contains glacially transported material. By contrast, Lawson (1979a) used a very restrictive definition to exclude any hint of reworking or addition of components that were not *directly* glacially deposited. Drewry (1986: 120) even stated that the genetic term "till" is no longer applicable and new nomenclature is necessary. One might think that such disparate views preclude any agreement about the definition of "till", let alone the development of a genetic classification. However, a considerable measure of agreement has been achieved by the

International Quaternary Association's (INQUA) Commission on Genesis and Lithology of Quaternary Deposits. A comprehensive classification of tills has emerged that has satisfied the majority of INQUA correspondents (Dreimanis 1989). The INQUA classification represents the broadest consensus concerning glacial sediments at the present time and the terrestrial elements are adapted in this book (Table 1.2). The factors considered in the INQUA classification are primarily the formational and depositional processes, the general environment of deposition and the position in relation to glacier ice (Table 1.2).

In the genetic depositional classification of till there are two main categories. *Primary tills* are formed mainly by direct release of debris from the glacier and deposited by primary glacial processes, namely meltout, lodgement, sublimation or during deformation induced by the glacier. *Secondary tills* are the products of resedimentation of glacial debris that has already been deposited by the glacier, with little or no sorting by meltwater. Within these categories several varieties of till have been documented (Table 1.2), although these represent end members in a continuous spectrum of depositional types (Dreimanis 1989).

Melt-out till is deposited by a slow release of glacial debris from ice that is not sliding or deforming internally.

Lodgement till is deposited by plastering of glacial debris from the sliding base of a moving glacier by pressure melting and/or other mechanical processes.

Sublimation till is till released by the sublimation (direct transition of ice to the vapour state) of debris-rich ice (Shaw 1989). It requires long-term extremes of cold and aridity for its formation, so its development is restricted to Antarctica.

Deformation till comprises weak rock or unconsolidated sediment that has been detached by the glacier from its source, the primary sedimentary structures distorted or destroyed, and some foreign material admixed (Elson 1989). This term suffers from having various other meanings, not recognized by INQUA, so it should be used with care.

Squeeze flow till is the result of squeezing or pressing of till by the weight or movement of glacier ice.

Flow till may be derived from any glacial debris upon its release from glacier ice or from a freshly deposited till, in direct association with glacier ice. Redeposition is accomplished by gravitational slope processes, mainly by gravity flow, and it may take place ice-marginally, supraglacially or subglacially, and subaerially or subaquatically.

These process terms may conveniently be used in combination with the terms for position, e.g. "supraglacial meltout till". Many other terms for till and combinations have been used, for which Dreimanis (1989) has provided a comprehensive review.

Recognition of lodgement till, meltout till and flowtill has long been a matter for debate, and the most useful criteria are tabulated in detail by Dreimanis (1989: Appendix D). Genetic terms for lithified sediments may have the suffix "-ite".

1.2.5 Genetic classification of glaciomarine sediments

In terms of preservation potential and volume, glaciomarine sediment are vastly more important than terrestrial glacial sediments, yet prior to the late 1970's little was known about the contemporary environment and sediment classifications were not based on direct observations of the processes (e.g. Harland et al. 1966). Since then, however, glaciomarine environments have received considerable

13

attention, notably those in Antarctica, Svalbard and Alaska, and several simple classifications have been adopted. For example, a genetic classification arising from wide-ranging American studies on the Antarctic continental shelf includes the following three main categories (Anderson et al. 1980, 1983):

Basal till - deposited on the shelf by grounded glaciers (therefore better grouped with terrestrial sediments).

Compound glacial marine sediment - resulting from a combination of ice-rafting (from icebergs and ice shelves) and normal marine sedimentation.

Residual glacial marine sediment - the product of ice-rafting coupled with bottom current activity that is sufficiently strong to winnow silts and clays.

To these categories may be added a fourth, *sediment gravity flows*, described by Wright & Anderson (1982) from Antarctica and by Miall (1983) from the Early Proterozoic Gowganda Formation.

From the perspective of position in relation to the ice margin of grounded tidewater glaciers in Svalbard, Boulton (1990) identified three main zones of sedimentation:

Inner proximal zone - sedimentation sufficiently high to inhibit benthic life and bioturbation is rare (0-7km from the ice margin).

Outer proximal zone - where benthic life and bioturbation are common (7-60km), and sedimentation is influenced by suspended and ice-rafted components.

Distal zone - outer fjord or outer shelf where suspended sediment concentrations are much less, where upwelling of deep water along the continental margin occurs, and where the sea bed is affected by waves and subject to reworking and erosion of finer materials.

Deep drilling combined with seismic investigation on the Antarctic continental shelf has yielded important information about depositional processes in a temporal context (Barrett et al. 1989), Barron, Larsen & Shipboard Scientific Party 1989, 1991). This has led to a classification that is related to the proximity of the sediment source (Hambrey et al. 1989, 1991), and depends on an assessment of the relative importance of rain-out deposition, ice-rafting and biogenic activity. In this case the 'proximity' is less to do with the ice (much of which may not be delivering sediment to the sea), than with the main source of sediment, such as in ice streams. This classification, together with sediment gravity flow is used here.

Waterlain till - sediment which is released from floating basal glacier ice and accumulates on the sea bottom without being affected by winnowing processes. In character it resembles basal till, from which it may be difficult to distinguish without clast orientation measurements. Francis (1975) introduced the term "waterlain till" in preference to the semantically incorrect waterlaid till of Dreimanis (see Dreimanis 1979). Dreimanis inferred the same sort of depositional processes as summarized above, but texturally it included both stratified and unstratified varieties of diamicton.This term has been adopted for Antarctic sediments (Hambrey et al. 1989 et seq.), but Dreimanis (1989) has subsequently advocated abandoning the term. It is retained here, however.

Proximal glaciomarine sediment - sediment which is composed principally of debris released from floating glacier ice and icebergs, and which has been affected by winnowing processes. A biogenic component in the form of shelly fauna and diatoms may also be present. *Distal glaciomarine sediment* - sediment which is principally of marine origin, such as suspended sediment and biogenic material, with a minor iceberg-rafted component (<1% ice-rafted material).

A comprehensive provisional classification of glaciomarine processes and sediments was compiled for INQUA by Borns & Matsch (1989), following the general principles of the terrestrial classification. It draws attention to the wide range of ice margin types and depositional processes, but does not define them or discuss how genetically different sediments may be distinguished, so its use at this stage is premature.

1.2.6 Genetic classification of glaciolacustrine sediments

Glaciolacustrine sediments embrace all material derived directly or indirectly from a glacier that is in contact with an enclosed standing-water body. Characteristically, they comprise sediments which have a direct glacial and iceberg component, and can be classified in the same way as those in the glaciomarine environment: waterlain till, proximal and distal glaciomarine sediment, and sediment gravity flows. They also include deltaic deposits and rhythmites, notably *varves* or *varvites* (lithified). A varve may be defined as a sedimentary bed or lamina or sequence of laminae deposited in a body of still water, and representing one year's accumulation. A varve comprises a thin pair of graded glaciolacustrine layers, seasonally deposited (usually by meltwater streams) in a glacial lake or other body of still water in front of a glacier. The varve normally includes a lower summer layer consisting of relatively coarse-grained, light-coloured sediment (usually sand or silt) produced by rapid melting of ice in the warmer months, which grades upward into a thinner winter layer, consisting of very fine-grained (clayey), often organic, dark sediment slowly deposited from suspension in quiet water while the streams were ice-bound. Counting and correlation of varves have been used to measure the ages of Late Quaternary glacial deposits. The term was introduced by De Geer in 1912 (Swedish: *varv*). Although characteristic of glacial lakes, not all glacial lakes have varves, and not all varves are glacial. Non-glacial examples have been described, for example, in the papers published in Schlüchter (1979). Furthermore, some varve-like sediments form in fjord settings, but are not annual. Glaciolacustrine sediments may include material released directly from ice, by ice-rafting.

1.3 Facies analysis of glacigenic sediments

For most of the period in which glacigenic sediments have been studied, it has been customary, both with regard to Quaternary and pre-Quaternary sediments, to apply genetic terms (such as till) uncritically. This approach has often led to confusion, because tills and tillites are such varied deposits that they have meant different things to different people. As a result, many palaeoenvironmental reconstructions have been based on inadequate data, or on data presented in a way that others cannot use.

It is now widely recognised that sequences of glacigenic sediments need to be examined objectively and, for this, facies analysis, as developed for other branches of sedimentology, is the most important approach. Some advocates of the facies approach have tended to be dismissive of other methods of analysis, but these too are needed if one is to gain a clear understanding of the mode of deposition and palaeoenvironment.

1.3.1 The facies concept

The concept of facies has been used ever since it was recognised that features found in particular rock units were useful for interpreting the environment of deposition and for predicting the occurrence of mineral resources. (Reading 1978, gives a useful summary).

A sedimentary facies is a body of sediment or rock with specified characteristics, namely colour, bedding, geometry, texture, fossils, sedimentary structures and types of external contacts. The term "facies" has been used in many different senses, for example in the strictly observational sense, in the genetic sense, and in an environmental sense. However, a facies should ideally be a distinctive rock that forms under certain conditions of sedimentation, reflecting a particular process or environment. Facies may be subdivided into *subfacies* or grouped into *facies associations* (Reading 1978), or considered on a regional scale in terms of *facies architecture*.

Here, glacial facies refer to the different sediment types one finds in a glacial environment, and which are interpreted as till, glaciofluvial, glaciolacustrine and glaciomarine deposits. Grouped together we have terrestrial glacial facies associations, glaciomarine facies associations and so on.

Lithology	Bedding characteristics	Bedding geometry	Sedimentary structures	Boundary relations
Diamict(on/ite)	Massive	Sheet	Grading: normal	Sharp
Gravel	Weakly stratified	Discontinuous	reverse	Gradational
Sand(stone)	Well stratified	Lensoid	coarse-tail	Disconformable
Mud(stone)	Laminated	Draped	Cross-bedding: tabular	Unconformable
	Rhythmic lamination	Prograding	trough	
	Wispy stratification		Lonestones (dropstones)	
	Inclined stratification		Clast-supported	
			Matrix-supported	
			Clast concns.: layers	
			pockets	
			Ripples	
			Scours	
			Load structures	
			Mottling (=bioturbation)	

Table 1.3. Principal descriptive criteria used in defining lithofacies in glacigenic sequences.

1.3.2 Glacial sedimentary facies

Examination of contemporary glacial environments indicates that sedimentary facies are varied and related in an often complex manner. Identification of these facies in Quaternary sequences broadly is relatively straightforward, but in detail, for example, it may be difficult to distinguish different types of glacigenic

16

sediment. Glacial, fluvial, aeolian, marine, lacustrine and mass flow process account for the wide variety of facies present. In pre-Pleistocene sequences the frequent lack of three-dimensional exposure of strata often makes it difficult to determine the precise depositional environment and the degree of direct glacial influence in a particular facies. It is therefore important to undertake first a descriptive facies analysis, using the criteria listed in Table 1.3, and applying non-genetic terms like diamict (on/ite), and only then interpret them. There are many instances where authors have interpreted their sediments without providing adequate descriptions.

1.3.3 Interpretation of facies

Briefly, the principal descriptive facies are listed below, with comments on how each may be interpreted:

(a) Massive diamict with striated, predominantly angular to subrounded stones - basal till deposited subglacially by meltout or lodgement, or waterlain till deposited by steady rain-out of debris from the base of floating glacier ice without rewoking, or till that has been subject to gravity flowage, either subaerially or subaquatically.

(b) Diamict with deformation structures (folds, thrusts, faults, convolutions) - deformation till formed by push or overriding by the glacier, or till that has undergone gravity flowage.

(c) Weakly stratified diamict with some microfossils and shells - deposited from the base of a floating glacier in a proximal glaciomarine or glaciolacustrine setting, with some reworking by bottom currents.

(d) Massive diamict with a greater or lesser degree of sorting (e.g. weak grading), sometimes underlain by soft sediment scour marks - subaquatic slumping of unstable till deposited on a slope (e.g. continental slope).

(e) Massive breccia with angular stones, sparse fine material - supraglacial till derived from rockfall and deposited by melting of underlying ice.

(f) Laminated mud/silt with outsize stones and marine fossils - clastic marine sediments with ice-rafted dropstones mainly from calved icebergs.

(g) Rhythmites with regular sand/silt or clay couplets, occasionally with outsized stones - varves (lacustrine) or tidally controlled laminae (marine) with dropstones.

(h) Graded laminae (mud/sand) or beds (sand/gravel) of sporadic origin, for example from turbidity currents in lakes or the sea.

(i) Stratified sands, cobbles and boulders, moderately well sorted and with subrounded to rounded stones, often with trough cross-bedding, ripples, mud-drapes - recycled till, transported and redeposited by running water (glaciofluvial) usually in a subglacial, proglacial and subaquatic environment; typically braided stream faciesor subaquatic outwash; alternatively, sand and gravel horizons may represent lag deposits, resulting from removal of fines from glacial sediments in the marine environment.

(j) Stratified and non-stratified, well-sorted silts - aeolian deposits (loess) formed as a result of wind erosion and transport (deflation) of outwash plains and till-covered areas.

(k) Facies cutting across the bedding, such as diamict beds with upper reworked, sorted parts preserving wedge-shaped features, polygons, stripes and circles - periglacial phenomena generally indicating permafrost conditions.

(l) Any of the above facies with beds that have been down-faulted, folded, and locally depressed, and subsequently overlain by other sediments, are the result of differential melting of buried stagnant glacier ice.

1.3.4 Lithofacies coding

In descriptive facies analysis it is common to use shorthand notation, e.g. *Dm* for massive diamict. Eyles et al. (1983) established a formal lithofacies code in order to allow rapid description and visual appraisal of field sequences or drill cores containing diamictons or diamictites, from which environmental interpretation can then be undertaken. This lithofacies approach had already been applied to braided stream deposits by Miall (1977). Examples of its application are to the early Proterozoic Gowganda Formation of Ontario (Miall 1983) and to the Cenozoic Yakataga Formation of Alaska (Eyles & Lagoe 1990).

The approach of Eyles et al. (1983) was strongly criticised by Karrow (1984), Dreimanis (1984) and Kemmis & Hallberg (1984) since, on its own, it is too restrictive, and regards other modes of till study (such as fabric and granulometric compositional analyses) as of secondary importance. The code also adds an interpretative letter, which deflects from the objective nature of the approach. Furthermore, lateral variations have not received adequate attention.

Independently of Eyles et al. (1983), Fairchild & Hambrey (1984) used a more simple facies abbreviation for shorthand descriptive notation of glacigenic lithofacies. They preferred this to be informal, each study being expected necessarily to generate its own scheme. It was argued that lithofacies analysis alone was insufficient to establish the precise mode of deposition of all glacigenic sediment, and at least some of the other methods outlined in Section 1.4 need to be employed. A flexible approach, involving fresh appraisal of each sequence, allows lithofacies to be defined to suit the sequence concerned. Complex codes are useful as abbreviations for illustrative purposes, but tend to hinder the conveyance of information to the reader.

1.3.5 Facies associations and facies models

The next step to considering lithofacies individually is to examine how they relate to one another, namely the study of facies associations. Some facies are mutually exclusive.

For example the terrestrially deposited massive till may be associated with aeolian siltstones fluvial sands and gravels, but not with widespread rhythmites containing dropstones and fossils. The variety of facies represented in a facies association reflects the advance and recession of glaciers in terrestrial, lacustrine, intertidal or marine environments, or a combination of them.

Having considered the origin of facies and their spacial arrangement, one is then in a position to develop a pictorial representation of the palaeo-environment and the processes operating therein, that is to develop a sedimentary model. Such models have been developed for a range of glacial settings, some of which are discussed later. Models require a certain amount of generalization of the processes, since the main point of them is to examine how they apply to similar settings elsewhere.

1.3.6 Stratigraphic (or facies) architecture

Glacial geologists have long been concerned to establish the large-scale, three-dimensional geometry of glacigenic facies and their relationships to one another, especially in the ancient record. This approach is now used to define what in the last few years has become known, as stratigraphic or facies architecture. The concept was applied first to fluvial sequences, when it was realised that lithofacies logging provided only part of the environmental picture. Its application is particularly apt for thick, laterally extensive glaciomarine sequences, now aided significantly by the availability of seismic profiling across present-day continental shelves. Major advances have been made concerning the development of high latitude continental shelves under the influence of ice sheets by combining drilling and seismic surveying (see Cooper et al. 1991 and Hambrey et al. 1992 for reviews concerning Antarctica, and King in press with reference to northern latitudes) In contrast, most terrestrial successions are normally of limited areal extent, so architecture can be conceived on a small scale.

The large-scale three-dimensional architecture of sedimentary sequences reflects the organization of sedimentary environments in space and time. Each sequence demonstrates a range of interacting processes which, if repeated, will reproduce characteristic lithofacies associations.

Stratigraphic architecture of glacigenic sequences is a response to the interactions between four related phenomena: (a) the geometry of the crust (b) the spatial and temporal pattern of expansion and decay of the ice sheet, and its relation to the global glacioeustatic cycle; (c) isostatic response of the crust, and (d) patterns of ocean circulation. Stratigraphic architecture provides a framework whereby major global questions can be addressed, e.g. the response of sea level changes to glaciation. A stimulating review of such an approach has been provided by Boulton (1990).

1.3.7 Sequence stratigraphy

A recent development in the analysis of sedimentary basins has been the adoption of a concept that has been in use by the petrôleum industry since the 1970s, namely *seismic stratigraphy*. Companies have had to base their investigations mainly on seismic records in which the ages and character of the sediments are, as often as not, unknown, at least until exploratory wells have been sunk. *Sequence stratigraphy* has been a subsequent development of this approach and now provides a means of obtaining a high resolution time-stratigraphic framework in which to place lithological and sub-surface (well-log and reflection seismic) data. Proponents of the technique have argued that it allows one to undertake global correlations, although others have disputed these claims. Sequence stratigraphy was originally developed for interpreting marine seismic sections, but many articles applying the technique to onshore areas have now been published. However, sequence stratigraphy has not yet been applied to many glacial sequences, so there is much potential for understanding better the relationships of facies in space and time, and the link to sea level changes such as those determined globally by Haq et al. (1987).

Sequence stratigraphy is essentially the exercise of defining packages of strata that are bounded by unconformities. A *sequence* is defined as "a stratigraphic unit composed of a relatively conformable succession of genetically related strata. and bounded at its top and base by unconformities or the correlative conformities", the latter being called *sequence boundaries*. The development of sequence boundaries is related to eustatic sea level changes, and has allowed the development of global sea level curves, often referred to as "Vail-curves" after Peter Vail of Exxon who first developed the concept. These curves have been refined in subsequent papers, notably that of Haq et al. 1987). Developed principally for Mesozoic-Cenozoic sedimentary basins, the curves were explained in terms of changes in global ice volume, although for the Mesozoic Era at least the presence of ice sheets has not been substantiated. The reader is referred to Wilson (1991) for a helpful introduction to sequence stratigraphy.

1.4 Other methods of analysis of glacigenic sediments

1.4.1 Grain-size distribution

Diamictons of glacial origin have a broad range of grain sizes, with representatives in all classes, together with relatively subdued peaks or none at all. However, examination of the grain-size distribution is helpful in determining (a) the nature of the source material, (b) how it becomes modified during glacier flow and when it is deposited, and (c) the extent to which it is

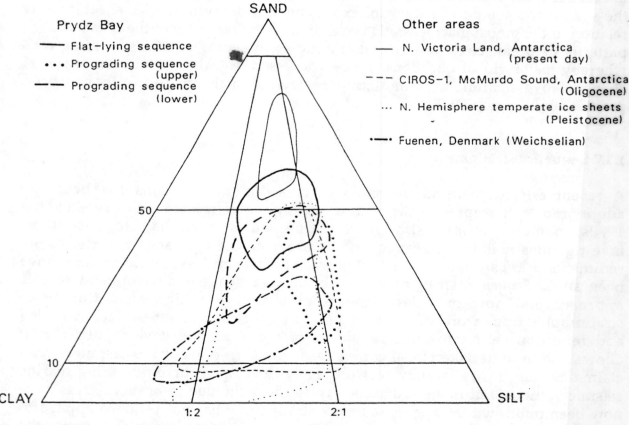

Fig. 1.3. *Ternary diagram to show the relevant proportions of sand, silt and clay in massive diamicts from different areas. Note that contemporary diamicts from Antarctica (Victoria Land) are sandy, whereas the Pleistocene examples from the Northern Hemisphere are more clay/silt-rich. The Prydz Bay and CIROS-1 samples are from older glacial sediments (mainly Oligocene) recovered in drill-holes on the continental shelf.*

modified following deposition. Examination of grain size can take place on all scales. For example, it is useful to estimate the proportion of material of gravel size (>2mm); this can be achieved visually using clast density charts. More common are analyses of the sand and finer fractions using a variety of methods ranging from basic sieving to the use of a sophisticated instrument like the SediGraph, which uses a laser beam to obtain the grain-size distribution.

Grain-size analysis has proved particularly useful in examining the processes that modify till after it has been deposited in the glaciomarine environment (e.g. Anderson & Molnia 1989, Barrett 1989a), where winnowing processes operate. With regard to terrestrial tills, individual sheets are remarkably uniform, but may vary greatly from one sheet to another (Sladen & Wrigley 1983). The regional variation between till sheets is related particularly to rock type and to the incorporation of pre-existing sediments by the glacier (Sladen & Wrigley 1983), or even to the thermal regime of the ice-mass (Barrett 1986). Figure 1.3 illustrates the grain-size distribution of various glacigenic sediments in a ternary diagram, including both terrestrial and continental shelf deposits from different climatic regimes.

1.4.2 Clast shape

A variety of parameters have been used for the analysis of particle shape in rocks (reviewed by Barrett 1980), so it is often difficult to compare one study with another. Perhaps the most useful for glacigenic sediments, because it links the sediments to the ice being transported, is the method presented Krumbein (1941) as applied by Boulton (1978). In this approach, Krumbein sphericity (based on ratios between the long, short and intermediate axes) is plotted against Powers roundness, estimated visually from shape charts. The shape charts also define fields of four different shapes: discs, spheres, blades and rods (Zingg shape) which also aid comparison of different glacial sediments. Boulton found distinct, but overlapping fields for supraglacial (rockfall debris), basal till and lodgement till in Spitsbregen and Iceland. This study has formed the basis for comparison with the glaciomarine environment of Antarctica (Domack et al. 1980, Kuhn et al.1993), and Baffin Island (Dowdeswell 1986), and with Precambrian glacial sediments from Svalbard (Dowdeswell et al. 1985) (Fig. 1.4). In most studies, fifty clasts have been measured at each site.

1.4.3 Clast fabric

Measurement of the orientation of the long axis of stones is one of the oldest and most widely used techniques employed in the investigation of diamicts, and has been extended to the smaller grains that are the only ones in sufficient numbers in cores. By comparison with glacial striations or flutes, it has been shown that diamicts commonly display a preferred orientation parallel to ice flow, hence they are useful in establishing regional patterns of ice movement. However, transverse orientations are also possible, such as in an end moraine, or the fabric may be disrupted by post-depositional flowage of the sediment.

According to the level of sophistication required, one of two methods may be used. The simplest is to measure the long axis orientation projected onto the horizontal plane. Data are grouped into convenient classes, e.g. 10°, 20°, 30° and plotted as a rose diagram. A simple statistical check using the Chi-squared test will establish the strength of the fabric at various confidence levels. This method is especially useful in material (e.g. cores) that does not lend itself to disaggregation but only permits one to take measurements on an exposed horizontal surface.

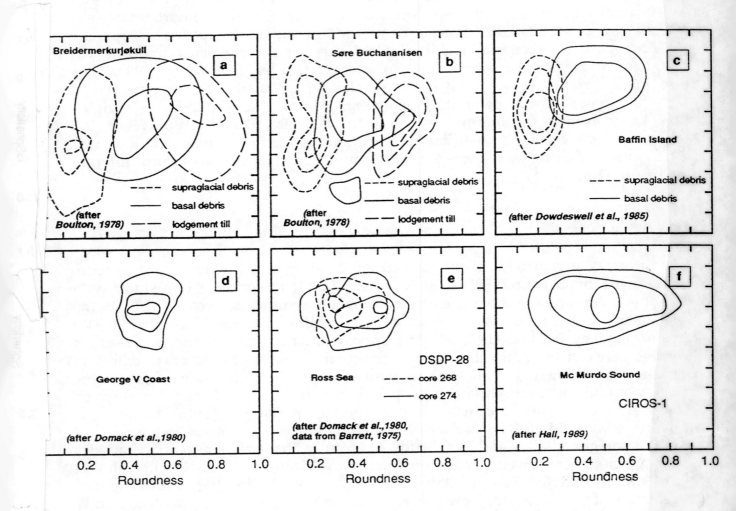

Fig. 1.4. Roundness - sphericity plots for Antarctic continental shelf clasts, compared with those denoting specific glacial transport paths from Boulton (1978). The Icelandic glacier is temperate; the Svalbard glacier is slightly cold and a little more analogous to the Antarctic glaciers. (From Kuhn et al. 1993; data from various sources).

The second method, which is preferable, is to measure the plunge of the stone axis as well as its orientation, to give the three dimensional view (Fig. 1.5). As with structural geological data, clast fabric data may be plotted on a stereographic projection, usually a Schmidt/Lambert equal area net (Phillips 1971). In this way it is possible to see whether there is an upglacier preferred orientation (the norm) or otherwise. Some recent studies (e.g. Mills 1977, Lawson 1979, Domack & Lawson 1985, Dowdeswell et al. 1985, Dowdeswell & Sharp 1986) have in addition undertaken an eigenvector/eigenvalue analysis to examine statistically the strength and value of the preferred orientations, for which computer programs are available. By plotting eigenvalues graphically, in combination with thestereographic projections it has proved possible to discriminate between basal melt-out till, lodgement till, sediment flows (flow tills) and waterlain tills, in order of declining fabric strength. It has even been possible to distinguish the variable effects of shearing within lodgement till, and between basal tills from highland and lowland locations.

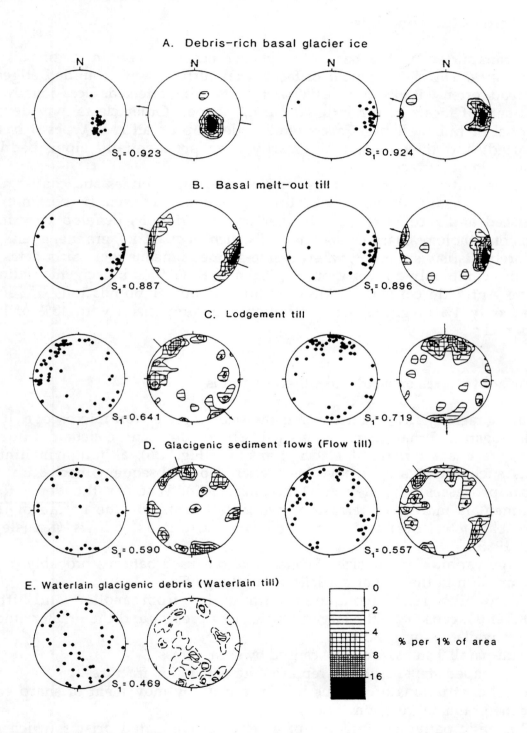

Fig. 1.5. Three-dimensional clast-orientation data from basal ice and a variety of glacigenic sediments, plotted on the lower hemisphere of a Lambert/Schmidt equal-area projection. The contoured and shaded diagrams represent the number of points per 1% of area. A, B and D are from Matanuska Glacier, Alaska; C from Skalafellsjökull, Iceland; E from Catfish Greek 'Till', Ontario (from Dowdeswell et al. 1985; from various sources).

There is little agreement on the number of clasts it is necessary to measure to obtain a valid preferred orientation. Lawson measured only 25, Mills 50, others as many as 300. The author has found that reproducible results are possible with 50 measurements, even where there is only a weak or low preferred orientation.

1.4.4 Surface features on clasts

Debris transported at the base of a glacier acquires certain distinguishing characteristics. The development of facets (flat surfaces with rounded edges) is widespread, often with as many as 80% of clasts being affected. Occasionally two parallel sets of facets give rise to "flat-iron" shapes. Other clasts may develop pentagonal or bullet shapes. Facets tend to develop on all rock types if basally transported, and they do not necessarily form preferentially along bedding, foliation or joint surfaces.

Striations and associated features, such as crescentic gouges and chattermarks develop on basally transported stones; they are especially common on subrounded and faceted clasts. Whether or not striations develop very much depends on lithology. Hard crystalline rocks such as quartzite, granite, gneiss and schist rarely display striations, whereas fine-grained igneous rock, carbonates and mudstones commonly do. For example, Kuhn et al. (1993) in glacigenic sediment from the Antarctic continental shelf found that out of populations of several hundred only 4% of gneisses had striations, in comparison with 43% of basic igneous rocks.

1.4.5 Surface features on sand and silt-sized grains

These are generally investigated using the scanning electron microscope. It has been demonstrated that quartz grains with sharp edges and conchoidal fracture patterns are characteristic of glacial transport (Fig. 1.6), although in lithified deposits such textures may have been altered by subsequent mechanical and chemical processes (Krinsley & Dornkamp 1973). Whether the depositional environment is marine or terrestrial has little bearing on the nature of these fracture characteristics, which may be summarized as follows (Krinsley & Funnell 1965):

(a) Large variation in the size of conchoidal breakage patterns probably caused by the variation in the size of particles ground together.

(b) Very high relief (compared with grains from aeolian and littoral environments), caused by the large particle sizes present and the greater amount of energy available for grinding.

(c) Semi-parallel steps, probably caused by shear-stress.

(d) Arc-shaped steps, probably representing percussion fractures.

(e) Parallel striations of various length, caused by movement of sharp edges against the grains in question.

(f) Prismatic patterns consisting of a series of elongated prisms which may represent cleavage and including a very fine grained background that may indicate recrystallization.

(g) Imbricated breakage blocks which look like a series of steeply dipping hogbacks

24

(h) Small scale grinding indentation - irregular markings which frequently appear with conchoidal blocks.

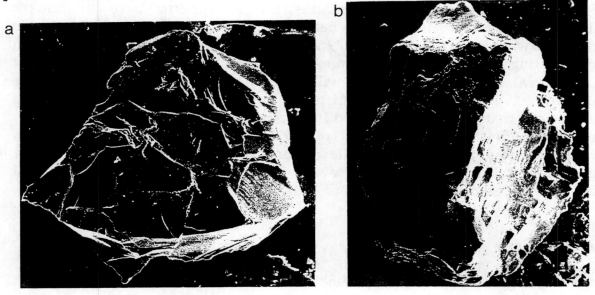

Fig. 1.6. Scanning electron microscope photographs of quartz grains from glacier ice at the margin of Grinnel Ice Cap, Baffin Island. (a) Sample from clean ice (supraglacial or englacial transport. (b) Sample from basal debris-rich ice. The longest axes in both cases meAsure approximately 0.8 mm. Photographs courtesy of J. A. Dowdeswell.

In general, these criteria are related to processes of slow intergranular attrition under stress. However, interpretation of these textures requires care. A detailed study of a contemporary glacial environment by Whalley & Kingsley (1974) indicated that no surface texture was present that could characterise any particular glacial sub-environment. The general variability of grain surfaces from whatever position in the glacier they were taken, made it impossible to establish the source and transport mechanism of a particular grain. Freshly weathered supraglacial debris, derived by rockfall, had many of the characteristics found in subglacially transported grains. Other processes, unrelated to glacial transport, can give rise to intergranular attrition, e.g. mudflows (Harland et al. 1966). Textures of quartz grains in till are also dependent on their nature in the parent rock, i.e. inherited characteristics (Whalley and Kingsley 1974). With regard to pre-Pleistocene tillites the usefulness of the technique is further limited as a result of surface texture modification during diagenesis, and few SEM studies of such rocks have proved useful.

Microscopic studies on garnet grains have proved to be of some use in providing evidence of glacial transport. Folk (1975) first recognised trails of crescentic marks on a garnet grain in till. These trails are of parallel orientation, are uniformly spaced and of similar size. On any one surface there may be several trails differing in width, length and orientation. They cannot be the result of random impact nor due to an orientation effect of the internal atomic lattice. The crescentic marks on garnets resemble other glacial chattermarks (although four orders of magnitude smaller); the latter are caused by the rhythmic release of strain by fracturing as a glacier moves over bedrock. It is thus thought that chattermark trails form on garnets when they are held fast in glacier ice and slowly grind past other grains or bedrock under great stress (Folk 1975). Chattermark trails have not been observed on other mineral grains; possibly the reason is that garnets are hard, while other grains or bedrock (including quartz) are subject to crushing.

Some 15% of garnets in Pleistocene tills in North America have chattermark trails, whereas as many as 33% have been documented in Late Palaeozoic tillites from Gondwanaland (Gravenor 1979). Chattermark trails are thus useful in discriminating between glacial and non-glacial diamicts. It has been found that the statistical chance that garnets will encounter the right conditions to be chattermarked increases with the distance travelled in englacial transport, a characteristic that has further assisted in interpreting the distribution and extent of ice in pre-Pleistocene sequences (Folk 1975, Gravenor 1979, 1980). Chemical etching can also produce crescentic features which might be mistaken for chattermark trails, but if there are several crescentic features in line, chemical action is considered unlikely (Gravenor 1981).

1.4.6 Mineralogy

The mineralogy of most clastic sediments tends to reflect the degree to which they have been transported and reworked. For example, the earlier-formed, high temperature minerals in igneous and high grade metamorphic rocks tend to be unstable at near surface temperatures and pressures. The order of stability, starting with the most stable, is as follows:

quartz, zircon, tourmaline
chert
muscovite
microline
orthoclase
plagioclase
hornblende, biotite
pyroxene
olivine

Although chemical weathering is now recognized as an important process in the ice-free areas of the Antarctic and Arctic, and may well have been instrumental in preparing the bedrock for glacial erosion , once ice is eroding and transporting fresh bedrock, this trend of mineral stability is no longer revealed. Chemical weathering , although it does occur at the base of a glacier, is of minor significance except in areas of carbonate bedrock. However, abrasion and grinding continually produces fresh rock which, if cemented in the ice-mass, is effectively preserved from subsequent chemical weathering. Supraglacial debris may be subjected to little more than frost weathering, and again the breakdown of even the most unstable minerals is slow. Till deposited after transport over long distances therefore contains a proportion of easily identifiable, fresh-looking feldspar grains, for example, whereas in other environments transport would lead to their being rapidly rapid rounded and chemically altered.

Clay mineral assemblages in tills, obtained by X-ray diffraction techniques, are also distinctive; clay size fractions with illite and chlorite are typical of glacial environments (Alley and Slatt 1976), unless containing a high proportion of preglacially weathered material. Analysis of the clay content of glacigenic sediments allows the proportions of the major clay minerals to be estimated, which in turn allows an assessment of the contribution of weathered *versus* unweathered material to be made. Such data have proved to be of particular value in deep-sea sediments bordering the Antarctic ice sheet, where the glacial component may not be immediately obvious (eg Ehrmann & Mackensen 1992). For example, the onset of continental East Antarctic glaciation around 36 m.y. ago is indicated by the change from smectite-dominated to illite- and chlorite-dominated assemblages, the latter indicating physical weathering under a cooler climate.

An exception to the absence of chemical alteration is where carbonate is present in the system. Work on ultra-thin sections on Precambrian tillites (Fairchild 1983 has shown that carbonate ground into a fine rock flour and deposited in a marine environment is prone to rapid recrystallization.

1.4.7 Geochemistry

Numerous analyses of whole-rock geochemistry of glacigenic sediments have been undertaken, but with rather unhelpful results - till typically has the geochemical character of a greywacke. However, one successful approach has been the application of major element chemistry to glacigenic and associated sediments in order to determine the pattern of climatic change. This approach was developed by Nesbitt & Young (1982) and applied to the diamictite-bearing Early Proterozoic Huronian Supergroup in Ontario. These authors devised a Chemical Index of Alteration (abbreviated to CIA with America's counter-espionage agency in mind!) which takes account of the fact that during chemical weathering feldspars degrade into clay minerals. Ca, Na and K are removed from the feldspars by soil solutions, so that the proportion of alumina and alkalis typically increases in the weathered product. The degree of weathering can therefore be quantified using molecular proportions:

$$CIA = [Al_2O_3/(Al_2O_3 + CaO^* + Na_2O + K_2O)] \times 100$$

where CaO^* is the amount of CaO incorporated in the silicate fraction of the rock. This allows one to contrast glacial and non-glacial sediments in the same basin.

The geochemical character of carbonates in glacigenic sediments can also be revealing. Five distinct types of carbonate have been recognized in glacial sediments by Fairchild et al. (1989) and Fairchild & Spiro (1990): precipitation from sea or lake water, with or without microbial action; from rocks ground up by glaciers; from rapid recrystallization of this "rock flour"; and from groundwater near saline lakes such as those in arid regions of Antarctica. These authors further demonstrated that proportions of different isotopes of oxygen (^{18}O, ^{16}O) provide an indication of palaeolatitude. This is based on the premise that present-day precipitation has increasingly lower proportions of the heavy isotope towards the poles.

In deep-sea sediments with or without a glacial component, in which the stratigraphic record is complete, the ratio between the heavy and light isotopes of oxygen in planktonic foraminifera may yield important climatic information that can be linked to global palaeotemperature and ice volume changes (Shackleton & Kennet 1975, Miller et al. 1987).

1.4.8 Palaeomagnetism

The magnetic characteristics of diamicts provide useful additional information concerning depositional processes (Eyles & Menzies 1983). In lodgement tills the magnetic particles are poorly aligned with respect to the earth's magnetic field as a result of shear dispersion during deposition. In comparison, diamicts formed as a result of deposition in the sea or a lake show a magnetic alignment within Earth's magnetic field with respect to azimuth and inclination. Magnetic orientation measurements are usually undertaken on small cubes or cylinders placed in a magnetometer in the laboratory.

With appropriate equipment, the magnetic fabric may be measured in the field, thus allowing ice flow directions to be determined rapidly without the time consuming work of clast fabric analysis. Palaeomagnetism is also useful in a stratigraphic context, since particular till units may have a distinctive fingerprint. For sequences that embrace reversals in Earth's magnetic field, palaeomagnetism may be used to provide temporal control on glacigenic sediments. However, this is of limited use in terrestrial sequences as deposition tends to be sporadic. Far better use may be made in this respect in deep-sea sequences.

1.4.9 Geotechnical properties

A variety of techniques used by engineers to assess the stability of glacigenic sediments may be mentioned briefly. These techniques allow certain geotechnical parameters to be determined and related to the geological processes responsible for them. The parameters include liquid limit, plastic limit, natural moisture content, shear strength and compressibility. The use of these parameters enables one to discriminate between, for example, different types of massive diamicts since a sheared lodgement till from which water has been squeezed will be tougher than a waterlain till. For a useful discussion of these parameters, the reader is referred to the review by Sladen & Wrigley (1983).

a) Evidence for terrestrial glaciation

Abraded surfaces
    Striated and/or polished surfaces
    Crescentic gouges; chattermarks
    Striated boulder pavements

Clast-rich beds with:
    Irregular thickness (usually, c. 50m)
    Lenses of sand/gravel (glaciofluvial)
    Depositional shear structures in massive diamict; otherwise structureless diamict
    Preferred clast orientation
Depositional fossil landforms, e.g. moraines, eskers

b) Evidence of glaciomarine/glaciolacustrine deposition

Massive to stratified beds, often 10's or 100's of metres thick, with gradational boundaries
Dropstones in stratified units
Random clast fabric
Slight sorting or winnowing at top of beds
Association with fossils
Association with rhythmites of varve or turbidite origin
Association with resedimented deposits (debris flows)

c) Evidence common to both environments

Variable clast lithologies
Poorly or non-sorted with wide range of clast sizes
Exotic (far-travelled) varieties of clasts
Fresh minerals
Constant mix of clasts over wide area common
Clast characteristics:
    Shape variable from angular to rounded
    Some striated and faceted surfaces
    Flat-iron/bullet-nosed shapes
    Calcareous crusts
    Fragile clasts
    Quartz grain textures; chattermarks on garnet grains

d) Other evidence of cold climate

Ice wedge casts
Fossil sorted stone circles, polygons and stripes
Fossil solifluction lobes
Association with lithified loess (loessite)

Table 1.4. principal criteria for establishing the glacial origin of diamict successions

1.5 Evidence of glaciation in the geological record

The recognition of glacigenic sediments in the rock record is of fundamental importance to palaeoclimatology and the reconstruction of palaeoenvironments. Their occurrence suggests harsher climatic conditions than the norm and, if their sedimentary characteristics are determined, it is possible to determine the presence, nature and extent of land, the direction of sediment transport and the characteristics of the depositional environment. As we have seen, till is an extremely varied material, but glacial environments are characterised by a variety of facies unique in the geological record. Environments of till deposition are varied, ranging from mountainous, to lowland, lacustrine, tidal and marine. Few sedimentary criteria in themselves are sufficient to allow one to infer a glacial origin for a particular deposit; however, the association of several distinctive features can indicate not only a glacial origin but whether it is marine or terrestrial (Table 1.4).

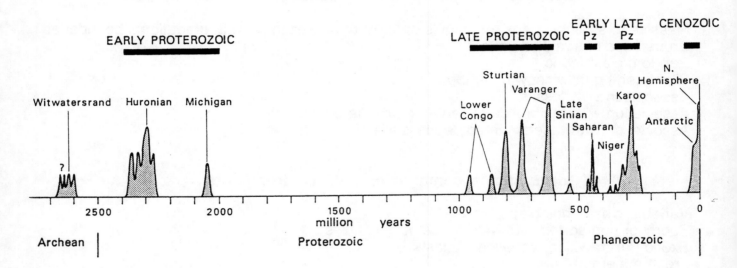

Fig. 1.7. Time plot in millions of years showing periods of glaciation and their names. A qualitative measure of the extent of Earth's surface affected is indicated by the heights of the peaks (after Hambrey 1992).

The recurrence of glacigenic sediments in the geological record suggests that ice ages are not as unusual as they have sometimes been made out to be (Fig 1.7: Table on inside front cover). Major ice ages are represented in Early Proterozoic and Neoproterozoic, Early and Late Palaeozoic and Cenozoic strata, while other geological intervals have not been devoid of glacial activity. Nevertheless, in terms of the bulk thickness of sediments, in the context of Earth's entire history, glacigenic sediments form but a small part, although their stratigraphic value is great, especially in non-fossiliferous rock. Notable are the wide-ranging correlations made possible using tillites as stratigraphic markers for Neoproterozoic time.

Because of their widespread nature, Cenozoic and especially Quaternary glacigenic sediments have attracted by far the greatest attention. Glacigenic sediments in various guises cover 8% of the Earth's land surface, including a third of Europe and a quarter of North America (Flint 1971). These sediments have created entirely new landscapes and have provided mineral-rich soils that have influenced vegetation, land use and the pattern of human settlement. In addition they have provided the bulk of the sand and gravel needed by the construction industry in much of Europe and North America.

Thus, the main focus has been on terrestrial sediments which are generally less than a few tens of metres in thickness, but in favoured locations, such as where deep basins have become filled, thicknesses may approach several hundred metres. By far the thickest accumulations of glacigenic sediment occur in marine areas, especially at the edges of continental shelves that bordered, or still border, the great ice sheets. Some examples of thicknesses of Cenozoic glacigenic sedimentary associations are given in Table 1.5.

Continent	Area	Country	Thickness (m)
North America	Great Lakes	USA	12
	Illinois		35
	Iowa		46-66
	Central Ohio		29 average 232 maximum
	New Hampshire		10 average 122 maximum
	Spokane Valley (Idaho/Washington)		335-442
	Fraser Delta		670
	Gulf of Alaska		5000
Europe	N. Germany	Germany	58 average
	Norrland	Sweden	4-7
	Denmark	Denmark	50
	Lubbendorf in Mecklenburg	Germany	470
	Grenoble	France	~400
	Heidelberg	Germany	397
	Imola, Po Valley	Italy	800
	East Anglia	UK	143
	Isle of Man	UK	175
	North Sea	-	920
Antarctica	McMurdo Sound	-	702 (min.)
	Prydz Bay	-	480 (min.)

Table 1.5.Typical thicknesses of drift sheets in lowland areas and continental shelves (from Flint 1971, and various other sources).

1.6 Extent of glacier ice today

The distribution and extent of ice on the Earth's surface are known principally from the work of the World Glacier Monitoring Service which has compiled a global inventory (in varying degrees of detail) of essentially all the relevant glacierized regions of the world (Table 1.6).

Continent	Region	Area (km^2)	Totals
South America	Tierra del Fuego/Patagonia	21,200	
	Argentina north of 47.5°S	1,385	
	Chile north of 46°S	743	
	Bolivia	566	
	Peru	1,780	
	Ecuador	120	
	Columbia	111	
	Venezuela	3	**25,908**
North America	Mexico	11	
	USA (including Alaska)	75,283	
	Canada	200,806	
	Greenland	1,726,400	**2,002,500**
Africa			**10**
Europe	Iceland	11,260	
	Svalbard	36,612	
	Scandinavia (incl. Jan Mayen)	3,174	
	Alps	2,909	
	Pyrenees/Mediterranean Mts.	12	**53,967**
Asia (with all Russia)	Commonwealth of Independant States	77,223	
	Turkey/Iran/Afghanistan	4,000	
	Pakistan/India	40,000	
	Nepal/Bhutan	7,500	
	China	56,481	
	Indonesia	7	**185,211**
Australasia	New Zealand	860	**860**
Antarctica	Subantarctic islands	7,000	
	Antarctic continent	13,586,310	**13,593,310**
GLOBAL TOTAL			**15,861,766**

Table 1.6. Distribution of glacierized areas of the world (from World Glacier Monitoring Service 1989).

The large ice sheets of Antarctica (85.7%) and Greenland (10.9%) together represent 96.6% of the world's total glacierized area (13,586,310 km^2). Of the remaining 3.4% (c. 550,00 km^2), about two-thirds comprise high latitude ice caps and ice-fields, and one third mountain glaciers. However, it is the latter which have impinged most directly on human activity, as a result of avalanches, debris-flows and outbursts of water, as well as, more positively, in providing water for hydroelectricity and irrigation.

The principal concern today is not from local glacier hazards, but whether human- induced greenhouse warming of the atmosphere will lead to melting of the polar ice sheets. The accuracy of the figures for area of the Antarctic ice sheet, given in Table 1.6, is less than that for the whole of Earth's other ice-masses, but the acquisition of more and better satellite data is gradually improving the picture. Current estimates suggest 80% of the world's fresh water is glacier ice, of which the greater part (30 million m^3) is in Antarctica.

If the Antarctic ice sheet were to melt totally, the sea level would rise 80 m (Drewry 1991). Melting of the Greenland ice sheet would add a further 7m. The contribution of the remaining glaciers would be a mere fraction of this.

1.7 Geological timescale

Several comprehensive geological timescales have been published in recent years, all of which differ in terms of the ages of period, epoch and stage boundaries. Of most relevant to this book is the dispute concerning the onset of the Quaternary Period. Internationally, the lower boundary is placed at 1.8 m.y., but most Europeans prefer one at 2.4 m.y., which corresponds to major cooling in the North Atlantic and the onset of ice-rafting. Within the Quaternary Period we have the Pleistocene and Holocene epochs, the latter beginning about 10,000 y. BP; these too have been much debated. In the earlier ice age record, the Proterozoic/Cambrian boundary has been of some significance, with datesof around 570 to 610 m. y. traditionally being assigned. Now this boundary has been placed much earlier at about 530 m.y, well clear of the Neoproterozoic glacial events.

For the purposes of this book, the International Union of Geological Sciences timescale of 1989 is used as far as possible (see inside front cover), although maintaining some level of consistency in discussions of older glaciations has not always been possible.

CHAPTER 2

GLACIER DYNAMICS

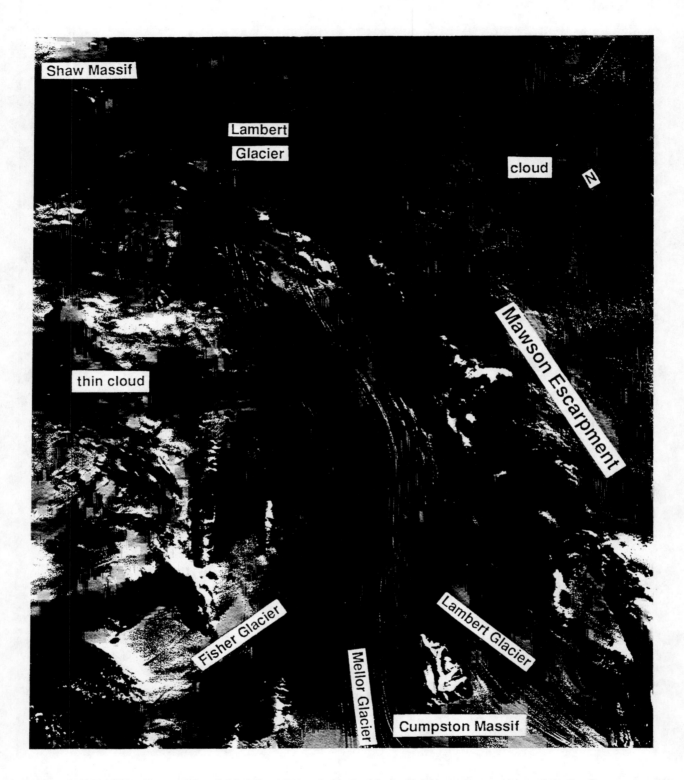

Fig. 2.1. Landsat image, dated 20 Feb. 1974, of the Lambert Glacier system, Antarctica. In the centre of the image is the strongly channelized, fast-flowing ice of three major flow units, the Fisher, Mellor and Lambert glaciers. Towards the top of the picture the Lambert Glacier passes into the floating Amery Ice Shelf, while to the left and right is relatively slow moving and thinner ice of the main ice sheet. The Lambert Glacier drains about a fifth of the East Antarctic ice sheet, and is the biggest valley glacier in the world, measuring 50 km across. The linear structures are medial moraines and longitudinal foliation (cf. Fig. 2.53). The distance from the top of the image to the bottom is c. 200 km. Image courtesy of the U.S. Geological Survey, Flagstaff, and previously published in Hambrey (1991).

2.1 Introduction

Understanding some of the complexities of glacier behaviour is essential in interpreting the origin of landforms, or sequences of glacigenic sediments, on the land surface or preserved in the geological record. In the past, many authors describing Pleistocene successions would have interpreted a till horizon within a multi-layered till sequence as representing a glacier advance during a well defined cold phase; the truth generally is more complex. For example, in a glacial environment, till is intimately associated with deposits that once would have been labelled interglacial. Furthermore, glacier advances are not necessarily synchronous nor related to climatic change in the short term. Indeed, some glaciers deposit most of their debris load during surges - catastrophic events related to dynamic instability of the ice - rather than to climatic change. Glacier dynamics have been reviewed in books by Paterson (1981) and Souchez & Lorrain (1991). Here, only those aspects that bear on landforms and sediments are considered here.

2.2 Formation and properties of glacier ice

Snow and ice are crystals with an hexagonal symmetry but otherwise occur in a wide variety of forms.

2.2.1 Derivation of glacier ice

Glacier ice is derived from (a) recrystallization of snow during diagenesis, (b) melting snow and refreezing at the surface to give superimposed ice, (c) freezing of rain water, and (d) condensation and freezing of saturated air to produce rime. In temperate regions, recrystallization of snow at the pressure melting point is the dominant source of glacier ice, whereas in north polar regions, where accumulation of snow is slight and summer melting occurs, superimposed ice (refrozen slush), is the main constituent. In the coldest and driest polar regions, notably the interior of Antarctica, no meltwater may be involved in diagenesis. Overall (c) and (d) are of little significance except locally.

2.2.2 Density of glacier ice

During the diagenesis of snow, its density increases progressively. Freshly fallen snow may have a density of as little as 0.05 gm cm^{-3}, snow that has survived one summer season (*firn*) 0.4 gm/cm^3, true glacier ice 0.83 to 0.91gm cm^{-3}, and pure ice (frozen water) 0.92 gm cm^{-3}. As the loosely packed, randomly orientated ice crystals with intervening air spaces in firn recrystallize in response to stress, the intercrystal air spaces are eliminated, to be replaced by crystals containing bubbles of air (Fig. 2.2). The density of ice, together with the size of ice crystals, thus increases with time and depth. In the interior of the Antarctic ice sheet this is a very slow process, and it may take thousands of years combined with burial under hundreds of metres of snow to achieve the same effects as just a few years and a few metres of burial in temperate regions.

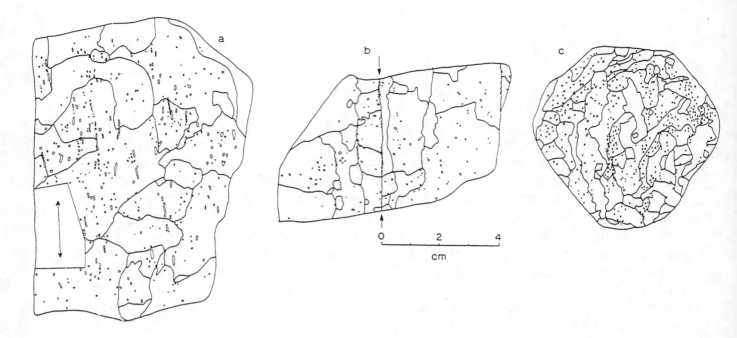

Fig. 2.2. Drawings of thin sections of foliated glacier ice showing irregular   shapes and interlocking nature of ice crystals and the relationship to entrapped air bubbles. (a) and (b) are from Charles Rabots Bre, northern Norway;   (c) is from the White Glacier, Axel Heiberg Island, Canadian Arctic. The arrows indicate the orientation of foliation in the ice.

2.2.3 Crystallography of glacier ice

Glacier ice is analogous to deformed rocks, although it is mono-mineralic. Recrystallization is indicated by changes in the orientation of the optic- or c-axes. However, the crystal shape is often very irregular, with an interlocking character (Fig. 2.2). In glaciers of temperate regions crystals may reach 25 cm in size. The c-axis is that on which a ray of incident-parallel light is transmitted normally. c-axes are measured using a universal stage or by examining the basal planes perpendicular to the c-axis, which are picked out when bubbly ice weathers. Glacier ice often shows a preferred orientation related either to stress or to the cumulative strain axes, so it is continually modified as it moves down glacier. Single maxima may develop perpendicular to the plane of shear when undergoing simple shear, but the pattern is often more complex, for example with 3 or 4 maxima clustered around the normal to the shear plane. Often c-axis preferred orientations are related to deformational structures, especially the layered structure called foliation (section 2.8.2). In zones of strong shear, large crystals break down into smaller crystals and generate a new foliation.

2.2.4 Deformation of a single ice crystal

The deformation of a single crystal has been studied experimentally by applying a constant stress to a single crystal orientated so that there is a component of shear-stress in the basal plane, and measuring strain as a function of temperature. When stress is first applied, ice immediately deforms elastically, followed by permanent deformation (*creep*), which continues as long as the stress is maintained. Deformation is thus possible under very low stresses, and it takes places in layers parallel to the basal planes of the crystal, a phenomenon called *basal glide*. Crystals can also deform if the basal plane is not favourably orientated, but then the applied stress needs to be much higher.

2.2.5 Deformation of polycrystalline ice

Under constant stress, the strain-rate initially increases with time, a characteristic known as *strain-softening*, but in polycrystalline ice many crystals are orientated so that they do not slip on the basal plane and they harden as the strain increases. If a constant stress is maintained until the total strain reaches a few per cent, the strain-rate may reach a steady value proportional to a power, n (typically 1.5 to 4), of the stress.

If the applied stress is great enough, polycrystalline ice effects undergo elastic deformation, followed by transient creep (strain-rate decreasing continually, i.e. strain-hardening), then constant strain-rate (secondary creep), then tertiary creep with the strain-rate increasing. Overall, higher stresses are needed to induce the same effect as in a single crystal, since most crystals are not orientated for basal glide in the direction of the applied stress.

Deformation in polycrystalline ice is accomplished by:
- movement of dislocations within crystals
- movement of crystals relative to one another
- crystal growth
- migration of crystal boundaries
- recrystallization

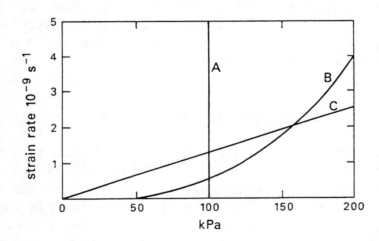

Fig. 2.3. *Graph showing the relationship between shear strain-rate and shear stress in deformable substances. Curve A represents Newtonian viscous flow, line B perfectly plastic material with a yield stress of 100 kPa or 1 bar, curve C is for glacier ice, defining Glen's flow law (after Paterson 1981).*

2.2.6 Flow law for polycrystalline ice

For glacier ice, a simple flow law has been derived which is widely used (Fig. 2.3). Many laboratory experiments have demonstrated the relationship between shear-strain-rate $\dot{e}_{xy}$ and shear-stress τ_{xy} over the range of stresses that are typical in normal glacier flow, that is up to 200 kPa (i.e 2 x atmospheric pressure or 2 bars). A generalized flow law (known as Glen's Law) that is applicable to glaciers

39

relates the effective shear-strain-rate $\dot{e}$ to effective shear-stress τ in the following manner (Nye 1957):

$$\dot{e} = A\tau^n$$

where n is a constant (discussed above), A depends on the ice temperature, crystal size and orientation, and impurities. The flow law is now reasonably well established, but widely differing values of A and n have been obtained in different experiments (e.g. 1.5 to 4 for n, with a mean of 3).

It is known that disseminated debris affects the way in which ice deforms, but in a manner that is not well understood. If the volume of sand in ice exceeds 10%, the creep-rate decreases with an increasing sand content. Below 10% there is no clear relationship (Hooke et al. 1972). Structural evidence sometimes supports this experimentally derived observation; in a debris-rich/debris-poor layered sequence, the debris-rich layer behaves as the more rigid (competent) material, creating structures called *boudins* (see Section 2.8.2).

Ice undergoes deformation for hundreds or thousands of years. Large total strains of 100% or more are encountered. The final crystal structure may reflect this history and not just the most recent stresses.

Deformation of boreholes and tunnels, and the spreading of an ice shelf under its own weight, broadly support the flow law. Instead of using the true flow law, some authors have assumed , for simplicity, that ice behaves as a pefectly plastic material. Such material does not start to deform until the stress reaches a critical value, known as the yield stress. Then the strain-rate becomes very large. Stresses in glacier ice often approach 200 kPa, which approximates to perfectly plastic behaviour with a yield stress of *c*. 200 kPa.

2.3 Morphology of glaciers

Glaciers range in size from tiny ice-masses only a few hundred metres across to the huge ice sheet of Antarctica which covers 13.6 million km^2. The global distribution of glacier ice is summarized in Table 1.7. Various names are applied to the wide range of glacier types (Table 2.1). At the largest end of the range is the Antarctic ice sheet, a composite of three dome-like ice sheets (Table 2.2). The bulk of the ice lies over East Antarctica, which is an elevated landmass divided by sub-sea level basins bearing ice up to 4776 m thick (Fig. 2.4). The mountainous backbone of the Transantarctic Mountains separates this ice sheet from the West Antarctic ice sheet which fills a marine basin, reaching a maximum thickness of over 4000 m and a depth below sea level of 2555 m. Both the East and West Antarctic ice sheets are divisible into several drainage basins (Fig. 2.4). Dynamically separate is the much smaller ice sheet that covers the mountainous spine of the Antarctic Peninsula. The combined total of Antarctic ice accounts for 91% of the world's freshwater ice and 85% of its freshwater. Huge floating slabs of ice, called ice shelves, buttress the East and West Antarctic ice sheets, notably the Ross and Filchner--Ronne ice shelves. There are also many other smaller ice shelves in Antarctica, but the Arctic only has small examples, in northern Ellesmere Island and Severnaya Zemlya. There appears to be a climatic control on the formation of ice shelves, the climate in the north being barely severe enough.

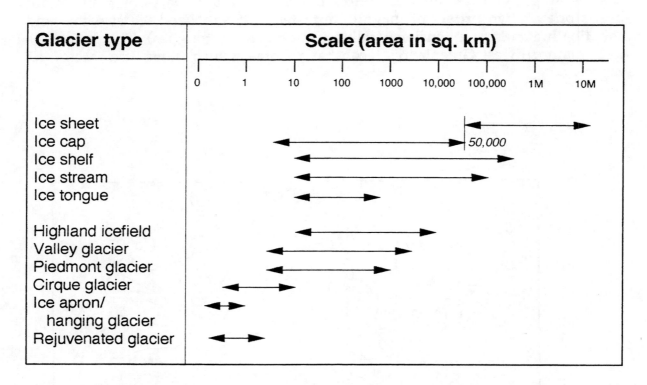

Glacier type	Scale (area in sq. km)

Ice sheet
Ice cap
Ice shelf
Ice stream
Ice tongue

Highland icefield
Valley glacier
Piedmont glacier
Cirque glacier
Ice apron/
    hanging glacier
Rejuvenated glacier

Table 2.1. Classification of glaciers according to their size (given in terms of area), shape, and relationship with the surrounding and underlying topography.

Region	Area (km²)	Volume (km³)	Average thickness (m)
East Antarctica	10,353,800	26,039,200	2,565
West Antarctica	1,974,140	3,262,000	1,700
Antarctic Peninsula	521,780	227,100	510
Ross Ice Shelf	536,070	229,600	430
Filchner-Ronne Ice Shelf	532,200	351,900	660
TOTALS	**13,918,070**	**30,109,800**	

Table 2.2. Dimensions of the various components of the Antarctic ice sheet (after Drewry 1983).

The only other ice sheet is that which covers most of Greenland (area: 1.7 million km²). This too is dome-shaped, filling a basin rimmed by coastal mountains to a depth of more than 3000 m, from which ice flows in many places as outlet valley glaciers (Fig. 2.5). The Greenland ice sheet accounts for 8% of the world's freshwater ice. The ice shelves and ice streams of Antarctica and Greenland presently produce the majority of icebergs and are responsible for the bulk of the ice-rafted debris that enters the oceans. Whereas ice sheets are largely slow moving (typically tens of metres per year), certain parts of them attain speeds one or two orders of magnitude faster in channelized ice streams, and their floating marine extensions (called ice (or glacier) tongues). The fastest ice stream is northwest Greenland's Jacobshavn Isbrae, which flows at a rate of 4.7 km yr⁻¹.

(a)

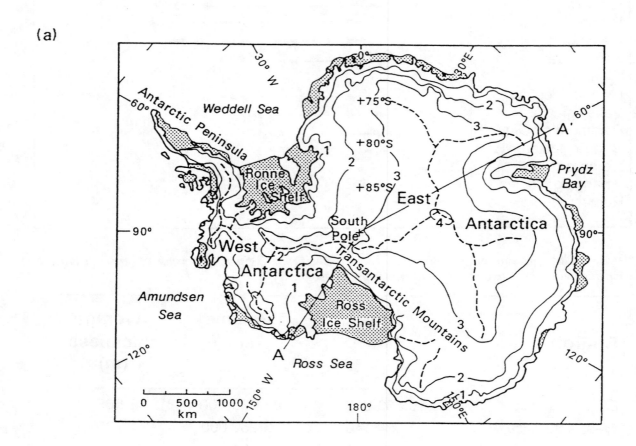

(b)

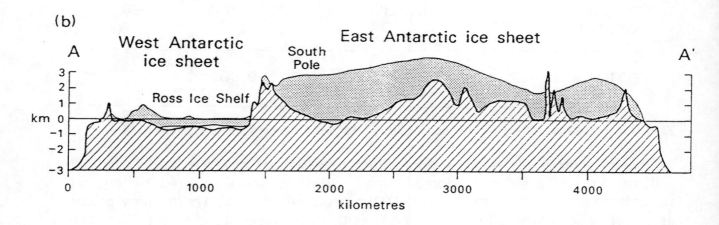

Fig. 2.4. *The Antarctic ice sheet: (a) ice drainage basins (after Drewry 1983); (b) cross-section drawn from ice surface and subsurface maps in Drewry (1983).*

Fig. 2.5. *Vestfjordgletscher, situated at the head of Scoresby Sund, East Greenland is one of the major outlet glaciers of the Greenland ice sheet seen in the background of this photograph.*

Fig. 2.6. *The largest valley glacier in the European Alps, the Grosser Aletschgletscher flows 22 km from the 4000 m peaks of the Jungfrau and Mönch in the background. Note the small ice-dammed lake of Märjelensee at the bottom of the photograph.*

The remaining 1% of the world's ice occurs in many different topographic settings. Ice caps are smaller varieties of ice sheets, defined arbitrarily as covering more than 50,000 km² (Armstrong et al. 1973). They are characterized by relatively slow radial ice flow. They are common in areas where the mountains have not been fully dissected and plateau remnants survive. Many examples occur in sub-Arctic and Arctic regions. Highland ice-fields are extensive areas of undulating ice, partially reflecting the form of the underlying bed, and through which mountains project as nunataks. Outlet valley glaciers descend from these ice-fields in all directions. Extensive highland ice-fields occur in the St Elias Mountains on the Yukon/Alaska border, the Queen Elizabeth Islands in the Canadian Arctic, Svalbard and Patagonia.

Fig. 2.7. The terminus of the grounded tidewater glacier, Nordenskiöldbreen, Spitsbergen. Sea ice in the foreground has been buckled into pressure ridges as a result of forward movement of the glacier in winter.

Valley glaciers emanate from either ice sheets or highland ice-fields, or are self-contained features, fed by ice accumulating on surrounding mountains (Fig. 2.6). The largest are several hundred metres thick, and they flow at rates of a few hundred metres a year. The longest, the Hubbard Glacier in southern Alaska, flows for some 100 km from highland ice-fields into a coastal fjord. Those entering the sea are tidewater glaciers, and there are two types: those that are

44

grounded on the sea bed (Fig. 2.7) and those that float (Fig. 2.8), the former being the most common. Compared with land-based valley glaciers, tidewater glaciers in fjords undergo far more pronounced fluctuations. Many subpolar and polar regions are characterized by tidewater glaciers. Some valley glaciers, especially those that originate from highland ice-fields, spread out into piedmont lobes if they leave the confines of the mountains. Small piedmont glaciers are common in the high Arctic, but the largest, the Malaspina Glacier of southern Alaska, measures 70 km across.

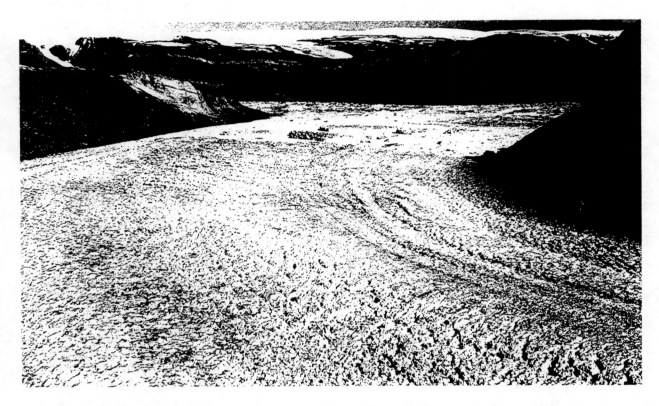

Fig. 2.8. Aerial view down the heavily crevassed floating Daugaard-Jensen Gletcher, East Greenland. Note the tabular icebergs and ice-filled head of Nordvestfjord (the world's longest fjord) in the background.

On a smaller scale are cirque (or corrie) glaciers. They are the result of accumulation under steep cliffs on mountain flanks, eventually creating an armchair-shaped hollow which is gradually overdeepened, because of rotational ice flow. Cirque glaciers develop preferentially on the lee side of a mountain mass where wind-blown snow accumulation is greatest. The best developed cirque glaciers occur in areas of moderate relief, especially high, dissected plateaux, as in Scandinavia and, formerly, Britain.

In contrast, high-relief areas such as alpine terrain, are characterized more by hanging glaciers or ice-aprons, which are small masses of ice clinging to precipitous rock slopes. Occasionally, they shed ice-avalanches of catastrophic proportions.

Lastly, rejuvenated or regenerated glaciers occur where ice-avalanche debris accumulates at a rate faster than it melts, below an ice-mass that is perched on the edge of a cliff above. They are frequently conical in form, reminiscent of scree slopes below gullies.

2.4 Cold and warm glaciers

The temperature distribution (thermal regime) of a glacier is of fundamental importance to its dynamics, both in terms of the way the ice deforms and with regard to the rôle that meltwater plays in lubricating the bed. For example, ice *below* the pressure melting point (cold ice) deforms less readily than ice *at* the pressure melting point (warm ice), to the extent that under a given stress, ice at 0°C deforms at a rate a hundred times faster than ice at -20°C.

Although transitions between cold and warm ice in the same glacier are common, it has become customary to classify glaciers thermally in the following manner:

• Warm (or temperate) glaciers: those in which ice is at the pressure melting point throughout except for a surface layer, c. 10--15m thick, that is subject to annual cooling in winter.

• Cold glaciers: those in which a substantial proportion of the ice is below the pressure melting point. There are two main types: those in which all the ice is below the melting point, so is dry and frozen to the bed, and those that undergo substantial surface melting in summer or are warmed to the melting point from beneath by geothermal heat. These are often referred to respectively as 'polar' and 'subpolar' glaciers, but these terms are misleading and are best avoided; even in high polar regions, glaciers undergo substantial melting, while some cold ice is present in temperate latitudes where the ice has originated from a high altitude.

The thermal regime of a glacier influences the rôle meltwater plays in the development of erosional and depositional landforms and sediments. In warm glaciers, water flows relatively freely and tends to reach the bed quickly, forming a subglacial channel and a water-film that facilitates sliding. As a consequence, recycling of material by glacial streams is important. In cold glaciers with surface melting, water tends to migrate towards the margins rather than to the bed, but sliding may take place if the ice is thick enough to allow geothermal heat to raise the temperature of basal ice to the melting point. Where ice is frozen to the bed, erosion of bedrock or deposition of sediment is inhibited and no glaciofluvial sediments form. The thermal regime of glaciers entering the sea (e.g. ice shelves, ice tongues) and the resulting icebergs also strongly influences sedimentation.

2.5 Glacier hydrology

Meltwater from glaciers plays a vital rôle in the processes of erosion, incorporation of debris into the ice and erosion. Furthermore, meltwater is a valuable resource, providing the means of generating hydroelectricity and of irrigating the land. Glaciers act as natural storage reservoirs, retaining water in winter, and releasing it in summer when it is most needed for irrigation. More destructively, meltwater bursting from ice-dammed or subglacial lakes, or when mixed with loose sediment to create a debris-flow, has caused much damage in several parts of the world, including Iceland, the Andes and the Himalaya. In the Quaternary and older geological record, catastrophic release of meltwater from glaciers has created impressive landscapes and shifted huge volumes of sediment. More detailed accounts of glacier hydrology are given by Paterson (1981) and Drewry (1986).

2.5.1 Sources of meltwater

Surface melting is by far the most important source of water in temperate and cold glaciers, and is supplemented by runoff from melting snow along the valley sides. Solar radiation is the principal energy source for this process. Rainfall is also able to contribute significantly to melting on these types of ice-mass. Melting rates on the lower reaches of a typical alpine glacier are 10 m yr^{-1}, whilst 2 m yr^{-1} is more typical of cold glaciers. In many parts of Antarctica and on the Greenland ice sheet, surface melting is also prevalent, but may not be the major source of water. In these areas, as well as beneath other cold glaciers of sufficient thickness, geothermal heat is an important generator of meltwater. Water is also produced by frictional heating as ice slides over its bed, and from the effects of ice being rapidly strained, especially close to the bed.

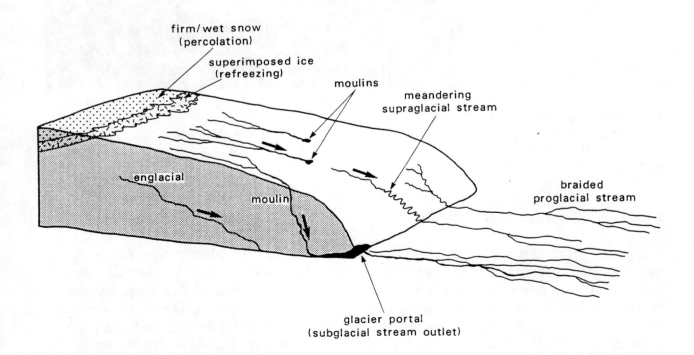

Fig. 2.9. Schematic 3-D cross-section through a glacier, illustrating the supraglacial, englacial and subglacial water-flow routes in a temperate glacier.

47

Fig. 2.10. Deeply incised meandering stream flowing towards the margin of Gornergletcher, Switzerland.

2.5.2 Water flow through a glacier

The most noticeable aspect of water movement in the glacier system is the manner in which it flows over the surface, but water flow englacially and subglacially is equally important (Fig. 2.9). The supraglacial drainage system resembles that developed on bedrock or consolidated sediment. Dendritic stream patterns are common. Channels tend to be straight or meandering and are often incised to a depth of several metres (Fig. 2.10). Straight channels are frequently controlled by supraglacial moraines or ice structures like foliation. Meandering channels may occur on exceptionally steep slopes; those on Charles Rabots Bre in Norway are found on slopes with an inclination of nearly 40°.

Supraglacial streams tend to enter an englacial position or reach the glacier bed via vertical shafts, *moulins*, which may drop sheer for several tens of metres. Frequently, moulins develop at sites of weakness in the ice, such as healed crevasses. On cold glaciers supraglacial streams commonly flow towards and alongside the margins, and the resulting torrent may prove difficult to cross (Fig. 2.11). Most discharge from cold glaciers is at the margins, whereas in temperate glaciers water normally emerges from a glacier portal at the lowest point on the bed. Otherwise, meltwater may spread out as a thin film across the bed.

Fig. 2.11. Lateral meltwater stream flowing along the flank of an arm of the cold Wordie Gletcher, northern East Greenland. The stream constantly undercuts the ice margin, causing frequent collapses and temporary blockages. Note the ice debris stranded on the banks to the left.

Water channels at the base of a glacier are of two principal types, named after the authors who examined the physical principles of their development. *Röthlisberger channels* (Röthlisberger 1972) are pipes incised into the ice above the bed. They are convex-up in cross-section, as may readily be seen from the form of the glacier portal at the snout, and they can be several metres high. In contrast, the smaller *Nye channels* are incised into bedrock (Nye 1973). In cross-section, these features have vertical sides and a rounded or flat bottom in cross-section. They tend to develop roughly parallel to ice flow but can spiral around obstacles. Nye channels belong to the family of *p-forms* that are described in Section 3.3.2. Well defined channels do not form beneath all temperate glaciers, however. Instead, thin films form, which facilitate sliding, but which slow down the throughput of water.

A certain amount of meltwater percolates down through snow and firn, although much of it may refreeze as ice-layers and ice-glands, thus inhibiting further downward flow. Where firn passes down into ice, the bulk of water percolation ceases, and the firn may become saturated in summer. However, a small volume of water can also penetrate through glacier ice itself, apart from through the more obvious crevasses and moulins. Ice tends to melt preferentially at crystal boundaries, and narrow veins, a fraction of a millimetre across, develop, allowing water to pass downwards. This type of intergranular flow has been described by Nye & Frank (1973). The vein network is rarely stable, because shearing of the ice tends to sever the interconnecting veins.

During the course of a summer melt season, the hydrological system undergoes considerable modification, as dye-tracing tests have shown. The combination of minimal meltwater production and ice deformation results in closure or partial closure of the previous summer's channels. Thus the glacier is able to hold back meltwater in the early part of the melt season, resulting in high basal water pressures. As the summer progresses, the channels become more open so that, towards the end, flow is uninhibited and basal water pressures are low. In stagnant ice the channels do not close up, but progressively enlarge each summer. The end result may be a body of ice, riddled with moulins, supraglacial and englacial channels that bear a strong resemblance to the landforms of limestone solution, hence the term *glacier karst*.

Not all meltwater that reaches the bed of a glacier emerges at the snout. Some of it is forced into the underlying sediment such as till, which thereby is easily deformed and facilitates glacier flow (section 2.7.3). Some water may flow into the bedrock if it is permeable or well jointed.

During periods of rapid meltwater production, englacial and subglacial streams may be under considerable pressure. Air bubbles may implode, creating a shock that can facilitate erosion of bedrock or sediment. This process, known as cavitation, is well known from studies of dam failure where small leaks may rapidly evolve into serious breaches. Occasionally, there is visible manifestation of water under high pressure. Outlet streams may emerge with considerable force, while at the surface fountains, or more commonly springs, may be observed.

Fig. 2.12. Ice-dammed "Between Lake", formed at the confluence of the White (left) and Thomson (right) glaciers on Axel Heiberg Island, Canadian Arctic (cf. Fig. 2.22). The photograph was taken during the jökulhlaup of 1975, the lake already having lowered several metres from the clearly defined water-marks on the ice to the left and right.

2.5.3 Ice-constrained lakes

Lakes associated with glaciers are of four main types:

- supraglacial lakes
- ice-dammed lakes
- subglacial lakes
- proglacial lakes

Lakes are important because they commonly fill up during the summer until they become unstable and water bursts out from them as *jökulhlaups*. Ponding of water by cold ice is widespread in the Arctic, but is also important in some temperate regions.

Supraglacial lakes are typical early melt-season features and are generally no more than a few metres deep. Many gradually empty as the drainage network opens up. In contrast, ice-dammed lakes fill slowly as the melt-season proceeds until the hydrostatic head is sufficient to allow water to flow under the ice, creating a high-discharge event (a jökulhlaup) that lasts typically for a day and leaves many icebergs stranded (Figs 2.12). However, not all lakes burst out annually.

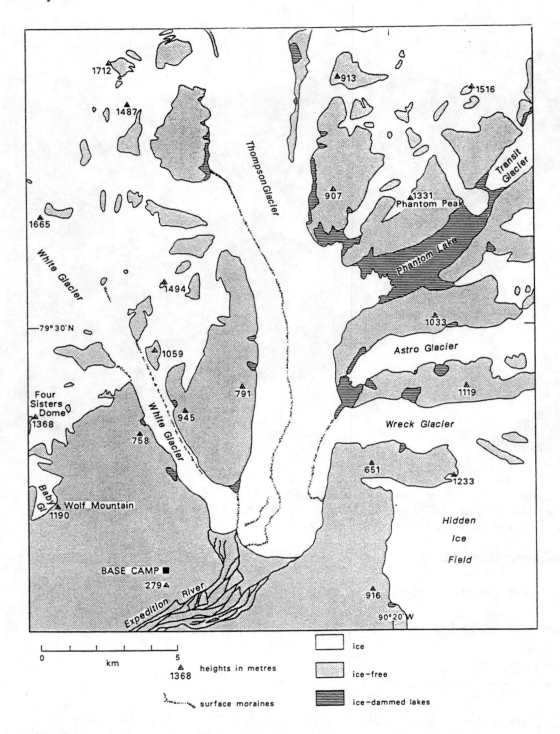

Fig. 2.13. Ice-dammed lakes of various types associated with Thomson Glacier and its tributaries, Axel Heiberg Island, Canadian Arctic.

Ice-dammed lakes form (a) in the notch where two valley glaciers join, (b) where a relatively ice-free side valley enters the main glacier valley, (c) where a side glacier enters and blocks off drainage in the main valley (Fig. 2.13). Ice-dammed lakes, if not properly monitored, can be very hazardous; many people have died during jökulhlaups in Switzerland and South America.

Subglacial lakes are known from Iceland where high heat flow from volcanically active areas can melt large volumes of ice, as beneath the ice cap of Vatnajökull, the type-area for the generation of jökulhlaups that devastate the coastal plains. Much larger subglacial lakes, identified in radio-echo sounding records occur beneath the East Antarctic ice sheet, the largest approaching 8000 km² in area (Drewry 1986).

Proglacial lakes occur at the front of a glacier, the ice sometimes terminating as a cliff. Lakes of this type tend to grow as a glacier recedes, unless the rate of sediment filling is too great. Proglacial lakes are generally stable unless a lower-level outlet route can be found up-glacier.

2.5.4 Water discharge from glaciers

Discharge readings have been taken below the snouts of many glaciers, especially where hydroelectricity is being generated. The pattern of discharge provides a good indication of the nature of water flow through the glacier. At any one time, discharge has two main components which are often out of phase (Fig. 2.14), a variable component related to weather conditions and an underlying base-flow component. Discharge has a marked diurnal variation (Fig. 2.14b), especially if the weather is sunny. The maximum discharge may be twice as much or more of the minimum discharge. The marked diurnal changes have the effect of producing a flood event every 24 hours. This behaviour is an important consideration for walkers who, having crossed a gentle stream in the early morning, may have to face a dangerous torrent in the afternoon. Base-flow volumes change much more slowly. Base flow continues throughout the winter in many cases, maintained by frictional processes or geothermal heat.

On a seasonal basis, glaciers release the bulk of a year's precipitation of snow as meltwater during the few warm months of summer. The seasonal variations normally bear little relation to precipitation; rather runoff is under a strong thermal influence. Several distinct periods of runoff on outwash plains below glaciers during the course of a year have been identified (Church & Gilbert 1975):

(a) Break-up of winter river-ice is represented by the spring snow melt. First, saturation of the snowpack occurs, generating slush; then water begins to flow over channel-ice, leading to the break-up of the winter-ice on the river bed. Drainage may be well developed beneath the snow before it becomes visible.

(b) The *nival* (snow-melt) flood follows with the establishment of a connected drainage network. Much of the stored water in glaciers is released at this stage to give anomalously high flows in comparison with the daily melt.

(c) By late summer, runoff continues more or less in accordance with the degree of meltwater generation, reflecting the clearance of internal and subglacial channels in the ice.

(d) The freeze-back period of autumn follows the cessation of melting on the glacier, although draining of some channels and groundwaters is maintained. Freezing in Arctic areas may be rapid but large areas of *Aufeis* (sheets of vertically orientated ice crystals) may develop as water continues to flow from springs or from beneath the glacier into subzero temperatures. Temperate glaciers may maintain a low steady rate of flow throughout the winter.

During stages (b) and (c) above, or even at other times, the general discharge pattern may be interrupted by jökulhlaups, which produce sharp runoff peaks. Storm precipitation as rain may produce a similar effect. The discharge curve for a jökulhlaup typically shows an initial steep rise and an almost instantaneous cut-off (e.g. Whalley 1971, Theakstone 1978).

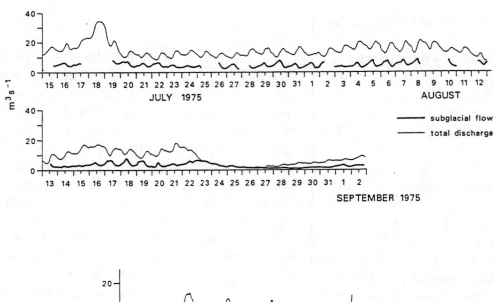

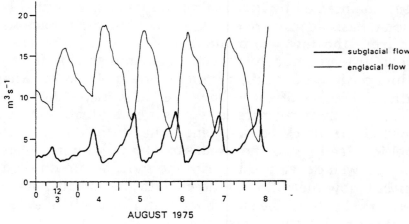

Fig. 2.14. Water discharge calculated from electrical conductivity   measurements in the Gornera, the stream from the Gornergletscher in Switzerland (after Collins 1977). (a) Curves showing total discharge and the proportion of discharge routed through subglacial conduits for the period 15 July-2 September 1975. (b) Diurnal variations of the discharge components routed through the englacial and subglacial networks during sustained ablation 3-8 August 1975. Note the out-of-phase character of the peaks and troughs of the two components.

High amplitude floods are also associated with surges. This is a response to sheetflow of water beneath the ice, after which the glacier settles back down on its bed and the flood stops abruptly.

2.5.5 Meltwater as a geological agent

The geomorphological and sedimentological products of glacial meltwater are described in Chapters 3, 4 and 5. Here it is sufficient to mention that meltwater is both a powerful erosive agent, permitting the development of sub-glacial gorges, Nye Channels and other 'p-forms', and a medium for transporting and depositing large volumes of sediment. Furthermore, material carried in solution is now regarded as important in certain circumstances, especially in areas of carbonate bedrock where both rapid dissolution of rock flour and various processes of secondary precipitation of calcium carbonate can occur (Fairchild et al. in press).

2.6 Glacier fluctuations in response to climate

Factors affecting the behaviour of the snout of a glacier are complex and numerous, but changes in climate are the most significant. The 'state of health' of a glacier, or its mass balance, is dependent on how much snow is preserved in its accumulation area and how much ice is lost in the ablation area during the summer season. The movement of a glacier is dependent on the characteristics of the supply and loss of material. If loss exceeds supply, such as during a climatic amelioration, the whole glacier will become thinner; if the reverse is the case, the glacier will thicken in either the accumulation area (through the build up of firn) or the ablation area (through reduced melting) or both.

2.6.1 Mass balance

Changes in the mass of a glacier from year-to-year, and the characteristics of these changes spatially, are studied in order to determine the balance between gains and losses, that is the glacier's mass balance. Mass-balance studies are made on valley glaciers throughout the world, and on some there are continuous records going back over thirty years, for example Storglaciären in Sweden since 1946, Hintereisferner in Austria since 1952, South Cascade Glacier in Washington State, USA since 1957, and White Glacier on Axel Heiberg Island, Canada since 1957 (Paterson 1981).

Mass balance reflects the difference between net gains (accumulation) and losses (ablation) in a given year. A positive mass balance refers to an excess of accumulation over ablation, and for a negative mass balance the reverse is true. Accumulation is represented by a variety of processes that eventually lead to the formation of ice; direct snowfall, avalanches, freezing of meltwater and slush, and the development of rime. Ablation embraces those processes which result in loss of material from the glacier: melting and the subsequent runoff of water, calving of icebergs into lakes or the sea, erosion by streams, evaporation and wind erosion.

In most cases, mountain glaciers have a clearly defined accumulation area towards their upper end, and an ablation area towards the snout. However, the influence of wind on the distribution of snow may complicate this pattern. In Antarctica the picture is rather different, since almost all ablation is by calving or through melting in contact with sea water. In some arid interior parts of Antarctica, ice streams may be subject to net ablation but, on approaching coastal areas where precipitation is greater, net accumulation may take place. Clearly, direct measurement of the mass balance of the Antarctic ice sheet is fraught with uncertainty, but as the resolution and extent of satellite data improve, continuing efforts are being made to assess the state of health of that continent's ice-cover. Series of satellite images have already been used to demonstrate the rapid recession and disintegration of an ice shelf in the Antarctic Peninsula under the influence of changing oceanographic conditions (Doake & Vaughan 1990). Further work of this nature, combined with surface observations, is of vital importance if we are to understand the links between ice volume and sea-level changes.

In mass-balance terms, the surface of a glacier may be divided into a number of zones, separated by distinct lines (Table 2.3). As summarized by Paterson (1981), the zones are best developed in a cold glacier descending from high altitude, but not all glaciers have all zones. The dry snow zone is developed only on the highest mountains of Alaska and the Yukon, and in the interiors of the Greenland and Antarctic ice sheets. On some cold glaciers all the accumulation is in the form of superimposed ice, whereas on many temperate glaciers this form of ice may be negligible.

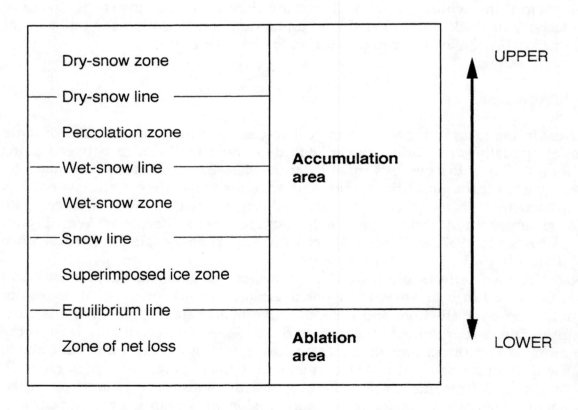

Table 2.3. Subdivision of a glacier surface into its constituent mass balance zones.

2.6.2 Effect of mass balance changes on flow

Glaciers continually grow and decay in response to climatically controlled mass-balance changes. The appearance and disappearance of the great Antarctic and

deformation along a 10 km flowline in the Barnes Ice Cap, Baffin Island, the strain ellipses showed compression normal to the particle path in the upper part of the ice cap (pure shear, or non-rotational strain), but gradual rotation of the ellipse towards parallelism with the particle path with increasing depth.

Fig. 2.19. Regelation of ice at the sole of the Glacier de Tsanfleuron, Switzerland, observed in a cave near the snout. Successive regelation layers with basal debris have built up in this vertical section which is about 20 cm high.

2.7.2 Basal sliding

Whereas internal deformation affects all glaciers, not all slide on their beds. Nevertheless, all warm glaciers and many cold glaciers do slide, the latter in response to geothermal heat flux from bedrock if the ice is sufficiently thick to act as an insulating blanket from the ambient air temperature. Extensive areas of basal ice in the Antarctic and Greenland ice sheets are at the pressure melting point and slide. Most fast-flowing valley glaciers and ice streams attain their high velocities because basal sliding is the major component of the total velocities. The basal sliding component has been recorded for several glaciers, examples being 50% for the Grosser Aletschgletscher in Switzerland, 90% for the Blue Glacier in Washington State (both valley glaciers); 9% for Vesl Skautbreen, a cirque glacier in Norway; and 0% for the Meserve, a glacier frozen to its bed in Antarctica. The highest percentages are associated with rapid melting or heavy rainfall, when the bed is lubricated. Basal sliding is largely responsible for the bulk of erosion, transport and deposition of debris taking place.

Basal processes have been subject to several rigorous theoretical analyses, notably by the physicists Weertman, Lliboutry and Nye (Paterson 1981: ch. 7), but the precise mechanisms are still far from fully understood. Three main processes have been recognized at the bed of the glacier. First, enhanced basal creep occurs in the lowest few metres of the glacier, a process possible in ice at any temperature, but one that is more effective in warm than in cold ice. This process allows ice to flow around an obstacle such as a boulder on the bed, and determines the direction in which basal debris is transported. A second mechanism, involving pressure melting and regelation, occurs at the ice/rock interface. When ice moves over a bump in the bed, pressure melting occurs on its upstream side, the water flows over and around the bump to the lower-pressure zone, where it refreezes and forms regelation ice, possibly in a cavity, and at the same time loose debris becomes attached to the glacier sole (Fig. 2.19). The critical maximum size of the bump is debatable, but a figure of 1 m has been postulated. This process has been observed under many active glaciers. A third mechanism involves slip over a water layer. Water at the base of a glacier not only 'lubricates' the bed but also smooths out the smaller irregularities. A layer of water only a few millimetres thick could increase the sliding velocity by up to 100%. Much of this water may originate from supraglacial meltwater streams reaching the bed *via* moulins and crevasses. Direct observations in subglacial cavities beneath temperate glaciers confirm the validity of these mechanisms, but in addition suggest that the motion is jerky (e.g. Theakstone 1967, Vivian & Bocquet 1973).

From the above it will be apparent that the bed roughness and the rôle of debris will have an important influence on the sliding velocity. If the bed is too rough because of bedrock irregularities, debris between ice and rock, and debris within the basal ice, the basal sliding velocity is appreciably reduced. The difficult problem of developing a theory taking these factors into account has yet to be solved.

2.7.3 Deformable beds

The 'effective bed' beneath a moving ice-mass may not necessarily be the ice/bedrock or ice/sediment interface, since the material beneath the ice may also be deforming. In recent years it has become widely recognized that large areas beneath glaciers and ice sheets are underlain by unconsolidated sediment which fundamentally affects the dynamic behaviour of the ice-mass (Boulton & Hindmarsh 1987). In essence, the deforming sediment forms part of the flowing ice-mass, and the 'effective' bed is the surface below which there is no forward motion.

If the base of the ice is at the pressure melting point, the underlying sediment may be saturated with water and thus be prone to deformation, especially if pore-water pressures are high. Boulton (1979) examined deformation in till beneath Breidamerkurjökull in Iceland, and found that this process accounted for 90% of the forward movement of the glacier, even though the deforming layer was only about 1/2 m thick. Displacement was about 1/2 m after 10 days. The strongest deformation appeared to be in the upper part of the unit (Fig. 2.20).

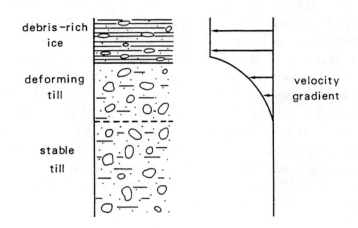

Fig. 2.20. Velocity vectors parallel to flow and associated displacements in the upper part of a deformable substrate of till below the Breidamerkerjökull, Iceland. Shape of profile generalized. Displacement about 0.5 m after 10 days (after Boulton 1979).

A similar process has been inferred for one of the Antarctic's major ice streams, named 'B' (Alley et al. 1989). It is conceivable that for large parts of ice sheets and valley glaciers, especially where flowing fast, deformation in subglacial sediment is the dominant component of flow. A direct result of this process is that substantial volumes of till may be transported beneath the ice by shear within the deformable layer. In the case of Ice Stream B, which merges with the Ross Ice Shelf and other ice streams, the end result of the process is inferred to be deposition at the grounding-line which in effect represents the end of the debris 'conveyor belt', giving rise to *diamict aprons* (a feature described from elsewhere in Antarctica by Hambrey et al. 1991) or in a genetic sense *till deltas* (Alley et al. 1989).

The bed may also deform if the sediment is frozen to the ice, as when a glacier overrides permafrostfrost. In such cases blocks or large 'rafts' of frozen sediment may be incorporated into the base of the ice-mass. This frozen sediment may deform in a manner that resembles the flow of ice (the flow laws are similar), becoming folded and fractured in the process.

2.7.4 Long- and short-term variations of glacier flow

Glaciers show major changes in surface velocity through time. On the longer timescale of several years or more, velocities change in response to climatic factors. Velocities increase in response to gains in mass, following several years of positive mass balance. By the same token, a succession of negative mass-balance years results in declining velocities and, if severe, to eventual stagnation of the ice. Exceptions to such behaviour occur in the special case of surge-type glaciers (Section 2.9), or when a subglacial volcanic eruption takes place.

Velocity variations on a seasonal or daily timespan are largely related to meltwater production and subglacial water pressures which controls the rate of basal sliding. Many studies have been undertaken on valley glaciers (see Paterson

1981: ch. 7). Typically, during the winter, when meltwater production largely ceases, ice-flow tends to close drainage channels. During early summer the first meltwater is forced into small cavities and water pressure becomes high. Thus ice-flow tends to reach a peak at this time, even though meltwater production does not reach a peak until a month or so later, but by this time the cavities and channels have opened up and water pressure is low. Similarly, sliding velocities reach a minimum a few months after the time that runoff ceases and water pressures are lowest. As an example, measurements over a 2-year period of the temperate valley glacier, the South Cascade, showed the lowest sliding velocity in November (120 mm day^{-1}) and the highest velocity in June (220 mm day^{-1}), an 83% difference. The creep velocity (comprising internal deformation) in this case, by contrast, varied only from 40 to 50 mm day^{-1}, a 25% difference. Minimum and peak discharges occurred in March/April and July respectively.

Velocities also vary diurnally. The bulk of surface melting is due to solar radiation, so the supply of meltwater to the bed and associated basal water pressures and sliding velocities is highly dependent on the weather. In clear weather, velocities may vary as much as 100%, with peaks in mid-afternoon and minima in early morning. On cloudy or rainy days the diurnal fluctuations are largely suppressed.

2.8 Structure of glaciers and ice sheets

Glacier ice is like any other type of geological material in comprising strata that progressively deform to produce a wide range of structures. The end product is a metamorphic rock that in temperate glaciers has deformed close to the melting point, and the original structures may be totally obliterated.

Two main categories of structures occur in glaciers (Table 2.4):
• primary structures which are result of deposition or accretion of new material.
• secondary structures which are the result of deformation.

2.8.1 Primary structures

The dominant structure in the first category is sedimentary stratification, which is the result of accumulation of snow year by year (Fig. 2.21a). Most snow turns into firn, then coarse, bubbly ice, but summer surfaces are often indicated by a refrozen melt-layer of bluish ice and dirt. Regularly layered ice is the product, although periods of excessive ablation, which remove several or many previous layers, give rise to unconformities. Within the snow pack, downward percolation of meltwater takes place during the summer. Refreezing of the meltwater creates ice-layers, and ice-lenses or pipe-like structures called ice-glands.

Another type of primary structure, regelation layering, is the result of pressure melting and regelation processes at the base of a glacier. It is often called basal foliation, but is not strictly a deformational structure. Commonly, it is parallel-laminated and has variable amounts of basal debris associated with it.

Where ice is subject to collapse, especially in the neighbourhood of ice cliffs and crevasses, ice debris may reconsolidate to give a breccia. This structure may be welded together by trapped snow or by the refreezing of meltwater.

64

Category	Structure		Type of deformation
Primary structures	Sedimentary stratification Unconformities Ice layers, lenses, glands Regelation layering (basal foliation) Ice breccia		No deformation
Secondary structures	Folds Foliation Boudinage Crevasse traces (tensional veins) Shear zones		"Plastic" deformation
	Ogives Closed fractures:	 normal faults strike-slip faults thrusts crevasse traces	"Brittle" deformation
	Open fractures	transverse crevasses marginal crevasses longitudinal crevasses splaying crevasses *en echelon* crevasses basal crevasses *Bergschrund* *Randkluft*	

Table 2.4. Classification of structures in glacier ice.

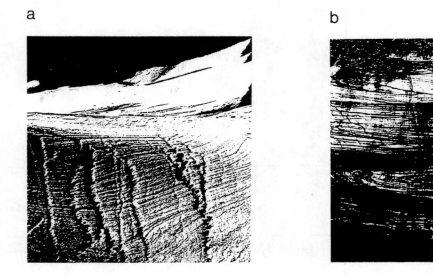

c

Fig. 2.21. *Structures in glacier ice: (a) Sedimentary stratification, deformed by flow and exposed on a 30-40° slope on Charles Rabots Bre, Okstindan, Norway. Each layer represents a year's accumulation of snow above the firn line. (b) Isoclinal z-fold associated with thrusts (the dirty layers) in the terminal cliff of Thomson Glacier, Axel Heiberg Island, Canadian Arctic. The wavelength of the fold is estimated to be about 5 m. (c) Typical well-developed longitudinal foliation in the White Glacier, Axel Heiberg Island, Canadian Arctic; note the supraglacial melt-stream flowing parallel to the ice.*

2.8.2 Secondary structures resulting from ductile deformation

Secondary structures include a range of features that can be classified broadly as the products of plastic flow and brittle fracture. Folds on all scales are commonly observed on the glacier surface and they involve both primary and secondary layered structures (Fig. 2.21b). Huge folds many kilometres across are well known from the piedmont glaciers of Alaska, notably the Malaspina and Bering (Post & LaChapelle 1971), while small folds only a few centimetres across can be distinguished in foliated ice. Normally, these folds fall into the *isoclinal* or *similar* categories, but occasionally *parallel* folds occur, reflecting more 'brittle' folding (Hambrey 1976). Refolded folds are quite common. Recumbent folds are often present in basal ice, and they reach several metres in amplitude in cold glaciers. Hudleston (1976) described fine examples from the Barnes Ice Cap.

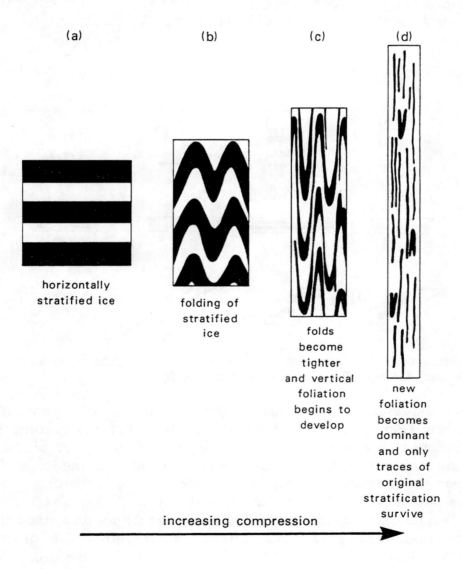

(a)　(b)　(c)　(d)

horizontally
stratified ice

folding of
stratified
ice

folds
become
tighter
and vertical
foliation
begins to
develop

new
foliation
becomes
dominant
and only
traces of
original
stratification
survive

increasing compression

Fig. 2.22. Development of foliation by folding and transposition of earlier layers (stratification or another foliation). Note how the increasing compression from A to B is accompanied by a marked change of overall shape. The foliation develops out of those parts of the original layers that are thinned the most; its final orientation is totally different from that of the original layers.

Foliation is a layered structure, comprising coarse-bubbly, coarse clear (blue) and fine (white) ice (Allen et al. 1960), that forms during flow (Fig. 2.21c). Coarse ice has crystals in the range of 1--15cm, whereas fine ice crystals are usually less than 5 mm in diameter. Foliation is also defined by elongated air bubbles, discrete planes of shear and flattened ice crystals. Foliation generally develops from pre-existing layers, notably stratification and the traces of former crevasses (Hambrey 1975, Hooke & Hudleston 1978). Folding of the layers and progressive attenuation of the folds allows the new structure to develop on the limbs of the sheared folds. Ultimately, only isolated fold hinges may remain. This process is known as transposition and is depicted in Figure 2.22; it is also a characteristic feature of metamorphic rocks such as schists and gneisses. Rarely, foliation appears to be a totally new structure, forming a sort of cleavage that has an axial-planar relationship to similar folds. Foliation develops essentially under two basic deformation regimes - pure shear and simple shear - which affect the pre-existing structural inhomogeneities in different ways (Fig. 2.23).

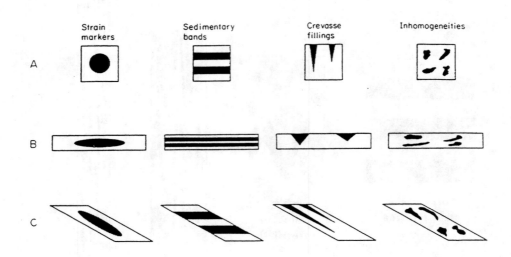

Fig. 2.23. The development of foliation from pre-existing structural   inhomogeneities under pure shear and simple shear regimes (after Hudleston & Hooke 1978).

Pure shear is represented by normal compression and equal expansion in the direction of flow (Fig. 2.23, rows A to B). So the structures become simply flattened, a process typical of upper and middle parts of the glacier. Simple shear results from the strong velocity gradient between bedrock and the ice, so is most prominent near the margins and the bed. In essence, the structures gradually rotate towards parallelism with the direction of flow (Fig. 2.23, A to C). The same effect can be demonstrated by drawing a circle or other shapes on a stack of cards, and smearing them out gradually. Foliation is therefore the product of cumulative strain (Hambrey & Milnes 1977, Hooke & Hudleston 1978).

Most foliation falls into one of two main geometrically arranged categories: longitudinal and transverse. Longitudinal foliation is normally steeply dipping and it occurs at the margins of a glacier, or in association with medial moraines. In some glaciers, where ice from a wide accumulation basin feeds into a narrow tongue, longitudinal foliation may extend across the entire width and be predominantly parallel to flow (Hambrey & Müller 1978). Isoclinal folds may have axes parallel to the foliation, and the entire structure may be the result of isoclinal folding of primary stratification and eventual transposition. Arcuate foliation originates in regions of transverse crevasses or ice-falls, forming first as crevasse traces (described below) before being compressed in a longitudinal direction below the ice-fall and deformed by flow into an arcuate pattern with the apex down stream. The geometry of arcuate foliation resembles that of a set of nested spoons (Allen et al. 1960). As the foliation moves downglacier, its upglacier dip declines from near vertical to a few tens of degrees near the snout. In both longitudinal and transverse foliation, the total strain ellipse is parallel to the structure, but in the former case this has been achieved mainly through simple shear, but in the latter by pure shear.

Boudinage structures are sausage-shaped features that result from compression of a multi-layered sequence with layers of contrasting competence (or ductility) (Fig. 2.24). They are widespread in glacier ice (Hambrey & Milnes 1975) and two main types may be distinguished: competence-contrast boudinage and foliation boudinage. The former type is evident in ice with distinct layers of coarse and fine ice, or in dirty and clean ice. Fine ice and dirty ice often appear to behave as the more competent (less ductile) material; the latter conflicting with the behaviour of debris-bearing ice in the lab. Foliation boudinage is the more common type, being intimately associated with longitudinal foliation. Commonly, the foliation has an asymmetric appearance, suggesting formation in a rotational strain regime.

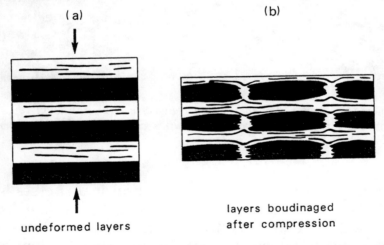

Fig. 2.24. *Schematic diagram showing the development of boudinage structures in a sequence containing layers of contrasting ductility or competence. The boudins form in the more competent layers, the less competent material on either side flowing into the spaces between the boudins (after Ramsay 1967). In glaciers, different ice types, for example fine-grained and coarse-grained, have different ductilities.*

2.8.3 Secondary structures resulting from brittle deformation

A wide variety of fractures result from brittle failure of ice, mainly in the upper 30–50 m of a glacier. Foremost among them are crevasses which are usually classified according to their orientation with respect to the ice-flow direction: transverse, marginal, longitudinal, splaying and *en echelon* (Meier 1960) (Fig 2.25). Exceptions are the little-studied basal crevasses, while a *bergschrund* is the crevasse which separates the flowing ice at the head of a valley from the ice that

69

adheres to the rock above. Crevasses may range from a few metres in length to hundreds of metres in a large valley glacier. Some crevasses in Antarctica are over 100 km long. Crevasses vary in width from a few millimetres to several tens of metres. Some of those in Antarctica are large enough to swallow large over-snow vehicles.

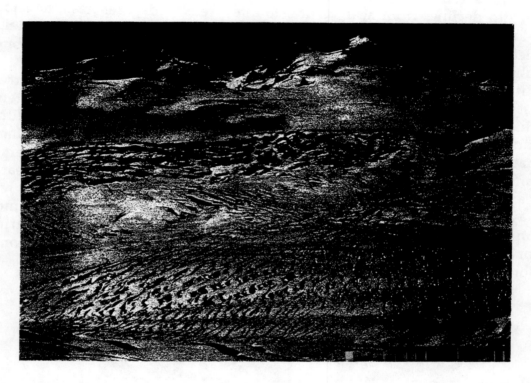

Fig. 2.25. Accumulation area crevasses of the mainly transverse and en echelon types in the complex upper basin of the Fox Glacier, New Zealand; bergschrunds can be seen in the uppermost reaches of the glacier.

Crevasses form only where at least one of the principal stresses is tensile, and where this stress is greater than the tensile strength of the ice. Crevasses normally open in the direction of this stress (Meier 1960), but, if the ice is passing through a changing stress regime or contains structural inhomogeneities, this may not be strictly true. Similarly, the stresses needed to initiate a fracture may not always be the same.

The depth of crevasses is dependent on the plastic flow of ice at the bottom. In temperate glaciers measured depths of crevasses have rarely exceeded 30 m, observations supported by theory, yet mountain literature abounds with examples of crevasses "hundreds of feet deep". Cold glaciers, on the other hand, may be considerably deeper. Crevasses may also attain a greater depth if they are water-filled (Robin 1974), in which case they can penetrate to the bed of the glacier.

Ice-falls are steep zones in a glacier where the entire surface is broken up by crevasses of many orientations. Such zones are difficult or impossible to traverse safely on foot (Fig. 2.26).

Crevasse traces is the name given to a wide family of initially vertical layers that form parallel to or extend from open crevasses (Hambrey 1975). One type is the product of brittle fracture, without separation of the two walls. Away from the fracture plane, ice recrystallizes adjacent to the plane under continued extension (Hambrey & Müller 1978) creating a feature analogous to tensional veins in rocks (Durney & Ramsay 1973). A second type of crevasse trace is the result of freezing of meltwater in a crevasse and its subsequent closure. These usually occur as prominent blue ice-layers, whereas those of the first type are more subtle. Crevasse traces survive ablation remarkably well, cropping out all the way to the snout in many valley glaciers, suggesting that deep-seated fracture may occur to the bed in both cold and temperate glaciers.

Fig. 2.26. The Hochstetter Icefall of Mount Cook, feeding the Tasman Glacier of New Zealand. The photo was taken prior to a major rockfall that obliterated much of the ice in this view (cf. Fig. 2.33).

Clean-cut but unopen fractures may sometimes be observed in glaciers. Normal faults, with the down-glacier wall dropped down by up to a few metres, form in ice-falls or either crevasse regions, and are visible in the walls of crevasses or along the sides of the glacier. Strike-slip faults, showing lateral displacements of up to a metre, are readily seen at the surface of the glacier and may be associated with bending of the ice, due to shear prior to fracture. In both cases distinctive features need to be displaced, in order to recognize these movements.

Thrusts, alternatively referred to as thrust faults or shear planes, have been observed along the margins and near the snouts of many glaciers. They are associated with strong compression in the ice, such as where the glacier impinges

71

on an obstacle or where the ice is slowing down. Although small thrusts may be observed in many temperate glaciers, they are best developed in cold ones. Surge-type glaciers also have many thrusts. The high Arctic is particularly well endowed with thrusts and their development is facilitated by:

• the rotation by flow of transverse crevasse traces into an attitude (up-glacier dip of *c*. 45°) that promotes displacement along the pre-existing planes of weakness;

• the transition from a basal sliding regime of thick ice to one in which the ice is frozen to the bed, such as near the snout where the rôle of geothermal heat is lessened under thinner ice.

Thrusts play an important rôle in the recycling of basal debris (as discussed in Section 2.10.2) and they tend to be associated with recumbent folds.

2.8.4 Secondary structures of composite origin

A final type of secondary structure is referred to as *ogives*, which are alternating light and dark bands, or waves, extending in arcuate fashion across the glacier surface below some ice-falls (Fig. 1.1). Each pair of light and dark bands represents a year's movement through the ice-fall, and where they extend to the snout they are useful measures of the transit-time of ice through the glacial system. Various hypotheses have been proposed for ogive development, but the most likely is that of Nye (1958) who suggested that the ice is thinner in the ice-fall in summer because of ablation, and collects more dust, thereby creating the darker layers and troughs. In contrast, in winter, snow fills the crevasses, the ice thickens, so creating the light bands and wave crests. Close examination of ogives in Bas Glacier d'Arolla, Switzerland, by the author indicates that they are composed of typical arcuate foliation, suggesting that these structures are genetically related. The darker layers thus represent more intensive crevasse and crevasse trace formation in summer.

2.8.5 Structural patterns in glaciers and ice sheets

Most structural glaciological work has been undertaken on cirque or valley glaciers, and several detailed maps, combining air-photograph information with ground measurements, have been published. In some cases, we can see how stratification evolves into longitudinal foliation, in others how arcuate foliation develops out of transverse crevasse fields. Foliation is generally strongest near the margins of a glacier and where associated with a medial moraine. In cirque glaciers, stratification may be the dominant structure throughout, but in many valley glaciers this structure is soon obliterated.

Satellite imagery provides a means of deciphering the surface structure of the Antarctic and Greenland ice sheets. Much Antarctic ice is discharged into the Southern Ocean via ice streams and ice shelves, and many show supposedly flow-parallel features which have been described as flowlines (e.g. Crabtree & Doake 1980, Swithinbank 1988). By comparing these Landsat images with air photos and ground observations of exposed ice, we have been able to interpret these linear structures as longitudinal foliation resulting from the probable passage of stratified ice into confined channels, thereby subjecting it to isoclinal folding (Reynolds & Hambrey 1988, Hambrey & Dowdeswell in press). Much can be inferred from Landsat-derived structural maps about flow dynamics and sediment-transfer paths. Figure 2.1 shows one of the largest ice-drainage systems in Antarctica, the Lambert Glacier system.

In conclusion, although glacier ice is rheologically simpler than most rocks, being mono-mineralic, deforming solely under the influence of one driving force, gravity, the structural pattern may be exceedingly complex. However, because ice deforms at a rate that can be measured and allows estimates of total strain to be made, glaciers can serve as models for deformation deep within the Earth's crust. For example, the same range of structures occur in glaciers as in mountain belts such as the Alps, even though the former deform ten orders of magnitude faster (Hambrey & Milnes 1977).

Structures in glacier ice tell us much about the nature of deformation on both short and long timescales. They can also be used to infer flow paths of debris, and explain the disposition of debris on, within and below the ice.

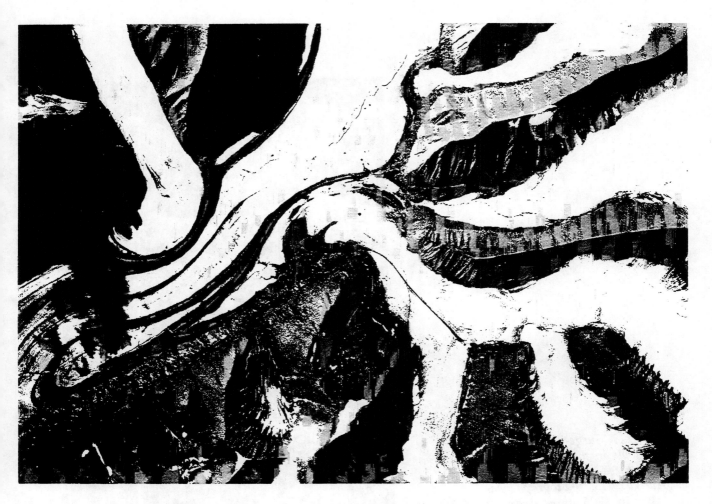

Fig. 2.27. Surge-type glacier complex in Spitsbergen (Battybreen) during a period of quiescence (1969). Irregular pulses of ice flow give rise to their distinctive looped and tear drop-shaped medial moraines. Vertical air photo No. RC8 S69 1466, courtesy of Norsk Polarinstitutt.

2.9 Glacier surges

Glacier surges are a form of exceptionally rapid flow that occurs for a short period (typically a few months) in certain glaciers, and is preceded and followed by longer intervals (typically many years) of quiescence when the ice is relatively stagnant. According to Sharp (1988a), about 4% of glaciers are of the surge type, but their distribution is restricted to certain areas, suggesting that their environmental setting may have some influence (Raymond 1987, Clarke, G. C. 1991). Both temperate and cold valley glaciers may surge, and it has even been suggested that large parts of the Antarctic ice sheet may also surge, a scenario which, if true, could have globally catastrophic consequences in terms of sea-level and climatic changes. Surge-type glaciers have two distinct zones, the reservoir and the ice receiving area, which do not coincide necessarily with the normal accumulation and ablation areas.

Fig. 2.28. Surge-induced thrusts and folds in the tongue of Variegated Glacier, southern Alaska. The photograph was taken in 1986, three years after the glacier's latest surge. The roughness of the ice surface is a relict of the intense crevassing induced by the surge.

2.9.1 Characteristics of surge-type glaciers

Glacier surges and surge-type glaciers have been studied in detail in several locations, including Alaska, the Yukon, the Pamirs and Svalbard. The processes involved during a surge event are therefore quite well known, but what actually triggers the surge and what is the nature of the bed over which the ice flows are the subjects of much debate. Excellent reviews of glacier surges have been provided by Kamb et al. (1985), Raymond (1987) and Sharp (1988a, b).

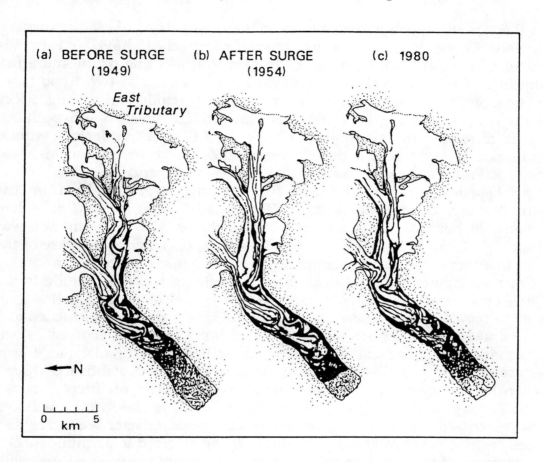

Fig. 2.29. Evolution of moraines (black) on Susitna Glacier, Alaska Range, Alaska following a surge in 1951-52 (from Clarke, T.S. 1991).

When a glacier surges, it displays several features: (a) its surface becomes very heavily crevassed; (b) ice velocities increase ten- or a hundred-fold (to 5 km yr^{-1}, for example) and fluctuate wildly; (c) large volumes of ice are transferred from the upper receiving basin to the lower part of the glacier, resulting in a net reduction of the surface gradient and leaving ice stranded at the higher levels; (d) the snout often (but not always) responds with a rapid advance, typically a few kilometres over a few months; (e) large volumes of turbid meltwater are released; (f) quiescent phase ice-structures, such as longitudinal foliation and medial moraines, are deformed into arcs and loops, resulting in the teardrop-shaped moraines that typify such glaciers (Fig. 2.27); (g) widespread thrusting in the ice occurs as the surge front, in the form of a topographic bulge, passes through (Fig. 2.28).

In contrast, during quiescence, the crevasses ablate away and the ice becomes slow- moving or stagnant, although ice velocities gradually build up to a new surge in the reservoir area. In the ice-receiving area the ice ablates largely by down-wastage, and a drainage network with distinctive potholes commonly develops. A series of surge cycles may give rise to several sets of teardrop-shaped medial moraines, which give the glacier a complex and fascinating appearance that changes dramatically during each successive surge (Fig. 2.29).

2.9.2 Mechanisms of glacier surging

Although there is still much uncertainty as to the mechanisms of glacier surging, comprehensive investigations on a number of glaciers have led to the development of several hypotheses, and the dismissal of some of the more far-fetched ideas that were prevalent up to fifteen years ago. The two hypotheses most favoured currently are dependent on a water-controlled triggering effect, whereby the subglacial drainage system is destroyed. However, one hypothesis, developed from studies on the temperate Variegated Glacier in Alaska, assumes a rigid substrate of bedrock; the other, from studies of the partially cold-based Trapridge Glacier in the Yukon, requires that the bed is deformable.

Rigid bed hypothesis. The rigid substrate model arises principally from the investigations made during the surge of Variegated Glacier in 1982-83, when measurements of borehole deformation revealed that *c.* 95% of ice motion was by basal sliding (Kamb et al. 1985). During quiescence, drainage at the base of the glacier, as in other non-surging temperate glaciers, takes place via a tunnel system, water emerging at the snout through a single portal. The prelude to the surge is the closure of the tunnel system, trapping meltwater and causing an increase in the basal water pressure. As a result, friction at the bed is lessened and the rate of basal sliding increases, allowing separation of the glacier from its bed and the formation of linked cavities controlled by bedrock irregularities. Surge velocities are attained once a linked cavity system has been established (Kamb 1987). When the linked cavity network develops into series of interconnected tunnels, water may be discharged more efficiently often in the form of a flood, and the surge ceases. In reality, the surge of Variegated Glacier was irregular, with pauses and increases in velocity, the latter being referred to as mini- surges. The final cessation of the surge, in July 1983, was abrupt and was associated with a huge outburst flood of turbid water that drastically altered the morphology of the outwash plain (Kamb et al. 1985). By this time there had been wholesale transfer of ice towards the glacier snout, but the surge front ceased moving when within 1 km of the terminus, and the surge did not manifest itself in the form of a rapid advance of the glacier snout.

Deformable bed hypothesis. An alternative mode of surging has developed from the idea that ice may rest on a bed of soft sediment which is readily deformed (section 2.6.3). Trapridge Glacier has been monitored continuously for over 20 years in the expectation that a surge is imminent, and it is thought that the glacier rests on a bed of deformable sediment which facilitates the initiation of a surge (Clarke et al. 1984). Unlike the Variegated Glacier, the Trapridge has a largey cold thermal regime, and in this respect is probably more typical of Arctic glaciers. The build-up to a surge is marked by the development of a wave-like bulge in the lower reaches of the glacier. The bulge marks the boundary between warm-based ice up stream and cold-based ice down stream, and it has progressively moved down glacier during the period of observation at a rate of some 30 m yr^{-1}. As with the Variegated Glacier model, a surge in a glacier of the

Trapridge type also begins with the destruction of the subglacial drainage system. However, in the latter case, the drainage system occurs within permeable sediments beneath the ice, but under certain circumstances the effective permeability may be reduced. Progressive thickening of the ice leads to an increase in basal shear-stress, which has the effect of reducing the permeability of till. Water pressure therefore builds up and the sediment is transformed into a slurry, the glacier then being able to flow at an enhanced velocity, which further weakens the sediment (Clarke et al. 1984). In this model, the surge terminates because the redistribution of ice leads to a lowering of the basal shear-stresses to a level that allows the subglacial sediment to return to a permeable state. The Trapridge Glacier type model requires a large supply of readily erodible material to create a thick-enough layer of subglacial sediment. Thus, soft bedrock and high rates of tectonic uplift may provide the necessary geological controls on the location of surge-type glaciers that behave in the manner envisaged for Trapridge Glacier.

Both the mechanisms discussed above seem plausible and they may not be mutually exclusive, even for the same glacier.

2.9.3 Do ice sheets surge?

From the standpoint of the stratigraphic record it is more important to know whether the Antarctic ice sheet is prone to surging, since it has been speculated that the Wisconsinan ice sheet may have behaved in this fashion (Clayton et al. 1985). Furthermore, it has been suggested that a major surge of the Antarctic ice sheet could so seriously disrupt ice/ocean interactions that Northern Hemisphere glaciation could once again be initiated (e.g. Hughes 1975). Theoretical evidence in favour of surging has been offered from the Lambert Glacier--Amery Ice Shelf system which drains about a fifth of the East Antarctic ice sheet (e.g. Budd & McInnes 1978, Allison 1979), but this is countered by a variety of approaches to interpreting the Landsat imagery of the glacier system by others (Robin 1979, McIntyre 1985, Hambrey & Dowdeswell in press). More convincingly, some ice streams feeding into the Ross Ice Shelf from the West Antarctic ice sheet, have surge characteristics. Ice Stream B (underlain by a deformable bed) is currently flowing at more than 800 m yr^{-1} whereas its neighbour, Ice Stream C is practically stagnant with a velocity of only 5 m yr^{-1} (Whillans et al. 1987). However, buried surface crevasses in Ice Stream C, suggest a behaviour 250 years earlier that resembled that of B (Shabtaie & Bentley 1987). The contrasting behaviour thus inferred may be related to large-scale surging, but with a periodicity of hundreds of years. However, to date, there are no reports that any parts of the Antarctic ice sheet have, in fact, surged. The fact the cold ice caps with composite basins in Svalbard do surge (Solheim 1991), suggests that the phenomenon could occur on a larger scale in Antarctica.

2.10 Debris in transport

Glaciers act as conveyor belts for large volumes of eroded material. Unlike most other erosional depositional systems, the thickest sediments accumulate farthest from the source and, furthermore, even the most distal sediments contain coarse bouldery material. The manner in which sediment is transported out of the glacier system is largely dependent on the thermal regime. Cold, sliding glaciers carry a heavy basal debris load which is deposited as till. In contrast, highly dynamic temperate glaciers have relatively little basal debris but a high supraglacial load, most of which is modified by meltwater soon after deposition.

Following Boulton (1978), we can conveniently consider the movement of debris in relation to its transport path through the glacier. Understanding of these paths provides a better basis for interpreting the depositional processes and deriving the source areas of the sediment.

Fig. 2.30. *In this photograph of Breithorn, Switzerland, steep tributary glaciers are bounded by rockfall-derived lateral moraines. As the glaciers enter the main trunk glacier, the Gornergletscher, flowing from left to right, the lateral moraines join to form medial moraines.*

2.10.1 High-level transport

Debris carried at a high level in a glacier is derived from the following sources (partly after Drewry 1986):

- rockfall from adjacent mountain slopes
- avalanche debris which includes snow, ice, soil and rock in various proportions from adjacent slopes
- debris-flows from adjacent slopes

- wind-blown dust (sand-grade and finer)
- volcanic ash
- marine salts and microflora derived from sea spray
- extraterrestrial, such as meteorites
- pollutants from human sources
- aerosols and gases.

Fig. 2.31. Small medial moraine in the Glacier de Saleina, Switzerland, illustrating the characteristic nature of rockfall-derived debris. In the background is a serrated arête with a relatively low col formed as a result of breaching of the arête at the head of the glacier.

Of these, rockfall is volumetrically by far the most important component on the surface of most valley glaciers. It tends to accumulate as lateral moraines, forming medial moraines where ice-flow units combine (Fig. 2.30). Commonly, these merge towards the snout, giving rise to a totally debris-covered surface. Rockfall debris is largely derived from frost-shattering processes and has a predominantly angular character (Fig. 2.31). Boulton (1978) has recorded the following characteristics, based on detailed examination of clast shape using Krumbein's (1941) sphericity/roundness chart, and grain-size distribution:

- clasts have low roundness (very angular and angular) and very variable sphericity (Fig. 1.4), and show little modification during transport;
- grain size analyses indicate a predominance of clasts coarser than 1ø and a deficiency of fines, with little modification down glacier.

Rock falling onto snow near the head of the glacier may be buried and may re-emerge below the equilibrium line. Rock falling on the surface near the equilibrium line will remain on the surface. In both cases the debris will be passively transported, unless it is able to reach a marginal or basal position by falling down crevasses or moulins.

Large-scale collapse of cliffs may turn free-falling rock into a fluid-like flow, especially if mixed with snow and ice, and the end product may be quite different from typical debris derived from rockfall. An impressive landslide took place in

New Zealand in the early hours of 14 December 1991, when a buttress leading up to the summit of Mount Cook (3,764m) collapsed (Chinn et al. 1992) (Figs 2.32, 2.33). The rock- and ice-avalanche descended 2,720 m, obliterating the Hochstetter ice-fall and part of the lower Tasman Glacier where it rode 70 m up the far moraine wall, 7.3 km from the source area. An estimated 14 million m³ of rock and ice were involved, and being mixed with snow on its descent behaved as a fluid that may have reached a velocity of 600 km hr⁻¹. In the process the height of Mount Cook, New Zealand's crowning peak, was reduced by 20 m. The resulting deposit was a pulverized mass of rock debris in which the originally angular shapes had been rounded, providing a marked contrast with material resulting only from free-fall. Although many years may separate such events in alpine regions, individual glaciers may show signs of several major rockfalls, albeit on a somewhat smaller scale than that at Mount Cook.

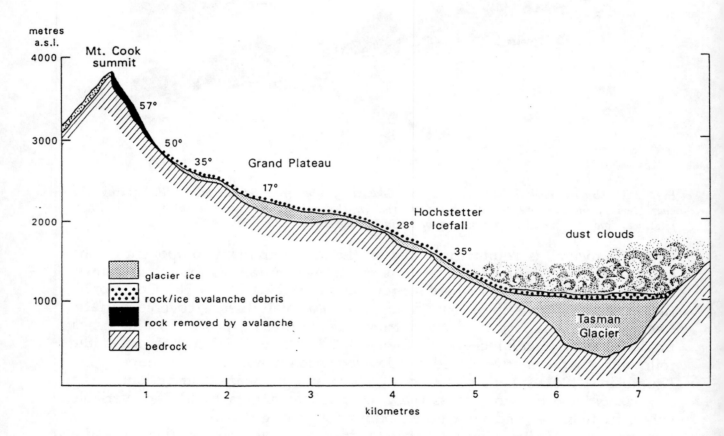

Fig. 2.32. Rock and ice avalanche from Mount Cook, New Zealand that fell onto the Tasman Glacier, 14 December 1991 (after Chinn et al 1992).

Supraglacial streams may remobilize the rockfall material and sort it. Pockets of such debris accumulate in hollows and, when the channel is abandoned, the debris retards ablation, so leading to the development of dirt-cones.

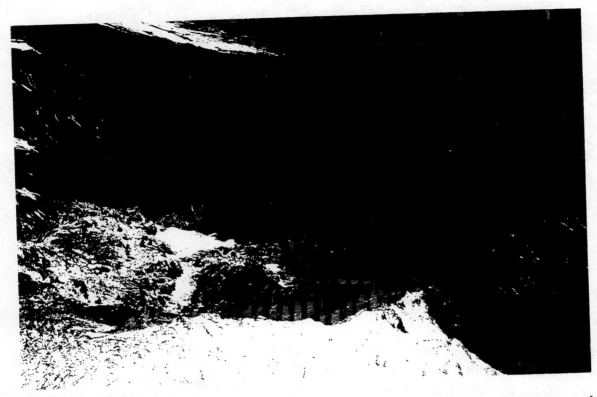

Fig. 2.33. *Aerial view of the Mount Cook avalanche site (outlined) two months after the event that reduced the height of the mountain by 20 m. On the lower left is the Hochstetter Icefall already having absorbed much of the debris into crevasses. Below, the Tasman Glacier flows from left to lower right. The avalanche has obliterated the already debris-covered glacier surface and extended a considerable way up the lateral moraine on the far side.*

2.10.2 Basal transport

Erosion at the bed of a glacier comprises various processes: crushing, fracturing and abrasion, which combine to produce a sediment very different from that carried on the surface. Debris eroded in this fashion is initially transported in a basal zone of traction, where particles frequently come into contact with the glacier bed and are retarded, so that large forces are imparted on both the particle and the bed.

Incorporation of debris into the basal ice is intimately associated with pressure melting and regelation (section 2.7.2). Pressure melting on the upstream side of a bump is followed down stream by refreezing of the released water, thereby allowing a thin layer a few millimetres to a few centimetres thick to be added as regelation ice to the glacier sole. Any loose debris, ranging in size from clay to boulders, is incorporated into the regelation layer, and concentrations of debris may reach 50% or more by volume. Beneath a temperate glacier, regelation ice rarely builds up to more than a metre or so, as it soon melts again as the ice moves down valley. In contrast, cold glaciers sliding on their beds build up dirty regelation layers (Fig. 2.34) as a succession that attains a thickness of several metres and persists down valley. Regelation ice-layers often appear foliated, like normal glacier ice, but derivation from water can often be determined using isotopic compositional contrasts (Souchez & Lorrain 1991).

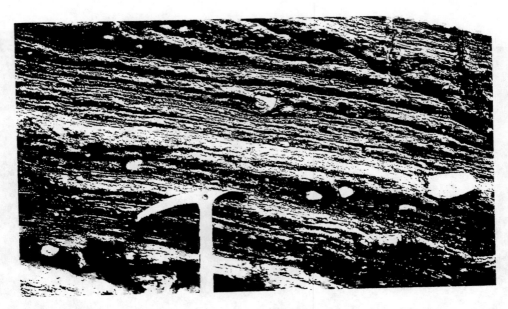

Fig. 2.34. Debris-rich basal ice formed as a result of regelation; Taylor Glacier, Dry Valleys, Antarctica.

A larger fragment (a cobble or boulder) may be incorporated into the glacier sole by ice deforming around it. It will be removed from its position, because shear within the basal ice results in rotation of the stone. Although a common process beneath sliding glaciers, it may also take place beneath cold-based ice to a limited extent.

Boulton (1978) demonstrated that clasts carried in basal ice (the zone of traction) have a significantly higher roundness (predominantly subrounded and subangular) and slightly higher sphericity than those carried in high-level transport (Fig. 1.4). In addition, the boulders tend to have relatively smooth faceted surfaces, although sharp-edged fractures on otherwise smooth boulders are common. Striated abrasional facets are also common. Clasts that are flattish and do not roll readily attain typical flat-iron shapes with a strong preferred orientation of striae. Clasts that are more spherical and roll easily develop striations with many intersecting sets. Lithology has some bearing on whether a clast develops striae or facets.

Boulton (1978) also recorded that in basal debris the grain-size distribution is polymodal, but overall is depleted in the coarse fraction and enriched in the fine fraction in comparison with debris undergoing high-level transport. The high proportion of fines is the result of comminution of larger particles in the zone of traction, a process characteristic of a crushing mill, and one that produces the rock flour that gives glacial meltwater its milky appearance.

A further distinction can be made between the shape of stones carried in the basal zone of traction (and subsequently deposited as melt-out till) and that of boulders deeply embedded in lodgement till and subsequently modified by overriding ice.The latter often acquire a smooth striated, often bullet-nosed up-glacier termination and a sharply truncated down-glacier termination. The striae form a single, slightly diverging set. Overall, embedded clasts in lodgement till show an even greater degree of roundness than debris from the zone of traction.

Fig. 2.35. Debris-rich basal ice, thrust to the surface of the lower White Glacier, Axel Heiberg Island, Canadian Arctic. Note the predominantly subrounded and subangular character of the stones (cf. rockfall debris in Fig. 2.32).

2.10.3 Debris incorporated into an englacial position

A certain amount of debris reaches an englacial position from the bed of the glacier and may even emerge at the surface (Fig. 2.35), a process that operates in both warm and cold glaciers, but especially in the latter. Deformational processes in the ice responsible for this include thrusting and folding.

Thrusting is a particularly common mechanism in cold glaciers, especially where a cold snout, frozen to the bed, is pushed from behind by sliding ice. Debris-rich basal ice may rise into the body of the glacier along a thrust. Many thrusts are related to transverse crevasse traces which have rotated into a orientation favourable for reactivation. Other planes of weakness may also promote thrust formation from the bed. Larger bodies of sediment may be incorporated into the ice where shear-stresses within the glacier propagate both into the underlying sediment and the sediment in front. The advancing glacier itself shows prominent thrusts and shear zones and the author suggests these may extend downwards onto the frozen sediment (Fig. 2.36). A "sole thrust" (in the manner of deformation in rocks in mountain belts) may develop in the sediment beneath and in front of the ice. As the ice advances, new thrusts develop progressively more forwards than the previous ones. Structural geological concepts have been applied to thrusting in push moraines by Croot (1987), but they apply equally well to the processes going on in the ice itself.

A second tectonic mechanism of debris incorporation involves folding of dirty basal ice. The folds tend to have limbs that are subparallel to the bed, and the hinges are perpendicular to the direction of ice flow. Hudleston (1976) has investigated such folds in the Barnes Ice Cap and demonstrated that they can form only when perturbations in flow occur. Repeated folding may cause the regelation layer to rise to a high level within the glacier.

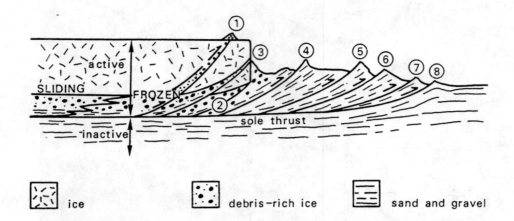

Fig. 2.36. Sketch to show how large masses of frozen subglacial sediment may be incorporated into an englacial position by thrusting. The thrusts extend down through the ice into the frozen sediment, there being no sliding at the ice/sediment interface. Forward movement of the ice leads to the development of a "sole thrust" from which upward-curving subsidiary thrusts develop progressively in front of each other.

The transfer of debris from the surface to an englacial position is associated with the opening of crevasses, a process which may be very effective in the short term (several years). With reference to the Mount Cook landslide referred to above, the Hochstetter ice-fall had, within just a few weeks, swallowed almost all the surface debris that had initially obliterated it (Fig. 2.33). Supraglacial streams may also be effective in transporting debris into the body of the ice, especially when they enter a moulin.

CHAPTER 3

GLACIAL EROSIONAL PROCESSES AND LANDFORMS

Fig. 3.1. *The Matterhorn, viewed from Allalinhorn, Switzerland. This classic horn has been eroded by glaciers on three sides. Two of these glaciers are visible here, divided by the steep Hörnli Grat, the arête by which the first ascent of the mountain was made (Photograph courtesy of Jürg Alean).*

3.1 Introduction

Glacial erosion has its most profound effect in mountain environments (Fig. 3.1), where material stripped off is rapidly transported to lower-lying regions. However, glacially eroded lowland areas are by no means uncommon. Large areas of the Laurentian Shield of North America, and those parts of the Sahara influenced by Ordovician glaciation, are lowland areas exhibiting significant glacial erosion.

3.2 Processes of glacial erosion

Glacial erosion comprises several processes: abrasion, fracture of structurally homogenous rock, fracture of jointed rock, each followed by entrainment of debris into the basal zone of the glacier, and meltwater erosion. Frost shattering of valley-side cliffs steepened by glacial erosion and the removal of debris supraglacially must also be taken into account.

Abrasion involves ice containing debris overriding bedrock. The debris scores the bedrock to produce fine-grained material (grain size generally <100 μm), as well as generating striations and other features discussed below (section 3.3.2). On the basis of laboratory work and direct observations beneath glaciers, the principal factors affecting glacial abrasion appear to be the presence and concentration of basal debris, the sliding velocity of the glacier on its bed, and the transport of debris towards bedrock, so continually renewing the abrasive surface. Other factors affecting the rate and type of abrasion include:

- ice thickness (generally the thicker the glacier the greater the pressure exerted on the bed),
- the presence of water at the glacier base (high water pressure has the effect of buoying up the glacier, so reducing the effective pressure, although this is counteracted by increased basal sliding),
- the relative hardness of basal debris in transport compared with bedrock,
- the size and shape of particles embedded in the ice,and
- the effectiveness of removal of the eroded debris, especially by meltwater.

Under certain circumstances, ice (or debris in the ice) exerts sufficient force on homogeneous bedrock to cause it to fracture. Crescentic fractures are one indication of this, apparently sheared boulders are another. Fracture can also occur by 'pressure release' after the glacier has disappeared (Lewis 1954). In effect, after rapid erosion, newly exposed rock is in a stressed condition, and joints may develop, commonly resulting in exfoliation of large sheets of rock on steep valley sides. Additionally, subaerial freeze/thaw in the presence of meltwater promotes fracturing, after the development of jointing in response to dilation (Harland 1957).

Rocks that are already jointed before the arrival of a glacier are particularly susceptible to erosion. Stratified, foliated and faulted rocks are also likely to be exploited more than structureless rock. The joints may or may not have been exploited by weathering before the onset of glaciation; if not, freeze/thaw beneath a glacier, the base of which is close to the pressure melting point, is a potential mechanism for loosening the blocks. The downglacier side of *roches moutonnées* is a favourable locality for block removal (Carol 1947). Once loosened, large blocks (*rafts*) may be detached from the bed.

Once material has been detached from the bed, it may be incorporated into the basal zone of the glacier by one of two processes. Small particles are incorporated by regelation, especially on the downstream side of bedrock obstacles, while larger blocks are picked up by the ice deforming around and surrounding them. For a glacier of simple shape, the efficacy of glacial erosion and transport reaches a peak near the equilibrium line.

3.3 Landforms resulting from erosion by glaciers

3.3.1 Modes of erosion

The most dramatic effects of direct glacial activity are erosional, and are represented by such major landforms as glacial troughs, cirques and alpine forms (horns and arêtes), medium-scale features such as *roches moutonnées*, and small-scale features such as polished and striated pavements, with their associated markings. Excellent detailed descriptions of these features, with many examples, are given by Embleton & King (1975), Sugden & John (1976) and Prest (1983).

As ice flows over an obstacle it tends to abrade an upglacier-facing surface, thereby smoothing it. In contrast, surfaces that face downglacier are subject to several mechanisms, including initial bedrock fracturing, loosening and displacement of rock fragments, and incorporation of those fragments into the base of the sliding glacier; such surfaces are rough, and are commonly referred to as plucked or quarried. Drewry (1986) has provided an extended quantitative treatment of these processes.

Features which are solely the result of abrasion are referred to as streamlined, and they include lowland bedrock areas which have been scoured, glacial troughs, watershed breaches, domes, whaleback forms, grooves, "p-forms", striations and polished surfaces. Other forms, which are the result of both abrasion and rock fracture, include trough-heads, cirques, cols, *roches moutonnées* and *riegels*; these are only partly streamlined. Another group of small-scale features is the result of rock crushing, whereby stones embedded in the base of moving ice subject the bed to sporadic impact; these are referred to as friction cracks and are non-streamlined forms. A final group of landforms are in effect residual, reflecting what is left after a combination of abrasion and fracturing by ice, and frost-shattering and hillslope movement, has attacked an elevated mass of rock. Such landforms are characterised by horns, arêtes and nunataks.

A landform classification based on one provided by Sugden & John (1986, p. 169), but modified according to the processes discussed by Drewry (1986), is given in Table 3.1. The following discussion treats each group in turn, more or less according to scale, the largest first.

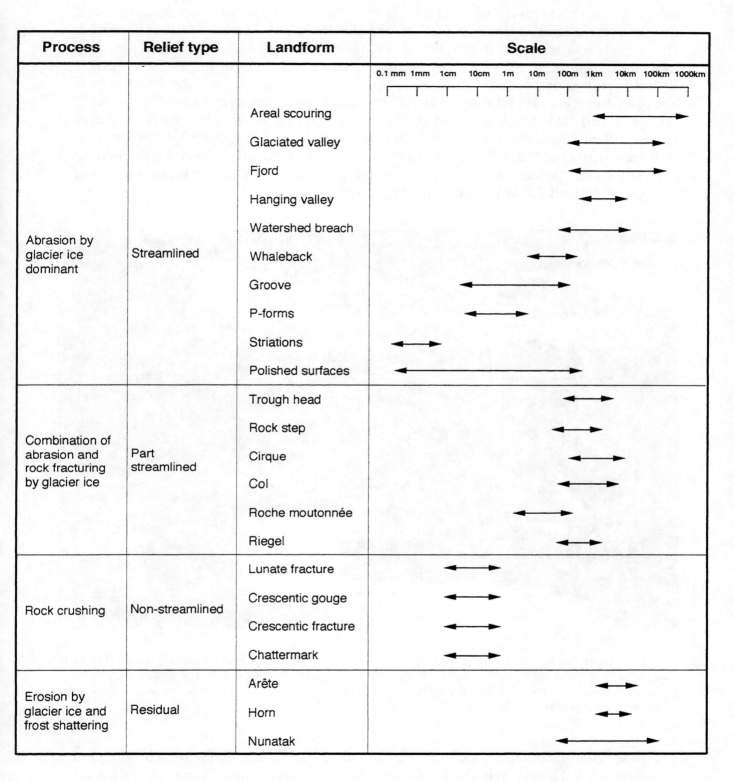

Process	Relief type	Landform	Scale
			0.1 mm 1mm 1cm 10cm 1m 10m 100m 1km 10km 100km 1000km
Abrasion by glacier ice dominant	Streamlined	Areal scouring	
		Glaciated valley	
		Fjord	
		Hanging valley	
		Watershed breach	
		Whaleback	
		Groove	
		P-forms	
		Striations	
		Polished surfaces	
Combination of abrasion and rock fracturing by glacier ice	Part streamlined	Trough head	
		Rock step	
		Cirque	
		Col	
		Roche moutonnée	
		Riegel	
Rock crushing	Non-streamlined	Lunate fracture	
		Crescentic gouge	
		Crescentic fracture	
		Chattermark	
Erosion by glacier ice and frost shattering	Residual	Arête	
		Horn	
		Nunatak	

Table 3.1. Classification of glacial erosional landforms according to process of formation, relief form and scale.

3.3.2 Glacial abrasion

Areal scouring
Some lowland areas bear abundant signs of glacial erosion on a regional scale, such as large parts of the Laurentian and Baltic shields. A low amplitude but

irregular relief, sometimes referred to as "knock and lochan" topography after the heavily scoured landscapes of the North-West Highlands of Scotland, characterizes such terrain. In this area, ice-masses flowed over, and scoured, Lewisian basement gneisses, occasionally leaving the summits of the younger Precambrian Torridonian sandstone peaks as isolated nunataks. Bedrock is exposed nearly everywhere, with *roches moutonnées* (see below) or rocky knolls (knocks) and lake (lochan)-filled depressions. Structural features, such as joints, faults, dykes and steeply inclined strata were selectively exploited by the ice. This type of landscape, which has a relief of about 100 m, must have been formed beneath a sliding ice sheet. Such conditions probably exist beneath large parts of the Antarctic and Greenland ice sheets today.

Fig. 3.2. *A sandstone tor of periglacial origin in the Kennar Valley area, Victoria Land, Antarctica. This area was once covered by ice as indicated by scattered erratics, but there was little erosion, probably because the ice was frozen to the bed. In contrast, deep erosion took place in the background where an ice stream was sliding on its bed.*

Sometimes associated with areally scoured landscapes are sets of parallel grooves and flutes, which in some respects resemble fluted moraine in predominantly depositional environments (ch. 4). Genetically they may be linked, in the sense that obstacles to flow may initiate a flute. Good examples are to be found eroded in the Precambrian basement gneisses of the Laurentian (Canadian) Shield, notably in northeastern Alberta (Embleton & King 1975).

In contrast to these areally scoured regions, some areas of low or moderate elevation may have experienced little glacial erosion, even though inundated by an ice sheet. In such cases only isolated erratics on a deeply weathered surface with fluvial or periglacial forms will indicate the former presence of ice. Ice-free areas in Antarctica illustrate well the effectiveness of glacial erosion where ice is channelled into valleys, but also its rather protective nature on the intervening plateau remnants (Fig. 3.2), even though erratics indicate that ice covered the entire region. It is likely that the thinner ice over the plateau was frozen to the bed and slow moving, protecting it from erosion. Much of the present-day ice-cover, too, is frozen to the bed, and erosion is minimal.

a

b

Fig. 3.3. Contrasting forms of glaciated valleys. (a) A true U-shaped profile, the Yosemite Valley in California. The near-vertical crag on the left is El Capitan, 1500 m high. (b) A more typical glaciated valley, Glen Rosa, Isle of Arran, Scotland, with a more open (parabolic) cross-section. Note the small breached watershed, cirque and horn named Cir Mhòr at the head of the valley.

Glaciated valleys

Glacial troughs, a term which embraces both valleys and fjords, are probably the most spectacular manifestations of glacial erosion. They are predominantly the product of abrasion as a result of ice being strongly channelized, but rock fracturing and plucking downstream of smoothed faces is common until iregularities have been removed. It is widely recognised (Flint 1971), however, that most troughs follow the line of existing river valleys, but the glaciers act as overdeepening, widening and straightening agents.

Many glacial troughs are eroded by valley glaciers descending from high mountains, such as where cirque glaciers extend towards lower ground, as in the European and New Zealand Alps, the Western Cordillera of North America, the Himalayas and other temperate areas today. Other glacial troughs develop beneath, or extend from, ice sheets or ice caps, where ice-streaming is prominent, as beneath the Greenland and Antarctic ice sheets, and other Arctic or sub-Arctic ice caps. In the case of East Greenland, the landscape is characterised by deep incisions into plateau-like terrain with striations to the top of the cliffs, while the plateaux themselves bear little evidence of erosion. The incisions form a complex dendritic pattern, depending on the alignment of preglacial river valleys, or a "chocolate tablet-like" arrangement if controlled by intersecting bedrock faults. Modification involves overdeepening, the development of trough-heads, steps, riegels and basins, the steepening of the valley sides and truncated spurs.

In cross-section, glaciated valleys are often described as U-shaped. Well-known valleys which approach a strictly U-shaped profile include Lauterbrunnental in Switzerland and the Yosemite Valley in California (Fig. 3.3a). These valleys feature flat bottoms, reflecting infilling by hundreds of metres of glaciofluvial gravels or lake sediments. The lower rock and scree slopes end abruptly against this flat floor. However, truly U-shaped profiles are exceptional, and most glaciated valleys have much less steep sides (Fig. 3.3b). An "ideal" glaciated valley, in mathematical terms, has a parabolic profile, according to the formula:

$$y = ax^b$$

where the x and y axes correspond respectively to the depth of the valley and half the width respectively. Most ideal troughs have an exponent (b) of about 2. On the other hand many troughs have asymmetrical profiles commonly related to differences in bedrock hardness. In the English Lake District, the Buttermere valley has a steep southern and southwestern flank comprising rocks of the hard Borrowdale Volcanic Group and intrusive syenite, whereas the opposite flank with its gentler, less craggy aspect is made of softer mudstones of the Skiddaw Group. The steeper reaches of many alpine-type glaciated valleys are characterized by a V-shaped profile. This may be due more to vigorous erosion by subglacial meltwater under high pressure than by abrasion by ice. Even many parabolic profiles are interrupted by V-shaped notches in the valley floor.

In contrast to the normally smooth cross-sections, the long profiles of glaciated valleys are extremely irregular, as a result of uneven overdeepening by the ice. Overdeepening is controlled by many factors, the most important of which are the presence of constrictions in the valley or the entering of a tributary glacier into the main stream, both of which result in accelerated ice flow and enhanced erosion; also important are changes in bedrock geology and structure. As a result, many glaciated valleys have multiple rock-basins filled by lakes. Often these valleys (and fjords) are overdeepened below the general level of the continental shelves offshore. Many examples exist in western Scotland; Britain's deepest lake, Loch Morar, is 315 m deep but its surface is only 15 m above sea level. Some of the largest ice-scoured hollows, now filled by the lakes, lie near the perimeter of the Alps, the deepest being Lago di Como (410 m) in northern Italy. Many of these lakes are deepened further by being dammed by material scooped out of the hollows and dumped as end moraines. Most glaciated valley lakes begin to fill rapidly with glaciolacustrine and glaciofluvial material, but they become much more stable when this source is cut off. Gravel fans from a side valley may grow out into a lake, while suspended sediments will also gradually fill the lake (ch. 6).

In the geological record, reports of glaciated valleys are rare, but a few examples have been described from the late Palaeozoic sequences in Gondwanaland (South Africa and Antarctica).

Fjords

The term *fjord* is of Norwegian origin (spelt *fiord* in North America and New Zealand) and, in a morphological sense, refers to a glacial trough the floor of which is below sea level. Fjords are actively forming today in Antarctica, Greenland, the Canadian Arctic Islands, Svalbard, Novaya Zemlya, southern Alaska, Chile and South Georgia, where tidewater glaciers reach the sea. Fjord coastlines from earlier glacial phases are well developed also in Norway, Scotland, Iceland, British Columbia and New Zealand.

Genetically, fjords and glaciated valleys are similar, and frequently pass into one another. They possess many common characteristics, of which the most striking is the depth to which they have been overdeepened, illustrating the effectiveness with which glaciers can erode, irrespective of the level of the sea.

Like glaciated valleys, fjords have a parabolic profile, although many more of them approach a true U-shape with vertical sides, particularly where they are cut in hard crystalline rocks beneath former ice sheets, such as in Greenland, southwestern Norway, Baffin Island, South Island of New Zealand (Fig. 3.4). Greenland and Norway have particularly long fjords, many over 100 km long, and the depth of some, from crest to sea bottom, attains 3000 m. Probably the world's biggest fjord is the Lambert Graben in East Antarctica, but it is filled by an ice stream that drains a large part of the East Antarctic ice sheet. If this ice were to disappear a fjord c. 800 km long, 50 km wide and 3 km from crest to floor would become exposed. These exceptional dimensions reflect the far longer period of glacierization in East Antarctica (at least 40 m y) than in other parts of the world, as well as the softer bedrock that occurs on the floor of the Lambert Graben compared with the adjacent land.

Many fjords have a characteristic long profile which relate to most intense glacial erosion at their inner ends, as it is there that the ice in contact with the bed is at its most dynamic. Thus fjords deepen quickly from their headward ends and then progressively shallow seawards, although this general trend is often interrupted by the presence of distinct basins. The mouth of a fjord is often marked by a sill with shallows, rocky reefs or even islands. The best-known example is Norway's Sognefjord, which has a maximum depth of 1308 m, but at the entrance the fjord is only 3 km wide and 200 m deep. The sills may be capped by morainic debris, although the depth of this material in relation to the eroded basins is small.

93

Fig. 3.4. Fjords in Greenland and New Zealand. (a) The world's longest open fjord Nordvestfjord which, with Scoresby Sund, stretches for 300 km from the open sea. An active tidewater glacier at the head (cf. Fig. 2.10) discharges tabular icebergs which can be seen heading towards the open sea. Rotten (greyish) sea ice floats on the fjord surface in the foreground. (b) The head of New Zealand's best known fjord, Milford Sound, with the horn, Mitre Peak (1692 m) on   the left. Glaciers last occupied this fjord in the late Pleistocene Epoch, although mountain glaciers (right) still survive. Successive phases of downcutting are evident from the change in slope of the fjord walls.

Recorded fjord depths may only partly reflect the depth to which glacial erosion has occurred, since large volumes of sediment often fill the overdeepened basins, sometimes to depths of several hundred metres, as in Glacier Bay, Southern Alaska today (ch. 7).

Less spectacular, but nonetheless renowned for their scenic beauty, are the sea lochs of western Scotland. Although with gentler slopes and a relief rarely exceeding 1200 m, these fjords show many of the same characteristics of the

larger fjords, namely multiple basins, deeper inner parts and rocky sills. In many cases, such as Loch Morar and Loch Maree, the sills have been raised above sea level since deglaciation as a result of isostatic rebound, creating freshwater lochs separated from the sea by short rapids.

Some fjords show progressive stages of downcutting, and even the upper part of the former river valley profile may be preserved. A 'U' within a 'U' also may be apparent (Fig. 3.4b). Major landforms associated with fjords include trough-heads, hanging valleys and breached watersheds.

Fjord systems are as varied in terms of their spacial arrangement as glaciated valley systems. Some have linear trends related to faults, as in East Greenland and Scotland. Other fjord systems that were fed by several equally important glaciers, such as Glacier Bay in Alaska, have a dendritic arrangement . Fjord systems with sinuous branches are common in southwestern Norway, although in detail the fjord walls are smooth, and only the spurs truncated.

Breached watersheds

The action of a valley glacier, when it grows sufficiently to spill out of its constraining trough as a diffluent ice stream, leads to downward abrasion of the lower cols. These cols may themselves attain a parabolic form in cross-section and eventually be so deeply eroded that the main trunk glacier is diverted. It may be, however, that the effect of diffluence of ice over a col is simply the transfer of some ice from one valley to an adjacent one.

Fig. 3.5. A near U-shaped breached watershed in East Greenland, perched high on a precipitous rocky ridge. The main glacier flowed down Nordvestfjord, behind the ridge from left to right, and some ice spilled over this col joining another ice stream that had overridden a low-level col to the left.

Most glaciated mountain regions demonstrate glacial diffluence. The British Isles have many examples, some of which are well described by Embleton & King

95

(1975). A good example connects Glen Sannox and Glen Rosa on the Isle of Arran (Fig. 3.3b). The former, with its own valley glacier, was partially blocked at its mouth by a major glacier from the Central Highlands flowing south down the Firth of Clyde. As a result some of the ice accumulating at the head of Glen Rosa spilled south over a col into Glen Rosa, as indicated by abraded rocks on the col and its overdeepened character. Glacial diffluence on a widespread scale is evident in the Grampian Highlands. A well-known example is the Lairigh Ghru, a high pass that divides the 1300-1400 m high Cairngorm Plateau into two distinct parts. Here ice in the upper Dee valley was restricted by ice to the south, and thus overspilled the col to the north, lowering the col by an estimated 230 m to 840 m. Many other examples in Scotland, Wales and the Lake District have been reported.

Watershed breaching is also a feature of fjord landscapes, and notable examples occur in southwestern Norway and East Greenland (Fig. 3.5). The latter area provides several excellent examples of watersheds having been cut down to near the level of the base of the main trunk glacier.

When diffluence attains a level at which all cols are being used to discharge ice the term glacial transfluence is used. The process is well illustrated by past and present ice sheets, and frequently represents a shift in the ice divide inland of the original mountain divide. This happened in Scandinavia and Scotland, as revealed by the transport of erratic boulders eastwards up and over the respective mountain divides. The Antarctic demonstrates transfluence on a vast scale. Here the East Antarctic ice sheet impinges against the Transantarctic Mountains at a level of around 4000 m and discharges ice towards the Ross Ice shelf over buried cols. Some of the biggest and fastest-flowing ice streams in Antarctica are cutting through the Transantarctic Mountains, including the Beardmore and Axel Heiberg glaciers which respectively provided routes to the South Pole for Scott and Amundsen in 1911/1912. Many parts of the Greenland ice sheet are behaving in a similar fashion. Greenland is more-or-less fringed by mountains which, even though over 3000 m high in places, are only partly able to constrain the ice sheet, and many glaciers spill out seawards through breaches in the coastal mountains. Another area of extensive ice-cover, the Icefield Ranges of Alaska and the Yukon, demonstrates the same process of watershed breaching, as ice escapes seawards to the south, through ranges over 4000 m high.

Hanging valleys
These are a characteristic feature of glaciated mountain landscapes, although not exclusively confined to them. The hanging valleys were occupied by tributary glaciers that were not as effective as eroding downwards as the main trunk glaciers. Thus one finds glacially shaped side-valleys ending abruptly against the steep wall of the valley (Fig. 3.6) or fjord (Fig. 3.7), and this provides the site for waterfalls to form. Cirques are a type of hanging valley, but normally the term is restricted to side valleys that had valley glaciers in their own right. One of the most famous examples of a hanging valley landscape is the Yosemite of California, where the Ribbon and Bridalveil falls both drop several hundred metres over vertical cliffs into the main valley (Fig. 3.3a). Most fjord regions have fine hanging valleys, as do high-alpine regions. Even relatively low highland areas like those of Britain have well developed hanging valleys, though seldom do they have sheer rock faces below.

a

b

Fig. 3.6. Hanging valleys. (a) The Steall Waterfall forms a graceful cascade as it drops into Upper Glen Nevis, Scotland. (b) This hanging valley terminates at the vertical northern flank of Milford Sound, New Zealand, creating a fine waterfall.

Domes and whaleback forms

If a glacier meets an obstruction, it may not be able to abrade it totally, but leaves it as an upstanding, smoothed rock hillock. There are two types, domes and whaleback forms which are totally the product of abrasion. On a similar scale, *roches moutonnées* (section 3.3.3) represent a combination of abrasion and plucking.

Domes are relatively unusual features and are best developed in areas of homogeneous bedrock which, after being eroded by the ice, is subject to *exfoliation*. This process involves peeling off, like the skin of an orange, curved slabs of rock, so presenting a smooth, streamlined form to the next ice readvance. Among the finest examples of glacially eroded domes are those cut into granite in the Yosemite National Park. Some of them are nearly symmetrical and reach heights of several hundred metres (Fig. 3.7).

Fig. 3.7. A series of ice-abraded domes of granite rising above the glacial Tenaya Lake basin, Yosemite National Park, California. A number of erratics litter the shore and shallows of the lake.

Whaleback forms, which have also been referred to as rock drumlins, are a good example of a streamlined landform, and are typically tens to a few hundred metres long, and from less than one to tens of metres high. Length:width ratios are commonly 1:2 to 1:4, and they are orientated parallel to flow. In this respect they are similar in shape and orientation to drumlins made of drift and, indeed, they may occur as peripheral members of drumlin fields. Clear examples of streamlined whaleback forms are rare, but Embleton & King (1975) and Sugden & John (1976) have provided examples from Scotland, Iceland, and the Baltic and Canadian shields.

On a larger scale, spurs and interfluves may be tapered and smoothed. The Finger Lakes of New York State provide one such area. All these forms carry the small-scale markings described below.

Striated, polished and grooved surfaces

Striations are among the most common features of glacial erosion. They are finely cut, U-shaped grooves on the surface of bedrock that has been scored by stones in the base of a sliding glacier (Fig 3.8a). Individual lines are sometimes a metre or more long occurring in parallel. Glacial striations are quite varied in form. If the cutting tool is rotated, so presenting a new cutting edge, the resulting striations appear to step sideways or lie *en echelon*. Many striations are asymmetric, blunt and deep at one end, tapering at the other end (Fig 3.8b). These are sometimes known as "nail-head" striations and they usually indicate ice flow towards the tapering end, but, on surfaces sloping up-glacier, gouging may give rise to the opposite pattern. Some striated surfaces have *rat-tails* (Fig. 3.8c), which are minor ridges extending downstream from knobs of more resistant rock which have protected the more easily striated material on either side. Although uncommon, rat-tails are good indicators of the sense of direction of ice flow as opposed simply to its orientation.

Fig. 3.8. Small-scale features of glacial abrasion. (a) Well-preserved striations made by Late Palaeozoic glaciers on early Proterozoic dolomite, near Douglas, Karoo, South Africa. Although these striations display strong parallelism, a curving set of "tram-lines" indicates occasional variations in ice-flow direction. Note the diamictite of the Dwyka Formation at top right. (b) Short nail-head striations on a steep rock face that flanked the McBride Glacier, Glacier Bay, Alaska in the 1970's. Ice flow was towards the right; note how the striations score progressively deeper into the bedrock until the abrading tool flips out. These nail-heads formed at a later stage of erosion than the dominant gently-inclined set   of regular striae. (c) Rat-tails in dolomite, developed down-glacier of resistant chert concretions which protect the bedrock from erosion, Wordie Gletscher, East Greenland. Ice flow was from bottom to top of the picture (d) Ice-abraded boulder pavement at the base of the Late Palaeozoic Dwyka tillite, Elandsvei, near Tweifontein, Karoo Basin, South Africa. Note the regular nature of the striations, formed as the ice flowed from right to left.

Striations commonly show a wide range of orientations in a small area. On uneven bedrock they reflect all the irregularities of basal flow of the glacier. Even on flat surfaces, rotation of the gouging stones results in marked deviations from the mean flow direction, while different recognizable sets may reflect longer-term changes in the directions of flow. Therefore, it might be thought that striations are of little use in reconstructing the mean directions of movement of former glaciers and ice sheets. Nevertheless, such directions have been obtained successfully in many cases. Measurement of a large number of striations in a small area facilitates statistical treatment and a graphical indication of mean ice flow (especially if supported by other evidence), even if individual striations are orientated up to 90° from the mean.

Fig. 3.9. Late Palaeozoic striations and related features formed in a thin veneer of soft sediment that lay on quartzite bedrock, Elandsvei, South Africa. (a) Ploughed ridge in front of pebble (above coin) with groove behind; flow was from right to left. (b) Flowage of soft sediment from top to bottom of picture has created small lobate features that partially obliterate the groove extending to the right of the head of the hammer.

Minute scratches in large quantities give rise to a polished appearance to the rock, although they may be visible only under a magnifying glass. The degree of polishing depends on the fineness of the abrading material. Fine-grained rocks with little bedding, foliation or jointing, are the most suitable for the generation and preservation of polished surfaces. Coarsely crystalline and easily weathered rocks are the least likely to bear well-preserved striations and polished surfaces. Exceptionally hard rocks, such as quartzites, are also unlikely to acquire many striations. Sometimes till itself may be subjected to glacial erosion, resulting in bevelling of the stones embedded in the till and the development of a striated boulder pavement (Fig. 3.8d). Even boulders in glaciomarine sediment may become striated if they are overrun by grounded ice (section 8.5.6). On exposure to air, striations and polish are often soon lost through weathering. They tend to survive best if subsequently buried rapidly by till. In this way, striations dating from 10-20,000 years ago may be well-preserved, as in the highland areas of Britain. In the older geological record, striated surfaces are often exposed beneath tillite horizons and provide convincing evidence of former glaciation. Impressive exhumed striated pavements of late Palaeozoic age are widespread in southern Africa (Fig. 3.8a) and other Gondwana continents. Ordovician and late Precambrian pavements are found in the Sahara; others of late Precambrian age are known from places as far apart as China and Norway.

Another type of striation arises when ice flows over bedrock that has a thin film of mud or sand. Such striations are recognizable by plough marks associated with stones pushing the soft sediment (Fig. 3.9a), and by slumping of the sediment into the groove (Fig. 3.9b).

As indicators of glaciation, striations have certain limitations. They are insufficient in themselves since other nonglacial agencies can give rise to striations. For example, floating lake, river or sea ice with embedded stones is capable of scratching a rock, although the resulting striations tend to be shorter, more irregular in orientation, and are found only in limited areas.

Debris-flows, both subaerial and submarine, can striate the surfaces over which they are flowing; volcanic debris-flows are particularly capable of this. Avalanches can also scratch a surface. Tectonic lineations sometimes closely resemble glacial striations. All these types of striations, however, tend to show a greater tendency for parallelism than do those formed by glaciers.

Striations are gradational with large furrows or grooves carved not only out of soft sedimentary rocks but also granites and gneisses. Grooves themselves may even be regarded as gradational into glacial valleys. Many grooves appear to be simply enlargements of single striations. Ice may occupy such grooves and further result in localized abrasion. Grooves attain depths and widths of a few metres, and lengths of several hundred metres (Fig. 3.10a,b). Many have overhanging walls which themselves are striated. Spectacular examples on islands in western Lake Eire have been described by Goldthwait (1979) and grooves 30 m deep and 1.5 km long have been reported in the Mackenzie River Valley in Canada (Flint 1971, p. 89).

"Plastically moulded (or p-) forms"
Some glaciated rock surfaces exhibit complex, smooth forms which are frequently known as "p-forms", features which have been the subject of considerable argument. Various origins have been proposed for these enigmatic features: to (a) the normal processes of glacial abrasion, (b) the movement of

saturated till at the base or sides of a glacier, and (c) the action of meltwater, especially under the high pressures that may exist in places beneath a glacier . All three processes may in fact be responsible for the different features within this group. The best preserved p-forms have been described from areas of resistant igneous and high-trade metamorphic terrains. Norway is particularly well endowed with p-forms, where they occur along deeply incised glacial valleys and fjords. p-forms are also to be found on some of the western isles of Scotland, e.g. Mull and the Garvellachs (see Ch. 3).

Fig. 3.10. Grooves resulting from prolonged glacial abrasion. (a) Steep-sided groove, measuring about 3 m deep and 8 m across formed under the Quaternary Laurentide ice sheet, Whitefish Falls, near Sudbury, Ontario. (b) Open, wave-like grooves formed under the Late Palaeozoic ice sheet that covered Gondwanaland, near Douglas, Karoo, South Africa. Note that in both cases the grooves themselves are striated.

On the basis of work in northern Norway, Dahl (1965) classified p-forms according to their geometry. Those that are likely to be more the product of glacial abrasion than any other process are:

Cavetto forms are channels cut on steep rock faces and orientated parallel to the valley sides. They may be up to 0.5 m deep, have sharp edges, and the upper part may overhang. Striations and crescentic gouges occur within them. Such linear features may well be the product of abrasion.

Grooves occur on flat open surfaces (as mentioned above), but have rounded edges. These features, often cut in extremely hard rocks, are also likely to be the result of glacial abrasion.

Other p-forms include *Sichelwannen, curved* and *winding channels* (alternatively known as *Nye channels*. These are described in section 5.3.1 as the predominant process is likely to be meltwater erosion.

Fig. 3.11. Trough-head in Gasterntal, Berner Oberland, Switzerland. The snout of the glacier Kanderfirn rests at the top of the heavily plucked rock face and the Little Ice Age lateral moraine is visible on the right.

3.3.3 Abrasion and rock fracture combined

Trough-heads and valley-steps

Many glaciated valleys and fjords terminate abruptly inland at steep, rocky faces called trough-heads or trough-ends. They mark the position where overdeepening in the longitudinal profile has occurred, but their origin is obscure. It has been suggested (Sugden & John, 1976, p. 184) that a trough-head represents a switch from sheet flow of ice (as at the edge of an ice cap) to channelized flow, combined with a change in the basal ice conditions to one that promotes slip. However, trough-heads are morphologically similar to valley-steps and the latter occur within alpine valleys where ice-flow is channelized throughout.

The trough-head or valley-step has a heavily "plucked" appearance, but ice-abrasion marks may be found on the less steep, downvalley-sloping surfaces on the craggy face itself, as well as on the crest of the trough-head. It is probable that some trough-heads and valley-steps were related to original breaks in slope caused by the cropping out of harder rock, but that the step migrates upvalley and is enhanced by a combination of abrasion and rock fracturing. As the ice flows over a vertical face it loses contact with the bed and creates a cavity. Here freezing and thawing processes may assist in the loosening of blocks. The ice regains contact with the bed lower down and abrades it, and the process is repeated down the cliff. The combination of these processes may help to perpetuate these features, but insufficient work has been done on their morphology to characterize adequately the processes resopnsible for their formation.

Some of the most spectacular trough-heads occur at the heads of fjords in southwestern Norway. Several branches of Sognefjord have good examples, the best known being at the head of Aurlandsfjord where a branch of the Oslo-Bergen railway spirals down about a thousand metres altitude from Myrdal to the village of Flåm on a short sediment fan close to fjord level. Another impressive example lies at the head of Gasterntal in the Bernese Oberland of Switzerland (Fig. 3.11). In Britain, fine trough-heads, but on a smaller scale, occur at the head of the Buttermere Valley in the Lake District and at the head of Loch Avon in the Cairngorms. In the USA, the Yosemite Valley terminates in a trough-head.

Valley-steps are more common than trough-heads. Many valleys in the Alps have several rock steps, each of which occurs where the valley narrows. In Wales, a series of valley-steps originating in a cirque occur on Snowdon, and there is another fine example in the Cuillins of Skye in Scotland.

Both trough-heads and valley-steps have been likened to large-scale roches moutonnées (below) and it is probable that all these forms are the result of common processes.

Riegels

This German term is given to the rock barrier that extends right across the valley, either holding back a lake or, when breached, as an upstanding transverse ridge (Fig. 3.12). The barrier may show limited signs of abrasion on the upglacier and much fracturing on the downglacier side. Riegels usually form where a band of resistant rock crosses the valley.

Cirques

Of all the landforms of glacial erosion, cirques are among the most fascinating, and they have long been regarded as one of the surest indicators of past glacial activity. Although the term cirque embraces a broad family of landforms, in its most characteristic form it resembles an armchair-shaped hollow high up on the mountainside (Fig. 3.13). The widespread occurrence of cirques (the term is French, but used internationally), is reflected in a wide variety of local names. In Britain the term *corrie* (from the Gaelic, *coire*) is normally used, or *cwm* (pronounced koom) in a Welsh context. Other terms are *Kar* (Germany), *botn* (Norway) and *nisch* (Sweden).

104

Fig. 3.12. Riegel or transverse rock barrier breached by the river at the right. The ice in the foreground is Franz Josef Glacier, Southern Alps, New Zealand.

Cirques are invariably present above the sides, and at the heads of glacial troughs, or even close to sea level in some glaciated coastal mountains. Cirques are an extremely varied landform in terms of both size and shape. They may be as little as a few hundred metres wide (Fig. 3.13), yet the largest (the Walcott cirque in Victoria Land, Antarctica) is 16 km wide and has a headwall 3 km high (Flint 1971: p. 133). Cirques in the steep terrain of Alpine regions tend to slope outwards and to be poorly developed. In areas of less pronounced relief they commonly contain lakes (tarns), filling rock basins or dammed by moraines. However, the length:height ratios from the lip of a mature cirque to the top of the headwall is remarkably constant, ranging from 2.8:1 to 3.2:1. Thus for the Western Cwm of Everest it is 3.2, for the Blea Water corrie in the English Lake District, a fraction of its size, it is 2.8 (Manley 1959).

Following analysis of the form of numerous Scottish cirques, Haynes (1968) found that their longitudinal profiles can best be described by logarithmic curves of the form:

$$y = k \, (1-x)^{e-x}$$

where x = the distance from headwall to lip, y = the depth of the cirque from the headwall to the basin, and k is a constant. The k values reflect how well the basin is developed. At the ends of the range of k values, k = 2 is characteristic of a deep cirque with a tarn and a steep headwall; whereas k = 0.5 is typical of a relatively open cirque with a gently inclined headwall (Fig. 3.13).

The detailed form of cirques is closely controlled by the structure of the rock, in particular jointing and bedding. A complicating feature, however, is that many cirques are composite in nature, with small cirques formed at a later, less intensive stage, of glacierization within the major feature. Many British mountain areas have cirque-within-cirque forms. A good example is Coire Bà, one of the largest in Britain, cut into the east face of the Black Mount in the Grampian Highlands. Here, several small cirques are incised into the headwall of the main cirque. In the Cairngorms, Coire an Lochain Uaine is a double cirque that forms part of a whole complex of cirques around the headwaters of the River Dee.

Fig. 3.13. A well-formed small cirque, Addacomb Hole, cut into Skiddaw Group metasediments (Ordovician). Formerly occupied by a morainedammed tarn, the cirque is now dry; the gullies cutting the moraine are clearly visible in the photograph.

Some cirques have a composite long-profile, representing a number of cirques at different elevations, known as cirque stairways. One of the best examples in Britain is on Snowdon. Here, the large cirque of Cwm Llydaw with its overdeepened basin containing a tarn, has a precipitous headwall, breached part way up by a smaller cirque, Cwm Glaslyn, also with a tarn. Above the latter, there is a small incipient cirque directly below the summit of Y Wyddfa. The stream connecting them falls over steep cliffs, the headwalls of the successive cirques. In full glacial times ice formed a continuous stream with ice-falls. At less intense phases, the higher cirque may have supplied the lower ones with avalanche debris, so for a time at least two of them may have contained separate glaciers, but equally likely is that each represents a different level of glacierization Another possibility is that accumulation may have been controlled by the wind so that the greatest accumulation was at the foot of headwalls irrespective of altitude.

One of the few studies of a cirque glacier is described in a classic monograph edited by Lewis (1960). The investigators examined the relatively simple Vesl-

Skautbreen in the Jotunheim of Norway. A tunnel was hand-dug through the ice by a team of volunteers to the cirque headwall in two places and a comprehensive three-dimensional picture of the velocity distribution and structure was obtained. The dominant structure is annual stratification which, as it moves down through the glacier, changes from a surface-parallel down-glacier dip to an up-glacier dip near the snout. Both this and the velocity measurements demonstrate that ice flow has a rotational slip component, thereby enabling the greatest erosion to occur directly below the level of the equilibrium line. This rotational component of flow is thus held to be responsible for the overdeepening that ultimately creates a basin for a tarn.

The initiation of a cirque is a matter for debate, but accumulation of snow on leeward slopes, especially in depressions (nivation hollows) is likely. Once successive years of snowfall have accumulated and turned to ice, flow takes place, and downward and backward erosion commences. As the floor forms, it is abraded, while dilation of the bedrock and freeze-thaw, resulting in rock fracturing and block removal, takes place at the headwall.

Roches moutonnées

Features similar in size to whaleback forms but which are only streamlined on the side facing up-glacier are known as *roches moutonnées*, a term introduced in 1787 by De Saussure because of their similarity to *moutonnées*, the wavy wigs that were in fashion at the time (Embleton & King 1975: p. 152). They are much more common than whaleback forms and have an asymmetric form as a result of glacial plucking along joints on their downstream side; they are particularly well-developed in jointed crystalline rocks. Rock fracturing, induced by the impact of ice-embedded stones, is probably also important on the crest of a *roche moutonnée*.

Large areas may be covered by *roches moutonnées*; they provide one of the most useful criteria for determining the former directions of ice flow. One of the most striking characteristics of glacially eroded terrain is the contrast between the smooth appearance of the eroded surface looking in the direction of former ice flow and the craggy appearance looking upstream. In many areas, the form and size of *roches moutonnées* are related to the other structures in the rocks besides joints, e.g. foliation, faults, dykes and alternating hard and soft lithologies.

The development of *roches moutonnées* is probably related to pre-existing hillocks. The suggestion by Carol (1947), following observations beneath the Oberer Grindelwaldgletscher in Switzerland, that plucking was initiated by meltwater freezing in joints in a zone of lower pressure beyond the crest of a protrusion in the glacier bed, is supported by observations beneath various other temperate glaciers and theoretical studies of basal sliding. *Roche moutonnée*-like forms, thought to be the result of the same processes, often occur as partially streamlined bosses on steep hillsides.

Most *roches moutonnées* range from a few tens to a few hundreds of metres in length, and several to tens of metres high. There are also smaller and very much larger ones in some areas. In parts of Scandinavia basement rock hills some hundreds of metres high are commonplace in the more subdued terrain. A well-known large *roche moutonnée* is the Lembert Dome in the Yosemite National Park; it rises about 250 m above the Tuolumne Meadows (Fig. 3.14).

Low-lying areas which have been heavily scoured, such as the Baltic and Canadian shields, and the North-West Highlands of Scotland display numerous, closely spaced *roches moutonnées* as well as small rock-basins. Many of these do not have a consistent shape because of complex structural controls such as variably orientated faults and foliations.

Fig. 3.14. Large (250 m high) roche moutonné in granite, Lembert Dome, Yosemite National Park, California. The smooth, abraded upstream face to the right contrasts with the steep, plucked face to the right.

In the geological record good examples of *roches moutonnées* can occasionally be found. Abraded bedrock surfaces in the Neoproterozoic sequence of Mauritania display a number of well-developed *roches moutonnées*, and others may be found beneath the Late Palaeozoic Dwyka Tillite of South Africa.

Crag-and-tail features
Crag-and-tail features (also called lee-side cones) are composite forms, being the result of both erosion (the crag) and deposition (the tail, comprising lee-side till). The upglacier craggy end is often steep and rough, while the tail is smooth and reminiscent of a drumlin. Britain's best known feature is Castle Rock in Edinburgh.

3.3.4 Rock crushing

Many striated and polished rock surfaces also show a variety of other small-scale features, normally a few centimetres in plan (but occasionally up to 2m) and of crescentic appearance. They have commonly been referred to collectively as friction cracks, but this is a poor term since Drewry (1986), among others, has demonstrated that they are the result of rock crushing under the repeated impact of debris in basal ice on a segment of bedrock until failure occurs.

There are several types of crush features, including lunate fractures, crescentic gouges, crescentic fractures and chattermarks (Fig. 3.15). They tend to lie with the concavity pointing either in or against the direction of flow. The principal fractures in these features tend to dip in a down-glacier direction. However, some authors have argued that this is an unreliable criterion for determining ice movement directions, although crescentic gouges do seem to be reliable. As with striations, care must be exercised in using crescentic fractures as precise indicators of glacier flow. These features, notably chattermarks, are also good indicators of the jerky nature of glacier flow, and the associated stick-slip behaviour.

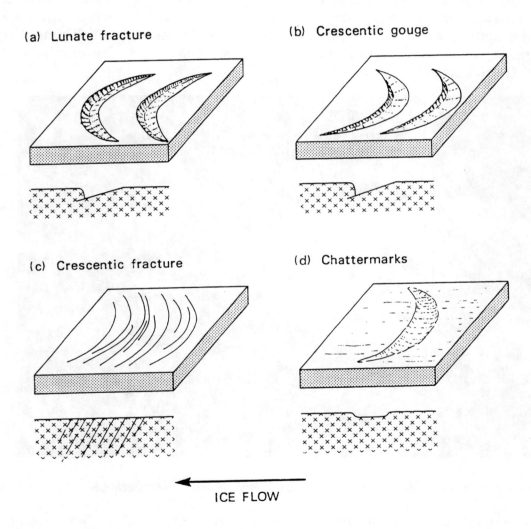

Fig. 3.15. *Small-scale features of glacial erosion with cross-sections, formed largely as a result of rock crushing (based mainly on Embleton & King 1968: Fig. 6.3).*

3.3.5 Erosion by ice combined with frost action

In mountainous terrain, valleys and cirques may progressively extend backwards as a result of erosion by the ice. Frost-shattering is also an important process, and the glaciers facilitate the removal of the debris. Progressive destruction of a mountain mass in this manner generates a family of landforms, which are essentially relict features. Two cirques which are being enlarged and approaching each other may eventually cut through the intervening ridge to produce a narrow, serrated ridge or *arête*, breached in places by cols. Three or more cirques eroding backwards against a single mountain mass may eventually meet and leave behind a pyramidal peak or *horn* (Fig. 3.1). Both horns and arêtes are common in areas that have undergone prolonged or active glacial erosion. If an ice sheet competely inundates these features at a later stage, then they may be worn down into dome-like peaks and rounded ridges.

Arêtes

The term *arête* is French (*Grat* is the German equivalent), and is applied to the knife-edge ridges that are so abundant in the Alps. Arêtes are mainly the result of backward erosion by adjacent cirque or valley glaciers, during which the rock dilates and fractures along joints. Frost-shattering facilitates the development of a jagged ridge, on which upstanding pinnacles are referred to by climbers as *gendarmes* (policemen).

Fig. 3.16. Narrow ice-draped arêtes on Mount Tasman, Southern Alps, New Zealand.

Arêtes are widespread in alpine terrains all over the world. The Western Cordillera in Canada and Alaska as well as the European and New Zealand Alps, the Andes and Himalayas all have classic landforms (Fig. 3.16). The western parts of the Britain and Norway have well-developed arêtes arising from late glacial cirque activity despite being under an ice sheet for much of the Pleistocene Epoch. Those well-known to British fell walkers for providing sporting routes to the summits include the ridge of Crib Goch on Snowdon (Wales), Striding Edge on Helvellyn in the Lake District, and the Carn Mòr Dearg ridge on Ben Nevis (Fig. 3.17) and the Cuillin ridge on the Isle of Skye (Scotland).

Fig. 3.17. Curving arête leading from Carn Mor Dearg towards the summit of Ben Nevis (off picture to the right) in Scotland. The range of hills in the background, the Mamores also comprises arêtes, together with cirques.

Cols

Mention has already been made of one special type of col - the watershed breach which is the product of abrasion as ice escapes from its confining trough. Many cols, however, are simply the result of backward erosion of adjacent cirque or valley glaciers by rock fracturing, and lowering locally of the intervening arête (Fig.2.32). A col may thus be a low notch, approachable only by ascending the steep headwall of the adjacent cirques. The arête leading down to the col may similarly be precipitous. Many Alpine passes are of this nature. In Britain erosion has only proceeded to this degree in relatively few places. The Cuillin Hills on Skye provide the best examples.

Horns

Backward erosion by three or more cirque glaciers in combination with frost-shattering, may lead to the intersection of the cirque headwalls, thus isolating an upstanding mass of rock, called a horn, a term loosely derived from the German for a peak. A horn represents the stage at which all the original smooth highland has been eroded. The horn is epitomized by the Matterhorn (Fig. 3.1) which stands astride the Swiss-Italian border. Three nearly symmetrical ridges rise steeply towards a pyramidal summit , each face carrying a steep cirque glacier. Other well-proportioned horns are Mount Assinboine in the Canadian Rockies, Mount Aspiring in New Zealand and K2 in the Karakorum Range in Pakistan. Most horns, however, lack this symmetry due to the uneven backward erosion of cirques. Some horns even have four or more arêtes leading up to their summits. Few British peaks acquire the status of a horn, but examples are Schiehallion in the Grampian Highlands and Cir Mhòr on the Isle of Arran (Fig.

3.3b). The Cuillins on the Isle of Skye have steeper sides but less well developed individual forms.

Nunataks

Rock outcrops ranging from less than a kilometre to hundreds of kilometres across that are surrounded by ice are known as *nunataks*, a term derived from the Inuit language. They include the last relics of mountains that have been subjected to valley glacier, cirque glacier or ice sheet erosion, as well as intensive frost shattering. At the other extreme, nunataks may be represented by entire mountain ranges that have only been affected by erosion on their flanks, and the intervening surface, though heavily frost-shattered, may preserve its preglacial form. Both end-members and everything in-between exist in Antarctica, the most extensive development being represented by the Transantarctic Mountains which stretches across the entire continent. The Greenland ice sheet has numerous nunataks towards its periphery, where it begins to spill through the coastal mountains, and nunataks are common in areas that have highland ice-fields such as Ellesmere Island and Spitsbergen. Smaller nunataks are characteristic also of heavily glacierized alpine regions.

3.4 Structural geological controls on glacial erosion

The processes of glacial abrasion, rock fracture and block removal are facilitated if there are weaknesses in the bedrock. Major structures, such as faults, commonly provide the main discharge routes from ice caps and ice sheets. However, the fault-controlled valleys may well have been in existence prior to glaciation and floored by deeply weathered material that subsequently was easily picked up by the ice. As already described, one of the Antarctic's major discharge routes, the Lambert Glacier, flows in a deep graben, bounded by parallel faults. Ice streams in the Scottish and Scandinavian ice sheets flowed along faults. Distinct preferred trends of lochs and are visible in western Scotland where a satellite image covering the Highlands (Fig. 3.18) shows both a series of NW-SE grouping and a major structure, the Great Glen Fault, running along Loch Linnhe and Glen More. Another good example is in East Greenland where two sets of intersecting faults, one running east-west, the other NW-SE, are now followed by fjords, isolating individual fault blocks.

On the other hand, some glacial valleys may not show any clear relationship to bedrock structure. The English Lake District is a case in point: the radial drainage system here was developed on a dome of Late Palaeozoic and younger rocks that developed in early Tertiary time. Ice continued the downcutting below an unconformity into folded early Palaeozoic rocks without adapting to the differently orientated structures in the lower strata. This phenomenon is known as superimposed drainage.

Landscapes of areal scouring display many close links between landform and bedrock structure. Complex assemblages of ribbon-lakes, small straight valleys, gullies and *roches moutonnées* all show a close relationship to bedrock foliation and cross-cutting faults, and the land surface may give the appearance of being strongly dissected. The early Proterozoic and Archaean rocks in the Lewisian basement of North-West Scotland and matching rocks in the Laurentian shield of Canada often contain ribbon-lakes and gullies following these structures.

112

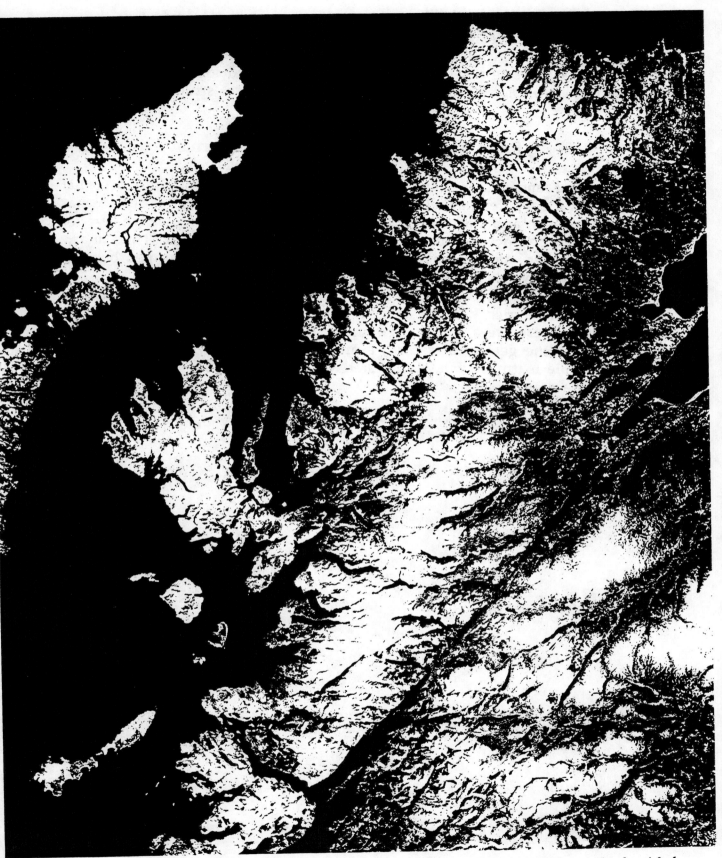

Fig. 3.18. Landsat image of the Scottish Highlands. The troughs (valleys and fjords) are clearly picked out by snow on the adjacent high ground. Many of these lineaments are structurally controlled, especially faults, and thus were subjected to preferential glacial erosion (Published by permission of...................).

Cirques are not necessarily located in areas of structural weakness, but their detailed form is often controlled by bedding foliation or jointing. Backwalls tend to be steep if there is a set of exploitable, near vertical joints. Similarly, the detailed form of *roches moutonnées* reflects the same sort of structural feature.

3.5 Preservation potential of erosional forms

Most erosional landforms are known from land areas and one might expect that their preservation potential is low, as erosion often continues after the ice has receded. However, as a glacier recedes it commonly deposits a protective layer of till on scoured bedrock and protects it from subaerial weathering. Thus, many Pleistocene eroded surfaces have been found where till resting on bedrock has been freshly removed.

On the longer timescale, erosional forms have less chance of survival, unless they are rapidly buried beneath a thick sedimentary pile in a subsiding basin. Nevertheless, there are a surprising number of striated pavements and *roches moutonnées* in the geological record, examples having been mentioned above, but larger-scale features such as glaciated valleys, cirques or horns, have only rarely been recognized.

CHAPTER 4

ENVIRONMENTS OF TERRESTRIAL GLACIAL DEPOSITION

4.1 Introduction

This chapter concentrates mainly on the environments where till and associated sediments are deposited, processes which are more pronounced in the lower part of a glacier system. It has been said that till is more variable than any other sediment known by a single name; partly because of this, there is, as yet, little agreement as to a suitable definition (see Ch. 1). In this section the various characteristics of tills and related terrestrial deposits, together with the landforms resulting from deposition, are considered.

4.2 Processes of terrestrial glacial deposition

Until comparatively recently, glacial geologists interpreted the mode of deposition of tills on the basis of their examination of Pleistocene deposits, and rarely considered the processes actually occurring in present day ice sheets, with the result that many misconceptions arose. The position has altered dramatically over the past twenty years or so, following many detailed studies of the processes involved, notably by Boulton in the 1960s and 1970s (reviewed by Sugden & John 1976), and others subsequently (reviewed by Dreimanis (1989). A comprehesive account of glacial depositional processes is given by Sugden & John (1976: Ch. 11). Although modern studies have drawn attention to the complexity of glacial environments, the better understanding obtained of subglacial processes allows us to make meaningful interpretations of Pleistocene and pre-Pleistocene glacigenic sequences. The relation of the most common types to a receding, wet-based glacier is shown in Figure 4.1.

4.2.1 Subglacial lodgement

The zone of erosion beneath a typical glacier is followed downwards by a transition to a depositional regime beneath the actively moving ice, usually as the glacier is losing its erosive capacity (Fig. 4.1). Pressure-melting allows material to be released from the dirty basal ice, particularly on the upstream side of bumps. Together with basal shear within the ice/debris mixture, which facilitates renewal of the debris supply, this process leads to debris being plasteres onto the bed, and progressively built up. This process, known as lodgement, occurs both on bedrock and older till surfaces, and leads to the filling of bedrock irregularities, so smoothing out the glacier bed. Some till may be squeezed into cavities downstream of a bump. Lodgement occurs beneath both advancing and receding glaciers, and the details are now well established from studies by Boulton (1970, 1971) on glaciers in Spitsbergen. A common characteristic of lodgement till is the presence of shear structures formed as the ice overrides the unconsolidated deposit. Where preserved in older deposits, the pattern of such structures is potentially useful as an palaeo-iceflow indicator. Shear-stress also results in the orientation of stones in the till developing at an angle of 45° to the bed (parallel to the plane of maximum shear-stress), but, as the material continues to deform, stones rotate so that they tend to approach parallelism with the bed, thus providing an additional means of establishing the direction of flow in old lodgement tills.

116

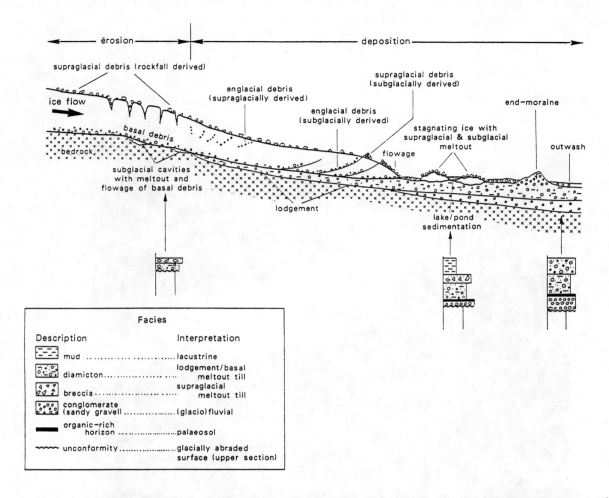

Fig. 4.1. Cross-sectional view of a typical temperate glacier to illustrate the various types of glacigenic sediment deposited as the ice recedes.

Deposition of lodgement till may be strongly affected by the transfer of debris-rich basal ice into an englacial or even supraglacial position by movement along shear or thrust zones within a glacier undergoing compressive flow (Section 2.10). Strictly speaking, the subsequent release of such deposits is by the melt-out process.

The rate of subglacial till deposition can be considerable, and 6m per century is regarded as reasonable (Sugden & John 1976: p. 218). Contrary to the belief of glaciologists up to about 20 years ago, it is now known that much of the Antarctic and Greenland ice sheets are at the pressure melting point at the base. Thus, the deposition of thick tills and tillites on a continental scale as recorded in Pleistocene and earlier successions is not difficult to explain.

4.2.2 Melt-out and flowage

Melting out of debris can occur either subglacially or supraglacially, a process that is particularly active at the margins of glaciers and ice-sheets where the ice is practically stagnant (Fig. 4.2). Geothermal heat is largely responsible for the deposition of subglacial melt-out tills. If the water is able to escape without disturbing the sediment, the resulting till may preserve traces of the original relationship of debris to the ice structures. Meltout till is commonly deposited in unstable situations and is thus prone to flowage.

Fig. 4.2. Deposition of subglacial meltout till and reworking by meltwater at the snout of the Vadrec da Morteratsch, Switzerland. Note the predominantly subrounded and subangular character of the clasts.

During the past 50 years, frequent reference has been made to the possibility of flowage of saturated till at the base of a glacier. For example, Gripp (1929) suggested that some englacial bands were the result of till being squeezed up into basal crevasses, a process Mickelson (1971) has observed in action, and Hoppe (1957) has explained certain types of hummocky moraines as a result of the same process. More recently, Boulton (1971, 1972) has noted that flowage can occur where thick unfrozen subglacial till exists beneath temperate ice. Basal crevasses are not particularly common beneath glaciers, as it happens, unless the ice is at an advanced stage of decay, but there are other spaces into which till can be squeezed, in particular hollows downstream of obstacles, abandoned subglacial stream channels and moulins. Stones in wet lodgement till may undergo reorientation if subsequently affected by flowage and the final fabric may be very different from till that has not been affected. Features of "plastic" deformation, particularly folds, may aid the recognition of such till.

Debris at the surface of a glacier may have a basal derivation, especially near the margins of cold glaciers and ice sheets (Fig. 2.35) or, in the case of valley glaciers, it may have accumulated as a result of subaerial weathering of rock faces. This debris melts out in summer as a result of ablation. A layer of debris generally retards melting of dirty ice in comparison with clean ice, so the resulting ice surface tends to acquire debris ridges. As the debris is lowered it tends to undergo sliding and reorientation, and rarely reflects the original fabric in the ice after deposition as till. A thick layer of supraglacial debris tends to be unstable and, as with basal melt-out till, often capable of flowage.

Regardless of whether the sediment is deposited supraglacially or subglacially, the process is particularly common near the snouts of receding or stagnating glaciers. The process has been particularly well studied by Boulton (1968, 1971) at the margins of various Spitsbergen glaciers and by Lawson (1981, 1982) at the snout of Matanuska Glacier in Alaska (Fig. 4.3a). The principal factors affecting debris flowage are the gradient of the bed on which it rests, whether it be ice or the glacier bed itself, the bed roughness, the amount of meltwater available for enabling the debris to become more fluid, and the fabric of the debris itself. Saturated till may flow on the gentlest of slopes. According to Boulton, three types of flow can be distinguished which, in decreasing order of rate of movement, are:

(a) mobile flow, in which stones may tend to settle towards the bottom producing a crude sorting pattern (Fig. 4.3b);

(b) semi-plastic flow, which often begins by slope failure along an arcuate slip-face (Fig. 4.3c) and is perceptible as a slow-moving lobate tongue (Fig. 4.3d); boulders may sink, fold structures may form, and washing by meltwater may produce laminations;

(c) downslope creep, which involves less water and is not perceptible to the eye. Although the till has a fabric that is subject to alteration and may acquire a weak foliation, it is unlikely to develop fold structures. Such till, in contrast to other types of flow till, is compact and massive.

Strictly, a "flow till" is resedimented and some authors (e.g. Lawson, 1979) have argued that the term should be abandoned. The often too subtle differences between such sediments and "non-disaggregated" tills arguably makes such a step premature.

4.2.3 Sublimation of debris-rich ice

Sublimation is the process of ablation whereby ice is vapourized directly without passing through the intervening liquid phase. Some authors have claimed it to be of negligible importance, but in the cold, arid parts of Antarctica, such as the Dry Valleys of Victoria Land, where temperatures rarely exceed 0°C, it is a common process (Shaw 1977). Here, debris-rich ice ablates from the surface downwards, producing a loose *sublimation till* which inherits the foliated structure from the ice (Fig. 4.4). Outside Antarctica the process is rare, however.

Fig. 4.3. Glacigenic sediment flowage in close proximity to Alaskan glaciers. (a) *General view of the snout of the Matanuska Glacier, illustrating a proglacial area that has been almost totally disturbed by flow processes.* (b) *The debris-covered snout of Matanuska Glacier illustrating mobile flow in the foreground; a slurry of fine material to the bottom of which coarser material settles.* (c) *"Semiplastic flow" often begins with slope failure within till as here, a few hundred metres from the snout of Matanuska Glacier.* (d) *Semi-plastic flows commonly terminate in a lobate tongue; here, failure of a till-covered slope has occurred, the debris flowing out across the surface of Orange Glacier.*

120

Fig. 4.4. Formation of sublimation till from debris-rich basal ice, Taylor Glacier, Victoria Land, Antarctica. The darker material is the dirty ice, and the lighter coloured cap is crumbly sublimation till that is inheriting the foliated structure from the ice.

4.2.4 Ice-push

During the winter months, when ablation ceases, it is common for even a generally receding glacier to advance a short distance and push up a small ridge of till and fluvioglacial material (Fig. 4.5). A succession of minor advances in winter during a phase of general recession produces a series of annual push moraines (e.g. Worsley 1974). More spectacular push-moraines often occur at the snouts of advancing glaciers, but are the product of glaciotectonic deformation (Section 4.2.6).

4.2.4 Subglacial chemical processes

The importance of chemical processes involving calcium carbonate beneath terrestrial glaciers was first noted by Ford et al. (1970). Other chemical processes involving silica have also been identified. It is now known that comminution of carbonate rock particles creates a rock flour that is highly reactive chemically. Carbonate is thus readily dissolved and precipitated (Fairchild 1983, Souchez & Lemmens 1985). Isotopic studies beneath Swiss glaciers in limestone areas have revealed that chemical action can be important at the ice/bedrock interface. Souchez & Lemmens (1985) and Sharp et al. (1989) found that calcite was formed by precipitation, as patchy coatings and cornices up to a few centimetres thick on polished and striated bedrock. Some of the coatings themselves were found to be striated (Fig. 4.6). Also, calcite-coated pebbles and thin limestone crusts were present in recently deposited basal tills.

Fig. 4.5. Small (2 m high) push-moraine from the preceding winter's advance at Steingletscher, Switzerland. Note the person on the ice-abraded bedrock with p-forms in the background.

Calcite is released when meltwater refreezes. Partial freezing is accompanied by concentration of salts in the residual water, eventually resulting in saturation. During freezing, water in equilibrium with the growing ice is progressively impoverished in heavy isotopes, notably oxygen-18. Sliding over protuberances on the bed results in pressure melting on the stoss side and refreezing on the lee side, thus leading to carbonate deposition. Fixed bedrock protuberances give fluted or furrowed coatings on the lee side. The ^{18}O isotopic composition indicates that the origin of the initial water which gives the precipitate is the basal ice-layer, not true glacier ice.

Fig. 4.6. Chemically precipitated calcite coatings, with striations and cornices, on limestone bedrock, Glacier de Tsanfleuron, Switzerland. Ice-flow was from left to right.

Calcium carbonate coatings on pebbles and calcite layers in Neoproterozoic tillites have been reported from China (Wang Yuelun et al. 1981), the Sahara (Deynoux 1980) and Svalbard (Hambrey 1982). Such carbonate coatings may form in a variety of settings, including subglacially, proglacialy anf pedogenically; or arise later during burial diagenesis or metamorphism (Fairchild et al. 1993).

4.2.6 Glaciotectonic deformation

Glaciotectonic deformation is now accepted by many researchers to be an important process, but until the publication of the book by Aber et al. (1989), this was not reflected in the general literature. The possibility that glaciers could deform shallow crustal rocks and sediments was recognized well over a century ago by Lyell during studies in Norfolk, England and the Italian Alps.

123

Glacitectonic processess can be recognized throughout glaciated terrains where sedimentary bedrock and thick drift deposits occur. Strictly speaking, glaciotectonic processes involve erosion as much as deposition; their discussion in this chapter relects the fact that they are most evident in nearly contemporaneous sedimentary sequences. The application of structural geological principles to the study of glaciotectonism has revolutionized our understanding of the resulting products.

Glaciotectonic deformation takes place in a variety of settings - in front of a glacier, beneath the ice margin or under the middle part of a thick ice sheet. Glaciotectonic deformation operates in any topographic setting, both during advancing and recessional phases, and involves all manner of sediment types in frozen, saturated and dry states.

The principal factors that are important for the genesis of glaciotectonic phenomena are as follows (from Aber et al.: Table 9.1):

- lateral pressure gradient,
- elevated groundwater pressure,
- ice-advance over permafrost,
- ice advance against a topographic obstacle,
- lithological boundaries in the substrate,
- surging of ice-lobes,
- subglacial meltwater erosion,
- damming of proglacial lakes,
- thrusting in front of the ice,
- compressive flow with basal drag,
- shearing of fault blocks up into the ice.

Glaciotectonic deformation takes place when the stress transferred from the glacier exceeds the strength of the material beneath or in front. The material may be subject to both brittle and ductile deformation, typical structures being thrusts and folds respectively. Sometimes, the glaier detaches a sediment-mass or slab of bedrock along a plane of décollement, incorporating it into the body of the ice before depositing it (cf. Section 2.10.3; Fig. 2.37). The structure of the glaciotectonically deformed sediment is strikingly similar to that of a mountain belt, such as the Alps or the Rockies, but on a much smaller scale.

4.3 Terrestrial glacigenic facies and facies associations

4.3.1 Boundary relations and geometry of glacigenic units

Most till is deposited during glacial recession after the ice-mass has already eroded bedrock or the underlying unlithified sediments. Thus, tills and tillites commonly unconformably overlie striated bedrock surfaces, or more rarely boulder pavements. These surfaces are irregular, but the basally deposited till tends to fill the hollows and smooth out the relief. The top of a basally deposited till may be quite regular and in distinct contrast with supraglacial till lowered onto it. The upper surface of the latter may, however, be extremely hummocky, reflecting uneven down-wastage of the ice, and the whole picture may be confused by till flowage. Overlying sediments may be fluvial or aeolian, forming a distinct break and tending to smooth out the surface. Within the glacigenic facies are intimate associations with fluvial materials which occur as lensoid or channel-like bodies. Till may be spread over many hundreds of square kilometres as a deposit varying in thickness from a few metres to tens (or exceptionally a few hundreds) of metres, as a result of accretion during successive advances. Entire sequences may become disrupted and structurally complex during glaciotectonic processes, and the final geometry will reflect large-scale movements of material en masse.

4.3.2 Sedimentary structures

Terrestrial tills characteristically lack well defined bedding or lamination. However, lodgement tills may have a distinct fissility or foliation, resulting from shearing as the glacier overrides the unconsolidated material. Exceptionally, meltout tills have a weak 'bedding' acquired from debris-rich dirty layers parallel to basal foliation in the ice. Flow tills, as already discussed, may be subjected to slight grading as the heavy stones settle during flowage and, in addition, bear evidence of slump folding with axial planes parallel to the depositional surface and fold axes normal to the flow direction. Differential loading of till onto other soft sediments may result in convolutions and injection phenomena, although these features are rare.

Till, of course, is rarely deposited in isolation from other sediments. Glaciofluvial deposits, for example, may be preserved as lenses or channels within a till unit and may have structures which are associated with normal braided river deposits, such as cross-bedding. Lacustrine sediments or aeolian sands may also form lensoid features within the till sequence.

4.3.3 Facies

The principal lithofacies in the terrestrial glacial environment are the product of direct glacial, fluvial, lacustrine and aeolian processes (see also Chs 5 & 6). Of the sediments release directly from the ice, diamicton is the principal lithofacies, especially in its massive form. generally, it is interpreted as lodgement, meltout or flow till, depending on the relationships with adjacent beds. Also common are breccias, derived from angular, supraglacially transported rockfall material. These facies may be intimately associated with conglomerates and sandstones of fluvial origin, laminated muds of lacustrine origin and cross-bedded sands of aeolian origin.

In general, terrain of alpine character produces more supraglacial debris with predominantly angular clasts than more subdued terrain. On the floors of many alpine valleys, occupied by temperate glaciers, there is so much reworking by meltwater, that little till can survive, the bulk of the sediment being glaciofluvial. The thickest diamicton sequences occur in the Arctic, where cold glaciers carry a much greater basal debris load than their alpine counterparts.

4.3.4 Facies associations

Unlike many glaciomarine sedimentary records, terrestrial sequences tend to be very incomplete. If a wet-based glacier readvances over sediment released earlier, recycling, removal and incorporation into the new deposits frequently occurs. Thus, not all successive advances are preserved, and the dominant facies tend to reflect the final recession of the ice. Complex diamict sequences may arise from melt-out processes with or without flowage. At the base of a glacier, melt-out may begin after lodgement has ceased, e.g. during glacial recession, so that the till becomes progressively younger upwards, whereas from the top the earliest released debris will lie at the top and so become younger downwards. Once the ice has disappeared, the one sequence will be superimposed upon the other. Thus, different types of till may be interstratified after the ice has disappeared, leaving behind a multi-layered till sequence relating to one glacial phase.

It is also apparent that flowage almost certainly complicates the picture, and recognition of the depositional characteristics of a typical till sequence will require very close scrutiny, and even then it may not be possible to make a meaningful interpretation.

4.3.5 Facies architecture

Terrestrial glacigenic sequences are characterized by their complexity and rapid vertical and lateral facies changes. Although there are few places where extensive vertical sections of Quaternary age demonstrate the two-dimensional geometry of facies, let alone the three-dimensional architecture, we can demonstrate the expected disposition on the basis of our understanding of the depositional processes. However, better exposed pre-Pleistocene terrestrial glacigenic sequences may aid our assessment of facies architecture.

4.4 Physical, chemical and mineralogical characteristics

4.4.1 Grain size distribution

Tills and tillites are characterized by extremely poor sorting, containing all size classes from clay to boulder (Fig. 4.7).

Grain-size distribution characteristics of till are dependent on whether it is supraglacially or basally derived. Triangular plots of the sand, silt and clay fractions have been presented for a variety of lodgement and supraglacially derived tills from North America and Britain (Sladen & Wrigley 1983). Material at the base of a glacier is subjected to attrition and breakdown into clay-size particles, and so is finer grained than supraglacial material, but there is a degree of overlap (Fig. 4.8a).

The nature of the terrain also influences the grain size distribution. Data from three types of terrain are depicted in Figure 4.8b. Shield terrains (1) and glaciated valley terrains (2) yield coarse, sandy, clast-rich tills, whereas glaciated sedimentary lowlands (3) yield tills which are rich in fines, especially where lacustrine and marine sediments have been incorporated into the tills.

The degree of glacial transport also influences grain size distribution in tills. Most tills have a bimodal or polymodal distribution, the latter tending to reflect the greater distance of transport. The composition of the principal rock type making up the till also influences the grain-size distribution. Coarse-grained igneous and metamorphic rocks, such as granites and gneisses, and sandstones tend to give rise to sandy tills, whereas the matrix of fine-grained sedimentary rocks is dominantly clayey or silty.

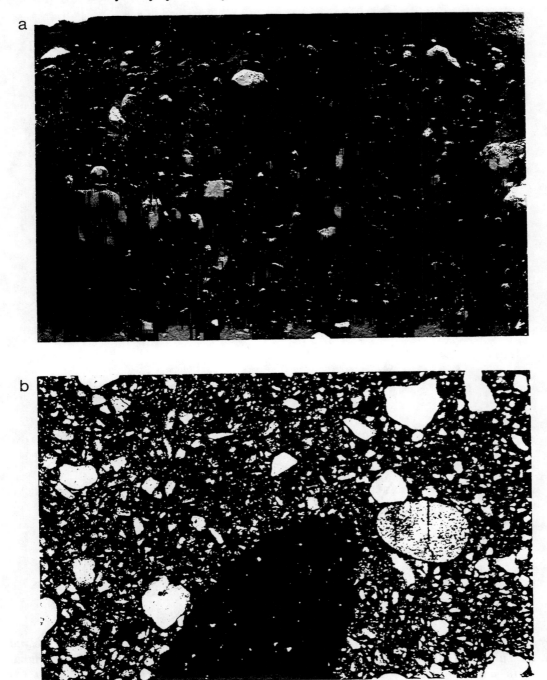

Fig. 4.7. Poorly sorted character of massive diamicton, interpreted as lodgement till. (a) Till deposited by the Devensian ice sheet, Loch Lurgainn, North West Highlands of Scotland. (b) Matrix of a lodgement tillite of Neoproterozoic age, Wilsonbreen Formation, Svalbard. The dark background is clay grade, most of the white grains are silt grade, and the larger fragments are of sand. The field of view horizontally is approximately 5 mm.

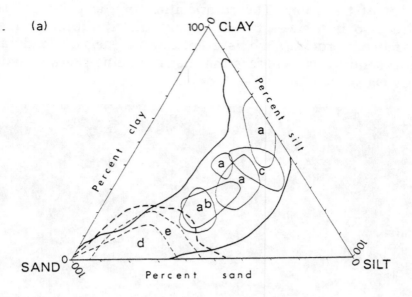

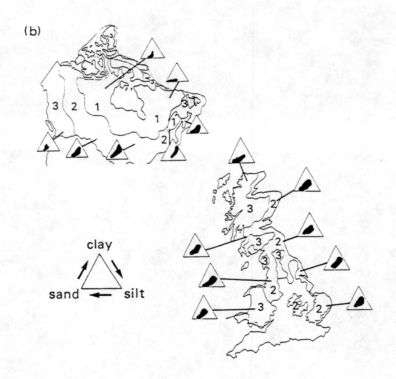

Fig. 4.8. *Grain-size distribution in the sand and finer fractions of tills from North America, Iceland and Britain. A. Thin solid lines - envelopes for lodgement tills from (a) Ohio, (b) Northumberland, (c) Ontario; thin dashed lines are envelopes for coarse-grained supraglacial diamicts from (d) Iceland, (e) Scotland. The heavy solid and dashed lines enclose all tills in these categories respectively. B. Grain-size distribution envelopes for lodgement tills from (1) shield terrain, (2) sedimentary lowlands of subglacial terrain and (3) glaciated valleys. (From Sladen & Wrigley 1983).*

Grain-size distribution is also influenced by the thermal regime of the ice-mass, since the availability of meltwater facilitates the grinding down of the coarser particles. In Antarctica today, where the supply of meltwater is limited, tills tend to be sandy, in contrast with those associated with northern hemisphere glaciers where meltwater is abundant (Fig. 4.9).

Fig. 4.9. *Basal till in a cold polar setting, Schirmacher Oasis, Dronning Maud   Land, Antarctica. This till comprises high-grade metamorphic clasts in a sandy matrix; there is little silt and finer material.*

4.4.2 Shape and surface features of clasts

A characteristic of till, unusual in other types of sedimentary rock, is that it contains stones of both local and far-distant (exotic) origin, usually well mixed and apparently homogeneous over wide areas. Many studies have been made on the provenance of stones in till but, as yet, few analyses have been made of pre-Pleistocene tillites. For Quaternary deposits, studies of boulder trains of a particular lithology in some cases have the potential for tracing ore bodies, although this method of prospecting is still in its early stages of development (Shilts 1976). The picture of apparent homogeneity of till is more complex than casual observation suggests. A statistical analysis of the Catfish Creek Till in Ontario (May & Dreimanis 1976) found that the matrix of a till is inherently more homogeneous than the stones, and that the lowest part of a till unit, containing locally derived material, is in fact so variable that it should be excluded when characterising the entire unit.

The shape of stones in till varies from rounded to angular, but normally subangular to subrounded varieties predominate. Debris falling onto a glacier surface generally is shattered and angular and subsequently is transported passively in a supraglacial or englacial position (Fig. 4.10a). Basal debris is subjected to abrasion and rotation, so that rounding of the irregularities takes place, and the stones have a broader range of shapes with subangular and subrounded being dominant (Fig. 4.10b). These characteristics are inherited in lodgement till (Fig., 4.10c). Reworking by subglacial meltwater leads to further rounding (Fig. 4.10d). Boulton (1978) has shown that roundness combined with sphericity permits one to determine the means of transport and deposition. Three distinct sphericity/roundness fields can be recognised for supraglacially derived boulders in high-level transport (i.e. those derived from rockfall), boulders from the zone of traction and boulders embedded in lodgement till (Fig. 1.4).

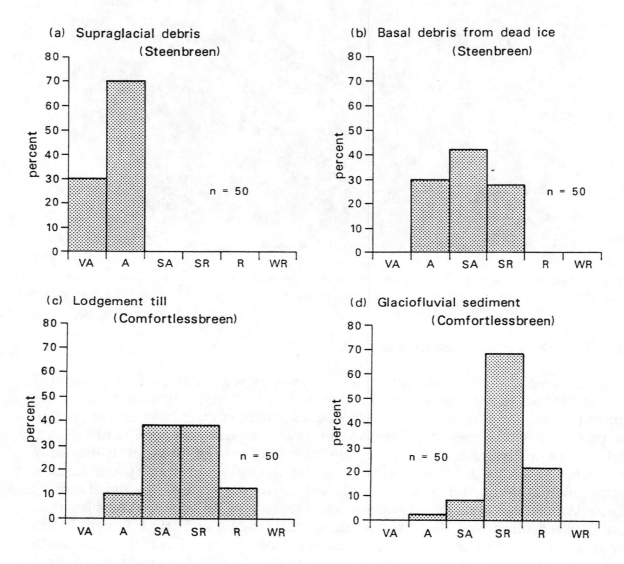

Fig. 4.10. *Range of shapes according to Power's roundness categories, in a variety of proglacial settings, Engelskbukta, Svalbard. (a) Supraglacial debris, Edithbreen; (b) debris from buried stagnant basal ice, Steenbreen proglacial area; (c) lodgement till from fluted ridge, Comfortlessbreen; (d) glaciofluvial material, Comfortlessbreen outwash area.*

Apart from roundness and sphericity, stones in basal till often have specific shapes which indicate their glacial origin, in particular faceted, pentagonal and flat-iron forms (e.g. Wentworth 1936, Flint 1971, Boulton 1978). Many stones are clearly more pointed at one end than the other. Lithology does not appear to play an important part in determining the roundness of stones, since a study by Dowdeswell et al. (1985) of carbonate and crystalline clasts in a Spitsbergen tillite has shown no detectable difference in the shapes of the different lithological populations. However, there are variations in Zingg shapes: bedded sedimentary or well foliated metamorphic rocks tend to produce blade, plate and flat-iron forms, partly because they tend to slide during traction, whereas gneisses and igneous rocks tend to be more rounded, as they are subject to rolling at the base of the glacier. Boulton (1978) and Sharp (1982) have further noted that the larger boulders (>0.5m) which are deeply embedded in lodgement till have a characteristic form: a smooth, bullet-nosed up-glacier end, and a truncated down-glacier one (Fig. 4.11). In contrast, boulders loosely embedded in, or resting on, the till surface have no such clearly defined form, and although they may be faceted, they lack the streamlined form of embedded boulders. Tills often contain a proportion of angular stones such as shales which would not normally survive extended transport in other media. Supraglacial or englacial transport enables such "fragile stones" to be preserved.

Fig. 4.11. Bullet-nosed boulder embedded in lodgement till near the now-stagnant margin of Burroughs Glacier, Glacier Bay, Alaska. Ice movement was from left to right.

Striations are commonly regarded as a distinctive feature of till stones (Fig. 4.12). However, supraglacially or englacially transported debris will have none, while perhaps only a relatively small proportion of basally transported stones will have striations. The proportion of stones with striations varies greatly (Dowdeswell et al. 1985), but generally it is of the order of 5-20%. Striations may occur on any part of a stone, but are more common on facets or on the surface of a stone projecting above lodgement till that has subsequently been overridden (Sharp 1982). Striations may be orientated completely at random, cross-cutting one another, such as when the stone has been subject to rolling, or may form sub-parallel sets, as for example when little rotation has taken place (Fig. 4.12). The most regular striations occur on boulders embedded in the till; these form a single set, diverging slightly down-glacier (Boulton 1978) or they have a consistent orientation (Sharp 1982). Lithology has a marked influence on whether a stone is striated or not (Section 1.4.4).

Fig. 4.12. Striated clast from the proglacial area of Aavatsmarkbreen, Spitsbergen. A single prominent set of parallel striae is cut by striae of many orientations. To the right is a prominent crescentic gouge indicating ice movement to the right. The stone is about 30 cm long.

In addition to facets and striations, some stones may bear other small-scale features of glacial abrasion of the type that are often found on abraded bedrock surfaces, such as crescentic fractures and chattermarks (Fig. 4.12). A feature, previously mentioned (Section 4.2.4), in areas of carbonate bedrock, is the presence of thin egg-shell-like muddy or calcareous coatings on some stones in tillites. The Chinese Sinian tillites have good examples of these, and there it has been suggested that, during glacial sedimentation, clay minerals were deposited and dissolved calcareous matter was precipitated. Transport of till stones by meltwater soon results in rounding and the removal of the glacial markings; rolling on a stream bed may give rise to percussion marks instead.

4.4.3 Clast fabric

It has been known since the work of Miller in 1850 (Flint 1971) that stones in till often have a preferred orientation, and that this is parallel to the direction of ice movement (where known from other evidence). However, only in the last thirty years or so, has this property been used, to any extent, in palaeoenvironmental reconstructions. In most tills and tillites a preferred orientation can only be determined by detailed measurements, but occasionally is it visible to the naked eye. On the other hand, in some situations, till fabric measurements reveal stones orientated transverse to the direction of ice movement.

Fabric studies are useful for determining the direction of glacier flow during deposition, possible with greater accuracy than from features resulting from abrasion especially if the stones are not equidimensional. The shape of stones, in particular roundness, sometimes has a marked effect on their orientation. For example, in lodgement till the long axes of the well-rounded shorter stones tend to be aligned parallel to the flow direction, whereas longer less rounded and wedge-shaped stones are transverse to flow (Embleton & King 1968: 308). Thus, because of the fact that both parallel and transverse fabrics can arise during glacial deposition, other sedimentological evidence must be examined before assessing the significance of a particular fabric. Sliding, shearing and post-depositional flowage can all determine the final fabric of a till. In general terms, we may expect the fabric of till to reflect both the position of debris in the ice during transport and the mode of deposition.

Figure 1.5 illustrates on stereographic projections (cf. Phillips 1971) the three-dimensional orientation of clasts in a variety of situations. Typically, clasts in basal ice show a strong preferred orientation. For example, Lawson (1979 b) at the snout of the Matanuska Glacier in Alaska found that in the debris-rich foliated ice from which basal melt-out tills were derived, the stone orientations plot as a single strong maximum, generally reflecting the local direction of flow, and lying in the plane of the foliation which itself was parallel to the glacier bed (Fig. 1.5, row A).

The basal melt-out till derived from the Alaskan glacier retained more-or-less the same fabric characteristics as the source ice (Fig. 1.5, row B) although if the pebbles were inclined in the ice, their angles of repose decreased on deposition. Supraglacial melt-out till also preserved the original fabrics of the ice except there was some dispersion of pebbles. Modification of the supraglacial till is often pronounced after deposition.

In lodgement till, the preferred orientation of stones tends to be parallel to the direction of local glacier movement, as one would epect. In addition it is common for stones to dip upglacier, rather in the manner of pebble imbrication in stream beds. Lodgement tills tend to have rather broad, even girdle-like maxima fabrics on a sterographic projection (Lawson 1979, Dowdeswell et al. 1985) (Figure 1.5, row C). On the other hand, under certain stress conditions, such as strong compressive flow, stones may become orientated transverse to flow (Dreimanis 1976). Boulton (1971) found similar strong fabrics in Spitsbergen, but with transverse-to-flow orientation fabrics in zones of compressive ice flow and parallel fabrics in zones of extending flow.

In the case of Matanuska Glacier, flowage of till is the dominant process, to the extent that only 5% of the till was unaffected. These flow tills (or sediment flows as Lawson preferred to call them) are varied in character and have complex fabrics ranging from fairly strong alignment parallel to flow where water content was high, to weak multi-maxima fabrics as the water content decreases, although this pattern was not always consistent (Fig. 1.5, row D).

All these fabrics contrast markedly with the pattern resulting from the settling of basal glacial debris through a water colums, which tends to show a random distribution within the horizontal plane (Fig. 1.5, row E).

Most fabric studies have been undertaken using pebbles and cobbles, but it is also possible to make use of elongate grains of sand size under the binocular microscope. This technique only permits us to derive two-dimensional data, but nevertheless is especially useful for studies of core material (see Ch. 8). Comparisons between the orientations of sand grains, pebble and cobbles, and boulders in a lodgement till in Switzerland demonstrate that reproducible fabric results are obtainable regardless of grain size.

4.4.4 Mineralogy of the matrix

Mineralogical analyses of terrestrial tills and tillites are relatively unusual. However, the technique has proved especially useful for tracing economic mineral deposits. This can be done, for example, by establishing palaeo-ice flow directions from the distribution of heavy minerals (e.g. garnets) in the matrix or in ore-bearing boulders. Glaciers appear to disperse material so that the concentration of minerals, elements or rock types reach a peak close to the source with an exponential decline in the direction of transport (Shilts 1976). The detection of such dispersal tails is a key feature in mineral exploration in till-covered areas. An interesting example of successful mineral exploration with limited resources is demonstrated by the Geological Survey and mining companies in Finland. In that country the populace have been encouraged to send by post free-of-charge any rock of exceptional appearance, with sizeable rewards for good samples. From 10,000 samples mailed each year, 5% lead to field trips and 0.25% to detailed study, and 10 mines have been established as a result. The resourceful Finns also use dogs to sniff out sulphide boulders and so define the dispersal pattern of ore-bearing rocks (Kujansuu 1976).

4.4.5 Geochemistry

The bulk composition of the matrix of till or tillite resembles that of greywackes, which it often is in a mineralogical sense (Pettijohn 1975 p. 176-178). These materials are normally rich in alumina, iron, alkaline earths and alkaline metals. Tills in limestone and dolostone regions have a matrix that is rich in CaO, MgO and CO_2. Tills deposited in an aqueous environment differ from lodgement and ablation tills because of the effects of sorting by currents and removal of the fines. Consequently, the matrix of such tills is more sandy and thus depleted in Al_2O_3, iron and K_2O, and higher in SiO_2.

4.4.6 Characteristics of quartz and heavy mineral grains

Scanning electron microscopy (SEM) on quartz or garnet grains enables different transport mechanisms or sedimentary environments to be distinguished on the basis of surface features on sand- and silt-sized particles, many of which are illustrated by Krinsley & Doornkamp (1973). Further details are given in Section 1.4.5.

4.5 Landforms in terrestrial glacial depositional environments

A wide range of depositional features, notably moraines and ice-contact deposits, are associated with glaciers and ice sheets, many of which are important in elucidating the palaeogeographical and palaeoclimatic patterns of a glaciated region. Their value in this regard is naturally of most significance for Quaternary events, since relics of glaciation survive at the surface, but the study of glacial and related landforms has proved to be an unreliable method of establishing a glacial chronology. In the earlier geological record such landforms are rarely preserved or at least are unrecognisable. However, there are exceptions, such as in "Ordovician" Sahara, and where landforms are preserved, we have a powerful means of determining paleoenvironmental conditions.

Various classifications of the landforms arising from terrestrial glacial deposition and associated ice proximal processes have been proposed. Here we adopt a classification based on the relationship of the landform to its source within or at the margins of the glacier, and to the direction of ice flow. The classification incorporates shape, sediment assocations and structures, and is based to some extent on a scheme published by Sugden & John (1976) and on a classification generated by INQUA (Goldthwait 1989) (Table 4.1). However, it should be borne in mind that the specific landforms described belong to a complete spectrum of forms. A particular feature may not therefore necessarily fit exactly into any one of these categories. Excluded from this discussion is a consideration of glacitectonic landforms which are a combination of erosion, folding and faulting, and deposition of bodies of bedrock and sediment. These forms are discussed in Section 4.6.

"Moraine" has been variously defined but here we follow Sugden & John (1976: 214): in defining it as an accumulation of glacial or glacier-worked sediments having an independent topographical expression. A moraine usually is made up of till, but there are exceptions, such as where ice-push has occurred

A number of excellent texts provide comprehensive summaries of the various glacial depositional landforms (e.g. Embleton & King 1968, Flint 1971, Sugden & John 1976), hence this brief account is intended simply to convey an impression of the wide range of features associated with such environments.

4.5.1 Debris forms on the glacier surface

In a geomorphological sense, these forms are largely emphemeral since, although they are the result of deposition, they have not finally been deposited. Nevertheless, they are striking elements of the glacial terrestrial environment.

Lateral moraines on the surface of a glacier are the result of intermittent rockfall onto marginal ice from ice-eroded cliffs. Since active glaciers tend to be convex in cross-section, debris can accumulate in a well defined band between the rock wall and the ice, where it will be subject to grinding and comminution during glacial movement. Thus lateral morainic debris commonly shows characteristics of both subglacial and supraglacial debris. Many lateral moraines, particularly those associated with cold glaciers, are ice-cored. These typically comprise a debris layer 2 m thick, resting on inactive glacier ice.

Lateral moraines may facilitate access to the middle of a glacier that is heavily crevassed at its margins. Although loose, unstable and uncomfortable to walk on, the boulders frequently bridge the crevasses and the debris cover encourages the ice to melt back at lesser angles.

Rockslides are common in steep mountain terrain due to the oversteepened slopes and severe frost action. Earthquakes in some areas, e.g. Southern Alaska, New Zealand, Andes, Caucasas, Himalayas, promote instability and rockslides have been known to blanket entire segments of a glacier. Post & La Chapelle (1971) illustrated one such rockfall on Sherman Glacier arising from one of the most powerful earthquakes ever recorded (in 1964). This blanketed the ice to such an extent that it retarded ablation and caused the glacier to readvance. Another rock avalanche is described in more detail in Section 2.10.1. Most rockslides form lobes of angular debris extending across the glacier perpendical to flow. With time they become deformed as they pass downglacier.

Table 4.1. *Classification of glacial depositional landforms according to position in relation to glacier and ice-flow direction. Glacier surface features before the ice has totally melted are included, but glaciotectonic landforms are excluded since these are the product of both erosion and deposition. Scale refers to maximum linear dimension*

Position in relation to glacier	Relation to ice flow	Landform	Scale
			1cm 10cm 1m 10m 100m 1km 10km 100km
Supraglacial; still actively accumulating	Parallel	Lateral moraine	
		Medial moraine	
	(Transverse)	Shear/thrust moraine	
		Rockfall	
	Non-orientated	Dirt cone	
		Erratic	
		Crevasse filling	
Subglacial during deposition	Parallel	Drumlin	
		Drumlinoid ridge	
		Fluted moraine	
		Crag-and-tail ridge	
	Transverse	De Geer (washboard)) moraine	
		Rogen (ribbed) moraine	
	Non-orientated	*Ground moraine:*	
		till plain	
		gentle hill	
		hummocky ground moraine	
		cover moraine	
Supraglacial during deposition	Parallel	Moraine dump	
	Non-orientated	Hummocky (or dead ice/disintegration) moraine	
		Erratic	
Ice marginal during deposition	Transverse	*End moraines:*	
		terminal moraine	
		recessional moraine	
		Annual (push) moraine	
		Push moraine	
	Non-orientated	Hummocky moraine	
		Rockfall	
		Slump	
		Debris flow	

Large isolated angular blocks of rock called *erratics*, sometimes the size of a small hut, may fall into the ice surface. Their size is dependent on the bedding and jointing characteristics of the rock. Igneous and high-grade metamorphic rocks with well-spaced joints tend to form the biggest blocks, whereas well-stratified or fissile rocks rarely produce large boulders.

Medial moraines occur on active glaciers, sometimes extending through to the bed, particularly where the debris originates from a spur between two valley glaciers (Fig. 4.13). Such debris will generally be angular and frost-shattered. Medial moraines may not appear on the surface until the snout is approached (Fig. 4.14), reflecting either early burial of the debris in the accumulation area or the presence of a rock mass below the ice surface in the ablation area. Medial moraines usually form elevated ridges on the glacier because the debris preferentially protects the ice from ablation, but a thin patchy cover of debris might occupy a linear depression because of enhanced ablation when scattered debris is in contact with ice. Most medial moraines are straight, but they may become folded due to compressive flow if the ice spreads out laterally as a piedmont glacier. Contorted (looped) moraines result from glacier surging (Fig. 2.28). Near the snout of an alpine glacier, medial moraines tend to merge with lateral moraines, forming a complete debris cover a metre or more thick. This may sometimes be stable enough to allow small plants or even trees (in Alaska) to grow on the slow-moving or stagnant ice.

Fig. 4.13. Medial moraines on the surface of a valley glacier, north of Nordvestfjord, East Greenland. These are derived from lateral moraines where flow units combine.

Linear ridges of debris on the glacier surface on the glacier surface may also result from ice deformational processes, especially shearing or thrusting of the debris-rich basal layers towards the surface in zones of longitudinal compression. These *shear* or *thrust ridges* may extend for several metres, are often arcuate, dipping upglacier if transverse, or towards the middle of the glacier if parallel to the sides. Such debris has a clear basal imprint including striated stones. The actual mechanism of debris incorporation is disputed, but simple "movement of debris up shear planes" as originally proposed by Goldthwait (1951) has been

137

considered by some to be untenable (Weertman 1961, Hooke 1968). A combination of basal freezing-on of debris, and movement of debris-rich ice along flowlines (Hooke 1968), or along thrusts developed along pre-existing structural weaknesses in the ice (Hambrey & Müller 1978), is more likely for steady-state glaciers, particularly if the relatively thin terminal part of a glacier is frozen to its bed, as in the case of a typical cold glacier (Section 2.10.3). Thrusting, with displacement rates of 0.1m/hr have been observed near the margin of Variegated Glacier during its 1982-83 surge, so it is likely that debris-bearing basal ice can be conveyed along thrusts to the surface if strain-rates are sufficiently high (Sharp 1985a).

Fig. 4.14. The snout of Gåsbreen, Hornsund, Spitsbergen, showing (a) medial moraines emerging from beneath the ice having followed englacial transport paths until this point, as a result of early burial in the accumulation area; (b) terminal moraine complex with a core of dead glacier ice.

Another form of debris mound is the *dirt-cone* which can occur anywhere on the glacier surface, most commonly on otherwise bare ice in the ablation area or on snow. The dirt-cones result from debris accumulating in pools in supraglacial streams; diversion of drainage followed by ablation combined with the protective influence of the debris cover, allows a mound to grow. Dirt-cones can attain several metres in height, and once formed be relatively persistent features on the ice surface. Even so, the veneer of debris is superficial, usually no more than a few centimetres thick. Dirt-cones on snow are less permanent; they result from slurries of debris flowing onto the glacier during the spring melt.

Crevasse-fillings are features with a consistent orientation and represent zones where debris was washed into otherwise clean crevasse by surface meltstreams. They are characterized by their parallelism to crevasse traces (Section 2.8.3). A special type of crevasse filling is associated with surge-type glaciers as on Eyabakkajökull, Iceland (Sharp 1985a,b) These really represent the intrusion of dykes of finer material from subglacial diamicton into crevasses as the glacier sinks to its bed after a surge.

4.5.2 Landforms formed subglacially, parallel to ice flow

Landforms which form at the glacier bed parallel to the direction of ice-flow are streamlined, and probably reflect a complex interplay between deposition and erosion of the unconsolidated sediment. These "bedforms" provide key evidence for understanding the processes that operate beneath glaciers and ice sheets, and are useful in palaeo-environmental reconstruction. However, the processes responsible for the creation of subglacial bedforms are still not well understood, and they have been subject to major controversy. The debate has centred around (a) those who advocate subglacial bed-formation by deformation of soft sediments by moving ice; and (b) those who suggest subglaial flooding on a catastrophic scale over specific areas of the substrate. Although it could be argued that these bedforms are erosional features, they are included in this chapter since they comprise unconsolidated sediments that may eventually be preserved in the stratigraphic record.

Drumlins are one of the most distinctive features of glacial depositional environments, and are particularly common where broad valley glaciers or ice sheets were flowing relatively fast. Arguably, they are the most intensively studied glacigenic landform, partly because of their value in reconstructing ice flow directions, but more especially the light they ought to throw on processes operating beneath large ice-masses. Major contributions in the field of drumlin research in recent years are contained with the volumes edited by Menzies & Rose (1987,1989) , while Embleton & King (1975) have given a thorough review of earlier work with many examples. The term drumlin is derived from *druim*, a Gaelic term for a mound or rounded hill.

Drumlins come in a great variety of shapes and sizes, for example ellipsoidal, egg-shaped, and irregular multiple ridges. The degree of elongation is variable (normally in the range 2.5:1 to 4:1, though exceptionally as much as 60:1) and often they have a long low tail. They occur singly, or in fields of hundreds, whence they give rise to the expression "basket of eggs" topography, on account of their resemblence to birds' eggs, which have a prominent asymmetry-blunt end facing upstream. The larger drumlins reach 50 m or more in height and 20 km in length, but others might be only 2 m high and 10 m long. Drumlins are part of a continuum of glacier bedforms that extend into flutes (Rose 1989).

Drumlins are widely, but sporadically developed in the depositional zones of the last great ice sheets of the northern hemisphere. The largest drumlin field, containing some 10,000 individuals, is probably that in central-western New York State. It occupies an area of 225 by 56 km between Lake Ontario and the Finger Lakes, and was deposited by the Wisconsinan Laurentide ice sheet. In the British Isles, extensive drumlin fields occur in southern Scotland, northwest England, parts of Wales, and in northern and western Ireland (Fig. 4.15); these were formed by the main Devensian ice sheet. Elsewhere, in Europee the extensive drumlin fields deposited by Weichselian ice sheets occur in Sweden and Finland, while smaller fields, related to the encroachment of Würmian glaciers onto the lowlands, are to be found in Switzerland and Germany.

139

Fig. 4.15. Drumlin field, SE Cumbria, England, viewed towards the Howgill Fells (left background) and the Pennines beyond and to the right. Ice flow was towards the right. Note the steeper upglacier ends of many drumlins (in shadow). (Photograph: BPE-86 with permission of Cambridge University Collection of Air Photographs.

The composition of drumlins is highly variable. Usually the dominant facies is massive diamicton, interpreted as till. Some observations have shown that the preferred orientation of clast long axes is parallel to the drumlin long axis, and hence to ice flow, but in some cases the preferred orientation varies throughout the drumlin, though tending to point towards the long axis. Some drumlins have a rock core and others comprise coarse, stratified glaciofluvial deposits. These sediments are commonly highly deformed, with folds overturned inthe direction of ice-flow; in such cases subglacial bed deformation is likely to be the dominant process.

A number of hypotheses have been proposed to acocunt for drumlins, according to a variety of subglacial conditions (Menzies 1989):

(a) Moulding of previously deposited material within a subglacial environment in which a limited amount of subglacial meltwater activity occurs.

(b) Formation resulting from textural differences in subglacial debris due to dilantancy (ability to contract in volume), pore-water dissipation, localized freezing, or localized basla ice-flow patterns;

(c) Formation due to the efects of active basal meltwater carving cavities beneath an ice-mass and subsequently infilling the space with a variety of stratified sediments, or by meltwater erosion of *in situ* sediments, Some of the major drumlin fields (e.g. Livingstone Lake, Saskatchewan) have been explained as the product of catastrophic meltwater floods beneath the large Pleistocene ice sheets, since the pattern and geometry of these features bears a close resemblance to bedforms such as beds of ripples or barchan dunes, which are the product of turbulent water-flow in other situations (Shaw et al. 1989).

Whichever of these mechanisms is correct, there is increasing evidence to suggest that deformation of unconsolidated, slurry-like material beneath a glacier is an important process (Section 2.7.3), Beneath fast-flowing glaciers, localized perturbations within the slurry may eventually lead to the development of at least some groups of drumlin-like landforms.

Drumlinoid ridges or *drumlinized ground moraine* are elongated, cigar-shaped ridges, and spindle forms (Fig. 4.16). Like drumlins, they are the product of ice streamlining, but under basal ice conditions that were unsuitable for discrete drumlins to form.

Fig. 4.16. Drumlinoid ridges and fluting, south of Thelon River, District of Keewatin, Northwest Territories, Canada. Ice flow was towards the bottom of the picture. (Photograph NASPL T301L-223 with permission of the Department of Energy, Mines and Resources, Canada.).

Fluted moraines or *flutes* represent an end-member of streamlined forms that form on fresh lodgement till surfaces. However, flutes may occasionally appear on surfaces of other material, for example glaciofluvial sand and gravel. Generally they appear as large furrows about a couple metres in wavelength, reminiscent of ploughed ground (Fig. 4.17), but megaflutes 100 m wide, 25 m high and 20 km long have been reported in Montana, USA. Superimposed flutes on megaflutes have been described from the proglacial area of Austre Okstindbreen, northern Norway by Rose (1989). Many flutes extend downstream of large boulders embedded in the lodgement till. There is usually a moderately strong preferred orientation of clasts parallel to the flute, although there may be some divergence from the linear trend on the flanks of the flute. Embedded boulders may attain an abrupt stoss side and gentle downstream form (Boulton 1978). Many mechanisms have been suggested, but most authors accept that at least some flutes are formed by the squeezing of a saturated till into the hollow formed by ice as it moves over a large boulder embedded in the till.

Fig. 4.17. *Fluted moraine in the proglacial areas of glaciers in Okstindan, northern Norway. (a) General view of fluted lodgement till at Austre Okstindbreen, the glacier snout lying to the right of the proglacial lake. (b) Close up view of a large flute at Vestre Okstindbreen; ice flow away from camera.*

Boulder beds with striated upper surfaces (*boulder pavements*) are massive, matrix-supported diamicts with moderate to poor sorting. Gravel clasts are mostly rounded and subrounded, and have long axes preferentially aligned parallel to the striae on the top surface of the pavement. Lodgement of basal debris, derived from previously winnowed till, boulder beaches of fluvial deposits, is thought to be responsible. Selective lodgement of boulders occurs downglacier when basal thermal conditions change from cold/freezing to warm/melting. Excellent examples have been described from the late Palaeozoic Dwyka Formation of South Africa (Visser & Hall 1985).

4.5.3 Landforms formed subglacially, transverse to flow

The origin of transverse moraines formed subglacially has received much less attention than longitudinal streamlined forms, even though they are often closely associated. Transverse moraines have a variety of names reflecting that of its discoverer, its type locality or its mode of formation.

Rogen or *ribbed moraines* are large-scale transversely orientated, somewhat irregular, ridges, typically 10-20 m high, 50-100 m wide and 1-2 km long. They were named after a lake in Sweden, and have been recently reviewed by the originator of the term, Lundqvist (1989). The ridges are often slightly arcuate and concave up glacier. Frequently, irregular cross-ribs link up three or four of the transverse ridges, enclosing small lakes and boggy hollows (Fig. 4.18). One of the most characteristic features of the ridges is their gradual transition into drumlins.

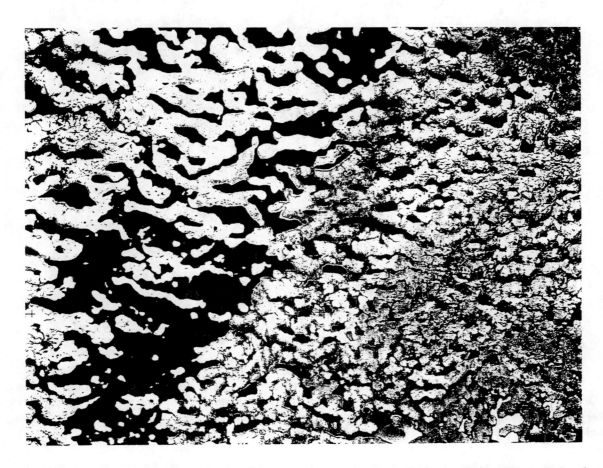

Fig. 4.18. Vertical aerial photograph of Rogen moraine, west of Kaniapiskau River, Labrador, Canada; the ridges are 15-20 m high. Photograph NAPL A11441-121, with permission of Department of Energy, Mines & Resources, Canada .

The moraines are composed of a clast-rich diamicton, and sediments laid down by water. Commonly, a collection of large boulders sits on top. Measurements of the long axes of clasts indicate a preferred transverse orientation. The association of Rogen moraines with streamlined sediments is illustrated, for example, by their crests being fluted or drumlinised. Rogen moraines may also pass imperceptibly into drumlins in a downglacier direction.

The following hypotheses concerning the origin of Rogen moraines have been discussed by Lundqvist (1989):

(a) Deposited as marginal moraines, forming a complex of end-moraines.

(b) Deposited as subglacial moraines, as emphasized by modern literature, formed under thick ice away from the ice-front, especially in a transitional zone between warm- and cold-based ice where the ice is under compression.

(c) Formation by active ice, as a result of tectonic processes within the ice, i.e. subglacial folding of debris-rich layers, or stacking of thrust-slices of debris-rich ice against obstacles to glacier flow, followed by meltout.

(d) By filling of open crevasses by supraglacial debris.

(e) By filling of basal crevasses by subglacial debris.

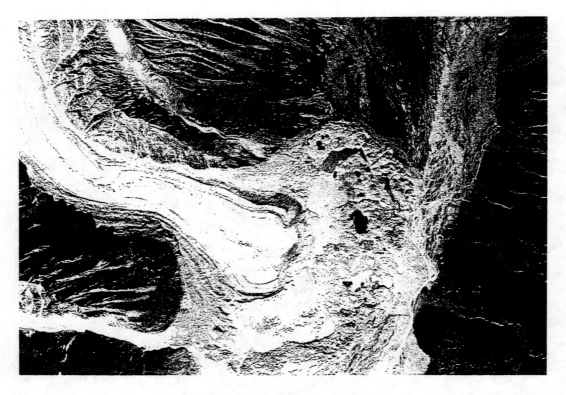

Fig. 4.19. Vertical aerial view of the snout-area of Roslin Gletscher, East Greenland, showing chaotic ice stagnation features formed following a surge. Hummocky moraine with intervening lakes, and thrustmoraines are the dominant features, and show a weak transverse trend. (Photo: Geodetic Institute for Greenland, Copehagen).

Lundqvist's most-favoured hypothesis for the formation of Rogen moraines is a combination of some of these processes, and he notes a genetic affinity with drumlins. In areas of extending flow near the ice-margin, drumlins form at obstacles to the flow. Where the flow becomes compressive, drumlin formation is incomplete, and only crescent-shaped ridges form. These forms are preserved only in the central part of a glaciated area, where the ice stagnated. Stagnant ice deposits are often superimposed on Rogen moraines. The importance of shearing/thrusting mechanisms upglacier of the snout in the creation of Rogen moraines has been stressed by Sugden & John (1976) and Bouchard (1989).

144

Moraines from surge-type glaciers. Transverse moraines that are the product of surging have been described from Eyabakkajökull in Iceland by Sharp (1985b). Here, broad , sedimentologically complex ridges up to 25 m high are present. The processes held to be responsible for their formation include: thrusting, gravity flow of glacigenic sediment, debris-dyke formation, glaciofluvial activity and wastage of buried stagnant ice. Moraine complexes of surge-type glaciers often appear as chaotic, hummocky moraines with numerous kettle holes between, but inspection of aerial photographs of such areas often reveals a clear linear trend to the hummocks (Fig. 4.19). In North America a hummocky supraglacial complex extends over half a million square kilometres in Alberta, Saskatchewan and Manitoba. Here the relief of these features attains 100 m (Paul 1983). Towards the southern limit of the Wisconsinan (last) ice sheet, it is conceivable that hummocky moraines were formed following a surge of part of the Laurentide ice sheet over the Mid-West United States. Rapid down-wastage followed the advance (however caused) and blocks of ice may have become detached from the parent mass. Certain areas show superimposed transverse ridges which could represent crevasse-fillings, creating an assemblage of forms that is characteristic of surge-type glaciers in Iceland as described by Sharp (1985b). However, this concept of surge-type deposition for the Mid-West moraines is highly controversial.

Other types of transverse moraines include De Geer and sublacustrine moraines; they are discussed in Chapter 6.

4.5.4 Non-orientated landforms, formed subglacially

Landforms of this nature are commonly referred to somewhat loosely as ground-moraine - areas of non-lineated, smooth to hummocky drift cover, especially of basal melt-out and lodgement till, with minor subglacial outwash lenses and a thin cover of supraglacial melt-out or flow tills. The various types of ground moraine, from thickest to thinnest are:

Till-plains: nearly flat or slightly rolling and gently inclined plains; mostly consisting of a thick till cover, often in multiple layers of varying composition, completely masking bedrock irregularities.

Gentle hills: mounds of till resting on detached blocks of bedrock.

Hummocky ground-moraine (or *dead ice/distingegration moraine*): a chaotic area of hills of basal till, flow till with minor reworked sediments associated with small streams and ponds. Supraglacial debris may drape the basal till.

Cover moraine: a patchy, thin layer of till revealing the bedrock topography, either entirely (veneer) or only partly (blanket).

Till-plains are normally associated with lowland glaciation. Large flat-lying areas underlain by till are widespread in Europe and North America. Much of the Cheshire Plain, the Vale of York, and East Anglia in Britain has flat sheets dominated by till, often 100 m thick. The plains of North Germany and lowlands north of the Alps have a mantle of till. In North America, till-plains are widespread in the States of the Mid-West. Hummocky moraines are common in both lowland and mountain regions and is a form of ablation moraine. Hummocky moraines are the result of downwastage of a glacier rather than recession, especially when the ice has ceased to be active. They are often genetically linked to supraglacial hummocky moraines, where basal debris is elevated to an englacial and supraglacial position, and mixed with supraglacial debris. It is normal for the hilly topography to be controlled by supraglacial meltout processes, rather than basal meltout, so these forms are discussed more thoroughly in Section 4.5.6. One of the most instructive papers on the genesis of these features was by Boulton (1967), who described processes at the margins of a glacier in Spitsbergen.

Hummocky moraine is a characteristic feature of the southern limits of the last great ice sheets of North America and northern Europe, where they attain a relief of tens and occasionally a few hundred metres. Large areas of hummocky moraine suggest that the ice sheets wasted rapidly. It is unlikely that supraglacial debris was much involved; the only potential source is from the bed by means of shear processes anyway, and such material is indicative of active ice, and generally limited to the snout position. In the British Isles, one of the largest supraglacial hummocky moraine complexes was formed when a major ice stream flowing down the Irish Sea impinged against the west flanks of the Pennines in Cheshire and Staffordshire (Paul 1983).

Hummocky moraines in highland areas are a feature of down-wastage of valley glaciers. They can be observed to form in almost any terrestrial glacial environment today. Particularly fresh-looking forms, with a relief of several tens of metres in valleys of the English Lake District and the Scottish Highlands, were deposited during the rapid wastage phase of the last valley glaciers and ice-fields around 10,000 years B.P. (Fig. 4.20).

Some erratics belong to this class of landform. Where large blocks have been detached from the bed of a glacier they may stand proud of the general level of the till plain.

Fig. 4.20. Late glacial hummocky moraines in Glen Derry, Cairngorms, Scotland.

4.5.5 Supraglacial landforms, parallel to ice flow

Features of this nature include lateral moraines and moraine dumps, the latter derived from medial moraines; these have already been mentioned in the context of features present on the surface of moving ice. However, the long-term preservation potential of lateral moraines is relatively small due to valley-side collapse after ice recession. That of medial moraines is even less because of ice front processes and glaciofluvial action.

Lateral moraines are some of the most impressive features of contemporary glacial mountain environments (Fig. 4.21)., especially above and downstream of those glaciers in the Alps, Scandinavia, the Western Cordillera of North America, and elsewhere which advanced strongly during the Little Ice Age of around 1700-1900 AD.

Fig. 4.21. A fine pair of lateral moraines along the flanks of Vadrec da Tschierva, Graubunden, Switzerland. Note the extent to which material has been accreted away from the valley sides, and the characteristic small valleys between the moraine crests and adjacent hillsides.

Lateral moraines form from the steady supply of frost-shattered debris that falls down onto the glacier margin from the cliffs above, and from debris that is rubbed between the glacier and the valley side. They are therefore composite supraglacial and subglacial features. The debris reflects both the angular nature of rockfall debris and the subrounded/subangular nature of basally (or marginally) transported stones. The latter type often show typical glacial abrasional markings, such as striations and faceted surfaces. The effect of comminution of debris is such that lateral moraines generally have a mud-grade matrix. This is manifested in the steep inner faces of lateral moraines when ice recedes rapidly, the clay binding the slope to some extent together, although the effect of rainwater and irregular dislodging of embedded boulders tends to produce parallel runnels (Fig. 4.21). It is deceptively difficult and often hazardous to climb the inner faces of fresh lateral moraines. In contrast the outer face of a lateral moraine, especially a Little Ice Age one, is stable and often well vegetated. This slope, facing the adjacent hillside, may have its own stream and small lakes at its foot .

Lateral moraines in Alpine terrain often grow towards each other by accretion of basal or marginal, rather than rockfall debris. In places such moraines may become somewhat detached from the valley sides (Fig. 4.21). The glacier may thus be more constrained by its moraines than by the steep valley sides. Advancing glaciers may break through the old moraines below the crest or dislodge boulders over the crest.

Lateral moraines which cling to the mountainside rapidly undergo collapse. After a few hundred years, only a low angle bench or a slight break in slope may be all that is left of a lateral moraine. However, even in Britain, which lost its glaciers 10,000 years ago, a discerning eye may be able on hillsides to pick out subtle notches that represent lateral moraine positions.

Fig. 4.22. Lateral moraines alongside Glacier de Corbassière, Valais, Switzerland. The outermost one (left) is probably the Little Ice Age moraine; the other three are ice-cored.

Another type of lateral moraine is ice-cored, and is a feature that is best developed in the polar regions where melting of a large mass of ice takes hundreds of years, although some good examples occur in Alpine terrain (Fig. 4.22). Ice-cored moraines may also comprise basal and supraglacial debris. Dynamically, they are separate from the main glacier; in fact, the ice is probably dead. The veneer of debris is typically 1-2 m thick, and dark wet scars indicate where slumping off the sides has occurred. Alongside cold glaciers, marginal streams may flank both sides of the moraine, one associated with glacial meltwater, the other with snow melt and rainwater.

Moraine dumps rarely survive the recession of a glacier. Near the snout, the debris is often spread over a broad zone and several medial moraines merge together to form a blanket, rather than distinct ridges. Even if a prominent medial ridge is left when the ice recedes, glaciofluvial action will tend to destroy it as the river braids constantly migrate across the valley floor.

4.5.6 Non-orientated landforms, formed supraglacially

As mentioned in Section 4.5.4, hummocky moraines are often of composite origin, involving the displacement of basal debris towards the glacier surface. However, in contrast to lowland ice sheets and cold glaciers which yield hummocky moraines of largely basal origin, mountains glaciers generate topographically similar features principally as a result of down-wastage of a thin supraglacial debris cover, derived principally from rockfall.

Hummocky moraine begins as unstable debris lying on a substantially debris-covered glacier surface, usually near the snout where the ice is relatively inactive. Uneven ice wastage gives rise to a chaotic hummocky appearance, and recycling by flowage of saturated debris and by local supraglacial and englacial meltwater streams is a common process. Sometimes, meltwater has carved out of the ice large tunnels and caverns an assemblage of forms referred to as glacier karst by analogy with limestone regions. The relief in areas of hummocky moraine may reach several tens of metres. The process of hummocky moraine formation may be be observed directly on many glaciers, for example the Unteraargletscher in Switzerland and the Tasman Glacier in New Zealand. Ultimately, the dead glacier ice is reduced to areas of ice-cored moraine and the final decay of ice leaves behind a series of hummocky moraines.

Mention should be made of *supraglacially-derived erratics*. These angular blocks may have fallen onto clean ice and not be linked with any surface moraine, so are isolated when the ice recedes. Occasionally, such erratics stand on pedestals if the underlying surface has since been eroded away, especially in areas of carbonate bedrock such as the Pennines of northern England where chemical weathering has taken place.

4.5.7 Ice-marginal landforms, transverse to flow

These forms are essentially represented by a range of *end moraines* which document the stages when a glacier remained stationery but active, or more especially when it advanced. They form perpendicular to ice motion and consist of belt of ground higher than the general level of the valley floor, often sweeping in a convex down-valley arc to join lateral moraines. End moraines are compositionally varied and complex, comprising any of the sediments that form in the proglacial area of a glacier.

A *terminal moraine* is a broad arcuate belt of debris that formed around an ice lobe during the period in which the ice was at its of maximum extent. It is frequently hummocky, pitted, comprises irregular short ridge crests, and contains high mounds of distinctive lithology derived from point sources. A long halt during a general phase of recession leads to the formation of a *recessional moraine*.

Both these sorts of moraine have a steep outerface when fresh (20-30°), whereas the inner slope is generally irregular and hummocky, and of relatively low angle (10-20°); they tend to have a convex down-glacier form in valleys, and even in lowland areas they are frequently lobate. In regions such as the Alps, well-preserved terminal moraines are relatively rare, due to destruction by meltwater. Lower land bordering the Alps and similar regions have well-formed moraines, deposited by glaciers of the last ice age, notably those which enclose the outlets of the large lakes in northern Italy. In North America, a fine set of terminal moraines lies to the east of the Teton Range in Wyoming. These were created by a piedmont glacier that spread out from the foot of the range and scooped out the lake basin which is now enclosed by a prominent ridge. One of the tallest terminal moraines (430 m) was created by the very dynamic Franz Josef Glacier when it debouched onto the coastal plain to the west of the Mount Cook range in New Zealand. Terminal moraines associated with the great ice sheets do not have such pronounced relief, but are impressive because of their lateral extent; northern Germany, Poland and Denmark have fine examples. The southern margin of the Laurentide ice sheet in North America is marked by a zone up to 600 km wide, comprising a terminal-recessional moraine complex, extending across the continent to the Rockies in the latitude of, and south of, the Great Lakes. North of this zone is a belt of ice disintegration features, the hummocky moraines, to which reference has already been made.

Dumping of supraglacial debris is a common mechanism in the creation of terminal and recessional moraines, since during an advance the snout is normally very steep and debris accumulates as a heap at the foot of the ice slope. Release of the basal debris load in one place for a long period will result in an irregular terminal moraine composed principally of subangular and subrounded stones, of which a proportion are striated. The process of thrusting, already described, may create distinct ridges. This latter mechanism is important at the margins of sub-polar ice margins today, and may have been significant in the creation of the end moraines of the last glaciation.

Other processes involved in terminal moraine formation include meltout, flowage and squeezing out of saturated till. Ice-cored end moraines may also develop, especially in the polar regions (Fig. 4.14). Clast fabric studies have tended to indicate a preferred orientation parallel to the moraine crest. Often terminal and recessional moraines are destroyed by meltwater during subsequent recession, and those that survive may have relatively subdued forms. Alternatively they may simply be represented by boulder belts, scattered outsize boulders (up to the size of a hut), the small debris having been washed away.

Rather more emphemeral features are *annual push moraines*, which are the result of small winter readvances pushing up a series of small, closely-spaced ridges a metre or so high, during a general period of recession. They are a common feature at the snouts of the relatively less dynamic temperate glaciers of Norway, and of smaller glaciers in alpine areas (Fig. 4.5).

The rather special forms of end-moraine that result from glaciotectonic processes are described in Section 4,6, However, it is conceivable that many more,

150

supposedly "conventional' end-moraines may turn out to be the product of glaciotectonism on further investigation.

4.5.8 Non-orientated ice-marginal and proglacial landforms

This group of landforms includes hummocky moraines and the product of mass movement of glacigenic sediment. The genesis of hummocky moraines is normally associated with supraglacial or basal meltout processes (as described in Section 4.5.4), but continues if irregular heaps of debris fall from the face of an ice-mass into the ice-marginal zone, or if dead ice becomes detached from the parent ice body.

Mass-movement processes which result in discrete landforms are rockfalls, debris-flows and slumps. Rockfalls are induced by (a) the oversteepening of valley sides by glacial erosion and (b) the development of joints parallel to the valley side as a result of pressure release following recession of the ice (Harland 1957), a type of exfoliation. Such rockfalls can sometimes fill a valley bottom and numerous instances occur in alpine valleys. Debris-flows and slumps most commonly involve tills. The high silt and clay content allows till-covered hillsides fail easily, especially when wet. Debris-flows commonly occur as distinct lobes, often several metres thick. Slumps are marked by scars of fresh debris and a rucked surface and bulge below. These processes are a day-to-day occurrence on land recently vacated by the ice, and may continue to affect hillsides thousands of years after the ice has receded.

4.6 Glaciotectonic landforms

Glaciotectonic structures are commonly manifested at the land surface as distinct landforms. Many Late Pleistocene and younger features display their original morphology, but others may have been buried by later depositional processes. Glaciotectonic landforms comprise a variety of hills, ridges and plains, all of which are constructed wholly or in pert of bedrock or drift, or both. The definitive work on glaciotectonic landforms (Aber et al. 1989) uses a five-fold classification. They are described below in order of decreasing topographic prominence.

4.6.1 Hill-hole pair

This feature represents a combination of an ice-scooped basin and a hill of ice-thrust, often slightly crumpled material, of similar size. The principal morphological features include: (a) an arcuate or crescentic outline of a hill, convex in the downglacier direction (b) multiple, subparallel, narrow ridges following the overall arcuate trend of the hill; (c) an asymmetric cross-profile, with steeper slopes on the downglacier side; (d) topographic depression on the upglacier side of the hill, covering an area approximately equal to that of the hill. The size of the hill-hole pairs varies from 1 km² to 100 km², with relief ranging from 30 to 200 m. Good examples of hill-hole pairs are situated at Wolf Lake in

151

Alberta, Herschel Island in the Yukon and at the contemporary ice margin of Eyabakkajökull in Iceland.

4.6.2 Large composite-ridges

These are the most typical and distinctive of all glaciotectonic landforms and are composed of large slices of up-thrust and commonly contorted sedimentary bedrock that is generally interlayered and overlain by large amounts of glacial drift. Large composite-ridges are up to 200 m high, 5 km wide and up to 50 km long, and have an arcuate form. Individual ridges within these complexes are typically several hundred metres high and up to 100 m wide. Elongated lakes often form in the valleys between the ridges. The *inside* of the ridge complex, generally demarcates the maximum position of the ice. Large composite-ridges usually involve considerable disruption of pre-Quaternary bedrock, perhaps to depths of 200 m below the surface. The folds and thrust blocks are stacked up in piggy-back fashion. Large composite-ridges are gemetrically similar to mountain belts that were the product of thin-skinned tectonic movements.

The best known site is Møns Klint in Denmark, where large thrust-blocks of Cretaceous chalk and drift are exposed in coastal cliffs up to 143 m high, with ridges extending inland. This complex was formed during the late Weichselian glaciation, as a result of the expansion of the Fennoscandian ice sheet. Similar ridge complexes have been described from Dirt Hills and Cactus Hills, Saskatchewan, and Prophets Mountains in North Dakota.

4.6.3 Small composite-ridges

Small composite ridges are smaller scale versions of the ridges described above, and they generally have a relief of less than 100 m. Many are composed only of unconsolidated Quaternary strata and are thus more susceptible to erosion. The terms *push-moraine* and *Stauchmoränen* (from the German) are commonly applied to such ridges. They are the product of thrusting of unconsolidated proglacial deposits, including glaciofluvial and marine sediments, especially in areas of permafrost (Fig. 4.23). Good examples are to be found in the Canadian Arctic and Svalbard. On Axel Heiberg Island, the snout of advancing Thomson Glacier is pushing forward a small composite-ridge that is 45 m high, 2.1 km long and 0.7 km wide, comprising almost entirely glaciofluvial sediments (Kalin 1972) (Fig. 4.24).

The forward part of a push moraine-complex usually comprises stacked sheets of glaciofluvial sand and gravel, forming ridges with steep outer faces and gently inclined inner faces, with no trace of till, so the sediment must have been thrust up in a frozen state in front of the ice margin. Inner parts of the complex comprise sheets of sand and gravel, but also till, glaciolacustrine or glaciomarine sediments. Recumbent folding is evident, especially in the less competent muddy materials.

Composite-ridges seem to be preferntially associated with surge-type glaciers. Many small composite-ridges have been mapped as conventional end-morianes, but for these a glaciotectonic origin cannot be ruled out. A continuous spectrum occurs between the two types of composite-ridge.

Fig. 4.23. Advancing front of Thomson Glacier, Axel Heiberg Island, Canadian Arctic. To the right a push-moraine complex of thrust fluvial gravels is forming. This complex was described in detail by Kalin (1971).

4.6.4 Cupola-hills

Isolated hills or a jumbled group of hills with no obvious source-depression, but having the general characteristics of ice-thrust masses, occur on bothe large and small scales. The most common form is the cupola-hill, which has an internal structure similar to that of composite-ridges, but which shows signs of having been subsequently overridden by the ice. Cupola-hills consist of deformed glacial and interglacial deposits, plus detached blocks ('floes') of older strata or bedrock, overlain by basal till which truncates the older strata. Cupola-hills have a dome-like form with long, even slopes, varying from near-circular to elongated ovals. They range from 1 to 15 km in length, and from 20 to at least 100 m in height. Cupola-hills are common in regions having a soft substratum that was affected by ice coming from different directions. Good examples of cupola-hills, described by Aber et al. (1989), occur on the islands of Møn and Langland in Denmark and at Martha's Vineyard in Massachusetts.

4.6.5 Megablocks and rafts

Large pieces of glacially transported bedrock, buried in drift are called megablocks or rafts. Although they are typically less than 30 m thick, their lateral dimensions

153

often exceeds 1 km². Transport distances may be great, for example a group of rafts in east-central Poland orignates in Lithuania, over 300 km to the NE. Rafts are generally removed from the bedrock along bedding planes, which act as planes of detachment when the ice overrides them, especially if the ground is frozen.

Perhaps the biggest raft, covering an area of about 1000 km² is found at Esterhazy, Saskatchewan, and many other rafts occur throughout the plains of southern Alberta. Rafts of Cretaceous chalk, measuring up to about 100 m in length are well exposed in coastal cliffs in Norfolk, England. These show signs of internal tectonic deformation including thrusting and gentle folding. A dolomite example from the Neoproterozoic Port Askaig Tillite in the Garvellachs, Scotland, and knicknamed 'The Bubble', is tightly folded.

4.6.7 Diapirs, intrusions and wedges

Various kinds of soft-sediment deformation features are widely known from glacigenic sequences, and range in lateral dimensions from a few centimetres to over 100 m. Intrusions include all those structures resulting from the injection or squeezing of one type of material into another. The intruded material is usually clay- or silt-rich sediment, whereas the host material may be of almost any composition if soft. Intrusions develop in a subglacial, water-saturated condition with intergranular movement as the main means of deformation. Structures may be grouped as those which originate from below (diapirs, dykes etc.) and those which are forced down from above (wedges and veins). Wedge structures may be misinterpreted as periglacial features such as ice wedges or thermal crack-fillings. Good examples of intrusions have been described by Aber et al. (1989) from Kansas, western Norway and Denmark.

4.7 Preservation potential of terrestrial glacigenic sediments

Compared with many non-glacial sedimentary sequences, terrestrial glacigenic deposits tend to be relatively thin. In highland areas their potential for preservation is low because of the efficacy of reworking processes, such as by streams or mass-movement, following ice recession. In less confined lowland areas, sheets of till may have a chance to build up to a thickness of several tens of metres, and remain largely undisturbed by non-glacial processes. On the other hand, subsequent ice advances tend to remove the evidenc of the previous depositional phases, and the earlier record of glacier fluctuations will be largely destroyed.

On a geological timescale of tens and hundreds of millions of years, even a somewhat sparse terrestrial record may be preserved from lowland areas, since the decay of ice sheets leads to a rise in sea level, thus burying the glacial sequences with marine sediments. Thus, although unimportant in volumetric terms, terrestrial glacigenic sediments are documented from all the main phases of Earth's glacial history. Probably the best examples are from the late Palaeozoic glaciation of Gondwanaland. However, for assessing the ancient glacial climatic record, we need to turn to glaciomarine sequences, since it is these that provide the thicker and more continuous sedimentary sequences, and are less likely to have been eroded.

In contrast to the sediments, glacial depositional landforms do not commonly survive intact to be incorporated in the sedimentary record. Indeed many moraines and ice contact features disappear when the glacier itself has gone, because they depend on the ice for support and are subject to reworking by mass-movement, fluvial action, periglacial processes, etc. Some streamlined forms, in theory, have a better chance of survival in the sedimentary record than some less dynamically produced forms because they may quickly become buried as a result of sedimentation associated with ice wastage. However, to recognise such forms in sedimentary sections can be difficult, and there are only a few instances of depositional landforms being identified in ancient glacigenic sequences.

CHAPTER 5

GLACIOFLUVIAL PROCESSES AND LANDFORMS

5.1 Introduction

In many areas, more sediment is transferred out of the glacier system by meltwater than directly by the ice. Huge quantities of sediment may be involved, but the rate at which it is moved is strongly dependent upon temperature, which in turn is a function of the season and the time of day. Often, one may obtain a good impression of the power of a meltstream from the noise of boulders rolling along the bed of the stream. The milky appearance of the water arising from suspended mud (*rock flour*) is another indication of the efficacy of such streams in removing sediment from a glacier (Fig. 5.1). Meltwater streams are also powerful agents of erosion, especially beneath the ice where the water may be under high pressure.

Fig. 5.1. *The Copper River, Alaska, one of the world's largest glacial meltwater streams. Laden with fine-grained glacial sediment, the stream has a milky appearance as it flows past and undercuts, the terminus of Childs Glacier.*

The rôle of water within, alongside and under the ice, and the character of runoff, has been discussed in Chapter 2. Here, we briefly examine erosional and sedimentary processes and the rôle they play in the development of landforms. Emphasis is placed on processes and landforms in close proximity to the ice margin, rather than in areas distant from the source glacier. For more extensive accounts of glaciofluvial processes and landforms, several works may be recommended: Rice (1972), Embleton & King (1975), Jopling & McDonald (1975), Sugden & John (1976), Miall (1977, 1983) and Drewry (1986).

Numerous examples of glaciofluvial sequences occur in the geological record. Of particular note are the Neoproterozoic and late Ordovician deposits of the Taoudeni Basin of West Africa (Deynoux & Trompette 1981a,b), and the late Palaeozoic of the northern Karoo Basin in South Africa (von Brunn & Stratten 1981).

5.2 Glaciofluvial processes

5.2.2 Meltwater erosion

Glacial meltwater at the base of a glacier, together with its sediment load, is often a major instrument of erosion. The importance of meltwater erosion and transport increases progressively towards the snout, thus assuming a dominant rôle below the equilibrium line. Water at the base is primarily derived as a result of supraglacial melting, reaching the bed via crevasses and moulins (glacier potholes). Some water is derived as a result of basal melting and some as runoff from the valley sides, but generally the latter components are much smaller.

Water at the base of the glacier flows in four main modes: (a) in tunnels cut into the ice, manifested at the snout by a semi-circular or semi-elliptical basal stream outlet or portal; (b) by sheetflow at the ice/bedrock interface; (c) by draining into the underlying sediment, allowing it to deform more easily, and (d) in tunnels incised into the bed, including both bedrock and unconsolidated sediment.Water is often under high pressure at the base, so facilitating erosion of the ice as well as the bedrock or sediment beneath. The hydroelectric power companies have long recognized the problem of water under high pressure in tunnels, and the process of *cavitation*, i.e. the growth and collapse of air bubbles, resulting in the generation of shock waves. The same effect is evident beneath glaciers but is compounded by the abrasive quality of suspended sediment.

Meltwater stream patterns are, to a great extent, related to the temperature of the ice. In temperate glaciers, most water finds its way to the base and appears at the snout near the lowest point. Lateral meltstreams are absent. In cold glaciers the path of meltwater is dependent on whether the glacier is frozen to the bed, but most water flows near the surface, with discharge taking place *via* lateral meltstreams, subglacial meltstream systems normally being poorly developed.

Ice-dammed lakes, on their release, also play a significant rôle in meltwater erosion. Typically, they build up during the ablation season until a critical level is reached. This may occur when the water has a sufficient head to force its way beneath the glacier, and lift up the ice bodily, thus releasing a catastrophic flood of water, which flushes out the system. Alternatively, the lake may overflow and cut a gorge into the glacier surface or spill over a neighbouring col. Subglacial lakes have also been recorded. In sedimentological terms, the consequences of catastrophic drainage from ice-dammed and subglacial lakes are frequently dramatic.

5.2.3 Transport of sediment and solutes

Sediment carried by meltwater streams may be categorized as suspended load and bed-load. The suspended load has been monitored near the snouts of many glaciers, and a clear relationship of major peaks with discharge has sometimes been demonstrated. For example, the glacier Erdalsbre in southwestern Norway indicated peaks of suspended sediment concentration coinciding with high discharge events (Østrem 1975). Peak values of sediment concentration exceeded 2000 mg/litre at 10-15 cubic metres per second, whereas typical low values for the same summer season were 100-200 mg/litre and 5 cubic metres per second. However, many of the smaller variations in suspended sediment content do not obviously match variations in discharge. At some glaciers, the peak sediment concentration may slightly precede the maximum discharge, probably reflecting the pick-up of sediment during the phase of rising discharge. In other cases, the peak discharge may coincide with a minimum of suspended sediment if the glacier bed is relatively clean and has insufficent loose material to be incorporated into the stream.

The bed-load in a stream comprises material that is rolled or bounced ("saltated") along the bed. Hydro-electric power schemes need to trap such debris, using grids over which the larger material rolls, but which allows the water with suspended sediment material through. Norwegian and Swiss engineers have gradually improved the efficiency of these traps, but nevertheless there is still considerable wear on tunnels and turbine blades from the suspended sediment.

The relative proportions of bedload and suspended sediment are highly variable; bedload is greater in steeper outwash areas than on gently inclined ones. Overall, it is common each year for the bulk of sediment discharge to take place during a few days of high magnitude flood events.

The rôle of meltwater as an agent of chemical solution of the bedrock has often been underestimated. Meltwater with dissolved carbon dioxide forms a weak acid that is capable of dissolving limestone or dolomite bedrock. Extensive investigations of these processes have been undertaken on the Glacier de Tsanfleuron in Switzerland (Sharp 1989). Selective mineral solution may also occur, for example of feldspars in igneous and metamorphic rocks, or the breakdown of micas into clay minerals, so weakening the rock as a whole. Mineral-rich waters may leave thin deposits of calcite or silica during regelation, especially on the lee side of small bedrock bumps in front of receding glaciers (Hallet 1976). Solutional furrows and extensive films of precipitated calcite indicate that an almost continuous film of meltwater exists beneath such glaciers.

5.2.3 Deposition of glaciofluvial sediments

The deposition of meltwater-transported sediment takes place in a variety of settings:

• Supraglacially, where the finer fraction of rockfall-derived material accumulates in crevasses, pools and streams on the glacier suface.
• Englacially, where surface debris is washed down moulins into the internal drainage system of a glacier.
• Subglacially, either in tunnels cut into the base of the glacier by basal meltwater, or in channels cut into subglacial till or solid bedrock beneath the ice.
• Ice marginally, where streams of both glacial and non-glacial origin are forced to flow along the ice margins where the glacier has a convex-up cross-profile.
• Proglacially, firstly in an ice-frontal position especially in the zone between the snout and the higher ground of an end-moraine complex, and secondly downvalley as the main meltwater stream spreads across the valley floor or plain.

Of the above, the first two are unimportant in terms of the final sedimentary product. Subglacial, ice-marginal and proximal proglacial sedimentary processes give rise to a variety of landforms (Section 5.3) that are frequently modified or destroyed soon after deposition, but which nevertheless are sometimes preserved in the geological record. By far the most important setting for the development and preservation of glacial meltwater sedimentary sequences is downvalley of the glacier, away from the immediate influence of marginal ice processes, as on braided outwash plains.

5.3 Glaciofluvial landforms

As noted earlier in this chapter, glacial meltwater is capable both of transporting large quantities of sediment and creating a wide variety of depositional landforms, and eroding bedrock or unconsolidated sediment to a considerable depth. These depositional and erosional processes take place subglacially, ice-marginally and proglacially, and the resulting landforms may be classified accordingly (Table 5.1). However, the forms described represent a continuum of features, some of which are the result of both erosion and deposition.

5.3.1 Landforms resulting from subglacial meltwater erosion

The most important features are a range of channels orientated approximately parallel to the former ice flow direction, and incised into the sub-ice bed, whether bedrock or unconsolidated sediment such as till or fluvial material. Some of these forms develop as a result of a combination of ice-abrasional and meltwater-erosional processes, and are called p-forms. Where abrasion is dominant they are described in Chapter 3.

It is known from studies of the hydrology of temperate glaciers that water tends to gather into discrete streams, especially towards the topographically lowest part of a glacier cross-sectional profile. In narrow valley glaciers, there may be just one major stream running down the ice-flow centreline, where stresses in the ice tend to be least, but in broad valley glaciers, or lobes of ice of the piedmont type, there may be a series of channels, interlinked in a complex pattern similar to braided channels, although the floors of the channels may be discordant with one another.

The largest features resulting from subglacial meltwater erosion are *tunnel valleys*. Several examples associated with the last ice sheet over northern Europe have been described (Ehlers & Gibbard 1991). In East Anglia in England, steep-sided, deep valleys have been cut in chalk and associated bedrock, and are normally filled with glacial meltwater sands, gravels or finer material; some till may also be present. Often, tunnel valleys show a reverse gradient, for example the so-called *Rinnen* of Denmark, northern Germany and Poland. Subglacial streams under high pressure are responsible for these features. Tunnel-valley fills associated with a late Pleistocene grounded tidewater ice-margin in the Irish Sea Basin have been documented, both in coastal exposures near Dublin and in seismic profiles offshore (Eyles & McCabe 1989). The channels are steep-sided, are stacked one on top of another, and measure 10 m in depth and up to 2 km in width.

Other spectacular channels are *subglacial gorges* cut into solid bedrock. Many such channels are extremely narrow in relation to their depth, several metres compared with tens of metres. They tend to be deepest and narrowest where the main valley narrows or steepens, and they cut through riegels and valley-steps in a manner which suggests that the stream was trying to smooth out the longitudinal gradient.

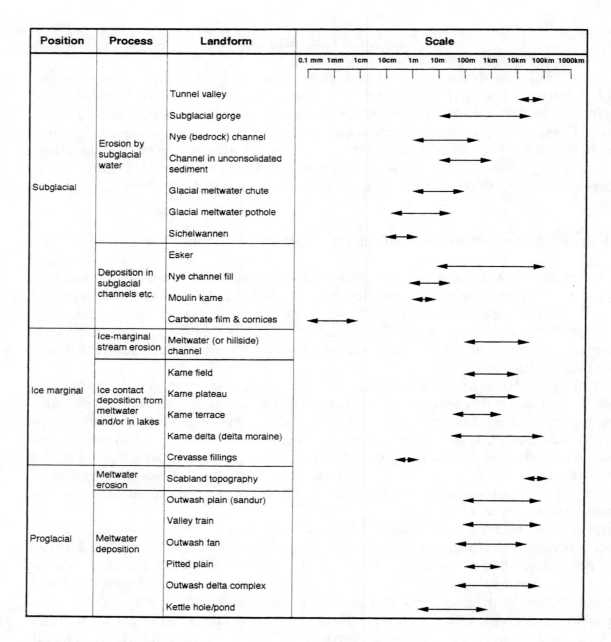

Position	Process	Landform	Scale
			0.1 mm 1mm 1cm 10cm 1m 10m 100m 1km 10km 100km 1000km
Subglacial	Erosion by subglacial water	Tunnel valley	
		Subglacial gorge	
		Nye (bedrock) channel	
		Channel in unconsolidated sediment	
		Glacial meltwater chute	
		Glacial meltwater pothole	
		Sichelwannen	
	Deposition in subglacial channels etc.	Esker	
		Nye channel fill	
		Moulin kame	
		Carbonate film & cornices	
Ice marginal	Ice-marginal stream erosion	Meltwater (or hillside) channel	
	Ice contact deposition from meltwater and/or in lakes	Kame field	
		Kame plateau	
		Kame terrace	
		Kame delta (delta moraine)	
		Crevasse fillings	
Proglacial	Meltwater erosion	Scabland topography	
	Meltwater deposition	Outwash plain (sandur)	
		Valley train	
		Outwash fan	
		Pitted plain	
		Outwash delta complex	
		Kettle hole/pond	

Table 5.1. Classification of glaciofluvial erosional and depositional landforms according to position in relation to the glacier margin. The scale range-bars refer to the maximum linear dimension of the landform.

The cross-sectional shape of a subglacial gorge is generally irregular, especially where the downstream gradient is high, but in gently inclined reaches, a flat bottom with vertical sides is more typical. Since gorges commonly exploit weaknesses in bedrock, such as faults or dykes, they are often straight.

Gorges are cut down by subglacial meltwater under high pressure, with cavitation playing an important rôle. They may also be moulded by ice, as indicated by striations, which fills the void left as the subglacial stream cuts downwards. A good example of a still partially ice-filled gorge is that at the snout of the present-day Unterer Grindelwaldgletscher in Switzerland. Numerous other subglacial gorges occur throughout the Alps, Scandinavia, and on a smaller scale the British Isles. Fine gorges tens of metres deep in Britain are to be found near Loch Broom in northwest Scotland (the Corrieshalloch Gorge) and in Glencoe; gorges occur in many other parts of Scotland, the Lake District and North Wales.

In some areas, channels may be many kilometres long, but lack a consistent downstream gradient. Instead, they continue over rises in the bedrock even those having an amplitude of more than 100 m, the water having been able to flow uphill because it was under high pressure. Following recession of the ice, *ribbon lakes* may be left filling the lower parts of the channel system. Channels may also be cut in unconsolidated till or other forms of glacial drift. They, too, are frequently deep in relation to their width, and near vertical sides are common. Typically they attain depths of several metres.

Fig. 5.2. Nye channel of late Pleistocene age cutting into diamictite of Neoproterozoic age, Garvellachs, Scotland.

Networks of subglacial channels, cut into both bedrock and drift have been described from many parts of Europe and North America. In Britain, the Southern Uplands and Pennines are well-endowed with such features.

On an order of magnitude smaller scale, there are bedrock channels called *Nye* (or *N-*) *channels*. These have vertical or undercut sides, are commonly sinuous (even meandering) and contain small pools (Fig. 5.2). Nye channels are especially common in carbonate bedrock, and here chemical erosion may largely be responsible for their formation. An exceptionally fine set of Nye channels have been described from the Glacier de Tsanfleuron in Switzerland (Sharp et al. 1989), and late Precambrian examples have been documented from Greenland (Moncrieff & Hambrey 1988).

Channelized meltwater flow tends to be associated with relatively low sliding rates. Subglacial channels of all sizes, once initiated, tend to be self-perpetuating and may carry most of the water, except in times of flood. In contrast, subglacial tunnels cut in the ice and referred to as *Röthlisberger* or *R-channels* are emphemeral as the creep of ice is constantly trying to close them, and often does so each winter. By the same token, creep of ice may encroach on bedrock channels at times of low discharge, allowing striations to develop, but subsequent meltwater activity tends to re-open them.

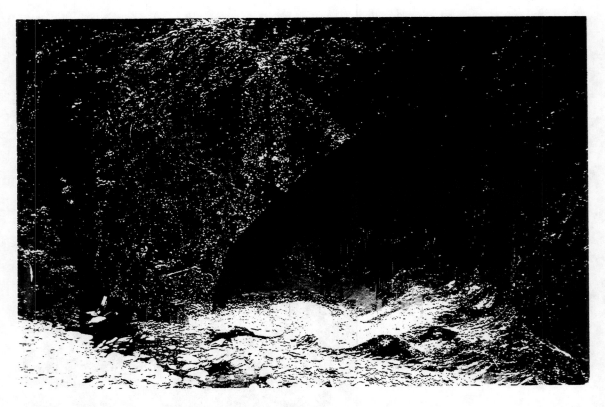

Fig. 5.3. Large potholes bored into gneiss in the valley wall below Franz Josef Glacier, New Zealand. The height of this feature is about 3m.

Sometimes, channels run directly down steep rock-slopes. These are *meltwater chutes*, and represent the places where marginal meltstreams descend to a lower position in the glacier.

Apart from channels, subglacial meltwater may create a variety of small landforms in bedrock. *Potholes* are circular shafts, ranging from a few centimetres across to over 10 m in diameter and up to 20 m deep, cut into bedrock either in the valley floor or the valley side (Fig. 5.3). Internally, the shaft reveals a spiral structure and the bottom often has a collection of well-rounded boulders and cobbles. Potholes are best developed in hard crystalline rocks, and fine examples occur in Scandinavia and the Alps, both on the valley floor and along the sides. In a sense, potholes represent a plunge-pool, strongly constrained by the ice, developing especially beneath a moulin that extends to the base of a glacier. In some circumstances, potholes are created by the cavitation process associated with water under high pressure. *Bowls* are a less well-developed form of pothole.

Of the various types of p-form, only *Sichelwannen* (German for "sickle-shaped troughs"), which are crescent-shaped depressions and scallop-like features on a hard bedrock surface (Fig. 5.4), have a predominantly meltwater origin, with cavitation playing an important rôle. *Sichelwannen* may not be symmetrical, but their axis always coincides with the direction of ice-flow with the horns pointing forwards. They range in size from a metre to 10 m in length.

Fig. 5.4. Bedrock scallop (Sichelwanne) formed by meltwater erosion combined with glacial abrasion on bedrock near Columbia Glacier, Alaska.

5.3.2 Landforms resulting from subglacial meltwater deposition

The presence of channels and cavities in and below ice-masses provides a ready site for the deposition of glaciofluvial material. The range of landforms formed subglacially passes imperceptibly into subaerial ice-contact forms and as a result the use of particular terms has not been consistent. In general, the term *esker* can be applied to most subglacial linear features (though some form ice-marginally, whereas the terms *kame* and *kame terrace* (Section 5.3.4) refer to ice-contact deposits formed adjacent to the snout and along the flanks of a glacier respectively). Terminological difficulties and many examples of such landforms have been thoroughly treated by Price (1973) and Embleton & King (1975).

Fig. 5.5. The proglacial area of the partly tidewater glacier named Comfortlessbreen, Spitsbergen. In the foreground is a welldefined sand and gravel esker, forming a ridge up to 4m high. Another esker, largely reduced to a low sinuous bank runs approximately parallel to the ice cliff in the background. To the right is an abandoned outwash fan-formed from a subglacial meltwater stream when ice was in contact with the gravel plain and discharge was at a higher topographic level than at present.

The term *esker* is of Irish origin and, nowadays, is applied to elongated ridges formed subglacially, and commonly normal to a glacier snout. Eskers consist of stratified glaciofluvial deposits, especially sand and gravel (Fig. 5.5). Eskers may be sinuous or straight; they can run uphill, and sometimes they bifurcate or are beaded. When first formed they are steep-sided, characteristically having a flat or slightly arched top. Some of the largest eskers occur in central Sweden and Finland where they extend for a few hundreds of kilometres. They rarely exceed 700 m in width and 50 m in height and, commonly, are an order of magnitude smaller than this. The internal structure of an esker frequently comprises sand and gravel with arched bedding, dipping out from the centre. Graded bedding (both normal and reverse), cross-bedding, climbing ripples, load structures, slump folds, faults and parallel lamination in clays may be other sedimentary structures present (Banerjee & McDonald, 1975). The coarser material is well-rounded and sorted, and clast-fabric studies have indicated a strong preferred orientation parallel to the ridge and dipping downstream. Palaeocurrents are broadly parallel to the esker trend. Eskers are generally formed subglacially by deposition from meltwater streams, sometimes by the total or partial blocking of subglacial Röthlisberger channels as discharge declines in late summer, as a result of which debris is not easily carried through the system. Eskers develop

166

best where ice is active but nevertheless receding, whence they can be seen to extend beyond the glacier snout. Eskers also form under stagnant ice but then they lack continuity. Some eskers may form by deposition from sub- or englacial streams as they emerge from a receding glacier into ponded water, and pass into kames.

Since eskers are associated most commonly with active ice, they tend not to be very stable features. They are further subjected to meltwater erosion as the glacier recedes. Some areas of Finland and Sweden have eskers in abundance. Here, lakes lie between the esker networks and the ridges themselves have been used effectively as road routes through otherwise difficult terrain. Reports of eskers in the older geological record are rare, but the Ordovician sequence in the Sahara records a number of well-preserved examples (Beuf et al. 1971).

In addition to eskers formed in Röthlisberger channels cut in the ice, sediment may also fill Nye channels cut into unconsolidated drift or bedrock beneath the ice. In this way, it is possible to find trough cross-bedded sand and gravel bodies enclosed by diamict or bedrock.

Rather rare landforms, *moulin kames,* are associated with debris accumulating at the bottom of moulins and are mounds several metres high or more. They are rather emphemeral features.

5.3.3 Ice-marginal landforms resulting from stream erosion

This type of landform is represented by meltwater channels perched high up on the valley sides and trending in the same direction as the former ice flow. As at the bed of a glacier, ice-marginal meltwater streams may erode both the ice and the adjacent sediment or bedrock. However, as noted in section 5.2.2 the two types of channel form under different glacier thermal regimes. Whereas subglacial channels are predominantly associated with temperate glaciers, *ice-marginal channels* tend to occur along the flanks of cold glaciers since the sub-zero ice-temperature prevents downward movement of meltwater through the glacier. Ice-marginal channels are common in the Queen Elizabeth Islands of Arctic Canada (Fig. 5.6), Greenland and Svalbard. Distinguishing between the different types of channel facilitates refinement of palaeoclimatic reconstructions. Marginal channels are often in direct contact with the ice (Fig. 2.11), or they lie between an ice-cored lateral moraine and the valley side, or both. It is thus not uncommon to find channels running essentially across a hillside after a glacier has receded. Complex channel cross-sections may result, since during each ablation season the marginal stream may melt its way beneath as it flows alongside the glacier, and then has its flow disrupted as the undercut ice collapses. Some hillsides might show a succession of meltwater channels at different levels, perhaps intimately associated with lateral moraines and having a form that mimics successively lower ice-levels (Fig. 5.6). Some channels may terminate abruptly if the water has found a direct route or *chute* straight down the hillside beneath the glacier.

Overflow channels, as their name suggests, are cut by marginal streams overflowing low cols at or below ice-surface level. The water in such cases is often dammed as lakes before incision. Many such channels of Pleistocene age were identified throughout glaciated temperate regions, but at a time before the characteristics of subglacial drainage were known. Thus, re-examination of these

Pleistocene channels indicates that many are not overflow channels at all, but subglacial ones.

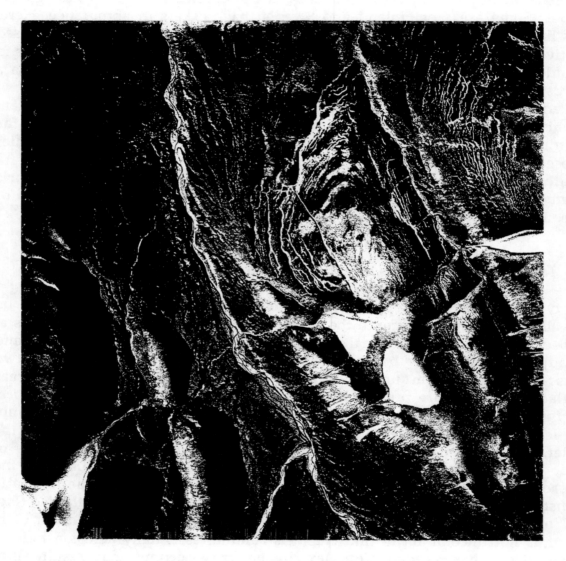

Fig. 5.6. Vertical aerial photograph illustrating abandoned ice-marginal meltwater channels, SW Ellesmere Island, North West Territories, Canada. The channels mark successive positions of the glacier tongue as the ice receded southwards up the main valley. (Photograph NAPL A16780-69, with permission Department of Energy, Mines and Resources, Canada).

5.3.4 Depositional features derived from meltwater in contact with the ice margin

The area immediately in front and along the side of a glacier is sometimes characterized by a wide range of landforms, deposited by meltwater, often in proglacial lakes. These forms are *kames* and related features. They are frequently associated with eskers, and the range of transitional features has led to confusion

168

concerning their definition. Broadly, the nomenclature of Embleton & King (1975) and Sugden & John (1976) is followed here.

The term *kame* is of Scottish origin and refers to a broad group of flattish-topped, ice-contact glaciofluvial landforms. They have no specific orientation with respect to the ice flow direction. They occur singly, as isolated hummocks, as broader flat plateau areas (*kame plateaux*), or as broken terraces, usually in a proglacial setting. Where a large area is covered with numerous discrete kames, the term *kame field* is sometimes used. Kames consist of well-sorted, stratified sand and gravel. They vary from a few hundred metres to over a kilometre in length, and are an order of magnitude less than this in width. They form in direct contact with the ice, so when the glacier recedes, their up-valley faces are prone to slumping and collapse. Kames were formed in abundance at the margins of the last great ice sheets, especially in Canada and Scandinavia. Their development was favoured particularly by the huge volumes of meltwater that were released during the rapid recession of the ice margins. Embleton & King (1975) have described a number of examples from Scotland, that were associated with decaying Southern Uplands ice.

Fig. 5.7. Kame terrace in Adams Inlet, Alaska formed during the recession of the Glacier Bay glacier system within the last century. Note that the former, steep ice-contact slope is now undergoing slumping.

169

Kame terraces are distinct from kames in that they form parallel to the ice-flow direction from streams running along the flanks of a stable or slowly receding ice margin (Fig. 5.7). Their composition, however, is similar to that of kames, comprising coarse, bedded gravel and sand with planar bedding or cross-stratification if deposited by streams, or fine-grained parallel-laminated clay, silt and fine sand if deposited in an ice-marginal lake. Kame terraces slope downvalley at an angle that approximates that of the former ice-level. They may also tilt gently up towards the adjacent hillside, especially if the material is derived from a side valley. As the ice recedes, kame terraces are prone to collapse. In favourable circumstances, ice still-stands allow a succession of terraces to form at different levels.

Related to kames, but occurring on a much larger scale, are *delta moraines* (also known as *kame deltas*). These features form transverse to the direction of ice-flow in a marginal position where a series of subglacial streams enter a proglacial lake or the sea. Delta-moraines may thus be intimately associated with eskers. In addition to typical glaciofluvial material, they also contain glacial debris derived from the ice itself. For example, ablation debris on the surface of the ice may fall directly onto the delta-top. Delta-moraines require the ice position to have remained relatively stable for a long period. Probably the largest delta-moraine complexes are the three "Salpausselkä moraines" of Finland associated with a lake over the southern Baltic Sea region that was dammed by the Fennoscandian ice sheet (see Ch. 6).

Where glaciofluvial processes substantially rework the deposits of an end-moraine complex, the resulting feature is a *kame moraine*. The process is especially pronounced between lobes of the glacier. Here, bedded gravel and sand accumulates, and the sediment surface develops *kettle holes* as a result of the melting of stagnant ice beneath. The orientation of individual kame sediment accumulations is random within the context of an end-moraine system.

A last group of minor landforms are *crevasse-fillings*. They are composed of stratified drift which entered the crevasses via supraglacial meltstreams. They are linear features a few tens of metres long, but may be of any orientation, depending on the type of crevasses that were filled.

5.3.5 Landforms resulting from meltwater erosion in a proglacial setting

Streams emerging from the snout of a glacier may be highly erosive, although less so than where they are flowing subglacially. Their typical high debris load and high velocity leads to channel-switching, and constant erosion and redeposition of proglacial sediment. Sometimes water collects in proglacial or ice-marginal lakes which, as the melt-season progresses, tend to overflow through *spillways*. The outlet streams, especially if the lake is held back only by unconsolidated sediment, may rapidly cause downcutting and lowering of the water level. Nowadays, large proglacial lakes are rare, but they seem to have been abundant near the southern limits of the great Pleistocene ice sheets. A good example of a spillway is the now abandoned major channel that formed when glacial Lake Agassiz, lying to the northwest of the Great Lakes, overflowed southwards into the Minnesota River and eventually the Mississippi.

Proglacial drainage patterns in front of the former ice sheets responded to fluctuations of the ice margin, and major drainage diversions and the creation of new channels have been the result. In Britain, the best known example was the diversion of the River Thames during the Anglian glaciation from a route from the Chilterns to East Anglia, to its present course through where London now stands.

Fig. 5.8. Valley train below Mt. Cook (centre-right) extending southeastwards from debris-covered Tasman Glacier (right) and Müller Glacier (centre). Note the partial superimposition of the Müller braided river system on the Tasman system as it enters the main valley.

5.3.6 Depositional landforms derived from meltwater in a proglacial setting

The landforms which result from meltwater after it leaves a glacier are not directly influenced by ice, except where stranded ice has been buried. However, in terms of volume, the amount of sediment may greatly exceed that directly deposited by the ice, at least in areas with temperate glaciers.

Braided river systems, known as *outwash plains* or *sandar* (from the Icelandic; singular *sandur*), develop down-valley of glacier snouts that terminate on land. In the strictly Icelandic sense, *sandar* are laterally unconstrained, and in their type area at the southern margin of the ice caps in southeastern Iceland their width may be as great or greater than their length. They are characterized by numerous active braids across most of the plain, but during jökulhlaups the whole area may be flooded.

In mountain areas, a braided river system or *valley-train* may extend across the whole width of a valley, mountains rising sharply from the valley floor. Good examples occur in Alaska and the Southern Alps of New Zealand (Fig. 5.8). In areas of post-glacial uplift, successively higher levels of a braided river system are abandoned on terraces as the river cuts downwards, the most recent active outwash plain being narrower than prior to uplift. In relation to a receding

171

glacier, a braided system may occur within the terminal moraine, often in association with proglacial lakes. A single channel may emerge from the moraine, and a new braided-stream system evolve outside the moraine. Glaciofluvial activity may rework much of the till and cover it almost completely. *Braided outwash-fans* develop where river systems, constrained by valleys, debouch onto lowlands beyond mountain ranges. Fans often coalesce, forming a broad succession of fans along a mountain front. Many examples may be cited, but the best known lie to the north of the European Alps.

A common characteristic of braided river plains is the presence of water-filled pits called *kettles* or *kettle holes* (Fig.5.9). Often, these are steep-sided and shaped like a cone when freshly formed. They are the result of the burial of dead glacier ice, either as a remnant of a glacier left behind during recession or as a detached block of ice swept downstream during a flood, in both cases being subject to burial by glaciofluvial material. Slow thaw of the ice undermines the gravel above, causing it to cave in and gather water. If an outwash plain is littered with numerous kettle holes the term *pitted plain* is often used.

Fig. 5.9. Kettle hole, actively forming in the outwash plain below Casement Glacier, Alaska.

Braided streams generally develop in response to marked fluctuations in discharge, and as such are not confined to glacial environments; for example ephemeral streams in deserts are commonly braided. (For comprehensive reviews of braided river environments see Miall 1977, 1983). In glacial environments, braided streams form mainly in response to seasonal variations of flow from glaciers. In winter, discharge from a glacier is frequently reduced to a mere trickle, but in early summer, when ice melt is combined with snow melt, the entire valley floor may be washed by the flood and much debris is transported. This represents an annual switch from empty channels to bank-full discharge. As a result distinct sedimentary structures (bed-forms) develop.The

172

reader is referred to a modern textbook of sedimentology for an account of these forms.

In coarse sediments, the development of primary structures is less well known. Large dune-like features of gravel have been left behind after catastrophic lake drainage. Pebble- or boulder-ridges, transverse to flow in outwash areas, are also known.

On a larger scale, all sediment types give rise to the channel *bars* that are so typical of braided glacial outwash streams (Figs. 5.10, 5.11). Bars may be relatively stable and may only shift during floods or when an entire channel migrates.

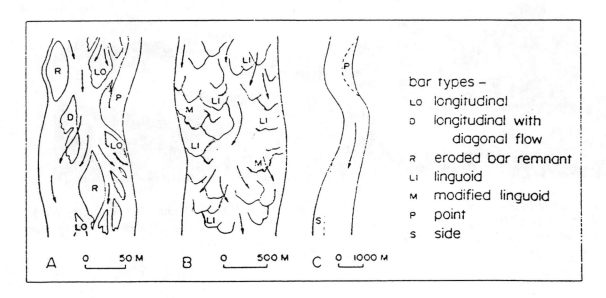

Fig. 5.10. Principal types of bar in braided river channels. (Miall 1977).

In general, the range of sedimentary structures reflects the distance from the glacier, permitting us to distinguish proximal environments comprising featureless or crudely parallel-bedded or discontinuously bedded gravels, from distal environments with a wide variety of structures including cross-bedding in sand bars, and dunes and ripples as channel bed-forms. In the proximal environment, there is a wide range of grain sizes, except in certain backwater locations, due to rapid deposition. Clast imbrication is common (Fig. 5.12) and disk-shaped clasts are tilted upstream by up to 30°. Palaeocurrent directions give a good approximation to the mean trend of channels and bars (Rust, 1975). Melting of associated ice, particularly if it has been buried, will produce various high-angle faults and soft-sediment deformation structures (McDonald & Shilts 1975), as well as generating a pitted surface on the plain.

In more distal areas, foreset bedding develops on the distal, prograding ends of some bars where they build out into deeper water. Festoon structures, parallel and oblique foreset-lamination, and trough-fill structures are also common. The sediments fine upwards and downstream on linguoid bars if the depositional history is simple.

Periodic flooding may distribute material over a wide area and lead to channel migration. Deposits associated with floods are (a) channel fills, (b) *levées* which are best developed in sandy river beds, and (c) surface veneers which comprise fine gravel to sand, the latter often being deflated to leave a lag-gravel, or having a rippled top.

Fig. 5.11. Longitudinal bars with diagonal flow in the braided river below, Fox Glacier, New Zealand (cf. Fig. 5.10). Note the generally coarse, gravelly material of the upstream ends of the bars. A point bar is also present, adjacent to the rock wall in the background.

Fig. 5.12. Upstream-dipping cobbles and boulders (imbrication) in the main channel below Franz Josef Glacier, New Zealand.

The texture of braided river deposits changes markedly downstream (Fig. 5.13). The size of material decreases downstream, but this is more noticeable in the channel-beds than on the surface of the outwash plain generally. The increase in roundness initially is very rapid. Sorting by lithology also occurs, resistant rocks becoming more prominent downstream. There tends to be little lateral mixing of sediments. Stone orientation is normally weakly parallel to flow, but sometimes it is transverse. In fine-grained sediments, several characteristic grain-size distributions have been recorded, e.g. a log-normal distribution, a distribution truncated at the fine or the coarse ends of the spectrum, or a bimodal distribution.

Braided-river deposits may accumulate to depths of a few hundred metres in favourable locations, and little evidence of direct glacial deposition may survive.

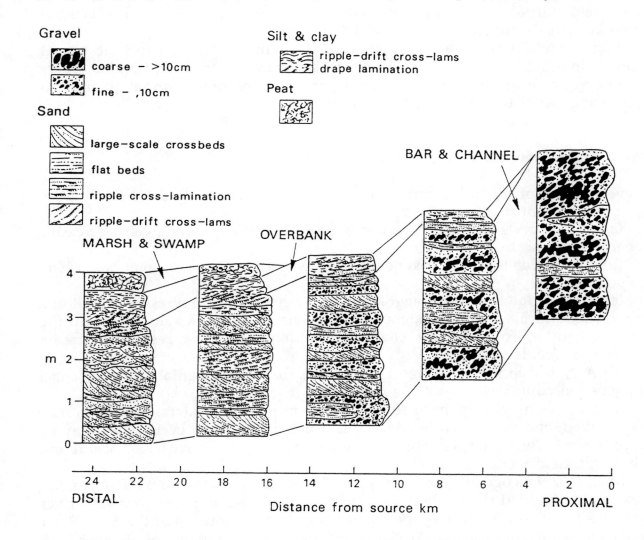

Fig. 5.13. Downstream variation in facies and sedimentary structures. Bar and channel sequences do not fine upwards, but are capped by a finer overbank facies that become more important in a down-fan direction (Boothroyd and Ashley 1975).

175

5.4 Glaciofluvial facies and facies associations

5.4.1 Lithofacies and their interpretation

The lithofacies of braided glacial river systems have been classified and presented as a generalized scheme by Miall (1983), based on the Scott River in Alaska, the Donjek and Slims rivers in the Yukon, and the nonglacial Platte River in Colorado. Lithofacies are subdivided into three main groups: gravel, sand and fines, and qualified by terms for internal textures and sedimentary structures (Table 5.2).

Lithofacies in eskers are predominantly gravel and sand, the fines having been removed by the turbulent flow that characterizes subglacial streams. These facies become disturbed as the ice is removed. A good example of the range of facies in an esker was described by Sanderson (1975).

Kames and kame terraces, comprising glaciofluvial material include gravel, sand and mud, with similar sedimentary structures to those in braided rivers. However, like eskers, they also show evidence of deformation or collapse following removal of the supporting ice-mass.

5.4.2 Braided river facies associations

Several types of repetitive vertical sequences may be envisaged for the braided river environment (Miall 1977):

(a) A flood cycle: a superimposition of beds formed at progressively decreasing energy levels.

(b) A cycle due to lateral accretion: a cycle generated by side or point-bar growth is possible.

(c) A cycle due to channel aggradation: this cycle would represent the fill of a channel or local channel system. Waning energy-levels would occur during sedimentation, followed by channel abandonment as a result of stream migration (*avulsion*).

(d) A cycle due to channel reoccupation: an abandoned, partially filled channel may be reoccupied as a result of avulsion.

Such cycles may range from 15 cm to 60 m in thickness, lateral variations may be marked, and in a given braided stream deposit all these cycle types may be represented. Thus interpretation of braided river sequences requires careful and thorough field work.

Miall (1978) recognized six main types of facies association in gravel- and sand-dominated braided rivers, three of which are applicable to glacial outwash river systems (Fig. 5.14). The *Scott*-type facies association is named after the Scott River outwash fan in southeastern Alaska which was studied in detail by Boothroyd & Ashley (1975). These authors described five distinct vertical profile sequences which, with increasing distance downstream are:

(a) Upper fan - coarse gravel facies, well imbricated downstream

(b) Upper midfan - also gravel but finer, with thin flat beds and large-scale trough cross-bedded sands. Sand-wedge deposits are important.

(c) Lower midfan - increasing sand in relation to gravel, and large-scale festoon cross-beds increasing in abundance.

(d) Lower fan braided facies - planar cross-bedding (resulting from slipface migration of bars) and abundant climbing-ripple lamination (also known as ripple-drift cross-lamination), resulting from bar-surface deposition.

(c) Lower fan meandering facies with abundant large-scale trough cross-bedding and planar to tangential cross-bedding.

Overbank deposits are absent on the upper fan, but increase in importance down-fan. The grain size in overbank silty climbing-ripple lamination and draped lamination decreases down-fan.

Facies	Sedimentary structures	Interpretation
Gravel: massive, matrix-supported	None	Debris-flow deposits
Gravel: massive or crudely bedded	Horizontal bedding, imbrication	Longitudinal bars, lag deposits, sieve deposits
Gravel: stratified	Trough cross-bedding	Minor channel fills
Gravel: stratified	Planar cross-bedding	Linguoid bars or deltaic growths from older bar-remnants
Sand: medium to very coarse; may be pebbly	Solitary or grouped trough cross-bedding	Dunes (lower flow regime)
Sand: medium to very coarse; may be pebbly	Solitary or grouped planar cross-bedding	Linguoid transverse bars, sand-waves (lower flow regime
Sand: very fine to coarse	Ripple marks of all types	Ripples (lower flow regime)
Sand: very fine to coarse; may be pebbly	Horizontal lamination, parting or streaming lineation	Planar bed flow (lower and upper flow regimes)
Sand: fine	Low angle (<10°) cross-bedding	Scour-fills, crevasse-splays, antidunes
Erosional scours with intraclasts	Crude cross-bedding	Scour-fills
Sand: fine to coarse; may be pebbly	Broad, shallow scours including cross-stratification	Scour fills
Sand	High angle, planar cross-stratification, horizontal lamination, shallow scours	Aeolian deposits
Sand, silt, mud	Fine lamination, very small ripples	Overbank or waning flood deposits
Silt, mud	Laminated to massive	Backswamp deposits
Mud	Massive with freshwater molluscs	Backswamp pond deposits
Mud, silt	Massive, desiccation cracks	Overbank or drape deposits
Silt, mud	Rootlets	Seatearth
Coal, carbonaceous mud	Plants, mud-films	Swamp deposits
Carbonate	Pedogenic features	Soil

Table 5.2. Lithofacies types in the braided river depositional environment (summarized from Miall 1983).

177

The *Donjek* type of Miall (1978) (Figure 5.14) is the most varied facies association, containing anything from 10-90% gravel. Marked fining-upwards cycles on several scales are present, the thicker ones reflecting either sedimentation at different topographic levels within the channel system, or successive events of vertical aggradation followed by channel switching. The differentiation of the channel-interfluve system into distinctive topographic levels is thus a prominent feature of this type of braided river system. Bar gravels dominate the lower, most active channels, while sand and pebbly sand occur on the higher elevations, and mud may be present in abandoned areas of the outwash plain or in interfluve ponds.

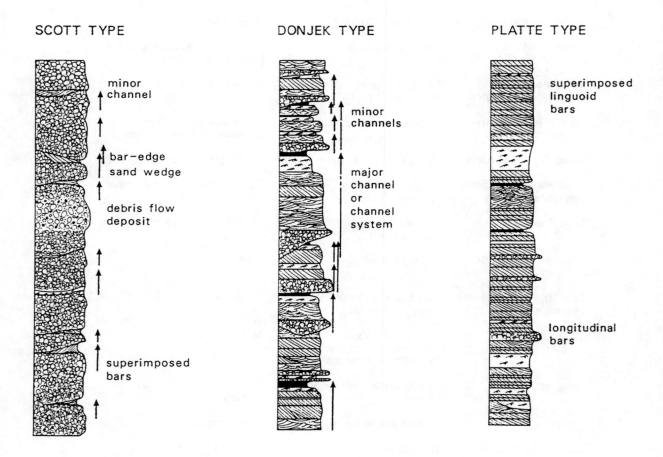

Fig. 5.14. Hypothetical facies association in the most common types of glaciofluvial sequences (Miall, 1983).

A third type of association, dominated by sand, is the *Platte* type. Although named after a nonglacial river, it is typical of the lower reaches of a sandur. Runoff is spread between numerous shallow distributaries, the topographic differentiation between channels and interfluves being less than for the *Donjek* type, and average channel depth and slope is smaller. Sandy bed-forms dominate, especially sand-waves, linguoid bars and dunes, resulting in a sequence that is largely planar cross-bedded. Minor gravel and fine, overbank deposits may be present, and cyclicity is rarely observed.

The distal reaches of an outwash river, or where wind-blown material (loess) is abundant, may give rise to a fourth, but little studied, facies association, the *Slims* type, named after a river in the Yukon. Low-relief bars and channels are characteristic, and the facies are mainly massive, laminated and ripple cross-laminated sandy silts.

In a different climatic regime, namely that of the maritime Arctic of Spitsbergen, which is under the influence of permafrost, Bryant (1983) has provided an account of the areal distribution of facies and landforms in a proglacial braided river valley. Comparative studies in southern Britian found many similarities with the Spitsbergen facies associations, allowing the inference that some glaciofluvial sequences were formed under periglacial conditions.

5.4.3 Facies architecture

The overall architectural styles of modern sandur or ancient outwash sequences have received little attention. Thus the examination of environmental changes recorded by such sequences has rarely been attempted. The geometry of glaciofluvial horizons at different levels within a grounded-ice depositional sequence generally requires good sections that are both vertically and laterally extensive, for which the pre-Quaternary record offers the best prospects.

The architecture of a typical esker is complex. The feature itself snakes across the surface, typically a braided-river plain. Dedding rises and falls gently as observed in a longitudinal profile, while in cross-section it is often gently arched and faulted. The geometry of late Ordovician esker systems have been described from the Sahara by Beuf et al. (1971). The architecture of kames and kame terraces has features which may also be found in braided-river and glaciolacustrine systems, but on a smaller scale.

5.5 Preservation potential of glaciofluvial sediments

Glacial outwash complexes commonly accumulate to depths of several hundred metres in major glacial troughs in temperate regions, as in Alaska or New Zealand. Otherwise, they form extensive fans beyond mountain ranges over lowland areas. Many fluvial sequences have survived into the rock record, especially where deposited in fault-bounded basins; the same should be true of glaciofluvial sediments. Thus, compared with sediments released directly from ice on land, they have a markedly better preservation potential. Glaciofluvial sequences in the rock record may be more abundant than might at first be apparent, especially if direct evidence of glaciation is missing.

In contrast to braided-river complexes, eskers, kames and kame terraces are commonly subject to erosion once the glacier has receded, and only a few have survived to become part of the rock record.

180

CHAPTER 6

GLACIOLACUSTRINE PROCESSES

AND SEDIMENTS

Fig. 6.1. Ice-dammed lake at the snout of Austerdalsisen, Svartisen, Norway. Ice at the right has blocked off the low col. The lake has been artificially lowered by tunnelling, to prevent damaging jökulhlaups.

Fig. 6.2. Proglacial lake below Mount Cook, New Zealand, resulting from recession of the debris-covered Hooker Glacier (behind iceberg) into a deeper basin. Note the suspended sediment in the water.

6.1. Introduction

As discussed in Chapter 2, lakes are associated with glaciers in several different ways. Lakes in direct contact with glacier ice are referred to as glacial lakes and may form along the glacier margins, especially along its flanks, or in front of the snout (ice-marginal and proglacial lakes respectively) (Figs. 6.1 and 6.2). Vertical, calving ice-cliffs are characteristic of glaciers that flow into lakes. Some lakes form subglacially, or from melting out of ice buried under outwash. Other lakes that are not in direct contact with an ice-margin may still be influenced by a glacier. In such cases, a braided glacial river may connect the glacier to the lake and provide the principal sediment source; these are not strictly glacial lakes, however. Glacial lakes form in both temperate and polar climatic regimes. Some of those in Antarctica (Fig. 6.3) are highly saline and remain perpetually frozen over, yet are quite warm at depth.

Many of the largest glacial lakes of today occupy overdeepened basins dating back to the last glacial period. Such lakes are important sediment traps, and sequences a hundred or more metres thick have been preserved. The Northern Hemisphere ice sheets were bordered in many places by lakes hundreds of kilometres across, and these too have trapped much sediment.

Fig. 6.3. Saline lake in front of Taylor Glacier, Victoria Land Antarctica. Salt crystals form on the surface of the lake-ice which is a permanent feature, although because of strong thermal stratification the deeper waters may be warm. Sedimentation in such lakes is characterised by a high proportion of evaporitic minerals.

Glacial lake processes are varied and give rise to complex assemblages of facies. They reflect input of material from subglacial or ice-marginal streams both as bed-load, which leads to the formation of deltas. Some sediment, carried in suspension, may settle over the entire lake floor. In addition, subaquatic gravity flow and littoral processes may operate and, in arid areas, aeolian input is also important. In many respects, the processes in glacial lakes resemble those in fjords, the latter being described in the following chapter, but there are differences due to the way in which suspended sediment interacts with the lake waters.

This chapter focuses briefly on the processes and products of glacial lakes, and emphasizes their importance during the Pleistocene Epoch. Embleton & King (1975) have provided more details about specific glacial lakes, while processes and products have been thoroughly aired in the volume edited by Jopling & McDonald (1975), and by Drewry (1986) and Ashley (1989).

6.2 Pleistocene-Early Holocene glacial lakes

Glacial lakes were far more extensive at the margins of the Northern Hemisphere ice sheets than they are today, and were subject to wide and rapid fluctuations. As a result, deposits and landforms associated with them are widespread. In Europe, the largest was the Baltic ice-lake which, around 10,500 years BP, stretched for some 1,200 km along the southern margin of the Weichselian ice sheet over Scandinavia. An equally extensive group of lakes developed in North America as the margin of the last ice sheet there, the Wisconsinan, receded. The most important of these, innundating nearly a million square kilometres of Manitoba, Ontario, Saskatchewan, North and South Dakota, and Minnesota, has been named proglacial Lake Agassiz (Teller 1985). This lake was the result of damming of rivers flowing into Hudson Bay by glacier ice as it receded northwards. The lake came into existence about 11,700 years BP. At its greatest height, it overflowed southwards into the Minnesota River valley and on into the Mississippi River. As glacier-recession took place, outlets to the east, into the Lake Superior basin developed. Occasional catastrophic bursts, during which up to 4,000 km^3 of water were released with a year or two, created channels into Lake Superior. By 8,500 years BP drainage bypassed the Great Lakes to the north, and by 7,500 years BP the lake was finally drained. The widespread sediments that accumulated in Lake Agassiz included various types of varve and laminated muds, resting on late-glacial till.

6.3 Physical character of glacial lakes

The pattern of sedimentation in a glacial lake, to a large extent, is controlled by density differences within the water mass. Density is controlled by temperature (being greatest at +4°C), the concentration of dissolved salts, and the amount of sediment in suspension. Most lakes possess a thermally controlled density stratification which varies during the course of a summer season. Typically, in summer, a well-mixed layer of low-density warmer water develops at the top of the water column, and there may be a sharp decrease in temperature at its base. In the autumn, as the surface waters cool and become denser, they sink, allowing them to eventually overturn completely. Mixing may also be induced by wind and waves.

The rôle that the input of glacial meltwater has on mixing depends on the position of the glacial stream outlet with respect to water-level. If there is a significant difference in density, the sediment-laden meltwater may maintain its integrity as a plume. Commonly, sediment plumes, being denser, sink to the bottom of the lake as an *underflow* (Fig. 6.4). The descending water often behaves

184

as a turbidity current, so giving rise to graded, rhythmically stratified sediments spread over the entire lake basin floor. If a low-density (e.g. a clear) subglacial stream enters a lake containing suspended sediment, it rises to the surface and becomes an *overflow* (Fig. 6.4). Turbulent-exchange between inflowing water and the lake waters may be limited unless the density contrasts are low. A third type of flow, an *interflow*, may also develop in some lakes.

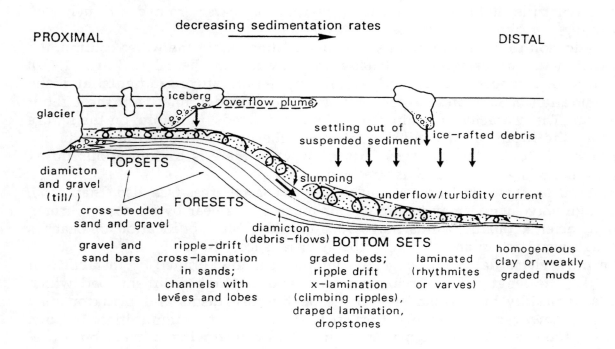

Fig. 6.4. Processes and sedimentary products in a glaciolacustrine setting.

6.4 Glaciolacustrine sedimentary processes

Glaciolacustrine processes operate in lakes that are in direct contact with the ice margin, and (according to some workers) those that are connected to the glacier by a short river. Here we are concerned mainly with the former, in which deposition comprises the following elements (Fig. 6.4):
   • direct deposition from glacier ice;
   • deposition from subglacial rivers that usually enter the lake below water level;
   • sedimentation from suspension;
   • sedimentation from gravity flows;
   • lake shore sedimentation:
   • biogenic sedimentation; and
   • evaporitic mineral sedimentation.
Several processes and the resulting features are common to fjords, and are dealt with more thoroughly in Chapter 7. The most characteristic facies in a glacial lake are deltaic deposits and lake bottom deposits (varves and other rhythmites).

In a glaciolacustrine delta situation, through which most glaciolacustrine sediment passes, deposition occurs as topsets, steeply-dipping foresets and thin bottom-sets (Church & Gilbert 1975). Foresets are formed by avalanching, and to some extent by slumping. Bottom-sets are typically rhythmites, resulting from the transport of sediment by turbidity currents combined with the settling-out of suspended sediment; if annual cyclicity is recognizable, they are referred to as *varves* (or *varvites* for the lithified equivalent). Vrve-like sediments, linked to tidal-processes, are found in fjord settings (Ch. 7), and care is needed in distinguishing between the two types.Turbidity underflow is not a continuous event, and may be significant on more than one occasion or not at all. In contrast to lakes, the salt-water in fjords allows flocculation, and varves may not form. In glaciolacustrine deltas generally, redistribution of coarse-grained sediments by slumping is a significant process.

In addition to the existence of suspended matter within the water column, the formation of varves requires density stratification in the lake (Sturm 1979). Density differences are caused mainly by temperature-gradients and the concentration of salts and suspended matter. Stratification may vary throughout the year. The character of laminated sediments depends, therefore, on the nature of both the input of suspended matter and the nature of stratification in the water body. Ideal clastic varves form only when there is discontinuous influx during periods when the lake is stratified.

Littoral processes in lakes resemble those in coastal areas (see also Chapter 7), although wave-action may be limited for much of the year by ice-cover. Strong, downglacier katabatic winds may cause considerable modification of beaches through wave action and shifting icebergs or lake-ice onshore.

In most glacial lakes the biogenic component is low, limited by instability of the lake, the length of the freezing season, and the high sediment load which tends to rapidly bury organisms. Some benthic organisms and plankton may survive, however, and even gather to form quite rich communities. In such cases sedimentary structures may be obliterated by burrowing animals, but trace-fossils may be preserved.

Some lakes in Antarctica, notably those in the Dry Valleys of Victoria Land, show abundant signs of biochemical activity in the form of benthic microbial mats and stromatolites, mineralized to calcite, growth being possible down to a few tens of metres' depth (Parker et al. 1981, Wharton et al. 1982). Most of these lakes differ from most glacial lakes in receiving only limited meltwater, in having no outlet, and in remaining frozen at the surface all year round. These lakes have developed strong thermal stratification, and because solar radiation has been preferentially absorbed by the lower saline part of ice-covered lakes, they have become surprisingly warm at depth (e.g. 26°C in Lake Vanda). A variety of evaporitic minerals accumulate, including glauberite, halite and carbonate. These lake-salts are derived from weathered bedrock and are concentrated by surface-freezing and ablation on the top surface of the lake-ice, or by evaporation if the lake surface melts.

6.5 Landforms resulting from glaciolacustrine deposition

6.5.1 Deltas

Deposition of delta-fronts generally takes the form of avalanching and, to a lesser extent, slumping of coarse material, against a background of sedimentation of suspended fine material (Fig. 6.2). Some delta-fronts are straight, especially where meltwater channels switch from one side of a valley to the other. Where sediment is delivered from a single or narrowly confined group of channels, delta -ronts are arcuate and often overlapping; they are little modified by current activity. Climbing-ripple lamination and draped lamination are characteristic of sedimentary-sequences formed at delta-fronts (Gustavson et al., 1975); Climbing-ripple lamination is a useful indicator of rapid sedimentation during high-discharge events.

Fig. 6.5. Glaciolacustine delta, Adams Valley, Glacier Bay, Alaska, comprising 200m of sand and gravel. The delta was formed as a result of the post-little Ice Age recession of the Glacier Bay complex as it dammed this side valley.

An example of a well-exposed glaciolacustrine delta, which formed earlier this century is well exposed in the Adams River valley, Glacier Bay, Alaska (Goodwin 1984). The 200 m-thick sequence comprises rhythmically bedded clays and silts with ice-rafted debris and diamictons, overlain by glacial outwash sands and gravels (Fig. 6.5). Another good example is a Weichselian glaciolacustrine delta that has been described from north Sjælland, Denmark (Clemmensen & Houmark-Nielsen 1981). Here the foresets, which are centimetres to decimetres in thickness, are composed of coarse-grained conglomerates (matrix-supported) to sands, with slight angular unconformities. Pebbles are scattered throughout. Sedimentary structures include parallel lamination, megaripple cross-bedding, climbing-ripple lamination and isolated scour-and-fill structures. The sediments contain a fine fraction, which represents the background sedimentation from suspension.

6.5.2 Delta-moraines

A delta-moraine forms where an ice-front remains stationary for a considerable time in a lake or the sea. They are the product of glaciofluvial deposition immediately in front of the ice margin. Frequently, sediment is fed into the lake subaquatically *via* esker channels. Fyfe (1990) has described what is probably the longest delta-moraine system exposed on land, the Salpausselkä Moraines which extend for some 600 km across Finland (Fig. 6.6). Facies range from boulder gravels to muds, and a wide range of sedimentary structures are present. However, diamicts are uncommon. Fyfe was able to explain the marked contrasts in the form and stratigraphy along the length of the moraine in relation to variations in the water-depth and the nature of the subglacial drainage system:

• Large individual deltas with braided tops ("ice-contact deltas") which built up to water level were the product of conduit-focused sedimentation.
• Lower, narrower coalescing fans of finer material ("grounding-line fans") were formed at the grounding-line by sediment fed from a distributed drainage system.
• Small, laterally overlapping subaqeous fans derived from unstable subglacial conduit systems, occurred where marginal water depths were greatest.
The differences in the subglacial conduit system is believed to be related to the reduction in the basal shear-stress as the water deepens, lowering the surface ice profile and destabilizing the subglacial stream network

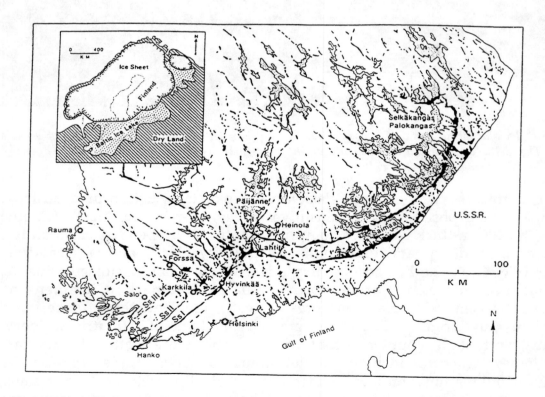

Fig. 6.6. Delta-moraine complexes formed in the Baltic Ice Lake at the southern periphery of the Fennoscandian ice sheet around 10,500 to 10,200 years BP. The three complexes are named Salpausselkä (Ss on figure) I, II and III. Note their association with eskers (Fyfe 1990).

188

6.5.3 De Geer moraines

De Geer moraines (also called washboard or sub-lacustrine moraines) are a group of landforms formed subglacially, transverse to ice-flow but, unlike Rogen moraines (Section 4.5.3), they form some way behind an ice-margin that calves into the lake, especially in broad, open depressions (Sugden & John 1976). De Geer moraines are a succession of discrete, narrow ridges, ranging from short and straight to long and undulating. The ridges are more delicate than those of Rogen moraine, although they are occasionally linked by cross-ribs. The ridges rarely exceed 15 m in height, and their spacing may be up to 300 m. They are composed of till with a cap of boulders, while lenses of sand and other stratified water-lain deposits, including varves, occur between the ridges.

The origin of these features is unclear, but may be linked to accumulation of sediment where the base of the glacier decouples from the bed of the lake, allowing debris to accumulate as a ramp between well-grounded and floating ice. Each ramp ceases to develop when the ice thins sufficiently to break off as an iceberg. Another model concerning the development of De Geer moraines, based on Finnish studies, suggest that local surging phases of the Fennoscandian ice sheet may have led to flexuring of the basal ice zone, so as to create basal crevasses. Susequent lowering of the ice-mass is considered responsible for squeezing highly saturated sediment which is then left as ridges as the ice melts (Zilliacus 1989).

6.5.4 Shorelines or strandlines

Many glacial lakes are subject to wide fluctuations, especially if ice-dammed. Hence beaches may develop at different levels, sometimes resulting in a staircase-like arrangement of shorelines or strandlines. Shorelines become evident during the draining of a lake, and a number of shorelines associated with late Pleistocene ice-dammed lakes survive.

Among the best-known preserved shorelines are the Parallel Roads of Lochaber in Scotland. Glens Roy, Gloy and Spean all have glacial lake shorelines, recognized by Agassiz around 1840 and described in detail by Jamieson in 1863; they figured prominently in the presentation of evidence of glaciation in Britain (Peacock & Cornish 1989).

There are three sets of *parallel roads*, at 260, 325 and 350 m, each related to cols over which ice-dammed water flowed (Fig. 6.7), first of all as progressively higher cols became blocked by ice, then as it receded. These events took place during the latest glacial phase in Scotland, the Loch Lomond Stadial about 10,000 years BP. The last glacial lake, at 260 m, covered an area of 73 km^2 and had a volume of 5 km^3, which is thought to have drained catastrophically as the ice-dam failed (Sissons 1979). Many of the lacustrine sediments are deformed, and some beaches have been displaced by faulting, probably as a result of an earthquake that occurred in response to crustal adjustment as the lake emptied. Deltas which formed in the lake from side-valley streams were soon dissected by the same streams after the lake had emptied.

Fig. 6.7. Glen Roy, Lochaber, Scotland showing three prominent "parallel roads" or ice-dammed lake shorelines.

6.6 Glaciolacustrine facies

From a knowledge of the lithology, geometry and sedimentary structures, most glaciolacustrine sediments may be interpreted either as deltaic sediments or lake bottom sediments, and classified as topset, foreset and bottom-set beds (Fig. 6.4). The relative importance of the different facies reflects the rôle played by direct deposition from subglacial or proglacial streams, the importance of deposition from grounded or floating ice, and the fluctuations of overflow, interflow and underflow conditions.

6.6.1 Deltaic sediments

Dissection of deltaic sequences after glacial lakes have drained has often provided useful cross-sections for lithofacies investigations (Fig. 6.8a). Deltaic sediments with both topsets and foresets are commonly referred to as *Gilbertian deltas,* a type described in the late 19th C from Pleistocene Lake Bonneville, centred on Utah, USA. Topset beds are typically sand and gravel, and are the result of braided stream deposition (Fig. 6.8b). In a Danish Pleistocene example, Clemmensen & Houmark-Nielsen (1981) found clast-supported conglomerates, with frequent imbrication and stones up to 20 cm in diameter, forming irregular sheets, interbedded with thinner sands with horizontal lamination or rare semiplanar or trough cross-bedding. Channel structures may also interfere with the upper part of the foreset beds. In a late Holocene glacial lake sedimentary complex at Austerdalsisen, Norway, Theakstone (1976) described large sets of

190

cross-bedding up to 8 m thick, with foresets at a moderate angle (c. 20°) (Fig. 6.8a). Lowe-angle foresets are known from other glaciolacustrine deltas.

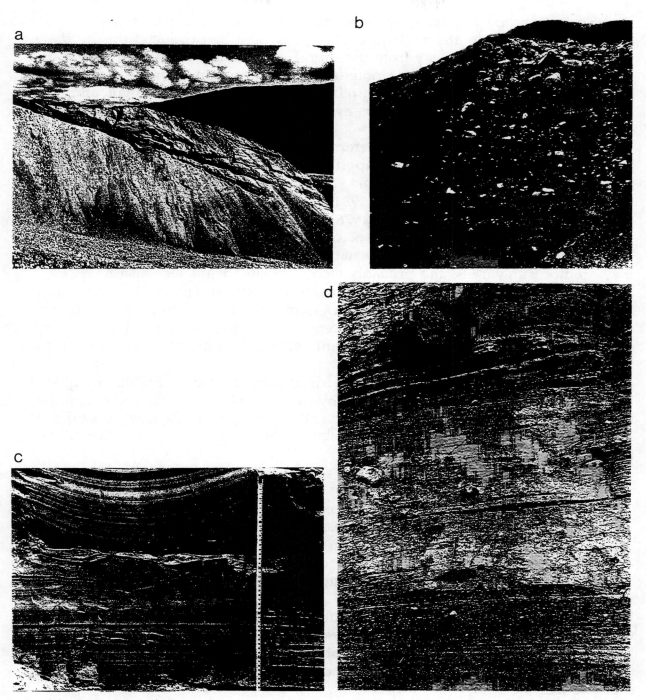

Fig. 6.8. Typical lithofacies in now-exposed glaciolacustrine sequences: (a) Cross-bedded sand and gravel forsets in a 30m section through the delta. (b) Sand and gravel topsets in a glaciolacustrine delta, upper Glen Roy, Scotland. The height of the section is about 4m. (c) Bottom set sand and silt rhythmites of complex origin, Austerdalsisen, Norway. The lower set is relatively undisturbed, but at the top there is a disturbed zone showing convolute lamination resulting from overpressurization of saturated sediment. The top 'synclinal' structure is probably the result of slumping with the mass being detached from more stable sediment beneath. (d) Stratified diamicton of Pleistocene age, Taylor Valley, Victoria Land, Antarctica. During expansion of the East Antarctic ice sheet, Taylor Valley was blocked at its mouth allowing a lake to form. This facies was derived as a result of rain-out of glacial debris from icebergs, accompanied by reworking by lake-bottom currents. Larger ice-rafted blocks have penetrated the stratification.

6.6.2 Lake bottom sediments

On the bottom of the lake, sedimentation from turbidity currents may be discontinuous during the melt-season. *Varves* may be produced on the distal prodelta slope and thin away from the source, with nearly all the thinning in the coarse summer layer. Climbing-ripple lamination is common in the foreset beds of the prodelta slope. Varves contain numerous graded laminae and rare climbing-ripple laminations, perhaps even isolated ripples in the coarse summer layer (Gustavson et al. 1975).

In the specific case in Denmark referred to above, the bottom-set beds contain low-angle climbing ripples, draped lamination, parallel-laminated clay, sigmoidal lamination, wave-ripple lamination and disturbed lamination (Clemmensen & Houmark-Nielsen 1981). The bottom-sets of the Austerdalsisen glaciolacustrine delta (Theakstone 1976) contain evenly bedded, laminated fine silt and clay, and some fine sand. In places, the laminae are draped over obstacles. The sediments are rhythmically alternating, in part due to colour variations, but they are not true varves, and probably were deposited from suspension (Fig.6.8c). Deformational structures occur in laminated bottom-set sediments (Theakstone 1976) including convolute lamination, flame structures, normal faults and thrusts with displacements of a few centimetres, and recumbent folds. Such structures are due to loading or slumping, although some may result from the melting of buried ice-blocks.

Correctly identified, varves may be diagnostic of sedimentation in glacial lakes, especially if laminae are pierced by dropstones. Varves comprise couplets of silt and clay , representing summer and winter accumulations of sediment respectively, and on the basis of their relative thickness they can be classified as follows (Ashley, 1975):

Group I: clay thickness > silt thickness
Group II: clay thickness = silt thickness
Group III: clay thickness < silt thickness

Group I varves do not occur as graded beds, but rather as two distinct layers, both the silt/clay and clay/silt contacts being fairly sharp. The silt unit is not as a whole normally graded, but consists of laminae containing minute graded beds. The clay unit, however, does show a generally decreasing grain size upwards; its upper surface may be uneven as a result of bioturbation. Clay accumulates unhindered throughout the year, but at periods of high discharge there is an influx of silt in turbidity currents. Such deposits are distal to the main sediment input source.

Group II varves are also couplets. The clay layer is graded, the silt layer shows multiple graded laminae and small-scale cross-lamination and erosional contacts. These sediments form in an intermediate position between the delta and the distal portions of the lake.

Group III varves show a considerable variation in thickness of the silt layer, whereas the clay thickness remains relatively constant, suggesting two different processes. A sharp contact occurs between the clay and silt layers. The silt layer shows laminations graded on the microscopic scale, but as a whole it does not always fine upwards. Clay layers, however, always fine upwards, suggesting that flocculation is not very important. This group is closely associated with deltas, i.e. they are formed in a proximal position. Graded beds are due to turbidity currents and the erosional contacts suggest current action. Where varves or varvites contain clasts bigger than the size of the layers, and reveal dropstone structures, these sediments provide some of the best evidence of glacial conditions (ice-rafting in a glaciolacustrine environment) in the rock record.

6.6.3 Rain-out diamict and sub-aquatic gravity flows

In lakes with a high input of ice-rafted material another facies, diamict, may be dominant (Fig. 6.8d). In a study of the lake deposits at Scarborough Bluffs, Lake Ontario, Eyles & Eyles (1983) documented the following lithofacies: diamict, sand, mud and minor gravel. This study is important beacuse it demonstrated the importance of diamict deposition in sub-aquatic situations, and provided an explanation for the common gradual passage from diamict into rhythmites. The Scarborough Bluffs sequence was regarded as typical of a large, glacially influenced lake-basin. Both massive and stratified diamicts are present; they consist of clayey silts with variable proportions of sand and gravel. Sedimentary structures include lenses, starved ripples, sand balls and pillows, intraformational breccias, undeformed laminated muds as interbeds, and conformable loaded contacts and diamict balls. The subaquatic origin of the diamicts was not accepted by some authors, but the sedimentary structures suggest that the diamicts formed by a variety of mechanisms, but notably by rain-out from icebergs and a floating ice margin, with fine material falling out of suspension, or by resedimentation by subaquatic gravity flows. Reworking by bottom currents led to the more gravelly concentrations. The other workers have argued that the bulk of the diamict was a lodgement till, but whichever explanation is correct, there is no doubt that in many diamict sequences, rain-out and sub-aquatic gravity flowage are important processes.

6.7 Glaciolacustrine facies associations

The assemblages of facies present in the glaciolacustrine environment have been described by Shaw (1975). In proximal glaciolacustrine successions (i.e. delta situations) he identified the following facies: gravel, cross-bedded sand, flat-bedded sand, cross-laminated sand, alternating sand, silt and clay, parallel laminated sediment and diamicton. Fining upwards successions were found to result from ice-front recession. Flat-bedded sands comprise multi-storied channel deposits. Increased deposition at the distributary mouth was found to result in the formation of bars covered by rippled surfaces. At the bottoms of deepwater channels, which breached distributary mouth-bars, dunes with cross-stratified cosets were present. Upward-fining trends, were sometimes interrupted as a result of channel migration. The finer grained facies were interpreted as overbank deposits.

In the late Palaeozoic Karoo tillite sequence of South Africa, Visser (1983a) described a vertical succession of facies that can be explained in terms of a glaciolacustrine retreat sequence. The lake bed is represented by heterogeneous diamictite, deposited during the last glacial advance over the lake floor. Next follows deformed siltstone with diamictite lenses and sandstone beds, showing evidence of debris-flowage, dropstone activity and underflows, and indicating that there was floating ice on the lake. On top lies varved shale, then rhythmite still suggesting the presence of ice, but not in contact with the lake. Finally the sequence is capped by black carbonaceous mud, indicating a marine transgression.

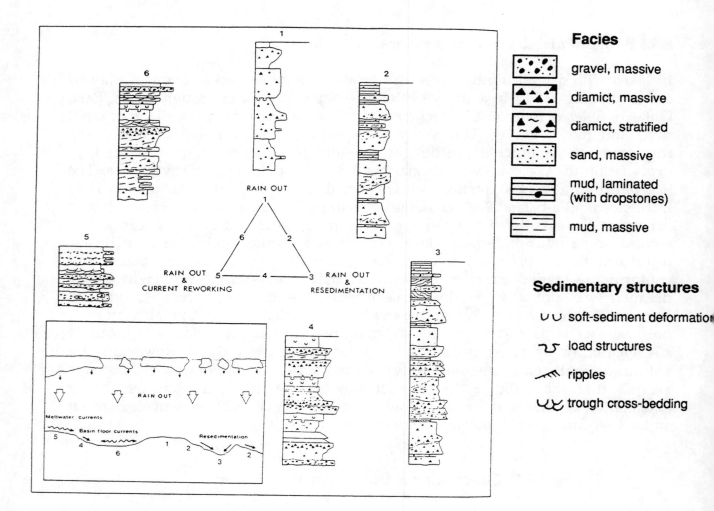

Fig. 6.9. Depositional model for diamict sequences in glacial lakes (as well as massive settings) subject to substantial influxes of fine-grained suspended sediment and ice-rafted material. The principal processes and the resulting facies associations are illustrated in these idealized profiles. The numbers in the triangle refer both to the sections and the schematic diagram in the inset. (adapted from Eyles & Eyles 1983).

The Scarborough Bluffs sequence, referred to above, provides a good illustration of a glacially influenced lake-bottom association. Vertical profiles logged by Eyles & Eyles (1983) show considerable variability in detail but, broadly, a dominance of diamictons, with lesser amounts of laminated sand and mud. By following the sequence laterally, Eyles & Miall (1984) showed that three main facies associations could be traced for several kilometres: a rain-out/resedimentation diamict association, a turbidite basin association and a deltaic association. Eyles & Eyles (1983) summarized the main processes operating and the resulting facies associations in a lake basin of this type in a facies model (Fig. 6.9); this model also applies to glaciomarine settings.

194

Rather different facies associations are associated with arid evaporating glacial lakes in Antarctica. Modern facies investigations are limited, but an analogous late Precambrian succession has been described by Fairchild et al. (1989) from Spitsbergen. In addition to the more normal lake facies of rhythmites, diamictites and sandstones, there are well-developed stromatolitic carbonates, which are transitional with the rhythmites. Carbonate is also a major component of the clastic facies, and evaporitic mineral pseudomorphs are common (Fig. 6.10). It is interesting to note that the common association of diamictites with carbonates and stromatolites, formerly used to discredit the late Precambrian glacial hypothesis, can now be viewed in the context of a modern Antarctic glaciolacustrine environment and, one might suspect, that other ancient sequences are of a similar nature.

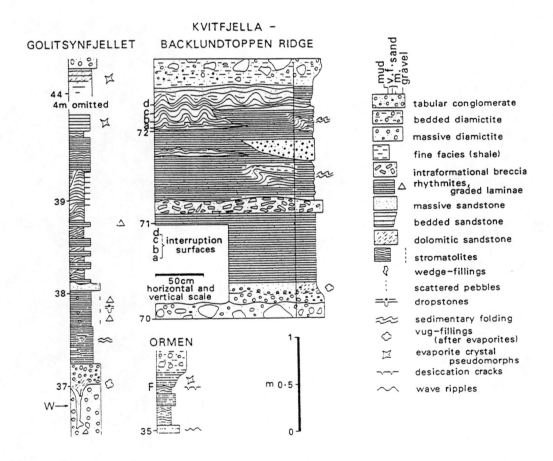

Fig. 6.10. Detailed logs through a glaciolacustrine facies association of the "Antarctic type" in which biochemical processes are important. Neoproterozoic Wilsonbreen Formation, NE Spitsbergen. The Kvitfjella-Backlundtoppen ridge section also shows the geometry of the facies as traced laterally. The vertical scales are in metres, measured from the base of the formation (Fairchild et al. 1989).

CHAPTER 7

GLACIOMARINE PROCESSES

AND SEDIMENTS

PART I: FJORDS

7.1 Introduction to the glaciomarine environment

The glaciomarine environment is here taken to include all marine areas in which glaciers reach the sea and influence sedimentation, ranging from fjords (Fig. 7.1) to those areas affected by iceberg drift and far-distant from the source glaciers. Until the early 1970s, our knowledge of glaciomarine environments was limited, principally because of the difficulty of undertaking systematic investigations in ice-infested waters. Nevertheless, glaciomarine sediments today are being deposited widely on contemporary continental margins, and form a substantial part of the middle and late Cenozoic stratigraphic record in these areas.

Fig. 7.1. Grounded tidewater glacier, Comfortlessbreen, at the head of the fjord Engelskbukta, Spitsbergen. In the foreground is a sandy pebble beach, derived as a result of reworking of subglacial material.

Interest in glaciomarine environments blossomed through the 1970s and 1980s and continues to flourish today. Several volumes containing papers devoted to glaciomarine sediment have been published in the last ten years (e.g. Andrews & Matsch 1983, Molnia 1983, Dowdeswell & Scourse 1990), as well as several review and numerous case-study papers. In addition, the text-book by Drewry (1986) covers the glaciomarine environment thoroughly, though much significant work has been published since. This interest has developed mainly because we now have sophisticated research vessels equipped for effective sea-bottom sampling, coring and drilling, and undertaking detailed seismic, bathymetric and oceanographic studies. Even so, some of the most significant advances in understanding fjord environments has come about from low-budget studies from small rubber boats in Alaska. The now-considerable body of sedimentological and geophysical data from glaciomarine environments, combined with advances in the understanding of the dynamics of ice-masses that reach the sea, have now given us a far better understanding of glaciomarine processes than fifteen years ago. However, many processes remain unclarified by direct observation and we look forward to the time when data might be obtainable directly from beneath glacier ice floating in the sea. A recent promising development has been the use of a remotely operated vehicle to investigate grounding-line processes beneath the floating tongue of Mackay Glacier, Victoria Land, Antarctica and tidewater glaciers in Alaska.

In contrast to the present-day glaciomarine environment, many Pleistocene and older glaciomarine sequences are well exposed, and the geometry of such sediments may be seen much more clearly in good onshore sections. Numerous studies have been made, reflecting the common occurrence of glaciomarine sediments in all the main periods of Earth´s glacial history (Frakes 1979, Hambrey & Harland 1981, Anderson 1983, Frakes et al. 1992). Indeed, because of their higher preservation potential compared with terrestrial sequences, much can be learned from ancient environments about current processes.

The glaciomarine environment is exceedingly complex and reflects not just glacial and marine processes, but inputs from biogenic sources, and from rivers and the wind. A general scheme for all glaciomarine environments to illustrate the main sediment sources, stores, pathways and processes leading to the deposition of glacigenic material on the sea floor is given in Figure 7.2.

It is convenient to consider glaciomarine environments according to two main geographical settings:

(1) Fjords (this chapter), in which sedimentation is influenced by tidewater or floating glaciers, rivers and streams, together with slope and marine processes.

(2) Continental shelf and the deep ocean (Ch. 8) in which sedimentation is dominated by grounded ice margins, floating glacier tongues, ice shelves and open-marine processes.

7.2. Fjord environments with active glaciers

All fjords by definition have, at some stage in their evolution, been influenced by glaciers. Today, 25% of fjords still have active glaciers. A variety of glacier regimes may be identified:

(a) *Alaskan regime* - highly dynamic, grounded, temperate glaciers, characterised by rapid sedimentation that is, or has been, facilitated by rapid tectonic uplift. These glaciers are found along the coast of the Gulf of Alaska, in British Columbia and Chilean Patagonia.

(b) *Svalbard regime* - dynamic, grounded, slightly cold glaciers, terminating in relatively shallow fjords (<200m deep), in which sedimentation is influenced by large amounts of meltwater during a short summer season. This type is dominant in Svalbard, parts of the Canadian Arctic and the Soviet Arctic.

(c) *Greenland regime* - dynamic, floating, cold glaciers in deep fjords (>200m). These are typically outlet glaciers from the Greenland ice sheet or the ice caps and highland ice-fields of Ellesmere Island and Baffin Island in the Canadian Arctic.

(d) *Antarctic maritime regime* - dynamic, cold, mainly grounded glaciers extending to near the mouths of short fjords with limited rock exposure and restricted surface melting. They are characteristic of the northern Antarctic Peninsula.

(e) *Antarctic arid regime* - sluggish, very cold, floating glaciers extending nearly to the mouth of fjords and fronted by fast-ice that may survive for several seasons without breaking up; surrounded by large areas of ice-free ground dominated by aeolian processes.

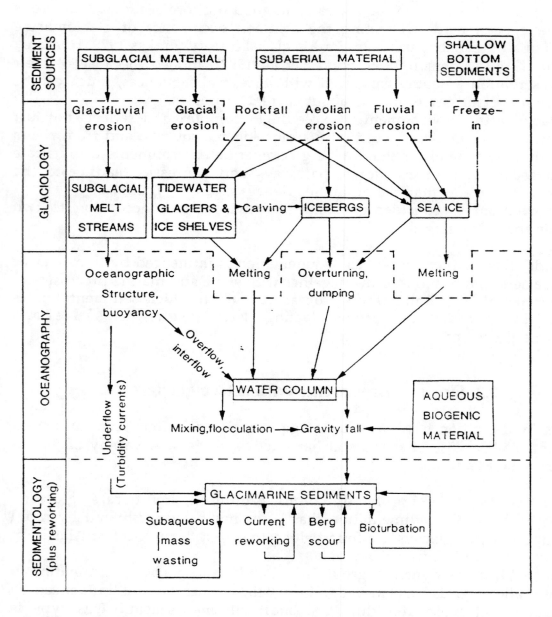

Fig. 7.2. The complex system making up the glaciomarine sedimentary environment, showing sediment sources, processes and environments of deposition (Dowdeswell 1987).

Sediment source	Debris type	Glaciological regime				
		Alaska	Svalbard	Greenland	Antarctic maritime	Antarctic arid
Direct from glacier	Subglacial debris	• •	• • •	• • •	• •	• • •
	Supraglacial debris	• •	•	•		•
	Glaciofluvial discharge	• • • • •	• • •	• • •	• • • •	•
	Supraglacial discharge	•	•			•
Marine	Icebergs	• •	•	• • •	•	•
	Sea ice		• •	•		•
	Biogenic	•	•	•	•	•
Non-direct, glacial terrestrial	Fluvial	• •	• • •	• •		•
	Mass movement	• •	•	•		•
	Wind	•	•	•		• • •

Low   High

•   • •   • • •   • • • •

Table 7.1. *Relative importance of sedimentary inputs in different fjord regimes.*

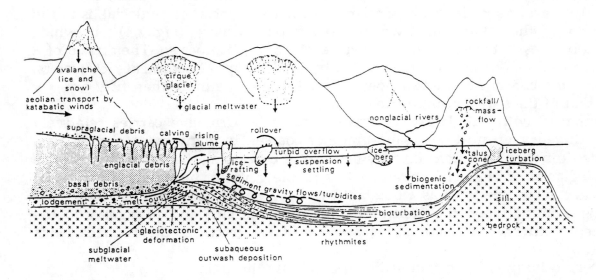

Fig. 7.3. *Sediment sources and processes operating in a fjord influenced by a grounded tidewater glacier.*

Sediments in fjords are supplied from numerous sources (Fig. 7.3). Glaciers themselves provide ice-contact deposits, glaciofluvial deposits and iceberg-rafted debris. In addition, there is a marine input, including suspended sediments and biogenic material, whilst the land provides fluvial, rockfall, aeolian and gravity-flow debris. The relative importance of sedimentary inputs according to different glaciological regimes is summarized in Table 7.1.

7.3. Processes controlling fjord sedimentation

7.3.1. Deposition from glaciers

Ice contact processes
The terminal position of a grounded tidewater and floating glacier tongue fluctuates seasonally, normally advancing in winter and receding in summer. The snout is generally subject to extensional stresses and is therefore heavily crevassed. Such glaciers are therefore more dynamic than their terrestrial counterparts. Icebergs calve from vertical ice cliffs, shoot up out of the water from below the waterline, or are launched subhorizontally away from the ice-cliff. Supraglacial and englacial debris falls into the water through meltout, calving or shaking. Although prolonged observations and sampling in proximity to such ice-cliffs is hazardous, a number of impressive studies have been made in recent years, notably adjacent to Alaskan glaciers, as summarized for example by Powell & Molnia (1989), Powell (1990) and Syvitski (1989).

The rate of sedimentation from ice at the glacier terminus depends on:

- the volume of ice being melted
- the debris content of the ice
- movement of the ice front

Sedimentation is strongly influenced by the nature of discharge from glacial streams into more dense marine waters. Submarine discharge of sediment-laden water from grounded tidewater glaciers, whether subglacial or englacial, is in the form of a jet, the behaviour of which depends on discharge (Fig. 7.4), but which invariably rises to the surface, sediment falling out as it does so. The arrival of a jet at the surface of the fjord is indicated by ´boiling up´ of sediment-rich waters, in which the concentration of suspended matter may be more than fifty or sixty times that of the surrounding waters.

Compared with subglacial streams, supraglacial drainage carries relatively little sediment and most is of sand-grade or coarser. In many cases, supraglacial streams do not enter a fjord directly because of crevassing, but find their way to join subglacial or englacial streams. On suitable glaciers, however, supraglacial meltwater enters the fjord as a bouyant overflow, as do non-glacial or proglacial streams. Alternatively, the meltwater falls over the cliff and plunges into the stratified marine waters below. Most sediment falls directly to the sea bed at this point (Fig. 7.3).

Iceberg calving is important sedimentologically through:

- the release of loose supraglacial sediment; this varies according to the transverse velocity profile of the glacier and the location of supraglacial debris;

• generating waves; large waves may be generated several times a day and in Alaska major calving events can devastate shorelines up to 100 km from the glacier (it is not advisable to camp close to the beach where calving glaciers are active!); waves can also lift ice-rafted material above the high-tide mark;

• controlling the position of the ice front in combination with the forward movement of the glacier.

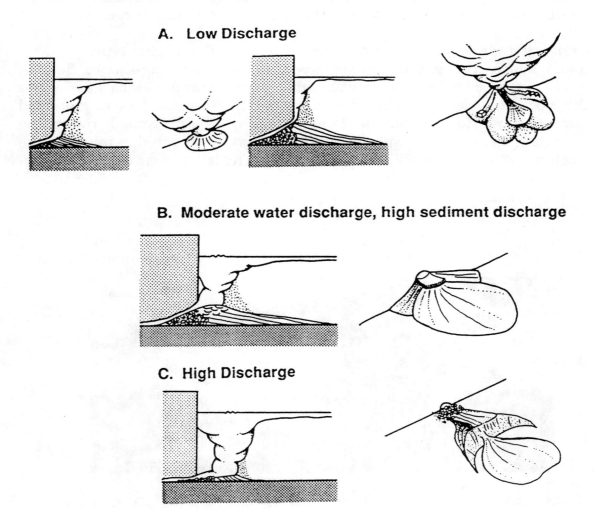

Fig. 7.4. *The variability of the sediment plume according to differences in discharge and the types of sediment fan associated with these differences (Powell 1990).*

The contribution of the various processes to the total sediment yield has been estimated for Coronation Glacier, an outlet glacier of the Penny Ice Cap on Baffin Island, as follows (Syvitski 1989):

86% glaciofluvial discharge
9% supraglacial dumping
3.7% Subglacial deposition
0.8% ice-rafted englacial deposition

The total sediment output from this glacier amounts to nearly two million tonnes a year, a figure which may be regarded as typical of many medium-sized cold glaciers. In Alaska the sediment budgets are considerably higher, but in Antarctica much less.

The style of sedimentation may change significantly if the glacier surges (Elverhøi et al. 1983) because of:

- substantially increased meltwater discharge;
- a radical change of the subglacial meltwater channel system to sheet-flood;
- erosion and reworking of glaciomarine sediments during a surge.

Many tidewater glaciers in Alaska and the Arctic are of the surging type.

Little is known concerning sedimentation at the grounding-line of floating fjord glaciers where ice loses contact with the bed. Submarine discharge of water probably occurs as a jet and ice contact processes may resemble those of grounded tidewater glaciers. However, the sediment plume will be prevented from rising by the floating ice, and the suspended sediment may be thoroughly mixed in the water column if, as seems likely, there is strong circulation near the grounding-line.

Fig. 7.5. Debris-laden iceberg, a typical illustration of ice-rafting processes, Columbia Glacier, Prince William Sound, Alaska.

Although volumetrically unimportant, the recognition of iceberg-rafted stones or dropstones (Fig. 7.5) in stratified sediments is one of the most important criteria for establishing a glacigenic origin of a rock sequence, especially in the pre-Pleistocene record. Compared with deposition at the grounding-line, deposition from floating ice (icebergs and sea ice) is relatively minor in most fjords (Fig. 7.3). The main controls on sedimentation are:

- the concentration and distribution of debris in the source glacier;
- the residence time of an iceberg in a fjord;
- the volume of ice calved;
- the rate of iceberg drift;
- the rate of iceberg melting;
- the amount of wave action, thereby influencing the number of overturning events.

The dominant control on the amount of debris in the parent ice-mass is its thermal regime (Dowdeswell & Murray 1990). Temperate glaciers have basal debris-rich layers of the order of centimetres to a few metres in thickness. In cold glaciers the debris-rich layers are commonly repeated by ice-tectonic processes, such as folding and thrusting, especially in the transition zone from basal sliding to a frozen bed. At the snouts of such glaciers, the debris-rich zone may be several metres thick, although sediment is unevely distributed. Glaciers in the highest latitudes, in which meltwater is limited, entrain debris by overriding a frontal apron composed of ice and debris, creating debris layers several metres thick. Surge-type glaciers incorporate debris to such an extent that the debris layer may reach ten metres in thickness; this largely arises from ice-tectonic processes, combined with flow over overpressurized basal sediments. In most cases the debris is parallel to an ice foliation, and may account for 50% by weight of the debris-rich ice.

Estimation of sedimentation rates from icebergs in fjords is difficult, even where the total glaciomarine sedimentation rate is known. However, a two-dimensional model developed for tidewater glaciers in fjords by Dowdeswell & Murray (1990) suggests (within an order of magnitude) that the rate for Alaskan fjords of 14 mm/yr is typical, representing 0.2-0.7% of the total glaciomarine sediment, and that in Svalbard the rate is 1 mm/yr, that is 0.1-0.3% of the total.

Sea-ice formation and break-up is an important process in the fjords of polar regions. During freezing, debris on the shore may be incorporated into the ice, later to be carried offshore under the influence of tidal currents. Material up to cobble size may be readily transported, creating a problem of discriminating iceberg-rafted from sea ice-rafted sediments, However, sediments derived from beaches will generally be better sorted and rounded.

7.3.2. Marine processes

Water circulation
The distribution of sediment in a fjord is largely controlled by circulation in the upper part of the water column, even in deep fjords. The juxtapositon of different water masses, namely fresh clean water, fresh turbid water and sea water, gives rise to water stratification. Stratification is most marked in summer when runoff and sediment transfer to the fjord is greatest.

A fjord often has a surface layer of fresh water derived from glacier streams, melting snow, ice and rain, most of which generally flows out of the fjord. Much of the sediment delivered to the continental shelf outside the fjord is transferred during just a few weeks when discharge is at its peak. Sediment may only be discharged from some longer fjords every few decades. Downfjord changes to the cross-sectional shape of the basin affect water-flow velocity; for example, constrictions cause flow to accelerate and reduce sedimentation. Surface layer velocity is affected by the tides.

Inflows of turbid freshwater plumes in areas of high precipitation and rapid ice recession, such as Alaska, may be dense enough to flow down to, and along, the fjord floor. Evidence for these underflows comes from coarse-grained sediments within 0.5 km of glaciers in Muir Inlet, Alaska. Such activity could have been important elsewhere when ice caps and glaciers were receding rapidly up fjords. These underflows can inundate an animal community on the bottom of the fjord or discourage its establishment.

The sporadic renewal of denser deep water is an important process because it prevents stagnation and oxygen-depletion of the fjord waters. When stagnation does occur, organic matter is preferentially preserved, and minerals typical of reducing environments, such as pyrite (iron sulphide), may form. Deep-water flow also may sweep away fine sediments on the fjord sill (if present) and redeposit them in the inner basin. Another factor is that, depending on its turbidity, deep water may either flush out a fjord (if with low turbidity) or add sediment.

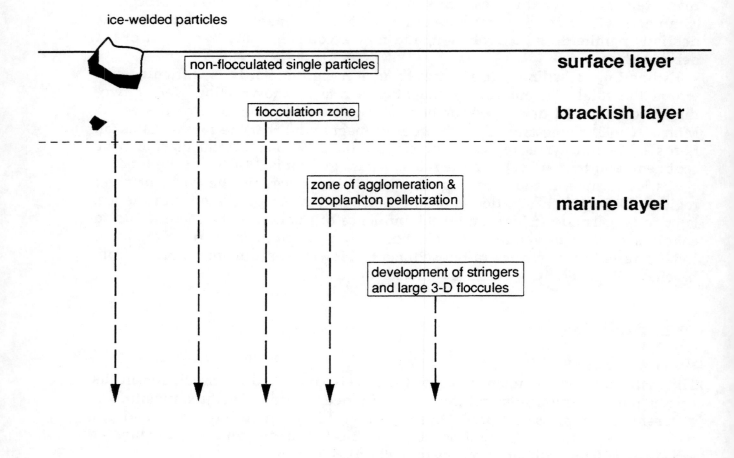

Fig. 7.6. Processes responsible for enhanced particle settling in fjords (adapted from Syvitski 1989).

Most sedimentary particles enter a fjord as single grains, and the rapidity with which they are deposited is mainly a function of size. However, enhanced particle settling may result from a number of different processes occurring at different levels in the water column (Fig. 7.6):

• flocculation: fine particles are attracted to each other when the normally repulsive electrostatic forces on their surface are neutralised by saline waters, for example near the ice front;

206

- agglomeration: attachment of sediment grains to each other by organic matter;
- pelletization: the result of zoöplankton ingesting sedimentary particles and egesting them as faecal pellets which then sink rapidly; suspended particles may settle as discrete layers or stringers;
- deposition of ice-welded pellets of mud or till derived from the glacier.

Tides

Reworking of sediments by tides is an important process in some fjords. Those fjords that have at their mouth a shallow sill may be prone to a variety of tide-related disturbances to the water. The sea-floor sediment, especially on the sill itself, may be winnowed as a result; the fines are transported either seawards to the shelf or into the fjord basin, leaving a relatively coarse, sandy and gravelly lag deposit. Even where no sill exists, tidal currents may be powerful enough to rework the fjord floor, carrying out of the system fine sand and mud from the inner parts of the fjord.

Waves

Wave action is mostly dissipated at the fjord sill, but some fjords, such as those in Greenland are open to the full fetch of the ocean. Calving of icebergs can also generate large waves. Where waves have strongly influenced shorelines, nearshore zones of sand and gravel are often followed offshore by a zone of stony lag sediment and washed bedrock, and then by poorly sorted sandy to muddy sediment. In Alaska, earthquake-generated waves may cause considerable devastation (Section 7.3.3).

Submarine slides and gravity flows

These processes are common as a result of overloading or oversteepening of slopes, especially as a result of high sedimentation rates near an ice margin. Part of a failed mass may develop into a fast-flowing turbid gravity flow, which on slopes is confined to gullies, but beyond spreads out over the fjord basin, filling hollows in the bed, as water fills a pond.

Biogenic processes

Like many other natural water bodies, fjords provide a habitat for a variety of marine organisms, which may contribute significantly to the sedimentary pile on the fjord floor. In Spitsbergen, diatom-blooms in springtime contribute to seafloor sediments in Kongsfjorden, while in southeastern Alaska prolific, bi-annual diatom blooms are reflected in the silica-content of the bottom-sediment. Diatoms are also a major component of the sediments in the outermost parts of Antarctic fjords. Minor biogenic components are represented by radiolaria and silicoflagellates; rare calcareous species of planktonic foraminifera may also occur.

Larger organisms are molluscs, crustaceans and worms, which usually are sparsely disseminated through the sediment. Nevertheless, they play an important rôle in mixing up (bioturbating) the sediment and disrupting or even destroying the stratification,.

Earth´s rotation

Sedimentation patterns in fjords are partly controlled by the Coriolis force, which is the effect of the earth´s rotation on a freely moving water-mass. In the

Northern Hemisphere, sediment-laden streams entering a fjord are commonly deflected towards the right-hand shore, and to the left in the Southern Hemisphere. Tongues or wedges of sediment tend to pile up along the respective sides of the fjord as a result. The effect increases in a poleward direction.

7.3.3. Influences on fjord sedimentation from land sources

In temperate and slightly cold glaciers, the dominant direct influence on fjord sedimentation from land that is ice-free near the shore, is from rivers and streams (Table 7.1). Often these rivers emanate from glaciers that terminate some distance from the sea and so carry abundant suspended sediment. In a typical Greenland fjord, rivers are locally important, but relatively insignificant in terms of the overall length of the fjord. In maritime Antarctica, most of the ground other than rocky cliffs is glacier-covered, and independent streams are rare. In arid parts of Antarctica, such as the Dry Valleys region of Victoria Land, stream activity does occur, but is emphemeral, and the main influence is from wind action on till-draped hillsides.

Mass-movements in the form of rockfalls, slumps or debris-slides are common, although they are usually small and localized. Alaska is exceptional in this respect, because, periodically, some of the world´s most powerful earthquakes generate huge rock slides. Perhaps the most dramatic event took place in Lituya Bay in July 1958, during a powerful earthquake along one of the region's major faults, the Fairweather. A landslide of 400 m³ of rain-soaked rubble was dislodged from a steep headwall, sheared off ice at the snout of the tidewater Lituya Glacier, and then rode on an air cushion up to an elevation of 525 m on the other side of the fjord, leaving in its wake rock stripped bare of soil and forest. A huge wave raced down-fjord at a speed estimated at 250 km/hr, ripping away vegetation as it went. One fishing boat was swept out of the fjord, over the exposed moraine at the mouth, into the open ocean, in spite of which, the two persons aboard the vessel managed to survive (US National Parks Service 1983). Even so, the total sediment input to these fjords from these slides represents only a relatively small proportion of the total sediment budget.

7.4. Patterns and rates of sedimentation in fjords

The patterns and rates of sedimentation are highly variable and depend upon whether tidewater glaciers or floating tongues terminate in a fjord, or only supply debris over land through meltwater streams, as well as upon the various processes described above. In tidewater glacier environments coarse gravel and sand tends to accumulate close to the glacier, but may occur patchily elsewhere, such as where diamictons have been winnowed by currents or where debris has fallen out of icebergs. Subglacial meltwater streams issuing from the glacier, which in many fjords provide the bulk of the sediment, deposit gravel, sand and mud, frequently draping diamicton that was deposited earlier. Ice-proximal mud tends to be laminated as a result of variations in discharge and tidal currents, often occurring as graded couplets. Mud extends from the glacier cliff into the distal parts of the fjord basin, and beyond if there is no sill, but ceases to be laminated because of low sedimentation rates, flocculation and bioturbation (Powell 1990).

Laminites have also been recorded from proximal settings in Antarctic fjords. In the maritime Antarctic Peninsula, glaciers are mainly grounded as far out as the mouth of the fjords they occupy, so relatively open bays are typical (Domack 1990). In such cases, although there are few visual signs of meltwater, sedimentation is inferred to be from suspended sediment in meltwater and sediment gravity flows issuing from the ice-bed interface. However, no plumes of sediment have been observed, so a process involving a combination of basal melting, and tidal pumping and flushing of subglacial sediment is believed to take place. The direct tidal signal is masked by low sedimentation rates and bioturbation. The organic content is low in these proximal sediments, but increases away from the ice margin. In the direction of the outer bay, the biosiliceous component increases markedly, and on the continental shelf diatom ooze with a minor amount of poorly sorted sand and gravel is dominant.

Laminites in colder Antarctic fjords, for example Ferrar Fjord in Victoria Land, form a very minor fraction of the sediments so far investigated. This suggests that in these areas sedimentation from subaquatic stream discharge is minimal, and that the mud fraction that occurs in both the ice-proximal and ice-distal sediments is mainly derived from basal melt-out sediment, and carried away in suspension more-or-less continuously.

An important feature of fjords is that major variations in sedimentation occur through time, both diurnally and seasonally, as well as at random, factors which contribute to the overall complexity of the sedimentation pattern. Many fjords have more than one distinct basin. In a typical fjord in Svalbard, the sediments are thickest in the innermost basin near the glacier; here over 100 m of till and compacted glacigenic sediment have accumulated. Outside the inner basin, the fjord has between 20 and 60 m of sediment, consisting principally of till or ice-front or surge deposits (Elverhøi et al. 1983). The fjords in southern Alaska, described by Powell & Molnia (1989), have much greater quantities of sediment (several hundred to over a thousand metres) owing to rapid crustal uplift, a process which is responsible for one of the most active erosion systems in the world.

In both these cases the fjord sediments are young, probably dating mainly from the last glaciation. In contrast, a fjord and glacial lake sequence drilled to basement in Ferrar Fiord, Antarctica, comprises a sedimentary record extending back into the Pliocene Epoch, at least 4 m.y. ago; however, despite the relatively long time interval represented, there is only 166 m of sediment (Barrett & Hambrey 1992).

In fjords lacking direct contact with a glacier, meltwater retains its dominant rôle, but much of the coarser material is trapped in prograding deltas or fans before, or as it enters, the fjord. Beyond these deltas, mud is the dominant sediment, derived mainly from the turbid waters of glacier-fed streams.

The rates of sedimentation in glacier-influenced fjords are exceedingly variable (Table 7.2). Glacier Bay in Alaska has yielded the most exceptional values with some 9 m yr^{-1} in its inner part, although 0.5 m yr^{-1} is perhaps a more typical figure close to most glacier margins in temperate and cold regimes. In the more distal reaches of a fjord sedimentation rates are an order of magnitude less. It has been demonstrated, using data from a number of different glaciological regimes (Boulton 1990), that sedimentation-rates decline logarithmically with distance from the ice-front.

Fjord	Relation to ice front	Sedimentation rate	Source
McBride Inlet, Glacier Bay, Alaska	Grounding line Submarine outwash fan	>13 m/yr 5 m/yr	Powell & Molnia 1989
Muir Inlet, Glacier Bay, Alaska	Inner fjord	9 m/yr	Molnia 1983
Glacier Bay, Alaska	Outer fjord	> 4.4 m/yr	Powell 1981
Kongsfjorden, Svalbard	Inner fjord Central fjord	50-100 mm/yr 0.4 mm/yr	Elverhøi et al. 1983
Van Mijenfjorden, Svalbard	Inner fjord	15 mm/yr	Elverhøi et al. 1983
Coronation Fjord, Baffin Island, N.W.T.	1 km 10-30 km 15 km	400 mm/yr 2-9 mm/yr 3.6 mm/yr (theoretical)	Syvitski 1989

Table 7.2. Some sedimentation rates in glacier-influenced fjords.

7.5 Textural characteristics of fjord sediments

The sub-gravel sized fraction of sediments has been investigated in a number of fjords. Boulton (1990) has focused on the grain-size distribution of the sediment plume from Kongsvegen in Svalbard and the resulting sediments in Kongsfjorden. A series of sediment traps indicates a skewed grain size distribution with a prominent peak at the sand-silt boundary. As the coarser material settles out first, the peak shifts down-fjord progressively into the silt mode. Fjord-bottom sediments show the same trend. For comparison, Boulton also showed the grain-size distribution for an iceberg and the resulting sediment, which proved to be similar; in this case, the sand component was even greater than that in the early plume sediments, but there was no pronounced peak.

Elverhøi et al. (1983) have examined the grain-size distribution of other facies in Kongsfjorden. The trend from poorly sorted basal tills, through somewhat muddier glaciomarine deposits on slopes and sills, to the muds of glaciomarine origin in the distal basins is clear.

The grain-size distribution of the principal facies in the sequence of Ferrar Fjord (Barrett & Hambrey 1992), demonstrates the poorly-sorted nature of the dominant facies, diamictite, although a pronounced peak at the sand-silt boundary is evident in some samples. The well-sorted nature of some stratified sands is evident, supporting the intepretation of an aeolian origin.

The roundness/sphericity characteristics of clasts within fjord sediments are similar to those in terrestrial glacigenic facies. Dowdeswell (1986) investigated the shapes of the various components associated with a fjord glacier on Baffin Island. The results were similar to those for debris in transport as recorded from terrestrial locations (*cf.* Figs. 1.4). There are few data on the fabric and surface markings of clasts in the gravelly sediments of fjords.

210

7.6 Facies analysis of fjord sediments

The characterization of facies in modern fjords has largely followed a process-orientated approach (e.g. Powell 1981, Powell and Molnia 1989, Elverhøi et al. 1983). Objective designation of lithofacies in the manner advocated by Eyles et al. (1983) has only been attempted infrequently. Much of our understanding concerning fjord facies and processes has come from the temperate glacier-fed Alaskan coast. The principal lithofacies, in so far as they can be gleaned from the literature, are listed and interpreted in Table 7.3. (Note that in this book the term 'lithofacies' is used differently from that in the papers describing Alaskan fjords, but in a manner more compatible with normal sedimentological treatment (e.g. Eyles et al. 1983). Contrasting examples from Antarctica and from the older geological record are also described below.

Lithofacies	Interpretation
Diamicton	a. Lodgement till b. Subglacial and meltout till c. Subaqueous sedimen gravity flows d. Silt and clay from meltwater streams + iceberg debris

Gravel (poorly sorted)	a. Subaqueous outwash: grounding-line fan
	                                         ice-contact delta
	                                         fluviodeltaic complex
	b. Gravity flow
	c. Lag deposit from winnowed diamicton

Gravel (well sorted)	a. Beach b. River mouth bar
Sand (poorly sorted)	a. Subaqueous outwash b. Gravity flow
Sand (well-sorted)	a. Beach b. River mouth bar
Mud with dispersed clasts	Meltstream-derived silt + clay, with minor ice-rafted debris ("iceberg zone mud")
Mud	a. Meltstream-derived silt + clay b. Tidal flat sediment
Rhythmites (laminated) sand and mud)	Deposition from underflows generated from meltwater streams, and influenced mainly by tides (cyclopels and cyclopsams)

Table 7.3. Principal lithofacies in Alaskan-type fjords and their interpretation.

7.6.1 Facies associations forming today in Alaskan fjords

Fjord lithofacies often occur in different associations. For example, a morainal bank comprises a chaotic mixture of diamicton, gravel, rubble and sand (Powell 1989) and a grounding-line fan comprises some, or all of these in different proportions (*cf.* Section 7.6). In a study of tidewater glaciers in Glacier Bay, Powell (1981) recognized five facies associations:

Association I: Facies of a rapidly receding tidewater glacier with its ice front calving in deep water (Fig. 7.7). Interpretative facies deposited close to the ice front consist of reworked subglacial till, subglacial stream gravel and sand, scattered and dumped coarse-grained supraglacial debris, and low ice-push (or De Geer) moraines from winter readvance.

Fig. 7.7. Rapidly receding McBride Glacier, Glacier Bay, Alaska. The water depth at the grounded ice cliff is over 100 m.

Association II : Facies of a slowly receding tidewater glacier with its ice front calving predominantly in shallow water. Ice-front recession has been retarded or stopped by a side-wall construction or a ramp in the fjord floor, but calving is still able to proceed apace. The resulting interpretative facies are ice-proximal, coarse grained, morainal-bank deposits, with an ice-contact delta or grounding-line fan (Section 7.7). Subaquatic gravity flows are common down the bank foreslope producing intertonguing sand layers within the more distally deposited iceberg zone mud (Fig. 7.8).

Association III: Facies deposited by a slowly receding or advancing tidewater glacier, rarely calving into shallow water. When ice fronts recede or advance into shallow water they often terminate in a protected bay and lose ice by surface-melting. Lateral streams dominate the environment close to the ice front. The resulting facies is an iceberg-zone mud, deposited both close to the ice-front away from stream outlets, and in more distal locations. Large ice-contact deltas, comprising gravel to mud facies which are structureless at stream outlets, develop along the ice-cliff and further down the fjord. These pass laterally into muddy sand and interlaminated sand and mud in the prodelta area and beyond, to which is added ice-rafted debris depending on the amount of supraglacial debris available.

Fig. 7.8. Ice-contact delta in front of Riggs Glacier, Glacier Bay, Alaska. At high-tide this delta is under water and the glacier prone to minor calving.

Association IV: Facies of a turbid outwash fjord. In this case, the glacier is terrestrial and produces a large outwash delta that progrades into the fjord. The resulting facies are coarse-grained fluvial deposits on the delta surface. The slope comprises sand and gravel which intertongues with marine glacial outwash mud in more distal areas. Very little ice-rafted debris is present, since only small icebergs are introduced *via* the meltstreams. Sand-silt rhythmites of turbidity current origin cover the floor of the fjord.

Association V: Facies of a shallow-water environment, distant from the ice-front. This association comprises tidal-flat muds and braided-stream and beach sands. These areas may be uplifted isostatically and disturbed by stranded icebergs. Eventually, this association acquires a cover of vegetation, which may be buried during a subsequent glacial readvance.

This classification underestimates the importance of temporal and spatial variations that result from changes in the position of the glacier terminus with respect to the depositional site. In many Alaskan fjords, where rapid recession is taking place, the compexity of the facies associations is even greater (Molnia 1983).

7.6.2 Plio-Pleistocene facies associations from an Antarctic fjord

The main facies in the Ferrar Fiord succession of Antarctica are massive to well stratified diamictite, mudstone and sand, and minor rhythmite (Barrett & Hambrey 1992). This association of facies is clearly different from that in fjords where meltwater is dominant. There is little mud such as might be associated

associated with a meltwater plume, or gravel typical of subaquatic outwash. Sedimentation was principally the result of:

Facies	Sedimentary features	Interpretation
Massive diamictite	Non-stratified; clasts uniformly distributed; horizontal shear surfaces; occasional striated clasts	Lodgement till; less commonly basal meltout till or waterlain till
Weakly stratified diamictite	Weak, wispy stratification in matrix, and dispersed clasts, including dropstones	Proximal glaciomarine/ glaciolacustrine sediment derived from basal debris melted out near the grounding-line; some winnowing
Well-stratified diamictite	Well-developed stratification in matrix and dispersed clasts, including dropstones	Proximal to distal glaciomarine/ glaciolacustrine sediment; strong bottom-current reworking
Massive sandstone	Fine to very fine, uniform, unstratified; a few beds are graded	Aeolian or supraglacial sand deposited directly or washed off floating ice; graded beds may be turbidites
Stratified sandstone	Fine to very fine; occasional fine horizontal mm-cm laminae	As above, but with intermittent deposition
Mudstone	Moderately and weakly stratified to unstratified; little sand and few gravel-sized clasts	Sedimentation from suspension well beyond ice margin, originating from basal debris close to the grounding-line; little from sea or berg ice
Rhythmite	Alternation of well-sorted, very fine sand and clayey silt layers from a few mm to a few cm thick; no dispersed gravel clasts	Sedimentation from intermittent underflows of fines winnowed from basal glacial debris with background sedimentation of aeolian sand; some may be varves, no ice-rafting

Table 7.4. Features and interpretation of lithofacies in the cored CIROS-2 fjord sequence, Ferrar Fiord, Victoria Land, Antarctica (Barrett & Hambrey 1992).

- lodgement of till as ice advanced across the fjord floor;
- release of waterlain till from a floating ice tongue near the grounding-line; and
- deposition of sand that had blown onto lake ice.

Clast provenance and Quaternary geomorphological studies indicate that the glaciers which created the fjords tended to recede during glacial maximum, when the main East Antarctic ice sheet expanded and blocked the entrance to the fjord, creating a lake. Conditions may have been cold enough for a semi-permanent ice-cover to be maintained on the lake, and discourage melting, so the main sediment source, other than direct deposition from ice, was aeolian.

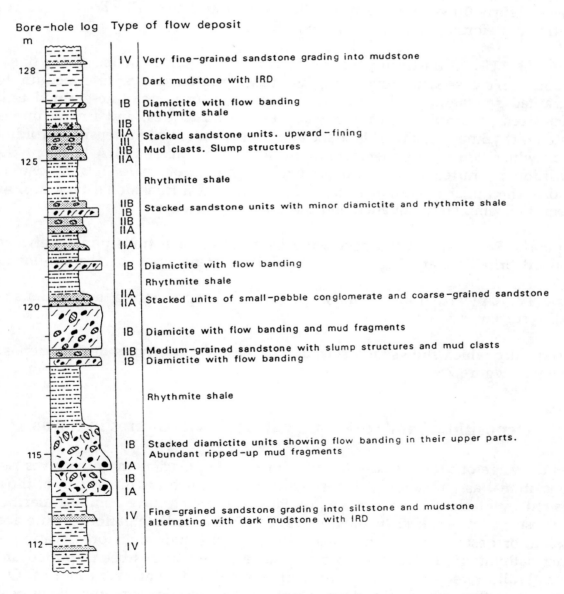

Bore-hole log
m

Type of flow deposit

IV — Very fine-grained sandstone grading into mudstone

Dark mudstone with IRD

IB — Diamictite with flow banding
Rhthymite shale

IIB
IIA — Stacked sandstone units. upward-fining
III
IIB — Mud clasts. Slump structures
IIA

Rhythmite shale

IIB
IB — Stacked sandstone units with minor diamictite and rhythmite shale
IIB
IIA

IIA

IB — Diamictite with flow banding

Rhythmite shale

IIA
IIA — Stacked units of small-pebble conglomerate and coarse-grained sandstone

IB — Diamicite with flow banding and mud fragments

IIB — Medium-grained sandstone with slump structures and mud clasts
IB — Diamictite with flow banding

Rhythmite shale

IB — Stacked diamictite units showing flow banding in their upper parts. Abundant ripped-up mud fragments
IA
IB
IA

IV — Fine-grained sandstone grading into siltstone and mudstone alternating with dark mudstone with IRD

IV

Fig. 7.9. Part of the late Palaeozoic Dwyka Formation, Kahari Basin, Southern Africa, recovered in a borehole, illustrating the relationship between different types of flow deposits (Visser 1983b).

215

7.6.3. Submarine debris-flow facies associations from the late Palaeozoic sequence of South Africa

The importance of subaquatic debris-flowage close to the grounding-line was recognized by Visser (1983b) from studies of a well-exposed sequence of late Palaeozoic age in the Kalahari Basin of southern Africa. Visser distinguished several types of flow based on lithology and texture, derived from material deposited near the grounding-line of a floating glacier. These flows occurred either as individual or stacked beds, and are described and interpreted as follows:

Type I (IA): diamictite, massive, medium-coarse, clast-rich individual beds about 1 m thick. The diamictite grades into clast-supported conglomerate. (IB): deformed argillaceous diamictite with thin, clay-rich laminae or flow-banding. There is an upward-transition to rhythmite-shale. These sediments represent cohesive debris-flows with laminar flow characteristics. IB shows features indicative of deformation of a late stage in the flow.

Type II: Upwards-fining conglomerate and sand beds. IIA: Basal conglomerate/coarse sandstone at the base, passing upwards first into medium to fine-grained sandstone with mud clasts, then into an upper fine-grained unit characterized by ripple marks and water-escape structures, and finally into a mudstone/rhythmite shale. This association represents a high-density turbidity flow, in which complete liquefaction occurred. IIB is similar to IIA but contains in addition ice-rafted debris and angular fragments of shale in a mudstone instead of the upper fine-grained unit. This represents a transitional type of flow between the completely liquefied and the cohesive varieties.

Type III: Reversly graded sandstone beds. These are transitional between cohesive debris-flows and high-density turbidity currents or, in part, grain-flow.

Type IV: Upwards-fining sandstone-siltstone, representing a low-density turbidity current.

A sequence which shows an association of these various types of flow deposit is shown in Figure 7.9.

7.7 Depositional and soft-sediment erosional features in fjords

The wide range of processes associated with receding glaciers in fjords gives rise to depositional assemblages and erosional features which are distinct from those deposited by glaciers and meltwater on land, but which have some similarities with those in large lakes. In most cases, continued sedimentation during ice-recession, or destruction during ice-advance will preclude their preservation as distinct bathymetric features. Only rarely have these features been uplifted and eroded to allow ready inspection of their internal sedimentary structures. Our knowledge is thus based to a large extent on the investigations that have been made in Alaskan and Arctic fjords and a variety of geophysical techniques, such as seismic profiling and side-scan sonar. The forms described below are part of a continuous spectrum of features, the first three together apparently being analogous to delta-moraines in lacustrine settings.

7.7.1 Morainal banks

Morainal banks form by a combination of lodgement, meltout, dumping, push and squeeze processes when a glacier terminus, grounded in water, is in a stable or quasi-stable state (Powell 1983, Powell & Molnia 1989). (Note that the American term *morainal* in this context is well-established in the literature, so is used in preference to the British adjectival form *morainic*.) These banks are often intimately associated with submarine outwash. Depositional processes are further complicated by the partial lifting of the ice-margin from the bed as the tide comes in, and by tidal-pumping, allowing subglacial meltwater to influence growth of the bank. Compositionally, morainal banks are composed of pockets of diamicton enclosed in poorly sorted sandy gravel (with clasts up to boulder size) or gravelly sand. The fore-slope of the bank is affected by resedimentation processes such as sliding, slumping and gravity flowage. The up-glacier side of the bank may also collapse once the glacier has receded.

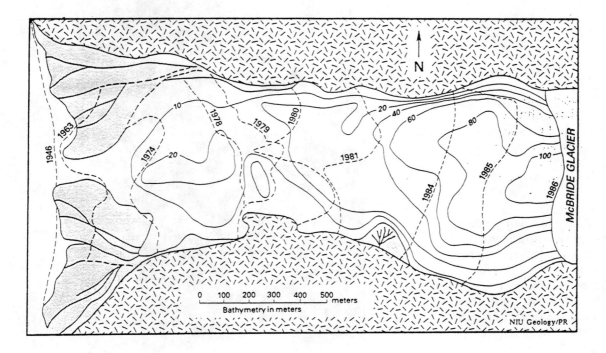

Fig. 7.10. McBride Inlet, Glacier Bay, Alaska, showing the morainal bank at the entrance (now exposed) and a younger morainal bank (between the 1978 and 1980 glacier fronts) that is capped by submarine outwash sediments. A small delta on the south side occurs near the 1984 position (Powell & Molnia 1989).

In McBride Inlet, a short arm of Glacier Bay in Alaska, Powell (1983) observed that a morainal bank formed between 1978 and 1980 (Fig. 7.10). Within this period, 1.2 million cubic metres of sediment were deposited at a rate of 400,000 cubic metres a year. Since the early 1980's the top surface has been partly exposed at low tide. The flanks of the bank slope at 15° (foreslope) and 9° (backslope).

217

7.7.2 Grounding-line fans

Also known as subaqueous (but more properly subaquatic) or submarine outwash-fans, grounding-line fans extend from a glacier-tunnel that discharges subglacial meltwater where a glacier terminates in the sea (Powell 1990) (Fig. 7.11). The bulk of the sediment is released very rapidly from meltwater which issues as a horizontal jet at, or near, the sea floor during full-pipe flow (Fig. 7.4). This outwash contains a range of particles from coarse gravel to mud, the coarser fraction being deposited almost immediatelty. The rest of the fan system includes sediment gravity flows, slumps and deposits from the plume of turbid meltwater as it rises above the bed. Much of the fine material is carried away to more distal settings in the sediment plume.

Fig. 7.11. Grounding-line fan developing in front of Harriman Glacier, Prince William Sound, Alaska. This type of feature, developed close to or just below water level, represents a platform that facilitates the advance of the glacier.

The complexity and nature of grounding-line fans, which are difficult to investigate in modern settings, is demonstrated by parts of the late Palaeozoic Dwyka Formation in South Africa (Visser et al. 1987; Fig. 7.12) and the late Palaeozoic Wynard Formation in Tasmania (Powell 1990). In the latter, six main facies were recorded:

• Stratified conglomerates with rounded to subrounded clasts and trough cross-bedding. This facies was interpreted as representing a recession of the glacier.

218

- Clast-rich stratified diamictite, with subangular to subrounded clasts, many of which are striated, and with striated cobble pavements, representing a quasi-stable terminus.
- Deformed pebble conglomerate and coarse sandstone with flow structures resulting from loading onto the diamictite and ice push as the fan prograded, followed by flowage as the ice support was lost when recession took place.
- Trough cross-bedded sandstone and conglomerates with deformation and injection of diamictite resulting from ice push during continued progradation of the fan.
- Cross-bedded sandstone and poorly sorted pebble-boulder conglomerates, with a few striated clasts, also formed during progradation.
- Stratified diamictite with 5-50% subangular to subrounded clasts, including sandstone rafts incorporated during soft-sediment deformation.

Fig. 7.12. Grounding-line fan complex in the late Palaeozoic Dwyka Formation, Kransgat River, Karoo Basin, South Africa. (a) Massive diamictite at the base, overlain by interbedded sandstones and diamictites that show signs of slumping. (b) Close-up view of slumped sandstone and diamictite. (c) Soft-sediment sandstone dyke penetrating diamictite, the result of rapid burial and fluidization of an overlying sandstone layer. (d) Stratified diamictite with large ice-rafted boulder, representing more distal facies compared with (a).

219

Geometrically, the better stratified parts of the fan have a barchanoid (concave down-flow) form.

Grounding-line fans which develop during sustained advances are not well-preserved because of reworking. Nor are fans that form during rapid recession well preserved because there is insufficient time for them to build up. Like morainal banks, they are best-developed if the ice margin is quasi-stable.

7.7.3 Ice-contact deltas

Developing out of grounding-line fans, when a glacier becomes more-or-less stable for tens of years or more, are tidally influenced fan-deltas (Powell & Molnia 1989, Powell 1990), alternatively known as proglacial fan-deltas (Fig. 7.8). The fan grows by direct addition of sediment from submarine outwash-streams and from sediment gravity-flows at the delta-front. Mass-flows elsewhere along the glacier front may contribute to the growth of the fan. As the glacier pushes against the delta, the sediment is subjected to folding and thrusting, processes which tend to increase the height of the delta (Fig. 7.4). With continued readvance, the glacier rides up the back of the fan and deposits lodgement till, and perhaps also forms typical morainal bank accumulations. If the terminus then recedes, basal ice may become buried. Tunnels, through which streams discharge, become inclined upglacier and prone to blockage, promoting occasional outbursts of sediment-charged water. During recession much of the coarser load is dumped on the reverse side of the fan, and backsliding and collapse of sediment is likely. The larger a fan grows, the more stable it becomes, providing a platform over which subsequent advance can take place (Fig. 7.11).

In Alaska' ice-contact deltas form extremely rapidly. A good example is that formed during the recession of Riggs Glacier in Glacier Bay (Fig. 7.8). In 1979, the glacier terminated in about 55 m of water, but by 1985 the delta-plain had aggraded above high tide mark and extended 100 m further into what had been 32 m of water in 1981; a total of one million cubic metres of sediment had thus accumulated in only four years.

7.7.4. Fluviodeltaic complexes

Where a river or stream enters the fjord directly, especially if carrying a large bedload from a glacier upstream, a delta with a braided top may develop (Fig. 7.13). During peak melting in early summer the bed-load comprises material up to cobble-size, which is deposited both as a topset and a foreset, together with finer gravel and sand. Channel-filling and switching gives rise to trough cross-bedding on a scale of tens of metres wide and several metres deep. Active channels may extend down the delta slope, and extend onto the fjord floor; levées and terraces may also develop along the sides of the channels. One such channel in Glacier Bay (Queen Inlet) is 32 m deep and up to 259 m wide, extending for 10.5 km (Syvitski 1989). Slumping is also a common process on the delta slope. Finer material will mostly be carried away in suspension, but some may be deposited between distributary channels in intertidal areas. The top of the delta front may be reworked by waves and tides, creating beach ridges. An example of a fluviodeltaic complex in Svalbard is illustrated in Figure 7.14a. The principal facies recorded, together with their interpretations are, from top to bottom, as follows:

Fig. 7.13. Fluviodeltaic complex in Kejser Franz Josef Fjord, East Greenland (foreground). Note the plume of sediment, derived from a glacially influenced river flowing from the right.

(a) sandy gravel: forms top-set of stratified pebbles and cobbles, mostly of subrounded and subangular shapes (Fig. 7.14b) - fluvially deposited on the delta top;

(b) sand, gravelly sand and sandy gravel - cross-bedded on a metre scale; fluvially derived, subaquatically deposited delta foresets (Fig. 7.14c);

(c) sand: graded coarse to fine sand or silt beds a few centimetres to decimetres thick, with dispersed gravel lonestones (Fig. 7.14d) - turbidity current-derived sediment with ice-rafted dropstones;

(d) mud: massive blue clayey silt and clay with whole or broken mollusc shells and dispersed stones - fjord bottom mud of suspension origin with minor ice-rafted debris.

(e) diamicton: massive gravelly sandy mud with shell fragments with up to 30% gravel clasts - waterlain or flow till deposited on the fjord floor.

7.7.5 Proglacial laminites

Laminites have been studied most thoroughly in Alaskan fjords, where they lie beyond the morainal banks and ice-contact deltas, but still in a proximal position in relation to the glacier (within 1-2 km) (Powell & Molnia 1989). Laminites are produced by deposition of suspended sediment from streams, tidal and wind-generated currents, subaquatic slumps and gravity flows. Variations in stream discharge combined with the effect of tidal currents frequently produce rhythmically laminated sediments called *cyclopsams* (graded sand-mud couplets) and *cyclopels* (silt-mud couplets) (Mackiewicz et al. 1984).

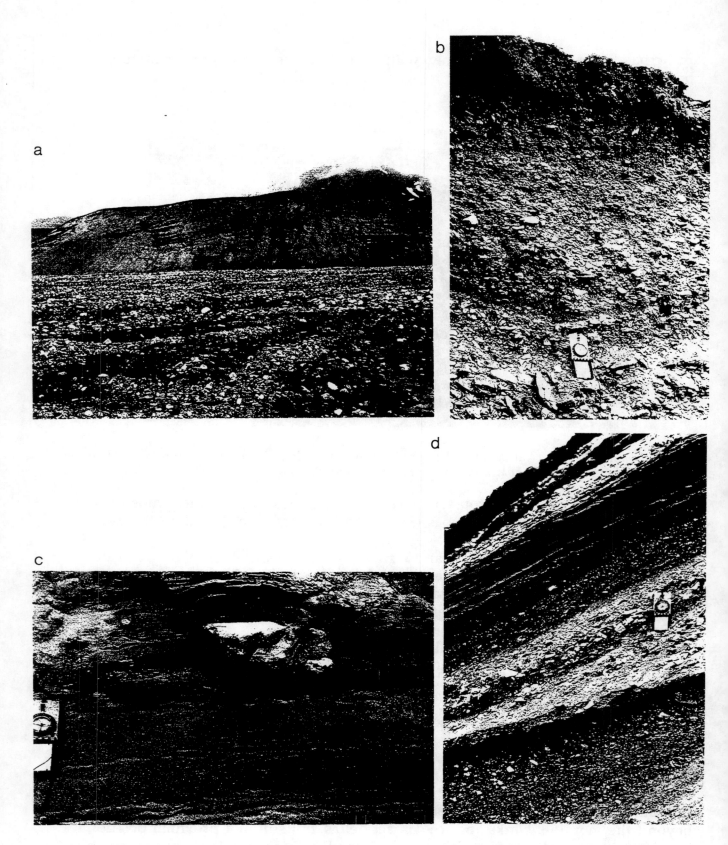

Fig. 7.14. *Dissected, elevated Weichselian fluviodeltaic complex at Engelskbukta, Spitsbergen, showing characteristic facies (a) general view of delta, showing low-angle foresets dipping seawards (to the right); (b) poorly sorted sandy gravel, forming a top-set about 1 m thick; (c) outermost forsets in delta, comprising interbedded sand and gravel; (d) gently dipping, well-sorted, stratified sand with ice-rafted clast.*

222

Experiments using traps in McBride Inlet, Glacier Bay (Cowan & Powell 1990) have shown that cyclopsams and cyclopels are derived from turbid overflow plumes, but that the deposition as laminae is related to tides, and two graded couplets are produced each day. The sand flux is greatest during low tide, and lowest near high tide. Small-scale gravity flows may interrupt the laminite sequence. Other laminae may be produced as a result of variations in discharge on a diurnal basis. Pleistocene cyclopsams and cyclopels have been recognized in other fjord areas with lower tidal ranges than Glacier Bay, ranging from temperate latitudes (e.g. Whidbey Island, Washington State; Domack 1984) to high Arctic latitudes such as Ellesmere Island and Svalbard (Fig. 7.15), although these may be daily or seasonal couplets, rather than tidal ones.

Fig. 7.15. Fjord-bottom sand-silt rhythmites (probably tidally related), associated with tidewater Comfortlessbreen, Engelskbukta, Spitsbergen. These sediments have been lifted above sea-level by large-scale glaciotectonic processes.

Cyclopsams are interbedded or interlaminated with coarser outwash sediments in the ice proximal zone. Cyclopels also form in a proximal position during periods of low discharge or in quiet areas adjacent to the glacier front. Beyond 1-2 km from the terminus, laminites are lost because all the silt has already settled out and sedimentation rates are much reduced, allowing bioturbation and flocculation to take place.

In the older geological record, tidal rhythmites have been reported, for example, from the early Proterozoic Gowganda Formation in Ontario, Canada (Mustard & Donaldson 1987), and the Neoproterozoic Elatina Formation, South Australia (Williams 1989). Many other ancient laminites in glacial sequences are probably of this type.

Fig. 7.16. Latest Proterozoic fjord-bottom rhythmites with abundant ice-rafted material, Henan Province, China.

An important additional component to laminites is often represented by *dropstones* (or *lonestones* if a non-genetic term is preferred), providing one of the most reliable criteria for recognizing glacial influence in the geological record (Fig. 7.16). Dropstones of ice-welded diamicton or mud may also be present; these are known as till-pellets, although some of them may reach boulder size. Dispersed clasts frequently exceed the thickness of the laminae, and show disruption of the laminae beneath, with draping of laminae over the top.

7.7.6 Fjord bottom sediment complexes

Further from the glacier, and interstratified with the proglacial laminites, is homogeneous mud derived from glacial rock-flour. This mud contains variable amounts of iceberg-rafted debris and therefore may grade into diamicton (more than about 1% gravel). Such a deposit has been referred to as *bergstone mud* (Powell & Molnia 1989), and its extent depends on the paths taken by icebergs. Dropstone structures are not visible in these homogeneous sediments.

7.7.7 Beach and tidal-flat features

Beaches are best developed close to areas where there is, or has been, an abundant supply of subglacial outwash sediments. Often beaches consist of coarse gravel, becoming sandy downdrift of the outwash streams. Beaches also form in sheltered rocky embayments. Spits commonly develop across river outlets, and these may be backed by lagoons and marshes (Fig. 7.17).

Tidal-flats are generally of fine sand and mud, often dotted with iceberg-rafted debris. Tidal-flats are common in areas of quiet water, towards the margins of river outlets; they often form on top of deltas after the main distributaries have moved elsewhere.

224

Fig. 7.17. Spits and lagoons, containing reworked glacial and glaciofluvial sand and mud, Engelskbukta, Spitsbergen. Comfortlessbreen in the background.

Beaches and tidal-flats in glacier-influenced fjords show a range of features arising from iceberg action as the tides rise and fall, although in the polar regions features produced by icebergs may be difficult to distinguish from those resulting from sea ice. Icebergs rapidly break down into *bergy bits*, ice-blocks several metres across. Tidal movements may drag an iceberg across the beach or tidal-flat, creating a groove with levées (Fig. 7.18a). Grooves run up and down the beach, or across it if long-shore drift takes place. Often grooves are irregular in direction and cut across one another. Icebergs may also be rocked by waves, creating wallows, and others may push up ridges as the tide comes in (Fig. 7.18b). If icebergs are not well grounded, then various bounce-, chatter- and roll-marks may form. Some icebergs become stranded in intertidal zones, and on sandy or muddy shores may become buried, creating so-called iceberg rosettes or a pitted surface when they melt (Fig. 7.18c). Dirty icebergs, comprising basal ice, may also become buried, leaving behind diffuse pods of till or mud in the beach deposit.

If strong tidal-currents and waves winnow out fine material from the surface of boulder- or cobble-rich diamictons in beaches, icebergs may compress the boulders into the matrix, creating a flat surface or type of gravel pavement (Fig. 7.18d). Fine, irregular striations may form as the ice moves over the pavement.

7.7.8 Iceberg turbate deposits

Icebergs may become stranded on shoals, disturbing the sediment and creating an *iceberg turbate*, although there seem to be few descriptions of such features in the literature. Since shoal deposits are typically winnowed diamicts, the turbate will have gravel as the dominant component. If it is interbedded with other facies, iceberg groundag may be evident in the form of deformation structures. If the icebergs contain debris, the resulting sediment will be of mixed composition and have a disorganised fabric, often building up into irregular ridges.

Fig. 7.18. Features formed on beaches in iceberg-influenced fjords. (a) Grooves on sandy gravel beach created by icebergs from Columbia Glacier, Prince William Sound, Alaska. (b) Push-ridges on beach formed by icebergs at high-tide, Columbia Glacier, Alaska. (c) Iceberg rosettes, pits resulting from burial of sea or glacier ice by beach material, followed by slow melting of the ice, Engelskbukta, Spitsbergen. (d) A gravel pavement formed by ice-compaction and tidal-current winnowing of shoreline sediments, Prince William Sound, Alaska.

7.8 Stratigraphic architecture in fjords

To the author's knowledge, no systematic studies, combining deep-drilling through fjord sediment to bedrock and seismic surveys to determine the stratigraphic architecture of a modern fjord, have been undertaken, although it has been possible to gain an appreciation of the structure of bottom sediments in some Arctic fjords using geophysical techniques combined with shallow coring.

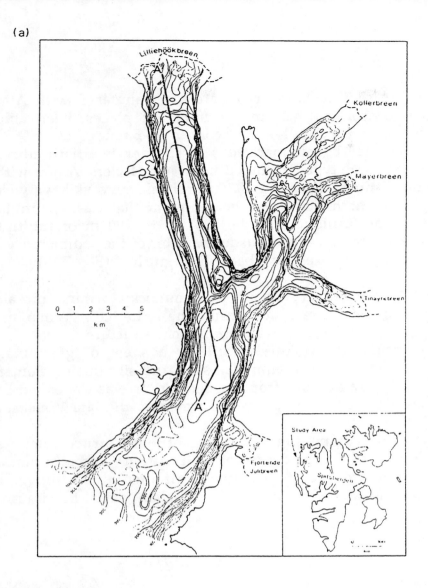

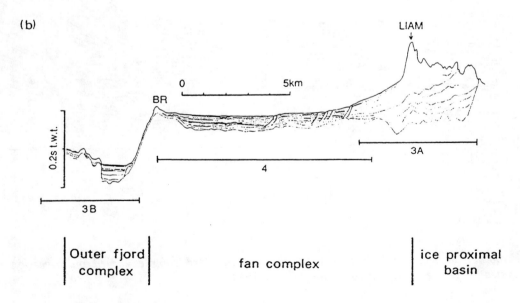

Fig. 7.19. Krossfjorden, NE Svalbard. (a) Bathymetric map (20 m interval contours), showing multiple basins and termini of tidewater glaciers influencing sedimentation. (b) Acoustic profile (3.5 kHz record) and interpretation illustrating seismic stratigraphic architecture in Lilliehöökfjorden, the NW arm of Krossfjorden, Spitsbergen. LIAM is the Little Ice Age moraine and BR represents a bedrock ridge and break of slope at the mouth of this arm of Krossfjorden (Sexton et al. 1992).

227

In Arctic fjords, where sedimentation rates are low compared with Alaska, high resolution acoustic profiling has, in a number of places, been able to penetrate to bedrock. As an example, the work of Sexton et al. (1992) in Krossfjorden, Spitsbergen may be summarized. Krossfjorden is a fjord currently influenced mainly by non-surge type, grounded tidewater glaciers. Along with its branches, it has several distinct basins (Fig. 7.19a). Seismic surveys have yielded details of the stratigraphic architecture of the sediment that has accumulated since the fjord system was fashioned by the ice. Up to 180 m of unlithified sediment above bedrock is present. The acoustic character of the sedimentary pile is variable, and three main complexes have been distinguished (Fig. 7.19b):

• Ice-proximal unit, interpreted as subaquatic , hummocky terminal moraines, formed annually during a general recession, inside the Little Ice Age maximum; the material is a fine-grained diamicton with no internal structure.
• Fan complex, comprising structureless till or a sheet of glaciomarine sediment at the base, overlain by well-laminated sand and silt; this glaciomarine fan comprises material that rained out from suspension or was deposited from debris-flows or turbidity currents; the complex formed as the late Weichselian glaciers receded up-fjord.
• Outer fjord complex, comprising a blanket of both structureless and laminated material (depending on the degree of bioturbation), and a hummocky till above bedrock, similar to the ice-proximal unit mentioned above; this complex formed during and following the late Weichselian glacial maximum.

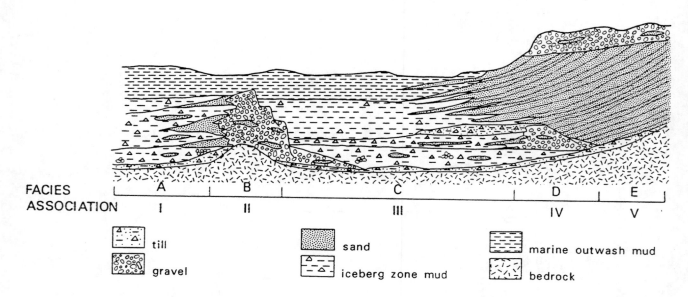

Fig. 7.20. Hypothetical facies architecture in a glaciomarine-fjord complex based on investigations of Alaska tidewater glaciers (simplified from Powell 1981).

Surveys of some Baffin Island fjords confirm the broad pattern found in Spitsbergen fjord basins of a glaciomarine mud-draped till, but in addition illustrate the importance of resedimentation and Coriolis effects in piling up masses of sediment to one side of the fjord (Syvitski 1989, Boulton 1990). In Alaskan fjords, the thickness and internal structure of sedimentary sequences have also been determined using high resolution profiling methods. However, the sedimentary facies that make up the several hundred metre thick sequences are not known directly, but Powell (1981) has developed a model of an hypothetical sedimentary sequence (Fig. 7.20) that may arise from the progressive development of the five facies associations he defined (Section 7.6.1), and which illustrates at least a little of the complexity of the stratigraphic architecture that may develop in a fjord.

CHAPTER 8

GLACIOMARINE PROCESSES
AND SEDIMENTS

PART II. CONTINENTAL SHELVES
AND DEEP SEA

8.1. Introduction

High-latitude continental-shelf areas under the influence of glacigenic processes are particularly sensitive to changes in climate . Ice sheets which border these continental shelves, not only control sea-level, but also generate cold bottom water that influences the oceans throughout the world. The problems and cost of access to the continental shelf glaciomarine environment - represented first and foremost by Antarctic ice shelves, outlet glaciers, ice tongues and ice cliffs - have so far limited the acquisition of knowledge concerning these areas. Nevertheless, a considerable body of data has been obtained during the past decade, and reviews have stressed the importance today of glaciomarine processes on high-latitude continental shelves (e.g. Andrews & Matsch 1983, Dowdeswell 1987, Drewry 1986, Hambrey et al. 1992).

Ten per cent or possibly much more of the world´s oceans and continental shelves receive glaciomarine sediment today (Drewry 1986). Around Antarctica, icebergs influence sedimentation, albeit to a volumetrically minor degree, up to a distance of several hundred kilometres from the continent (Fig. 8.1a). The northern iceberg limit extends to the southern tips of South America, South Africa and New Zealand, and there were reports during the 19th century of icebergs creating a hazard to shipping around Cape Horn and the Cape of Good Hope. Because of their favoured preservation potential, continental shelf sediments are the most common of pre-Pleistocene glacigenic sequences, and many detailed accounts have been published in recent years.

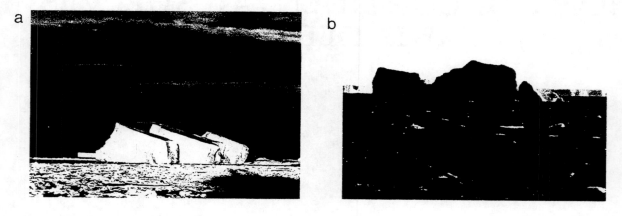

Fig. 8.1. Decaying icebergs off Dronning Maud Land, East Antarctica: (a) Tilted, tabular iceberg grounded on the outer continental shelf. (b) Iceberg in advanced stages of decay close to the coast. Note the prominent debris-band, probably originally incorporated at the base of the ice-mass, and the ice shelf in the background.

One of the earliest and most stimulating attempts at reconstructing the glaciomarine environment, typified by Antarctica and the margins of the Pleistocene ice sheets of the Northern Hemisphere, was that of Carey & Ahmad (1961). The model developed by these authors was widely used, but was founded on few field data and it seriously overstressed the rôle of ice shelves as providers of sediment to the marine environment. Recent reconstructions are founded on a much better database, although this is still inadequate in view of the wide range of glaciological conditions that occur at the margins of marine ice-masses. Relatively poorly known are the modes of debris transport by ice to the continental shelf, the thermal regime of the transporting glaciers, the rates of sediment supply, the importance of meltwater, the rôle of water depth and ocean currents, the character of sediments deposited beneath floating ice and their modification by marine processes, and the rôle of sea ice. Nevertheless, we now have a good idea of the range of processes that influence sedimentation in these settings, and they form the basis of this chapter.

8.2. Transport of glacigenic sediment out of the ice-sheet system

8.2.1. Nature of sediments arriving at the grounding zone

The volume and distribution of sediment within ice-masses are largely controlled by ice dynamics and thermal regime (see Drewry & Cooper 1981, Drewry 1986, Dowdeswell 1987 for more detailed reviews). As noted in Chapter 7, the thickness of the basal debris-rich zone varies according to glacier type. This debris zone is derived as a result of erosion at the base of a sliding ice-mass and entrainment by regelation of dirty ice. Ice-masses which are predominantly cold tend to have thicker basal debris layers than do temperate glaciers; commonly this layer is 10 m or more thick where terminating on land, but it is not known for certain whether similar thicknesses prevail where ice-masses enter the sea. From coring and radio-echo sounding of ice sheets, the thickness of the basal layer tends to be of the order of 1% of the total thickness, so some Antarctic glaciers may have as much as 100 m of basal debris when they reach the grounding-line. Sediment-rich ice occurs in discrete bands (Fig. 8.1b), but the concentration is variable, ranging from 0.01% to 70%, or about 1% on average, showing an irregular but approximately exponential decrease upwards. This distribution and density pattern is important from the point of view of determining the rate of sedimentation once the ice-mass begins to float.

Other sediment sources, volumetrically, are relatively minor around most ice margins that border continental shelf areas. Exceptions are the few instances in Alaska where glaciers terminate in the open sea; here supraglacial debris is also important.

8.2.2. Ice-ocean sediment transfer routes

Excluding lodgement, meltout and shearing of the debris layer beneath the grounded ice sheet, debris is transferred out of the Antarctic ice-sheet system in three ways (Drewry & Cooper 1981).

Via ice shelves. On entering deep water, grounded ice on a broad front decouples from the bed and floats out over the sea, where it undergoes creep-thinning as a result of acceleration following the reduction of basal and lateral drag (Fig. 8.2). For sedimentation, it is important to consider mass balance and particle paths. Most ice shelves receive considerable snow accumulation, so particle paths are inclined *downwards* as the accumulation increases seawards. If there is net basal melt, particle paths intersect the bottom of the ice shelf, and any debris in transit in the basal layer is likely to be melted out prior to calving at the ice front. So sedimentation today is important at the inner margins of a continental shelf near the grounding-line, but of little significance on the outer shelf and in open

233

ocean. However, freeze-on of oceanic ice beneath some ice shelves is important. Beginning close to the grounding-line, freeze-on may prevent all of the basal and englacial debris from being released. The remaining debris is therefore transferred to the ice shelf edge and deposited in the open sea from icebergs.

Fig. 8.2. *Embayment in the Brunt Ice Shelf, off Halley Station, East Antarctica. RRS Bransfield anchored against fast ice.*

Via ice cliffs. Where grounded Antarctic ice calves directly into the sea from an ice margin that is grounded in relatively shallow water, ice cliffs form (Fig. 8.3). Wave action at the foot of the cliffs and the slow rate of advance of the cliffs promote sedimentation close to the ice edge prior to calving. In the Arctic, extensive ice cliffs are found only on Nordaustlandet in the Svalbard group of islands.

Fig. 8.3. *Grounded ice cliff near Barnes Glacier, Ross Island, Antarctica, with winter fast-ice at its foot.*

Outlet glaciers and ice streams. Inland ice draining towards the coast is channelled into ice streams or outlet glaciers, the latter being constrained by rock walls. Such glaciers generally have a high rate of discharge and may extend out into the sea as ice tongues. Observations of calving icebergs from such glaciers show that they retain a basal debris layer up to 15 m thick, with variable amounts of supraglacial debris. The sediment-release process is similar to that under ice shelves, but in addition they produce an almost continuous supply of small, debris-rich icebergs. The bulk of the glacigenic sediment in the Southern Ocean today probably comes from outlet glaciers.

	Debris type		
Glaciological regime	**Basal**	**Englacial**	**Supraglacial**
Outlet glaciers			
via mountains	• • • •	• •	• •
ice stream	• • • •	•	•
Ice cliffs			
near mountains	• • •	• •	• •
ice sheet edge	• •	•	•
Ice shelf			
from ice sheet	• •	•	•
via mountains	• • •	• •	•

Sediment quantity:

trace	small	moderate	substantial
•	• •	• • •	• • • •

Table 8.1 Variations in the location and relative proportions of different types of entrained sediment, according to type of ice-mass (after Drewry 1986).

The relative importance today of these three modes of transfer of ice from Antarctica have been estimated to be as follows (Drewry & Cooper 1981):

Ice shelves	62%
Ice cliffs	16%
Outlet glaciers and ice streams	22%

However, from a sedimentological point-of-view, there are significant differences in the locations and volume of sediment produced by these different glacier types (Table 8.1).

Ice shelves and outlet glaciers respond rapidly to oceanographic and climatic changes. Their grounding-lines thus migrate frequently across the continental shelf, producing thick complexes of strongly diachronous sediment (Drewry & Cooper 1981).

(a) Ice shelf in recessed state

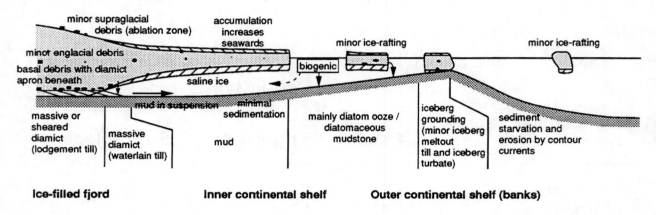

(b) Ice shelf advanced to edge of continental shelf

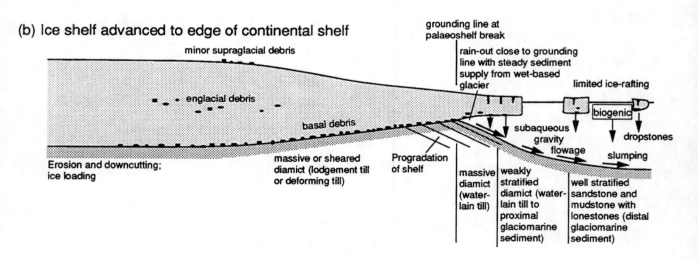

Fig. 8.4. Ice dynamics, sediment sources and sedimentary processes, products and their interpretation at the margin of the Antarctica ice sheet. (a) Ice shelf with grounding-line at the inner part of the continental shelf, as at the present day. (b) Ice shelf having grounded on the continental shelf and advanced across it to the continental shelf break, where it becomes decoupled from the bed (glacial maxima).

8.3. Processes controlling sedimentation on glacier-influenced continental shelves

The range of sediment transport paths, processes of deposition and resulting sedimentary facies are summarized in Figure 8.4 for an ice shelf in recessed and advanced conditions.

8.3.1. Delivery of sediment to the glaciomarine environment

Ice dynamics

As much as a half of the Antarctic ice sheet was considered by Zotikov (1986) to be at the pressure melting point at its bed, a view supported by clear evidence from radio-echo sounding echo of subglacial lakes beneath the thicker parts of the ice sheet (Oswald & Robin 1973). Under these conditions, basal sliding is bound to generate a debris layer. Although this basal debris layer is normally only visible in a few places around the margin of the continent (Drewry 1986), it is believed to be widespread in the interior of Antarctica. Deposition of the bulk of this sediment in the marine environment is by subglacial melting on the continental shelf beneath an extended grounded ice sheet or close to its grounding-line. Melting of icebergs provides a more limited source of debris in the open ocean.

Another mechanism of transport of basal debris to the marine environment has been identified beneath ice streams hundreds of metres thick draining the West Antarctic ice sheet and feeding the Ross Ice Shelf (Shabtaie & Bentley, 1987; Alley et al., 1989). Both theory and seismic velocity data support the view that the fast-flowing ice streams are moving over a water-saturated bed of till (or perhaps previously deposited glaciomarine sediment) and deforming it, thereby transporting it seawards. Transport of this material ends as the ice stream merges with the ice shelf and achieves buoyancy, but at the same location melting of the underside of the ice stream by currents beneath the ice shelf releases additional basal debris. These processes of debris transport and release are followed by gravity flow and slump processes to create, beneath the inner margin of the ice shelf, a prograding complex of glacially derived sediment. Alley et al. (1989) described the resulting feature as a *till delta*, although the non-genetic term *diamict apron* is preferred here. This type of prograding wedge, but on a much larger scale, has been recognized in seismic records from the Antarctic shelf for some time (Hinz & Block 1983), and is now considered by some to be characteristic of ice sheet deposition (Cooper et al. 1991). It is conceivable that this process of diamict apron formation was the primary mode of continental shelf progradation when glaciers reached the shelf break, as well as being, on a smaller scale, the main mechanism for deposition on the inner shelves today (Hambrey et al. 1992).

Debris can also be transported into the marine realm after falling or being blown onto the glacier surface. Rock debris and wind-blown sand are widely observed on the surface of outlet glaciers passing through the Transantarctic Mountains in southern Victoria Land (Barrett et al. 1983), and are probably equally common in other localities where ice passes through coastal ranges or where nunataks are exposed. In only a few places, however, is supraglacial debris sufficiently abundant to be evident in aerial photographs or satellite images, and it is probably unimportant as a proportion of the total sediment, except where the sediment supply from other sources is exceptionally low. However, aeolian sediment from these sources can be spread as far as the northern limit of icebergs because it is carried on top of, and within the ice. Such debris can be readily recognized from its well sorted fine-grained nature (aeolian sand) and angular shape (scree).

In the Arctic, rockfall and aeolian debris on icebergs derived from glaciers and ice caps terminating at the open coast (e.g. northeastern Svalbard) are of minor importance. Only in Alaska, where a few glaciers approach the open coast, is supraglacial transport locally important. However, during Pleistocene glacial periods, when ice descended from the lofty mountains to the open coast on a broad front, the transfer of supraglacial sediment was highly significant, not only in Alaska, as demonstrated by very thick late Cenozoic sequences that have been tectonically uplifted, but also off northwest Europe, British Columbia and the southern Andes.

In summary, the main source of glacial debris for the open marine glacial regime in Antarctica is attributed to outlet glaciers and ice streams (Drewry & Cooper 1981, Orheim & Elverhøi 1981). However, most of this debris is probably deposited close to the grounding-line, from which position it may be further distributed (and modifed) by gravity flows or currents. Debris deposited from icebergs is volumetrically minor compared with that originating from other sources. This, too, has the attributes of basal glacial debris. Such debris may occur far from the source if the iceberg turns over before the basal debris has melted out. Ice-rafted debris in the stratigraphic record can provide important evidence of the existence of floating ice at sea level.

Glacial meltwater input.
Sediment in suspension and bedload can enter the marine environment directly from supraglacial and subglacial meltwater channels in tidewater glaciers, floating ice tongues and ice shelves, or from terrestrial streams. In the Antarctic the latter are small and rare, with maximum discharges of a few cubic metres per second. According to Zotikov (1986), subglacial meltwater is generated at a rate of several cubic km/year beneath the Antarctic ice sheet and its outlet glaciers. However, this volume is insufficient or the water inadequately channelized to generate visible turbid sediment plumes in surface waters around Antarctica. Some subglacial transport of fine-grained sediment is suspected from the abundance of terrigenous sediment in deep-water mud accumulating around the continent today. In contrast, meltwater and associated suspended sediments have been observed immediately seawards of some glaciers at the northern tip of the Antarctic Peninsula, where temperatures are significantly higher (Griffith & Anderson 1989).

Glacial meltwater is much more significant in the Northern Hemisphere, where cold ice fronts are grounded on the continental shelf. For example, the ice caps on Nordaustlandet in Svalbard, which together provide the longest continuous sea-frontage of any Northern Hemisphere ice-mass (120 km), reveal a number of discrete subaquatic stream outlets, as indicated by sediment plumes.

Sea ice and river ice
Terrigenous debris entrained into sea ice occurs in littoral environments by freeze-on, as described in Chapter 7. This process is facilitated if extensive shallow areas exist, as is the case in the Arctic, but in the Antarctic such

238

conditions are relatively rare. Ice-rafting of littoral debris that becomes anchored to the bottom of sea ice, and of fine sediment that may be incorporated in sea ice when turbid water freezes, is of major importance in the Arctic Ocean and surrounding seas, in areas far removed from any glaciers. Indeed, muds derived from sea ice are thought to form about 80% of the late Cenozoic sediments in the central Arctic Ocean (Clark & Hanson 1983).

Sea ice also plays a rôle in the sedimentation of diatom ooze in Antarctica. Diatom blooms occur beneath the sea ice as it breaks up in summer, but the diatoms become trapped in the ice when new platelet ice forms beneath the floes. Often the ice floes are strongly coloured, giving rise to so-called *brown ice*.

Arctic rivers discharge large volumes of ice into the ocean during the spring break-up, allowing river gravels and sands to be ice-rafted off shore. To distinguish these types of ice-rafting sediment from iceberg sediment is of major importance in assessing the palaeoclimatic significance of dropstones in sediments, for example around the Arctic Basin.

Sea and river ice-rafted debris has been reported from a number of pre-Pleistocene sequences, even for times when there is no other evidence of glaciation, e.g. the Jurassic and Cretaceous of the Canadian Arctic, Spitsbergen, Siberia and Australia (Frakes & Francis 1988).

The transport of aeolian sediment onto sea ice and its subsequent release into the sea is an important processes in some polar areas; depositional rates of up to 1 mm yr^{-1} have been recorded in McMurdo Sound, offshore from the Dry Valleys in Antarctica.

Biogenic material

The polar seas are biologically rich in those areas where the sea ice breaks up in summer. Upwelling of nutrient-rich bottom waters and low clastic sediment input, facilitates the production of a diverse fauna and flora. Siliceous ooze is the principal component around Antarctica, where a zone several hundred kilometres wide, made up largely of diatom fragments with a minor ice-rafted terrigenous component, rings the continent. To the north lies calcareous ooze comprising coccoliths and foraminifera, while the transitional zone coincides with the boundary between cold Antarctic waters and the warmer oceans to the north (the zone being referred to as the Antarctic Convergence). The importance of diatomaceous sediment has been also been recognized near the Antarctic coast. For example, Domack (1988) found a predominance of diatom ooze just outside the fjords of the Antarctic Peninsula, as did Dunbar et al. (1989) in McMurdo Sound. The dilution of the ooze by terrigenous sediment depends on the proximity of outlet glaciers. On the other hand, areas such as the inner Weddell Sea which have a more permanent and relatively dense sea-ice cover, have little diatomaceous sediment. Some diamictites dating back to Oligocene time, recovered from the deep drill holes, also contain significant amounts of diatoms (Barrett 1989; Barron, Larsen & Shipboard Scientific Party 1989). Nearly pure oozes with only minor ice-rafted material are present in many places near the Antarctic coast. The distribution of siliceous biogenic material on the inner Antarctic shelf is strongly influenced by marine currents, and transport is generally in a westerly direction.

The distribution of diatomaceous sediments in the Arctic is less clearly defined, and to a large extent reflects the permanancy of the sea ice. For example, the relatively ice-free coasts of western Svalbard have abundant diatomaceous sediment, whereas the Arctic Ocean, with its near-complete sea-ice cover, does not.

Other biogenic components (e.g. foraminifera, calcareous nannofossils, radiolaria, bryozoa, siliceous sponges, invertebrate fossils) form only a fraction of a per cent of the total sediment, but their abundance along with diatoms demonstrates the high productivity of waters in polar regions.

8.3.2 Ice-ocean interactions

The nature of the interaction between glaciers and the sea on continental margins is relatively poorly known. In particular, the processes operating just seawards of the grounding-line, where tidal pumping may be important, and beneath floating ice-masses, are poorly known, and glacial sedimentologists await the development of new technology that will allow investigation of these inaccessible areas by remotely operated vehicles or drilling through the ice. The most important ice--ocean interactions have been outlined by Drewry (1986) and Dowdeswell (1987), and are summarized as follows.

Rates of ice cliff melting
The rate of debris production from grounded ice cliffs is dependent on the ice velocity, the temperature difference between the water mass and the ice front, the velocity of any current, and the length of contact between the ice front and the water mass. For ice margins terminating in shallow water, surface warming of the sea in summer accounts for the loss of significant amounts of ice.

Melting and freezing at the base of ice shelves
The thermal regime at the base of an ice shelf is important to glaciomarine sedimentation because it controls whether all or most of the basal debris is released at the grounding-line through melting, or whether some of it is transferred to the open ocean by being protected by the freezing-on of a layer of oceanic ice. Ice floating in water is prone to melting because heat is transferred from the warmer liquid into the ice. The process beneath an ice shelf is facilitated by water circulation in the form of tidal currents and meltwater plumes. The geometry of the underside of the ice also helps, since this slopes downwards towards the land; ice shelves typically thin from around 1000 m near the grounding-line to 200--500 m at the seaward limit. The most vigorous circulation is near the grounding-line, thought to be in the form of convective turbulent flow linked to ´tidal breathing´, so it is here that most debris melts out.

In certain circumstances, basal freeze-on occurs. If deep-flowing saline water close to the freezing point rises as the grounding-line is approached, its freezing point is raised due to lowering of pressure, and may freeze as it makes contact with the underside of the ice shelf. Towards the outer part of the ice shelf, however, currents advect warmer water from beyond the ice shelf and undermelt occurs. Therefore, the basal freezing process is most active within a short distance of the grounding-line.

From observations on the Ross and Amery ice shelves, it is known that sedimentation is highly sensitive to the rate of melting. Thus ice shelves represent barriers to the movement of basal glacial sediment from the continent offshore. Basal freeze-on is known to take place beneath a number of ice shelves, including the Ross Ice Shelf (Zotikov 1980), the Amery Ice Shelf (Budd et al. 1982) and the Filchner-Ronne Ice Shelf (Thyssen 1988). These freeze-on zones may be over a 100 m thick, and represent as much as a third of the total ice shelf thickness (Fig. 8.4a). However, ice cores have demonstrated an absence of rock debris at the boundary between glacier and oceanic ice in cores through both the Amery Ice Shelf (Site G-1, Morgan, 1972) and the Ross Ice Shelf (Site J-9, Zotikov et al. 1980), supporting the view that the bulk of the basal debris has indeed been lost near the grounding-line.

Iceberg calving

Changes in oceanographic conditions are often the principal factors affecting long-term iceberg production from floating glacier tongues and ice shelves. A rising sea level causes the grounding-line to lift and move inland and the ice-mass to become unstable. Increased iceberg production may therefore herald a time of marked climatic warming and rapid recession of the ice shelf, a point that has to be taken into account when interpreting the climatic record from sediments with an ice-rafted component. An illustration of this is that the climatic warming in the Antarctic Peninsula this century is leading to the rapid break-up of the more maritime ice shelves and a huge increase in iceberg production. On the other hand, ice shelf and outlet glacier calving may not be related to oceanographic or climatic changes at all. Many ice shelves tend to advance at rates up to 1 km/yr over tens or even hundred of years before calving huge icebergs are calved. Recent examples include tabular icebergs more than 100 km long calved from the Filchner and Ross ice shelves.

Grounded glaciers also respond to oceanographic changes, but are more closely influenced by climatic factors and mass balance. Icebergs in such cases are usually small, tens of metres across and irregular in shape, usually spalling off the cliff face, rather than producing tabular icebergs. These glaciers also increase their iceberg production during rapid recession. Temperate glaciers that once extended across the shelf of the Gulf of Alaska often behaved in this manner during the Pleistocene Epoch.

Iceberg melting rates, fragmentation and overturning

The rate of sediment release by iceberg melting depends on the melting rate, the distribution of sediment within the icebergs, and the number of overturning and fragmentation events. Debris melting out on the surface of an iceberg may remain there until an overturning event occurs, or it flows off as a slurry or is washed off by meltwater or rain, processes which cause a pulse in sedimentation. In contrast, debris on the steep sides and base of an iceberg may melt out steadiliy. These processes are summarized in Figure 8.5, which documents the loss of debris from a hypothetical iceberg.

A tabular iceberg is relatively stable if its width exceeds its thickness, but melting rounds off the sharp edges, while waves undercut the sides at the waterline making the cliffs prone to collapse and produce bergy bits; as a result the iceberg becomes unstable. In an investigation of more than 2000 icebergs in the eastern Weddell Sea, Orheim (1980) found that about 100-150 km offshore only 15% had not overturned.

Rates of melting of icebergs have not been studied directly, but estimates of 10-50 m/yr have been made for clean Antarctic icebergs by Budd et al. (1980). Dirty ice melts more slowly than clean ice, but nevertheless a dirty ice-layer at the base of an Antarctic iceberg will be lost within a few years. However, any internal debris layers may survive until the iceberg finally disintegrates; normally, this is south of the Antarctic Convergence (50°-60°S), but a few icebergs drift into the warmer waters beyond.

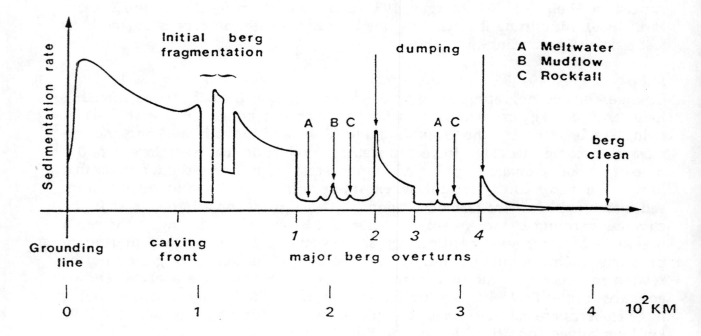

Fig. 8.5. Pattern of debris release from an iceberg derived from an ice stream, and resulting sea-floor sedimentation-rate. Note the various effects of calving and overturning of icebergs, the contributions from meltwater, mudflow and rockfall activity (Drewry 1986).

Iceberg drift paths
The general pattern of iceberg drift is controlled by ocean currents. Thus icebergs can move in a totally different direction to sea ice, which is more influenced by the vagaries of the wind. Around Antarctica, evidence from satellite imagery indicates that icebergs within 100-200 km of the coast tend to move in a westerly direction in the Eastwind Drift, so remaining close to the continent for up to five years. Thus most debris is released in this coastal girdle. Those icebergs which move further offshore come under the influence of the Westwind Drift and move back eastwards. The Antarctic Convergence represents an effective northern limit for most icebergs. The average speed of Antarctic icebergs, over a year or more, ranges from 0.11-0.2 m/sec.

Northern Hemisphere icebergs are insignificant in size compared with their Antarctic counterparts. Most are trapped in fjords, but a number of distinct iceberg routes are known. The East Greenland Current flows south from the Arctic Basin and receives icebergs from major calving glaciers in northeastern Greenland. Most icebergs become caught up in the northbound West Greenland Current and are transported into Baffin Bay. Here they mix with icebergs that calve from the large outlet glaciers of West Greenland and the small glaciers in the Canadian Arctic. A current on the west side of Baffin Bay then carries them southwards through the Davis Strait into the North Atlantic. Some reach the Grand Banks off Newfoundland, and one sank the Titanic in 1912 with a loss of 1503 lives. Small icebergs from Svalbard and the Soviet Arctic islands find their way into the Arctic Basin, while larger icebergs are occasionally produced by the Ward Hunt Ice Shelf in northern Ellesmere Island. One such iceberg (ice island

242

T3) has been tracked since 1947, and has followed an irregular, but roughly clockwise spiralling track through the Arctic Ocean.

8.3.3. Oceanographic processes

Oceanographic processes include those that are not just restricted to glaciomarine environments. However, the interaction of glacier ice and sea water influences the salinity and temperature, and thereby controls the location, rate of deposition and character of glaciomarine sediments. A fuller discussion for the Antarctic is given in Dunbar et al. (1985), but the links between oceanography and sedimentology are not well known at present.

A number of processes may be considered to be particularly important in the glaciomarine environment (see Dowdeswell 1987 for further information):

Meltwater inflow and water stratification
In the Arctic or subarctic, water stratification arises when meltwater from the land or glaciers enters the sea, but on the open coast, except when the winter sea-ice cover has yet to melt, currents and wind tend to mix the water. In the Antarctic there are only a few signs of meltwater flowing into the sea.

Sediment suspended in meltwater
Sediment plumes are generated where a subglacial stream discharges into the sea. Since a significant proportion of the Antarctic ice sheet is sliding on its bed, suspended sediment derived from meltwater must be inferred at grounding-lines where ice streams and outlet glaciers reach the sea and begin to float as glacier tongues or ice shelves. Sediments with abundant terrigenous mud in proximal and distal glaciomarine settings may be derived from this source. However, the process itself has not been documented.

Tides
Tides are important at grounded ice cliffs for similar reasons as in fjords (Chapter 7), although they show a less pronounced imprint on the sediment. Additionally, tides control circulation beneath floating ice-masses, facilitating melting at the grounding-line and flushing out of suspended sediment.

The effects of sea ice on circulation
When sea ice forms, it leaves the underlying water preferentially enriched in brine. This brine-rich water forms a layer beneath the ice, but because it is more dense it becomes unstable. This leads to overturning and mixing of the water layers. During ice break-up, less dense fresh water is generated, inhibiting mixing and producing a relatively stable layer. Sea ice also minimises the effect of wind on the sea, and also inhibits mixing.

In Antarctica, brine-enriched waters accumulate on the western sides of the wider ice shelves and in depressions in the coastline. Together with surface winds, the density contrast leads to a clockwise circulation in large embayments.

Depth of water
The physical character of the Arctic and Antarctic continental shelves differ greatly. The northern shelves are typical of most shelves in more temperate

latitudes in being relatively shallow and affected by a wide range of coastal processes. The Antarctic continental shelf, however, is unique in having great depth (often more than 500 m), rugged bathymetry, landward tilt and almost total glacier-ice cover at its rim. These characteristics are the result of crustal (*isostatic*) loading by the thousands of metres of ice on the continent. The near-absence of ice-free land and the presence of deep water immediately offshore in the higher polar areas of Antarctica mean that there is no wave-dominated coastal zone. Thus beaches are uncommon and the rôle of sea ice in transporting sediment is much less important than in the Arctic or sub-Antarctic.

Bottom currents
Near the Antarctic continental shelf break, some of the denser shelf water flows along and down the slopes, mixing with and modifying the deep water or producing *Antarctic Bottom Water*. In some areas, such as the cold Weddell Sea, the Antarctic Bottom Water is a powerful current, winnowing and in places deeply eroding the sea floor sediments, as well as having a major influence on global ocean circulation.

In the Arctic, upwelling of warmer deep waters leads to the creation of *polynyas*, areas of relatively ice-free ocean in winter, such as the Northwater located between Ellesmere Island and North Greenland.

8.3.4. Processes of reworking

Modification of glacigenic sediments by processes characteristic of all continental margins may be a significant factor in explaining particular facies distributions. Especially important are the following:

Current activity
Bottom currents can redistribute the finer fraction (sand and mud) of glaciomarine sediments, especially if it has been made less cohesive by bioturbation, leaving behind a lag deposit composed principally of gravel. Other bed forms, such as current ripples, or even megaripples and sand waves form in shallow depths (as on Arctic shelves) where currents are strong, and wave-induced structures also occur. Currents also transport biogenic material into zones incapable of supporting life on their own (e.g. beneath the innermost parts of ice shelves).

Subaquatic mass-movement
Reworking by subaquatic mass-movement of sediment on unstable slopes includes a continuous spectrum of processes according to the amount of deformation undergone by the sediment as it moves:

• *Sliding*: displacement of the sediment mass along a slip plane, but accompanied by little internal deformation.
• *Slumping*: displacement involving internal folding but not disaggregation of the sediment mass.
• *Debris flowage*: movement of a mass of sediment as a slurry involving total reorganisation and mixing of the sediment particles, ripping up clasts from the substrate.

244

• *Turbidity flowage*: sediment carried in suspension, settling out to produce graded beds.

Subaquatic mass-movement is considered to be a common process on the Antarctic continental shelf, which is relatively rugged and has slopes up to 15° (Wright & Anderson, 1982, Wright et al. 1983). Unsorted ice-rafted debris may be transformed into sorted turbiditic sands in this way.

Mass-movement is probably even more important on the Antarctic continental slope (although less steep), and may be intimately associated with rain-out of glacial debris when the grounding-line reaches the shelf break. There is also seismic evidence, in at least one place (Prydz Bay) to suggest that a considerable width (c. 10 km) and depth (c. 300 m) of the Antarctic shelf, probably composed of diamict has suffered a rotational slide (Hambrey et al. 1991).

In comparison, ice-influenced Northern Hemisphere shelves are relatively subdued, and mass-movement processes are mainly limited to the continental slope, the troughs which extend across the shelf, and deltas bordering the coast. Continental slope slumping is well seen in seismic profile across the northwest British (Stoker 1990) and Barents Shelf margins (Boulton 1990, Vorren et al. 1989, 1990).

Fig. 8.6. Tabular icebergs stranded on the eastern Weddell Sea continental shelf, illustrating the extent to which sea-floor sediment may be subject to iceberg scour. Coats Land and the East Antarctic ice sheet are in the background.

Scour by icebergs

The importance of iceberg scouring (or ploughing or gouging) on both northern and southern high latitude shelves has recently been documented, following the deployment of geophysical techniques, such as side-scan sonar and bottom topography. Scour takes place when moving icebergs come into contact with the bed. In Antarctica many tabular icebergs have a draught of 500 m and cause considerable disruption of the sediment if they become grounded (Fig. 8.6). Northern Hemisphere icebergs are much smaller, but here the continental shelves are shallower. Iceberg scouring and sub-surface deformation are potentially serious hazards to bed installations, such as hydrocarbon platforms, pipelines and cables.

Scour by sea ice
In contrast to icebergs, sea ice scouring is confined to shallow water (Drewry 1986). Multi-year ice may reach an average thickness of 3 m, but the inverse sides of pressure ridges can scour the sea bed to depths of 30 m.

Bioturbation
Reworking of sediment by benthic organisms is common in both proximal and distal glaciomarine sediments. For example, homogenization of once stratified diamicton or mudstone may be a major process, and can create confusion concerning interpretation of the depositional process. Bioturbation commonly is indicated by mottling of the sediment, and by distinct burrows. Much of the structureless sandy mudstone and diamictite in cores from McMurdo Sound has been interpreted as once stratified sediment that has been bioturbated; in this case some shelly fossils of the organisms that caused the bioturbation are visible.

Compaction and recycling of sediment by ice readvances
As a grounded ice-mass advances across the continental shelf the sediment becomes overpressurized. Water may be lost and the sediment becomes stiff (geotechnically *overconsolidated*), i.e. having a greater shear strength and bulk density than might be expected for its depth. Many polar shelf areas are covered by overconsolidated Pleistocene diamicts, indicating deposition from grounded ice, for example the Barents Shelf (Solheim et al. 1990) and Antarctica (Elverhøi 1984). Sometimes, it is previously deposited glaciomarine sediment that has been loaded. In thick glacial sequences, like that in Prydz Bay, Antarctica several overconsolidated horizons of diamictite are underlain by geotechnically and seismically defined unconformities.

Each new ice advance is potentially capable of incorporating older material. In particular, recycling of biogenic material may result in incorrect biostratigraphic sediment zonation and correlation of the sedimentary sequences. However, on a time span of tens of millions of years this does not appear to have been a significant problem for dating either the McMurdo Sound or Prydz Bay cores.

8.4 Physical, chemical and mineralogical characteristics

8.4.1 Sediment texture

Sediment texture is usually expressed in terms of grain size distribution, a parameter which in glaciomarine environments depends on:

(a) the size distribution of the source sediment, such as in basal, englacial or supraglacial debris, or on subaquatic glaciofluvial discharge or sedimentary gravity flowage;

(b) the energy level of the depositional environment as demonstrated by wave-grading and seaward fining of shallow marine settings.

Several studies have been made of the sand and finer grain size distribution on high latitude continental shelves, and a selection of distributions for a variety of facies are illustrated in Figure 8.7. The poorly sorted nature of most of these facies is a reflection of the source area, and a common origin from basal debris-rich ice for most of the sandy mud and muddy sand facies. In the drill-core CIROS-1, reworking processes are indicated by the massive sand facies, which is a sedimentary gravity flow. Similarly, massive sand from the continental slope of Prydz Bay represents a deposit that has been reworked by currents.

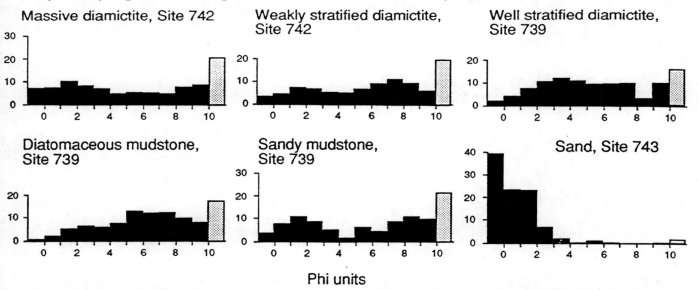

Fig. 8.7. *Grain size distributions for glacigenic sediments recovered from the continental shelf of Antarctica, (a) during Leg 119 of the Ocean Drilling Program to Prydz Bay, East Antarctica (from Hambrey et al. 1991).*

Investigations by Anderson et al. (1980) have shown that plots of mean grain size *versus* sorting allow one to discriminate between basal tills deposited by grounded ice on the shelf, glaciomarine sediments derived from ice rafting, and winnowed glaciomarine sediments (Fig. 8.8).

Compared with unmodified glaciomarine sediments so far collected from Antarctica, diamicts on the Alaskan shelf (Eyles & Lago 1990) are texturally similar, although there is a suggestion that the clay component is more prominent. It is useful to compare the compositions of glacigenic samples from different areas, using ternary diagrams (triangular plots), in order to compare the thermal regime of the glaciers which supplied sediment to the glaciomarine environment (Barrett 1989, Hambrey et al. 1991). Figure 1.3 demonstrates that Prydz Bay diamicts fall approximately between typical middle latitude Pleistocene tills (Denmark) and present clay deposits from polar glaciers in Antarctica. A higher proportion of fines suggests meltwater is an important component of the system. The Prydz Bay, and more especially the CIROS-1 samples suggest meltwater has played an important part during deposition of these predominantly Oligocene sediments. However, care must be taken in interpreting such data, as other factors, for example changing bedrock source material, may also influence the grain size distribution of the sediment.

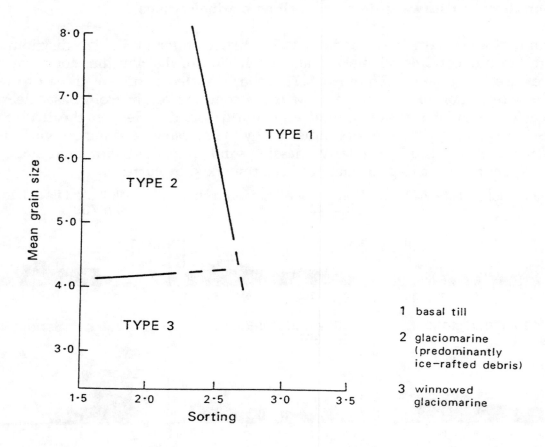

Fig. 8.8. Fields illustrating mean grain size versus sorting for representative types of Antarctic glaciomarine sediment from the Ross, Weddell and Bellingshausen seas and George V Coast. (Summarized from Anderson et al. 1980).

8.4.2 Clast shape

The shapes of gravel clasts in glaciomarine sediments are inherited from those in the transporting medium (cf. Fig. 1.4). Glaciomarine sediments do not show the modified character that arises during deposition by lodgement. Thus the basal debris that reaches the sea contains a mixture of rounded to angular clasts with a strong modes in the subangular and subrounded categories, suggesting basal ice transport. However, some more southerly parts of Antarctica have a high proportion of angular and subangular clasts which, in the absence of any potential supraglacial source, may indicate erosion by subzero ice rather than by sliding (Fig. 8.9).

No systematic differences in clast shape are evident between ice-proximal and ice-distal settings, nor between different glaciological settings (ice shelves, grounded ice walls). Neither does there appear to be any lithological control on the shape of clasts; a comparison of quartzite, granite, schist and metavolcanic rocks shows that each lithology has similar roundness-sphericity distributions (Kuhn et al. 1993).

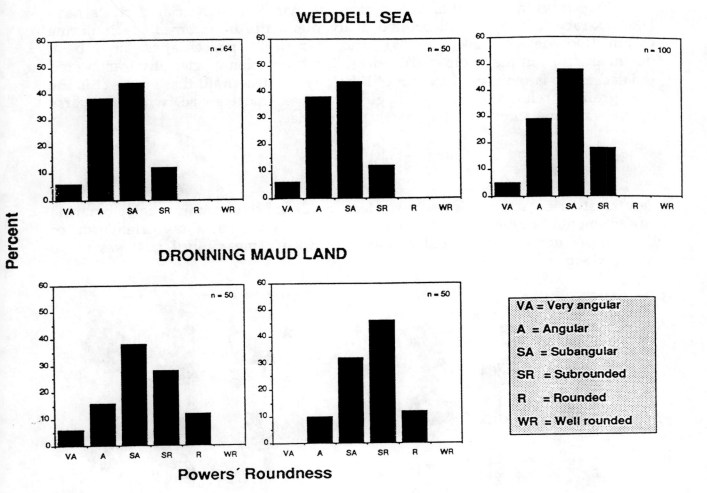

WEDDELL SEA

DRONNING MAUD LAND

VA = Very angular
A = Angular
SA = Subangular
SR = Subrounded
R = Rounded
WR = Well rounded

Powers' Roundness

Fig. 8.9. Roundness characteristics of gravel clasts in sediments from the continental shelf of Antarctica and the deep sea. Note the more angular nature of material from the Weddell Sea compared with that from the shelf off Dronning Maud Land. In the absence of any likely supraglacial source the angularity may be related to fracture of bedrock by ice, accompanied by little, if any, basal sliding.

8.4.3 Surface features on gravel clasts

Stones carried at the base of a sliding glacier acquire a variety of surface features through contact with the bed. The most common feature in glaciomarine sediments is faceting, which at its best developed is evident in the form of flat-iron shapes (two parallel facets). Other clasts may be modified into bullet shapes. The important diagnostic feature of glacial transport, striations, occur on a widely varying proportion of clasts. Domack (1982) recorded striae on 12% of clasts from piston cores off the George V Coast, Antarctica; Barrett (1975) observed that 10% of stones sampled at Deep Sea Drilling Project sites in the Ross Sea were striated; and Hall (1989) recorded that 60% of stones extracted from the CIROS-1 core in

McMundo Sound. In contrast, the Prydz Bay cores yielded only a handful of striated clasts despite over 1000 m of drilling through glacigenic sediment (Barron, Larsen & Shipboard Scientific Party 1989). That surface markings are clearly dependent on lithology is indicated by data from the Lazarev Sea (Kuhn et al. 1993).

In Precambrian sediments in Svalbard, as many as 18% striated clasts have been extracted from a well-bedded shaly diamictite of inferred glaciomarine origin (Dowdeswell et al. 1985), and significant percentages have been documented from many older sequences. The proportion of glacially transported striated clasts is mainly a function of lithology because in all these cases it is the fine grained relatively soft lithologies that bear most striae, whereas coarse igneous and metamorphic clasts rarely carry striae.

8.4.4 Clast fabric

Few clast orientation studies have been made in contemporary glaciomarine environments because coring techniques normally do not allow orientation of the sample nor do they yield enough pebbles for meaningful analyses to be undertaken.

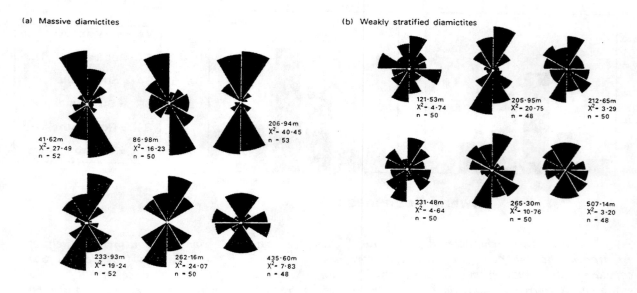

Fig. 8.10. *Typical fabrics plotted as rose diagrams of sand-sized and larger material, from diamictites in the Miocene-Oligocene CIROS-1 drillcore, McMurdo Sound, Antarctica. Boxed figures include depth below sea floor, Chi-squared values (significant at the 95% level for values >11) and number of measurements; the latter is a measure of strength of preferred orientations. First row illustrates massive diamictites of lodgement till (significant preferred orientation) and waterlain till (no preferred orientation). Second row is for weakly stratified diamictites deposited from floating ice (from Hambrey 1989).*

Tests by the author on terrestrial sediments indicate that the orientation of sand-sized grains is not significantly different from that of pebbles and large clasts. Thus grain orientations can be used effectively for aiding the interpretation of a glacigenic sediment recovered from the sea-bed. Ideally, three-dimensional measurements would be preferred to allow thorough statistical analysis, but this is rarely possible, and most measurements to date apart from Domack's have been 2-dimensional.

250

Figure 8.10 illustrates two-dimensional data plotted as rose diagrams from massive and weakly stratified diamictites in the 702 m long CIROS-1 drill-core in the Antarctic. The data are also amenable to Chi-squared statistical treatment in order to assess whether a preferred orientation is apparent at different significance levels. Massive diamictites interpreted as lodgement till deposited on the sediment floor show a strong preferred orientation as do their terrestrial counterparts. Other massive diamictites considered to be waterlain tills have a random fabric, as do most stratified diamictites, as would be expected from settling through water. Random waterlain fabrics may be modified by slumping or currents but these processes have not yet been fully investigated.

Obtaining samples from weathered pre-Pleistocene glaciomarine sequences exposed on land has proved to be a better prospect for obtaining clast fabrics, especially in three dimensions. Such studies confirm the patterns obtained for grains in more recent glaciomarine environments.

8.4.5 Geochemistry

Few systematic geochemical investigations of glacial sediments have been undertaken, and those that have, demonstrate somewhat mixed success in interpreting palaeoenvironments. The difficulty lies in interpreting analyses of sediments in which there are several genetically distinct components (Fairchild et al. 1988). Nevertheless, the potential value of whole-rock geochemistry has been demonstrated by various authors (e.g. Frakes 1975, and Nesbitt & Young 1982, and by Fairchild & Spiro (1990) with respect to the isotopic composition of carbonates in glacial sediments.

Environments of deposition of terrestrial and marine glacigenic deposits are sufficiently varied to lead to differences in sedimentary geochemistry. Although glacial regimes are not characterized by active chemical weathering because of the low temperatures, the proportions of dissolved elements is substantial - rock flour particles are easily attacked because they have a large surface area. On land, oxygenated sediments dominate, whereas subaquatic sediments are more reducing and frequently strongly deficient in oxygen. Thus one would expect elements which are easily precipitated in the presence of oxygen, e.g. iron and manganese, to be preferentially concentrated in terrestrial tills, but depleted in waterlain tills and glaciomarine sediments. This is borne out by the limited amount of data available. Norwegian terrestrial tills have iron abundances of 5-11%, whereas Antarctic glaciomarine sediments 2-5%.

In contrast some metals such as Cu, Pb, Zn and V tend to form solids more readily in reducing environments and, although there are data only for Cu and V, this indeed appears to be true.

Biogenically influenced chemical sedimentation
Following ice recession, the rôle of chemical sedimentation varies considerably. For example, in the late Palaeozoic intracratonic basins of South Africa and Brazil, dark organic and siliceous shales accumulated, as well as in the Precambrian Adelaide geosyncline. Other post-glacial sequences show regionally extensive pink, orange, cream dolostones with, or followed by, stratiform baryte and chert (e.g. the Neoproterozoic of Kimberley in Australia, the Grand and Petit Conglomerate in Zaïre, and in Mauritania and Svalbard).

251

The accumulation of siliceous shales and chert may reflect concentrations of silicon in basin waters as a result of dissolution of this element from silicate minerals in glacial meltwaters. This process is often related to acid volcanism, for example in the Neoproterozoic Tindir Group of Alaska in which thick chemical sediments are associated with glacigenic sediments. The silicon may be released when waters become sufficiently acid or alternatively removed from solution by siliceous micro-organisms.

Stratiform baryte deposits are often associated with sedimentary phosphates and may therefore form by precipitation from seawater, such as during a transgression.

Carbonate isotope geochemistry and cementation
Until recently, most carbonates were interpreted as warm-water deposits, a view that led some authors (e.g. Schermerhorn 1974) to dismiss many supposed Precambrian tillites as non-glacial on the grounds that diamictites associated with carbonates or with a carbonate matrix could not be glacial. However, cold-water carbonates have been found in a number of modern polar marine settings. For example, in the Barents Sea clastic carbonates rest on glacigenic sediments (Bjørlykke et al. 1978) and in the Weddell Sea, bioclastic carbonate sediments form an integral part of glaciomarine sediment (Elverhøi 1984). Much of the sea floor bordering the East Antarctic ice sheet in the eastern Weddell and Lazarev Seas was found during a cruise of the German icebreaker FS *Polarstern* in 1991 to be supporting well-developed bryozoan colonies; often these were anchored on ice-rafted debris. However, the analogy made between bioclastic carbonate accumulations in the Barents Sea and carbonates associated with Proterozoic glaciomarine sequences is not necessarily a valid one, because skeletal organisms can precipitate calcite from waters undersaturated for carbonate, whereas cyanobacteria cannot (Fairchild & Spiro 1990).

The suggestion that carbonate precipitation from marine waters could occur when sea ice formed and increased the salinity, and the water subsequently warmed (Carey & Ahmed 1961) has not been substantiated. Nevertheless, a few percent of carbonate occur in many Antarctic glaciomarine sediments today and probably plays a rôle in cementation. Cementation by carbonate is a major feature of the Oligocene CIROS-1 core in McMurdo Sound (Bridle & Robinson 1989) and also occurs in a number of diffuse horizons in the contemporaneous Prydz Bay sequence (Barron, Larsen & Shipboard Scientific Party 1989), even though there are no obvious detrital or biogenic sources for the carbonate. The presence of extensive carbonates in Proterozoic glaciomarine sequences may suggest that the waters were oversaturated, but this hypothesis remains to be tested. The use of carbonate isotopes ^{18}O and ^{13}C in interpreting the palaeoenvironmental signficance of glaciomarine sediments is a promising new line of investigation that has already been proved its value in interpreting Quaternary terrestrial tills and pre-Pleistocene glaciolacustrine sediments (Fairchild & Spiro 1990).

8.4.6 Mineralogy

Determination of the mineral fraction of the silt and larger fraction of glaciomarine sediment can assist establishing its provenance, as it can for terrestrial sediments. Minerals of this size are not substantially altered in the marine environment before they are buried. New minerals may form however, especially in deep ocean basins influenced by ice-rafting where sedimentary rates are low, e.g. manganese oxide coatings have been found on ice-rafted clasts in the South Indian Ocean (Barron, Larsen & Shipboard Scientific Party 1989) and in the Weddell Sea during the 1991 cruise of FS *Polarstern*.

Clay minerology has not often been undertaken on glaciomarine sediments. However, the technique can provide important clues concerning the climate that prevailed on land before and during the input of glacial sediment to the marine environment. Various Antarctic studies, e.g. Anderson et al. (1980) in the Ross Sea, Elverhøi & Roaldset (1984) in the Weddell Sea, Hambrey et al. (1991) in Prydz Bay have indicated that illite, chlorite and smectite are the most common clay minerals, the first two of which are considered typical of marine sediments in high latitudes. Kaolinite is generally the product of warm, moist environments, but was found in large quantities in the early Oligocene part of the glacigenic sequence drilled during Leg 119 of the Ocean Drilling Program in Prydz Bay, reflecting erosion of a kaolinite-rich source rock and deep chemical weathering of the rock prior to glacial transport. The same signal of kaolinite input to the cold waters of the southern Indian Ocean was identified in well-dated cores taken from the Kerguelen submarine plateau (Ehrmann 1991) and the comparison with Prydz Bay provides clear evidence of the onset of large-scale glaciation on East Antarctica.

8.5 Erosional features on the continental shelf and slope

Various erosional phenomena, mostly associated with grounded ice or subglacial meltwater, may be found on continental shelves at the present day. Some are similar to those occurring on land, but others are unique.

8.5.1 Submarine troughs

These are common bordering areas that have, or are being subject to, intense glacial erosion. Narrow troughs are continuations of fjords and essentially are genetically the same. Good examples occur in the NW British, Norwegian and Greenland continental shelves. Broad troughs with gently sloping sides on continental shelves have also been interpreted as the product of glacial erosion, well-developed examples include the Filchner depression (or Crary Trough) in the Weddell Sea, Antarctica which is over 1100m deep in places, up to 200 km wide and extends across the continental shelf for 400 km beyond the present Filchner Ice Shelf limit (as well as a further 600 km beneath the ice shelf). It is closed by a sill at a depth of about 600 km (Fütterer & Melles 1990). On the Barents Shelf, Bjørnøyrenna deepens from 200 m to 500 m towards the shelf edge through a distance of 600 km; its width is 200 km or more (Fig. 8.11).

Troughs, many tens of metres wide, and tens of metres deep detected in high-resolution seismic records in the Ross Sea (Anderson & Bartek 1992), associated with major erosional surfaces have also been interpreted as the product of glacial erosion. They are similar in scale to modern troughs and have been carved by ice streams. Sometimes, an *ice stream boundary ridge* is developed between postulated ice streams, giving rise to a system of ridges and troughs; these may also be accretionary features.

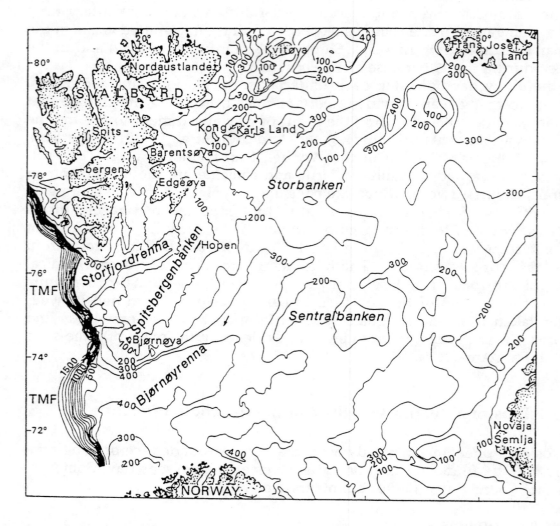

Fig. 8.11. Bathymetry of the Barents Shelf, showing troughs/glaciated valleys (Storfjordrenna and Bjørnøyrenna) and trough-mouth fans (TMF) at the western edge of the shelf (Solheim et al. 1990).

Valleys cutting into the continental slope are also a feature of the Gulf of Alaska coast (Carlson et al. 1990). They are often continuations of fjords and continental shelf troughs. These valleys are erosional features, incised into the lithified strata that underlies the shelf. They have a U-shaped cross-section and concave longitudinal cross-sections that commonly shoal at their seaward limit. Diamicton drapes the walls of the valleys and the upper continental slope beyond. The shelf-valleys end abruptly at the shelf edge.

8.5.2. Tunnel valleys

U-shaped channels, measuring some 3 km wide and 100 m deep have been detected in high resolution seismic records from the Ross Sea continental shelf, Antarctica (Anderson & Bartek 1992). These have been interpreted as large-scale subglacial meltwater channels, analogous to tunnel valleys on land. Similar features, have been identified on the Scotian Shelf off Canada (Boyd et al. 1988).

8.5.3 Iceberg and sea ice scours

The processes that create major erosional features on continental shelves influenced by iceberg drift are discussed in Section 8.3.4. Icebergs have variable keel geometries and the scours reflect this geometry. Most common are multiple, parallel scour channels bordered by levées. In the Northern Hemisphere iceberg scouring occurs as far south as the Labrador Shelf and Grand Banks off Newfoundland. Here, scours reach dimensions of 20-100 m in width and 2-10 m in depth, and one extending for more than 60 km has been reported. Both recent and relict iceberg scours have been described from the shallower parts of the Barents Shelf (Solheim 1991), where they have an average relief of 2-5 m and widths of 20-80 m (Fig. 8.12). Relict scours also occur in shelf areas that were affected by Pleistocene ice-masses west of the British Isles and Norway.

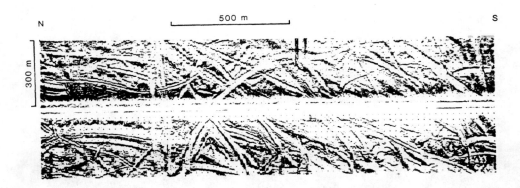

Fig. 8.12. Iceberg ploughmarks on the floor of the Barents Sea in the vicinity of Bråsvellbreen, Nordaustlandet revealed by side-scan sonar techniques (photograph courtesy of Anders Solheim, Solheim 1991).

Wherever the Antarctic shelf has been investigated, numerous scour marks up to 25 m deep and 250 m wides, are visible (Elverhøi 1984, Lien et al. 1989). It is difficult to imagine the deformation that might occur when mega-icebergs (100 km long) become grounded on shallow banks. Not surprisingly, Elverhøi (1984) maintained that in areas of present and former scour activity, the detailed stratigraphy of glaciomarine sequences may be disrupted.

Apart from linear scours, pits formed by wallowing of icebergs and iceberg turbates have been observed from the Grand Banks off Newfoundland. Pleistocene iceberg scour marks have occasionally been reported (though mainly from lakes), but there are very few references to them in the older geological literature, yet they must be present in continental shelf sequences. Woodworth-Lynas & Guigné (1990) has examined contemporary and Pleistocene scours in three dimensions, in order to assess the nature and depth to which deformation takes place beneath the keel.

Various names have been applied to sea ice scours, including plough marks, scores, grooves and furrows. Sprag or jigger marks are formed by uneven movement across the sea floor (reviewed by Drewry 1986). Sea ice scours have a rather different regional distribution compared with that of iceberg scours, and in the Arctic are much more widespread than iceberg scours. Although iceberg and sea ice scours are morphologically similar, the latter tend to be of much lower relief and with finer, more closely spaced grooves. In Antarctica sea ice scours are rare because of the deep water.

8.5.4 Slope valleys

Slope valleys are groups of gullies forming a dendritic pattern on the continental slope, and have been described from the Gulf of Alaska (Carlson et al. 1990). The smoothness of the valley sides suggests that there is a cover of sediment, dumped there when the glacier tongues occupied the shelf valleys. Gullies merge to form small canyons on the lower slope and small submarine fans occur at the outlets of some canyons. These gullies are believed to have formed as a result of erosion by sediment gravity flows derived from terminal moraines directly at the shelf edge or directly from the glacier face as it was attacked by waves. The dendritic pattern on the Gulf of Alaska slope has been modified by tectonic compression on the rising continental margin.

Fig. 8.13. Striated boulder pavement formed by glacier "touch-down" onto a winnowed surface of rain-out diamicton, Yakataga Formation, Middleton Island, Gulf of Alaska.

8.5.5 Boulder pavements

Striated boulder pavements may form either at the base of grounded ice sheets or on intertidal flats where sea ice abrasion takes place. In glacial settings boulder pavements result from (a) the progressive accretion of boulders around an obstacle, (b) sub-glacial erosion of older sediment, or (c) by development of a lag deposit in a marine setting. It is the last process which concerns us here.

Boulder lag deposits on relatively shallow continental shelf areas are the product of wave and tidal-current winnowing of diamicton especially when the sea level has been lowered. They become abraded by glacier ice as it advances across the shelf (Fig. 8.13). In the most comprehensive investigation of submarine boulder pavements, undertaken by Eyles (1988) on the early Pleistocene Yakataga Formation in the Gulf of Alaska, it was envisaged that boulder-bearing diamict was deposited by rain-out of debris from icebergs or from an ice shelf. Advance of an ice shelf, partially in contact with the bed, resulted in the boulders acquiring flat, striated tops, but no significant reworking of the bed took place. Rising sea level led to rapid ice shelf recession and an immediate resumption of diamict accumulation.

In contrast to shallow-marine boulder pavements, those formed on land and associated with lodgement till deposition show preferred orientation of boulders and typical bullet-nosed shapes, as well as a more varied facies association.

Boulder pavements have occasionally been reported in the older geological record, e.g. in the Late Paleozoic glacigenic sequence in Brazil and the late Precambrian sequences of Greenland and Svalbard. Some of these, no doubt, could be reinterpreted in the light of the Eyles (1988) model.

8.6 Depositional features on the continental shelf and slope

The application of geophysical techniques including high-resolution and multi-channel seismics and side-scan sonar, have expanded our knowledge of the depositional forms on ice-influenced continental shelves in the past decade. Some have been interpreted in the light of supposedly analogous features on land, but other forms have no direct counterparts on land.

8.6.1 Fluviodeltaic complexes on the continental margin

Proglacial fan deltas topped by sandur plains and building out from the coast in proximity to piedmont glaciers occur along the Gulf of Alaska, in Iceland, Greenland and Svalbard. The largest are associated with the Bering and Malaspina glaciers in Alaska. The plains are characterized by gravelly longitudinal bars with cross-bedded sands and gravels. The coastline areas are dominated by waves, and river-mouth bars and spits extend several thousand metres in the direction of the prevailing onshore drift. Lagoons often occur behind the beach ridges, and sometimes are directly influenced by the ice. Prodelta sediments are sandy to depths of about 50 metres and up to 15 km offshore, and pass into clayey silts further out onto the continental shelf.

Coastal fluviodeltaic complexes forming more distally from the source glaciers tend to have more stable and abandoned parts, allowing marshes to develop with organic muds, and channels to be filled by the products of estuarine processes. One example is the Copper River in southern Alaska which probably has the largest delta dominated by glacial meltwater in the world. The braided delta top is subjected to aeolian processes and large sand dunes have developed in places. The shoreface is complex comprising, in a seaward direction, marginal islands, a breaker bar, middle shoreface sands, lower shoreface sands and muds, and prodelta to shelf muds.

8.6.2 Delta-fan complexes

Detected by high-resolution seismic profiling, a delta-fan complex is the result of deposition beneath, and close to, the grounding-line of an ice sheet a long way from the open sea (Anderson & Bartek 1992). The complex is essentially a prograding sequence with prominent lamination illustative of glaciomarine deposition, as well as gravity-flow material derived from basal glacial debris. A good example, several hundred metres thick, occurs on the outer shelf of the Ross Sea. Its upper surface is elevated compared with the level of continental shelf behind, thus forming a bank on which icebergs are prone to grounding. Delta-fan complexes have also been identified in the middle of the continental shelf.

8.6.3 Subglacial deltas

Subglacial deltas are also prograding features, but lack stratification in the proximal parts. They grade into laminated deposits in the delta bottom-set deposits, and downlapping onto glacial erosional surfaces (Anderson & Bartek 1992). They are believed to form by "conveyor-belt" recycling of soft subglacial sediment, and provide a platform over which the ice may advance, according to the model of Alley et al. (1989). These features, too, have been identified in the Ross Sea.

8.6.4 Till-tongues

During seismic surveys of the eastern Canadian and Norwegian continental shelves, King & Fader (1986) discovered wedge-shaped bodies characterized by acoustically indecipherable reflections, interfingering with stratified glaciomarine sediments, and connected to large offshore moraines (Fig. 8.14). These authors inferred that the wedges comprised subglacially-derived till, and they named the features *till-tongues*. They inferred that the features formed at an oscillating ice margin where the ice became buoyant at the grounding-line. King et al. (1991) modified the till-tongue model to include a proglacial apron comprising a series of sediment gravity flows. Till-tongues have also been documented from seismic record in the Ross Sea, Antarctica, where they rest on erosion surfaces, thus providing a means of identifying glacial advances in seismograms (Anderson & Bartek 1992).

8.6.5 Trough-mouth fans

Trough-mouth fans are major features of continental margins across which large ice streams drained in the past, and they consist of arcuate fans of prograding sediment. The large ones, such as the 400 km wide Bear Island (Bjørnøya) Trough-Mouth Fan on the Barents Sea margin (Fig. 8.11) and the 300 km wide Prydz Trough-Mouth Fan in East Antarctica, reflect sedimentation from major marine-based ice streams at the grounding-line or at the calving limit when they reached the continental shelf break. Little is known of the facies in trough-mouth fans, although drilling through prograding sediments down to 100 m below the sea floor at the mouth of Prydz Bay yielded a sequence of unconsolidated massive diamicton. This diamicton was believed to represent advanced ice, the sand indicating reworked sediment deposited during interglacials (Barron, Larsen & Shipboard Scientific Party 1989). Seismic records indicate that slumping occurs frequently.

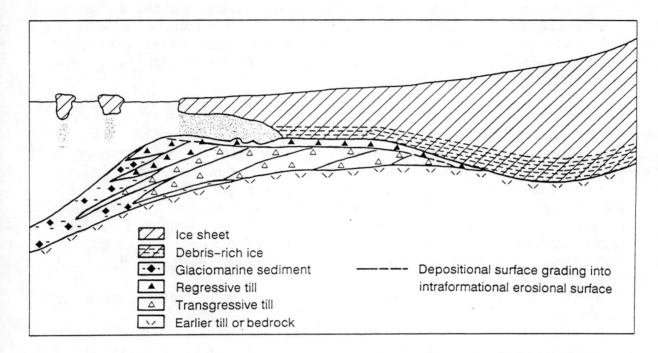

Fig. 8.14. Schematic representation of till-tongue associations formed under both transgressive and regressive regimes (modifed from King et al. 1987).

The sediments on the top of a smaller trough-mouth fan, at the mouth of Isfjordrenna Spitsbergen, comprise an upper part of proximal glaciomarine mud capped by a gravel lag, and a lower part with dense, massive unfossiliferous diamicton deposited by grounded ice (Boulton 1990). Frequent slope failure is evident here also, but mainly as small-scale mass-movements. The accretion of trough-mouth fans is aided by erosion and shelf-overdeepening at the inner parts of the continental shelf.

8.6.6 Diamict(on/ite) aprons

These are features (named by Alley et al. 1989 as till deltas) that have been inferred, rather than directly observed, to form at the grounding-line of ice shelves, floating glacier tongues or at ice margins grounded to the edge of a steep slope, such as the continental shelf break. The processes involved in their formation have been described earlier (Section 8.3.1). They comprise diamicton released at the break in slope by a combination of shearing of saturated till at the base of the ice, rain-out at the grounding-line and slumping of this sediment down the slope. They develop, not from point sources, but from a continuous line along the break in slope. They may form the bulk of the material in trough-mouth fans, but are not confined to them. The prograding sequence in Prydz Bay,

East Antarctica, is a good example (Section 8.9.1). Diamict aprons are forming extensively today only at the grounding-lines of ice shelves and floating glacier tongues in Antarctica, well back from the continental shelf break.

8.6.7 Shelf moraines

It has long been known that prominent ridges at the edge and in the middle of high latitude continental shelves are probably end-moraine complexes. King et al. (1987, 1991) have investigated such features created by the Fennoscandian ice sheet using seismic techniques on the mid-Norwegian continental shelf. They identified three types of moraine complex:

- *Linear moraines*. These are composite features consisting of (a) a lower stacked till-tongue succession, resulting from subglacial deposition near the grounding-line under conditions of advance, and (b) an unconformably overlying upper part formed during recession. Linear moraines are distinct ridges running for hundreds of kilometres at the shelf edge.
- *Tabular moraines* form in an intermediate-shelf position. They are irregular in plan view, with lateral dimensions as great as 70 km and a relief of 50-75 m. They have both abrupt and diffuse boundaries with the surrounding topography. On Trænabanken there are as many as five such complexes inside the linear moraine.
- *Hummocky moraines* are less extensive, but have a relief that may be more pronounced (25-100 m).

Both tabular and hummocky moraines are thought to have a common origin, formed where active ice was thinning and becoming buoyant over a broad zone in the form of a grounding zone, rather than a grounding-line. Deposition may be enhanced by a simultaneous rise in sea level. Both these types of ridge were prone to iceberg scour during the period of recession of the ice sheet.

Submarine end-moraine complexes have also been identified in seismic profiles on the northern Hebrides and West Shetland continental shelves (Stoker & Holmes 1991).

8.6.8 Flutes and transverse ridges

Largely depositional forms, resulting from a grounded ice-mass sliding on or deforming the sea bed on the northwestern Barents Sea, have been investigated using geophysical methods. They are represented by a system of parallel grooves and ridges, comprising stiff diamicton (Fig. 8.15). In several places the groove-and-ridge system is associated with short straight to arcuate ridges which run approximately perpendicular to the strongly linear pattern (Fig. 8.15). By comparison with terrestrial depositional landforms the grooves and ridges are interpreted as a set of flutes, described in Section 4.5.2.

The flutes are often 100 to 500 m long, but lengths over more than 1 km have been recorded. Their widths range from 1-15 metres, with 4-8 metres being typical. The fluted sea floor covers an area of around 4000 km^2 in water depths of between 160 and 300 m in northernmost Bjørnøyrenna (Fig. 8.11). They are parallel to the inferred ice flow direction down the submarine trough.

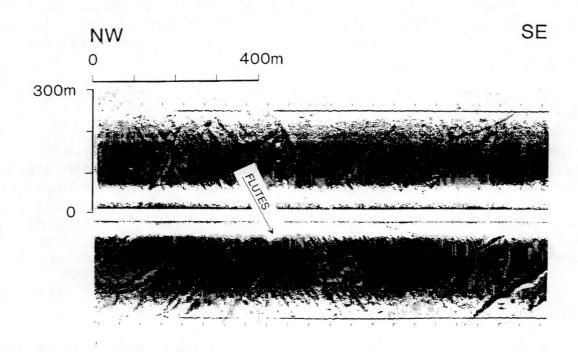

NW   SE

0   400m

300m

0

FLUTES

Fig. 8.15. Flutes and superimposed transverse (De Geer) ridges on the floor of the Barents Shelf revealed by side-scan sonar (Photograph courtesy of Anders Solheim; Solheim et al. 1990).

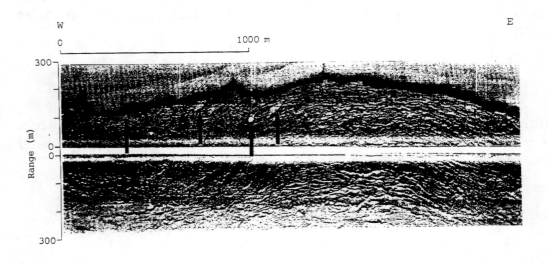

W   E

0   1000 m

300

Range (m)

0
0

300

Fig. 8.16. Side-scan sonar image of discontinuous, arcuate ridges, formed subparallel to the front of surge-type Bråsvellbreen, northern Barents Sea. Possible iceberg-impact features close to the ice margin are arrowed (Photo courtesy of Anders Solheim; Solheim 1991).

The flutes run across the associated transverse ridges, and through the gaps between ridge segments. The transverse ridges vary in length from 100-500 m, and in width from 15-30 m. They reach heights of 8 m. The ridges are intepreted as De Geer moraines, formed by seasonal ice push, at or close to the grounding-line.

In the special case of marine-based surging glaciers, Solheim (1991) described from Bråsvellbreen, Nordaustlandet, Svalbard a distinct suite of sea floor morphologies, associated first with rapid advance and then stagnation of strongly fractured ice. Numerous discontinuous arcuate ridges, that formed subparallel to the ice margin during the surge, extend for several hundred metres across the surge zone (Fig. 8.16). A set of linear ridges forming a rhombohedral pattern is also developed.

8.5.9 Mass-movement features

Oversteepened slopes with poorly consolidated diamicton and glaciomarine sediment are unstable, so submarine slides, slumps and gravity flows are common in the glaciomarine environment, as discussed in various contexts already. These features have a hummocky, sometimes lobate morphology, frequently having a relief of several metres and extending laterally for hundreds of metres. Larger collapse features on a scale of tens of kilometres occur occasionally at glaciated continental margins.

8.6.10 Raised beaches

These are common features of glacioisostatically uplifted areas bordering continental shelves. Raised beaches consist of well sorted sand and gravel with rounded clasts, originally supplied mainly from glacial or fluvioglacial sources. The best preserved raised beaches are of Holocene age, and good examples occur in northwest Europe, Canada, the USA, and the sub-Antarctic islands. Pre-Holocene raised beaches have commonly been modified or obliterated by later glacial erosion.

8.7 Sediment distribution patterns

A brief outline is given here of the areal distribution of glacigenic sediment in a number of continental shelf settings that today are directly or indirectly under the influence of ice.

8.7.1 Seawards of Antarctic ice shelves

Weddell Sea
Sediment patterns in the Weddell Sea are variable and related to bathymetry and current activity (Fütterer & Melles 1990). In front of the Filchner-Ronne Ice Shelf, the grain size distribution of surface sediments is related to water depth and is interpreted to be mainly current-controlled (Haase 1986). Very well-sorted fine to medium sand and moderately well-rounded pure sand occur to the west of Gould Bay in water depths of about 250 m. The high current velocities required (up to 20 cm sec^{-1}) to produce such sediments are tidal (Haase 1986), as confirmed by one current measurement of 40 cm sec^{-1} perpendicular to the ice front (Robin 1979). As water depth increases towards the west, the mud fraction increases, and therefore slower currents are inferred. In addition, the increase in the gravel component towards the Antarctic Peninsula illustrates the enhanced rôle of

iceberg rafting. A higher proportion of sand on the eastern flank of the Ronne Trough is related to a deep current flowing south under the ice shelf. Sub-bottom seismic profiling north of the Ronne Ice Shelf often failed to penetrate the gravelly floor. Elsewhere, the sediments are interpreted as having been compacted by Pleistocene grounded ice.

In the Crary Trough (also known as the Filchner Depression), the surface sediments are mainly depth-dependent and consisting of pebbly and sandy muds of glaciomarine origin. The western flank of the trough shows a deepening trend from sand to mud along its the axis. On the eastern flank the same fining-downwards trend is revealed but the sediments are gravelly (diamictons) on the upper continental slope and near the Filchner Ice Shelf edge. Grain-size distribution is controlled by current circulation and input of ice-rafted debris. Sands indicate the effect of currents which winnow out the fine material. The surface sediments are unconsolidated and are mainly up to 1m in thickness, underlain by overconsolidated diamicton (interpreted as a basal till of late Wisconsinan age). Unlike many other parts of the Antarctic margin, the biogenic component forms only a minor proportion of the total sediment.

Ross Sea
Biological productivity in the Ross Sea is much higher than in the Weddell Sea, and this is reflected in the proportion of silicon derived from diatoms in the sediments. Dunbar et al. (1985) have shown that the surface sediments in the Ross Sea are mixtures of unsorted ice-rafted debris, siliceous-biogenic material (diatom fragments), calcareous shell debris, terrigenous mud transported in suspension. Terrigenous mud makes up up to 50% of the surface sediment along the eastern part of the Ross Ice Shelf front, while to the west this is mixed with 10-50% of biogenic silica. An interesting feature of the ice-rafted debris distribution is that it increases in an offshore direction. The low concentration near the calving front suggests that most basal meltout occurs near the grounding-line, while on the outer shelf debris-bearing icebergs from valley glaciers are more frequent.

8.7.2 Continental shelf bordering mountain terrain with temperate glaciers

The coast of British Columbia and Alaska todayis a region characterized by a continental shelf flanked by high, tectonically active mountains, which still support many tidewater glaciers. Analogous settings are probably common in the older geological record, notably those of Neoproterozoic age. The distribution of superficial sediments on the northern Gulf of Alaskan coast (Carlson et al. 1990) seawards of the major fjords and glaciofluvial systems is illustrated in Figure 8.17. In the simplest case, a graded shelf sedimentary profile appears to be developing today, sand accumulating in the nearshore zone grades outwards into clayey silt in mid-shelf, the principal source being sediment brought to the coast by glacial meltwater. This Holocene facies association wedges out near the outer shelf, where diamicts, consisting of gravelly sand, muddy gravel and pebbly mud crop out extensively; these diamicts are probably of Pleistocene age.

Tectonic activity in places has uplifted the banks (e.g. Tarr Bank, Middleton Island and Kayak Island, Fig. 8.17), periodically permitting the fine sediment to be winnowed from the diamict and creating cobble and boulder-lag deposits, which are subject to abrasion during a subsequent ice-advance (Section 8.5.5). The modern glacially-formed submarine troughs or sea-valleys have concentrations of relict coarse sediments adjacent to their mouths near the outer shelf, and may represent end moraine complexes. These troughs are significant sedimentary traps, and seismic reflection profiles show accumulations of modern sediment tens to hundreds of metres thick.

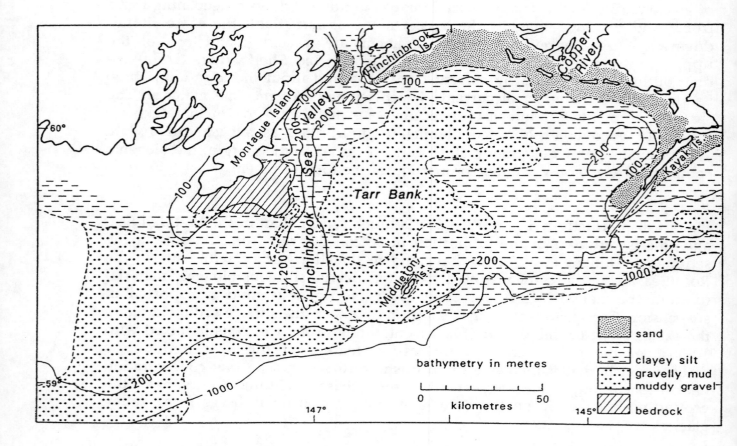

Fig. 8.17. Sediment distribution map of the Gulf of Alaska continental shelf and upper slope (after Carlson et al. 1990).

The continental shelf-edge and upper slope is covered mainly by gravelly sandy silts, or diamicts (with up to or more than 25% gravel) and silty sands. The sediment was probably deposited at the shelf edge during one of the most recent glacial stages when sea level was comparatively low, either as glaciogenic mud with dropstones or as a result of resedimentation of basal tills.

8.7.3 Continental shelf partly influenced by cold glaciers

The Barents Shelf is one of the broadest continental shelf areas in the world and during the Pleistocene Epoch was covered by ice. Sedimentation today is influenced by cold glaciers on the rugged Svalbard archipelago in the north, and much of the northern shelf is covered by sea ice in winter. The surface sediments of the western shelf have been mapped by the Norwegian Polar Research Institute (Elverhøi 1984).

264

Late Weichselian blue-grey glaciomarine sediments are overlain transitionally by olive grey mud that is partly derived from suspended matter of glacial meltwater origin. In the north the mud is pebbly as a result of iceberg- and sea ice-rafting. On Spitsbergenbanken biogenic material (shell fragments and barnacles) make up 60-90% of the surface sediments as a result of current winnowing. On the southern and southeastern slopes of Spitsbergenbanken the olive-grey mud is missing and a cobble lag, resulting from strong bottom current activity, is present.

8.8 Sedimentation rates

Sedimentation rates on continental shelves are generally several orders of magnitude less than in fjords (cf. section 7.5). In the polar regions estimates for Holocene sedimentation include 0.02 - 0.05 mm yr^{-1} for the Weddell Sea (Elverhøi & Roaldset 1983) while 0.02 - 0.07 mm yr^{-1} is indicated for the Barents Sea (Elverhøi 1984).

Over longer time intervals rates have been determined for the CIROS-1 core in McMurdo Sound. The lower, Early Oligocene part of the core gives an average sediment rate of 0.2mm yr^{-1}, whereas the upper part (mainly late Oligocene - early Miocene) is 0.04 mm yr^{-1} (Harwood et al. 1989). The sharp difference may reflect a greater tendency to uplift of the neighbouring Transantarctic Mountains during the earlier phase. The longer term average sedimentation rate for progradation of the outer Prydz Bay continental shelf is 0.008-0.012 mm yr^{-1}, although large parts of the sequence are missing and may have been eroded.

In contrast to these polar continental shelves, sedimentation rates are at least an order of magnitude higher on the cool temperate Gulf or Alaska Shelf. From the thickness and age estimates given by Eyles & Lagoe (1990), an average sediment rate of 1 mm yr^{-1} is obtained.

In the deep ocean-basins, sedimentation tends to be much more continuous, but overall the rates tend to be of the same order of magnitude. For example the Ocean Drilling Program recorded 0.013 mm yr^{-1} at Site 645 in southern Bay and 0.03 mm yr^{-1} at Site 745 in the southern Indian Ocean, much of the accumulation being terrigenous material of glacial derivation.

8.9 Sedimentary facies in glaciomarine environments

Although the range of processes operating in the modern glacimarine environment has now been widely investigated (section 8.3), little is known of the nature of sedimentary sequences accumulating in such environments (Eyles & Lagoe 1990). However, the older geological record provides excellent opportunities for studying thick sequences of glaciomarine sediment. The best exposed and by far the thickest late Cenozoic glaciomarine sequence occurs along the Gulf of Alaska. Here 5 km of glaciomarine strata, belonging to the Yakataga Formation, have been progressively uplifted by tectonic collision of the Pacific Ocean and North American plates and span 5-6 million years (Eyles & Lagoe 1990). Even longer records (35-40 million years) have been recovered from drill-holes on the Antarctic continental shelf, although these show a number of hiatuses (Barrett 1989, Cooper et al. 1991b, Hambrey et al, 1992). Numerous glaciomarine successions of pre-Cenozoic age have also been documented. Late Precambrian strata provide several excellent glaciomarine successions. Andrews & Matsch (1983) have provided a bibliography of papers describing glaciomarine sequences, and Anderson (1983) has summarized the main occurrences in the geological record, based on the global survey compiled by Hambrey & Harland (1981).

8.9.1 Sedimentary structures in glaciomarine sediments

Glaciomarine sediments demonstrate a wide range of sedimentary structures reflecting their composite origin. Sediment deposited subaquatically from ice by meltout, without reworking, is a non-sorted, homogeneous diamicton, although over several metres there may be slight variations in the content of gravel (Fig. 8.18a).

As soon as these sediments are subject to bottom currents, mud may be preferentially winnowed out leaving rippled and scoured surfaces (Fig. 8.18b) and giving rise to stratified diamicton (proximal glaciomarine sediment) with bedding typically on a centimetre-scale. Wispy bedding and signs of loading are common, reflecting the high water content (soupiness) of the sediments (Fig. 8.18c,d).

Bedding is often punctured by dropstones (Fig. 8.18e) over which subsequent sedimentary layers are draped. Many elongate dropstones stick in the sediment vertically, if it is not too soupy. Some dropstones occur as clasts of diamict or glacial mud (till pellets) and are assumed to have been dropped in as frozen lumps; usually they are up to a few centimetres in diameter, although some are bigger. Stronger currents remove sand as well as mud, leaving a gravel lag, which has a gradational relationship with the underlying undisturbed sediment but is followed sharply by later sediments. If sedimentation has been continuous, then diamictons often show gradational boundaries from stratified to non-stratified varieties as well as into less ice-influenced muddy sands and sandy muds. Successions may be tens of metres thick and lack distinct bedding hiatuses.

Recycling of diamictons by mass-movement processes generates new sedimentary structures. Slumping of stratified diamicton gives rise to isoclinal folds. In slides, partial disaggregation of beds may produce rootless folds detached from the original bed. Submarine debris-flows commonly affect glacigenic sediments. They tend to be represented by beds ranging from a few centimetres to several metres in thickness (Fig. 8.18f). The end product may still be a massive diamicton, difficult to distinguish from the original material. However, the flow normally has a sharp top and bottom, contains clusters of several gravel clasts, ripped-up from the bed over which it passes and occasional faint inverse or normal grading.

If mass-movement of diamicton develops into a turbidity flow, the end product shows normal grading, usually of the "coarse-tail" type (i.e. in which larger clasts may occur throughout the graded bed but in smaller quantities towards the top). Both debris-flows and turbidity flows may load the underlying sediment causing it to deform in a convolute manner or to penetrate the debris-flow as a sedimentary dyke. Sandy and muddy glaciomarine sediments of more distal origin are subject to the same reworking processes as diamictons; their structures are typical of continental margins generally, with the added influence of floating ice. The glaciomarine environment may yield rhythmically laminated

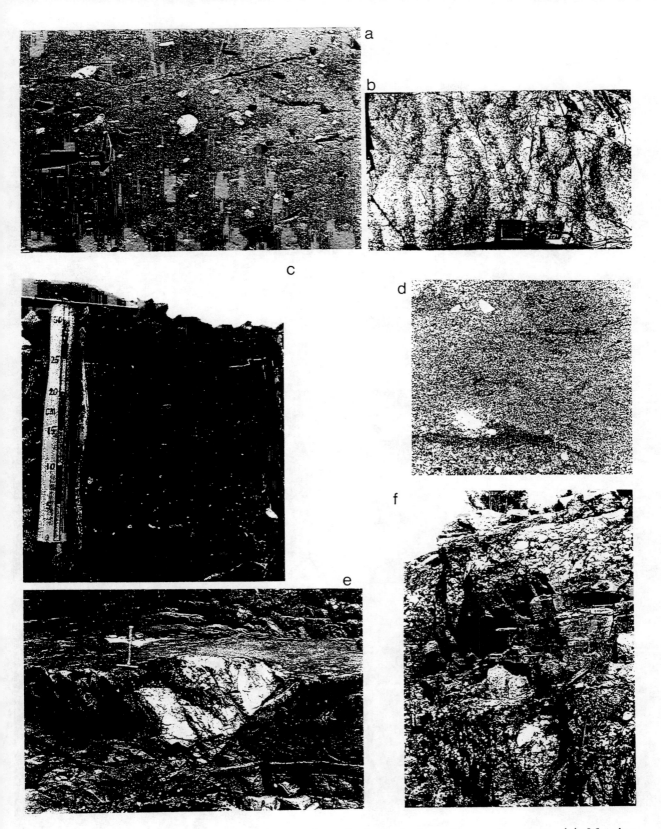

Fig. 8.18. Features in sediments in the continental shelf glaciomarine environment. (a) Massive diamictite, interpreted as waterlain till, Neoproterozoic Ulvesø Formation, East Greenland. (b) Wave ripples and winnowed surface on top of diamictite bed, Neoproterozoic Wilsonbreen Formation, NE Spitsbergern. (c) Clast-poor diamictite with wispy, diffuse stratification, the result of slight winnowing of "soupy" waterlain till followed by slumping, Wilsonbreen Formation, NE Spitsbergen. (d) Thin-section of rock shown in (c), showing variable grain size distribution, including clay-rich wispy beds (dark) and diffuse winnowed silty beds (bottom and top). (e) Large dropstone of pegmatite in sandy turbidites, top of the Dalradian succession at Macduff, Banffshire, Scotland (age: Ordovician or Neoproterozoic).(f)Matrix-supported conglomerate horizons interpreted as sub-aquatic debris-flows, interbedded with siltstone of glaciomarine origin, Whitefish Falls area, Ontario, Canada.

267

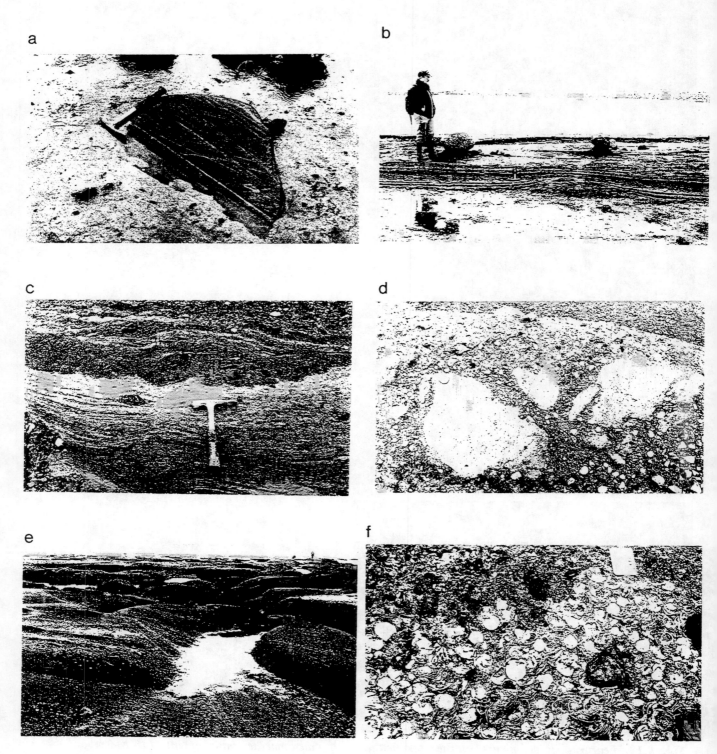

Fig. 8.19. *Some typical facies in the glaciomarine environment, Plio-Pleistocene Yakataga Formation, Middleton Island, Gulf of Alaska. (a) Massive diamict with striated clast, resulting from rapid deposition of mud together with ice-rafting. (b) Stratified diamict formed where some sorting by bottom currents of material as in (a) had taken place. (c) Stratified diamict with prominent gravel lag horizon. (d) Deformed masses of sand, the largest 0.5m in diameter enclosed within diamict facies, resulting from dismemberment of beds during loading. (e) Large-scale cross-bedding in a sand and gravel member at base of the Yakataga Formation; these represent slope deposits, reworked from glacial material on the shelf. (f) Coquina bed, formed during a phase of little sedimentation, except for the dropstones, when current winnowing was taking place.*

sediments, although these are much less common than in fjords. Normally the laminae are graded, representing turbidity current deposits, and they can range from proximal to distal.

Ice advancing on its bed across glaciomarine sediment, such as on a continental shelf, may abrade gravel lag deposits or accumulations of ice-rafted gravel to give a boulder pavement. If the underlying sediment is partially consolidated it may fracture and slickensides might form on the fault planes. More intensive glaciotectonic deformation may occur to depths of tens of metres, manifested in thrusts and overturned folds and the incorporation of rafts of underlying sediment occasionally more than a hundred metres long. In this respect, these structures are similar to those of advancing ice on land.

8.9.2 Lithofacies in the glaciomarine environment

Lithofacies are normally described in terms of the following: diamicts, gravel, sand, mud and combinations thereof, and their lithified equivalents. Subfacies are designated according to whether the rock is massive, stratified, or bioturbated, or has other distinguishing features. These lithofacies are then interpreted in the context of the whole sequence and by using other parameters (section 8.5). Massive diamict (Fig. 8.19a) can variously be interpreted as a lodgement till, deformable till beneath ice, waterlain till, or a debris-flow, or as a distal glaciomarine mud with ice-rafted debris that has been bioturbated or deposited in a quiet-water basinal setting. Weakly to well-stratified diamict (Fig. 8.19b) represents the transition to sediment increasingly dominated by marine processes, notably reworking by currents in which the influence of ice-rafting becomes clearer (i.e. proximal glaciomarine). Muddy sands to sandy muds demonstrate the transition to distal glaciomarine with ice-rafted debris becoming relatively less important. Other facies represent the processes that rework the sediment after deposition, notably mass-movement (Fig. 8.19c,d), downslope channeling (Fig. 8.19e), and by the addition of a biogenic component in the form of both macrofossils and microfossils such as diatoms (Fig. 8.19f).

Lithofacies have now been described from numerous Cenozoic and older sequences. The Antarctic continental shelf displays a wide range of facies, including sediments released directly from glacier ice, which are summarized and interpreted according to proximity of ice and water depth in Table 8.2. Relative proportions have also been calculated from which it may be noted that diamictite, sandstone and mudstone predominate. Similar glaciomarine facies occur elsewhere, although of course the biogenic influence varies according to age and was negligible in the Proterozoic Era.

The range of facies that accumulate on the continental slope during periods when an ice sheet is grounded to continental shelf edge has been investigated on the Norwegian and Barents continental slopes using a combination of shallow coring and seismic techniques (Yoon et al. 1991). Both diamictons and mud lithofacies, containing abundant signs of down-slope movement have been examined. These are listed in Table 8.3, together with their interpretations.

_____

Table 8.2. Facies recovered in deep drill-cores taken from the Antarctic continental shelf, and their interpretation (summarized from Hambrey et al. 1991).

Facies	Description	Interpretation
Massive diamictite	Non-stratified muddy sandstone or sandy mudstone with 1-20% clasts; occasional shells and diatoms.	Lodgement till (with preferred orientation of clast fabric) or waterlain till (random fabric).
Weakly stratified diamictite	As massive diamictite, but with wispy stratification; bioturbated and slumped; partly shelly and diatomaceous.	Waterlain till to proximal glaciomarine sediment.
Well-stratified diamictite	As massive diamictite with discontinuous and contorted stratification; occasional dropstone structures; abundant diatoms and shells.	Proximal glaciomarine/ glaciolacustrine sediment.
Massive sandstone	Non-stratified, moderately well-sorted to poorly sorted sandstone, with minor mud and gravel component; loaded bedding contacts.	Nearshore to shoreface with minor ice-rafting in distal glaciomarine setting; better sorted sands with loaded contacts are gravity flows; associated with slumping.
Weakly stratified sandstone	As massive sandstone, but with weak, contorted, irregular, discontinuous, wispy, lenticular stratification; brecciation, loaded contacts and bioturbation.	Nearshore with minor ice-rafting in distal glaciomarine setting; better sorted sands are gravity flows; associated with slumping.
Well-stratified sandstone	As massive sandstone, but with clear, but often contorted stratification.	Nearshore with minor ice-rafting in distal glaciomarine setting; some slumping.
Massive mudstone	Non-stratified, poorly sorted sandy mudstone with dispersed gravel clasts; intraformational brecciation and bioturbation; dispersed shells and shell fragments.	Offshore with minor ice-rafting in distal glaciomarine setting; some slumping or short-distance debris flowage.
Weakly stratified mudstone	As massive mudstone but with weak, discontinuous, sometimes contorted stratification defined by sandier layers; bioturbated.	Offshore to deeper nearshore with minor ice-rafting in distal glaciomarine setting; slumping common.
Well-stratified mudstone	As massive mudstone, but with discontinuous, well-defined stratification, with sandy laminae; syn-sedimentary deformation and minor bioturbation.	Deeper nearshore with minor ice-rafting in distal glaciomarine setting; some slumping.
Diatomaceous ooze/diatomite	Weakly or non-stratified siliceous ooze with >60% diatoms; minor components include terrigenouss mud, sand and gravel.	Offshore, with minor ice-rafting in distal glaciomarine setting.
Diatomaceous mudstone	Massive mud or mudstone with >20% diatoms and minor sand.	Offshore with sedimentation predominantly influenced by ice-rafting and underflows in distal glaciomarine setting.
Bioturbated mudstone	As massive mudstone, but stratification highly contorted or almost totally destroyed by bioturbation.	Offshore to deeper nearshore with minor ice-rafting; extensively burrowed.
Mudstone breccia	Non-stratified to weakly stratified, very poorly sorted, sandy mudstone intraformational breccia with up to 70% clasts; syn-sedimentary deformation and minor bioturbation.	Offshore to deeper nearshore slope-deposits with minor ice-rafted component, totally disrupted by debris flowage.
Rhythmite	Graded alternations of poorly sorted muddy sand and sandy mud; stratification regular on a mm-scale; dispersed dropstones.	Turbidity underflows derived from subglacial source, with ice-rafting in a proximal glaciomarine setting.
Conglomerate	Non-stratified to weakly stratified, poorly-sorted, clast- to matrix-supported sandy conglomerate; normal and reverse grading evident; clasts up to boulder size; intraclasts of mudstone frequently incorporated; loading and other soft-sediment features present.	Slope debris-flows derived directly from proglacial glaciofluvial material, or from subaqueous discharge from glacier; well-defined beds may be fluvial.

Facies	Characteristics	Depositional process
Thick-bedded, disorganized mud	Coarse-grained clasts dispersed in fine matrix without internal organization; more than 1 m thick; bioturbation minimal	Debris-flowage
Thin-bedded, disorganized mud-clast mud	Abundant mud clasts and rock fragments randomly scattered in mud matrix; bioturbation slight to common; variable thickness (1-10 cm)	Vertical settling of ice-rafted debris
Silt-clay couplet	Couplet composed of a basal silt unit and an overlying clay unit; basal silt thinly laminated and clay unit homogeneous; individual unit < 5 cm thick; bioturbation restricted to upper part; facies boundaries well-defined	Fine-grained turbidity current
Indistinctly laminated mud	Poorly sorted mud showing irregular and discontinuous laminae a few decimetres thick; bioturbation minimal	Downslope bottom-current with high sediment fallout rate
Layered mud	Irregular alternation of thin (a few mm) silt-rich and silt-depleted mud layers; bioturbation slight to common	Deep-sea contour current, heavily laden with fine-suspended sediments
Indistinctly layered mud	Poorly sorted mud, exhibiting indistinct, discontinuous layering and discontinuous trains of horizontally orientated coarse grains; thickness variable; bioturbation common; boundaries sharp or gradational	Deep-sea contour current
Bioturbated mud	Poorly sorted mud, intensley disturbed by bioturbation; primary structure absent except for diffuse banding; facies thickness variable; boundaries poorly defined and irregular	Hemipelagic sedimentation; contour current
Deformed mud	Mechanically deformed mud showing shear-lineation, crenulated laminae, wispy laminae, microfault and swirl structure	Slumping/sliding

Table 8.3. Facies recorded in shallow cores taken from the continental slope off northern Norway and the SW Barents Shelf (summarized from Yoon et al. 1991).

8.9.3 Glaciomarine facies associations

Facies associations are normally assemblages of various lithofacies in vertical sequences of strata that display a common theme, e.g. glaciomarine. As such they reflect environmental changes through time. However, it is necessary to consider how these facies change in an areal sense, since this will explain the vertical associations. Thus a typical present-day transition from waterlain till to distal glaciomarine is best represented by a recession sequence in a vertical profile, with massive diamictons at the base passing up into mudstones with dispersed stones.

Figure 8.20 is an idealized example of the type of facies association that may develop as a temperate glacier influences and then advances across a continental shelf, but many other associations may arise according to the thermal and hydrological characteristics of the glacier.

Many detailed records of glaciomarine environments have been provided by the geological record. Five examples, from tectonically contrasting settings are summarized below.

Continental facies at a convergent plate margin
The Plio-Pleistocene Yakataga Formation provides exposures that are almost complete for hundreds of metres vertically and laterally along the uplifted rim of the Gulf of Alaska. The Middleton Island sequence provides a excellent illustration of the facies that are preserved when extremely dynamic temperature glaciers fed by heavy snow in mountains that are being rapidly uplifted, advance across the continental shelf and subsequently recede rapidly (Eyles & Lagoe 1989; Eyles & Lagoe 1990). Typical facies are illustrated in Figure 8.19.

Glaciomarine deltaic complex at the margin of a subsiding rift basin
The 702m long core obtained from the drill-site CIROS-1 in western McMurdo Sound, Antarctica has provided a unique record of glaciation extending back to earliest Oligocene time (36 m.y. ago). The dominant facies are diamictite, sandstone and mudstone with abundant flora and fauna (Table 8.2). The facies reflect deposition from temperate glaciers as the Transantarctic Mountains were being uplifted. An early phase of relatively deep water glaciomarine sedimentation ended with emergence (at about 370 mbsf), and was succeeded by a series of advances and recessions of grounded ice across the continental shelf (Fig. 8.21) (Barrett 1989).

Glaciomarine facies associated in a subsiding ensialic basin
Thick sequences of interbedded glacioterrestrial and glaciomarine sediments are characteristic of subsiding basin within regions of continental crust. The basins may be linear features, bounded by faults, with connections to the open sea. Much of the North Sea Pleistocene sequence, which reaches a maximum thickness of c. 920 m, occupies a linear trough extending NNW from the Dutch coast for several hundred kilometres.

The Neoproterozoic glacial era provides several superbly exposed sequences that were deposited in an ensialic basin, for example the East Greenland/NE Svalbard basin (Fairchild & Hambrey in press), and the Port Askaig Tillite in Scotland (Spencer 1971, 1985). Glaciomarine sediments are an important component in these mixed terrestrial-marine sequences.

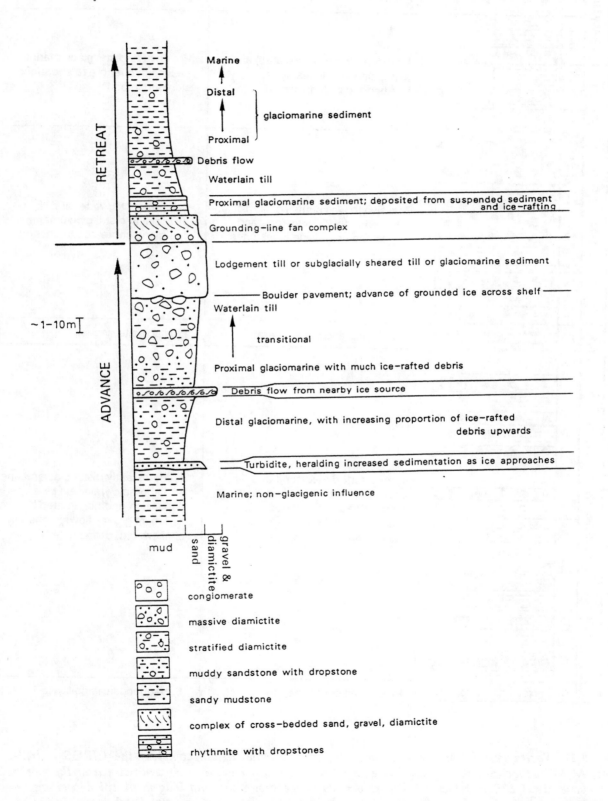

RETREAT

Marine

Distal

glaciomarine sediment

Proximal

Debris flow

Waterlain till

Proximal glaciomarine sediment; deposited from suspended sediment and ice-rafting

Grounding-line fan complex

Lodgement till or subglacially sheared till or glaciomarine sediment

Boulder pavement; advance of grounded ice across shelf

Waterlain till

ADVANCE

~1–10m

transitional

Proximal glaciomarine with much ice-rafted debris

Debris flow from nearby ice source

Distal glaciomarine, with increasing proportion of ice-rafted debris upwards

Turbidite, heralding increased sedimentation as ice approaches

Marine; non-glacigenic influence

mud | sand | gravel & diamictite

conglomerate

massive diamictite

stratified diamictite

muddy sandstone with dropstone

sandy mudstone

complex of cross-bedded sand, gravel, diamictite

rhythmite with dropstones

Fig. 8.20. Hypothetical succession formed as a result of the advance and recession of a temperate glacier across the continental shelf.

273

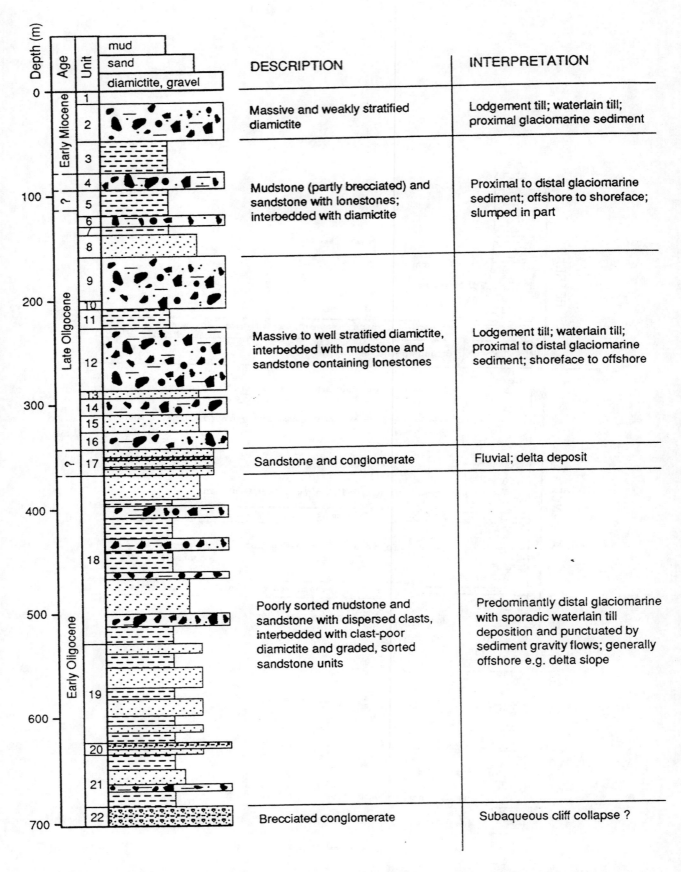

Fig. 8.21. *Facies association from a subsiding glaciomarine deltaic complex, the CIROS-1 drill-hole, McMurdo Sound, Antarctica. Note that the sequence comprises two distinct parts; the upper one above about 370 m which illustrates periods of ice grounding and lodgement till deposition on the continental shelf, and the lower part which is deeper water and mainly distal glaciomarine. A thin fluvial sequence between is the only indication of subaerial conditions in this core. (Hambrey et al, 1992).*

274

Subsiding passive continental margin dominated by marine ice sheet
Five sites, drilled by leg 119 of the Ocean Drilling Program on a 180 km long transect across the continental shelf of Prydz Bay to the continental slope, prove an insight into the development of a continental margin under prolonged influence (over some 40 m.y.) of a major ice sheet that flows into the sea (Barron, Larsen & Shipboard Scientific Party 1989; Hambrey et al. 1991). The principal facies are listed and intepreted in Table 8.2. Seismic records, together with borehole information indicate that there is (i) an upper, flat-lying sequence of grounded-ice and floating-ice sediments, with periods of compaction by ice-sheet loading indicated by overconsolidated sediments, and (ii) a lower prograding sequence, mainly composed of glaciomarine sediments.

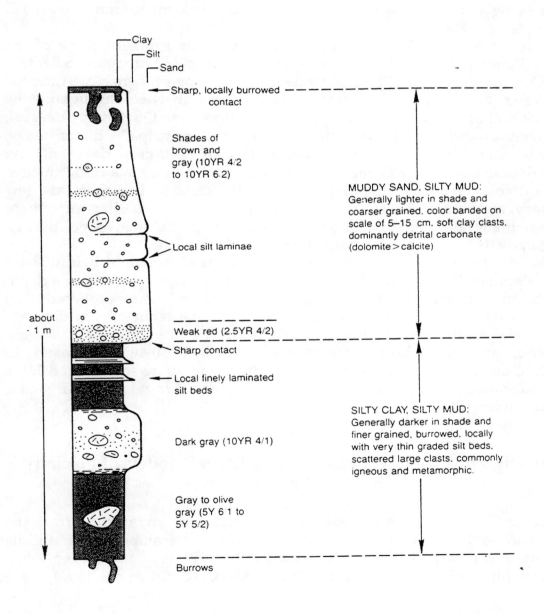

Fig. 8.22. Composite sketch of the general features of a 'typical' lithological cycle from ODP site 645, Baffin Bay. These cycles are believed to represent interglacial/glacial phases, the latter being indicated by the scattered clasts of ice-rafted material. (Note that the use of "mud" here differs from more normal usage). (Srivastava, Arthur & Shipboard Scientific Party 1987).

Ocean basin influenced by iceberg-rafting

Most of our knowledge concerning ice-influenced sedimentation in the deep ocean basins comes from the Ocean Drilling Program (ODP) and its forerunner the Deep Sea Drilling Project (DSDP), operated by the Americans on behalf of the international scientific community. Numerous sites have been drilled in the polar oceans, and a wide range of facies have been documented. The sediments generally comprise a mixture of pelagic and terrigenous components, although the relative proportions are extremely variable. On the one hand the bulk of the sediment may consist of diatom ooze (in the coldest waters or calcareous ooze (in warmer conditions), with a small ice-rafted component that reflects the geology of the onshore glacierized terrain. On the other hand, microfossils may form a minor component, with terrigenous mud, brought in by bottom currents or as turbidites being dominant. Again, larger fragments represent ice-rafted material. Two examples are given here.

In the first case, ODP Site 745 lies at the base of the southern slope of the Kerguelen Plateau in 4082 m of water (Barron, Larsen & Shipboard Scientific Party 1989; Ehrmann 1991). The site toaday is under the influence of Antarctic Bottom Water, derived from beneath the Antarctic ice shelves. Throughout the 215 m thick section, which spans the uppermost Miocene to Quaternary interval, both terrigenous and marine components are present, principally diatom ooze, and silt and clay. Much of the core shows a clear alternation of clayey diatom ooze and diatomaceous clay (sometimes with minor silt) on a scale of decimetres to metres. Pebbles and granules are scattered throughout, but there are few sedimentary structures. These alternations may reflect a greater supply of terrigenous material when the ice sheet was closer to the shelf edge, that is, reflecting cyclicity in the behaviourof the ice sheet .

ODP Site 645 lies low down on the continental slope off southern Baffin Island in a water depth of 2020 m. Here, the sea floor was penetrated by coring to a depth of 1147 m and an early Miocene to Quaternary record was recovered. The influence of ice is perceived above 753 m, in late Miocene sediments, in the form of quartz grains and small pebbles (dropstones). The predominant facies are muddy sand, sandy mud and detrital-carbonate mud (with 30-40% carbonate). As in the south, these form alternations on the decimetre to metre scale (Fig. 8.22), a cyclicity supposed to represent glacial periods with heavy ice-rafting, and interglacials without.

8.10 Stratigraphic architecture and depositional models of glacially influenced continental shelves

In this section, examples of continental shelf stratigraphic architecture from the Antarctic and Arctic are given (Fig. 8.23). Several other examples from middle latitudes have also been published, for example, off NW Scotland (Stoker 1990), S.E. Canada (King & Fader 1986) and the Gulf of Alaska (Carlson et al. 1990)

8.10.1 Antarctica

A number of features distinguish the Antarctic from other continental margins:

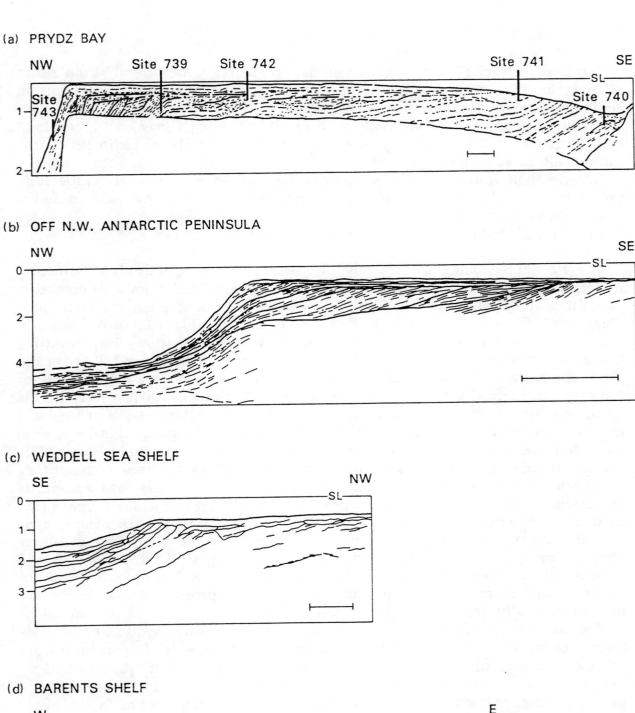

(a) PRYDZ BAY

NW   SE

Site 739   Site 742   Site 741

Site 740

Site
743

(b) OFF N.W. ANTARCTIC PENINSULA

NW   SE

Two-way travel time (sec.)

(c) WEDDELL SEA SHELF

SE   NW

(d) BARENTS SHELF

W   E

Plio/Pleistocene

Oligo/Miocene   Paleo/Eocene

Mio/Pliocene

Eo+Oligocene

oceanic crust   continental crust

Fig. 8.23. Stratigraphic architecture of polar continental shelves. Note the common occurrence of prograding wedges (of which at least the upper part is probably glacial), overlain by ice-dominated flat-lying sequences. (Summary diagram, based on various sources is from Hambrey et al. 1992).

277

(a) The Antarctic continental shelf has a greater water depth than other shelves, mainly because of ice sheet loading, with the result that bottom sediments remain unaffected by wave action.

(b) Shelf profiles indicate shallowing towards the shelf break as a result of greater glacial erosion of inshore areas, and accretion of glacial sediment on the outer shelf, in addition to loading.

(c) The shelf comprises a lower prograding sequence of normal compaction. Overlying this is a flat-lying sequence comprising over-compacted sediments with high seismic velocities.

(d) The shelf today is relatively starved of sediment.

The Prydz Bay shelf is probably the best-known high-latitude continental margin as the architecture of the shelf has been determined from a combination of drilling and seismic surveys (Barron, Larsen & Shipboard Scientific Party 1989, Stagg 1985, Cooper et al. 1991a,b, Hambrey et al.1991) (Fig. 8.23a). As noted in Section 8.4, a transect across the shelf shows a continuous but complex prograding sequence, covered by a flat-lying sequence. In the inner 120 km of the shelf, the prograding sequence is dominated by fluviatile-deltaic sediments of Mesozoic to Early Cretaceous age (Turner 1991). The outer 70 km of the shelf is dominated by a massive diamictite apron which attains a thickness of at least 400 m, and is interpreted as waterlain till deposited close to the grounding-line as the extended ice sheet decoupled from its bed at the palaeo-shelf break. The glacigenic part of the prograding sequence comprises discrete sedimentary packages, some of which are truncated at the top, whereas others pass landwards into parts of the flat overlying sequence. The drilled part of the shelf shows two distinct glacigenic sequences: a lower gently dipping one with strong syn-sedimentary or possibly glaciotectonic deformation at the base, and an upper steeper prograding sequence with numerous signs of slumping. Some sedimentary packages (not drilled) show signs of large-scale continental margin collapse and slumping. The topmost parts of the prograding sequence are glaciotectonically deformed and over-consolidated.

The flat-lying sequence, which overlies the prograding sequence, consists mainly of massive diamictite, interpreted largely as the result of deposition from grounded ice. Some of it is over-compacted, indicating that it was affected by loading, probably during a succession of ice advances. The flat-lying sequence generally thickens seawards from a few metres to around 250 m in the outer parts of the shelf, and illustrates successive stacking of separate units towards the shelf break. Each unit represents a distinct depositional phase. The flat-lying sequence is thickest beneath the banks of the outer shelf, and thinnest in a major channel that extends transversally across the shelf. Some of these separate units pass laterally seawards into prograding packages, including the topmost one.

Seismic data from other parts of Antarctica indicate broadly similar features to those in Prydz Bay. The continental shelf of the Weddell Sea is underlain by a prograding sedimentary sequence, the outer part of which is inferred to be glacigenic, truncated by repeated grounding of the Filchner Ice Shelf (Fig. 8.23c). On top of it is a flat-lying sequence (Hinz & Kristoffersen 1987) also of mainly glacial origin. The same holds for the Pacific margin of the Antarctic Peninsula (Larter & Barker 1989) (Fig. 8.23b). In both cases the uppermost sediments are over-compacted, and several ice advance and recessional cycles are in evidence. The margin of the Antarctic Peninsula differs from Prydz Bay in being

278

significantly steeper (14° compared with 4°), yet displays no evidence of slumping or collapse.

Model for the Cenozoic development of the Antarctic continental shelf
Based on the stratigraphic architecture of Prydz Bay, which is probably typical of many parts of the continental margin of Antarctica, a generalized model from the time of onset of glacierization at sea level has been developed (Hambrey et al.1992).

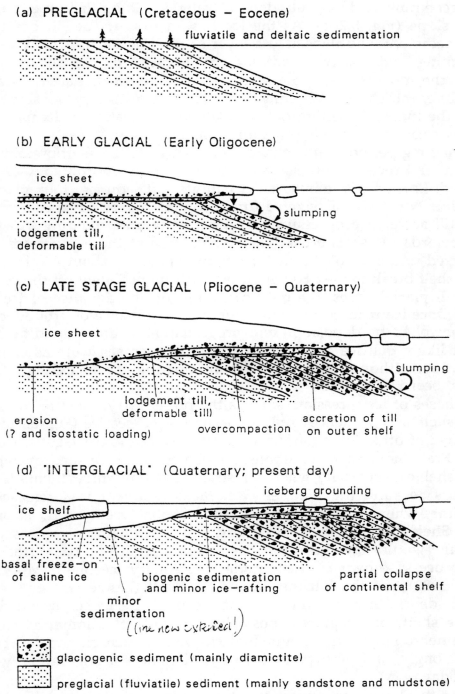

(a) PREGLACIAL (Cretaceous – Eocene)

fluviatile and deltaic sedimentation

(b) EARLY GLACIAL (Early Oligocene)

ice sheet

lodgement till, deformable till

slumping

(c) LATE STAGE GLACIAL (Pliocene – Quaternary)

ice sheet

erosion (? and isostatic loading)

lodgement till, deformable till)

overcompaction

accretion of till on outer shelf

slumping

(d) 'INTERGLACIAL' (Quaternary; present day)

iceberg grounding

ice shelf

basal freeze-on of saline ice

biogenic sedimentation and minor ice-rafting

minor sedimentation (ice now extinct!)

partial collapse of continental shelf

glaciogenic sediment (mainly diamictite)

preglacial (fluviatile) sediment (mainly sandstone and mudstone)

Fig. 8.24. *Model for the development of the Antarctic continental shelf, based on drill-hole and seismic data, as exemplified by Prydz Bay. This sequence of events may be inferred for continental margins generally that have been under the prolonged influence of ice.*

279

Prior to the onset of glacierization at sea level, the Antarctic continental shelf prograded under the influence of fluvial and deltaic processes (Fig. 8.24a). As the ice sheet developed, it isostatically depressed the land, and the shelf became flooded before the ice advanced across it. The thicker ice over the inner part of the shelf may have led to a reversal of the shelf gradient, and as grounded ice extended to the edge of the shelf, loading increased.

The grounded ice reached, and became decoupled at, the palaeoshelf-break by early Oligocene time, and began to deposit waterlain till on the upper parts of the continental slope (Fig. 8.24b). As the ice flowed over its bed it deformed the underlying sediment, and deposited a layer of till. The till itself may have acted as a deforming and erosive medium, and because of the shearing motion induced by the moving ice, some of it may have been displaced towards the grounding-line, thereby contributing to deposition on the palaeoslope. To this was added the material melting out of floating ice, creating a diamicton apron. Slumping was common on this prograding diamict apron.

During the long period characterized by advances to the shelf break with shelf progradation, and recedes with sediment starvation, the morphology of the shelf was modified. Erosion of the inner shelf with its soft sediment took place during glacial maxima, whereas till accreted further out as the ice advanced to the shelf break. The till accreted progressively as a flat-lying unit, building upwards and outwards towards the shelf break. The combination of these processes in Prydz Bay is reflected in the tilt of the shelf from an average of about 400 m below sea level at the shelf break to over 800 m on the shelf immediately off the coast.

During full glacial times, the only constraint on the advance of ice was the shelf break. Once ice reached this limit, subsequent advance of grounded ice was probably controlled by deposition and progradation at the grounding-line. It is not thought likely that ice shelves bordering the open sea were as extensive as those of today, although deposition from floating ice is evident from the sedimentary record (Fig. 8.24c).

During phases of ice recession, such as the present day, most of the Antarctic continental shelf was influenced only by icebergs and sea ice (Fig. 8.24d). Today may be typical of other 'interglacials' or periods of reduced ice-cover during the Quaternary Era. Under such conditions, ice floated as an ice shelf over the inner continental shelves, especially where constrained within embayments like Prydz Bay, the Ross Sea and the Weddell Sea. Most sediment was released at the grounding-line, where strong bottom melting probably occurred, as under the Amery Ice Shelf (Robin 1983). However, any sediment not released at the grounding-line would have been protected from melt-out by basal freeze-on, an important process, known to occur under the Amery Ice Shelf (Morgan 1972; Budd et al. 1982) and the Filchner-Ronne Ice Shelf (Lange & MacAyeal 1986) today. Such debris thus remained in the ice until the icebergs, which calved from the ice shelf, disintegrated, possibly hundreds of kilometres from their source. Sedimentation on the continental shelves in such circumstances would have been strongly, and perhaps exclusively, diatomaceous, as in many parts of Antarctica today.

In contrast to Pleistocene conditions, the earlier glacial record (Oligocene to Miocene and possibly Pliocene) was characterized by warmer (probably temperate) ice sheets, and the interglacials may have been cool-temperate with vegetation (e.g. the beech *Nothofagus*). Marine sediments may even have been deposited in a seaway across the middle of the continent (Webb et al. 1984). The preservation of interglacial muds on the shelf has not been as effective as that of diamictite, however.

8.10.2 Arctic

A northern hemisphere analogue of Prydz Bay is the south-west margin of the Barents Shelf (Vorren et al. 1989, 1990), but its glacial record is much shorter, with regional glaciation only starting about 0.8 m.y. ago (although older glaciomarine sediments may exist). According to Solheim & Kristoffersen (1984), the Barents Shelf was covered by ice sheets five to ten times during late Cenozoic time. The data comes from a combination of seismic data and a number of shallow cores and boreholes through sea bottom sediments. Broadly, the architecture of the shelf is characterized by a prograding wedge partly capped by a flat-lying glacial sequence. These sediments are underlain in the south by faulted and tilted sedimentary rocks mainly of Mesozoic age, while at the continental slope a wedge of supposedly Cenozoic age occurs. The shelf area is dominated by a broad submarine valley (probably glacially overdeepened), the Bjørnøyrenna (Bear Island Trough), which attains a depth of 500 m and which drained much of the ice that covered the Barents Shelf. At the edge of the shelf is a huge trough-mouth fan, similar in size to that in Prydz Bay and on which gravity sliding was an important process. A seismically-interpreted profile across the continental shelf and trough mouth fan is shown in Figure 8.23e (Vorren et al. 1990). Unit *TeE* is regarded as the glacigenic part of the sequence, and the regional unconformity at its base was ascribed to glacial erosion, and considered to represent the glacial-preglacial boundary. The age of this boundary was tentatively put at 0.8 million years when a change to larger ice volumes became apparent (Vorren 1990). Unit *Te* comprises four major subunits showing stepwise progradation of the palaeoshelf edge. Progradation of the continental shelf probably took place during glacial periods when the Barents Ice Sheet extended close to the shelf edge. Intervening interglacials were probably characterized by sediment starvation at and beyond the shelf edge. Assuming 0.8 million years is the age of the base of Unit *TeE* shelf progradation was about 30 km/million years, i.e. approximately ten times faster than in preglacial times (Vorren 1990).

Sediments from a borehole through the superficial layer on the shelf comprise overconsolidated glacigenic diamicts. Other facies may be related to proximal prograding diamict apron deposits, or derivation from meltwater streams associated with ice fed from the Scandinavian highlands to the south.

The geometry of the outer shelf units provides few clues as to the nature of the sediments. However, the suggestion that ice reached the palaeoshelf break on several occasions implies that the slope sediments also belong to a diamict-apron complex, a view supported by the Prydz Bay data.

8.11. Preservation potential of glaciomarine sediments

Deposition on continental shelves provides the best opportunity for glacigenic sediments to be preserved in the geological record. Hence most sequences from pre-Pleistocene glacial periods tend to be influenced, if not dominated, by continental shelf deposits. Shelf sequences generally show hiatuses, because of repeated advances of ice across the shelf. Erosion is most intense on the inner shelf where the ice is thicker and the early part of the glacial record may be lost. The outer shelf also loses part of its record as ice advances across it, but considerable build-up of diamictite from grounded ice and as glaciomarine sediment may occur there. Such areas provide a long but discontinuos record of sedimentation. The most complete direct record of glaciation occurs within the prograding sequences that underlie the outer shelf. These accumulate preferentially during periods when ice is grounded to the continental shelf-break. Interglacial periods may be represented by sediment starvation on the palaeoshelf-slope, and subject to current erosion, so that the sedimentary record shows a bias towards glacial conditions.

The tectonic setting of many high latitude continental shelves facilitates preservation of glaciomarine sediments. Many are slowly subsiding and allow hundreds, perhaps thousands, of metres of glacigenic sediment to accumulate, e.g. the Ross Sea, Weddell Sea and Prydz Bay in Antarctica. A rather different picture emerges from the Gulf of Alaska. Here several kilometres of sediment have also accumulated, but primarily because of the huge supply from the rapidly uplifting mountains bordering the Gulf. Continental shelf sequences have been thrust up to form spectacularly well-exposed outcrops along the coast and on offshore islands. In contrast to continental shelf areas, deep ocean sediments may preserve a continuous history of glaciation, but since such sediments lack a direct glacial component other than ice-rafted debris, the record may be difficult to interpret. Furthermore, since oceanic crust and associated sediments older than 200 m.y. have only been recorded in fragments, most having been subducted, the longer term preservation of deep sea glacigenic sediments is poor.

GLOSSARY OF GLACIAL TERMS

Ablation. The process of wastage of snow or ice, especially by melting

Ablation area/zone. That part of the glacier surface, usually at lower elevations, over which ablation exceeds accumulation.

Abrasion. The wearing down of rock surfaces by rubbing and impact of debris-rich ice.

Accumulation area. That part of the glacier surface, usually at higher elevations, on which there is net accumulation of snow, which subsequently turns into firn and then glacier ice.

Antarctic Bottom Water. A cold sea-bottom current, originating beneath ice shelves in Antarctica, and flowing northwards at depth to influence the world's oceans.

Areal scouring. Large-scale erosion of bedrock in lowland areas by ice sheets.

Arête (from French). A sharp, narrow, often pinnacled ridge, formed as a result of glacial erosion from both sides.

Aufeis (from the German). River ice that forms as a result of continued discharge of water from a glacier after the winter freeze-up has begun. It comprises continuous sheets of columnar crystals of ice.

Avulsion. The process of channel switching in a braided river system.

Axial planar relationship. Referring to parallelism with the axial plane of a fold structure.

Basal glide. Deformation of an ice crystal along discrete bands called basal planes.

Basal plane. The plane within the hexagonal ice crystal that is normal to the optic axis.

Basal shear stress. The force exerted by an ice-mass on its bed.

Basal sliding. The sliding of a glacier over bedrock, a process usually facilitated by the lubricating effect of meltwater.

Basket-of-eggs topography. Extensive low-lying areas covered by small elongate hills called drumlins (q.v.).

Bedrock flutes. Ridges, rounded in cross-section, formed parallel to the direction of ice movement.

Bergschrund (from the German). An irregular crevasse, usually running across an ice slope in the accumulation area, where active glacier ice pulls away from ice that adheres to the steep mountainside

Bergstone mud. Homogeneous sandy mud containing scattered gravel, deposited mainly from suspension and from icebergs in a glacier-influenced fjord or bay.

Bergy bit. A piece of floating glacier ice up to several metres across, commonly derived as a result of disintegration of icebergs.

Boudin (from the French for sausage). A sausage-shaped block of less ductile material separated by a short distance from its neighbours within a more ductile medium. Boudins normally form perpendicular to the maximum compressive stress.

Boulder bed. A bed of glacigenic sediment in which the concentration of boulders is exceptionally high (e.g. 50% or more).

Boulder clay. An English term for till (q.v.), no longer favoured by glacial geologists.

Boulder pavement. A concentration of striated boulders at the top of a bed of poorly sorted sediment that collectively have been planed off by overriding ice. Pavements are formed as a result of glacial abrasion on previously deposited till in a terrestrial environment, or by touch-down of ice onto the sea floor in a glaciomarine setting.

Braided outwash fan. A fan-shaped feature comprising glaciofluvial sediment, emanating from a terrestrial glacier.

Braided stream. A relatively shallow stream that has many branches that commonly recombine and migrate across a valley floor. Braided streams typically form downstream of a glacier.

Breached watershed. A short, glacially eroded valley, linking two major valleys across a mountain divide.

Breccia. Coarse, angular fragments of broken rock or ice, cemented or frozen together into a solid mass.

Brown ice. Brown-coloured sea ice, commonly found around Antarctica. The colour is due to finely disseminated diatom fragments.

Calving. The process of detachment of blocks of ice from a glacier into water.

Cavetto form. A channel cut into steep rock faces orientated parallel to the valley sides. They are the product of glacial abrasion and/or meltwater. Overhanging upper lips and striations are typical.

Cavitation. Growth and collapse of bubbles in a fluid (e.g. subglacial meltwater) in response to pressure changes. Bubble collapse generates shock waves which result in enhanced erosion of subglacial channels.

Channel bar. An elongate ridge of sediment between channels in a braided river system. It usually shows fining of sediment in a downstream direction.

Chattermarks. A group of crescent-shaped friction cracks on bedrock, formed by the juddering effect of moving ice.

Chute. A vertical groove in solid bedrock, formed as a result of meltwater erosion at the ice/bedrock contact.

Cirque (from French). A armchair-shaped hollow with steep sides and backwall, formed as a result of glacial erosion high on a mountainside, and often containing a rock basin with a tarn (q.v.) (c.f. **corrie, cwm**).

Cirque glacier. A glacier occupying a cirque.

Coarse-tail grading. A sedimentary bed in which the size of the largest clasts decreases upwards, while the size of the remaining material remains constant.

Col (from French). A high-level pass formed by glacial breaching of an arête or mountain mass.

Cold glacier. A glacier in which the bulk of the ice is below the pressure melting point, although ice at the surface may warm up to the melting point in summer, while ice at the bed may also be warmed as a result of geothermal heating.

Cold ice. Ice which is below the pressure melting point, and therefore dry.

Compressing flow. The character of ice flow where a glacier is slowing down and the ice is being compressed and thickened in a longitudinal direction.

Coriolis force. The inertial force associated with variation in the tangential component of the velocity of a particle. It results in an apparent deflection in the centrifugal force generated by the rotation of the Earth.

Corrie (from Gaelic *coire*). A British term for cirque (q.v.).

Cover moraine. A patchy, thin layer of till, revealing, in part, the bedrock topography.

Crag-and-tail. A glacially eroded rocky hill with a tail of till formed downglacier of it.

Creep. Permanent deformation of a material under the influence of stress.

Crescentic fracture. A crescent-shaped crack resulting from friction between debris-rich ice and bedrock; they commonly occur in groups aligned parallel to ice-flow direction.

Crescentic gouge. A crescent-shaped scallop, usually several centimetres across, formed as a result of bedrock fracture under moving ice.

Crevasse. A deep V-shaped cleft formed in the upper brittle part of a glacier as a result of the fracture of ice undergoing extension.

Crevasse-filling. A crevasse formed at the base of a glacier and filled with soft sediment.

Crevasse traces. Long veins of clear ice a few centimetres wide, formed as a result of fracture and recrystallization of ice under tension without separation of the two walls; these structures commonly form parallel to open crevasses and extend into them. Thicker veins of clear ice resulting from the freezing of standing water in open crevasses are also called crevasse traces.

Cumulative (also **total** or **finite strain**). The total amount of strain that a material (e.g. rock, ice) has undergone, usually in response to the prolonged application of stress.

Cupola hill. An isolated hill with no obvious source depression, but having the general characteristics of an assemblage of ice-thrust masses.

Curved (winding) channel. Sinuous channel cut into bedrock, usually by a subglacial stream under high pressure.

Cwm. The Welsh term for cirque (q.v.), also sometimes used more generally outside Wales.

Cyclopel. A graded silt-mud couplet (laminated) formed by tidal processes operating in a glacier-influenced fjord.

Cyclopsam. A graded sand-mud couplet (laminated) formed by tidal processes operating in a glacier-influenced fjord.

Debris-flow. A type of gravity flow involving the movement on inhomogeneous, unconsolidated material down a slope as a slurry. Total disaggregation of the material takes place, but little fine material is carried away in suspension.

De Geer moraines. A group of moraines formed subglacially, transverse to flow, beneath a glacier terminating in a lake. They have the form of discrete narrow ridges, which may be well spaced.

Delta-fan complex. A prograding sequence of glaciomarine sediment, gravity-flow material and basal glacial debris, formed at the grounding-line of an ice shelf or ice stream on the continental shelf.

Delta-moraine (also kame delta). A delta complex formed in an ice-frontal position, commonly below water-level, where subglacial streams enters a proglacial lake or the sea. Such features are composed of a mixture of glaciofluvial and glacial debris.

Diagenesis. The changes that occur in a sedimentary sequence following deposition (e.g. the alternation of snow to glacier ice or soft sediment to rock).

Diamict. A non-sorted terrigenous sediment containing a wide range of particle sizes. Embraces both **diamictite** (lithified) and **diamicton** (unconsolidated).

Diamict apron (cf. till delta). A prograding sequence comprising diamict and reworked sediments, formed just seaward of the grounding-line/zone at a break in slope on the sea floor.

Diffluence. The processes whereby ice in one valley overflows a col or group of cols into an adjacent valley.

Dilation. The reduction in volume of a material. Commonly occurs in bedrock following ice-removal, and leads to the development of cracks.

Dirt-cone. A thin veneer of debris, draping a cone of ice up to several metres high, formed as a result of the debris locally retarding ablation of the glacier surface.

Diurnal variation. Variations taking place on a day-night cycle.

Dome. A smooth, rounded boss of glacially abraded bedrock, commonly exceeding hundreds of metres in diameter.

Dropstone. A relatively large clast which falls through the water column into soft sediment, disrupting the bedding or laminae. Draping of sediment over the top of the clast subsequently occurs. In a glacial context dropstones are released from icebergs.

Drumlin (from Gaelic). A streamlined hillock, commonly elongated parallel to the former ice flow direction, composed of glacial debris, and sometimes having a bedrock core; formed beneath an actively flowing glacier.

Drumlinoid ridges (drumlinized ground-moraine). Elongate, strongly linear ridges, intermediate between drumlins and fluted moraine (q.v.).

Englacial debris. Debris dispersed throughout the interior of a glacier, derived either from the surface through burial in the accumulation area and through falling into crevasses, or from the uplifting of basal debris by thrusting processes.

Englacial stream. A meltwater stream that has penetrated below the surface of a glacier, and is making its way towards the bed.

Erratic. A boulder or large block of bedrock that is being or has been transported away from its source by a glacier.

Esker (from Gaelic). A long, commonly sinuous ridge of sand and gravel, deposited by a stream in a subglacial tunnel

Equilibrium line/zone. The line or zone on a glacier surface where a year's ablation balances a year's accumulation (cf. firn line). It is determined at the end of the ablation season, and commonly occurs at the boundary between superimposed ice (q.v.) and glacier ice.

Exfoliation. The process of removal of sheets of bedrock along joints parallel to the rock surface.

Extending flow. The character of ice flow where a glacier is accelerating and the ice is being stretched and thinned in a longitudinal direction.

Facies (sing. & plur.). A sediment type characterized by an assemblage of features, including lithology, texture, sedimentary structures, fossil content, geometry, bounding relations. **Lithofacies** refers to a particular lithological type.

Facies architecture. The large-scale two- or three-dimensional geometry of facies associations, usually on a basin-wide scale and revealed in seismic profiles.

Facies association. The grouping of facies into an environmentally coherent assemblage, e.g. a glaciomarine facies association, usually applied to vertical sedimentary logs.

Fault. A displacement in a glacier formed as a result of fracture of the ice without separation of the walls. It is recognised by the discordance of layers in the ice on either side of the fracture. A **normal fault** is a high-angle fracture resulting from the maximum compressive stress acting vertically and the intermediate and least compressive stresses horizontally. A **thrust fault** is a low-angle fracture in which the maximum compressive stress acted horizontally and the least compressive stress vertically. A **strike-slip fault** is a fracture showing sideways displacement where both maximum and minimum compressive stresses both acted horizontally.

Finite strain (see Total strain).

Firn (from German). Dense, old snow in which the crystals are partly joined together, but in which the air pockets still communicate with each other

Firn line. The line on a glacier that separates bare ice from snow at the end of the ablation season.

Fjord (from Norwegian) (**Fiord** in North America and New Zealand). A long, narrow arm of the sea, formed as a result of erosion by a valley glacier.

Fluted moraine (flutes). rounded, strongly linear ridges, up to a few metres in width and height, usually formed in association with lodgement till on land. Similar features have been recorded in shallow glaciomarine settings.

Foliation. Groups of closely spaced, often discontinuous, layers of coarse bubbly, coarse clear and fine grained ice, formed as a result of shear or compression at depth in a glacier.

Fold. Layers of ice that have been deformed into a curved form by flow at depth in a glacier. **Isoclinal folds** have parallel limbs and thickened hinges. **Similar folds** have a thickened hinge and thinned limbs. **Parallel folds** maintain a uniform layer-thickness around the fold. **Recumbent folds** are those with near-horizontal axes and are commonly associated with thrust-faults.

Floe (see **Raft**).

Gendarme (from the French for policeman). A pinnacle of rock on the crest of a narrow ridge, especially an arête.

Gentle hill. A mound of till resting on a detached block of bedrock.

Geothermal heat. The heat output from the earth's surface. This affects glaciers especially in the polar regions, by warming the basal zone to the pressure melting point.

Gilbertian delta. A delta produced by a stream with a high bed-load entering a quiet water-body. Such deltas have steep slopes, and avalanching causes the delta to advance and create a single cross-bedded set of sand and gravel.

Glacial debris. Material in the process of being transported by a glacier in contact with glacier ice.

Glacial drift. A general term embracing all rock material deposited by glacier ice, and all deposits of predominantly glacial origin deposited in the sea from icebergs, and from glacial meltwater.

Glacial period/glaciation. A period of time when large areas (including present temperate latitudes) were ice-covered. Numerous glacial periods have occurred within the last few million years, and are separated by interglacial periods (q.v.).

Glacial trough. A valley or fjord, often characterized by steep sides and a flat bottom, with multiple basins, resulting primarily from abrasion by strongly channelled ice.

Glaciated. The character of land that was once covered by glacier ice in the past (cf. *glacierized*).

Glacier. A mass of ice, irrespective of size, derived largely from snow, and continuously moving from higher to lower ground, or spreading over the sea.

Glacier ice. Any ice in, or originating from, a glacier, whether on land or floating on the sea as icebergs.

Glacierized. The character of land currently covered by glacier ice (cf. glaciated).

Glacier karst. Debris-covered stagnant ice, sometimes found at the snout of a retreating glacier, with numerous lake-bearing caverns and tunnels.

Glacier sole. The lower few metres of a (usually sliding) glacier that are rich in debris picked up from the bed.

Glacier table. A boulder sitting on a pedestal of ice, resulting from the protective effect of the rock mass on ablation of the ice surface during sunny weather.

Glacier tongue (see **ice tongue**).

Glacigenic sediment. Sediment of glacial origin. The term is used in a broad sense to embrace sediments with a greater or lesser component derived from glacier ice.

Glaciomarine sediment. A mixture of glacigenic and marine sediment, deposited more or less contemporaneously.

Glaciotectonic deformation (glaciotectonism). The process whereby subglacial and proglacial sediment and bedrock is disrupted by ice-flow. It is usually manifested in the form of distinct topographic features in which folds and thrusts are commonplace.

Glen's Flow Law. The empirical relationship which describes the manner in which ice deforms in response to an applied stress. First proposed by the British physicist John Glen following experimental studies in the early 1950's.

Gravity flowage. The process of transport of unconsolidated sediment down a slope, either subaerially or subaquatically. The term embraces debris-flow and turbidity currents at opposite ends of the spectrum.

Groove. A glacial abrasional form, with striated sides and base, orientated parallel to the ice-flow direction. Grooves are often parabolic in cross-section and up to several metres wide and deep.

Grounding-line (or grounding-zone). The line or zone at which an ice-mass enters the sea or a lake and begins to float, e.g. in the inner parts of an ice shelf or an ice stream.

Grounding-line fan. A subaquatic fan made up of material emerging from a subglacial tunnel where a glacier terminates in the sea. Comprises a complex range of facies.

Hanging glacier. A glacier that spills out from a high level cirque or clings to a steep mountainside.

Hanging valley. A tributary valley whose mouth ends abruptly part way up the side of a trunk valley, as a result of the greater amount of glacial downcutting of the latter.

Highland icefield. A near-continuous stretch of glacier ice, but with an irregular surface that mirrors the underlying bedrock, and punctuated by nunataks (q.v.).

Hill-hole pair. A combination of an ice-scooped basin and a hill of ice-thrust, often slightly crumpled material of similar size, resulting from glaciotectonic processes.

Horn. A steep-sided, pyramid-shaped peak, formed as a result of the backward erosion of cirque glaciers on three or more sides.

Hummocky (ground-) moraine. Groups of steep-sided hillocks, comprising glacial drift, formed by dead-ice wastage processes. Some hummocky moraines may be arranged in a crude transverse-to-valley orientation, and reflect thrusting processes in the glacier snout. Both terrestrial and marine types are known.

Ice age. A period of time when large ice sheets extend from the polar regions into temperate latitudes. The term is sometimes used synonymously with glacial period (q.v.), or embraces several such periods to define a major phase in earth's climatic history.

Ice apron. A steep mass of ice, commonly the source of ice avalanches, that adheres to steep rock near the summits of high peaks.

Iceberg. A piece of ice of the order of tens of metres or more across that has been shed by a glacier into a lake or the sea.

Iceberg turbate. Sediment disturbed and added to by icebergs grounding on a sea- or lake-floor.

Ice cap. A dome-shaped mass of glacier ice, usually situated in a highland area, and generally defined as covering < 50,000 sq. km.

Ice cliff (ice wall). A vertical face of ice, normally formed where a glacier terminates in the sea, or is undercut by streams. These terms are also used more specifically for the face that forms at the seaward margin of an ice sheet or ice cap and rests on bedrock at or below sea level.

Ice-contact delta. Delta formed from a subglacial stream where a glacier terminates in a standing body of water. The top surface is in the intertidal zone.

Icefall. A steep, heavily crevassed portion of a valley glacier.

Ice-marginal channel. A water-channel formed at the margin of a (usually cold) valley glacier, inside the lateral moraine, if present.

Ice sheet. A mass of ice and snow of considerably thickness and covering an area of more than 50,000 sq km.

Ice shelf. A large slab of ice floating on the sea, but remaining attached to and partly fed by land-based ice.

Ice stream. Part of an ice sheet or ice cap in which the ice flows more rapidly, and not necessarily in the same direction as the surrounding ice. The margins are often defined by zones of strongly sheared, crevassed ice.

Ice stream boundary ridge. A submarine ridge of bedrock or soft sediment defining the former boundary between two ice streams as they crossed a continental shelf.

Ice tongue (or glacier tongue). An unconstrained, floating extension of an ice stream or valley glacier, projecting into the sea.

Interflow. Flow in which water entering a lake or the sea is of the same density as the main body, resulting in ready mixing of the two.

Interglacial (period). A period of time, such as the present day, when ice still covers much of the earth's surface, but has retreated to the polar regions.

Internal deformation. That component of glacier flow that is the result of the deformation of glacier ice under the influence of accumulated snow and firn, and gravity.

Isotopes. Varieties of elements, all with identical chemical, but not precisely equal physical , properties.

Jökulhlaup (from the Icelandic). A sudden and often catastrophic outburst of water from a glacier, such as when an ice-dammed lake bursts or an internal water pocket escapes.

Kame (from the Gaelic). A steep-sided hill of sand and gravel deposited by glacial streams adjacent to a glacier margin.

Kame delta (see **delta moraine**).

Kame field. A large area covered by numerous discrete kames.

Kame-moraine. An end-moraine complex that has been substantially reworked by glaciofluvial processes.

Kame plateau. An extensive area of ice-contact sediments formed adjacent to a glacier that has not yet been dissected.

Kame terrace. A flat or gently sloping plain, deposited by a stream that flowed towards or along the margin of a glacier, but left above the hillside when the ice retreated.

Kettle (or **kettle hole**). A self-contained bowl-shaped depression within in an area covered by glacial stream deposits, and often containing a pond. A kettle forms as a result of the burial of a mass of glacier ice by stream sediment and its subsequent melting.

Kinematic wave. The means whereby mass-balance changes are propagated downglacier. The wave has a constant discharge, and moves at a velocity faster than the ice velocity. Kinematic waves are visible as bulges on the ice surface; when they reach the snout, the glacier is able to advance.

Knock-and-lochan topography (a Scottish term). Rough, ice-abraded, low-level landscape, comprising small hills of exposed bedrock, and rock basins with small lakes and bogs.

Lahar. Debris-flow consisting primarily of volcanic ash and lava boulders. Heavy rain and/or melting snow and ice during a volcanic eruption mixes with the loose deposits and forms fast moving tongues of slurry.

Laminite. A laminated sediment.

Large composite-ridge. A ridge, or series of ridges, in front of a glacier and comprising large slices of upthrust, contorted sedimentary bedrock and drift. They exceed 100 m in height.

Lee-side cone (see **crag-and-tail**).

Levée. A bank of sediment bordering a stream channel, constructed during period of bank-full discharge.

Linear moraine. A composite feature on a continental shelf, comprising till tongue (q.v.) sediments overlain by subaquatic ice recessional features.

Lithofacies (see **Facies**).

Little Ice Age. That period of time that led to expansion of valley and cirque glaciers world-wide, with their maximum extents being attained in many temperate regions about 1700-1850 AD or around 1900 in Arctic regions.

Lodgement. The process whereby basal glacial debris is "plastered" on to the substrate beneath an actively moving glacier.

Loess. Wind-blown sediment of silt-grade, often derived as a result of winnowing of fines from glacial outwash plains.

Lonestone. An isolated clast of pebble or larger size in a fine-grained matrix.

Lunate fracture. Moon-shaped crescentic fracture with the horns and steeper face at the downglacier end.

Mass balance (or **mass budget**). A year-by-year measure of the state of health of a glacier, reflecting the balance between accumulation and ablation. A glacier with a positive mass balance in a particular year gained more mass through accumulation than was lost through ablation; the reverse is true for negative mass balance.

Megablock (also known as a **raft**). A large piece of glacially transported bedrock, buried in drift. Such blocks are thin in relation to their linear dimensions, which may exceed 100 m.

Mixton(ite). a synonym for **diamicton(ite)**, but now largely redundant.

Morainal bank. A bank formed below water in front of a stable ice-front in a fjord or lake, as a result of lodgement, melting, dumping, push and squeeze processes.

Moraine. Distinct ridges or mounds of debris laid down directly by a glacier or pushed up by it. The material is typical till, but fluvial, lake or marine sediments may also be involved. Longitudinal moraines include a **lateral moraine** which forms along the side of a glacier; a **medial moraine** occurring on the surface where two streams of ice merge; and **fluted moraine** which forms a series of ridges beneath the ice, parallel to flow. Transverse moraines include a **terminal moraine** which forms at the furthest limit reached by the ice, a **recessional moraine** which represents a stationary phase during otherwise general retreat, and a set of **annual moraines** representing a series of minor winter readvances during a general retreat. A **push**

moraine is a more complex form, developed especially in front of a cold glacier during a period of advance.

Moulin. A water-worn pot-hole formed where a surface meltstream exploits a weakness in the ice. Many moulins are cylindrical, several metres across, and extend down to the glacier bed, although often in a series of steps.

Moulin-kame. A mound of debris up to several metres high that accumulated in the bottom of a moulin.

Nail-head striation. An asymmetric, relatively short scratch-mark, blunt at the downstream end and tapering towards the upstream end.

Nival flood. The flood that accompanies the phase of rapid snow-melt in spring/early summer.

Nunatak (from Inuit). An island of bedrock or mountain projecting above the surface of an ice sheet or highland icefield.

Nye channel (named after a British physicist). A channel cut into bedrock by subglacial meltwater under high pressure. Usually < 1 m across; commonly deeper than they are wide.

Ogives. Arcuate bands or waves, with the apex pointing down-glacier that develop in an icefall. Alternating light and dark bands are called *banded ogives* or *Forbes' bands*; *wave ogives* are a second type. Each pair of bands or one wave and trough represents a year's movement through the icefall.

Outwash plain. A flat spread of debris deposited by meltwater streams emanating from a glacier (cf. sandar).

Overconsolidated. A geotechnical term to indicate a more indurated sediment than would be expected from the depth observed below the surface. Overconsolidation may be due to loading by an ice-mass.

Overflow. A flow of water entering a lake or the sea, having a lesser density than the main body of water, and therefore remaining at the surface as a distinct layer.

Overflow channel. A channel cut through a hill or ridge by meltwater, as a result of ponding back by an ice mass.

Permafrost. Ground which remains permanently frozen. It may be hundreds of metres thick and only the top few metres thaws out in summer.

Piedmont glacier (from the French). A glacier which spreads out as a wide lobe on leaving the mountains or a narrow trough.

Pitted plain. A plain of glaciofluvial sediment with numerous lake-filled depressions (kettles), resulting from the melting of buried blocks of glacier ice.

Plastically moulded (or p-) forms. Smooth rounded forms of various types cut into bedrock by the erosive power of ice and meltwater under high pressure combined.

Plough-mark. (i) A groove or furrow caused by the impact and movement of icebergs across the sea or lake floor, or along a beach under the influence of tides. (ii) A groove formed by a stone at the ice/bedrock interface scoring soft sediment and pushing up a small ridge in front.

Polynya. An area of ocean with sea ice that remains relatively ice-free in winter, due to the upwelling to the surface of warmer water.

Portal. The open archway that develops when a meltwater stream emerges at the snout of a glacier.

Pressure melting. Enhanced melting resulting from the effect of ice impinging on bedrock, or by an object under stress impinging on ice.

Pressure release. The process of fracturing of bedrock that takes place as a result of the removal of stress as an overlying body of ice melts.

Pure shear (non-rotational strain). Deformation of a substance by extension normal to an applied compressive stress and extension parallel to it. This results in no change to the strained area, nor to rotation of the cumulative strain axes.

Raft (see **Megablock**).

Randkluft (from the German). The narrow gap that develops between a rock face and steep firn and ice at the head of a glacier.

Rat-tail. A minor ridge, parallel to striations, extending downglacier from a knob of more resistant rock.

Regelation ice. Ice which is formed from meltwater as a result of the lowering of pressure beneath a glacier.

Rejuvenated (or regenerated) glacier. A glacier which develops from ice avalanche debris beneath a rock cliff.

Rhythmite. A sedimentary unit comprising a repetitive succession of sedimentary types, e.g. mud/sand, irrespective of time and relative thickness. **Varve** is one specific type of rhythmite.

Ribbon lake. A long, narrow lake resulting from glacial erosion, and commonly containing multiple basins.

Riegel (from the German). A rock barrier that extends across a glaciated valley, usually comprising harder rock, and often having a smooth up-valley facing slope and a rough down-valley facing slope.

Rinnen (from the German) (see **Tunnel valley**).

Roche moutonnée (from French). An rocky hillock with a gently inclined, smooth up-valley facing slope resulting from glacial abrasion, and a steep, rough down-valley facing slope resulting from glacial plucking

Rock basin. A lake- or sea-filled bedrock depression carved out by a glacier.

Rock flour. Bedrock that has been pulverised at the bed of a glacier into clay- and silt-sized particles. It commonly is carried in suspension in glacial meltwater streams, which take on a milky appearance as a result.

Rockslide. An accumulation of rock debris resulting from a catastrophic rockfall. Sometimes during the fall, debris becomes airborne and accumulates as a pulverized mass, far distant from the source.

Rogen (ribbed) moraine (named after a Swedish lake). Large-scale, transversely orientated, irregular ridges of complex origin, including thrusting.

Röthlisberger channel (named after a Swiss glaciologist). A channel incised upwards into the base of a glacier by a subglacial meltstream.

Sandar (plur. **Sandur**) (from Icelandic). Laterally extensive flat plains of sand and gravel with braided streams of glacial meltwater flowing across them. They are usually not bounded by valley walls and commonly form in coastal areas.

Sea ice. Ice that forms by freezing of the sea (cf. ice shelf and icebergs which float on the sea).

Sedimentary model. A pictorial representation of the processes and sedimentary products that contribute to a particular sedimentary environment.

Sedimentary stratification. The annual layering that forms from the accumulation of snow, and is preserved in firn and sometimes in glacier ice.

Seismic stratigraphy. The application of seismic profiling techniques to the interpretation of the stratigraphy of inaccessible (subsurface) areas, notably offshore.

Sérac (from French). A tower of unstable ice that forms between crevasses, often in icefalls or other regions of accelerated glacier flow.

Sequence stratigraphy. The determination of the stratigraphy of sedimentary sequences by linking unconformities of temporal significance (sequence boundaries) to sea-level changes, thereby aiding regional correlation. A development of seismic stratigraphy, but applicable to well-exposed onshore sequences.

Shear (or thrust) ridge. A ridge or line of debris cropping out at the surface of a glacier, thrust up from subglacial position, especially near its snout.

Shelf moraine. A prominent ridge interpreted as an end-moraine complex, formed at the edge or in the middle of a continental shelf.

Shelf valley (see **Submarine trough**).

Sichelwanne. A crescent-shaped depression or scallop-like feature cut into bedrock, probably largely the result of meltwater erosion. The axis is coincident with the ice-flow direction, and the horns of the feature pointing forwards.

Sill. A submarine barrier of rock or moraine that occurs at the mouth of, or between rock basins in, a fjord.

Simple shear. Deformation of a substance by displacement along discrete (often closely spaced) surfaces or shear planes. The orientations of the principal axes of cumulative strain rotate as deformation proceeds. This type of deformation can be demonstrated by gradually smearing out a stack of cards.

Slope valleys. Groups of gullies forming a dendritic pattern on the continental slope associated with subglacial discharge at times when the ice advanced to the continental shelf edge.

Slumping. The process of downslope mass-movement whereby the internal organization of the mass of debris is not totally disaggregated. Deformation may be evident in the form of recumbent fold structures.

Small composite-ridge. Smaller scale versions (<100 m in height) of large composite-ridges (q.v.) and comprising mainly unconsolidated sediment thrust up into a series of ridges by a glacier.

Snout. The lower part of the ablation area of a valley glacier.

Snow. An agglomeration of precipitated ice crystals, most of which are star-shaped and delicate, forming a low density mass with a high air content

Snow swamp. An area of saturated snow lying on glacier ice. If movement is triggered on a slope a slush avalanche may develop. The snow swamp may not be visible until one steps into it.

Sole (see **Glacier sole**).

Sole thrust. The lowest thrust in a deformed sedimentary sequence which defines the plane below which no further displacements occur.

Strain. The amount by which an object becomes deformed under the influence of stress.

Strain-softening. The effect whereby, under a constant stress, the strain-rate increases with time.

Stauchmoräne (from the German). A collective term for ice-push ridges of the large and small composite-ridge types (q.v.).

Stress. The force applied to an object.

Striae. Linear, fine scratches formed by the abrasive effect of debris-rich ice sliding over bedrock. Intersecting sets of striae are formed as stones are rotated or if the direction of ice flow over bedrock changes.

Striated. The scratched state of bedrock or stone surfaces after the ice has moved over them.

Subaquatic. The state of being underwater; in this context commonly refers to glaciolacustrine and glaciomarine environments.

Subglacial debris. Debris which has been released from ice at the base of a glacier. It usually shows signs of rounding due to abrasion at the contact between ice and bedrock.

Subglacial delta (cf. **till delta, diamict apron**). Prograding wedge of glacial, marine and recycled debris formed seawards of the grounding-line of ice shelves or ice streams.

Subglacial gorge. A steep, often vertically sided gorge cut into bedrock by a subglacial stream under high pressure.

Subglacial stream. A stream which flows beneath a glacier, and which usually cuts into the ice above to form a tunnel.

Sublimation. The process whereby a material (e.g. ice) passes from the solid to the vapour state directly, without melting intervening.

Superimposed drainage. A drainage system unrelated to current bedrock structure, inherited from an earlier phase of development on once overlying bedrock, now removed.

Superimposed ice. Ice which forms as a result of the freezing of water-saturated snow. It commonly forms at the surface of a glacier between the equilibrium line and the firn line, and provides additional mass to the glacier.

Supraglacial debris. Debris which is carried on the surface of a glacier. Normally this is derived from rockfalls and tends to be angular.

Supraglacial stream. A stream that flows over the surface of a glacier. Most supraglacial streams descend via moulins (q.v.) into the depths or to the base of a glacier.

Surge. A short-lived phase of accelerated glacier flow during which the surface becomes broken up into a maze of crevasses. Surges are often periodic and are separated by longer periods of relative inactivity or even stagnation.

Tabular iceberg. A flat-topped iceberg that has become detached from an ice shelf, ice tongue or floating tidewater glacier.

Tabular moraine. Irregular morainic complexes formed in an intermediate continental shelf position.

Tarn. A small lake occupying a hollow eroded out by ice or dammed by moraine; especially common in cirques.

Temperate glacier (see **warm glacier**).

Thermal regime. The state of a glacier as determined by its temperature distribution

Thrust. A low angle fault, usually formed where the ice is under compression. Thrusts commonly extend from the bed and are associated with debris and overturned folds.

Thrust ridge (see **shear ridge**).

Tidewater glacier. A glacier that terminates in the sea, usually in a bay or fjord.

Till. A mixture of mud, sand and gravel sized material deposited directly from glacier ice.

Till delta. A prograding wedge of till that has been transported in conveyorbelt-like fashion to the grounding-line of an ice shelf or ice stream (cf. diamict apron).

Tillite. The lithified equivalent of **till**.

Tilloid. A till-like deposit, but of uncertain origin (a term rarely used today).

Till-pellet. A lump of till that was frozen when incorporated into the host sediment.

Till-plain. A nearly flat or slightly rolling and gently inclined surface, underlain by a near-continuous cover of thick till.

Till tongue. An inclined wedge of till and other material associated with the grounding-line of an oscillating marine ice margin.

Tongue. That part of a valley glacier that extends below the firn line.

Total strain (see **Cumulative strain**).

Transfluence. The large-scale breaching of a mountain-range by glaciers emanating from an ice sheet, with all cols being occupied by discharging ice.

Transverse ridge (see **De Geer moraines**).

Trim-line. A sharp line on a hillside marking the boundary between well vegetated terrain that has remained ice-free for a considerable time, and poorly vegetated terrain that until relatively recently lay under glacier ice. In many areas the most prominent trim-lines date from the Little Ice Age (q.v.).

Trough-head (or **trough-end**). The steep, plucked transverse rockface at the head of a glacial trough.

Trough-mouth fan. A large-scale prograding, arcuate fan at the mouth of a continental shelf trough. The fan extends out from the shelf-edge into the deep ocean basin.

Tunnel valley (Rinnen). A large subglacial, steep-sided channel cut into soft sediment or bedrock by meltwater. The channel may have a reverse gradient in places.

Turbidity current. A high-density, debris-laden current that flows down a slope. The resulting **turbidite** is a sediment characterized by graded bedding.

Unconformity. A discontinuity in the annual layering in firn or ice, resulting from a period when ablation cut across successive layers.

Underflow. A flow of water of greater density than the lake or sea into which it flows, thereby descending a slope as a bottom-hugging current or turbidity flow.

Valley glacier. A glacier bounded by the walls of a valley, and descending from high mountains, from an ice cap on a plateau, or from an ice sheet.

Valley train. A braided-river system that extends across the whole width of a valley between steep-sided mountains.

Varve. A sedimentary bed or lamina, or sequence of laminae deposited in a body of still water (usually a lake) and representing one year's accumulation.

Varvite. A lithified varve.

Warm (temperate) glacier. A glacier whose temperature is at the pressure melting point throughout, except for a cold wave of limited penetration that occurs in winter.

Warm ice. Ice which is at the melting point regardless of pressure. The temperature may be slightly below 0°C at the base of a glacier where the ice is under high pressure.

Whaleback. A smooth, scratched, glacially eroded bedrock knoll several metres high, and resembling a whale in profile.

BIBLIOGRAPHY

Aber, J. S., D. G. Croot, M. M. Fenton 1989. *Glaciotectonic landforms and structures.* Dordrecht: Kluwer.

Allen, C. R., W. B. Kamb, M. F. Meier, R. P. Sharp 1960. Structure of the lower Blue Glacier, Washington. *Journal of Geology* 68, 601--625.

Alley, D. W. & R. M. Slatt 1976. Drift prospecting and glacial geology in the Sheffield Lake - Indian Pond area, northcentral Newfoundland. In R. F. Leggett (ed), *Glacial till.* Royal Society of Canada Special Publ. No. 12, 249--266.

Alley, R. B., D. D. Blankenship, S. T. Rooney, C. R. Bentley 1989. Sedimentation beneath ice shelves - the view from Ice Stream B. *Marine Geology* 85, 101--120.

Allison, I. F. 1979. The mass budget of the Lambert Glacier drainage basin, Antarctica. *Journal of Glaciology* 22, 223--235.

Anderson, J. B. 1983. Ancient glacial-marine deposits: their spatial and temporal distribution. In B. F. Molnia (ed), *Glacial-marine sedimentation*, 10--92. New York: Plenum Press.

Anderson, J. B. & G. M. Ashley 1991. *Glacial marine sedimentation; Paleoclimatic Significance.* Geological Society of America Special Paper 261.

Anderson, J. B. & L. R. Bartek 1992. Cenozoic glacial history of the Ross Sea revealed by intermediate resolution seismic reflection data combined with drill site information. In *The Antarctic Paleoenvironment: A perspective on global change.* American Geophysical Union, Antarctic Research Series Vol. 56, 231--263.

Anderson, J. B. & B. F. Molnia 1989. *Glacial-marine sedimentation.* Washington: American Geophysical Union Short Course in Geology No. 9 (28th Int. Geol.Congress).

Anderson, J. B., D. D. Kurtz, E. W. Domack, K. M. Balshaw 1980. Glacial and glacial marine sediments of the Antarctic continental shelf. *Journal of Geology* 88, 399--414.

Anderson, J.B., C. Brake, E. W. Domack, N. Myers, R. Wright 1983. Development of a polar glacial-marine sedimentation model from Antarctic Quaternary deposits and glaciological information. In *Glacial-marine sedimentation*, In B. F. Molnia (ed), 233--264. New York: Plenum Press.

Andrews, J. T. & C. L. Matsch 1983. *Glacial marine sediments and sedimentation: an annotated bibliography.* Norwich: GeoAbstracts No. 11.

Armstrong, T., B. Roberts, C. Swithinbank 1973. *Illustrated glossary of snow and ice.* Cambridge: Scott Polar Research Institute, 60pp.

Ashley, G. M. 1975. Rhythmic sedimentation in glacial lake Hitchcock, Massachusetts-Connecticut. In Jopling, A. V. & B. C. McDonald (eds) *Glaciofluvial and glaciolacustrine sedimentation*, 304--320. Tulsa: Society of Economic Paleontologists and Mineralogists, Spec. Publ. 23.

Ashley, G. M. 1989. Classification of glaciolacustrine sediments. In: Goldthwait, R.P. and C. L. Matsch (eds), *Genetic classification of glacigenic deposits*, 243--260. Rotterdam: Balkema.

Banerjee, I & B. C. McDonald 1975. In Jopling, A. V. & B. C. McDonald (eds) *Glaciofluvial and glaciolacustrine sedimentation.* Tulsa: Society of Economic Paleontologists and Mineralogists, Spec. Publ. 23.

Barrett, P. J. 1975. Characteristics of pebbles from Cenozoic marine glacisal sediments in the Ross Sea (DSDP Sites 270--274) and the south Indian Ocean. In D. E. Hayes & L. A. Frakes (eds), *Initial Reports, Deep Sea Drilling Project, Leg 28*, 769--784. Washington: US Government Printing Office.

Barrett, P.J. 1980. The shape of rock particles' a critical review. Sedimentology 27, 291-303.

Barrett, P.J. (ed.) 1986. *Antarctic Cenozoic history from the MSSTS-1 drillhole, McMurdo Sound.* Wellington: DSIR Bull. 237, 174 pp.

Barrett, P.J. 1989a. Sediment texture. In P. J. Barrett (ed), *Antarctic Cenozoic history from the CIROS-1 drillhole, McMurdo Sound, Antarctica*, 49--58. Wellington: DSIR Bulletin 245.

Barrett, P.J. (ed) 1989b. *Antarctic Cenozoic history from the CIROS-1 drillhole, McMurdo Sound, Antarctica.* Wellington: DSIR Bulletin 245, 251pp.

Barrett, P. J. & M. J. Hambrey 1992. Plio-Pleistocene sedimentation in Ferrar Fiord, Antarctica. *Sedimentology* 39, 109--123.

Barrett, P.J., A. R. Pyne, B. L. Ward 1983. Modern sedimentation in McMurdo Sound, Antarctica. In R. L, Oliver, P. R. James & J. B. Jago (eds) *Antarctic Earth Science*, 550--555. Canberra: Australian Academy of Science and Cambridge: Cambridge University Press.

Barrett, P. J., M. J. Hambrey, D. M. Harwood, A. R. Pyne & P.-N. Webb 1989. Synthesis. In Barrett, P. J. (ed.) *Antarctic Cenozoic history from the CIROS-1 drillhole, McMurdo Sound, Antarctica*. Wellington: DSIR Bulletin 245, 241--251.

Barron, J., B. Larsen & Shipboard Scientific Party, 1989. *Leg 119, Kerguelen Plateau and Prydz Bay, Antarctica*. Proceedings of the Ocean Drilling Program, Vol. 119, Part A, 943pp.

Barron, J., Larsen, B.& Shipboard Scientific Party, 1991. *Leg 119, Kerguelen Plateau and Prydz Bay, Antarctica*. Proceedings of the Ocean Drilling Program, Vol. 119, Part B, Scientific Results.

Beuf, S., B. Biju-Duval, O. de Charpal, P. Rognon, O. Gariel, A. Bennacef 1971. *Les grès du Paléozoique inférieur au Sahara. Sédimentation et discontinuité. Evolution structurale d'un craton*. Paris: Edit. Technip.

Bjørlykke, K., B. Bue, A. Elverhøi 1978. Quaternary sediments in the western part of the Barents Sea and their relation to the underlying Mesozoic bedrock. *Sedimentology* 25, 227--246.

Boothroyd, J. C. & G. M. Ashley 1975. Processes, bar morphology and sedimentary structures on braided outwash fans, northeastern Gulf of Alaska. In Jopling, A. V. & B. C. McDonald (eds) *Glaciofluvial and glaciolacustrine sedimentation*, 193--222. Tulsa: Society of Economic Paleontologists and Mineralogists, Spec. Publ. 23.

Borns, H.W. & C. L. Matsch 1989. A provisional genetic classification of glaciomarine environments, processes and sediments. In: R. P. Goldthwait & C. L.Matsch (eds), *Genetic classification of glacigenic deposits* 261-266. Rotterdam: Balkema.

Bouchard, M. A. 1989. Subglacial landforms and deposits in central and northern Québec, Canada, with emphasis on Rogen moraines. *Sedimentary Geology* 62, 293--308.

Boulton, G.S. 1967. The development of a complex supraglacial moraine at the margin of Sørbreen, Ny Friesland, Vestspitzbergen. *Journal of Glaciology* 6, 717--736.

Boulton, G.S. 1968. Flow tills and related deposits on some Vestspitzbergen glaciers. *Journal of Glaciology* 7, 391--412.

Boulton, G.S. 1970. The deposition of subglacial and melt-out tills at the margins of certain Svalbard glaciers. *Journal of Glaciology* 9, 231--245.

Boulton, G.S. 1971. Till genesis and fabric in Spitzbergen, Svalbard. In R. P. Goldthwait (ed), *Till, a symposium*, 41--72. Columbus: Ohio State University Press.

Boulton, G.S. 1972. Modern Arctic glaciers as depositional models for former ice sheets. *Quarterly Journal of the Geological Society of London* 128, 361--393.

Boulton, G.S. 1978. Boulder shapes and grain-size distribution of debris as indicators of transport paths through a glacier and till genesis. *Sedimentology* 25, 773--799.

Boulton, G.S. 1979. Processes of glacial erosion on different substrata. *Journal of Glaciology* 23, 15--38.

Boulton, G.S. 1990. Sedimentary and sea level changes during glacial cycles and their control on glacimarine facies architecture. In Dowdeswell, J. A. & J. D. Scourse (eds.) *Glacimarine Environments: Processes and Sediments*. Geological Society Spec. Publ. No. 53, 15--52.

Boulton, G. S. & R. C. A. Hindmarsh 1987. Sediment deformation beneath glaciers: rheology and geological consequences. *Journal of Geophysical Research* 92 (B9), 9059--9082.

Boyd, R., D. B. Scott, M. Douna 1988. Glacial tunnel valleys and Quaternary history of the outer Scotian Shelf. *Nature* 333, 61--64.

Bridle, I. M. & P. H. Robinson 1989. Diagenesis. In Barrett, P. J. (ed.) *Antarctic Cenozoic history from the CIROS-1 drillhole, McMurdo Sound, Antarctica*. Wellington: DSIR Bulletin 245, 201--207.

Brodzikowski, K. & A. J. van Loon 1991. *Glacigenic sediments*. Developments in Sedimentology No. 49. Amsterdam: Elsevier, 674pp.

von Brunn, V & T. Stratton 1981. Late Palaeozoic tillites of the Karoo Basin of South Africa. In M. J. Hambrey & W. B. Harland (eds), *Earth's pre-Pleistocene glacial record*, 71--79. Cambridge: Cambridge University Press.

Bryant, A. D. 1983. The utilization of Arctic river analogue studies in the interpretation of periglacial river sediments in southern Britain. In: Gregory, K. J. (ed.) *Background to palaeohydrology*, 413--431: Chichester, England: John Wiley.

Budd, W. & B. J. McInnes 1978. Modelling surging glaciers and periodic surging of the Antarctic ice sheet. In A. B. Pittock, L. A. Frakes, D. Jenssen, J. A. Peterson, J. W. Zillman (eds) *Climatic change and variability: a Southern Hemisphere perspective*, 228--233. Cambridge: Cambridge University Press.

Budd, W., M. J. Corry, T. H. Jacka 1982. Results from the Amery Ice Shelf Project. *Annals of Glaciology* 3, 36--41.

Budd, W. F., T. H. Jacka, V. I. Morgan 1980. Antarctic iceberg melt rates derived from size distributions and movement rates. *Annals of Glaciology* 1, 103--112.

Cameron, D. 1965. Early discoveries XXII, Goethe - discover of the Ice Age. *Journal of Glaciology*, 5, 751--754.

Carey, S. W. & N. Ahmed 1961. Glacial marine sedimentation. In G. O. Raasch (ed), *Geology of the Arctic, Vol. 2*, 865--894. Toronto: University of Toronto Press.

Carlson, P. R., T. R. Bruns, M. A. Fisher 1990. Development of slope valleys in the glacimarine environments of a complex subduction zone, Northern Gulf of Alaska. In J. A. Dowdeswell & J. C. Scourse (eds), *Glacimarine environments: processes and sediments*, 139--154. London: Geological Society Special Publ. 53.

Carol, H. 1947. The formation of roches moutonnées. *Journal of Glaciology* 1, 57--59.

Chinn et al. 1992.

Church, M. & R. Gilbert 1975. Proglacial fluvial and lacustrine environments. In A. V. Jopling & B. C. McDonald (eds). *Glaciofluvial and glaciolacustrine sedimentation* 155--176. Tulsa: Society of Economic Paleontologists and Mineralogists, Spec. Publ. 23.

Clark, D. L. & A. Hanson 1983. Central Arctic Ocean sediment texture: a key to ice transport mechanisms. In B. F. Molnia (ed) *Glacial-marine sedimentation*, 301--330. New York & London: Plenum Press.

Clarke, G. K. C. 1987. A short history of scientific investigations on glaciers. *Journal of Glaciology, Special Issue Commenmorating Fiftieth Anniversary of the International Glaciological Society*, 4--24.

Clarke, G. K. C. 1991. Length, width and slope influences on glacier surging. *Journal of Glaciology* 37, 236--246.

Clarke, T. S. 1991. Glacier dynamics in the Susitna River basin, Alaska, USA. *Journal of Glaciology* 37, 97--106.

Clarke, G. K. C., S. G. Collins, D. E. Thompson 1984. Flow, thermal structure and subglacial conditions of a surge-type glacier. *Canadian Journal of Earth Sciences* 21, 232--240.

Clayton, L., J. T. Teller & J. W. Attig 1985. Surging of the southwestern part of the Laurentide Ice Sheet. *Boreas* 14, 235--241.

Clemmensen, L. B. & M. Houmark-Nielsen 1981. Sedimentary features of a Weichselian glaciolacustrine delta. *Boreas* 10, 229--245.

Collins, D. N. 1977. Hydrology of an Alpine glacier as indicated by the chemical composition of meltwater. *Zeitschrift für Gletscherkunde und Glazialgeologie* 13, 219--238.

Cooper, A. K., P. J. Barrett, K. Hinz, V. Traube, G. Leitchenkov, H. M. J. Stagg 1991a. Cenozoic prograding sequences of the Antarctic continental margin: a record of glacio-eustatic and tectonic events. *Marine Geology* 102, 175--213.

Cooper, A. K., H. M. J. Stagg, E. Geist 1991b. Seismic stratigraphy and structure of Prydz Bay, Antarctica: implications from ODP Leg 119 drilling. In J. Barron, B. Larsen & Shipboard Scientific Party. *Proceedings of the Ocean Drilling Program, Sci. Results, Vol.119*, 5--25. College Station, Texas.

Cowan, E. A. & R. D. Powell 1990. Suspended sediment transport and deposition of cyclically interlaminated sediment in a temperate glacial fjord, Alaska, U.S.A. In J. A. Dowdeswell & J. C. Scourse (eds), *Glacimarine environments: processes and sediments*, 75--90. London: Geological Society Special Publ. 53.

Crabtree, R. D. & C. S. M. Doake 1980. Flow lines on Antarctic ice shelves. *Polar Record* 20, 31--37.

Croot, D. G. 1987. Glacio-tectonic structures: a mesoscale model of thin-skinned thrust sheets? *Journal of Structural Geology* 9, 797--808.

Dahl, R. 1965. Plastically sculptured detail forms on rock surfaces in northern Nordland, Norway. *Geografiska Annaler* 47, 83--140.

Deynoux, M. 1980. *Les formations glaciares du Précambrien terminal et de la fin de l'Ordovicien en Afrique de l'Ouest. Deux exemples de glaciation d'inlandsis sur une plate-forme stable.* Tr. Lab. Sci. Terre, St.-Jerôme, Marseilles, 17(B).

Deynoux, M. & R. Trompette 1981a. Late Ordovician tillites of the Taoudeni Basin, West Africa. In M. J. Hambrey & W. B. Harland (eds), *Earth's pre-Pleistocene glacial record*, 89--96. Cambridge: Cambridge University Press.

Deynoux, M. & R. Trompette 1981b. Late Precambrian tillites of the Taoudeni Basin, West Africa. In M. J. Hambrey & W. B. Harland (eds), *Earth's pre-Pleistocene glacial record*, 123--131. Cambridge: Cambridge University Press.

Doake, C. S. M. & D. G. Vaughan 1990. Rapid disintegration of the Wordie Ice Shelf in response to atmospheric warming. *Nature* 350, 328--330.

Domack, E. W. 1982. Sedimentology of glacial and glacial marine deposits on the George V - Adélie continental shelf, East Antarctica. *Boreas* 11, 79--97.

Domack, E. W. 1988. Biogenic facies in the Antarctic glacimarine environment: basis for a polar glacimarine summary. *Palaeogeography, Palaeoclimatology & Palaeoecology* 63, 357--372.

Domack, E. W. 1990. Laminated terrigenous sediments from the Antarctic Peninsula: the role of subglacial and marine processes. In J. A. Dowdeswell & J. C. Scourse (eds), *Glacimarine environments: processes and sediments*, 91--104. London: Geological Society Special Publ. 53.

Domack, E. W. 1984. Rhythmically bedded glaciomarine sediments on Whidbey Island, northwestern Washington State and southwestern British Columbia. *Journal of Sedimentary Petrology* 54, 589--602.

Domack, E. W. & D. E. Lawson 1985. Pebble fabric in an ice-rafted diamicton. *Journal of Geology* 93, 577--592.

Domack, E. W., J. B. Anderson, D. D. Kurtz 1980. Clast shape as an indicator of transport and depositional mechanisms in glacial marine sediments: George V continental shelf, Antarctica. *Journal of Sedimentary Petrology* 50, 813--820.

Dowdeswell, J. A. 1986. Distribution and character of sediments in a tidewater glacier, southern Baffin Island, N.W.T., Canada. *Arctic & Alpine Research* 18, 45--56.

Dowdeswell, J. A. 1987. Processes of glacimarine sedimentation. *Progress in Physical Geography* 11, 52--90.

Dowdeswell, J. A. & R. L. Collin 1990. Fast-flowing outlet glaciers on Svalbard ice caps. *Geology* 18, 778--781.

Dowdeswell, J. A. & T. Murray 1990. Modelling rates of sedimentation from icebergs. In J. A. Dowdeswell & J. C. Scourse (eds), *Glacimarine environments: processes and sediments*, 121--138. London: Geological Society Special Publ. 53.

Dowdeswell, J. A. & M. J. Sharp 1986. The characterization of pebble fabrics in modern glacigenic sediments. *Sedimentology* 33,699--710.

Dowdeswell, J. A. & J. C. Scourse (eds) 1990. *Glacimarine environments: processes and sediments*. London: Geological Society Special Publ. 53.

Dowdeswell, J. A., M. J. Hambrey & R. T. Wu 1985. A comparison of clast fabric and shape in late Precambrian and modern glacigenic sediments. *Journal of Sedimentary Petrology* 55, 691--704.

Dreimanis, A. 1976. Tills: their origin and properties. In R. F. Leggett (ed), *Glacial till*, 11--49. Royal Society of Canada, Special Publ. 12.

Dreimanis, A. 1984. Lithofacies types and vertical profile models; an alternative approach to the description and environmental interpretation of glacial diamict and diamictite sequences. Discussion. *Sedimentology* 31, 885--886.

Dreimanis, A. 1989. Tills: their genetic terminology and classification. In R. P. Goldthwait, R.P. & C. L. Matsch (eds), *Genetic classification of glacigenic deposits*, 17--83. Rotterdam: Balkema.

Dreimanis, A. 1979. The problems of waterlain tills. In: Schlüchter, Ch. (ed), *Moraines and varves*, 167--177. Rotterdam: Balkema.

Drewry, D. J. 1983. *Antarctica: Glaciological and geophysical folio*. Cambridge: Scott Polar Research Institute.

Drewry, D. J. 1986. *Glacial geologic processes*. London: Edward Arnold, 276p.

Drewry, D. J. 1991. The response of the Antarctic ice sheet to climatic change . In C. Harris & B. Stonehouse (eds), *Antarctica and Global Climate Change*, 90--106. Cambridge: Scott Polar Research Institute & Belhaven Press.

Drewry, D. J. & A. P. R. Cooper 1981. Processes and models of Antarctic glaciomarine sedimentation. *Annals of Glaciology* 2, 117--122.

Dunbar, R. B., J. B. Anderson & E. W. Domack 1985. Oceanographic influences on sedimentation along the Antarctic continental shelf. In *Oceanology of the Antarctic Continental Shelf*, Antarctic Research Series 43, 291--312.

Ehlers, J. (ed) 1983. *Glacial deposits in North-West Europe*. Rotterdam: Balkema, 470pp.

Ehlers, J., P. L. Gibbard, J. Rose (eds) 1991. *Glacial deposits in Great Britain and Ireland*. Rotterdam: Balkema, 580 pp.

Ehrmann, W. U. 1991. Implications of sediment composition in the southern Kerguelen Plateau for paleoclimate and depositional environment. In J. Barron, B. Larsen & Shipboard Scientific Party. *Proceedings of the Ocean Drilling Program, Sci. Results*, Vol.119, 185--210. College Station, Texas.

Ehrmann, W. U. & A. Mackensen 1992. Sedimentological evidence for the formation of an East Antarctic ice sheet in Eocene/ Oligocene time. *Palaeogeography, Palaeoclimatology & Palaeoecology* 93,85--112.

Elson, J.A. 1989. Comment on glacitectonite, deformation till and comminution till. In: Goldthwait, R.P. and C. L. Matsch (eds), *Genetic classification of glacigenic deposits*. Balkema, Rotterdam, 85--88.

Elverhøi, A. 1984. Glacigenic and associated marine sediments in the Weddell Sea, fjords of Spitsbergen and the Barents Sea: a review. *Marine Geology* 57, 53--88.

Elverhøi, A. & E. Roaldset 1983. Glaciomarine sediments and suspended particulate matter, Weddell Sea shelf, Antarctica. *Polar Research* 1,1--21.

Elverhøi, A., O. Lonne, R. Seland 1983. Glaciomarine sedimentation in a modern fjord environment, Spitsbergen. *Polar Research* 1, 127--149..

Embleton, C. & C. A. M. King 1968. *Glacial and periglacial geomorphology*. London: Edward Arnold, 608pp.

Embleton, C. & C. A. M. King 1975. *Glacial geomorphology*. London: Edward Arnold, 583 pp.

Eyles, N. (ed) 1983. *Glacial geology*. Oxford: Pergamon, 409pp.

Eyles, C. H. 1988. Glacially and tidally-infuenced shallow marine sedimentation of the late Precambrian Port Askaig Formation. *Palaeogeography, Palaeoclimatology & Palaeoecology* 68, 1--25.

Eyles, C. H. 1988. A model for striated boulder pavement formation on glaciated shallow marine shelves, an example from the Yakataga Formation, Alaska. *Journal of Sedimentary Petrology* 58, 62--71.

Eyles, C. H. & N. Eyles 1983. Sedimentation in a large lake: a reinterpretation of the late Pleistocene stratigraphy at Scarborough Bluffs, Ontario, Canada. *Geology* 11, 146--152.

Eyles, N. & A. M. McCabe 1989. Glaciomarine facies within subglacial tunnel valleys: the sedimentary record of glacio-isostatic downwarping in the Irish Sea Basin. *Sedimentology* 36, 431--448.

Eyles, N. & M. B. Lagoe 1989. Sedimentology of shell-rich deposits (coquinas) in the glaciomarine upper Cenozoic Yakataga Formation, Middleton Island, Alaska. *Bulletin, Geological Society of America* 101, 129--142.

Eyles, C. H. & M. B. Lagoe 1990. Sedimentation patterns and facies geometries on a temperate glacially-influenced continental shelf: the Yakataga Formation, Middleton Island, Alaska. In J. A. Dowdeswell & J. C. Scourse (eds), *Glacimarine environments: processes and sediments*, 363--386. London: Geological Society Special Publ. 53.

Eyles, N. & J. Menzies 1983. The subglacial landsystem. In N. Eyles (ed), *Glacial Geology*, Pergamon Press, Oxford, 19--70.

Eyles, N. & A. D. Miall 1984. Glacial facies models. In R. G. Walker (ed), *Facies models*, 15--38. Toronto: Geological Association of Canada.

Eyles, N., C. H. Eyles, A. D. Miall 1983. Lithofacies types and vertical profile models; an alternative approach to the description and environmental interpretation of glacial diamict and diamictite sequences. *Sedimentology* 30, 393--410.

Fairchild, I. J. 1983. Effects of glacial transport and neomorphism on Precambrian dolomite crystal sizes. *Nature* 304, 714--716.

Fairchild, I. J. 1993. Balmy shores and icy wastes: the paradox of carbonates associated with glacial deposits in Neoproterozoic times. *Sedimentology Review* 1, 1--16.

Fairchild, I. J. & M. J. Hambrey 1984. The Vendian succession of north-eastern Spitsbergen: petrogenesis of a dolomite-tillite association. *Precambrian Research* 26, 111--167.

Fairchild, I. J. & M. J. Hambrey in press. Vendian basin evolution in East Greenland and NE Svalbard. *Precambrian Research*.

Fairchild, I. J. & B. Spiro 1990. Carbonate minerals in glacial sediments: geochemical clues to palaeoenvironment. In J. A. Dowdeswell & J. C. Scourse (eds), *Glacimarine environments: processes and sediments*, 241--256. London: Geological Society Special Publ. 53.

Fairchild, I. J., G. L. Hendry, M. Quest, M. E. Tucker 1988. Chemical analysis of sedimentary rocks, 274--354. In M. E. Tucker (ed) *Techniques in Sedimentology*. Oxford: Blackwells.

Fairchild, I. J., Hambrey, M. J., Spiro, B & Jefferson, T. H. 1989. Late Proterozoic glacial carbonates in NE Spitsbergen: new insights into the carbonate-tillite association. *Geological Magazine* 126, 469--490.

Fairchild, I. J., L. Bradby, B. Spiro 1993. Reactive carbonate in glacial sediments: a preliminary synthesis of its creation, dissolution and reincarnation. In M. Deynoux, J. M. G. Miller, E. W.

Domack, N. Eyles, I. J. Fairchild, G. M. Young (eds), *Earth's glacial record*, ***--***. Cambridge: Cambridge University Press.

Flint, R. F. 1971. *Glacial and Quaternary geology.* Wiley, New York, 892pp.

Flint, R. F., Sanders, J. E. & Rodgers, J. 1960. Diamictite: a substitute term for symmictite. *Bulletin Geological Society of America* 71, 1809--1810.

Folk, R. L. 1975. Glacial deposits identified by chattermark trails in detrital garnets. *Geology* 3, 473--475.

Ford, D. C., P. G. Fuller, J. J. Drake 1970. Calcite precipitates at the soles of temperate glaciers. *Nature* 226, 441--442.

Frakes, L. A. 1975. Geochemistry of Ross Sea diamicts. In D. E. Hayes, L. A. Frakes & Shipboard Scientific Party, *Initial Reports Deep Sea Drilling Project, Leg 28*, 789--794. Washington DC: US Government Printing Office.

Frakes, L. A. 1979. *Climates throughout geological time.* Amsterdam: Elsevier, 310pp.

Frakes, J. A. & J. E. Francis 1988. A guide to Phanerozoic cold polar climates from high-latitude ice-rafting in the Cretaceous. *Nature* 333, 547--549.

Frakes, L. A., J. E. Francis & J. I. Sykes 1992. *Climate modes of the Phanerozoic.* Cambridge: Cambridge University Press, 274 pp.

Francis, E. 1975. Glacial sediments: a selective review. In: Wright, A.E. & F. Moseley (eds). *Ice ages: ancient and modern.* Liverpool: Seel House Press, 43--68.

Fütterer, D. K. & M. Melles 1990. Sediment patterns in the southern Weddell Sea: Filchner Shelf and Filchner Depression. In U. Bleil & J. Theide (eds), *Geological history of the polar oceans: Arctic versus Antarctic*, 381--401. Dordrecht: Kluwer Academic Publishers (NATO/ASI Series C).

Fyfe, G. J. 1990. The effect of water depth on ice-proximal glaciolacustrine sedimentation: Salpausselkä I, southern Finland. *Boreas* 19, 147--164.

Garwood, E. J. 1932. Speculation and research in Alpine glaciology: an historical review. *Quarterly Journal of the Geological Society of London* 88, xciii--cxviii.

Glen, J. W. 1952. Experiments on the deformation of ice. *Journal of Glaciology* 2, 111-114.

Goldthwait, R. P. 1951. Development of end moraines in east-central Baffin Island. *Journal of Geology* 59, 567--577.

Goldthwait, R. P. 1979. Giant grooves made by concentrated basal ice streams. *Journal of Glaciology* 23, 297--307.

Goldthwait, R. P. 1971. Introduction to till, today. In Goldthwait, R. P. (ed.), *Till: a symposium*, Ohio State Univ. Press, Columbus, 3--26.

Goldthwait, R.P. 1989. Classification of glacial morphologic features. In R. P. Goldthwait, R.P. & C. L. Matsch (eds), *Genetic classification of glacigenic deposits*, 267--277. Balkema, Rotterdam.

Goodwin, R. G. 1984. *Neoglacial lacustrine sedimentation and ice advance, Glacier Bay, Alaska.* Ohio State University, Institute of Polar Studies Report 79, 183 pp.

Gravenor, C.P. 1979. The nature of the Late Palaeozoic glaciation in Gondwana as determined from an analysis of garnets and other heavy minerals. *Canadian Journal of Earth Sciences* 16, 1137--1153.

Gravenor, C.P. 1980. Heavy minerals and sedimentological studies on the glaciogenic Late Precambrian Gaskiers Formation of Newfoundland. *Canadian Journal of Earth Sciences* 17, 1131--1141.

Gravenor, C.P. 1981. Chattermark trails on garnets as an indicator of glaciaiton in ancient deposits. In M. J. Hambrey & W. B. Harland (eds), *Earth's pre-Pleistocene glacial record*, 17--18. Cambridge: Cambridge University Press.

Griffith, T. W. & J. B. Anderson 1989. Climatic control of sedimentation in bays and fjords of the northern Antarctic Peninsula. *Marine Geology* 85, 181--204.

Gripp, K. 1929. Glaciologische unde geologische Ergebnisse der Hamburgischen Spitzbergen-Expedition 1927. *Naturwiss. Verein in Hamburg, Abh. Geb. Naturw.* 22 (2--4), 146--249.

Grove, J. M. 1988. *The Little Ice Age.* London: Methuen.

Gustavson, T. C., G. M. Ashley, J. C. Boothroyd 1975. Depositional sequences in glaciolacustrine deltas. In: Jopling, A. V. & B. C. McDonald (Eds.). *Glaciofluvial and Glaciolacustrine Sedimentation*, 264--280. Tulsa: Society of Economic Paleontologists and Mineralogists, Spec. Publ. 23.

Haase, G. H. 1986. Glaciomarine sediments along the Filchner/Ronne Ice Shelf, southern Weddell Sea - first results of the 1983/84 Antarktis-II/4 Expedition. *Marine Geology* 72, 241--258.

Hall, K. J. 1989. Clast shape. In P. J. Barrett (ed), *Antarctic Cenozoic history from the CIROS-1 drillhole, McMurdo Sound, Antarctica*, 63--66. Wellington: DSIR Bulletin 245.

Hallet, B. 1976. The effect of subglacial shemical processes on glacier sliding. *Journal of Glaciology* 17, 209--221.

Hambrey, M. J. 1975. The origin of foliation in glaciers: evidence from some Norwegian examples. *Journal of Glaciology* 14, 181--185.

Hambrey, M. J. 1977. Foliation, minor folds and strain in glacier ice. *Tectonophysics* 39, 397--416.

Hambrey, M. J. 1982. Late Proterozoic diamictites of northeastern Svalbard. *Geological Magazine* 119, 527--551.

Hambrey, M. J. 1989. Grain fabric studies on the CIROS-1 core. In P. J. Barrett (ed), *Antarctic Cenozoic history from the CIROS-1 drillhole, McMurdo Sound, Antarctica*, 59--62. Wellington: DSIR Bulletin 245.

Hambrey, M. J. 1991. Structure and dynamics of the Lambert Glacier-Amery Ice Shelf system: implications for the origin of Prydz Bay sediments. In J. Barron, B. Larsen & Shipboard Scientific Party, Ocean Drilling Program Leg 119, Scientific Results, 61--76. College Station, Texas: Ocean Drilling Program.

Hambrey, M. J. 1992. Secrets of a tropical ice age. *New Scientist*, 42--49 (1 Feb. 1992).

Hambrey, M. J. & J. C. Alean 1992. *Glaciers*. Cambridge: Cambridge University Press, 206pp.

Hambrey, M. J. & J. A. Dowdeswell in press. Stable flow of the Lambert Glacier - Amery Ice Shelf system: structural evidence from Landsat imagery. *Annals of Glaciology*.

Hambrey, M. J. & A. G. Milnes 1975. Boudinage in glacier ice - some examples. *Journal of Glaciology* 14, 383--393.

Hambrey, M. J. & Milnes, A. G. 1977. Structural geology of an Alpine glacier (Griesgletscher, Valais, Switzerland). *Eclogae Geologicae Helvetiae* 70, 667--684.

Hambrey, M. J. & Müller, F. 1978. Ice deformation and structures in the White Glacier, Axel Heiberg Island, Northwest Territories, Canada. *Journal of Glaciology* 20, 41--66.

Hambrey, M. J., Milnes, A. G. & Siegenthaler, H. 1980. Dynamics and structure of Griesgletscher, Switzerland. *Journal of Glaciology* 25, 215--228.

Hambrey, M. J. & W. B. Harland (collators and editors) 1981. *Earth's pre-Pleistocene glacial record*. Cambridge University Press, 1004 + xv pp.

Hambrey, M. J., P. J. Barrett, K. J. Hall, & P. H. Robinson 1989. Stratigraphy. In: Barrett, P.J. (ed.) *Antarctic Cenozoic history from the CIROS-1 drillhole, McMurdo Sound, Antarctica*. DSIR Bulletin 245, 23--48, Wellington, New Zealand.

Hambrey, M. J., W. U. Ehrmann & B. Larsen 1991. The Cenozoic glacial record of the Prydz Bay continental shelf, East Antarctica. In J. Barron, B. Larsen & Shipboard Scientific Party, Ocean Drilling Program Leg 119, Scientific Results, 77--132. College Station, Texas: Ocean Drilling Program.

Hambrey, M. J., P. J. Barrett, W. U. Ehrmann & B. Larsen 1992. Cenozoic sedimentary processes on the Antarctic continental shelf: the record from deep drilling. *Zeitschrift für Geomorphologie*, Suppl.-Bd. 86, 77--103.

Haq, B. U., J. Hardenbol, P. R. Vail 1987. Chronology of fluctuating sea levels since the Triassic. *Science* 235, 1156--1167.

Harland, W. B. 1957. Exfoliation joints and ice action. *Journal of Glaciology* 3, 8--10.

Harland, W. B. 1964. Critical evidence for a great Infra-Cambrian glaciation. *Geologische Rundschau* 54, 45--61.

Harland, W.B. & K. N. Herod 1975. Glaciations through time. In A. E. Wright & F. Moseley (eds). *Ice ages: ancient and modern*, 189--216. Liverpool: Seel House Press.

Harland, W.B., K. N. Herod, & D. H. Krinsley 1966. The definition and identification of tills and tillites. *Earth Science Reviews* 2, 225--256.

Harwood, D. M., P. J. Barrett, A. R. Edwards, H. J. Rieck, P.-N. Webb 1989. Biostratigraphy and chronology. In P. J. Barrett (ed), *Antarctic Cenozoic history from the CIROS-1 drillhole, McMurdo Sound, Antarctica*, 231--239. Wellington: DSIR Bulletin 245.

Hayes, D. E. & L. A. Frakes & Shipboard Scientific Party 1975. *Initial Reports of the Deep Sea Drilling Project, Leg 28*. Washington DC: US Government Printing Office.

Haynes, V. M. 1968. The influence of glacial erosion and rock structure on corries in Scotland. *Geografiska Annaler* 50A, 221--234.

Hinz, K. & M. Block 1983. Results of geophysical investigations in the Weddell Sea and in the Ross Sea, Antarctica. *Proceedings of the 11th World Petroleum Congress*, London & New York: Wiley, 75--91.

Hinz, K. & Y. Kristoffersen 1987. Antarctica: recent advance in the understanding of the continental shelf. *Geologische Jarbuch* E37, 3--54.

Hooke, R. LeB. 1968. Comments on the formation of shear moraines: an example from south Victoria Land, Antarctica. *Journal of Glaciology* 7, 351--352.

Hooke, R. LeB. & P. J. Hudleston 1978. Origin of foliation in glaciers. *Journal of Glaciology* 20, 285--299.

Hooke, R. LeB., B. B. Dahlin, M. T. Kauper 1972. Creep of ice containing dispersed fine sand. *Journal of Glaciology* 11, 327--336.

Hoppe, G. 1957. Problems of glacial geomorphology and the ice age. *Geografiska Annaler* 39, 1--17.

Hudleston, P. J. 1976. Recumbent folding in the base of the Barnes Ice Cap, Baffin Island, Northwest Territories, Canada. *Bulletin Geological Society of America* 87, 1684--1692.

Hudleston, P. J. & R. LeB. Hooke 1980. Cumulative deformation in the Barnes Ice Cap, and implications for the development of foliation. *Tectonophysics* 66, 127--146.

Hughes, T. J. 1975. The West Antarctic Ice Sheet: instability, disintegration and the initiation of ice ages. *Review of Geophysics & Space Physics* 15, 1--46.

Imbrie, J. & K. P. Imbrie 1979. *Ice ages: solving the mystery.* London: Macmillan Press, 224 pp.

IUGS 1989.Global stratigraphic chart. *Episodes* 12 (2), insert.

Jopling, A. V. & B. C. McDonald 1975. *Glaciofluvial and glaciolacustrine sedimentation.* Tulsa: Society of Economic Paleontologists and Mineralogists, Spec. Publ. 23.

Kalin, M. 1971. The active push moraine of the Thomson Glacier. *Axel Heiberg Island Research Report, Glaciology* 4, 68 pp.

Kamb, W. B. 1987. Glacier surge mechanism based on linked cavity configuration of the basal water conduit system. *Journal of Geophysical Research* 92 (B9), 9083--9100.

Kamb, W. B., C. F. Raymond, W. D. Harrison, H. F. Engelhardt, K. Echelmeyer, N. Humphry, M. Brugman, T. Pfeffer 1985. Glacier surge mechanism: the 1982-1983 surge of Variegated Glacier, Alaska, *Science* 227, 469--479.

Karrow, P. F. 1984. Lithofacies types and vertical profile models; an alternative approach to the description and environmental interpretation of glacial diamict and diamictite sequences. Discussion. *Sedimentology* 31, 883--884.

Kemmis, T. J. & T. J. Hallberg 1984. Lithofacies types and vertical profile models; an alternative approach to the description and environmental interpretation of glacial diamict and diamictite sequences. Discussion. *Sedimentology* 31, 886--890.

King, L. H. in press. Till in the marine environment. Journal of Quaternary Science 8**,***--***.

King, L. H. & G. Fader 1987. Wisconsinan glaciation on the continental shelf - southeast Atlantic Canada. *Geological Survey of Canada Bulletin* 363, 72pp.

King, L. H., K. Rokoengen, T. Gunleiksrud 1987. Quaternary seismostratigraphy of the Mid Norwegian Shelf, 65°--67°30'N - a till tongue stratigraphy. *IKU Publ.* 114, 58pp.

King, L. H., K. Kokoengen, G. B. J. Fader, T. Gunleiksrud 1991. Till-tongue stratigraphy. *Bulletin, Geological Society of America* 103, 637--659.

Krinsley, D.H. & J. C. Doornkamp 1973. *Atlas of quartz and sand surface textures.* Cambridge: Cambridge University Press.

Krinsley, D. H. & B. Funnell 1965. Environmental history of sand grains from the Lower and Middle Pleistocene of Norfolk, England. *Quarterly Journal of the Geological Society of London* 121, 435--456.

Krumbein, W. C. 1941. Measurement and geological significance of shape and roundness of sedimentary particles. *Journal of Sedimentary Petrology* 11, 64--72.

Kuhn, G., M. Melles, W. U. Ehrmann, M. J. Hambrey & G. Schmiedl 1993. Character of clasts in glaciomarine sediments as an indicator of transport and depositional processes, Weddell and Lazarev Seas, Antarctica. *Journal of Sedimentary Petrology* 63, 477--487.

Kujansuu, R. 1976. Glaciogeological surveys for ore-prospecting purposes in northern Finland. In R. F. Leggett (ed), *Glacial till*, 225--239. Royal Society of Canada, Special Publ. 12.

Larter, R. D. & P. F. Barker 1989. Seismic stratigraphy of the Antarctic Peninsula Pacific margin: a record of Pliocene-Pleistocene ice volume and paleoclimate. *Geology* 17, 731--734.

Lawson, D. E. 1979. A comparison of the pebble orientations in ice and deposits of the Matanuska Glacier, Alaska. *Journal of Geology* 87, 629--645.

Lawson, D. E. 1981. Sedimentological characteristics and classification of depositional processes and deposits in the glacial environment. *US Army Cold Regions Research & Engineering Laboratory, Report* 81-27, 16pp.

Lawson, D. E. 1982. Mobilization, movement and deposition of active subaerial sediment flows, Matanuska Glacier, Alaska. *Journal of Geology* 90, 279--300.

Lien, R., A. Solheim, A. Elverhøi, K. Rokøngin 1989. Iceberg scouring and sea bed morphology on the eastern Weddell Sea Shelf. *Polar Research* 4,43--57.

Lewis, W. V. 1954. Pressure release and glacial erosion. *Journal of Glaciology* 2, 417--422.

Lewis, W. V. 1960. *Norwegian cirque glaciers*. Royal Geographical Society Research Series 4, 104 pp.

Lundqvist, J. 1989. Rogen (ribbed) moraine - identification and possible origin. *Sedimentary Geology* 62, 281--292.

Mackiewicz, N. E., R. D. Powell, P. R. Carlson, B. F. Molnia 1984. Interlaminated ice-proximal glacimarine sediments in Muir Inlet, Alaska. *Marine Geology* 57, 113--147.

Manley, G. 1959. The late-glacial climate of North-west England. *Liverpool & Manchester Geological Journal* 2, 188--215.

May, R. W. & A. Dreimanis, 1976. Compositional variability in tills. In R. F. Leggett (ed), *Glacial till*, 99--120. Royal Society of Canada, Special Publ. 12.

McCabe, A. M. & N. Eyles 1988. Sedimentology of an ice-contact glaciomarine delta, Carey Valley, Northern Ireland. *Sedimentary Geology* 59, 1--14.

McDonald, B. C. & W. W. Shilts 1975. Interpretation of faults in glaciofluvial sediments. In: Jopling, A. V. & B. C. McDonald (eds). *Glaciofluvial and glaciolacustrine sedimentation* 123--131. Tulsa: Society of Economic Paleontologists and Mineralogists, Spec. Publ. 23.

McIntyre, N. F. 1985. A re-assessment of the mass balance of the Lambert Glacier drainage basin, Antarctica. *Journal of Glaciology* 31, 34--38.

Meier, M. F. 1960. Mode of flow of Saskatchewan Glacier, Alberta, Canada. US Geological Survey Professional Paper 351, 70pp.

Menzies, J. 1989. Subglacial hydraulic conditions and their possible impact upon subglacial bed formation. *Sedimentary Geology* 62, 125--150.

Menzies, J. & J. Rose 1987. *Drumlin Symposium*. Rotterdam: Balkema.

Menzies, J. & J. Rose 1989. *Subglacial bedforms - drumlins, Rogen moraine and associated subglacial bedforms*. Sedimentary Geology Spec. Issue 62 (2/4).

Miall, A. D. 1977. A review of the braided-river depositional environment. *Earth Science Reviews* 13, 1--62.

Miall, A. D. 1978. Lithofacies types and vertical profile models in braided river deposits: a summary. In A. D. Miall (ed), *Fluvial Sedimentology*. Memoir Canadian Society of Petroleum Geology 5, 597--604.

Miall, A.D. 1983a. Glaciomarine sedimentation in the Gowganda Formation (Huronian), Northern Ontario. *Journal of Sedimentary Petrology*, 53, 477--491.

Miall, A. D. 1983b. Glaciofluvial transport and deposition. In N. Eyles (ed), *Glacial geology*, 168--183. Oxford: Pergamon.

Mickelsen, D. M. 1971. *Glacial geology of the Burroughs Glacier area, southeast Alaska*. Columbus: Institute of Polar Studies Report 40, Ohio State University, 149 pp.

Miller, K. G., R. G. Fairbanks, G. S. Mountain 1987. Tertiary oxygen isotope synthesis, sea level history, and continental margin erosion. *Paleooceanography* 2, 1--19.

Mills, H.H. 1977, Basal till fabrics of modern alpine glacier. *Bulletin Geological Society of America* 88, 824--828.

Mills, W. 1983. Darwin and the iceberg theory. *Notes and Records of the Royal Society of London* 38, 109--127.

Molnia, B. F. 1983a. Subarctic glacial-marine sedimentation: a model. In B. F. Molnia (ed) *Glacial-marine sedimentation*, 95--144. New York & London: Plenum Press.

Molnia, B. F. (ed.) 1983b. *Glacial-marine sedimentation*. New York & London: Plenum Press, 844 pp.

Moncrieff, A. C. M. 1989. Classification of poorly sorted sedimentary rocks, *Sedimentary Geology* 65, 191--194.

Morgan, V. I. 1972. Oxygen isotope evidence for bottom freezing on the Amery Ice Shelf. *Nature* 238, 393--394.

Mustard, P. S. & J. A. Donaldson 1987. Early Proterozoic ice-proximal glaciomarine deposition: The lower Gowganda Formation at Cobalt, Ontario, Canada. *Geological Society of America Bulletin* 98, 373--387.

Nesbitt, H. W. & G. M. Young 1982. Early Proterozoic climates and plate motions inferred from major element chemistry of lutites. *Nature* 299, 715--717.

Nye, J. F. 1952 The mechanics of glacier flow. *Journal of Glaciology* 2, 82--93.

Nye, J. F. 1957. The distribution of stress and velocity in glaciers and ice sheets. Proceedings of the Royal Society of London 239A, 113--133.

Nye, J. F. 1958. A theory of wave formation on glaciers. *International Association of Hydrological Sciences Publ.* 47, 139--154.

Nye, J. F. 1973. Water at the bed of a glacier. *International Association of Hydrological Sciences*, Publ. 95, 189--194.

Nye, J. F. & F. C. Frank 1973. Hydrology of the intergranular veins in a temperate glacier. *International Association of Hydrological Sciences*, Publ. 95, 157--161.

Orheim, O. & A. Elverhøi 1981. Model for submarine glacial deposition. *Annals of Glaciology* 2, 123--129.

Østrem, G. 1975. Sediment transport in glacial streams. In A. V. Jopling & B. C. McDonald (eds). *Glaciofluvial and glaciolacustrine sedimentation*, 101--122. Tulsa: Society of Economic Paleontologists and Mineralogists, Spec. Publ. 23.

Oswald, G. K. A. & G. de Q. Robin 1973. Lakes beneath the Antarctic ice sheet. *Nature* 245, 251--254.

Parker , B. C., G. M. Simmons, F. G. Love, R. A. Wharton, K. G. Seaburg 1981. Modern stromatolites in Antarctic Dry Valley lakes. *Bioscience* 31, 656--661.

Paterson, W. S. B. 1981. *The physics of glaciers*. Oxford: Pergamon Press, 380pp.

Paul, M. A. 1983. Chapter 3: The supraglacial landsystem. In N. Eyles (ed), *Glacial geology*, 71--90. Oxford: Pergamon.

Peacock, J. D. & Cornish, R. 1989. *Glen Roy area*. Quaternary Research Association Field Guide, Cambridge, 69 pp.

Pettijohn, F. J. 1975. *Sedimentary rocks*. New York: Harper & Row, 628 pp.

Phillips, F. C. 1971. *The use of stereographic projection in structural geology*. London: Edward Arnold, 90 pp.

Post, A. & E. R. La Chapelle 1971. *Glacier ice*. Seattle: University of Washington Press & The Mountaineers, 111 pp.

Powell, R. D. 1981. A model for sedimentation by tidewater glaciers. *Annals of Glaciology* 2, 129--134.

Powell, R. D. 1983. Glacial-marine sedimentation processes and lithofacies of temperate tidewater glaciers, Glacier Bay, Alaska. In: B. F. Molnia (ed) *Glacial-marine sedimentation*, 185--232. New York & London: Plenum Press.

Powell, R. D. 1984. Glacimarine processes and inductive lithofacies modelling of ice shelf and tidewater glacial sediments based on Quaternary examples. *Marine Geology* 57,1--52.

Powell, R. D. 1990. Glacimarine processes at grounding-line fans and their growth to ice-contact deltas. In J. A. Dowdeswell & J. C. Scourse (eds), *Glacimarine environments: processes and sediments*, 53--74. London: Geological Society Special Publ. 53.

Powell, R. D. 1984. Glacimarine processes and inductive lithofacies modelling of ice shelf and tidewater glacier sediments based on Quaternary examples. *Marine Geology* 57, 1--52.

Powell, R. D. & B. F. Molnia 1989. Glacimarine sedimentation processes, facies and morphology of the south-southeast Alaska shelf and fjords. *Marine Geology* 85, 359--390.

Prest, V. K. 1983. *Canada's heritage of glacial features*. Geological Survey of Canada Misc. Report. 28, 119 pp.

Price, R. J. 1973. *Glacial and fluvioglacial landforms*. Edinburgh: Oliver & Boyd.

Ramsay, J. G. 1967. Folding and fracturing of rocks. New York: McGraw-Hill, 568 pp.

Raymond, C. F. 1987. How do glaciers surge? A review. *Journal of Geophysical Research* 92 (B9), 9121--9134.

Reading, H. G. 1978. Facies. In H. G. Reading (ed), *Sedimentary environments and facies*, 4--14. Oxford: Blackwell Scientific Publications.

Reynolds, J. M. & M. J. Hambrey 1988. The structural glaciology of the George VI Ice Shelf, Antarctica. *British Antarctic Survey Bulletin* 79, 79--95.

Robin, G. de Q. 1974. Depth of water-filled crevasses that are closely spaced (letter). *Journal of Glaciology* 13, 543.

Robin, G. de Q. 1975. Ice shelves and ice flow. *Nature* 253, 168--172.

Robin, G. de Q. 1979. Formation, flow and disintegration of ice shelves. *Journal of Glaciology* 24, 259--271.

Robin, G. de Q. 1983. Coastal sites, Antarctica. In G. de Q. Robin (ed), *Climatic record in polar ice sheets*, 118--122. Cambridge: Cambridge University Press.

Rose, J. 1987. Drumlins as part of a glacier bedform continuum. In J. Menzies & J. Rose (eds). *Drumlin Symposium*, 103--116. Rotterdam: A. A. Balkema.

Rose, J. 1989. Glacier stress patterns and sediment transfer associated with the formation of superimposed flutes. *Sedimentary Geology* 62, 151--176.

Röthlisberger, H. 1972. Water pressure in intra- and subglacial channels. *Journal of Glaciology* 11, 177--203.

Rust , B. R. 1975. Late Quaternary subaqueous outwah deposits near Ottawa, Canada. In: Jopling, A. V. & B. C. McDonald (eds). *Glaciofluvial and glaciolacustrine sedimentation* 177--192. Tulsa: Society of Economic Paleontologists and Mineralogists, Spec. Publ. 23.

Sanderson, H. C. 1975. Sedimentology of the Brampton esker and its associated deposits: an empirical test of theory. In: Jopling, A. V. & B. C. McDonald (Eds.). *Glaciofluvial and Glaciolacustrine Sedimentation*, 155--176. Tulsa: Society of Economic Paleontologists and Mineralogists, Spec. Publ. 23.

Schlüchter, Ch. (ed) 1979. *Moraine and varves*. Rotterdam, Balkema, 441 pp

Sexton, D. J., J. A. Dowdeswell, A. Solheim & A. Elverhøi 1992. Seismic architecture and sedimentation in northwest Spitsbergen fjords. *Marine Geology* 103, 53--68.

Shabtaie, S. & C. R. Bentley 1987. West Antarctic ice streams draining into the Ross Ice Shelf: configuration and mass balance. *Journal of Geophysical Research* 92 (B9), 8865--8884.

Shackleton, N. J. & Kennett, J. K. 1975. Paleotemperature history of the Cenozoic and the initiation of Antarctic glaciation: oxygen and carbon isotope analyses in DSDP Sites 277, 279 and 281. In J. P. Kennett, R. E. Houtz & Shipboard Scientific Party, *Initial Reports of the Deep Sea Drilling Project, Leg 29*, 743--755. Washington DC: US Government Printing Office.

Sharp, M. 1982. Modification of clasts in lodgement tills by glacial erosion. *Journal of Glaciology* 28, 475--481.

Sharp, M. 1985a. Crevasse-fill ridges: a landform type characteristic of surging glaciers? *Geografiska Annaler* 67A, 213--220.

Sharp, M. 1985b. Sedimentation and stratigraphy of Eyabakkajökull - an Icelandic surging glacier. *Quaternary Research* 24, 268--284.

Sharp, M. 1988a. Surging glaciers: behaviour and mechanisms. *Progress in Physical Geography* 12, 349--370.

Sharp, M. 1988b. Surging glaciers: geomorphic effects. *Progress in Physical Geography* 12, 533--559.

Sharp, M., J. C. Gemmell, J.-L. Tison 1989. Structure and stability of the former subglacial drainage system of the Glacier de Tsanfleuron, Switzerland. Earth Surface Processes & Landforms 14, 119--134.

Sharp, R. P. 1988. Living ice: understanding glaciers and glaciation . Cambridge: Cambridge University Press, 225 pp.

Shaw, J. 1975. Sedimentary successions in Pleistocene ice-marginal lakes. In A. V. Jopling & B. C. McDonald (eds). *Glaciofluvial and glaciolacustrine sedimentation*, 281--303. Tulsa: Society of Economic Paleontologists and Mineralogists, Spec. Publ. 23.

Shaw, J. 1977. Till deposited in arid polar environments. *Canadian Journal of Earth Sciences* 14, 1239--1245.

Shaw, J. 1989. Sublimation till. In R. P. Goldthwait & C. L. Matsch (eds), *Genetic classification of glacigenic deposits*, 141--142. Rotterdam: Balkema.

Shaw, J., D. Kvill, B. Rains 1989. Drumlins and catastrophic subglacial floods. *Sedimentary Geology* 62, 62, 177--202.

Shilts, W. W. 1976. Glacial till and mineral exploration. In R. F. Leggett (ed), *Glacial till*, 205--224. Royal Society of Canada, Special Publ. 12.

Sladen, J.A. & W. Wrigley 1983. Geotechnical properties of lodgement till - a review. In N. Eyles (ed), *Glacial Geology* 184--212. Oxford: Pergamon Press.

Sissons, J. B. 1979. Catastrophic lake drainage in Glen Spean and the Great Glen, Scotland. *Journal of the Geological Society of London* 136, 215--224.

Solheim, A. 1991. The depositional environment of surging sub-polar tidewater glaciers. *Skrifter Norsk Polarinstitutt* 194, 97 pp.

Solheim, A., L. Russwurm, A. Elverhøi, M. Nyland Berg 1990. Glacial geomorphic features in the northern Barents Sea: direct evidence for grounded ice and implications for the pattern of deglaciation and late glacial sedimentation. In J. A. Dowdeswell & J. D. Scourse (eds), *Glaciomarine environments: processes and sediments*, 253--268. London: Geological Society Special Publ. 53.

Souchez, R. A. & M. Lemmens 1985. Subglacial carbonate deposition: an isotopic study of a present-day case. *Palaeogeography, Palaeoclimatology & Palaeoecology* 51, 357–364.

Souchez, R. A. & R. D. Lorrain 1991. *Ice composition and glacier dynamics*. Berlin: Springer-Verlag, 207pp.

Spencer, A. M. 1971. *Late Precambrian glaciation in Scotland*. Geological Society of London, Memoir 6, 100 pp.

Spencer, A. M. 1985. Mechanisms and environments of deposition of the late Precambrian geosyynclinal tillites: Scotland and East Greenland. *Palaeogeography, Palaeoclimatology & Palaeoecology* 51, 143–157.

Srivastava, S. P., M. Arthur & Shipboard Scientific Party 1987. Proceedings of the Ocean Drilling Program, Initial Reports 105. College Station, Texas: Ocean Drilling Program.

Stagg, H. M. J. 1985. The structure and origin of Prydz Bay and MacRobertson Shelf, East Antarctica. *Tectonophysics* 114, 315–340.

Stoker, M. S. 1990. Glacially-influenced sedimentation on the Hebridean slope, northwestern United Kingdom continental margin. In J. A. Dowdeswell & J. C. Scourse (eds), *Glacimarine environments: processes and sediments*, 349–362. London: Geological Society Special Publ. 53.

Stoker, M. S. & R. Holmes 1991. Submarine end-moraines as indicators of Pleistocene ice-limits off northwest Britain. *Journal of the Geological Society* 148, 431–434.

Sturm, M. 1979. Origin and composition of varves. In Ch. Schlüchter (ed), *Moraines and varves*, 281–285. Rotterdam, Balkema.

Sugden, D. & B. S. John 1976. *Glaciers and landscape*. London: Edward Arnold, 376 pp.

Swithinbank, C. W. M. 1988. *Antarctica*. In R. S. Williams & J. G. Ferrigno (eds) *Satellite image atlas of the glaciers of the world*. US Geological Survey Professional Paper 1386-B, 278pp.

Syvitski, J. P. M. 1989. On the deposition of sediment within glacier-influenced fjords. *Marine Geology* 85, 301–329.

Teller, J. T. 1985. Glacial Lake Agassiz and its influence on the Great Lakes. In Karrow, P. F. & Calkin, P.E. *Quaternary evolution of the Great Lakes*. Geological Association of Canada, Special Paper 30, 1–16.

Theakstone, W. H. 1967. Basal sliding and movement near the margins of the glacier Østerdalsisen, Norway. *Journal of Glaciology* 6, 805–816.

Theakstone, W. H. 1976. Glacial lake sedimentation, Austerdalsisen, Norway. *Sedimentology* 23, 671–688.

Theakstone, W. H. 1978. The 1977 drainage of Austre Okstindbreen ice-dammed lake, its causes and consequences. *Norsk Geografisk Tidsskrift* 32, 159–171.

Thyssen, F. 1988. Special aspects of the central part of the Filchner-Ronne Ice Shelf, Antarctica. *Annals of Glaciology* 11, 173–179.

Turner, B. R. 1991. Depositional environment and petrography of preglacial continental sediments from Hole 740A, Prydz Bay, Antarctica. In J. Barron, B. Larsen & Shipboard Scientific Party, Ocean Drilling Program Leg 119, Scientific Results, 45–56. College Station, Texas: Ocean Drilling Program.

US National Park Service 1983. *Glacier Bay: a guide to Glacier Bay National Park and Preserve, Alaska*. US National Park Service, Division of Publications, Washington DC, 129 pp.

Van der Meer, J. J. M. 1987. *Tills and glaciotectonics*. Rotterdam: Balkema, 270 pp.

Visser, J. N. J. 1983a. Glacial-marine sedimentation in the Late Paleozoic Karoo Basin, Southern Africa. In B. F. Molnia (ed) *Glacial-marine sedimentation*, 667–701. New York & London: Plenum Press.

Visser, J. N. J. 1983b. Submarine debris flow deposits from the Upper Carboniferous Dwyka Tille Formation in the Kalahari Basin, South Africa. *Sedimentology* 30, 511–524.

Visser, J. N. J. & K. J. Hall 1985. Boulder beds in the glaciogenic Permo-Carboniferous Dwyka Formation in South Africa. *Sedimentology* 32, 281–294.

Visser, J. N. J., J. C. Loock, W. P. Colliston 1987. Subaqueous outwash fan and esker sandstones in the Permo-Carboniferous Dwyka Formation of South Africa. *Journal of Sedimentary Petrology* 57, 467–478.

Vivian, R. & G. Bocquet 1973. Subglacial cavitation phenomena under the Glacier d'Argentière, Mont Blanc, France. *Journal of Glaciology* 12, 439–451.

Vorren, T. O., E. Lebesbye, K. Andreassen, K.-B. Larsen 1989. Glacigenic sediments on a passive continental margin as exemplified by the Barents Sea. *Marine Geology* 85, 251–272.

Vorren, T. O., E. Lebesbye, K. B. Larsen 1990. Geometry and genesis of the glacigenic sediments in the southern Barents Sea. In J. A. Dowdeswell & J. C. Scourse (eds), *Glacimarine environments: processes and sediments*, 269--288. London: Geological Society Special Publ. 53.

Wang Yuelen, Lu Songnian, Gao Zhenjia, Lin Weixing, Ma Guogan 1981. Sinian tillites of China. In M. J. Hambrey & W. B. Harland (eds), *Earth's pre-Pleistocene glacial record*, 386--401. Cambridge: Cambridge University Press.

Webb, P.-N., D. M. Harwood, B. C. McKelvey, J. H. Mercer & L. D. Stott 1984. Cenozoic marine sedimentation and ice volume variation on the East Antarctic craton. *Geology* 12, 287--291.

Wentworth, C. K. 1922. A scale of grade and class terms of clastic sediments. *Journal of Geology* 30, 377--390.

Wentworth, C. K. 1936. An analysis of the shape of glacial cobbles. *Journal of Sedimentary Petrology* 6, 85--96.

Weertman, J. 1961. Mechanism for the formation of inner moraines found near the edge of cold ice caps and ice sheets. *Journal of Glaciology* 3, 965--978.

Weertman, J. 1987. Impact of the International Glaciological Society on the development of glaciology and its future rôle. *Journal of Glaciology, Special Issue Commenmorating Fiftieth Anniversary of the International Glaciological Society*, 86--90.

Whalley, W. B. 1971. Observations of the drainage of an ice-dammed lake - Strupvatnet, Troms, Norway. *Norsk Geografisk Tidsskrift* 25, 165--174.

Whalley, W. B. & D. H. Krinsley 1974. A scanning electron microscope study of surface textures of quartz grains from glacial environments. *Sedimentology* 21, 87--105.

Wharton, R. A., B. C. Parker, G. M. Simmons, K. G. Seaburg, F. G. Love 1982. Biogenic calcite structures forming in Lake Fryxell, Antarctica. *Nature* 295, 403--405.

Whillans, I. M., J. Bolzan, S. Shabtaie 1987. Velocity of ice streams B and C, Antarctica. *Journal of Geophysical Research* 92 (B9), 8895-8902.

Williams, G. E. 1989. Late Precambrian tidal rhythmites in South Australia and the history of the Earth's rotation. *Journal of the Geological Society* 146, 97--111.

Wilson, R. C. L. 1991. Sequence stratigraphy: an introduction. *Geoscientist* 1, 13--23.

Woodworth-Lynas, C. M. T. & J. Y. Guigné 1990. Iceberg scours in the geological record: examples from glacial Lake Agassiz. In J. A. Dowdeswell & J. C. Scourse (eds), *Glacimarine environments: processes and sediments*, 217--234. London: Geological Society Special Publ. 53.

World Glacier Monitoring Service 1989. *World Glacier Inventory* IAHS (ICSI)-UNEP-UNESCO.

Worsley, P. 1974. Recent "annual" moraine ridges at Austre Okstindbreen, North Norway. *Journal of Glaciology* 13, 265--277.

Wright, R. & J. B. Anderson 1982. The importance of sediment gravity flow to sediment transport and sorting in glacial marine environment: Eastern Weddell Sea, Antarctica. *Bulletin of the Geological Society of America* 93, 957--963.

Yoon, S. H., S. K. Chough, J. Thiede & F. Werner 1991. Late Pleistocene sedimentation on the Norwegian continental slope between 67° and 71°N. *Marine Geology* 99, 187--207.

Zilliacus, H. 1989. Genesis of De Geer moraines in Finland. *Sedimentary Geology* 62, 309--317.

Zotikov, I. A., Zagorodnov, V. S., J. V. Raikovsky 1980. Core drilling through the Ross Ice Shelf (Antarctica) confirmed basal freezing. *Science* 207, 1463--1465.

Zotikov, I. A. 1986. The thermophysics of glaciers. Dordrecht: Riedel, 275 pp.

HABIA SERIES LIST

Hairdressing

Student textbooks

Hairdressing and Barbering The Foundations: The Official Guide to Hairdressing and Barbering at Level 2 REVISED 7e *Martin Green*

Begin Hairdressing and Barbering: The Official Guide to Level 1 3e *Martin Green*

Hairdressing and Barbering The Foundations: The Official Guide to Hairdressing and Barbering VRQ at Level 2 1e *Martin Green*

Professional Hairdressing and Barbering: The Official Guide to Level 3 7e *Martin Green and Leo Palladino*

The Pocket Guide to Key Terms for Hairdressing *Martin Green*

The Official Guide to the City & Guilds Certificate in Salon Service 1e *John Armstrong with Anita Crosland, Martin Green and Lorraine Nordmann*

The Colour Book: The Official Guide to Colour for NVQ Levels 2 & 3 1e *Tracey Lloyd with Christine McMillan-Bodell*

eXtensions: The Official Guide to Hair Extensions 1e *Theresa Bullock*

Salon Management *Martin Green*

Men's Hairdressing: Traditional and Modern Barbering 3e *Maurice Lister*

African-Caribbean Hairdressing 3e *Sandra Gittens*

The World of Hair Colour 1e *John Gray*

The Cutting Book: The Official Guide to Cutting at S/NVQ Levels 2 and 3 *Jane Goldsbro and Elaine White*

Professional Hairdressing titles

Trevor Sorbie: The Bridal Hair Book 1e *Trevor Sorbie and Jacki Wadeson*

The Art of Dressing Long Hair 1e *Guy Kremer and Jacki Wadeson*

Patrick Cameron: Dressing Long Hair 1e *Patrick Cameron and Jacki Wadeson*

Patrick Cameron: Dressing Long Hair 2 1e *Patrick Cameron and Jacki Wadeson*

Bridal Hair 1e *Pat Dixon and Jacki Wadeson*

Professional Men's Hairdressing: The Art of Cutting and Styling 1e *Guy Kremer and Jacki Wadeson*

Essensuals, the Next Generation Toni and Guy: Step by Step 1e *Sacha Mascolo, Christian Mascolo and Stuart Wesson*

Mahogany Hairdressing: Step to Cutting, Colouring and Finishing Hair 1e *Martin Gannon and Richard Thompson*

Mahogany Hairdressing: Advanced Looks 1e *Martin Gannon and Richard Thompson*

The Total Look: The Style Guide for Hair and Make-up Professional 1e *Ian Mistlin*

Trevor Sorbie: Visions in Hair 1e *Trevor Sorbie, Kris Sorbie and Jacki Wadeson*

The Art of Hair Colouring 1e *David Adams and Jacki Wadeson*

Beauty therapy

Beauty Basics: The Official Guide to Level 1 REVISED 3e *Lorraine Nordmann*

Beauty Therapy – The Foundations: The Official Guide to Level 2 VRQ 6e *Lorraine Nordmann*

Beauty Therapy – The Foundations: The Official Guide to Level 2 6e *Lorraine Nordmann*

Professional Beauty Therapy – The Official Guide to Level 3 REVISED 4e *Lorraine Nordmann*

The Pocket Guide to Key Terms for Beauty Therapy *Lorraine Nordmann and Marian Newman*

The Official Guide to the City & Guilds Certificate in Salon Services 1e *John Armstrong with Anita Crosland, Martin Green and Lorraine Nordmann*

The Complete Guide to Make-Up 1e *Suzanne Le Quesne*

The Encyclopedia of Nails 1e *Jacqui Jefford and Anne Swain*

The Art of Nails: A Comprehensive Style Guide to Nail Treatments and Nail Art 1e *Jacqui Jefford*

Nail Artistry 1e *Jacqui Jefford*

The Complete Nail Technician 3e *Marian Newman*

Manicure, Pedicure and Advanced Nail Techniques 1e *Elaine Almond*

The Official Guide to Body Massage 2e *Adele O'Keefe*

An Holistic Guide to Massage 1e *Tina Parsons*

Indian Head Massage 2e *Muriel Burnham-Airey and Adele O'Keefe*

Aromatherapy for the Beauty Therapist 1e *Valerie Worwood*

An Holistic Guide to Reflexology 1e *Tina Parsons*

An Holistic Guide to Anatomy and Physiology 1e *Tina Parsons*

The Essential Guide to Holistic and Complementary Therapy 1e *Helen Beckmann and Suzanne Le Quesne*

The Spa Book 1e *Jane Crebbin-Bailey, Dr John Harcup, and John Harrington*

SPA: The Official Guide to Spa Therapy at Levels 2 and 3, *Joan Scott and Andrea Harrison*

Nutrition: A Practical Approach 1e *Suzanne Le Quesne*

Hands on Sports Therapy 1e *Keith Ward*

Encyclopedia of Hair Removal: A Complete Reference to Methods, Techniques and Career Opportunities, *Gill Morris and Janice Brown*

The Anatomy and Physiology Workbook: For Beauty and Holistic Therapies Levels 1–3. *Tina Parsons*

The Anatomy and Physiology CD-Rom

Beautiful Selling: The Complete Guide to Sales Success in the Salon *Rath Langley*

The Official Guide to the Diploma in Hair and Beauty Studies at Foundation Level 1e *Jane Goldsbro and Elaine White*

The Official Guide to the Diploma in Hair and Beauty Studies at Higher Level 1e *Jane Goldsbro and Elaine White*

The Official Guide to Foundation Learning in Hair and Beauty 1e *Jane Goldsbro and Elaine White*

CENGAGE
Learning®

Professional Hairdressing and Barbering:
The Official Guide to Level 3,
Seventh Edition
Martin Green and Leo Palladino

Development Editor: Claire Napoli

Content Project Manager: Susan Povey

Senior Manufacturing Buyer: Eyvett Davis

Typesetter: MPS Limited

Cover design: HCT Creative

Text design: Design Deluxe

For product information and technology assistance,
contact **emea.info@cengage.com.**

For permission to use material from this text or product,
and for permissions queries,
email **emea.permissions@cengage.com.**

British Library Cataloguing-in-Publication Data
A catalogue record for this book is available from the British Library.

ISBN: 978-1-4080-7338-4

Cengage Learning EMEA
Cheriton House, North Way, Andover, Hampshire, SP10 5BE United Kingdom

Cengage Learning products are represented in Canada by
Nelson Education Ltd.

For your lifelong learning solutions, visit **www.cengage.co.uk**

Purchase your next print book, e-book or e-chapter at
www.cengagebrain.com

Printed in China by RR Donnelley
Print Number: 02   Print Year: 2015

Dedication to Leo Palladino BA

This seventh and fully revised edition of *Professional Hairdressing and Barbering* is dedicated to the memory of Leo Palladino, whose original chapters on the science and practice of hairdressing formed the basis of the *Official Guides to Hairdressing* as they are known today.

Leo's knowledge and understanding of his profession contributed to his renown and success in becoming the best-selling author in the UK and beyond on theoretical and practical hairdressing. His books have become and continue to represent the industry benchmark for learning in this industry.

During his stellar career in hairdressing, trichology and education, Leo lectured at college level, progressing to Head of Science at Gloucestershire College of Arts and Technology. He served on the Hairdressing Council as an advisor and toured the country with companies such as Wella.

As Chief Examiner for City & Guilds he had ultimate responsibility for setting syllabi and marking examination papers, and as a board member of Habia he set the industry standards as we know them today.

Throughout his life, Leo demonstrated an exceptional level of commitment both to his family and to his profession. His legacy lives on beyond a lifetime's dedication to the hairdressing industry.

Contents

Foreword ix
About Habia and VTCT x
Acknowledgements xi
Credit list xiii
About the author xv
Level 3 in Hairdressing and Barbering xvi
About the book xxi
About the website xxiii
Introduction by the author xxiv

PART ONE Customer-centred services

1 Consultation and Advice 4

Learning objectives 4
Introduction 5
The consultation 7
Analyse the hair, skin and scalp 15
Influencing factors and features 30
Make recommendations to clients 36
Advise clients on hair maintenance and management 38
Revision questions 42

2 Promote Services and Products 44

Learning objectives 44
Introduction 45

Salon services and products 46
Gain client commitment to using additional services
   or products 52
Revision questions 56

3 Customer Service 58

Learning objectives 58
Introduction 59
Service improvement 60
Implement changes in client service 69
Assist with the evaluation of changes in client service 71
Revision questions 73

PART TWO Hairdressing technical services

4 Creative Cutting 76

Learning objectives 76
Introduction 77
Maintain effective and safe methods of working when cutting 78
Preparation for cutting 84
Creatively restyle women's hair 91
Cutting and styling techniques 94
Provide aftercare advice 104
Revision questions 105

5 Creative Barbering 108

Learning objectives 108
Introduction 109
Maintain effective and safe methods of working when cutting 110
Preparation for cutting 113
Creatively restyle men's hair 124
Provide aftercare advice 131
Revision questions 132

6 Beards and Moustaches 134

Learning objectives 134
Introduction 135
Maintain effective and safe methods of working when cutting facial hair 136
Preparation for cutting 139
Create a range of facial hair shapes 144
Provide aftercare advice 147
Revision questions 149

7 Shaving 150

Learning objectives 150
Introduction 151
Maintain effective and safe methods of working when shaving 152
Prepare the hair and skin for shaving 157
Shaving 158
Provide aftercare advice 161
Revision questions 162

8 Creatively Style and Dress Hair 164

Learning objectives 164
Introduction 165
Maintain effective and safe methods of working when styling hair 166
Creatively style and dress hair 175
Provide aftercare advice 184
Revision questions 185

9 Creatively Dress Long Hair 186

Learning objectives 186
Introduction 187
Maintain effective and safe methods of working when dressing long hair 188
Creatively dress long hair 195
Provide aftercare advice 207
Revision questions 208

10 Hair Extensions 210

Learning objectives 210
Introduction 211
Maintain effective and safe methods of working when adding hair extensions 212
Plan and prepare to add hair extensions 218
Attach hair extensions 225
Cut and finish hair with extensions 232
Maintain and remove hair extensions 234
Provide aftercare advice 235
Revision questions 239

11 Colouring Hair 240

Learning objectives 240
Introduction 241
The principles of colour and colouring 242
Depth and tone 252
Lightening hair 256
Safe methods of working for colouring and lightening services 261
Skin and hair tests 268
Colour applications 275

Colouring problems 283
Provide aftercare advice 285
Revision questions 287

12 Colour Correction 288

Learning objectives 288
Introduction 289
Maintain effective and safe methods of working when
    colour correcting hair 290
Find out what is wrong 292
Plan and agree a course of action to correct colour 296
Provide aftercare advice 309
Revision questions 310

13 Perming Hair 312

Learning objectives 312
Introduction 313

Effective and safe methods of working when perming
    hair 314
Prepare for perming 320
Create a variety of permed effects 327
Alternative/creative winding techniques 332
Straightening hair 336
Neutralizing 338
Provide aftercare advice 343
Revision questions 344

14 Develop Your Creativity 346

Learning objectives 346
Introduction 347
Plan and design a range of images 348
The basics of good hair design 350
Produce a range of creative images 359
Revision questions 364

PART THREE Supporting management

15 Health and Safety 368

Learning objectives 368
Introduction 369
Check that health and safety instructions are followed 370
Make sure that risks are controlled safely and
    effectively 383
Controlling risks in the workplace 387
Revision questions 391

16 Personal Effectiveness 392

Learning objectives 392
Introduction 393
Contribute to the effective use
    and monitoring of resources 394
Consumer rights and legislation 401
Stock and stock control 403
Time and time management 408
Meet productivity and development targets 409
Revision questions 415

17 Promotional Activities 416

Learning objectives 416
Introduction 417

Contribute to the planning and preparation of promotional
    activities 418
Implement promotional activities 426
Advertising, PR and the press 429
Participate in the evaluation of promotional activities 432
Revision questions 434

Appendix 1: People's rights and consumer legislation 436

Appendix 2: Answers to revision questions 439

Appendix 3: Useful addresses and websites 441

Glossary 443
Index 446

Web Chapters

18 Facial Massage

19 Hair Patterns and Designs

Foreword

Martin Green, just like his legendary predecessor Leo Palladino, is incredibly passionate about his industry – both as a practitioner and as an educator.

Martin has dedicated his whole life to developing learners. Always at the forefront of technology and didactic learning, Martin demonstrates within these pages his range of skills and the depth of his knowledge of the hairdressing industry.

If you are truly committed to reaching the top of this highly fulfilling profession, make a smart move and invest in this superb study guide. As Director of Standards and Qualifications for Habia, I am delighted to add my endorsement to its scope and content, and I'm equally satisfied that it reflects the current standards and ethos of Habia.

Jane Goldsbro
Director of Standards and Qualifications
Habia

About Habia and VTCT

About Habia

Habia, the Hair and Beauty Industry Authority, is appointed by government to represent employers in the hair and beauty sector. Habia's main role is to manage the development of the National Occupational Standards (NOS) for hairdressing, barbering, beauty therapy, nails and spa. They are developed by industry for industry and represent best practice when achieving skills and knowledge for a particular job role.

Habia is also responsible for the development and implementation of apprenticeship frameworks and for issuing apprenticeship certificates, alongside providing information to employers on government initiatives that may affect the hair and beauty industry – be it educational, environmental or financial. A central point of contact for information, Habia provides guidance on careers, business development, legislation, and salon health and safety. Habia is part of SkillsActive, the Sector Skills Council that covers hair and beauty, sports and the active leisure sector.

About VTCT

VTCT is a government-approved awarding organization offering vocational qualifications across the hairdressing and beauty sector. It is the first non-unitary awarding body accredited to offer Principal Learning for the new Diploma in Hair and Beauty Studies. VTCT's full qualification package includes complementary therapies, sport and leisure, business skills, hospitality and catering. VTCT is involved in many new initiatives being introduced into the education system, including the embedding of general education skills and online assessment.

Acknowledgements

The author and publisher would like to thank the following contributors who have collaborated so closely with us to produce this book.

For constant partnership in reflecting industry standards:
Habia

For expert review and feedback on the material:
Louise Holm at Sparsholt College
VTCT

For providing images:

Alamy
Avlon
BaByliss PRO
Balmain Hair
Beauty Express
Brian Plunkett MIT: Trichocare
    Consulting Ltd
Cinderella Hair
Connect-2-Hair Ltd
Denman
Dr John Gray
Dr Seymour Weaver
E. A. Ellison & Co. Ltd
G. E. Betterton
Goldwell
Gorgeous PR
Habia

HMSO
HSE
iStockphoto
IT&LY
L'Oréal Professionnel
Paul Falltrick for Matrix
Phototake
Professor Andrew Wright
ProStyles
Redken
REM UK (Ltd)
Saks Hair & Beauty
Shutterstock
Wahl (UK) Ltd
Wella
Wellcome Images

The publisher would like to thank the many copyright holders who have generously granted us permission to reproduce material throughout this textbook. Every effort has been made to contact all rights holders, but in the unlikely event that anything has been overlooked please contact the publisher directly and we will happily make the necessary arrangements at the earliest opportunity.

For their help with the photoshoots:

Southampton City College   Ken Franklin

Claire Napoli   Nathan Allan

Tina Arey

And special thanks to:

The Vice Principal and staff of Kudos Hair and Beauty Salon at Southampton City College

Stylists

Tina Arey

Daniel Bartlett

Marie Bungay Hyland

Julie Burrows

Katie Munday

Michelle Parker

Models

Rylee Bean

Jennifer Delve

Rikki Drake

Mia Gordon

Reo Gordon

Chelsea Hope

Lee McMahon

Leeann Walsh

Credit list

About the author

Martin Green is a highly experienced hairdressing practitioner and college lecturer with over 40 years of experience in the industry. During that time, he has been a consultant for Habia where he was part of the original team that created the first industry standards. He has worked for awarding authorities such as the City and Guilds of London Institute, where he was a regional verifier for the south-west, and at the Vocational Training Charitable Trust where he wrote a wide range of assessment materials.

Martin is author of the *Official Guides to Hairdressing,* all published by Cengage. These include *Begin Hairdressing Level 1, Hairdressing and Barbering Level 2* and *Professional Hairdressing Level 3.* All are bestsellers in their domain.

Martin's energy and passion for his craft are keenly demonstrated through his being a unique blend of practitioner, teacher and author. His enthusiasm has been unyielding and he has shown a deep commitment to the development of e-learning and online resources as well as print materials.

Martin has won a national Joint Information Systems Committee (JISC) Hi5 award at advanced level for *Widening Participation in Education* in recognition of his achievements in this sector.

Level 3 in Hairdressing and Barbering

The Level 3 qualification provides the main industry-recognized qualification for salon apprenticeships and college-based courses. The structure of the Level 3 qualification requires the learner to complete a core set of mandatory units, and then make choices for a range of *mix and match* topics, based upon your own particular interests.

The different units from hairdressing or barbering have a varying amount of values or *credits*. This flexibility within the qualification and the Qualification Credit Framework (QCF) now allows the learner to choose between barbering and hairdressing at a diploma level or, alternatively, to accumulate individual units and credits towards lower levels of qualification within the QCF, such as a certificate or an award. This credit accumulation system may seem confusing at first, but it is explained in more detail further on.

An Introduction

Habia, the Hair and Beauty Industry Authority, is the representative organization responsible for defining the standards for our hair and beauty industry. The National Occupational Standards (NOS) that they produce are taken and used by awarding organizations such as; City & Guilds (C&G), Vocational Training Charitable Trust (VTCT) or ITEC to create the qualifications that you take part in. Therefore, in simple terms, Habia produces the standards that you work towards and ITEC defines the conditions and specifications against which you are assessed.

The NOS that are used in hairdressing or barbering have a common structure and design. Some of the units are specific to the industry such as cutting and styling, whereas others such as health and safety or personal effectiveness are common across many vocational sectors. This common structure allows for people to be credited with units of competence that are accepted in other industries too. In the past, this has always been a problem, as someone who qualified in one sector was not necessarily acknowledged as being competent in another, even though he or she had the training and experience.

Level 3 in Hairdressing and Barbering

The mandatory units are as follows:

Monitor procedures to safely control work operations

Promote additional services or products to clients

Provide hairdressing consultation services

Creatively cut hair using a combination of techniques

The optional units for Group 1 are as follows:

Colour hair using a variety of techniques

Provide colour correction services

Creatively style and dress hair

Creatively dress long hair

Develop and enhance your creative hairdressing skills

Create a variety of permed effects

Provide creative hair extension services

Provide specialist consultation services for hair and scalp conditions

Provide specialist hair and scalp treatments
(Note that this unit and the 'Provide specialist consultation services
for hair and scalp conditions' unit **must** be taken together.)

The optional units for Group 2 are as follows (only one unit can be taken from this group):

Contribute to the financial effectiveness of the business

Support client service improvements

Contribute to the planning and implementation of promotional activities

Units and learning outcomes

A unit relates to a specific task or skill area of work. It is the smallest part of a qualification and carries its own credit value, which you can build up to achieve a qualification.

A learning outcome describes in detail the skill and knowledge components of the unit that need to be completed.

For each unit, when all the outcomes have been achieved, a unit certification may be awarded. A Level 3 qualification is made up of a specific number of units required for the occupational area. Some of the units are mandatory (compulsory) and some are optional (not compulsory). All mandatory units must be achieved to gain the Level 3 qualification and a specified number of optional units must be selected to study in addition to the mandatory units to attain the qualification.

Unit title and learning outcomes (example Consultation Unit)

Unit title	Learning outcomes
Provide hairdressing consultation services	Identify client's needs and wishes
	Analyse the hair, skin and scalp
	Make recommendations to clients
	Advise clients on hair maintenance and management
	Agree services with your client

The NOS covers the learning outcomes in detail. They specify how each task is to be performed by listing the performance criteria. They also cover the circumstances, conditions or situations in which these actions must be carried out; this is called the range.

Performance criteria

The performance criteria are a list of essential actions. Although these may not be necessarily in the order in which they should be carried out, they do provide a definitive checklist of what needs doing. During assessment, these performance criteria form the specification of how a task must be performed.

Learning outcome	Performance criteria
Identify clients' needs and wishes	**By:** a) encouraging your client to express their wishes and views **By:** b) allowing your client sufficient time to express their wishes and views **By:** c) asking relevant questions in a way your client will understand **By:** d) using visual aids to present clients with suitable ideas to help them reach decision **By:** e) encouraging your client to ask about areas of which they are unsure f) Confirming your understanding of your client's wishes before making any service recommendations

Range

Range statements provide a number of conditions or applications in which the learning outcomes must be performed. Quite simply, they state under what particular circumstances, and on what occasions, or in which special situations the activity must take place.

Some example range statements – identifying which situations or circumstances need to be included when undertaking the task – follow:

Learning outcome	Range
Identify clients' needs and wishes	**By means of:** a) questioning **By means of:** b) observation **By means of:** c) testing

Essential knowledge underpinning the activity

When you do your work properly, you need to know what you are doing and why you are doing it. The terms theory, learning and principles generally refer to essential knowledge and understanding, in other words, what you **must** know.

Learning outcome	Essential knowledge and understanding
Identify clients' needs and wishes	**You should know:** how and when tests are carried out on hair and skin
	You should know: how the following factors limit or affect the services and products that can be offered to clients: ◆ lifestyle ◆ adverse hair, skin and scalp conditions
	You should know: current fashion trends and looks

At the point where a task's performance criteria and range have been covered and knowledge has been learned and understood, the task is carried out competently and a skill has been acquired.

Shared knowledge

Units and learning outcomes often share similar components. Therefore, some of the performance criteria used within a main outcome from one task is often similar to that used within another.

Example:

Maintain effective and safe methods of working when styling hair

Maintain effective and safe methods of working when cutting

Similarly, the knowledge that is essential to underpinning one learning outcome often features in another. This duplication may at first seem unnecessary, but it occurs because of the modular, stand-alone design of the units. (Remember, each individual unit can be awarded a certificate.)

This can be useful in terms of speeding up the learning process, as sometimes knowledge or skills learned in one activity are then directly applicable to other tasks. This is also useful when it comes to recording these learned experiences, because the knowledge learned in one situation can be quickly cross-referenced to other similar activities in your portfolio.

Assessment

Your competence – your ability to carry out a task to a standard – is measured during assessment. Your ability to carry out a task — performance evidence — will be observed and checked against the performance criteria. Therefore, your assessor will be watching to see how you carry out your work.

Sometimes it is not possible to cover all the situations that might crop up in one performance. Therefore, in that situation, your assessor might ask you questions about what you have done and how you might apply that in different circumstances. To help you get used to this, the activities that appear throughout this book contain many types of questions that you might be asked.

Your understanding and background knowledge of work tasks is also measured through questions asked by your assessor. Sometimes you might be asked to give a personal account of what you have learned. This could take the form of writing a sequence of events that need to be accomplished to complete the task satisfactorily. Other questions may ask you specifically about particular tasks; more often than not, these types of questions take the form of short-answer, or multiple-choice questions. Again, the activities covered within this book give plenty of examples and practice.

About the book

The common structure and design that exist within National Occupational Standards (NOS) are mirrored in many ways within this text. For the first time in the official hairdressing series, revisions and updates have been totally reworked to provide:

◆ the fastest possible navigation to the areas of learning that you need to know

◆ a book that covers all aspects of the professional qualification for both hairdressing and barbering standards

◆ a chapter structure that mirrors the industry standards in unit as well as outcome format

◆ a comprehensive glossary and index to help you find and understand key terminology.

How to use this book

You can use this book in a number of different ways:

1 You can use the revised chapter structure to cover complete units as you get to them within your training.

2 You can use the book as a quick guide and overview of the methods that you will be studying and the new practices that you are learning about.

3 You can use it as your stand-alone course guide covering the A–Z of hairdressing and barbering at professional Level 3.

The new format of this book will help you to read and use it more easily. Each chapter opens with a simple quick look at what you need to do and what you need to know, and then extends deeper by covering each aspect of your training using a comprehensive approach.

Throughout this textbook you will find many colourful text boxes designed to aid your learning and understanding, as well as highlighted key points. Here are examples and descriptions of each.

ALWAYS REMEMBER

Suggests good working practice and helps you develop your skills and awareness during training.

CourseMate video boxes highlight certain techniques available to view online.

HEALTH & SAFETY

Draws your attention to related health and safety information essential for each technical skill.

HAIR SCIENCE
Highlights the essential hair science knowledge needed for each unit.

ACTIVITY

Featured in all chapters to provide additional tasks for you to further your understanding.

TOP TIP

Shares the author's experience and provides positive suggestions to improve knowledge and skills in each unit.

Directional arrows point you to other parts of the book that explore similar or related topics, so you can expand your learning.

REVISION QUESTIONS

At the end of each chapter there is a useful revision section which has been specially devised to help you check your learning and prepare for your oral and written assessments.

Use these revision sections to test your knowledge as you progress through the course and seek guidance from your supervisor or assessor if you come across any areas that you're unsure of.

About the website

U sing our Hairdressing and Barbering Level 3 CourseMate alongside this textbook will provide a richly blended solution to your learning.

CourseMate is a highly interactive resource which brings course concepts to life. It is designed to support lecturers and students by providing a range of online resources, activities and video footage that perfectly integrates with classroom learning to fulfil the guided learning requirement for each unit.

For students:

◆ Searchable ebook

◆ Step-by-step videos

◆ Interactive multi-choice quizzes

◆ Interactive activities and games

For lecturers:

◆ Lesson plans

◆ PowerPoint slides

◆ Activity hand-outs

◆ Engagement Tracker tools to help you track students' progress and monitor their learning

For further information about CourseMate, please contact emea.fesales@cengage.com

Introduction by the author

Learning new things can be very enjoyable; on the other hand, it can be daunting. There is a reason for this that I ask you to consider: *Have you ever thought about the things that you enjoy doing and those that make you apprehensive? Are they not often the same?*

Rest assured that your feelings are completely natural, because it is human nature to feel good about things when they are going well, and equally natural to feel worried or concerned about the unknown.

When you apply this philosophy to learning, you experience a real sense of enjoyment at the point where you master new skills, after you have put so much effort into studying the theory and practice. The result is a deep feeling of personal and professional fulfilment – and that's why you have this book between your hands.

Learning is said to be a journey, and the first part of that journey, the commitment you make to your studies at the outset, is the most difficult. Remember those first few steps make the difference between being able to say: 'I can't do it' and 'I **can** do it now.'

I wish you every success for the future and hope that Level 3 is only a few short steps away for you.

Martin Green

Notes

PART ONE

Chapter 1
Consultation and Advice

Chapter 2
Promote Services and Products

Chapter 3
Customer Service

Customer-centred services

First impressions are vital in most walks of life. In the hair and beauty industry they obviously count for a great deal. How a client is greeted and served from the moment they walk through your salon doors is vitally important.

Your role is to find out what each client aspires to, and try to meet these needs and desires to their satisfaction – so that they will keep returning to *your* salon and even recommend your services to family and friends.

To become a true professional, you need the technical skills to provide those services. But you also need great listening and communication skills to inspire confidence in your clients – no matter what service they come to you for.

1 Consultation and Advice

LEARNING OBJECTIVES

◆ Be able to identify the client's wants and needs.

◆ Be able to analyse the hair, skin, and scalp.

◆ Carry out tests and make recommendations to clients based on your findings.

◆ Be able to advise clients on how they can maintain their own hair.

◆ Understand the salon's services, products and prices.

◆ Know your salon's policies and legal obligations.

◆ Be able to communicate professionally and provide aftercare advice.

KEY TERMS

acid	confidential	keratin	telogen
allergy alert test	contra-indications	monilethrix	temporal
alopecia	discolouration	occipital bone	traction alopecia
alpha keratin	fragilitis crinium	para dyes	trichologist
anagen	folliculitis	paraphenylenediamine (PPD)	whorls
arrector pili	graduation	referral	
beta keratin	hygroscopic	skin test	
catagen	incompatible chemicals	tariffs	

UNIT TOPIC

Provide client consultation

INTRODUCTION

All hairdressing services start at the same point, whether you are about to take on a difficult colour correction service or carry out a straightforward dry trim. The information that you need to do the job properly can only be found out during a thorough consultation. Spend the time with the client before you start in order to:

◆ develop a professional relationship and gain their confidence and trust

◆ find out what they want and discuss what it will look like

◆ see if their expectations are realistic in relation to their hair condition or length

◆ identify any limiting factors or contra-indications to the service

◆ discuss their options after you have made your evaluation

◆ negotiate a course of action and talk about the costs involved.

If you do all that, the rest should be easy.

Consultation and advice

PRACTICAL SKILLS

Find out what clients want by questioning and visual aids

Look for things that will influence the ways that you do your client's hair

Learn how to carry out a variety of tests

Make recommendations or alternatives based upon your findings

Provide aftercare advice to your clients

Make sure that you update the client's records after consultation

UNDERPINNING KNOWLEDGE

Know your salon's range of products and services

Understand the factors that can affect your hairstyling options

Know the contra-indications that can limit services and the possible causes for them

Know how to communicate effectively and professionally

ALWAYS REMEMBER

A professional distance is created during consultation that is not there at any other time. From the client's standpoint they are still on the outside, observing; remember, they can form an idea of the salon without any commitment. You will only get one chance to create a first, professional impression that lasts.

INDUSTRY ROLE MODEL

TOMMY VAN DER VEKEN Creative Director, Tommy's Hair Company

" Tommy's Hair Company has three salons in North Wales and Chester. We are five times winners of the British Salon of the Year 2 award, and are the proud recipients of the Most Wanted Business Thinker award on two occasions. My role within Tommy's is to develop the creative vision of the Tommy's brand and to provide ongoing progressive technical training and advancement to all members of our team.

With an active column of clients within each of our three salons, I still have an enormous passion for working with salon guests helping them to look and feel their best.

The consultation

During consultation a client begins to learn about your 'professional expertise', the breadth and depth of technical knowledge and subject specialism that you can offer, whereas the communication that takes place during routine services has a very different feel to it. The bond that develops during consultation sets the tone for an ongoing business relationship.

> " The consultation is a fact-finding mission, which allows you to identify your client's needs and wishes. This will help you to make the right recommendations for their service, and if undertaken successfully you will gain your client's trust immediately. Remember, a fantastic hairdresser conducts a great consultation every time – regardless of whether it's the client's first or fiftieth visit!
>
> *Tommy Van der Veken*

About you – the senior stylist

As a true professional you lead by example; you are responsible for the quality of service that you provide to your clients and, as a senior stylist, you are also a focal point to the other staff for back-up, support, advice and reassurance. Your experience is respected and your opinions are valued; the directions taken by other less senior staff may depend on your actions.

ACTIVITY

Putting consultation into practice

Client consultation can be practised within the salon/barber's shop with your work colleagues; the service lends itself well to role plays.

To develop these skills further, simulate your consultation skills by conducting a consultation for the following services upon each other.

Hairdressers	Barbers
Colour correction	Colour correction
Bridal hairstyle	Shaving
Perming	Perming
Hair extensions	Hair patterns/design

What are the signs of a true professional?

For those who join the hairdressing industry directly after leaving school, it is unlikely that they will know what is really required until they have had some experience of working with, and for, other people. But they can get a taster of this when they progress through Levels 1 and 2. When younger people find out how hard it can be, they can often have second thoughts. This is one of the main reasons why the industry has a high drop-out rate. Too

TOP TIP

Even if your regular customers are totally happy with the services you provide, take the opportunity to enhance the professional balance by always providing the consultation review. It will make them feel special and enhance the relationship with you.

ALWAYS REMEMBER

Effective communication relies upon listening to the client, hearing what they have to say, responding to them positively in how you reply and backing that up with the right body language.

many people start to train with an uninformed, unrealistic view of what customer service and working with the public involves.

People joining the industry later in life, perhaps wanting a career change, probably have a more realistic idea. Generally because they have acquired life skills over a long period of time, something that the younger people have not. Also they have all been clients themselves and so their perspective is very different. They have had the opportunity to observe and receive good customer service and will want to provide that same level of professionalism to others.

Identify clients' needs and wishes

Hairdressing is about relationships as well as technical skills – and the best relationships are created through good communication. People can often find it difficult to convey what they really mean; they use technical terms inaccurately, they see aspects and parts of styles in pictures that they like, but they cannot easily express these aspects to the stylist.

Your communication skills need to be well developed to find out what the client really wants. Too often people can find it hard to discuss personal issues or different aspects of their appearance.

Most people have no self-visualization; they do not have the ability to imagine what a particular style would look like on them. Also they would not want to expose any physical features or aspects with which they are not happy. The challenge is to sympathetically discover these features and, more importantly, to find ways of making the most of your client's appearance in the light of the real limiting factors that you can see from a professional point of view.

Good, effective communication

Good communicators use different skills in their daily routines. This involves the following essential skills.

Excellent listening skills This is the ability to hear and understand what the client is saying. It is particularly useful when the client has a friend or family member with them who keeps talking.

A good speaker Long pauses during conversations can often be uncomfortable but equally non-stop talking can lead to a lack of understanding. Knowing when it is right to speak or keep quiet is an invaluable interpersonal skill. During normal consultation you, the hairdresser, will be taking the lead. You will be asking questions, i.e. trying to gain enough information in the time available to make the right judgements. You will be weighing up what the client wants against the limitations arising from the analysis. You will be helping the client to agree on the various possible options and planning the necessary treatment.

'Reading' skills The ability to read situations, to understand what has been said or sometimes *not* said, is exceptionally useful. There are times when your client will look a certain way or say something that makes you think again. In these situations, your ability to read the situation, i.e. your perceptiveness and your response, may have a big impact on your long-term relationship.

Body language

We generally rely upon our ability to speak and hear as the 'natural' way of communicating. In fact, we use all of our senses to communicate. But are they used equally? Sight is probably at the top of the list as this is the fastest but often an unreliable way of drawing conclusions. In addition to using words we show our interest, attitude and feelings through physical expressions. Non-verbal communication (NVC) or body language is especially important. It can truly show what we are feeling, even if our words are saying something quite different!

We express our feelings with body language through eye contact, posture and general body positioning. So it is very important that we send the right message, particularly when dealing with clients and potential customers.

Eye contact

Always maintain eye contact when in conversation with your client. Where possible, maintain the same eye level. For example, when you carry out a consultation with the client and they are seated, sit beside the client or opposite them. Standing over or above them and looking down will convey a feeling of authority, and might appear as if you are trying to assert yourself and take control. This is threatening, intimidating and definitely the wrong signal to send to a potential client.

Physical contact

Most people are embarrassed by physical contact from someone they do not know well.

Posture, body position and gestures

Much has been written on the subject of body language and its effects. It is far too complex a subject to address in a few simple paragraphs. Posture, or composure of the body, is a form of body language in this context. Reading this form of communication is a skill that develops over time. However, there are a few basic rules that can help to convey the right message and create a good impression.

> " A consultation is the perfect opportunity to impress new people. So remember to be polite, courteous, welcoming, understanding and appreciative of the client's needs. Be professional at all times, use your training and knowledge to make well-informed suggestions and be confident in your abilities.
>
> *Tommy Van der Veken*

◆ Folded arms, and the crossing of arms on the chest, are considered to be protective gestures and can convey a closed mind or show defensiveness.

TOP TIP

Body zones – Proxemics

Do not crowd or appear over-familiar with your client. Imagine how you might feel if someone came up to you and got a little too close.

◆ Open palms, as a gesture supporting explanation or information, with hands at waist height and palms upwards, can indicate that the person has nothing to hide. This is interpreted as openness or honesty.

◆ Scratching behind the ear or the back of the neck while listening can indicate that the listener is uncertain or doesn't understand. Rubbing the nose while listening can indicate that you don't believe what you are hearing.

◆ Inspecting fingernails or looking at a watch is a plain and simple indication of boredom.

◆ Talking with your hand in front your mouth may lead the listener to believe you are not being honest. You're hiding yourself by your gestures.

◆ Shifting from foot to foot: this can indicate that you're worrying about getting found out! It also says that you would rather be somewhere else: to get away so that no guilty expressions are spotted.

These forms of communication are only an indication of feelings and emotions. In isolation they may not mean anything at all. However, taken together they can send a very clear message. Make sure that you send the appropriate signals and look interested, keen, ready to help and positive. Above all, show that you can listen.

Mirroring posture If you're in tune with the person you're speaking to you will often find you unconsciously mirror each other's body postures. So, if you rest your hand on your chin, the other person will follow you. If he or she leans forwards, you'll find yourself making the same move and so on.

This technique is used in interviews to help make the candidates feel at ease. Sometimes the technique is used as a subconscious sales tool, when sales representatives want us to part with our money.

Lying can be shown by:

◆ sweating

◆ excessive hand movements

◆ biting of fingernails

◆ chewing of the inside of the mouth

◆ drying up of the mouth

◆ lack of eye contact.

Effective body language during consultation

Remember, as a professional you should be attentive, positive and actively participating in the consultation process. You can show this in the way that you:

◆ sit to the front edge of your chair or stool, facing the client (not through the mirror)

◆ maintain eye contact and eye level, do not look down on the client

◆ acknowledge what has been said by nodding or confirming with a 'yes' at the appropriate moments.

Questioning techniques

There are three main types of question that you could use during your consultation. These are: open or closed questions, choice questions and feeling questions. These questioning styles are explained further on pages 11 and 12.

> The goal of the consultation is for you to find out exactly what your client wants and by asking open questions the client is more likely to open up and explain in greater detail their specific needs and wishes. You can then use your knowledge and experience to make recommendations based on their answers.
>
> *Tommy Van der Veken*

ACTIVITY

The illustration below shows the differences between questioning styles. Use this as a discussion point that you could use with your colleagues in role play. Try out different questioning techniques to see how they prompt for additional information then answer the following questions.

1 Which questioning style created the most response?
2 Which questioning style created the least response?
3 Which style do you think you would try to use in your consultations with clients?

Open questions These are good to use if you want to gain more in-depth information from the client. They start with 'who', 'what', 'when', 'why', 'where' and 'how'. Examples are:

◆ What products do you use on your hair?

◆ When did you last wash your hair?

◆ How often do you use the straightening irons on your hair?

Closed questions These are useful for a quick 'yes' or 'no' elimination response. Examples are:

◆ Have you had permanent colour on your hair before?

Open/closed questions

Open questions will result in explanations or descriptions. **Closed** questions will result in short answers such as 'yes' or 'no'.

Feeling questions

Feeling questions ask 'how do you **feel** about a fringe?' 'What do you **think** if I restyle it today?' 'Do you **want** more volume?'

Questioning styles

Choices questions

'Would you like to keep the parting in the centre, **or** would you like to try it on the side for a change?'

◆ Would you like me to do it in your lunch hour today for you?

◆ Are you against moving the parting from the centre?

Feeling questions   These types of questions focus upon the client and are good to use when you are trying to gauge a personal opinion or feeling. Examples are:

◆ How do you feel about taking the length back up to the shoulders?

◆ What do you think about changing the colour today?

Good consultation takes time and this can be a problem, particularly when you have a busy schedule. Feeling-type questions, or a variant on them, can be most useful here as these can be deployed in a persuasive way to *help* clients make choices.

Choices questions   These could be considered to be leading questions and are a good sales tactic when trying to pinpoint a few possible options. Examples are:

◆ Would you prefer the smaller intense leave-in treatment or the normal after-shampoo conditioner?

◆ Shall we blow-dry your hair and finish it in a different way? Perhaps curls or waves instead of straight?

◆ Shall we see if there is time today to do it all, or would you rather make a separate appointment for the highlights after the cut?

◆ Where would you like to wear the bulk of the hair? Piled on top, at the side or smooth on the sides and fixed at the back?

Whatever style of question you choose to use, you must make sure that you remember the following:

Identify the limitations or influencing factors   Some of these will crop up in your conversation, but the main influencing factors are going to arise during your visual inspection of the hair and scalp. So whether you do your inspection first or later will depend on the service that you are going to do, or how comfortable the client is with the consultation process.

Avoid misunderstandings

Misunderstanding is the main reason that clients get dissatisfied with hair! You must adapt your consultation style to suit the level of understanding of the client. Don't expect the clients to know what a reverse graduated bob is with razor-sliced and shattered edges. The client may like the sound of what you are saying but will not know or understand the techniques or methods involved. When it's done, they might hate it!

"

By doing a recap of the consultation with the client you can make sure they understand your vision and are happy with your suggestions on how to achieve the look they want. During the recap you can also ensure the client is aware of the cost, duration of the service and maintenance required with their finished look.

Tommy Van der Veken

Avoid using technical jargon; keep it for the training salon. You want the client to be happy with the final effect, so keep it simple. If a client uses our technical jargon with you, find out exactly what she means; it might not be what we mean by it.

Allow time for the client to consider your advice and don't rush them into making swift decisions. If they do feel rushed or that you are being 'pushy' you will be alienating them from the outset. That's a difficult position to be in as you will feel the tension during the service and that is an uncomfortable situation for a stylist.

You need all the facts before you start. So you might find that you need to ask a few questions before you make any examinations. Many people can be intimidated by having to walk into a salon in front of others. Be sensitive to their needs and watch their body language for any signs of defence and remember to be patient throughout the service.

Listen to their responses – hear what the client is saying. If there seems to be hesitation or reluctance, be positive and give reasons for your plan of action. If you cannot dispel the concerns, you will need another plan!

Confirm and agree – at any point where a decision is being made. Make sure that you confirm and summarize at all the points along the way. Only after you have both agreed a course of action can you proceed to the next stage.

ALWAYS REMEMBER

Being highly self-motivated affects others too. The way in which we conduct ourselves at work has a direct, positive impact on those around us. Conversely, demotivation spreads quickly!

ACTIVITY

Dealing with customer resistance

Most of the resistance from clients about services or products that you recommend is more to do with a lack of understanding rather than disinterest. Your professionalism enables you to explain what would be right for them and the benefits that they can provide.

Complete this activity by filling in the missing information in the table below. The first one has been started for you.

Consultation aspect	What might a client say in a defensive way?	What can you say to dispel resistance, uncertainty, hesitation or general confusion?
Recommending a retail product	No thanks, I have bought mousse before at the supermarket and it didn't work.	Well, that was unfortunate. What was it that you found a disappointment?
Recommending a new hairstyle		
Recommending a conditioning treatment		
Recommending further (unplanned) salon services		

Have empathy Clients don't come to the salon just when there is a celebration. Because hair grows at an even rate, they are likely to arrive when all sorts of things are going on in their lives and in yours too. It's easy to have sympathy. As hairdressers, we can

ACTIVITY

Hairstyles are strongly linked to emerging trends in the fashion industry.

So how would you keep abreast of the current fashions? Use the Internet, hair magazines and professional trade press to create a collection of work, consisting of the latest looks for reference in your portfolio.

Then, after creating the collection, complete the information in the table below.

	Describe the looks	How does it differ from previous looks?	Where did you find your source data/ information?
Current hair fashions			
New trends, i.e. emerging looks			

be sympathetic every day. We can all relate to the stories told by our fellow staff over lunch or during breaks – the stories about our clients in certain situations and circumstances. However, having empathy or the ability to put yourself in somebody else's shoes is quite difficult. Sincerity is an emotion that is easily read through your body language and so people can usually tell if you are being insincere.

The close bond that forms between client and hairdresser is often tested at more stressful times. From the client's point of view, the visit to the salon is not just about having their hair done; it can involve the 'feel-good' factor too. As hairdressers we must recognize this. The hairdresser's role seems to involve a number of aspects – a little bit of social worker, a touch of psychologist, a bit of entertainer and always a friendly listener. In order to do all these things, we need to have and show empathy. It is an extremely useful personal attribute for supporting and understanding others.

ALWAYS REMEMBER

Customer expectations

During consultation with your client, take a little extra time to find out what specific service expectations are required.

◆ What aspects of customer service does the client already receive?
◆ What aspects are missed out? From these, which would it be feasible to introduce?
◆ Would they be happy to pay for these additions?

TOP TIP

Motivation

Discuss with your fellow staff members the benefits of self-motivation and the impact that it has on the whole team. Then consider the disadvantages of demotivation and the impact it will have. Summarize and record your findings.

Who are we dealing with?

The analytical skills used during client consultation will, at a higher professional level, involve both objective and subjective thinking. These aspects are not examined at Level 2 when the stylist makes a number of observations, looking for influencing factors that control their thinking and arriving at a technically sound course of action.

At Level 3, after gaining more experience of working with people, we modify the way that we approach clients and can start to work intuitively; we subconsciously draw information from the client. It is a natural thought process that makes us look for patterns and work out the best way forward for that client. Obviously, different people will do this to a different extent and this gives us the professional edge. The artistic stylist who visualizes the perfect style for their new client but without any previous analysis, saying 'leave it to me', is taking a dangerous path. So what information do you need in order to get it right?

ACTIVITY

Standardizing a consultation within the salon

It is very easy to overlook small details during consultation and this may have an impact on the final results. So it can be helpful to devise a standardized consultation process for all salon staff.

1 What sorts of things would you need to include within your consultation system?
2 What other aids or information would you need for the system?
3 What sorts of things do you need to record?
4 How would you keep records of the consultations?

Leading the consultation

The biggest difference between consultation at Level 3 and Level 2 is that a Level 3 stylist must have a broader understanding of people. Experience enables you to combine improved communication skills with a wider range of technical skills. The Level 3 stylist (or someone with the same experience) is more likely to be able to handle a far wider range of people, personalities and their styling requirements. This level of professional communication requires:

◆ control

◆ responsibility

Taking control of the whole situation is essential. This will occur when the new client realizes that it is you who is going to conduct the consultation. However, you have to gain the trust of the client. This can only happen if you are prepared to shoulder the responsibility for the eventual outcome. So this is the opportunity to get it right and win over another client.

Analyse the hair, skin and scalp

You need a thorough understanding of all the hair and skin problems that could affect your choices when selecting a suitable course of action for your client. This section addresses these aspects.

Structure of hair

The cross-section taken through the hair lengthways shown in the diagram on page 16 gives a microscopic view of the three specific layers.

The cuticle is the outer layer of colourless cells which forms a protective surface to the hair. It regulates the chemicals entering and damaging the hair and protects the hair from excessive heat and drying. The cells overlap like tiles on a roof with the free edges pointing towards the tips of the hair. The number of layers is proportional to hair texture. Hair with fewer layers of cuticle is finer than coarser hair types, which

TOP TIP

Visual aids are an extremely useful way of conveying information.

What types of visual aid do you use during client consultation?

TOP TIP

Trust is hard earned. Gaining a customer's trust and loyalty takes time. Even when you have earned it, the bond remains fragile, so handle with care.

ALWAYS REMEMBER

You need to have the confidence to lead during consultation. It is your positive approach and outlook that will engage your client in the conversation and draw out the vital information.

Cross-section of hair

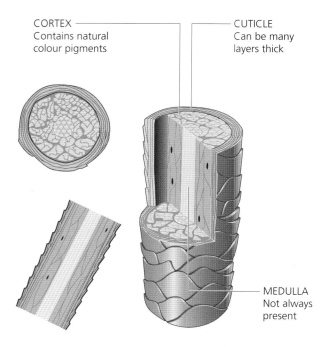

CORTEX
Contains natural
colour pigments

CUTICLE
Can be many
layers thick

MEDULLA
Not always
present

> A good consultation will analyse the client's hair, skin and scalp to identify the quality and quantity of their hair, any strong hairlines and growth patterns, and any potential factors which may limit or affect services and the choice of products.
>
> An excellent consultation will also consider the client's personality, fashion sense, age, height, weight, build, lifestyle, and most important of all facial shape. When pieced together you should have a complete picture of your client and be able to choose the perfect cut, shape and colour for them.

Tommy Van der Veken

have several layers. Hair in good condition has a cuticle that is tightly closed, limiting the penetration of moisture and chemicals. Conversely, hair that is in a dry or porous condition has damaged or partially missing cuticle layers. One simple indicator of cuticle condition relates to the time taken to blow-dry hair. Hair in good condition will dry quickly in proportion to the amount of hair on the head (density). The closely packed cuticle allows the dryer to chase the water from the hair shaft. Porous hair absorbs moisture and therefore takes far longer to dry and is unfortunately subjected to more heat, which exacerbates the problem.

The cortex is the middle and largest layer. It is made up of a long fibrous material which has the appearance of rope. If looked at more closely, each of the fibres is made up of even smaller chains of fibres. The quality and condition of these bundles of fibres will determine the hair's strength. The way in which they are bonded together has a direct effect upon curl and ability to stretch (hair elasticity). It is within this part of the hair that the natural hair colour is distributed. These pigments are diffused throughout the cortex and their colour(s) and rate of distribution determine the colour that we can see. It is also in this layer that both synthetic colours and permanent waves make the permanent chemical changes.

TOP TIP

A closed smooth cuticle is the most important sign of healthy hair. Healthy hair imparts shine, dries more quickly, is resistant to chemical treatments and holds styles and colours better than hair with a raised/damaged cuticle.

The medulla is the central, most inner part of the hair. It only exists in medium to coarser hair types and is often intermittent throughout the length. The medulla does not play any useful part in hairdressing processes and treatments.

Chemical properties of hair The bundles of fibres found in the cortex are made from molecules of amino acids. There are about 22 amino acids in hair and the molecules of each contain atoms of elements in different proportions. Overall, the elements in hair are in approximately these proportions:

◆ carbon: 50%

◆ hydrogen: 7%

◆ nitrogen: 18%

◆ oxygen: 21%

◆ sulphur: 4%.

The amino acids combine to form larger molecules, in the form of long chains called polypeptides, or, if they are long enough, proteins. One of the most important of these is keratin. Keratin is an important component of nails, skin and hair. This protein makes them flexible and elastic. Because of the keratin it contains, hair can be 'elastic' and can be stretched, curled and waved.

In hair, keratin forms long chains which coil up like springs. They are held in this shape by cross-links between chains. The three kinds of link are disulphide bridges (sulphur bonds), salt bonds and hydrogen bonds. Salt bonds and hydrogen bonds are relatively weak and easily broken, allowing the springs to be stretched out. This is what happens in curling. The normal, coiled form of keratin is called alpha keratin. When it has been stretched, dried and fixed into style, it is called beta keratin. The change is only temporary. Once the hair has been made wet or has gradually absorbed moisture from the air, it relaxes back to the alpha state. Disulphide bridges are much stronger but these too can be chemically changed, as in perming.

Physical properties of hair

Hair retains a certain amount of natural moisture which provides flexibility, allowing it to stretch and recoil. But hair that is dry and in poor condition is less elastic. Hair is hygroscopic: it absorbs water from the surrounding air. How much water is taken up depends on the dryness of the hair and the moistness of the atmosphere. Hair is also porous. There are tiny tube-like spaces within the hair structure and the water is drawn into these by capillary action, rather like blotting paper absorbing ink. Drying hair in the ordinary way evaporates only the surface moisture, but drying it over long periods or at too high a temperature removes water from within the hair, leaving it brittle and in poor condition. Damaged hair is more porous than healthy hair and easily loses any water, which makes it hard to stretch and mould. Curled hair returns to its former shape as it takes up water, so the drier the atmosphere, the longer the curl or set lasts.

TOP TIP

Hair in good condition has natural moisture levels that help to give it elasticity and shine.

Hair health and condition

Moisture levels within the hair are essential for maintaining good condition. We can see the evidence of this moisture from the shine that we associate with great-looking hair. 'Bad hair' denotes poor condition and the lack of shine is due to the unevenness of the hair's surface, i.e. the cuticle. A roughened cuticle surface is an indicator of either physical or chemical damage. Each of these states is difficult to correct. In mild cases of dryness, treatments can be applied to improve the hair's manageability and handling. In seriously damaged, porous hair, the hair has no way to resist the absorption of chemicals

and will be saturated by them. There are no long-lasting remedies for this, so regular reconditioning treatments are essential.

During a consultation the health and condition of the hair is your starting point. Whatever happens next should be a process of improving what went on before. A client will expect the service or treatment that you advise to be a step in the right direction. You will need to look for each of the following properties and aspects.

Features of hair in good condition

◆ Shine and flexibility.

◆ Smooth outer cuticle surface.

◆ Strength and resistance.

◆ Good elasticity (ability to stretch and return to original length).

◆ Good natural moisture levels.

Features of hair in poor condition

◆ Raised or open cuticle.

◆ Damaged, torn hair shaft.

◆ Split ends.

◆ Low strength and resistance.

◆ Over-elastic/stretchy.

◆ Dry, porous lengths or ends.

Physical hair damage is caused by

◆ Harsh or incorrect usage of brushes and/or combs.

◆ Excessive heat from styling equipment.

Chemical hair damage is caused by

◆ Incorrect over-timing of all colouring and perming treatments.

◆ Strengths of hydrogen peroxide.

◆ Over-bleaching and highlighting services.

◆ Excessive overuse of colouring products.

◆ Perm products that are too strong or over-processing.

◆ Chlorine compounds found in swimming pools.

Weathering

◆ Hair is also damaged by excesses of sunlight.

Good health

The normal and abnormal working of the body has a direct effect on the hair and scalp. Good health is reflected in good hair and skin. A balanced diet with plenty of fresh fruit and vegetables, and lots of water to drink, also contributes to good health.

◆ Chemicals and medication used in the treatment of disease take their toll on the hair and skin.

◆ Genetic factors affecting hair growth determine hair strength and texture.

◆ Women's hair is usually at its best during pregnancy.

◆ Deterioration of the hair and skin after giving birth is usually due to stress and tiredness.

The skin

Epidermis The epidermis is the surface of the skin. This outer protective layer of the skin is called the stratum corneum and is a hard, cornified layer, consisting of 15 to 40 layers of flattened skin cells or corneocytes; these constantly migrate up from deeper regions and fully replace themselves about once a month. The corneocytes are filled with keratin and a fatty lipid that make a barrier to prevent loss of water through the skin.

Beneath the stratum corneum lie keratinocytes which produce keratin and form the building blocks of the epidermis. In the same area, Langerhans cells scout for invading pathogens while melanocytes produce the pigment melanin that protects the skin from UV radiation. Merkel or nerve cells send messages via the nerve receptors to the brain to register sensation.

Dermis The dermis is the thickest layer of the skin. It is here that the hair follicle is formed. The dermis is made up of elastic and connective tissue and is well supplied with blood and lymph vessels. The skin receives its nutrient supply from this area. The upper part of the dermis, the papillary layer, contains the organs of touch, heat and cold, and pain. The lower part of the dermis, the reticular layer, forms a looser network of cells.

The subcutaneous fat lies below the dermis. It is also known as the subcutis or the hypodermis. It is composed of loose cell tissue and contains stores of fat. The base of the hair follicle is situated in this area.

> **TOP TIP**
>
> The skin is the largest organ of the body and if laid flat would cover an area of about 21 square feet. It is a protective barrier and is made up of many layers.

Layers of the epidermis

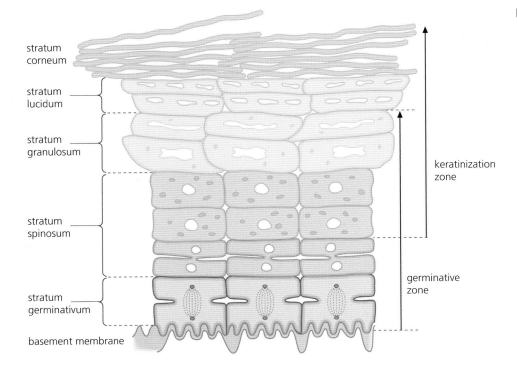

stratum corneum

stratum lucidum

stratum granulosum

keratinization zone

stratum spinosum

germinative zone

stratum germinativum

basement membrane

The hair in skin

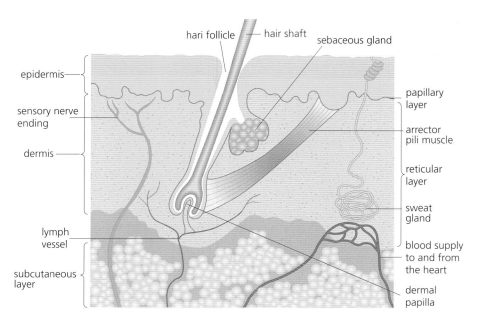

Subcutaneous tissue gives roundness to the body and fills the space between the dermis and muscle tissue that may lie below.

Hair follicle Hair grows from a thin, tube-like space in the skin called a hair follicle.

◆ At the bottom of the follicles are areas well supplied with nerves and blood vessels, which nourish the cellular activity. These are called hair papillae.

◆ Immediately surrounding each papilla is the germinal matrix, which consists of actively forming hair cells.

◆ As the new hair cells develop, the lowest part of the hair is shaped into the hair bulb.

◆ The cells continue to take shape and form as they push along the follicle until they appear at the skin surface as hair fibres.

◆ The cells gradually harden and die. The hair is formed of dead tissue. It retains its elasticity due to its chemical structure and keratin content.

Sebaceous gland The oil gland, or sebaceous gland, is situated in the skin and opens out into the upper third of the follicle. Oil or sebum is secreted into the follicle and onto the hair and skin surface.

Sebum helps to prevent the skin and hair from drying. By retaining moisture it helps the hair and skin to stay pliable. Sebum is slightly acid – about pH 5.6 – and it forms a protective antibacterial covering for the skin.

Sweat glands A sweat gland lies beside each hair follicle. These are appendages of the skin. They secrete sweat which passes out through the sweat ducts. The ends of these ducts can be seen at the surface of the skin as sweat pores. There are two types of sweat gland: the larger, associated closely with the hair follicles, are the apocrine glands; the smaller, found over most of the skin's surface, are the eccrine glands.

Sweat is mainly water with salt and other minerals. In abnormal conditions sweat contains larger amounts of waste material. Evaporation of sweat cools the skin. The function of sweat, and thus the sweat glands, is to protect the body by helping to maintain the normal temperature.

The arrector pili muscle The hair muscle, or *arrector pili*, is attached at one end to the hair follicle and at the other to the underlying tissue of the epidermis. When it contracts it pulls the hair and follicle upright. Upright hairs trap a warm layer of air around the skin. The hairs also act as a warning system; for example, you soon notice if an insect crawls over your skin.

Hair growth

Hair is constantly growing. Over a period of between one and six years an individual hair actively grows, then stops, the follicle rests, degenerates and the hair finally falls out. Before the hair leaves the follicle, the new hair is normally ready to replace it. If a hair is not replaced, then a tiny area of baldness results. The lives of individual hairs vary and are subject to variations in the body. Some are actively growing while others are resting. Hairs on the head are at different stages of growth.

Stages of growth The life cycle of hair is as follows:

◆ **Anagen** (growing stage) is the active growing stage of the hair, a period of activity of the papilla and germinal matrix. This stage may last from a few months to several years. It is at this stage of formation at the base of the follicle that the hair's thickness is determined. Hair colour too is formed in the early part of anagen.

◆ **Catagen** (preparing to rest) is a period when the hair stops growing but cellular activity continues at the papilla. The hair bulb gradually separates from the papilla and moves further up the follicle.

◆ **Telogen** (resting stage) is the final stage, when there is no further growth or activity at the papilla. The follicle begins to shrink and completely separates from the papilla area. This resting stage does not last long. Towards the end of the telogen stage, cells begin to activate in preparation for the new anagen stage of regrowth.

The new anagen period involves the hair follicle beginning to grow down again. Vigorous papilla activity generates a new hair at the germinal matrix. At the same time the old hair

TOP TIP

Hair texture

The more layers of cuticle that the hair has, the greater its resistance to absorbing moisture and chemicals. Therefore coarse hair in good condition can often take longer to perm than finer hair types.

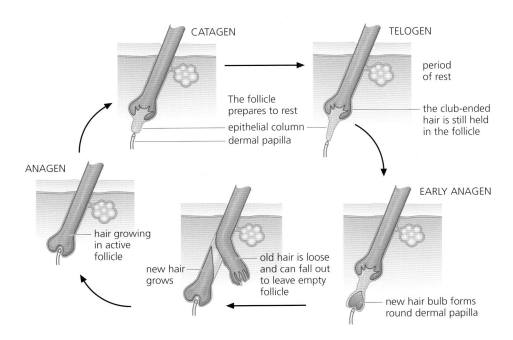

Stages of hair growth

is slowly making its way up and out of the follicle. Often the old and new hair can be seen at the same time in the follicle.

In some animals most of the hairs follow their life cycle in step, passing through anagen, catagen and telogen together. This results in moulting. Human hair, however, develops at an uneven rate and few follicles shed their hair at the same time. (If all hairs fell at the same time we would have bald periods.)

Hair texture

Individual hair thickness is referred to as hair texture and the main types are:

◆ very fine hair

◆ fine hair

◆ medium hair

◆ coarse hair.

The main differences between the hair textures relate to the number of layers of cuticle.

Hair and skin tests

There are different tests that you can carry out to help diagnose the condition and likely reaction of your client's skin and hair. These tests will help you to decide what action to take before, during and after the application of hairdressing processes. You will need to record all these results onto the client's record file.

Development strand test A strand test or hair strand colour test is used to assess the development of a colour while it is processing. It is carried out as follows:

1   Most colouring products just require the time recommended by the manufacturer – check their instructions.

2   Rub a strand of hair lightly with the back of a comb to remove the surplus colour.

3   Check whether the colour remaining is evenly distributed throughout the hair's length. If it is even, remove the rest of the colour. If it is uneven, allow processing to continue, if necessary applying more colour.

ALWAYS REMEMBER

Tests are a vital part of general hairdressing services and should not be missed out or ignored!

TOP TIP

Always follow the manufacturer's instructions when conducting/carrying out tests.

HEALTH & SAFETY

Health and safety: Allergy alert test (formerly known as patch test or skin test)

The allergy alert test is used to assess the reaction of the skin to chemicals or chemical products. In the salon it is mainly used before colouring. Some people are allergic to external contact of chemicals such as PPD (found in permanent colour). This can cause dermatitis or, in more severe cases, permanent scarring of skin tissue and hair loss. Some people are allergic to irritants reacting internally, causing asthma and hay fever. Others may be allergic to both internal and external irritants. To find out whether a client's skin reacts to chemicals in permanent colours, the following test must be carried out at least 48 hours prior to the chemical process.

Note: skin allergy testing is not just for new clients, it has now been found that clients can develop sensitivity to chemicals through prolonged use of the same or similar products. Therefore, testing for adverse reactions is essential and should be carried out regularly.

Colour test This test is used to assess the suitability of a chosen colour, the amount of processing time required and the final colour result. Apply the colour or lightening products you propose to use to a cutting of the client's hair and process as recommended.

Allergy alert test

1 Use a small amount of a dark, natural shade (as recommended by the manufacturer).

2 Clean an area of skin about 8 mm square behind the ear (or sometimes in the fold of the arm). Use a little water on cotton wool to prepare the area.

3 Apply a little of the colour to skin.

4 Ask your client to report any discomfort or irritation that occurs over the next 48 hours. Arrange to see your client at the end of this time so that you can check for signs of reaction.

5 If there is a positive response, i.e. a skin reaction such as inflammation, soreness, swelling, irritation, or discomfort, do not carry out the intended service. **Never ignore the result of an allergy alert or** skin test.

6 If there is a negative response, i.e. no reaction to the chemicals, then the service can be carried out as planned.

Warning: In recent years, there has been a growing number of successful personal injury claims made against salons where the necessary precautions have not been taken.

TOP TIP

Contra-indications

A contra-indication is something that signifies that a service CANNOT be carried out.

Allergy alert or skin test

ALWAYS REMEMBER

Incompatible chemistry

Henna is still widely used throughout the world as a hair and skin dyeing compound. In the UK people using natural henna will often add other ingredients, such as coffee, wine or lemon juice, to intensify the final colour. However, people in other countries also add compounds to henna; for example, in India and Turkey people sometimes add iron ore deposits which are crushed into the powder to increase the 'reddening' effect. If this mix subsequently comes into contact with hydrogen peroxide (either through colouring or perming) a chemical reaction will take place. In the exchange that takes place permanent damage and breakage will occur.

Test cutting In this test a piece of hair cut from the head is processed to check its suitability, the amount of processing required and the timing, before the process is carried out. The test is used for colouring, straightening, relaxing, reducing synthetic colouring, i.e. decolouring, lightening and incompatibility.

Test curl This test is made on the hair to determine the lotion suitability, the strength, the curler size, the timing of processing and the development. It is used before perming. This is done by winding test roller(s) up at the back of the head and then applying the perm solution, allowing it to develop and finally neutralizing.

Curl check or development test curl This test is used to assess the development of curl in the perming process. The test is used periodically throughout a perm by looking for the optimal 'S' development of the wound hair.

ALWAYS REMEMBER

Sensitivity and PPD

This test is used to assess the client's tolerance of chemicals introduced to the skin – PPD. The abbreviation stands for paraphenylenediamine (para dyes), the main ingredient within permanent colour that is a known irritant to skin and eyes and can cause an allergic reaction.

Incompatibility test

Perm lotions and other chemicals applied to the hair may react with chemicals that have already been applied, such as home-use products. The incompatibility test is used to detect chemicals/elements which could react with hairdressing processes such as colouring and perming. The test is carried out as follows:

1   Protect your hands by wearing gloves.

2   Place a small cutting of hair in a small dish.

3   Pour into the dish a mixture of **20 parts** of 6% (20 vol) hydrogen peroxide and **1 part** ammonium thioglycolate (general purpose perm solution). Make sure that you are not bending over the dish to avoid splashing the chemicals onto your face or inhaling any resultant released fumes.

4   Watch for signs of bubbling, heating or discoloration. These indicate that the hair already contains incompatible chemicals. The hair should not be permed, coloured or lightened if there are any signs of reaction. Perming treatment might discolour or break the hair and could burn the skin.

Elasticity test

Elasticity test

This test is carried out on a dry single hair and used to determine how much the hair will stretch and then return to its original position. It is an indicator of the internal condition of the hair's bonded structure and ability to retain moisture. By taking a hair between the fingers and stretching it you can assess the amount of spring it has. If the hair breaks easily, care needs to be taken before applying any hairdressing process and further tests are indicated – a test curl or a test cutting, for example. Natural healthy hair in good condition will be elastic and more likely to retain the effects of physical curling, setting or blow shaping longer. It will also take chemical processes more readily. Hair with little elasticity will not hold physical shaping or chemical processes satisfactorily.

Note: Always carry out an elasticity test on dry hair. Wet hair is weaker, as the hydrogen bonds and salt bonds are already broken, and testing the hair wet will therefore give you a false result.

Porosity test

The porosity test will assess the amount of damage to the cuticle layer and therefore the hair's ability to absorb moisture or liquids – another indicator of poor condition. If the cuticle is torn or broken, it will soon lose its moisture and become dry. It may be able to absorb liquids more quickly, but its ability to retain them is reduced. If the cuticle is smooth, unbroken and tightly packed, it may resist the passage of moisture or liquids. By running the fingertips through the hair, from points to roots, you can assess the degree of roughness. The rougher the hair, the more porous it will be and the faster it will absorb chemicals.

Note: If you are trying to *feel* the roughness of the cuticle layer, you have to run your finger and thumb against the lie of the cuticle layer, and the free edges always point towards the points/ends of the hair.

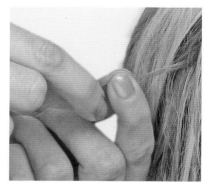

Porosity test

ALWAYS REMEMBER

Natural moisture levels

The natural moisture levels in hair play a significant part in the way that hair responds to treatments and styling. If the natural levels can be retained following perming, colouring and lightening, the client's hair will remain manageable, easier to detangle and able to hold thermal styling effects for far longer.

If those natural levels are depleted the hair becomes porous and will tangle easily, it is less manageable and will not be able to hold a set for long. Pre-chemical treatments help to reduce the hairs' moisture reduction.

Hair and scalp diseases, conditions and defects

An initial examination should be carried out before any hairdressing process occurs, so that any adverse conditions can be identified. Diseases of the hair and scalp may be caused by a variety of infectious organisms. However, not all hair and scalp conditions are dangerous; some non-infectious conditions can easily be addressed within the salon.

Infectious diseases

	Bacterial diseases				
	Condition	**Symptoms**	**Cause**	**Treatment**	**Infectious**
Folliculitis	**Folliculitis** Inflammation of the hair follicles	Inflamed follicles, a common symptom of certain skin diseases.	A contact bacterial infection, or due to chemical or physical action.	Medical referral to GP	Yes
Impetigo	**Impetigo** A bacterial infection of the upper skin layers	At first a burning sensation, followed by spots becoming dry; honey-coloured, crusts form and spread.	A staphylococcal or streptococcal infection.	Medical referral to GP	Yes
Sycosis	**Sycosis** A bacterial infection of the hairy parts of the face	Small, yellow spots around the follicle mouth, burning, irritation and general inflammation.	Bacteria attack the upper part of the hair follicle, spreading to the lower follicle.	Medical referral to GP	Yes
Furunculosis	**Furunculosis** Boils or abscesses	Raised, inflamed, pus-filled spots, irritation, swelling and pain.	An infection of the hair follicles by staphylococcal bacteria.	Medical referral to GP	Yes

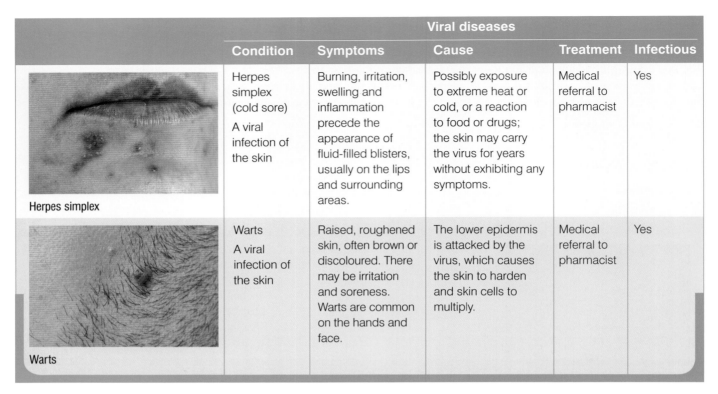

Viral diseases					
	Condition	**Symptoms**	**Cause**	**Treatment**	**Infectious**
Herpes simplex	Herpes simplex (cold sore) A viral infection of the skin	Burning, irritation, swelling and inflammation precede the appearance of fluid-filled blisters, usually on the lips and surrounding areas.	Possibly exposure to extreme heat or cold, or a reaction to food or drugs; the skin may carry the virus for years without exhibiting any symptoms.	Medical referral to pharmacist	Yes
Warts	Warts A viral infection of the skin	Raised, roughened skin, often brown or discoloured. There may be irritation and soreness. Warts are common on the hands and face.	The lower epidermis is attacked by the virus, which causes the skin to harden and skin cells to multiply.	Medical referral to pharmacist	Yes

Animal (parasitic) infestations					
	Condition	**Symptoms**	**Cause**	**Treatment**	**Infectious**
Head lice	Head lice (pediculosis capitis) Infestation of the hair and scalp by head lice	An itchy reaction to the biting head louse, 'peppering' on pillowcases and minute egg cases (nits) attached to the upper hair shaft close to the scalp.	The head louse bites the scalp feeding on the victim's blood. Breeding produces eggs, which are laid and cemented onto the hair shaft for incubation until the immature louse emerges.	Referral to a pharmacist	Yes
Scabies	Scabies An allergic reaction to the itch mite	A rash in the skin folds, often found between fingers and toes, in wrists and elbows, around the midriff, in the underarm area, or genital area; extremely itchy at night.	The itch mite burrows under the skin where it lays eggs.	Medical referral to GP	Yes

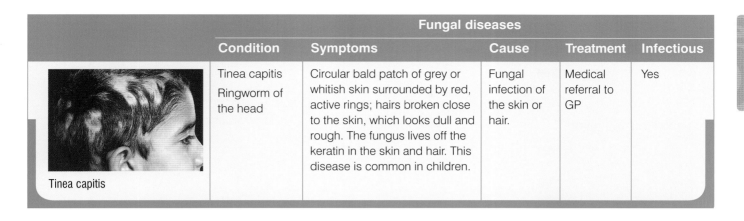

Fungal diseases					
	Condition	**Symptoms**	**Cause**	**Treatment**	**Infectious**
Tinea capitis	Tinea capitis Ringworm of the head	Circular bald patch of grey or whitish skin surrounded by red, active rings; hairs broken close to the skin, which looks dull and rough. The fungus lives off the keratin in the skin and hair. This disease is common in children.	Fungal infection of the skin or hair.	Medical referral to GP	Yes

Non-infectious diseases

Conditions of the hair and skin					
	Condition	**Symptoms**	**Cause**	**Treatment**	**Infectious**
Acne	Acne Disorder affecting the skin's sebaceous glands	Raised spots and bumps within the skin, commonly upon the face in adolescents.	Increased sebum and other secretions block the follicle and a skin reaction occurs.	Medical referral to GP	No
Dermatitis	Eczema and dermatitis In its simplest form, a reddening of the skin	Ranging from slightly inflamed areas of the skin to severe splitting and weeping areas with irritation and soreness.	Many possible causes, eczema often associated with internal factors, i.e. allergies or stress Dermatitis a reaction or allergy to external factors.	Medical referral to GP	No
Psoriasis	Psoriasis An inflamed, abnormal thickening of the skin	Areas of thickened skin, often raised and patchy. Often on the scalp and also at the joints (arms and legs).	Unknown	Medical referral to GP	No

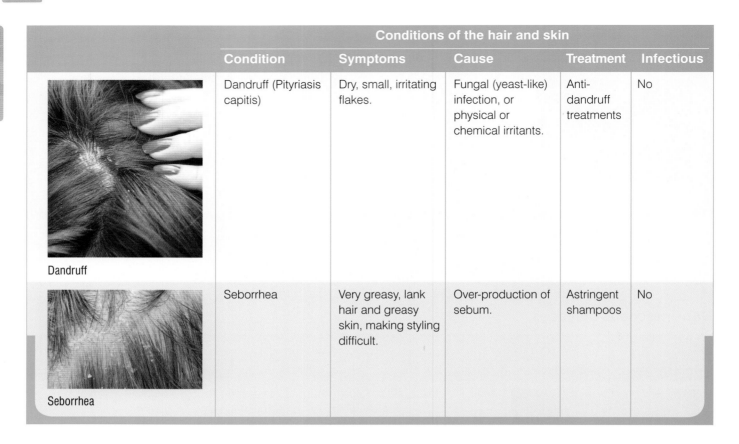

Conditions of the hair and skin					
	Condition	**Symptoms**	**Cause**	**Treatment**	**Infectious**
Dandruff	Dandruff (Pityriasis capitis)	Dry, small, irritating flakes.	Fungal (yeast-like) infection, or physical or chemical irritants.	Anti-dandruff treatments	No
Seborrhea	Seborrhea	Very greasy, lank hair and greasy skin, making styling difficult.	Over-production of sebum.	Astringent shampoos	No

Alopecia (hair loss)			
Alopecia areata	Alopecia areata	The name given to balding patches over the scalp. Often starts around or above the ears, circular in pattern ranging from 1–2.5 cm in diameter.	Trichological referral
Traction alopecia	Traction alopecia	Hair loss as a result of excessive pulling at the roots from brushing, curling and straightening. Very often seen with younger girls tying, plaiting or braiding long hair.	None

(Continued)

Alopecia (hair loss)

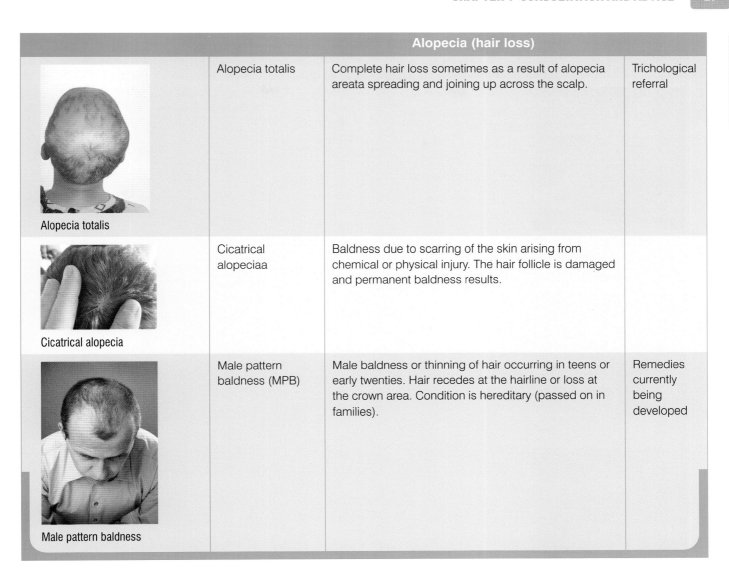

Alopecia totalis	Alopecia totalis	Complete hair loss sometimes as a result of alopecia areata spreading and joining up across the scalp.	Trichological referral
Cicatrical alopecia	Cicatrical alopeciaa	Baldness due to scarring of the skin arising from chemical or physical injury. The hair follicle is damaged and permanent baldness results.	
Male pattern baldness	Male pattern baldness (MPB)	Male baldness or thinning of hair occurring in teens or early twenties. Hair recedes at the hairline or loss at the crown area. Condition is hereditary (passed on in families).	Remedies currently being developed

Defects of the hair

	Condition	Symptoms	Cause	Treatment
Split ends	Split ends (Fragilitis crinium) Fragile, poorly conditioned hair	Dry, splitting hair ends	Harsh physical or chemical treatments	Cutting off or special treatment conditioners
Monilethrix	Monilethrix Beaded hair	Beadlike swellings along the hair shaft, hair often breaks at weaker points.	Irregular development of the hair forming during cellular production.	None

(Continued)

Defects of the hair			
Condition	**Symptoms**	**Cause**	**Treatment**
Trichorrexis nodosa Trichorrexis nodosa Nodules forming on the hair shaft	Areas of swelling at locations along the hair shaft, splitting and rupturing the cuticle layer.	Harsh physical or chemical processing	None, although cutting and conditioning may help
Sebaceaous cyst Sebaceous cyst Swelling of the oil gland	Bumps, lumps and swellings on the skin or scalp containing fluid, soft to the touch.	Sebaceous gland becomes blocked allowing a build-up of fluid to take place.	Medical referral
Damaged cuticle Damaged cuticle Broken, split, torn hair	Rough, raised, missing areas of cuticle; hair loses its moisture and becomes dry and porous.	Harsh physical or chemical processes	None, although cutting and conditioning may help

Influencing factors and features

The purpose of consultation is to arrive at a suitable hairdressing outcome through the process of analysis and evaluation. It should be done in a way that gives the client confidence in both the salon and in you.

It is important to study the 'complete picture' when you first meet the client so that you have as much information as possible when you come to advise the client.

During consultation you will be considering all the aspects of the client's physical features. Is there anything that doesn't work? Is the client happy with the existing style? They may point to areas which you feel may not be right, so you need to be able to express your technical appraisal in a clear, simple way without confusion. Avoid using any technical jargon or trade terms.

Consultation is personal and individual for each client and on every occasion. It is important to consider technical and personal image aspects:

TOP TIP

Always make and maintain eye contact with your client.

Be aware of your client's expressions and react appropriately.

- cutting and final shape of the hair
- volume or colour that will enhance the style
- finishing options of blow-drying or dressing the hair
- their hair type, hair growth, natural colour and face shape
- their personal image, lifestyle and personality
- the amount of time they can give to their hair.

ALWAYS REMEMBER

Look at the volume and quality of your client's hair. Remember that clients with fine hair want it to look bigger. Look at the proportion, partings and distribution of the hair.

How much natural movement has the hair got? Will it impede the styling plan?

Are there any strong growth patterns to contend with?

Expression Look for responses to your suggestions. You need to read the client's facial expressions and react to them appropriately.

Hair growth patterns

Hair is the frame for the face. The length, quantity, quality and texture of the hair all contribute to the total image. Fine hair often lacks body; most clients want fullness and volume which will last. Cutting methods to achieve this include volumizing techniques, where body can be created by using longer layers levelled in line with the client's baselines and face shape. You should also consider proportioning the hair weight. By setting hair or using light perms, you can create bulk and volume which give foundation to shape and style.

The hair's movement refers to the amount of curl or wave within the hair lengths. However, its growth pattern denotes the direction in which it protrudes from the scalp. Natural hair fall can be seen on wet and dry hair and strong directional growth will have a major impact on the lie of the hair when it is styled. So it is essential that it is taken into account during consultation.

Double crown The client with a double crown will benefit from leaving sufficient length in the hair to cover the whole area. If cut too short, the hair will stick up and not lie flat.

Nape whorl A nape whorl can occur at either or both sides of the nape. It can make the hair difficult to cut into a straight neckline or tight 'head-hugging' graduation. Often the hair naturally forms a V-shape. Tapered neckline shapes may be more suitable, but sometimes the hair is best left long so that the weight of the hair over-falls the nape whorl directions.

Cowlick A cowlick appears at the hairline at the front of the head. It makes cutting a straight fringe difficult, particularly on fine hair, because the hair often forms a natural parting. The strong movement can often be improved by moving the parting over so that the weight over-falls the growth pattern. Sometimes a fringe can be achieved by leaving the layers longer so that they weigh down the hair.

Widow's peak The widow's peak growth pattern appears at the centre of the front hairline. The hair grows upwards and forward, forming a strong peak. It is often better to cut the hair into styles that are dressed back from the face, as any 'light fringes' will be likely to separate and stick up.

Double crown

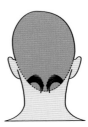

Nape whorl

Cowlick

Widow's peak

Face and head shape

The natural shape of the head, face and facial features are formed by the underlying bones of the skull. The contours of the head are its focal points and those on the side of the head are formed by the parietal and temporal bones. The shape of the back of the head and nape are formed by the occipital bones and the frontal bone forms the forehead shape.

The proportions of the hair mass and distribution in relation to the face and head are a vital element in choosing a suitable style for the client.

The face shape is made up of straight or curved lines, and sometimes a combination of the two. Straight, fine shapes appear angular and chiselled or firm and solid. They can be triangular, rectangular, square or diamond shaped. Curved-line shapes appear soft and may be round, oval, pear shaped or oblong. Shapes which have some straight and some curved lines are defined as heart shaped or soft square shaped.

To create a good balance, the hairstyle and face shape need to be compatible. So an angular haircut will not suit a soft, rounded face and a soft hair shape will not complement a chiselled face. Hair shape outlines can be made to look quite different from the front by simply changing a parting from side to centre. Side partings tend to make the face appear wider, while centre partings close down the width of a wide forehead.

An oval face shape suits any hairstyle. Round faces need height to reduce the width of the face. A centre parting can also help to reduce width. Long facial proportions are improved with short, wider hairstyles. Square-shaped faces need round shapes with texture onto the face to soften them. Longer lengths beyond the jaw line improve the balance and proportion.

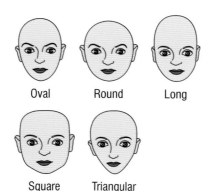

Oval   Round   Long

Square   Triangular

Different face shapes

ACTIVITY

Checklist

What are the shape, position and size of the client's ears?

What is the shape and size of the client's nose and mouth?

Are these features a major concern to your client?

What shape is your client's facial and head shape?

Are there any significant features that need to be accounted for?

Remember that small faces need 'opening up' while larger faces need narrower framing effects.

Ears, nose and mouth   Often ears are out of balance, which can affect the cut if you use them as a guide. Generally, large ears or even large lobes are accentuated by hair cut short or dressed away from the face. It is often better to leave hair longer over the ears unless it is an essential part of the style's impact.

Your client may wear a hearing aid and this could be a sensitive issue. Some clients wish to have all signs of an aid hidden, although others do not mind. You should discuss this sensitively with your client; they may feel too embarrassed to bring up the subject. The size of any hearing aid will need careful consideration when completing the total image.

The position, shape, size and colour of the nose and mouth are very important in the facial expression. The angles that are created can be softening or harsh and must not be ignored when the image is being planned. Hair shape and make-up can contribute to create the required effect.

Eyes We already know the importance of maintaining eye contact when communicating with the client, showing that you are listening to each other. But eyes play a major part in the selection of hairstyles too. Heavy fringes can accentuate and frame or they can obstruct vision.

Eye colour is a guide to the natural colouring of the client. This is a useful pointer when choosing hair colour. Eye shape is another element to be considered when choosing which style to create. Ideally, eye, head and face shape should all be complemented by the hairstyle and colour.

The eyebrows frame the eyes. Their shape, size and colour are all significant. A very harsh appearance is created if the eyebrows are removed; other effects are created by adding shapely lines. Eyelash and brow colouring help to balance facial features and give the eyes more definition.

Spectacles should also be considered when you are deciding on a hairstyle. Frame and lens colour, size and shape have a major impact upon styling and therefore must be taken into account.

ALWAYS REMEMBER

The eyes are the focal point of the face. Good images in magazines use this aspect to sell everything from clothes to hairdressing. Your client will be drawn to strong images, but are they wearable in everyday life? The majority of 'hair shots' use hair in ways across the eye line to create artistic impact. These snapshots of 'still life' are stimulating but may not be practical. You must make a point of this during your consultation.

What colour and shape are your client's eyes?

What are the eyebrows like?

Does your client normally wear spectacles?

TOP TIP

How long or short is your client's neck?

How wide or narrow is your client's neck?

Have you taken these features into consideration?

Neck and shoulders The length, fullness and width of the neck will affect the fall of the back and nape hair. Longer necks allow better positioning of long hair. They are complemented by high, neat lines; for example, mandarin collars or polo-neck tops. Short necks need to be uncluttered, with short hair and low collars. Long and thin necks are more noticeable with short styles and will therefore be better suited to longer hair around them. Shorter necks can be counterbalanced with height or upswept hair styles.

Body shape The body shape of the client needs to be considered. You need to balance the amount, density and overall shape of the hair to your client's physical body shape. This is particularly important if your client considers their shape or size to be a particularly important factor. For example, a small, clinging hairstyle would look wrong on a large body shape.

Lifestyle, personality and age

Remember that people are usually constrained in their hairstyle choices by what they do for a living or what they like to do in their spare time. Usually, people who work in environments where they have face-to-face contact with clients have to be more particular about the image they portray. This is a very important factor in style selection. Many clients need practical and manageable styles for work. Nurses, doctors and caterers, among others, may require styles which keep the hair off the face, or they may have to wear face and head coverings at work.

You should also find out whether the client does a lot of sport or exercise in their leisure time. If so, the hairstyle will have to be versatile and able to withstand a lot of washing and possibly heat styling. Dancers, athletes and skaters, among others, need hairstyles which will not get in their eyes and obscure their vision. Fashion models may require elaborate styles for special photographic or modelling sessions or displays. It is also helpful to consider how the style could create different effects when the client is going out. If you are styling for a special occasion, it is worth asking what dress will be worn. A beautiful outfit needs to be accompanied by an elegant hairstyle. However, this style will need to be altered for normal wear.

Character and personality can often override physical features when you are choosing a style for your client. A self-confident client will be able to wear looks that a more self-conscious client cannot and you need to take this into account. Is your client confident and outgoing or shy, timid and retiring, not wishing to stand out in a crowd? Is your client professional and business-like?

Also consider what age group your client falls into. There are a few basic rules that apply to people at certain ages but be careful not to stereotype people because of their age!

- ◆ **Children** – simple, practical shapes (although parents often try to suggest fashions).
- ◆ **Teenagers** – fashionable, trendy and willing to try new things.
- ◆ **Young people** – something suitable for work and leisure/going out.
- ◆ **Parents** – practical and attractive, often shorter styles.
- ◆ **Middle aged** – softening shapes to disguise wrinkles.
- ◆ **Senior age** – softening shapes.
- ◆ **Young business professionals** – fashionable cuts.
- ◆ **Older men** – simple, practical styles.

These are only general guides – there will always be exceptions to the rules.

Make-up   The way in which the client uses make-up may give a clue to her personality, dress sense and style. In business, a well made-up person is seen to be finished, groomed and in control. Lipstick reflects the colour and depth of the skin and eyes, creates light on the face and defines the lips. Shiny lipstick is most suitable for someone with a glitzy image. Matt lipstick might be more suitable for the quieter person, or when a natural look is required. Lip balm or gloss looks natural and keeps lips soft.

TOP TIP

What is the purpose of the style – fashion, special occasion?

Work, social and leisure pursuits are all factors to be considered.

Different styles suit certain age groups.

Will the final effect suit the client's lifestyle requirements?

TOP TIP

Make-up

How much make up is your client wearing?

What image is she trying to portray: natural, classic and business-like, dramatic or romantic?

Make-up can create a range of images: soft and kind, to bold and extrovert. It can make people feel different about their appearance and can boost confidence if it is properly applied.

Manageability Different styles need different amounts of commitment from the client once they leave the salon. You need to think about these factors when consulting with the client to make sure that they will carry on being happy with the style you both choose until it is time for their next visit.

Using visual aids in consultation

"Visual aids can help you to find out exactly what the client wants from you. Sometimes the client may not actually know what they want or may find it difficult to explain it to you. Using colour charts, magazines and books as points of reference during the consultation can be an invaluable way to better understand the client's wishes and help them reach a decision.

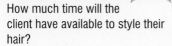
Tommy Van der Veken

Pictures Visual aids are essential! Use them and encourage clients to bring in pictures too.

'A picture paints a thousand words' – it may be a cliché but it is also very true. Pictures convey aspects of hairstyles or effects that are very difficult to put into words. They eliminate ambiguous technical jargon and establish a basis for things that you and the client can confirm.

Pictures are an immensely important visual aid and another form of language that hairdressers understand very well. One reason for this is their understanding of visual/spatial imagery. However, there is a vast difference between what the client sees in a photo and what the hairdresser will see.

Imagine that a client brings in a photo as an example for her own hairstyle. What aspects of the image does she find appealing? What is the client deriving from the image?

◆ mood

◆ attitude

◆ sex appeal

◆ colour contrasts

◆ fun 'night-clubbing/party' hairstyle

◆ perimeter shape.

But you, as the hairdresser, might see:

◆ a textured, above-the-shoulder-length bob

◆ no fringe – hair across the eyes

◆ unrealistic hair across the face

◆ false depiction of hair colour produced by photographic lighting/lighting effect

◆ limited flexibility for work/social wear.

From this example you can see that great images convey a lot more than hairstyles. The client will not understand the technical aspects of the style – she may just like it. You see something with a trained eye that is quite different.

Colour charts Colour charts are extremely useful for hairdressers. We rely on them every day. However, they are not always a very helpful medium for the client who may have very little ability for self-visualization.

The colour chart is a useful tool for hairdressers and a nice colouring book for clients. But unless you can help the client visualize the amount, intensity, density and saturation of the resultant colour, the visual aid will have little or no impact on the decision-making process of whether to buy colour or not.

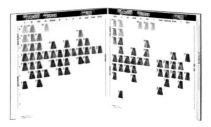

Colour chart

Computer-generated images The Internet is a valuable medium for conveying information about hair and hairstyles. Many salons have their own websites. The ease of using computer-generated images and uploading them to salon sites is proving a very cost-effective advertising medium. However, there is very little information technology used for graphical purposes within salons. Although many software packages are available for generating makeover effects, they are not much used at present. The constraints of cost, training and operator time are still barriers to salons that want to harness this technology.

Make recommendations to clients

Advising clients relies on a good knowledge of what is available, in order to match products and services to clients' requirements. Acquiring this knowledge is a continual process, as products are always being updated.

When you have completed your analysis and identified both your client's wishes and the factors that influence your styling options, you should be able to make your recommendations to the client.

TOP TIP

A colour chart is useful to show tones and hues, but don't expect your client to be able to visualize an all-over effect from a small sample.

TOP TIP

Try out products yourself so that you can pass on your experience to clients. Keep a note of what you've tried and what you did or did not like.

Other people's experiences are useful to know. You could do some market research on clients, friends or colleagues.

Specialist magazines often contain adverts and articles about new products on the market. Look at the features page to see what's new.

You will need to know:

◆ The ranges of services and products your salon offers.

◆ The prices and timings of those services and prices of the products available.

◆ Any services that need particular or special advice prior to implementation.

◆ Current fashion trends and looks.

◆ Any situations where referral is needed.

Services and products

You must know the ranges of services and products that your salon offers. In most situations, your salon will have a service or product that meets 99 per cent of the client's needs. It would be impractical or impossible to cover every type of treatment. You must be familiar with what your salon offers; you need to know what the costs are of these services/products and how long they will take.

Some services also need additional preparation or special conditions before they can be carried out. For example, your salon may do hair-pieces for hair-up work; some salons may do added clip-on hair extensions, whereas others offer the complete bonded extensions systems. In any of these cases it would be unlikely that all the necessary materials to do all of these services will be in stock. So you need to be able to tell your client exactly when those items will be available, the differences, benefits and pitfalls, how long it will take, how they will look after them and how much it will cost.

There are other special situations too. You may wish to offer a colouring service to a new client, but you know that you will have to conduct your skin test first. So you will need to tell the client what they can expect and also what you need to know and how they should contact you in the event of any adverse, contra-indications (in other words, a positive reaction to a patch test).

Current fashions

Your knowledge should be up-to-date and there are numerous ways in which you can do this:

◆ trade shows and exhibitions

◆ courses and seminars

◆ trade magazines

◆ fashion magazines

◆ TV and the Internet.

You should keep in touch with what is happening in the celebrity world. These well-known icons create fashion; they are prepared and groomed by personal stylists who are employed by production and promotional companies so that they are always in the eye of the public. People are stimulated by the entertainment industries and the media are always following the lives of celebrities from film, TV and music. The success of magazines such as *OK* and *Hello* is due to the attention they pay to these people's lives and what they are doing.

Your customers will expect you to be aware of what is happening and you will need to be ready to advise whether these new looks are going to suit them.

Referrals

There will be situations where the client's anticipated service cannot be provided.

This could be due to:

◆ adverse hair and skin problems

◆ your salon not offering that particular service or treatment.

ALWAYS REMEMBER

Keep abreast of all your salon's recent additions. If new products are introduced, then it's your job to find out: what they do; how they work; how long they take; and how much they cost.

TOP TIP

There is always another way of tackling a problem. When you can't provide what the client wants, always offer suitable alternatives.

See adverse hair and scalp conditions pages 25–30.

ALWAYS REMEMBER

If you are not sure what the problem is and it looks medical, refer the client to a pharmacist or their GP.

ALWAYS REMEMBER

If you need to refer clients to a trichologist, you can find a listing through the Trichological Institute on the Internet:

www.trichologists.org.uk.

TOP TIP

Make sure that records are found prior to consultation and used to check on previous client history. Update the records in line with your actions taken after consultation.

Specialist remedial referral Some hair and scalp conditions can be treated in the salon, but many more will require specialist, remedial attention. You need to know which ones are handled by the various specialists.

For example, a mother brings a child in for a haircut and the child obviously has nits and cannot be dealt with in the salon. You need to be sympathetic and not over-reactive. Tell the mother that the child is infected and explain what course of action is necessary. Explain the benefits of looking out for infestations on a regular basis and the signs that they would probably see, i.e. small whitish/grey nodules attached to the hair close to the scalp, generally around the back and nape, evidence of the louse itself or as peppering of brown speckles over the child's pillow at night and itching. But that is as far as your comment should go. You need to refer them to a pharmacist so that the mother can purchase a remedy and apply it at home.

A client with eczema may be aware of their condition, but you could be concerned that a planned service may aggravate the condition further; a referral to their doctor first, is preferable, if only to eliminate the concern that your planned service won't make the condition worse.

If you find a condition that you are not sure about, tell the client. Say, 'I'm not sure what the problem is but I do think you should get it checked out before we continue our planned services.'

Salon referrals You need to keep up-to-date with the salons in your area that offer special services that your salon does not cover. Services such as trichological analysis, hair extensions, hair transplants, wigs and hairpieces may be carried out locally in other salons. But remember, with referrals you must follow your salon's policy for external re-direction.

Advise clients on hair maintenance and management

What the clients really want is to be able to achieve the same finished effects that you do in the salon. They may not appreciate the amount of time that you put into your training as you just seem to make it happen so easily!

You already know the limitations created by adverse hair conditions, hair textures, types and tendencies. But, to the client, you are the enabler; you create the beautiful effects and now you need to pass on a little advice and a few tips on how the effect can be managed away from the salon between visits.

You need to:

◆ find out what the client is already doing with their hair

◆ explain the benefits modifying the care regimen in relation to the new effect

◆ advise them on the correct product regimen to maintain their hair at home

◆ tell them how to maintain their hair between salon visits.

What is the client already doing with their hair?

Your client will usually have their own routine for washing and styling their hair. This may be a simple affair: a client washes their hair because they think it needs it, puts a towel around their head and then runs a brush through it to finish it off! But generally the client already has a care regimen in place that is designed to meet their current needs. This could involve a treatment shampoo and conditioner and a variety of styling and finishing products which are all carefully used together with a range of hairstyling equipment.

Whatever care routine your client has, you need to be aware of what they are doing and how they are doing it. You should explain the benefits of modifying the care regimen in relation to the new effect. Remember this important phrase: 'People are reluctant to change'. Bear this in mind when you make any reference to changing from one care regimen to another. Your clients will be used to a routine that suits both their:

◆ lifestyle and

◆ existing hair requirements.

If you create a new effect, or provide a service that does not fit with their lifestyle, you will probably find out that it just doesn't work. Whatever you do for them, it must fit in with their job, their leisure interests and – most importantly – the time that they can afford to maintain it.

The other aspect relates to what they are already doing with their hair. If the products that they are already using conflict with what you are doing, you must make this known. What is the point of doing half a job? You are in partnership with your client. What you do and advise them to do, must be taken on board so that they can have the best possible chance of replicating the effect.

Advise clients on the correct product regimen to maintain their hair at home
Products are an essential part of the equation. What they use has a direct effect on the result. For example, if a client is used to a conditioning product for normal-frequent use, and you do a highlighting service for them, do you think that their current conditioner is up to the job? Similarly, if your client has been using defining waxes on their hair for some time because of the texture it achieved on their short hair, do you think it would still be appropriate if they had just had a full head of hair extensions? No, is the short answer.

All modern, fashionable hairstyles involve some form of product care and maintenance. Even casual, carefree styles need the right shampoos and conditioners. But most hairstyles are far more demanding than this. You need to advise your clients on the correct products for their hair, how to use them and the benefits they will receive from using them.

Explain how to maintain clients' hair between salon visits
The advice that you give clients on managing the effects themselves is crucial. If they can get a similar effect to that 'just left the salon' look, then you will have made a big difference.

Remember to cover the obvious things too. It can be easy to forget the basics when you are trying to explain how to make the most out of their hair. Often it is the simplest things that make all the difference and have a long-lasting effect on the general condition of the hair or how long the style will last. For example, a client who is always using very hot equipment to style their hair needs advice that their hair needs protecting from the heat or there won't be anything left to protect! Similarly, if their lifestyle involves a lot of sport, or going to the gym, then they are going to get hot; any moisture – whether environmental or from the body – is going to dampen the hair and alter the look.

ALWAYS REMEMBER

Be positive and suggest ways around problems and, if possible, find ways to avoid the negative effects altogether.

Agree services with your client

With all other parts of the consultation done, you are almost ready to complete the service. You have:

✓	Identified what the client wants – your initial part of the consultation had looked at what the client's expectations were by asking questions and using visual aids.
✓	Identified what the client's needs were – in light of what they wanted, you matched this against their perceived needs and what you saw as the actual needs of the hair following your analysis.
✓	Analysed the hair, skin and scalp – and looked for any adverse conditions, incompatibility issues and conducted or arranged for any necessary tests.
✓	Made your recommendations – based on what you found during the analysis and where this conflicted with client's expectations (because you thought they were unrealistic or they didn't realize that there were other issues to address) you have offered other suitable alternatives in a caring and sympathetic way.
✓	Provided your advice on how the look or effect can be maintained and recreated at home by using the right maintenance routines or purchasing the correct home-care product regimen.

So with all these things done; that must be everything? Not quite. You have made your assumptions based upon the client's needs; your professionalism has ensured that any misunderstandings were made clear. But without seeking confirmation from the client, you cannot continue.

The final part of the consultation will require the commitment from the client that confirms that:

✓	They are aware of what the agreed service entails (timings or special conditions).
✓	They have agreed to the services and/or products that you have negotiated.
✓	They understand the costs for the services and/or products and these have been considered acceptable.

And that you:

✓	Made sure that all records have been completed and are now up-to-date.
✓	Made the necessary appointment(s).
✓	Are now ready to carry out the agreed service.

Final notes

Service timings Different services have differing durations and timings and, in a busy salon, making the most of the available time in an appointment system is a difficult operation. Some special services like hair extensions can take many hours, whereas routine services like cut and blow-dry may take only 45 minutes. Whatever the service you must bear in mind that time is as valuable as money.

You need to be realistic when you provide that information to the client. Their time is precious; so you need to give them an accurate idea of how long things will take or what sorts of things are involved. If you are unable to give an exact time, don't make it up; give them a range that covers the minimum and the absolute maximum. Let them decide if that is suitable; if not, offer alternatives, there is always another day.

Service and product costs People don't like hidden costs. Hairdressing is built on trust, loyalty and, most importantly, repeat business. In other words, we want our clients to be satisfied with what we provide. We want them to return in the future, where we can build on the relationship further and develop a long-term, ongoing business arrangement, and this marks the difference between customers and clients.

With this in mind, you should always provide an honest and accurate cost for what has been agreed. If the client decides that she wants to add other things into the service package, that's fine – providing no misunderstandings occur.

Taking deposits You will also need to check to see if a deposit is needed. Often in situations where there is a large investment of salon or stylist's time, then other preparations need to be arranged in advance and it would be normal practice to take a deposit. This would be done at the time of making the appointment and would be refundable or deducted from the bill at the time when the service is finally carried out.

Client records Keeping accurate, up-to-date client records is essential for good salon management and to ensure clients receive a service appropriate for their individual needs. The records also give vital information should there be any subsequent client complaint or, worse, if there was any legal case against the salon.

Remember this is personal and private information, it must be treated as confidential and you have a duty to uphold the rights of your clients if you keep their personal information on file.

TOP TIP

Taking a deposit

Check what the policy is in your salon. What sorts of services need deposits in advance?

For more information see the Data Protection Act (1998) in the Appendix.

ACTIVITY

Calculating bills

1 Using your salon's price tariffs, calculate the cost for client X:

A full head of highlights (with lightening or high lift colour) on long hair with a reshape (trim) cut and blow-dry?

Now add to this an in-salon conditioning treatment.

2 What would be the costs of the following services and products in your salon for client Y?

A A restyle cut and blow-dry for a client with long hair.

B A retail-size shampoo and conditioner for dry damaged hair.

C A retail-size defining paste.

Now add A, B and C together to create the final bill for client Y.

Consultation checklist

◆ Listen carefully to what is requested.

◆ Use visual aids to assist the consultation process.

◆ Communicate the possible effects.

◆ Explain why certain effects are not possible.

◆ Give good reasons for suggested actions.

◆ Ensure that the client understands what is being said.

◆ Agree on a final and suitable course of action.

◆ Assure and reassure throughout.

◆ Make it clear if follow-up appointments are necessary.

◆ Carry out the agreed service or treatment.

◆ Encourage the client to rebook the next visit before she leaves.

◆ Maintain the client's goodwill and safety throughout the appointment.

◆ Record the details for future reference.

REVISION QUESTIONS

Q1.	The three stages of hair growth are anagen, _____ and telogen.	Fill in the blank
Q2.	The cortex is the outermost layer of the hair.	True or false?
Q3.	Select all that apply. Which of the following are infectious diseases?	Multi-selection
	Impetigo	☐ 1
	Scalp ringworm	☐ 2
	Alopecia	☐ 3
	Head lice	☐ 4
	Psoriasis	☐ 5
	Eczema	☐ 6
Q4.	The natural colour of hair depends on the amount of PPD within it.	True or false?
Q5.	Which of the following is commonly known as split ends?	Multi-choice
	Trichorrexis nodosa	○ a
	Monilethrix	○ b
	Tinea capitis	○ c
	Fragilitis crinium	○ d
Q6.	Dandruff is a condition of the scalp usually caused by fungal infection.	True or false?
Q7.	Which of the following tests are relevant to semi-permanent colouring services?	Multi-selection
	Skin test	☐ 1
	Strand test	☐ 2
	Curl test	☐ 3
	Incompatibility test	☐ 4
	Porosity test	☐ 5
	Test cutting	☐ 6

Q8. The active part of the root from which hair grows is called the _____ matrix.

Fill in the blank

Q9. Which face shape suits most hairstyles and lengths?

Multi-choice

Square   O   a

Oblong   O   b

Oval   O   c

Triangular   O   d

Q10. During consultation and hair analysis, any contra-indication found will not allow the planned service to be carried out.

True or false?

2 Promote Services and Products

LEARNING OBJECTIVES

◆ Know the salon's range of services, products and their prices.

◆ Be able to inform the clients about the salon's services and products.

◆ Be able to sell services and products to clients.

◆ Know your salon's policies and legal obligations.

KEY TERMS

benefits	in-salon promotions	point-of-sale
features	planogram	window displays

UNIT TOPIC

Promote additional services or products to clients

INTRODUCTION

The services and treatments provided by salons are continually revised and updated. We want clients to enjoy their salon experience and maintain the effects provided in the salon by purchasing the right products for use at home.

Therefore, by offering new or improved services and products your salon is able to:

◆ increase client satisfaction

◆ retain client loyalty for the future.

When we have satisfied our clients, we need them to return (and within a reasonable timescale) to maintain the financial and ongoing success of the salon. We can stimulate client's loyalty by encouraging them to:

◆ buy our other services, products and treatments

◆ return on a regular basis

◆ share their positive experiences with other people.

Promoting services and products

PRACTICAL SKILLS

Find out which salon services or products would be suitable for your clients

Be able to promote suitable services to your clients

Be able to promote suitable products to your clients

Find the right times to introduce different services and products to your clients

Provide aftercare advice to your clients and promote retail products

UNDERPINNING KNOWLEDGE

Know your salon's ranges of products and services

Understand the factors that can affect clients' willingness to try new services and products

Know how to match clients' needs to the benefits of the salon's services and products

Know how to communicate effectively and professionally

Recognize client buying signals or client disinterest

Salon services and products

Promoting ourselves first

It may sound unlikely, but good communication can far outweigh technical excellence. You could be an average stylist with reasonable communication skills and you will go far; unfortunately, a brilliant stylist with poor communication skills will not have much of a future in a salon setting! When it comes to creating success, communication is key.

INDUSTRY ROLE MODEL

CHRIS MOODY **Salon Director of Moodyhair and Global Platform Artist for Redken NYC**

"I have been hairdressing for 30 years. I manage and coach a team of stylists, as well as maintaining my own clients. I present and teach worldwide on behalf of Redken NYC, teaching in their London and New York academies, as well as presenting main-stage shows in the USA, Canada, Asia and the Far East. My passion lies in coaching and mentoring other stylists and hairdressing teachers to become the best they can be at their craft.

Good customer service involves being client focused. It is centred upon the needs of the client and is reflected in all of the routine salon operations: the telephone response times, salon refreshments and magazines, visually pleasing interiors, and the polite and friendly staff. It is your duty to be positive and helpful whenever you communicate with clients.

What is available?

Most salons have tariffs and point-of-sale material that provide information about the services and products they offer. This is a good way of giving information to clients as such materials can easily be seen and provide a talking point in the salon. However, printed material is not always at hand or available, particularly when services and product ranges are being continually updated.

TOP TIP

Try to put yourself in the client's position. How would you like and expect new services or products to be introduced and recommended to you?

> Use all the salon's services and products yourself. Have your hair regularly cut, coloured and treated, and always use the products you retail yourself on your own hair. That way you can give a personal testimonial on what you know and have experienced rather than what you read or heard. Practise what you preach!

Chris Moody

You need to be aware of changes and additions, so that you can provide current advice on available services or products that would benefit your clients. You also need to keep up-to-date on all the **features** and **benefits** of newly introduced services and products. If there are new additions in the salon that you do not know about, you will need to ask your manager or supervisor for more information.

Products are not wallpaper! A display of products should not be just a decoration used to cheer up a 'dingy' corner; it has far more important purposes. Retail displays are an expensive investment for the salon and stock on shelves represents a significant financial outlay. Retail products are an essential part of the salon's income; for many salons, a significant proportion of their overall annual turnover comes from retail sales.

There are hidden benefits to the purposes of retailing which you may not have thought about. The promotional displays play an integral part in supporting the salon's image. The salon *only* purchases products for use and resale that reflect the quality of services offered. The sale of retail products enables the clients to gain an extension of the salon experience in the products that they take home.

Features and benefits

In order for you to promote and ultimately sell your salon's services, treatments and products, you need to understand how clients feel and what they see. To do this you need to consider each service or product in terms of its features and benefits, and how these will meet the needs of your clients.

◆ 'Features' are the functions; i.e. what the service, treatment or product does.

◆ 'Benefits' are the results of the functions; i.e. advantages, what the service or product achieves.

For example, suppose you recommend a client to spend £11.50 on a conditioning treatment. Why should she do this? What are the features and benefits of the service?

◆ The features are that it re-conditions dry, damaged hair and is easy to apply.

◆ The benefits are that it improves the dryness and helps to smooth damaged lengths, thus improving handling, making the hair easy to manage and comb, enhancing the hair further with deep shine.

Therefore, these benefits justify the client spending £11.50 on her hair.

You need a good understanding of your salon's services and products so that you can recommend them with confidence to your clients. Knowing what will work and what will not for different clients are the signs of a true professional.

Inform the clients through recommendation

New clients coming into our salons are part of a very special group – maybe recommended by satisfied customers. Recommendation is a powerful form of communication. It does not work just among friends; it should be the natural way for you to open a conversation too.

The perfect time to talk about the variety of services and treatments available to clients is when they are in the chair. Recommendation is the simplest way of extending the range of services to our clients and enhancing the professional relationship.

> Never open a consultation with the words 'Is it the same again then?' Sit face to face and ask open questions: 'What have you liked?' or 'What challenges have you had?' or 'If we could improve one thing about your hair what would it be?' Really listen to the answers and suggest solutions based on what your salon offers and sells.
>
> *Chris Moody*

The most effective way of introducing clients to products that will benefit them is to use them on their hair in the salon. This style of introduction is a simple process of seeing:

◆ something that the client needs

◆ how it is used or applied

◆ the results that it achieves.

This style of promotion works by showing the product to the client and then by passing it to them so that they can look for themselves. By doing this the client can:

◆ hold it

◆ smell it

◆ discover how it works.

In-salon promotions

In-salon promotions are the most popular way of *sending messages* to the clients; it can become a topic of conversation while they are having their hair done. However, any form of salon activity costs money and that financial outlay has to be budgeted for.

The first part of the process is to find out what the promotion is about: if it is linked to a new product introduction, then there will be some form of support material to help inform the clients. Whenever a new product or range of products is introduced, the manufacturer will provide a range of point-of-sale material to help with its launch. The amount of supporting material tends to be distributed according to the levels of purchases made by the salon.

> Incentivize your regular clients to recommend their friends through vouchers, add-ons, special promotions for the client who recommends your salon and for their friend. Personal recommendations are the most effective way to gain new clients, and rewards like this turn regular clients into raving fans!
>
> *Chris Moody*

Implementation stages of product or service promotion

The following information provides you with a checklist for implementing an in-salon promotion.

Checklist

✓	Find out the budget
✓	Make individuals aware of the product's features and benefits
✓	Let staff know what their roles and targets are
✓	Conduct the promotion
✓	Evaluate the effectiveness of the promotion

Management will set a budget, which may include the purchase of the product range as well as the additional cost of setting up the promotional activities. Everyone should be informed in advance about the nature and purpose of the event. This provides staff with a clear idea of:

- what the newly introduced products do
- how they benefit clients
- how they are applied
- the cost of the products.

Normally, at the start of a promotion, there would be an introductory discount for a short period and this may tie-in with an advertising campaign, perhaps in magazines or even TV.

The most successful promotions include the following:

- Introductory discounts – only available for a set period at a lower price.
- Multi-buys – typically BOGOFs (buy one get one free) or get three for the price of two.
- Special offers – buy all three and get a free beach towel or scarf.

Staff must be made aware of the promotional plan, any incentives linked with the promotion and their personal targets for the introductory period. Each member of staff could have an individual role to play in the overall team plan. For example, the shampooist may ask the client if she wants to try the new conditioner on her hair at the basin. From this, a client gets the chance to gain a first impression/experience of the product.

The promotion continues at the styling position where display materials, e.g. show cards or leaflets, inform the client about the benefits from using the product; this can be reinforced by the stylist's recommendation and advice. Staff would also ask if the client noticed and liked the *smell* of the conditioner when it was applied, or if she can *see* and *feel* the difference that it has made to her hair.

Then finally, before the client leaves, she *connects* again with the promotion when she sees a well-put-together display in reception. The receptionist could now ask if she would like to add the product to her bill for her use at home.

The promotion should therefore be a whole team approach. This is particularly necessary in larger salons where individual staff can lose the continuity when clients are helped by several different individuals; it is important to reinforce the message to make a sale.

Using our senses

When a salon promotion is being implemented it has to be a *sensory event*. Effective selling is an experience for the client; it provides the purchaser with a variety of aspects that they can see, smell and touch.

Our senses provide the *channels* which influence people into making a purchasing decision. The more senses involved in the promotion, the greater the chance that the customer will buy.

At the end of the promotional period, the success and impact of the event can be evaluated through team discussion, review of sales reports and feedback from the clients on how the promotion was received.

Promotional materials

A typical product promotion will consist of:

◆ Point-of-sale merchandising – central island, open cabinet, shelf displays.

◆ Shelftalkers – printed promotional slips/cards fixed to/dangling from shelves: 'mobile' ones that bob or bounce deliver the best results.

◆ Eye-catching information displays – locate them to be seen at reception or centrally in treatment areas.

◆ Arrangement of popular lines at eye level with price details; use 'price watch' stickers.

◆ Linked displays, with money off and other special offers – such as first visit offers, loyalty schemes, recommend-a-friend discounts and promotional tie-ins with local stores.

Window displays

Window displays are an essential way of advertising services, their costs and displaying products. When maintained by the salon staff they can be very cost-effective, but if not done well they can look quite amateur. Larger companies, high street salon chains, and the national salons in shopping centres and malls always employ *visual merchandisers*. Usually the visual merchandising (VM) team will be provided through an external marketing company, but some of the larger hairdressing/barbering companies have their own VM departments.

Where companies like Toni and Guy, Regis or Headmasters have their own visual merchandising departments, they will plan marketing campaigns at head office. Then, each month, as they promote certain services or products, they will send out the **point-of-sale** materials and a **planogram** of how the displays should look.

The planogram is a colour plan of exactly how the promotion is to look from outside the salon and from inside. It shows the salon manager and staff exactly:

◆ what products are to be used within the display

◆ the amounts and sizes of products to be used

◆ where the products fit in to the display

◆ where the point-of-sale materials i.e. show-cards, brochures, leaflets and so on, are to be placed.

Website

The most popular and obvious way to find out about what a business has to offer is through its website. All salons have some form of Internet presence, some link with marketing companies to rent space in directory sites, and many have their own websites.

Typically, the larger the salon business, the greater impact its website will have. Booking online is a common feature and it allows potential clients to look for appointment availability, at times when they can make it. This frees-up valuable receptionist time and allows them to focus their customer service on clients walking through the door.

Gain client commitment to using additional services or products

Recognizing interest – buying signals

You need to be able to differentiate between genuine interest and a polite and friendly but negative response. Just because someone responds in a polite or friendly way is not an indication of 'I want to buy'.

Genuine customer interest is shown in three ways:

✓	The client will *ask* how that will benefit them, e.g. 'Would that be suitable for my type of hair?' or 'How does that work?'.
✓	The client will *show* their interest by taking the product and holding it for closer examination, or through their positive body language.
✓	The client picks up promotional materials and will *read* about the product.

So when you introduce or recommend the client to a new service or product, you will need to look for one of the signs above. If you do not see such a response, it could indicate a lack of interest from the client and you should be careful how you proceed. Do not push it too hard: even if you think it is the best product or service that your client could have. You do not want to be seen as 'pushy' or a 'commission seeker'. It will not help your professional standing and it may spoil your professional relationship too!

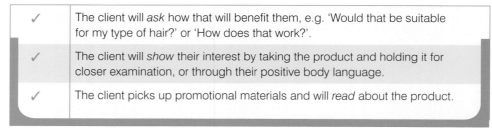

Never assume your clients don't want, need or can't afford to buy professional home care products. Tell your clients what you are using to create their hairstyle. At the end of the service remember to place the products on your station and say 'This is what you need and what has been used on your hair today.' It's not a hard sell; it's simply letting your client know what's on offer.

Chris Moody

Communication

There are many ways to inform clients about your salon's products and services. Establishing effective communication between you and the client is the most important aspect in determining success in your role. The relationship between stylist and client is built on quality of service, professional advice, trust, support and a listening ear. Good communication ensures productive and effective action. On the other hand, poor communication can lead to misunderstandings, misinterpretation and mistakes.

Verbal communication is what you say and what others hear, it should always be:

◆ clear to the listener

◆ brief, but what you say should create interest and initiate a response

◆ uncomplicated – your information should be easy to understand, avoiding the use of technical terms

◆ friendly, polite and courteous.

Questioning styles

◆ **Feeling questions** – these focus upon the client. Examples are: 'How does your hair feel now after using the new treatment today?' 'What do you think about the new colour range that we are using now?'

◆ **Open questions** – these are useful when you want the client to give you information. Examples are: 'What products do you use when you wash your hair at home?', 'How do you apply the colour when you do it at home?'

◆ **Closed questions** – these questions lead the client to give only simple yes or no responses and yield very little information. Examples are: 'Have you washed your hair with anything different lately?' 'Do you find the colour application at home is easy?'

> Good restaurants always tell you what the 'house specials' are that day; they give their guests information about all the choices they have. Let your clients know what else you can do. If they know you offer a quick, inexpensive highlighting service, they may decide to try it. If they don't know, they can't!
>
> *Chris Moody*

Non-verbal communication (NVC) is also known as *body language*. It can reveal what we really mean even if we are saying something different.

We express ourselves with body language through:

◆ posture

◆ gestures

◆ facial expression

◆ tone and manner of voice.

It is very important that we send and receive the right messages, particularly when dealing with clients and new customers.

ALWAYS REMEMBER

Genuine interest is expressed by the client wanting to know more about the products and services you offer.

For more on questioning styles, such as feeling questions, open and closed questions, see CHAPTER 1 Consultation, pages 11–12.

TOP TIP

Body zones proxemics

Do not crowd or appear over-familiar with your client. Imagine how you would feel if someone came up to you and got a little too close. What do you do? Immediately back off and go on the defensive.

ALWAYS REMEMBER

Good selling

◆ Listening, asking questions, showing interest.

◆ Using the client's name.

◆ Empathizing (putting yourself in the client's place), establishing a bond.

◆ Recognizing non-verbal cues (dilated pupils = 'I approve'; ear rubbing = 'I've heard enough').

◆ Identifying needs; helping clients reach buying decisions.

◆ Knowing your products/services.

◆ Highlighting the results or user benefits; demonstrating these where possible.

◆ Thinking positively, talking persuasively, projecting confidence and enthusiasm.

For more on physical behaviour that may contradict your verbal communication see CHAPTER 1 pages 8–11

ALWAYS REMEMBER

Bad selling

What you should avoid:

◆ Doing all the talking.

◆ Not listening, not 'hearing' unspoken thoughts, arguing.

◆ Interrupting – but never letting the clients interrupt you – thus losing an open opportunity for giving extra information.

◆ Hard selling, 'spieling' (working to a script).

◆ Threatening – 'You won't get it cheaper anywhere else', knocking the opposition.

◆ Manipulating – 'Oh dear, I'll miss my sales target'.

◆ Knowing nothing about the product.

◆ Treating 'no thanks' as personal rejection.

◆ Blinding clients with science.

◆ Staying mainly silent, waiting for an order.

◆ Insisting the client should buy the product.

TOP TIP

Excellent communicators are good listeners. They understand how to ask the clients the right questions and listen effectively to responses, building on the information given to them and then finally helping the client to buy.

ACTIVITY

The table below lists a variety of different personality traits that could be demonstrated by clients within the salon. Complete the missing information by stating how you would go about handling each of the types.

A client is showing	What signs would they be showing?	How would you handle this trait?
Anger		
Poor confidence		
Uninterest		
Interest in a product		
Disbelief in what you are saying		
Signs of being confused		

> Encourage all your clients to rebook as they leave the salon. Ask outright if they would like to book again. If your client won't rebook, offer to write a suggested date on a card and ask if you may call them the week before to arrange their appointment.
>
> *Chris Moody*

Consumer legislation

The rights of your customers must not be compromised; customers are protected from fraudulent and sharp practice by a variety of consumer legislation. For more information see Chapter 16 and Appendix 4.

ACTIVITY

The following table lists a number of situations that could occur. Complete the missing information for each one.

Scenario	Which legislation applies?	What action should be taken?
A client returns a pair of electrical straighteners, bought from the salon at the weekend. She says that they do not work.		
A client is dissatisfied with the colour that her hair has turned out. She feels it is too dark.		
A client is unhappy about a product that she has bought; she says it makes her scalp itch.		
A client feels that her 'special promotion' purchase of a hairdryer is a 'con', because she has seen the same item on the Internet at a cheaper price.		

REVISION QUESTIONS

Q1. A product's _____ are aspects of the product that state what it does.   Fill in the blank

Q2. A key feature of good customer service is being customer focused.   True or false?

Q3. Which of the following would be considered as an indication of positive communication?   Multi-selection

Avoiding eye contact   ☐   1

Smiling   ☐   2

Standing over the client and talking to them through the styling mirror   ☐   3

Talking with your hand covering your mouth   ☐   4

Sitting at the same level as the client, talking to them face to face   ☐   5

Sitting with folded arms   ☐   6

Q4. People have a comfort zone that is an invisible space around their body.   True or false?

Q5. Which of the following is a specific benefit of semi-permanent colour?   Multi-choice

It lasts well   ○   a

It's quick and simple to use   ○   b

It runs and stains easily   ○   c

It changes the hair's natural condition   ○   d

Q6. A 'shelftalker' is a salesperson.   True or false?

Q7. Which of the following are effective forms of promotion?   Multi-selection

Point-of-sale material   ☐   1

Eye-catching displays   ☐   2

Dusty shelves   ☐   3

Poorly trained staff   ☐   4

Clever clients   ☐   5

Eye-level product placement   ☐   6

Q8. Excellent _____ skills are an essential aspect of good communication.   Fill in the blank

Q9. Which legislation protects the clients from defective purchases?   Multi-choice

Data Protection Act (1998)   ○   a

The Sale of Goods Act (1979)   ○   b

Disability Discrimination Act (DDA 2005)   ○   c

The Prices Act (1974)   ○   d

Q10. If a client asks how a product would benefit them, then it is a sign of genuine interest.   True or false?

Notes

3 Customer Service

LEARNING OBJECTIVES

◆ Be able to use feedback to identify areas for improvement.

◆ Be able to make changes in customer services.

◆ Be able to help evaluate changes to customer services.

◆ Know how to support service improvements.

KEY TERMS

customer feedback
client questionnaires

Data Protection Act (1998)
unique selling point (USP)

UNIT TOPIC

Support client service improvements

INTRODUCTION

As hairdressers and barbers, you provide services to your clients on a daily basis; this involves talking about your client's hair, the way that they handle it and providing useful tips and advice. This close, loyal relationship provides you and the salon with a unique opportunity to gain valuable feedback and the client's experiences should influence the way that future services are provided.

This chapter considers what salons can do to improve their offering to customers. It means you will be able to support the salon management in evaluating the quality of existing services and helping to bring about changes to improve salon services for the future.

Customer service

PRACTICAL SKILLS

Find out what clients want by asking questions and providing questionnaires

Look for ways to improve customer services

Feedback information from customers to management

Make recommendations on how things can be improved

Help to implement changes to customer services

Support other staff and management through teamwork

UNDERPINNING KNOWLEDGE

Know your salon's ranges of products and services

Understand the factors within the salon that can affect customers' experiences and goodwill

Know how to collect and collate information for evaluation purposes

Know how to communicate effectively and professionally

Understand why changes to customer service are necessary

Know how to work with others to support change

Service improvement

Why change your services?

Quite simply, salons have to change because the demands of the clients continually change. Businesses cannot afford to *stand still*. Another salon can do what you do, offer the same products and may be able to offer the service at a more convenient time. Salons that stay the same lose clients and close down.

INDUSTRY ROLE MODEL

ESTHER VAN DER VEKEN   Business Director Tommy's Hair Company

“We have three salons in North Wales and Chester and have been recognized for our excellence in customer care, winning the British Hairdressing Business Awards Customer Care, and Front of House Team of the Year awards. My role within Tommy's is to put in place an evolving structured business strategy, which focuses on staff recruitment and retention, marketing and PR initiatives, staff education and training, and customer service improvements. Before entering the hairdressing industry with Tommy's

I was a professional model in the Netherlands. I had always admired the hairdressing industry for its creativity and passion towards education, so was thrilled when Tommy Van der Veken invited me to join him in his new venture – Tommy's Hair Company.

Salons need to improve their services because:

1 the customers expect it

2 other competition

3 they need to generate more business – increase sales

4 they want to improve productivity and effectiveness.

The diagram explains this in another way; here we see the factors that influence the need for change.

Unfortunately we do not know what our competitors are planning; we do not know what our customers will want next month and we do not know how much the business will benefit from improved services. However, we do know that: competition is not going to disappear; our clients will always continue to expect more; and that it will be beneficial for the business in the longer term. How big a benefit in terms of financial gain, better marketplace offering and perceived image, can all be measured later by evaluation, reports and **customer feedback**.

> Customer care is at the heart of everything we do within Tommy's. We refer to our clients as 'guests' and through our effective customer care programme we provide our salon guests with a sanctuary of care, which not only offers extremely high levels of service but delivers a consistent and enjoyable, positive experience.
>
> *Esther Van der Veken*

What is different about your salon?

In marketing speak a business *creates its own niche* in the marketplace by developing a **unique selling point (USP)** for its products or services. This USP sets a business aside from the other competitors; it makes the business different and is the main reason for customers walking through the door.

So if you look at the services and products your salon offers compared to others, what is different?

◆ Do you offer hair extensions or other services that your competitors do not?

◆ Does your shampoo work differently from others in the supermarket?

◆ Do you have flexible opening hours?

Remember that it is not necessarily what you offer that matters, it is the way in which it is offered.

> " Try to carve out a niche in the marketplace for yourself. Develop a unique selling point, which will make you more attractive to customers than your competitors. This could be a new or alternative colour service or express blow dry. There will always be something that you can do differently, which will surprise your clients and have them talking about you to their friends and family.
>
> *Esther Van der Veken*

The secret is in the *way* that salons conduct business with their clientele. Their customer service becomes their USP and is the reason why clients keep coming back.

Use feedback to identify potential client service improvements

In order to improve services you must first find out how your clients feel about the salon services you already offer. You have to gather feedback information so that some form of evaluation can take place. Then, depending on those findings, you can decide what to do. This can happen either through formal structured ways or informally during conversation with the client.

Customer feedback We learn a lot about our services from those people receiving them – or at least we should do. Customer feedback is extremely important. Client likes and dislikes are the first indicators of what works for them and the things that we should address. The feedback that we get from clients must shape what we do in the future. Without change and improvement, the salon will lose business.

Why do salons lose clients? Fact: It costs 20 times more to find a new customer than it does to retain an existing one.

Salons lose customers all the time and this can happen for a number of reasons, such as:

◆ moving house

◆ job moves

◆ friends' influence

◆ poor service

◆ boredom (the client wants a change).

TOP TIP

Your salon's new clients are generally somebody's former ones!

Moving house and **job moves** are unfortunate for business, but are sometimes necessary for clients and their families. This loss of clientele is an unfortunate fact of life and part of the acceptable percentage that the business has to be prepared for. It is also satisfying if clients keep in touch. Most salons will have a few clients who have moved away but cannot bear to think of forging new associations with another salon. We love to talk about it too, particularly when others cannot understand why they want to travel so far to have their hair styled.

Friends' influence can have a powerful impact on the business. When friends pressurize your clients to try their own hairdresser you have a two-to-one chance of retaining the custom:

1   The client likes the new experience and *you lose*.

2   The client does not like the new experience but is too ashamed to return and drifts off. *You lose again*.

3   The client does not like the new experience and cannot wait to get back to your safety. *You win*.

Then again, your clients should be recommending you. In this situation, you can win again. The power of personal recommendation is immense and should be encouraged. Many salons run 'recommend a friend' campaigns. This is a strong business builder so use it.

Poor service is self-explanatory and wholly unacceptable. This is the quickest way to lose clients and all those others that they meet.

Boredom is unfortunately the main reason why many salons lose clients. It is very hard to predict when boredom will set in, but, if you have a strong rapport with your client, you should be able to spot the signs. Typically, it will occur when you fall into the trap of regularly doing the same style repeatedly. This may not be your fault either. It is human nature to stick to what you know but it can be very damaging.

Fact: There are only a handful of styles that *truly* suit the client. Does that surprise you?

The above statement is based on a number of significant factors including physical proportions, client competence/ability and client lifestyle routines. Trying other styles is fine, but you will find that the client will gravitate back to the safe home ground of 'tried and tested' at some point in the future. The way to combat boredom is how you handle your client's styling preferences.

Product manufacturers do the same in marketing. *New and improved* means 'we have modified what you have been happy with before and given it a new slant'; this can apply to its formulation, packaging or both. In marketing terms, this is known as 'extended product life'. We can invent a new strategy and call it 'extended client life'.

> **TOP TIP**
>
> Personal recommendation is a very powerful marketing tool: if your clients recommend your skills and services to their friends, it is invaluable.

For advice on dealing with negative feedback, see page 72.

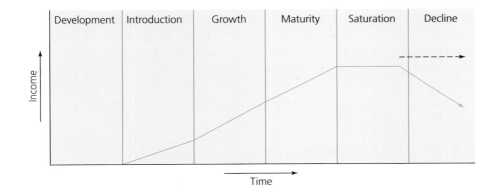

From the diagram, we can see the typical graph for a newly introduced product. Taking it through the development phase (similar to planning new services or treatments), we move into the introduction phase where sales start to increase following advertising and promotion. Moving forwards, further sales bring steady growth, which leads to the maturity phase. If continued interest is not stimulated, it will be followed by a fall-off in sales during the saturation phase. It is at this *boredom* point that the marketers reinvent the product with 'new and improved' in order to achieve an extended sales life of the product.

Collecting information from the client

You have a unique opportunity to find out about the service that you provide to clients by collecting information from them and evaluating it. This can be done openly or discreetly and is purely a matter of choice. However, if you do keep the collected written information on computer you will need to ensure that you keep this confidential and that the salon is not breaching the **Data Protection Act (1998)**.

> To ensure we are well informed of our guests' needs and wants we use several guest feedback methods including online surveys, courtesy calls, stylist/guest conversations and guest questionnaires. These are all invaluable tools for better understanding our guests and improving our guest services.
>
> *Esther Van der Veken*

There are many ways of sampling consumer opinion. In the salon it is often easier to talk to the clients as part of the general conversation that takes place during visits. If it becomes part of the salon's policy for collecting the feedback, a system for collating it in an organized way needs to be devised.

The simplest way to collect and evaluate feedback is to organize it into a table or spreadsheet. You also need to:

◆ Discuss with other staff the things that you should address.

◆ Ask your clients the same questions.

◆ Avoid closed questions that require simple yes or no answers.

◆ Record their responses.

Informal information collection

Verbally, during routine discussion Informal information collection happens every day during routine conversations with your clients. For example, you will be discussing something about their hair, the salon or salon staff. Regardless of the other topics that you chat about, this is an ideal time to learn about the experiences gained by the clients during their visits, but, unless you remember to back this up with some form of recording and documentation, much of what you hear is likely to be forgotten.

For example, if clients start saying:

1   'I didn't like the new conditioner; it left my hair all flat and lank.'

2   'That new styling mousse makes my scalp itch.'

3 'Have you got something like *Fashion Today* to read, instead of *Home and Garden*?'

4 'Do you have proper fresh ground coffee or just instant?'

Then this sort of feedback could be indicating that something is fundamentally wrong. The existing services do not meet the client's expectations.

Through non-verbal means People will often reveal their innermost feelings as they express themselves through their mannerisms, gestures and their body language. If a client is not happy about a certain part of the service they may not tell you directly, but you will get the general impression by what their body language is *telling* you.

See **CHAPTER 1** on body language, page 9.

Formal systems for collecting information

The simplest way of collecting useful data from the client would be through a client suggestion box or alternatively via **client questionnaires**. These are very useful systems but they do need a lot of preparation and planning. If not carefully focused they can create too much information that will:

◆ not be easy to evaluate

◆ be too diverse and general

◆ create too many issues to address.

This will make the survey almost useless.

Questioning styles Before looking at the method of collection, it would be useful to review the techniques of asking questions.

See **CHAPTER 1** Consultation, and **CHAPTER 2** Promote services and products.

The choice of questions used within your survey technique will depend on how you want to evaluate the data. However, the simplest way of indicating opinions is by an objective type of questioning, asking for straightforward 'yes' or 'no' answers. The drawback to this type of question is that it does not produce any additional information.

On the other hand, open style questions allow clients to elaborate in their answers. This will produce more information but it may be more difficult to quantify as the range of responses can be quite diverse and evaluation can be more complex.

Leading questions are *loaded* in their style, they force people to answer within a narrow band of options and provide easily quantifiable data for management. However, although they may gain the information that the salon wants, the clients' true feelings are not revealed.

Multiple choice questions use the Likert scale. For example:

Shampooing and conditioning

Q. Do you feel that the newly introduced, complementary head massage has been a useful addition to the backwash service?

Much better	Slightly better	No difference	Slightly worse	Far worse

ACTIVITY

Competitor comparison

You need to put your services and products to the test, which will mean making like-for-like comparisons between your salon and the competition. The table shows the types of information that you should address.

Service attributes	Worse (−3, −2, −1)	Same	Better (+1, +2, +3)	How vital is this attribute (essential, preferred, other)?
Service availability				
Range of services/treatments				
Ranges of products				
Communication standards				
Presentation/appearance				
Quality of work output				
Aftercare and advice				
Consultation				
Time allowed				
Refreshments + costs				
Payment options				
Salon image and location				

Much of this information is available through your clients from their feedback and general discussion, but there is a big difference between useful data and idle gossip.

Client surveys These are a particularly useful way of formally collecting client feedback. If you choose to keep the feedback anonymous, you may find that the client will provide a true perspective on the services she has experienced. If the survey is well constructed, it will be easy to quantify the feedback and sort the customer data.

Suggestion boxes These can provide lots of feedback but, with so many different comments, it may be tricky to attempt to address every client's innermost thoughts. However, it is worth remembering that this type of feedback does point to *real* client feelings and can give clear indications of clients' expectations.

Questionnaires These are very useful tools for collecting customer feedback. They can pinpoint customer feelings and provide sufficient data for analysis. This is because well-drafted consumer surveys are designed to 'steer' responses into clear, specific, quantifiable areas. They require careful thought in their construction and will often use the multiple-choice type of question as a way of channelling the answers. The favoured style of question for eliciting the strength of people's feelings about a specific issue is shown in the example above using the Likert scale.

Telephone survey Another way of gathering feedback is through telephone surveys, although they can be quite time-consuming to organize and implement during normal salon hours. Bearing in mind the amount of unsolicited calls that people receive at home, you may find that the clients consider this to be an intrusion of privacy and may respond to you accordingly. Many people do not like being telephoned in this way, so it may be your last option for obtaining feedback.

Salon website A website can be useful for getting clients to provide comments, at a time that suits them. If people are allowed more time to give their feedback then the resulting data may be more focused (and therefore more useful) than simply asking clients to answer questions while they are waiting in reception.

The feedback technique does not have to be too sophisticated or technically demanding. Larger salons and chains may have database-driven websites that can enable question and answer interactions; a simpler email response through a site's 'Contact Us' button will be fine for standard HTML-type websites.

> Customers now demand real value for money and want to experience honest genuine care every time they visit a salon. Never before has supporting client service improvements been so important! You have to be focused on providing genuine quality service, exceptional care and attentiveness, and dedicating more time to making your guests both look and feel special.
>
> *Esther Van der Veken*

What are your competitors doing?

The environment in which the business operates is constantly changing owing to many different economic and social factors. The original business plan is soon out of date and there is continuous competition from other salons in the area. The increased pressures that your competitors bring to bear will constantly undermine your salon's market position. **Your competition wants your customers**, just as much as you want theirs.

The success of your business depends upon increasing:

◆ service quality and value

◆ market share

◆ prices for services and treatments

◆ levels of staff abilities/skills.

Better understanding and knowledge about your competition are essential and competitor analysis is now big business. There are many specialist companies offering a wide range of 'undercover' services, like *mystery shoppers* (see page 68), designed to infiltrate and make comparisons between service quality levels. The types of information that competitors want will vary, but your pricing policy and service standards will be seen as particularly valuable.

You may have seen this at work already. How often does a bogus client contact the salon to find out the current costs for particular services or stylist availability? Even if you were sure that another salon is just trying to gain information, it would be difficult to refuse to talk to them. Pricing policy depends on many things, however, so in isolation the information is meaningless. Only when other aspects of service are considered can a far more useful customer comparison be made.

The pricing of salon services relates to:

◆ labour costs

◆ business fixed costs, i.e. rent, rates, light, heating, insurances

◆ costs of equipment, products and consumables

◆ expected profit margins

◆ the target (client) market.

So if these factors relate to the total viability of the business, what would happen if the information fell into the wrong hands? The only additional costs that they will incur are those involved in trying to attract your clients through marketing, promotion and advertising.

What do your competitors know about you?

1 Other businesses know what you should be paying your staff.

2 They know your business costs in relation to salon location (sources: letting agents, local authority, accountants).

3 They know what products you use, how much you pay for them and how much profit you make by selling them.

4 From your salon tariff, they know what sort of customer you are trying to attract.

There is not much that they don't know about you (if they can be bothered to find out).

The only thing that your competitors have to calculate is the profitability of the service and whether it is something they can match or even beat on price.

You need to put your services and products to the test, which will mean making like-for-like comparisons between your salon and the competition. The table in the previous activity on page 66, 'Competitor comparison', shows the types of information that you should address.

TOP TIP

Much of this information is available through your clients from their feedback and general discussion, but beware of the difference between feedback that provides useful data and that of cynical, idle gossip.

Mystery shopper The other popular method for collecting quality service information is the mystery shopper, or in this case 'mystery client'. In this method of collecting useful feedback, an unknown person has to go out and sample the services of your salon or a competitor. They are given a brief beforehand and will be looking for specific aspects of the salon service during their visit. They will pay at the end of the service as normal and will provide their feedback at some later time. They will not make themselves known to you at any part of the process and this is an ideal way of making comparisons between your offering and those of the competition.

Implement changes in client service

Example salon survey form

Salon survey – customer satisfaction

1. How well do you think that you were received when you arrived at the reception of our salon today?

Very well	Good	OK	Not well	Poorly
23	17	11	0	0

2. Does this reflect the usual way in which you are received at our salon?

Yes	No
40	11

Sharing the information

After collecting the feedback, the data needs to be organized with the number of respondents and the answers received. This could be done by creating a table or spreadsheet on the computer.

When the information has been collated, it may be useful to share it with all the salon team, particularly as this becomes an open forum and provides a basis for discussion. However, where responses could be targeted at individuals then sensitivity is required or alternatively, where there is negative feedback relating to individuals, it should be taken up with the person on an individual basis as part of their performance appraisal.

ACTIVITY

Care should be taken in quantifying data. In the example shown in the salon survey on customer satisfaction above, the two questions seem to be targeted at the quality of service received on entering the salon.

1 Is question 1 in the example a useful question?

2 Why?

3 Is question 2 in the example a useful question?

4 Why?

5 What does question 2 indicate if several people are responsible jointly for reception duties?

Identifying possible changes to services

When the data have been collected, organized and evaluated, there are two obvious reasons for making changes to services:

◆ Customers are dissatisfied with some part of the existing provision

◆ The salon wants to improve the existing provision.

If clients were dissatisfied with some aspects of the existing service, then it would be important to address and correct those issues. However, it is worth bearing in mind that sometimes the further improvements may not be feasible. When customer expectations exceed the salon's resources then a management decision is required. For example, imagine that a salon has always provided complimentary drinks to its clients in the past; then following a survey it has been noted that clients want a freshly ground cappuccino coffee instead. The salon management now has three possible options:

◆ They continue to offer the poorer quality, less expensive complimentary service as they have always done in the past.

◆ They decide to invest in an expensive, commercial ground coffee machine and choose to charge the clients for the service in the future.

◆ They invest in the new machine, continue to offer an enhanced complimentary service and profit margins fall.

You can see the dilemma: they now know that the client does not like their existing complimentary service, but does that mean that clients will be happy to pay for a premium service in the future? *Not necessarily!*

Other improvements to services may be far more obvious. If a client complains that a gown smells of perfume, it is both negligent and reflects poor service by the staff for not removing it from the salon and putting it in the laundry. Obviously, if services are not meeting the expectations of clients, something needs to be done straightaway. Sometimes existing services need improving and this means changing the salon's provision to enhance what is already offered.

> "
> Sharing customer care information and feedback is vital to salon success. We hold regular guest feedback meetings where we analyse and discuss all of the guest feedback we have received. By constantly analysing our customer care performance, we ensure that every member of our team is focused on delivering fantastic customer care.
>
> *Esther Van der Veken*

Presenting and sharing your ideas

Where you have found a useful contribution and enhancement for the salon's services, you need to prepare your case. It does not matter how small the changes seem to be, as they all have some value. However, you do need to justify your reasons for change; you must think about the impact that your suggestions will have on others and on the salon. Can the salon afford it? Or rather, to put it another way, can the salon afford *not* to do it?

Work out if you need the support of others first. If you need to gain the assistance of your fellow staff (as your ideas might directly affect what they already do) then you may find that you would be better off presenting your plan to them first. This is particularly useful as they may have views or opinions that could help you shape and hone down your final presentation to management.

In many situations, a change to services needs a supporting programme of training too. For example, if after a survey it was found that the salon should provide some sort of outreach service for weddings then those involved in delivering the service to the bride and her entourage at home would need to be proficient at all sorts of special occasion hair work.

When you have decided what changes need to take place, you need to make sure that all the staff are made aware of how improvements will be implemented. There are several ways of doing this (see below).

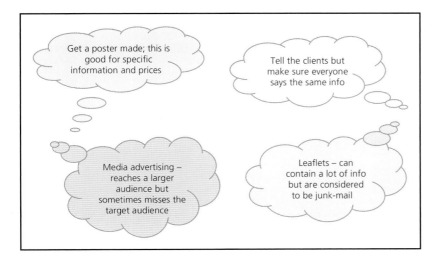

Be positive and explain changes

You may have gone through a lengthy process to find ways in which you can improve the client's salon experiences, but remember that people do not always welcome change. Generally, people tend to resist change and often, where a benefit is not immediately apparent to them, they may wonder why you have had to change things at all.

It is therefore important to explain any changes – you do not have to go into every detail, the main reasons will do. For instance you might say: 'You'll find that we have included a complimentary conditioning treatment with all our clients who have colour services, Mrs Smith. We have found that our clients were not taking up this particularly beneficial part of the service, so we now include it as part of a total service package.'

It is also important to be positive – if you are enthusiastic, your clients will be far happier about the changes themselves. If you seem less than positive then you are likely to find a general reluctance from everyone.

Assist with the evaluation of changes in client service

After the changes have been introduced, their impact will need to be evaluated. Feedback can be gathered informally from clients during routine salon services, but what you then do with this feedback is vitally important.

Ideally there should be a group discussion and someone – possibly you – should record the main outcomes of the team meeting so that documented information can be fed back to the management. Each member of the team will have varied feedback so be ready to take on-board comments from all the team.

ACTIVITY

With your fellow staff/students, discuss the various methods/systems of communication that your salon could use to find out what clients think and feel about the current services provided.

Negative comments

Not all feedback is positive and there will be times when new initiatives are not received well! It is not easy to give negative feedback, and no one likes criticism, but there are some constructive ways to go about it.

It is too easy to sound arrogant or flippant, so you need to be tactful in your delivery. If possible do not single out individuals; they will not like it and you will alienate yourself from the rest of the team. If possible, accentuate the positive; find ways of tempering and balancing the feedback with an equal amount of positive points as well as negative ones. Constructive criticism can be a useful learning experience; just think how supportive it can be when looking at someone's areas for improvement in appraisal.

If there is negative feedback coming your way, think about the ways in which you can adapt and change to accommodate the improvements. Try to be mature about it: it is impossible to get new strategies right on every occasion and there will be plenty of opportunities for the future. Remember to learn by any mistakes and then move onwards and upwards.

> You want your customers to have one thought in their mind when they leave your salon: 'I can't wait to go back!' Within Tommy's we don't sell services or products, we sell an *experience*. If you can work to that ethos the result is that your clients feel valued, appreciated and understood rather than feeling 'processed'.
>
> *Esther Van der Veken*

ALWAYS REMEMBER

Service improvement is the most important reason for making changes and taking a business in the right direction. Be motivated by it and make plenty of positive contributions towards it.

Salon and legal requirements

Before any information-collecting activities can be put into practice, make sure that there are no contraventions to your salon's policy or any breach of legal requirements.

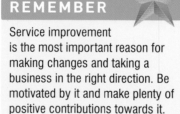

See CHAPTER 16 and APPENDIX 4 for legal requirements and restrictions on keeping and using personal information.

ACTIVITY

Customer service

With the help of your receptionist, you can conduct a simple survey to find out when salon customers want to come in. Create a simple graph displaying the days of the month along the bottom axis and a range of times (say 8.00 am to 8.00 pm) along the vertical axis. When customers call the salon, mark each point on the graph to show when they want to come in as opposed to what is available. Each plotting point could be colour coded to refer to different staff members. Run the exercise over a complete calendar month to see what happens.

◆ Are there any specific days that people would prefer?
◆ Are there any specific times that people would like?
◆ Do the findings of the survey highlight any shortcomings or inefficiencies with the present work systems and time scales?

REVISION QUESTIONS

Q1. A _____ question style prompts a simple *yes* or *no* response. Fill in the blank

Q2. A leading question forces people to answer within a narrow band of options. True or false?

Q3. What would provide strong reasons for changing or improving services? Multi-selection

Competition from other salons? ☐ 1

Manufacturer's sales pressure? ☐ 2

Customer feedback ☐ 3

Too much stock on the shelves ☐ 4

Good sales and turnover ☐ 5

Friends and family feedback ☐ 6

Q4. Client boredom leads to lost custom. True or false?

Q5. Which style of questioning provides the most effective way of finding out how strongly a client feels about salon services? Multi-choice

Open questions ○ a

Closed questions ○ b

Multiple choice (Likert scale) ○ c

Leading questions ○ d

Q6. Customer feedback has little importance in shaping a business's future. True or false?

Q7. Which of the following responses indicate that the newly introduced complementary head massage has been a useful addition to the salon's services? Multi-selection

Strongly agree ☐ 1

Agree ☐ 2

No opinion ☐ 3

Disagree ☐ 4

Strongly disagree ☐ 5

Q8. A formal collection of client information could be made through questionnaires or a _____ box. Fill in the blank

Q9. In multiple choice (Likert scale) question types, the middle option shows what sort of response? Multi-choice

A strong agreement ○ a

A mild agreement ○ b

Neither agreement nor disagreement ○ c

A mild disagreement ○ d

Q10. A negative feedback from a client questionnaire is a useful indicator. True or false?

PART TWO

Hairdressing technical services

Chapter 4
Creative Cutting

Chapter 5
Creative Barbering

Chapter 6
Beards and Moustaches

Chapter 7
Shaving

Chapter 8
Creatively Style and
Dress Hair

Chapter 9
Creatively Dress Long Hair

Chapter 10
Hair Extensions

Chapter 11
Colouring Hair

Chapter 12
Colour Correction

Chapter 13
Perming Hair

Chapter 14
Develop Your Creativity

Your expertise at Level 3 is what makes you a true hairdressing/barbering professional. The technical skills you need to operate successfully at this level are considerable, therefore the main section of this book is dedicated to the full range of advanced practices that you need to master.

From creative cutting to styling, perming and colouring, we present all the aspects of theoretical and practical training you need to study. Look at each client individually: their face shape, hair condition, posture, profession and personality. Get to know them and their hair intimately. Work safely and efficiently, communicate openly, and go the extra mile to make sure they leave your care looking and feeling amazing.

4 Creative Cutting

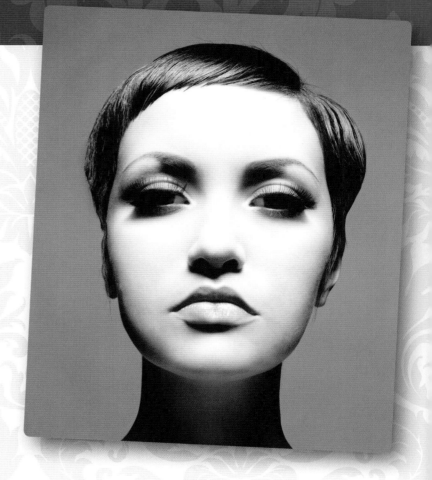

LEARNING OBJECTIVES

◆ Maintain effective and safe methods of working when cutting.

◆ Creatively restyle women's hair.

◆ Be able to work safely when cutting hair.

◆ Understand how to use a variety of techniques to create different effects.

◆ Be able to communicate professionally and provide aftercare service and advice.

KEY TERMS

brick cutting
clipper over comb
club cutting
concave
convex
corn rows

disconnection
dreadlocks
fading
freehand cutting
humid
point cutting (pointing)

scissor over comb
slicing
tapering
texturizing
total look

UNIT TOPIC

Creatively cut hair using a combination of techniques

INTRODUCTION

This chapter focuses on aspects of creative cutting and techniques that will enable you to construct a variety of fashionable styles. It will provide new alternatives to help you think more *laterally* about style design and visualization; it should give your existing skills a boost and help you to find new ways of applying them.

Creative cutting design involves deciding what needs doing, making the right choices, and selecting the visual images that will suit the client and enable them to see what is an achievable style for them.

Creative cutting

PRACTICAL SKILLS

Learn how to work with care, attention and safety at all times

Look for things that will influence the ways that you cut your client's hair

Learn how to work with different types of hair

Learn how to achieve a variety of creative effects and styles

Learn how to work with, or correct, simple faults and common problems when cutting

Provide aftercare advice to your clients

UNDERPINNING KNOWLEDGE

Identify how to work safely during cutting

Identify how different cutting techniques achieve different effects

Identify the common problems that can affect the way that you cut hair

Identify how the client's physical aspects and features can affect styling options

Identify how different cutting techniques achieve different effects

Identify how to communicate effectively and professionally

Maintain effective and safe methods of working when cutting

Building from the basics

As you have become more skilled in styling hair, your confidence and willingness to explore new ways of doing things develops into true professionalism. This is the result of hard work and *experience*.

INDUSTRY ROLE MODEL

ERIK LANDER   **Partner in Top Spot Hair Design, Corby, Guest Artist for Goldwell Hair Cosmetics, Habia Education Director**

I started working in the family salon at the age of 14 years and worked my way up, while honing my practical skills. After five years as a stylist, I had the opportunity to start delivering training at the Goldwell Academy in London, where I still work. I applied to become a member of the Habia skills team, at its inception, working my way through the ranks to become Habia's Education Director.

> Being a perfectionist is a massive advantage with precision cutting, as it's all about the detail. Everything evolves from the basic cuts and you need to perfect these before moving on to the more advanced techniques.
>
> *Erik Lander*

At Level 3, you are building from the basics and practising a range of techniques, gaining further experience, building up your confidence and producing a wider variety of new effects. This chapter considers the following essential aspects:

◆ reparation and maintenance

◆ client consultation

◆ identifying the influencing factors that affect style choice

◆ cutting techniques and accuracy

◆ good customer care.

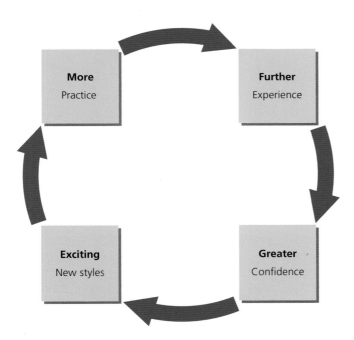

Preparing the client

Your salon will have its own policy and codes of practice for preparing clients and you must observe these. These policies might include a procedure for gowning and protecting the client from spillages or hair clippings, methods for preparing tools and equipment, and the general expectations for personal standards in relation to technical ability and hygiene. Even an experienced stylist never neglects the basics. Above all, you should remember the client's personal comfort and safety throughout the salon visit:

◆ **Cover up the client** – Always use a clean, fresh gown and place a cutting collar around the client's shoulders. Make sure that the gown is on properly and fastened around the neck. It should cover and protect the client's clothes and come up high enough to cover collars and necklines. Do not make the fastening too tight, but it should be close enough at least to protect the client's clothes and stop hair clippings from going down their neck, which is both uncomfortable while they are in the salon and irritating if they are returning to work or are busy for the rest of the day.

◆ **Make sure that the hair is clean** – You cannot cut hair well if it is loaded with hairspray or it has product build-up. If the client uses a lot of finishing products on their hair, you will need to make sure the hair is clean before you start. You should be able to comb hair freely during sectioning so that you can achieve the correct holding angles and cutting angles without tangles or binding.

◆ **Adjust the chair** – Client working height has a lot to do with your working posture and safety too. Client comfort should extend to the point where it makes the salon visit a welcome and pleasurable experience. Clients should not clutter the floor around the styling chair with bags, magazines and shopping. Anything that can safely be stored should be put away; it is not only a distraction, it is also a safety hazard.

◆ **Work position in relation to mirror** – The positioning of the client in front of the mirror is very important. Any angle of the head other than perpendicular to the mirror and the angle of the head to the seated position will affect the line and balance of the haircut.

All salon workstations have built-in foot rests and there are good reasons for this.

The foot rest:

1 Is there to improve the comfort for the seated client at any cutting height.

2 Helps balance the client and encourages them to sit squarely in front of the mirror.

3 Tries to discourage the client from sitting cross-legged.

4 Promotes better posture; it makes the client sit back properly in the chair.

ALWAYS REMEMBER

Salon chairs are designed with comfort and safety in mind; your client should be seated with their back flat against the back of the chair, their legs uncrossed and the chair at a height at which it is comfortable for you to work. You need to be able to get to all parts of the head, so the chair's height needs adjusting to suit the height of the client. Do not be afraid of asking the client to sit up: it is in their best interest too!

This is all critical for you and for the client in ensuring their comfort throughout, and so your task is not made more difficult. For example, if your client sits with crossed legs, it will alter the horizontal plane of their shoulders and this will make your job of trying to get even and level baselines more difficult.

TOP TIP

Always dry the client's hair well to remove the excess water before combing, blow-drying, or finger drying.

Hair drying Rough-dry the hair so that you work with damp hair throughout the cut. Dry off the client's hair so that they are not sitting with saturated hair; it is uncomfortable for them, as wet hair soon feels cold and will drip onto the gown and their clothes. Damp hair is required during cutting as it enables the natural tendencies, movements and directions of the hair to be seen. It means the wave movement and hair growth patterns can be considered.

Your personal hygiene

Personal hygiene is vitally important for anyone working in personal services. Your personal hygiene (or lack of it) will be immediately noticeable to every client you work with. Daily showering and regular brushing of teeth are essential for you, your colleagues and the clients, as body odour and bad breath are unpleasant in any situation. Other strong smells are offensive too: the smell of nicotine from smoking is very off-putting to the client, particularly if they are a non-smoker. Do not smoke when you are on duty.

HAIR SCIENCE: DID YOU KNOW?

We all carry large numbers of microorganisms inside us, on our skin and in our hair. These organisms, such as bacteria, fungi and viruses, are too small to be seen with the naked eye. Bacteria and fungi are seen through a microscope, but viruses are too small even for that.

Many microorganisms are quite harmless, but some can cause disease. Those that are harmful to people are called pathogens. For example, viruses cause flu and cold sores. Thrush and athlete's foot are caused by fungi, and bronchitis and impetigo are caused by bacteria. Conditions like these are infectious; they can be transmitted from one person to another.

The body is naturally resistant to infection; it can fight most pathogens using its inbuilt immunity system, so it is possible to be infected with pathogenic organisms without contracting the disease. When you have a disease, the symptoms are the visible signs that something is wrong. They are the results of the infection and of the reactions of the body to that infection. A doctor should always treat infectious diseases; non-infectious conditions can be treated with medication from the pharmacist.

Preparing the tools and equipment

Always get your scissors, razors, clippers, combs and sectioning clips ready beforehand – and that does not mean as the client arrives to sit in the chair from the basin! You must be prepared; your tools and equipment must be hygienically clean and safe to use. You would need to remove your combs from the Barbicide™ or sterilizer prior to use; these should be rinsed, dried and ready to use.

Plastic and rubber cutting tools

◆ Never use dirty or damaged tools. Germs can breed in the crevices and corners, and can cross-infect other clients.

◆ Clean or wipe all tools before disinfecting or sterilizing.

◆ Cutting collars should be washed and dried before use.

Other cutting tools

◆ Barbicide™ will corrode and damage metal tools like scissors and razors.

◆ Clippers must be checked for blade alignment and any damage before use.

Preventing infection

A warm, humid salon can offer a perfect home for disease-carrying bacteria. If bacteria can find food in the form of dust and dirt, they may reproduce rapidly. Good ventilation, however, provides a circulating air current that will help to prevent their growth. This is why it is important to keep the salon clean, dry and well aired at all times, and work areas free from clutter or waste items. This includes clothing, work areas, tools and all equipment. Some salons use sterilizing devices as a means of providing hygienically safe work implements. Sterilization means the complete eradication of living organisms. Different devices use different sterilization methods, which involve heat, radiation or chemicals.

UV sterilization

Ultraviolet radiation Ultraviolet (UV) radiation provides an alternative sterilizing option. The items for sterilization are placed in wall- or worktop-mounted cabinets fitted with UV-emitting light bulbs and exposed to the radiation for at least 15 minutes. If your scissors or combs are sterilized in a UV cabinet, remember to turn them over to make sure both sides have been exposed to the UV radiation.

Chemical sterilization Chemical sterilizers should be handled only with suitable personal protective equipment, as many of the solutions used are hazardous to health and should not be exposed to the skin. The most effective form of salon sterilization is achieved by the total immersion of the contaminated implements into a jar of fluid.

Autoclave The autoclave provides a very efficient way of sterilizing using heat. It is particularly good for metal tools, although the high temperatures are not suitable for plastics such as brushes and combs. Items placed in the autoclave take around 20 minutes to sterilize. (Check with manufacturer's instructions for variations.)

Your work position

The client's cutting position and height from the floor have a direct effect on your posture too. You must be able to work in a position where you do not have to bend 'doubled up' to do your work. Cutting involves many arm and hand movements and you need to be able to get your hands and fingers into positions where you can cut the hair unencumbered, without the risk of bad posture.

1 You should adjust the seated client's chair height to a position where you can work upright without having to over-reach on the top sections of their head.

2 You should clear trolleys or equipment out of the way, so that you get good all-round access (300°) around the client.

Working efficiently, safely and effectively

Working efficiently and maximizing your time are essential, so you must make the most of the salon's resources and take care in the way that you handle the equipment and products that you use. Always treat the salon's materials in the same way that you would look after your own equipment; aim to minimize waste, being careful about how much product you use.

You need to work in an orderly environment, with the materials that you need at hand and the equipment that you want to use in position and ready for action. Remember to keep an eye on the clock; you must remember that you have to work to time and that means providing the service in a commercially acceptable time. Your client has a schedule too. If their time constraints seem unreasonable or unrealistic, make sure that you tell them beforehand. For example, if they are expecting to be back within their lunch-hour and you need to restyle, cut, dry and finish long hair in that time, let them know how long it is likely to take.

TOP TIP

Look out for ways of doing things that can make your client's visit more comfortable and pleasurable. This is the first step in providing a better customer service.

Your working position and posture

Hairdressing, as you already know, involves a lot of standing. You should always adopt a comfortable but safe work position – although sometimes comfortable and safe are not necessarily the same thing.

ACTIVITY

Every salon has its own way of doing things. Use the following headings to note down your salon's practices in respect to:

◆ disposing of sharps
◆ conducting client consultation
◆ preparing clients for wet and dry cutting
◆ preparing and maintaining tools and equipment.

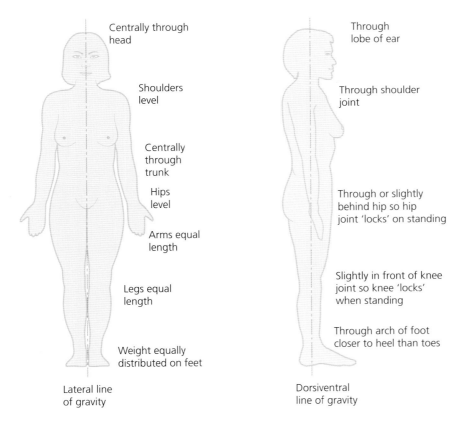

Centrally through head

Shoulders level

Centrally through trunk

Hips level

Arms equal length

Legs equal length

Weight equally distributed on feet

Lateral line of gravity

Through lobe of ear

Through shoulder joint

Through or slightly behind hip so hip joint 'locks' on standing

Slightly in front of knee joint so knee 'locks' when standing

Through arch of foot closer to heel than toes

Dorsiventral line of gravity

A naturally comfortable position for work should allow you to stand close enough to the styling chair without touching it; this should allow you to position your shoulders and torso directly above your hips and feet with your weight evenly distributed. You should not have to twist at any point as you can easily work around the chair or get your client to turn their head slightly towards you. You should wear flatter shoes rather than high heels for work, so that your body weight is comfortably supported on the widest parts of the feet. This will allow you to work longer without risk to injury or fatigue.

Always make a point of lifting your arms to check the working height for your client. If you have to raise your arms anywhere near horizontal during your work, you will find that your arms will start to ache very quickly. Make your adjustments to the styling chair, either up or down to suit your needs. (But remember to tell the client first!)

Preparation for cutting

Consultation

Effective communication with the client, as in any service, is essential for cutting hair. Consultation is not just a process that takes place before a service; it is a continual process of reconfirming *what* is taking place *while* it is happening. Therefore, during your discussions, you must determine what the client wants and weigh this against the limiting factors that will influence what you need to do.

> Consultation is key to the cut, if you get the consultation wrong the cut could be technically perfect but it is still wrong as it is not what the client asked for.

Erik Lander

You need to understand your clients fully and be able to negotiate and seek agreement with them throughout the service.

Using visual aids Visual aids such as style magazines, Internet downloads and trade publications are excellent sources for fashionable effects. Pictures are a universal language and this is a good way of demonstrating ideas and themes, showing clients what might work.

The haircutting style that you choose with your client should take into account each of the following points:

◆ face and head shape

◆ physical features and body shape, size and proportion

◆ hair quality, abundance, growth and distribution

◆ age, lifestyle and suitability

◆ purpose.

General styling limitations

The proportions, balance and distribution of the hairstyle will be a frame for the head and face. Therefore, you need to examine the head and face carefully, to identify if it is round, oval, square, heart-shaped, oblong or triangular. Only an oval face will suits all types of hairstyle, so all the others listed present some form of styling limitations; in other words, they become a factor influencing styling choice.

ALWAYS REMEMBER

Be sure to listen to your client's requests. Mistakes can be avoided if you achieve a clear understanding of what the client is asking for.

Physical feature	How best to work with it
Square and oblong facial shapes	Are accentuated by hair that is smoothed, scraped back or sleek at the sides and top. The harshness of lines and angles are made less conspicuous by fullness and softer movement.
Round faces	Are made more conspicuous if the side and front perimeter lengths are short or finish near to the widest part of the face. This is made worse if width is added at these positions too. Generally this facial shape is complemented by hair length beyond the chin.
Square angular features, jaw, forehead, etc.	Are improved with softer perimeter shapes; avoid solid, linear effects around the face. Shattered edges and **texturizing** will help to mask these features.
Flatter heads at the back	Are improved by graduation, creating contour and shape that are missing from having a flatter occipital bone.

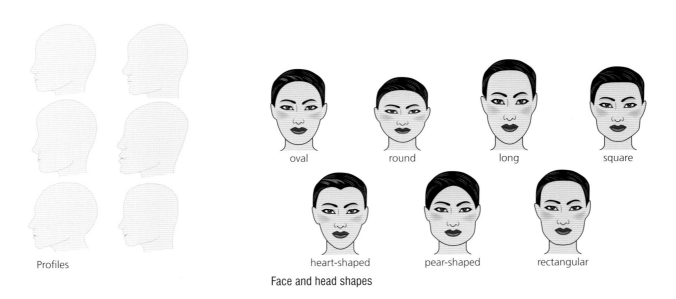

Profiles

oval   round   long   square

heart-shaped   pear-shaped   rectangular

Face and head shapes

Head, face and body physical features

Physical feature	How best to work with it
Prominent nose	Hair taken back away from the face accentuates this feature, while hair around the face and forehead tends to diminish the feature.
Square jaw line	Is softened by longer perimeter lengths either coming around and on to the face or styled with fullness. Conversely, shorter side lengths will have an opposite effect.
Protruding ears	Are better left covered rather than exposed. Sufficient hair should be left to cover and extend beyond their length.
Wrinkles around the eyes	Are made more obvious by hair being scraped back at the temples or straighter more angular effects.
Narrow foreheads	Are disguised by softer fringes and side partings, whereas they are made more obvious by hair taken back away from the face.

(Continued)

Physical feature	How best to work with it
Larger body shapes	The overall effect is balanced with longer, fuller hairstyles; they are made much worse by short, sleek, layered shapes.
Small faces	Can be overwhelmed if the hair is left long with a centre parting and no fringe.
Large faces	Are accentuated by short cropped or sculpted hair.
Shapes of glasses	There are so many available and are a definite fashion accessory. Generally, people will have had assistance in the selection and suitability of their frames. Therefore, they should not work against the hairstyle that you want to create, as they should already suit the shape and size of the face.
The way the head is held	Many people tilt their head to one side or forwards. Sometimes this is because they are tall and want subconsciously to reduce their height; sometimes they hide behind their hair because they lack confidence or they think it makes them more alluring. You need to look out for this natural posture but not make any comment other than asking whether they have any preference to finished lengths, fringes, and partings.

Reason or purpose for hairstyle

The reason or purpose for the hairstyle is a big factor in deciding what is suitable or not.

◆ A style suitable for a special occasion will differ from one for work. The requirements for competition or show work are quite different from those for general daily wear. However, versatility needs to be considered for everyone; people want styles that they can dress up or down.

◆ Some jobs have special conditions about hair lengths and styles; for example, people working in the armed services or police have to wear their hair above the collar while at work. Men have easily accommodated this by using clippers for very short styles. Women have either had to have short-layered styles or hair that is long enough to wear up and out of the way.

Quality, quantity and distribution of hair

◆ Good hair condition is an essential factor in great hairstyling. Some aspects of a hairstyle cannot be altered by cutting alone; for instance, if the hair is dry, dull and porous when the client enters the salon, it still will be when she leaves.

◆ Regular salon clients – the ones you tend to see more often than the others –tend to have something in common, e.g. difficult hair. It can be difficult for a number of reasons; it can be fine or unmanageable, lank and lacking volume or just not responsive to styling without force. Thin, sparsely distributed hair is always a problem: if there is not enough hair to get coverage over the scalp, then there is not a lot you can do about it. You should remember not to put too much texturizing into it; this will only make the problem more noticeable. Fine hair presents many problems too. Very fine hair is affected by dampness and quickly loses its shape. This type of hair always benefits from moisture-repelling styling products, so get your client in the habit of using them.

◆ Dry, frizzy hair can also be a problem, as the more thermo styling it receives, the more moisture is lost and the less it responds to staying in shape – in other

ALWAYS REMEMBER

The success of any hairstyle is based on the information that you get during the consultation. Be thorough: an extra five minutes spent discussing the final effect could make all the difference!

words, the harder it is to style. Dry, unruly thick hair needs to be tamed and most clients with this problem would like their hair to look smoother and shinier. Again, this is a conditioning issue and you need to deal with it before tackling the style. Sometimes this type of hair benefits from finishing products, so put them on as you finish and define the hairstyle.

◆ Very tight curly hair can be difficult, particularly if your client wants it to appear straight. It is possible to smooth and straighten hair, particularly when you use ceramic straighteners or thermal styling. However, keeping it straight is another matter, and you may want to consider other options.

◆ Cutting wavy hair presents some problems but not if it is looked at carefully before shampooing. Avoid cutting across the crests of the waves; you cannot change the natural movement in the hair so try to work with it.

◆ Straight hair, particularly if it is fine textured, can be difficult to cut. Cutting marks or lines can easily form if the cutting sections and angles are not right. Make sure that you only take small sections of hair and remember to crosscheck, at 90° to the angle in which you first cut, to avoid this happening to you.

Hair positioning, type, growth and tendency The perimeter outline formed by the hair in relation to the shape of the face is the first thing most people see. It is this effect that people often notice first. The complete hairstyle is based upon the frame that the hair creates for the face. How you 'fill in' the detail – the movement, direction, colour and placement – is down to your interpretation, understanding, technical ability and experience.

Hair growth direction and distribution should be a major consideration for what is achievable within a hairstyle. You need to make allowances for strong movement, high or low hairlines, natural partings, hair whorls, cowlicks, widow's peaks, and double crowns. Look for these before shampooing. The client cannot compensate for these, so when the hair is in need of washing, they will be plain to see. After the hair is washed the degree and strength of the feature can be seen and then you can reconsider how you will tackle it.

Style suitability

Style suitability refers to the effect of the hair shape on the face, and on the features of the head and body.

Aesthetically and artistically speaking, the client's 'hair will look right' when the hairstyle does one of two things. It *harmonizes*, i.e. suits the shape of the face and head – and is therefore a backdrop to an overall image; or it *contrasts*, i.e. it accentuates features of the face and head – by creating a prominent frame for the overall image.

Age As much as you would like to demonstrate your creative ability on everyone who walks through the salon door, bear in mind that some styles are inappropriate for certain clients. Beyond the physical aspects of style design, age does create some barriers to suitability.

Balance Balance is the effect produced by the amount, fullness and weight distribution of hair throughout the style. The opposite, imbalance, is lack of those proportions. Symmetry or symmetrical even balance occurs when the hair mass is distributed equally. Asymmetry or asymmetric effects occur when the overall shape

> **TOP TIP**
> When you choose a suitable hairstyle, always allow for the natural fall of the hair.

> For more on haircuts and age see **CHAPTER 1** Consultation, page 34.

does not have the same distribution on either side. However, both symmetrical and asymmetrical shapes can be balanced.

ACTIVITY

Complete the table below to say why each of these consultation factors are important and what might happen if these aspects are ignored.

Factors	Why is this important?	What will happen if this is ignored?
Client's requirements		
Hair growth patterns		
Hair texture		
Hair density		
Lifestyle		
Physical features		

Cutting tools

Scissors Scissors are and will always be the most important piece of hairdressing equipment that you will ever own. Your future income, popularity and success will rely upon this relatively inexpensive item. If you look after them, you will be surprised how long a single pair will last. Scissors can be used on either wet or dry hair and vary greatly in their design, size and price. Scissors should never be too heavy or too long to control; heavy scissors become cumbersome in regular use and if they are too long you will not be able to manipulate them properly for precision, angular work. Long blades are good for cutting solid baselines on longer hair, but a real nuisance for precise work around hairlines and behind ears!

To judge the balance and length of a pair of scissors, put your fingers in the handles as if you were about to use them. When the scissors are held correctly, the pivotal point should just extend beyond the first finger. This allows the blades to open easily and means that the thumb is in an ideal position to work them.

The more expensive scissors will often have one single blade that has small serrations throughout the length. This is really beneficial as this becomes the lower blade in cutting,

ALWAYS REMEMBER

Use quality combs: the comfort of quality combs and in particular cutting combs is the most important factor.

Your professionalism can be measured by the comb that you use. There is nothing worse than using cutting combs on clients, when each time you take a section you scrape and scratch the client's scalp. You will also find that, in regular use, if you persist in using cheap combs your hands will become sore, as the teeth will scratch you when you pass the comb into your hand on every section that you take!

grips the hair and stops it from being pushed away by the closing blades. Sharpening of this type of scissor blade is not recommended as this factory finish will be removed immediately.

Cutting comb Try to use only quality cutting combs. You will find that by spending only a little more you will get so much more out of them. The design of a cutting comb for hairdressing is different to that for barbering. The hairdressing cutting comb is parallel throughout its length, whereas the barbering comb is tapered.

There are two sorts of cutting comb. The first and by far the most popular has two sets of teeth: one end to the middle is fine and close together; the other end is wider and further apart. This allows you to do fine sections on fine hair and wider sections on coarser hair. The second type of cutting comb has uniform teeth throughout the length of the comb.

The length of cutting combs varies greatly, and what is best for you will depend on the size of your hand and what you can manipulate easily. The normal length of a cutting comb is around 15 cm but longer ones are now very popular and give a better guide for cutting baselines; these can be 2 or 3 cm longer. The quality of combs and the materials they are made from varies greatly. The best-quality combs are made from hard but flexible plastics and have the following properties:

◆ They are very strong but flexible; the teeth do not chip or break in regular use.

◆ They remain straight in regular use and do not end up looking like a banana after a couple of weeks.

◆ They are constructed by injection moulding and do not have sharp or poorly formed edges (as opposed to combs that are made from pressings and have flawed seams and tend to scratch the client's ears and scalp).

◆ They are resistant to chemicals making them ideal for cleaning, sterilization and colouring (as they will not stain).

◆ They have anti-static finishes that help to control finer hair when dry cutting.

Thinning scissors These can be used on dry or wet hair and can have either one or two serrated blades. These cutting surfaces will remove bulk or density from the hair, depending on the way in which they are used. This has two useful applications for cutting:

1 The tips of finely serrated scissors provide a quick way for texturizing the perimeter edges of hairstyles;

2 The whole blades can be used for removing weight (**tapering** or thinning) from sections of hair but closer to the head.

Thinning scissors with both blades serrated will remove hair more quickly than those with serrations on only one side; this is more noticeable on scissors that have broader notches in them as opposed to fine teeth.

Razors The open or 'cut-throat'-style razor used to be made from a single steel blade, which was hinged and closed into a protective handle.

The modern counterpart for this has disposable blades, which can be removed and disposed of safely in a sharps box. Razor cutting is done on wet hair and *always* with sharp blades. Because of the way in which razors are used, razoring should never be done on dry hair as it will pull and tear the hair even if the blades are new.

Electric and rechargeable clippers Clippers consist of a moulded, easy-to-handle body with a pair of serrated cutting blades.

TOP TIP

Never keep scissors in your pockets: it is unhygienic but, more importantly, it is a dangerous thing to do.

ALWAYS REMEMBER

New scissors are purchased in a useful protective case, so get into the habit of keeping them in it. This will make them easy to identify and will provide useful protection when they are carried around.

Clipper attachments Clippers have a range of plastic cutting attachments. These enable the clippers to be used safely in cutting wet or dry hair to a uniform layered length. These attachments vary the cutting depth:

◆ Grade 1 – extremely short, just leave a shadow of hair across the scalp.

◆ Grade 2 – very short allowing the skin to still be seen on all but very thick hair.

◆ Grade 3 – moderately short, the first grade length that enables the skin not to be seen on most but very fine hair types.

◆ Grade 4 – short, but a very popular length, leaving enough hair to resemble very short layering.

◆ Grades 5–8 – longer grades that produce layered effects that could be achieved by club cutting techniques.

ALWAYS REMEMBER

Avoid ingrowing hairs

Ingrowing hairs are an uncomfortable aspect of closely cut wavy or curly hair. When hair is razored or clippered close to the skin, say at the neckline, the action of the cutting implement tends to slightly pull the hair within the follicle. After the hair is cut, the hairshaft tends to retract back into the follicle giving a very close and smooth finish to the touch. Unfortunately, if the area is not kept thoroughly clean the open area of the follicle can become partially blocked with dead skin cells or dirt. This will change the growing direction of the hair within the follicle, often allowing it to find a new position within the skin. Bearing in mind that the average rate of hair growth is about 1.25 cm (12.5 mm) per month; if allowed to continue, an ingrowing hair will result. The physical effects of this will not be noticed until:

◆ It starts to create a raised area, often accompanied by a mild infected spot.

◆ It begins to itch as it causes irritation beneath the skin.

If the ingrowing hair is not exposed to the surface or removed from the skin, a longer-term, more serious condition/infection will result.

TOP TIP

Maintain your tools

Clippers – a little clipper oil should be applied last thing so that the lubrication can provide a protective coating overnight. Then, before use, they can be cleaned, any excess oil removed and made ready for the next day's use.

Scissors – carefully wipe over the blades at the end of the working day to remove any fragments of hair and then apply a little clipper oil to the pivot point to prevent any corrosion around the fastening screw. This will prolong their life and stop them from binding or getting stiffer to use.

ACTIVITY

Note: this activity should be carried out with the electric clippers removed from the power supply or rechargeable clippers 'run down'.

Knowing how to remove, replace, realign and maintain a clipper's blades is an essential part of hairdressing. Get your supervisor to show you how the lower blade-retaining screws are undone and removed. This will give you access to both cutting blades and the area below the armature (the vibrating arm that works them) for cleaning purposes.

When you have dismantled the blades, check them for signs of corrosion. If a rusted area exists, it will look like blackened areas around the blade edges. If blades have been allowed to get to this stage they should be replaced by new ones, as their ability to cut cleanly without friction has been greatly reduced.

With the blades stripped down, you can now use clipper oil to lubricate the two blades, wiping any excess away. Now you can replace the blades (the right way up) and partially re-tighten the retaining screws. Finally, readjust the alignment of the blades and tighten the screws. Check the alignment once more: the lower cutting blade should extend around 3 mm further than the upper cutting blade with the clippers adjusted fully forwards to the shortest cutting length.

Hand them back to your supervisor so that your maintenance can be checked.

Neck brushes, water sprays and sectioning clips Neck brushes will remove loose hair clippings from around the neck and face. Get used to passing the neck brush to your client when you are cutting dry hair as the small fragments are irritating when they fall onto the face. Neck brushes usually have synthetic bristles and these are easily washed and dried before they are used.

Water sprays are used for damping down dry or dried hair, to assist you in controlling the haircut. Stale water is unhygienic, so make sure that the water is emptied out and refilled on a daily basis.

Sectioning clips are made from hard plastics or thin alloys. They are used to divide the hair and keep bulk out of the way while you work on other areas. They are sterilized by immersing them into Barbicide™ solution for the manufacturer's recommended length of time.

Cutting checklist

✓	Make sure that the client is protected adequately first.
✓	Always gain agreement before attempting anything new or different.
✓	Make sure you consider the reasons and the purpose for the style.
✓	Assess the style limitations, hair problems or physical features.
✓	Avoid technical jargon or style names. If technical terms are used, always clarify in simple terms to avoid confusion.
✓	Do not just do the style if you think that it is wrong. If there are reasons why you think it will be unsuitable, you will be doing the client a big favour in the longer term if you tackle the issue straight away.
✓	Always give the client some advice on how to handle the style themselves.
✓	Give the client an idea of how long it will take.

Creatively restyle women's hair

Accurate sectioning

In order to be able to manage sizeable amounts of hair at any one time, you must organize and plan the haircut. This becomes automatic as you think:

◆ How do I go about this?

◆ Where do I start?

◆ What is the finish going to be like?

Being organized is about working in a methodical way; it is the way in which you habitually start at the same point. Then you divide and secure all the rest of the hair out of the way, finish that bit and then take down the next part to work on, and so on. Each part or section that you work on should be small enough for you to cope with, without losing your way and continuing on blindly. It seems a strange term to use, but 'blindly' is exactly the right word. If the sections are too deep or too wide, you will not be able to see the cutting guide that you need to work to. Accurate sectioning guarantees that every section is cut to the same length every time.

> Slow down, take your time and make sure that you take precise, clean sections so that you can clearly see your guideline. The guide must be followed at all times until you fully understand about the structure of a haircut and then you can start to work creatively using more freehand techniques.
>
> *Erik Lander*

Cut hair with natural fall

Being aware of and cutting with natural fall is extremely important. Looking for the directions of growth within the hair is an essential part of consultation as it is within the execution of the haircut. If you work with the natural fall, i.e. partings, nape hair growth, double crowns, and so on, you will be compensating for these anomalies and be able to produce an easier to manage result.

ALWAYS REMEMBER

Cutting baselines/perimeter outlines

The baselines will determine the perimeter of the hairstyle, or part of the style, and may take different shapes according to the effects required. A baseline is a cut section of hair, which becomes a cutting guide for the following sections of hair. There may be one or more baselines cut: for example, a graduated nape baseline may be cut, another may be cut into the middle of the hair at the back of the head. Other baselines may be cut at the sides and the front of the head.

Symmetrical: the baseline for evenly balanced hair shapes in which the hair is divided equally on both sides of the head. Examples are hairstyles with central partings or with the hair swept backwards or forwards.

Asymmetrical: the baseline to be used where the hair is unevenly balanced, for example where there is a side parting and a larger volume of hair on one side of the head, or where the hair is swept off the face at one side with fullness of volume on the other.

Concave: the baseline may be cut curving inwards or downwards. The nape baseline, for example, may curve downwards.

Convex: the baseline may be cut curving upwards and outwards – the nape baseline, for instance, may be cut curving upwards.

Straight: the baseline may be cut straight across, for example, where you wish to produce a hard, square effect.

TOP TIP

As previously covered in the section on preparation, always try to wash the hair, even if it is only a dry haircut. This way when you dry the hair off, you will be able to see the natural fall far easier and be able to work with it.

Controlling the shape

Three things control the quality and accuracy of a haircut:

◆ *The holding angle* – the angle at which the hair is held out from the head.

◆ *The cutting angle* – the angle at which the scissors, razor, etc. cuts the hair.

◆ *The holding tension* – the even pressure applied to sections of hair when it is held.

Cutting lines Also known as perimeter lines, these are the outline shape created when layered hair is held directly out with tension (perpendicular) from the head. The curves and the angle in relation to the head, determine the shape of the cut style. The main ones are:

◆ the contour of the shape from top to bottom

◆ the contour of the shape around the head, side to side.

Cutting guides These are prepared sections of hair that control the quality of the haircut. When the cutting guide is held, and first cut, it is to this length and shape that all the other following sections relate. In preparing this cutting guide you need to take all the client's physical features and attributes into consideration, i.e. eyes, eyebrows, nose, bone structure, head shape, neck length, hairlines, etc.

Cutting the lines and angles Comb the hair and hold the sections with an even tension. The tension ensures that accuracy is maintained throughout the cut and the position in which you hold and cut the hair determines the position the cut sections take when combed back on the head. The angles and lines of cutting depend on the different lengths required by the style. The first cutting line – the outer perimeter line – may be related to the nape (when starting at the back). The second cutting line – the inner perimeter line – depends on the different lengths required throughout the style.

ACTIVITY

The table below contains a list of terms used within cutting. Copy out the table and then:

(a) give an explanation of each term

(b) find examples or draw a sketch of the effects.

Cutting term	What does this term mean?	Examples of the effect
Graduation		
Fading		
Reverse graduation		
Disconnection		

Cross-checking the cut You should always cross-check at different points during the cut. This is done by taking and holding sections in a different plane, i.e. at right angles (90°) to the original cut sections. In other words, if you originally cut the hair in vertical meshes, your cross-check will be done with horizontal meshes.

Cross-checking provides a final technique for checking the continuity and accuracy of the haircut. Where you find an imbalance in weight or extra length that still needs to be removed, it provides you with the opportunity to create the perfect finish.

Dealing with cutting problems

A lack of attention during the cut or a missed detail or aspect during the consultation can lead to cutting mistakes. The variety of mistakes are too varied to cover, but they normally result in an imbalance in weight proportions or a difference in perimeter lengths.

If, on finishing a symmetrical cut, you feel that one side seems to be slightly longer than the other, you need to stop before taking anything off the apparently longer side. If it is obviously longer and the client has commented too, you need to put your comb and scissors down and recheck through the fingers.

Standing behind the client, you take a small piece of hair from the same position, on either side, with your forefingers and thumbs. Then slowly slide your fingers down either side until you get to the ends. Looking at the length through the mirror, see if the ends are at the same length either side; if they are, the cut is fine. If, however, there is a difference in both lengths, then you now have the chance to redress the balance.

Clients are very aware of their fringe length, if a fringe is taken too short, the client feels very exposed. But you can reduce the effect by reducing a solid fringe line by slightly *point cutting* to 'break up' the density and reveal a little skin of the forehead through the hair.

What bothers people is the stark contrast between the solid line of the fringe and the skin; this focuses even more attention on the area. Therefore the solution is to reduce this contrast by softening the demarcation line. (This technique of reducing obvious mistakes can be used throughout the perimeter of hairstyles, as it works in most cases.)

Finally, another popular cutting fault is caused by an imbalance of weight in layering on one side compared with the other. Again, if you find that the layer pattern on one side seems different from the other side, stop. You need to find out if the fault is because:

◆ one side is longer than the other

◆ there is a greater reduction in weight by texturizing on one side rather than the other.

If it is length then you can easily re-cut the longer side to match. But if it is due to weight reduction, you will see a 'collapse' in the overall style shape on the side that has had the greater amount of texturizing. You can then remedy the fault by further texturizing the hair from the thicker side.

Cutting and styling techniques

Club cutting

This is the method of cutting hair bluntly straight across. It systematically cuts all the hair, at an angle parallel to the first and middle finger, to the same length. It is the most popular technique and often forms the basis or first part of a haircut, before other techniques are used.

Club cutting is used in both layered and one-length cuts and is therefore a particularly suitable choice for maintaining or creating bulk and volume. It is an ideal way to cut finer hair types or for use with people who have sparser (less dense) hair.

Holding with an even tension.

Cutting with the scissor blades parallel to the fingers.

A completed 'club cut'.

Freehand cutting

Freehand cutting is mainly used on straighter hair for creating the profile or perimeter shape. The technique is used on straighter hair because curlier hair needs more control, through holding and tensioning, if you are to make an accurate cut. As the name suggests, freehand cutting relies upon one hand holding and combing the hair into position, and the other controlling the scissors to make the cut. More often than not, when cutting longer, one-length hair, the comb is used to create the guide for making the cut. This technique is more widely used in cutting fringes. Adults with fringes are particularly cautious about what the exact finished length should be. Therefore, it is easier to comb the length into position and create a profile shape that both suits the client and follows or covers the eyebrows. This would be guesswork if the hair were held between the fingers and cut, because the width of your fingers would obscure the exact length and position, you are trying to cut.

Freehand cutting using the comb to hold the hair.

Rechecking the accuracy of the line.

Always have an open mind to new cutting techniques, as there are so many different ways to achieve the result. If you don't achieve the desired effect then you need to start looking at your techniques and analyse why.

Erik Lander

Scissor over comb

This technique has been traditionally a barbering technique. In recent years, there has been a move in hairdressing generally towards easier-to-manage hairstyles, so therefore this technique is now widely used in hairdressing for cutting short styles on both men and women.

Scissor over comb cutting is ideal for producing contoured, layered shapes and close-cut, 'faded-out' perimeters. Faded or graduated perimeters have no set cut length (i.e. baseline); they rely upon the hairline profile and are graduated out from that into the rest of the hairstyle. The technique is used with either wet or dry hair as it uses the comb as a guide instead of the fingers.

Clipper over comb

Clipper over comb involves exactly the same technique as scissor over comb; instead of using scissors, the clippers whisk away the hair. Again, this technique enables you to cut hair far closer than you would be able to if you were holding it between your fingers.

Thinning

Thinning can be done with scissors or a razor and can be used for reducing or tapering bulk from thicker hair without reducing the overall length, or as a way of texturizing the profile of hairstyles to remove lines and angular shapes to create softer, faded, more ambiguous effects.

ALWAYS REMEMBER

Cutting combs used for scissor over comb need to be very flexible. You may want to buy a barbering, tapered comb as this makes cutting easier and quicker. This does not apply to clipper over comb techniques. Because of the weight and bulkiness of electric clippers, it is easier to use a standard parallel cutting comb.

TOP TIP

Hair grows at an average rate of 12.5 mm per month.

Point cutting with thinning scissors.

Removing the bulk from the ends.

Maintaining the same cutting angle throughout the technique.

Finally removing any definition to the previously club-cut lengths.

Razor tapering.

Removing weight from the ends.

The tapered effect.

Texturizing technique no. 1

Texturizing by **point cutting** or **pointing** is a technique where the angle of the cut changes to become almost parallel with the held hair. It is a way of reducing lines and bulk from the ends (1–2 cm) of the hair in order to create softer, more textured edges. It uses the point ends of the scissors and is more successful on straighter hair than wavy; it does not add any value to curly hairstyles at all. If curlier hair is point cut it can often make it more difficult for the client to manage: the hair would lose perimeter density and the curl would increase, making the hair fluffier.

Note the scissor position, parallel to the held lengths.

Point cutting to 'shatter' the club-cut lengths.

Continue the technique until the weight is removed.

Texturizing technique no. 2

Texturizing by **brick cutting** is similar to point cutting in that it only uses the point ends of the scissor blades, but is intended to remove 'fine chunks' of hair from the mid-length and nearer the root of hand-held sections. Its main advantage is when cutting shorter hair to create stiffer sections that support the outer hair perimeter shape. Put another way, it can produce volume while creating spikier edges.

Removing mid-length hair with the point of the scissors.

Change positions within the length to avoid any cutting lines.

Weight reduces quickly.

The tapered effect.

Texturizing technique no. 3

Texturizing by **slicing** is a technique for either very sharp scissors or razors. Slicing will produce a tapering effect in a hair section without reducing the overall length. It is always done with the hair held at an angle slightly downwards. The scissors or razor is introduced to the hair nearer the root and then, in one continuous and angled, downward motion, it takes a longer slice out and towards the ends of the held hair.

The finished effects will produce 'shattered' looks with irregular, tousled appearances that can be dressed with product to create texture and definition.

Positioning the scissors safely away from the fingers.

Scoop away the hair without totally closing the blades.

The final part of the movement.

Holding the section.

Offer the razor to the hair almost parallel to the angle of the hair.

Draw down the razor, through the held section.

Continue the action further down the section.

The tapering effect achieved by the razor.

Disconnection

Disconnection creates a demarcation area between two different levels. It is a deliberate style or styling feature that draws attention to the different levels and can be done on all one-length hair or layered effects. Typical examples for this would be a classic bob with a fringe; in this style, where the fringe stops, the sides continue down to the perimeter length creating a 'step' in the hairstyle. But this is not the only type of disconnection: a contemporary bob can have deliberate steps within the sides, this could be to draw attention to cheekbones or a jaw-line. Other examples of disconnection are regularly used on short, layered styles that have longer hair over-falling the shorter perimeters.

ALWAYS REMEMBER

One-length cuts, layering and graduation

These techniques create a comprehensive palette or toolbox of skills that are used to create all of the final cutting effects. You choose the *tools* to create the effect that you want.

A *one-length cut* is achieved by combing vertically down and cutting straight across. A *graduation* or *graduated cut* is achieved by creating a layered shape that is longer on the upper sections and shorter underneath.

A *reverse graduation* is achieved by creating a layered shape that is shorter on the upper sections and longer underneath.

> When cutting you must pay attention to how you comb the hair, e.g. combing it flat, full, left, right, up or down, as this affects how the hair will lie. Also, looking closely at how the hair lies will tell you if the hair is longer or shorter in certain areas, the outer contour or profile of the cut shape is critical for things such as balance, asymmetry, length and the overall style.
>
> *Erik Lander*

Combining the techniques

None of the techniques already mentioned are used in isolation by the creative cutter. A precision or artistic cutter will use a combination of these cutting disciplines to create a finished effect. As you become more experienced in your work you will probably find that you would need to use at least 30–50 per cent of these in each haircut in order to achieve satisfactory fashionable results.

Sources for more information on cutting

You need to keep abreast of what is happening in fashion and creative circles: style magazines, websites on the Internet, and music and TV are all sources for what is happening now, but that is only part of the story.

Fashion predictions and forecasts are created by clothes designers, and these are made well in advance of the current season. Each designer shows their collections up to 12 months in advance. This provides a good source of information for you, as the designers are always attempting to create a **total look** involving the hair and hair styling too.

Fashions have to start somewhere too. Every new theme can be tracked back to images, textures and differing cultures from the past. Historical collections in museums are a good source for new ideas, as well as the Internet. Look for the themes created by tribal and ethnic cultures too; many of these such as **dreadlocks**, **corn rows** and etched scalp patterns may be commonplace now but have their roots definitely cemented in the past.

ALWAYS REMEMBER

The most important factor in cutting hair

Very few people can do two things at the same time. You will have to learn very quickly that you need to hold a conversation with the client without losing your way and concentrating on the haircut. This is the biggest single cause of poor quality hairdressing. (If you cannot do this yet, tell your client that you need the time to focus on the task ahead.)

ACTIVITY

The table below contains a list of terms used within cutting. For each one give an explanation of the term and what effect they achieve within a haircut.

Cutting term	What does it describe?	What effect(s) does it achieve?
Pointing		
Chipping		
Slider/slice cutting		
Feathering		

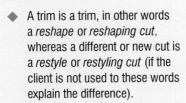

TOP TIP

Always ask how much the client wants cut off.

- A trim is a trim, in other words a *reshape* or *reshaping cut*, whereas a different or new cut is a *restyle* or *restyling cut* (if the client is not used to these words explain the difference).

- An inch to one person may be a centimetre to another. Do not be ambiguous; find out exactly how much needs to be cut off!

Checklists

Before the cut

✓	Communicate with your client and discuss the requirements.
✓	Use visual aids to help interpret the client's wishes or to show ideas and themes.
✓	Examine the hair – its type, length, quality, quantity and condition. Look for factors that influence the choice of style and cutting methods.
✓	Explain if there are any limitations that will affect the result.
✓	After your analysis, agree or negotiate with the client the suitable courses of action to take.
✓	Try to show the hair length to be removed.
✓	Discuss the time that will be taken and the price that you will charge.
✓	Proceed only when all checks have been made and the client has agreed to your proposals.
✓	Ensure that you choose the correct tools and techniques for achieving the variety of effects.

TOP TIP

Far more customers are dissatisfied as a result of the stylist not listening and taking too much off than because of poor or inaccurate haircuts.

During the cut

✓	After shampooing and towel drying, dry off the hair so that any previously masked tendencies can clearly be seen.
✓	Try to keep the hair damp but not saturated so that any newly added technical features can easily be seen.
✓	Take care with your precision or accuracy by checking each angle at which the hair is taken and held from the head.
✓	Create your baselines and guideline cuts first, so that there is continuity within the section patterning.
✓	When preparing baselines and guide sections, make sure that you attend to the features of your client's face and head. Use these as guides for accurate directions in the cut lines.
✓	Remember always that the first cuts you make often determine the finished shape of the style.

After the cut

✓	Cross-check each of the sections of the side, nape, top and front for accuracy and finish.
✓	Check the density, texture and features of the haircut.
✓	Position, place and mould the hair where necessary to see the shape clearly.
✓	When all the loose hair clippings have been removed and the client is prepared and comfortable, continue to blow-dry, set and finish the style.

STEP-BY-STEP: GRADUATED LAYERS – RESTYLE LONG (BEYOND THE SHOULDERS)

1 After greeting your client, conduct a full consultation to determine the style, length and angle of the layers that you will cut.

2 While the hair is wet, create a zigzag parting from recession to recession and secure on top with clips.

3 Secure the remaining hair either in clips or plaits.

4 Starting at the crown, comb the hair out following the head shape, 90 degrees to over directed as desired.

5 Take radial sections through the back and repeat the method.

Use a point cutting technique for seamless layering and ensure the length is shorter at the crown and longer at the perimeter.

6 Visually check the length of the back layers by holding the hair out from the same place on either side of the head.

7 Release the side sections and work in clean vertical sections using the back sections as a guide.

Adapt your technique to suit the hair type and texture, over direct if you wish to maintain length in the front.

8 Visually check the length of the front layers by combing the hair up from the same place on either side of the head, texturize as required.

9 Release all the hair and comb, use slicing techniques to visually blend layers.

10 To shape the hair at the sides, determine where the shortest length is going to sit and work downwards to connect to the base line. Direct the hair forwards if you wish to maintain length.

11 Apply styling products and blow-dry the client's hair into the desired style. Slicing, channelling and deep point cutting techniques can be used to soften and texturize the hair when it is dry.

12 Finished disconnected long layers.

CourseMate video: Graduated layers haircut

STEP-BY-STEP: CONTEMPORARY ASYMMETRIC CUT

1 After greeting your client, conduct a full consultation to determine the style, length and angle of the layers that you will cut.

Decide where the parting should lie, confirm this with the client and place it accurately.

2 Create a clean section from the parting just below the crown to the lower opposite side of occipital bone. (This will be the shorter side of the asymmetric style).

Section the rest of hair and clip it out of the way.

3 Cut the hair using a short graduation technique. If the cut is not graduated, adapt the technique to suit the style.

4 Take a radial section and use uniform layering to blend the graduation, adapting the technique to suit.

5 Take the front long section and over direct it to the back, removing the required length.

6 Use texturizing techniques to visually blend the style, channelling, point cutting and slicing.

7 Personalise the style when dry by shaping the fringe and perimeter, then cross check for accuracy.

8 The asymmetric cut is very striking when finished.

CourseMate video: Contemporary asymmetric haircut

STEP-BY-STEP: CLASSIC INVERTED BOB

1 After greeting your client, conduct a full consultation to determine the style, length and angle of the bob that you will cut.

2 Create a centre profile section featuring a diagonal forward section at the nape.

3 Section a vertical guide approximately one inch wide and hold it at a 45 degree angle to determine the degree of graduation.

4 Continue by cutting clean vertical sections throughout the nape.

5 Continue to work up to the occipital bone, as far up as the style requires.

6 Create a horseshoe section from temple to temple, checking the correct length by using already cut hair as a guide.

7 Determine the angle of the hair and place guide, dropping forwards to the front sections and angling as required.

8 Blend the weight line once the cut is dry using a deep point cutting technique, slicing or channel cutting.

9 The finished inverted bob is a real classic.

CourseMate video: Classic inverted bob haircut

Provide aftercare advice

Good service is provided through good advice and recommendation. The work that you do in the salon needs to be reflected by the client who can achieve and maintain the same effects at home.

Home and aftercare checklist

✓	Talk through the style as you work; that way the client sees how you handle different aspects of the look.
✓	Show and recommend the products/equipment that you use so that the client gets the right things to enable them to get the same effects.
✓	Tell the client how long the style can be expected to last and when they need to return for reshaping.
✓	Demonstrate the techniques that you use so they can achieve that salon hair look too.

> **TOP TIP**
>
> Make a point of talking through your styling techniques as you go as this is really useful to the client who gets useful advice on how to recreate a similar effect at home.

Styles need to suit their purposes and you need to bear this in mind when talking with the client. If the client's leisure routine or lifestyle is going to have an impact upon their hairstyle then you should make a point of telling them this. If they work in a hot, steamy kitchen then their hair is likely to go rather limp when they take their head covering off. Similarly, someone who does a lot of sport is constantly in and out of the shower, so his or her hairstyle needs to reflect this. Sometimes, if the hair is long it can easily be tied back out of the way, but if it is short you need to make some other suggestions.

Show and recommend the products and equipment you use

As you talk about the ways in which you have styled the hair, make a point of talking through the products that you have used as well. You know by experience the products that you would use on the client's hair to achieve different results and you also know the ones that you would avoid. Make sure to tell the client too, because they have not had the benefit of your training and they do not know.

So when you use a particular product, let them have a closer look. This way they get to see, smell and feel the product too and, subconsciously, this has a powerful effect on them. By doing this, you are involving the client in what you are doing and giving them a greater experience of the service. They will be able to see a direct link between the effects that you are achieving on their hair, with the added benefits of buying those particular products that will help them to recreate a similar effect.

Explain how routine styling tools can have detrimental effects

Only hair in good condition is easy to maintain. You know how difficult it is to make dry, damaged hair look good. Your clients can recognize the difference between good and poor condition and, given the choice, they will always choose hair that has lustre, shine, flexibility and strength.

So you need to warn them of the pitfalls of repeatedly using hot styling equipment. Ask if they use them at home and if they say that they use straighteners or tongs on a daily basis then tell them about the benefits of using heat protection sprays.

Demonstrate the techniques that you use Clients want to be able to recreate the effects that you achieve in the salon and this is your chance to show them how to do it. Show them how to do such things as correct combing, blow-drying or positioning of brushes. We have all seen the effects when these are not done properly, so make a point of giving them a few tips on how they can achieve a similar result for themselves and how long they can expect it to last.

> " You will get out of this industry exactly what you put in. It is the most amazing industry to work in, but it's a very fast moving one and this means that you have to constantly move forward and keep up to date with what's current. Remember you never stop learning: education, education, education!

Erik Lander

REVISION QUESTIONS

Q1.	Accuracy is achieved by _____ and cutting the hair at the correct angle.	Fill in the blank
Q2.	A razor should be used on dry hair.	True or false?
Q3.	Select the texturizing techniques from the following list.	Multi-selection

Club cutting	☐	1
Graduation	☐	2
Slice cutting	☐	3
Fading	☐	4
Point cutting	☐	5
Chipping	☐	6

Q4. Symmetrical styles produce outline shapes that are equally balanced.   True or false?

Q5. Which of the following is not a cutting term?   Multi-selection

     Cross-checking   O   a

     Thinning   O   b

     Free hand   O   c

     Free style   O   d

Q6. Disconnection is a term defining a continuous outline shape or layer patterning.   True or false?

Q7. Which of the following hair growth patterns does not affect the way that hair lays after it is cut?   Multi-selection

     Nape whorl   ☐   1

     Double crown   ☐   2

     Widow's peak   ☐   3

     Low hairline   ☐   4

     Cow lick   ☐   5

     High hairline   ☐   6

Q8. A _____ line or perimeter outline forms a guide line for a bob-shaped haircut.   Fill in the blank

Q9. Which of the following cuts would easily describe a disconnection?   Multi-choice

     Graduation in a long hairstyle   O   a

     Reverse graduation in a long hairstyle   O   b

     A fringe in a shoulder-length bob style   O   c

     Texturizing in a short cropped style   O   d

Q10. 'Personalizing' is the term that refers to any technique that is used to complete a style, tailoring it to the client's specific needs.   True or false?

Notes

5 Creative Barbering

LEARNING OBJECTIVES

◆ Be able to work safely when cutting men's hair.

◆ Be able to creatively restyle men's hair.

◆ Understand how to use a variety of techniques to create different effects.

◆ Be able to communicate professionally and provide aftercare advice.

KEY TERMS

blending

blunt cutting

disconnection

fading

graduation

male pattern baldness (MPB)

scissor over comb

UNIT TOPIC

Creatively cut hair using a combination of barbering techniques

INTRODUCTION

Over the past 20 to 30 years, contemporary men's hairstyling in Britain has been based around suitability and purpose rather than fashion. This has meant that the decisions that men make about their hair reflect what they do in their working lives, how they spend their leisure time and how much hair they have. However, this has started to change. Hair and hairdressing have always been an essential part of women's appearance and have now become a major interest for men too.

Creative barbering

PRACTICAL SKILLS

Learn how to work with care, attention and safety at all times

Look for things that will influence the ways that you cut your client's hair

Learn how to work with different types of hair

Learn how to achieve a variety of creative effects and styles

Learn how to work with, or correct, simple faults and common problems when cutting

Provide aftercare advice to your clients

UNDERPINNING KNOWLEDGE

Know how to work safely during cutting

Understand how different cutting techniques achieve different effects

Know the common problems that can affect the way that you cut hair

Know how the client's physical aspects and features can affect styling options

Understand how different cutting techniques achieve different effects

Know how to communicate effectively and professionally

Maintain effective and safe methods of working when cutting

Beyond basic barbering

Creative barbering at Level 3 uses all the cutting techniques available within the salon, so there is a lot to learn. Your experience with clients will enable you to offer far more than the basic 'quick cut' services and you should optimize these opportunities by using creative cutting techniques as a 'springboard' to a range of other exciting and profitable styling services.

INDUSTRY ROLE MODEL

KENYON YATES International Barbering Tutor, Assessor and Demonstrator

I left school at 15 with a very low opinion of my intelligence. I didn't consider barbering at all until my barber suggested it! He asked his boss if he would consider me as an apprentice. His boss took a long look at me and said 'No not really', so that made me decide – right I'll show you! In the early 1960s jobs were very easy to come by and I could have tried my hand at a job a day. However I stuck to it and gradually developed the skills. Then along came the Beatles and not only shook the world of music but blew barbering (artistically) out of sight for 30 or so years.

ALWAYS REMEMBER

A finished hairstyle always looks better if it is cut wet. Try to educate your own clients into booking for a wet cut at least.

◆ It will be easier for you to create new effects.

◆ Clients will be able to see a better, more professional result.

◆ You will generate a better professional service with your clients.

◆ It will raise expectations so that men do not expect merely a quick, cheap haircut.

◆ It is more hygienic for everyone concerned.

The preparatory outcome for this unit is very similar to other technical services, although there are variations:

◆ Specific differences for male and female clients.

◆ Policy at the barber's shop where you work.

◆ Different tools and equipment used.

As client protection is the first aspect that you must consider, a quick reference for cutting dry or wet is listed below along with the essential knowledge components.

Protect the client

Gowning the client This can take place before or after the consultation; it really depends upon the salon. In either event, client protection must always take place before starting the cut.

For dry cutting	Use a clean fresh cutting gown and put it on your client while he is sitting at the styling location. Make sure that the back is fastened and that any open, free edges are closed together, keeping any loose clippings away from the client's clothes. Place a cutting collar around his neck to ensure that any bumps or lumps in his clothing don't present any false, physical baselines for the haircut and that the collar edges fit snugly against the neck, so that there are no irritating hair fragments that will leave the client itching until he gets home.
For wet cutting	Gown the client as for dry cutting, but when your client is at the basin, place a clean fresh towel around his shoulders before positioning him back carefully and comfortably. Make sure that the basin supports the client's neck properly and that the flanged edges of the basin nestle comfortably on to the client's shoulders which are protected from any spills or seepage by a clean fresh towel.

As is the case with other technical services in hairdressing, it is essential that you work safely when cutting hair. In doing this, you must take the time to prepare and protect the client adequately.

This means:

◆ Finding all the equipment that you need, such as gowns, towels, combs, scissors, razor, clippers, etc.

◆ Checking that they are prepared for use, e.g. new blades for the razor, freshly laundered towels and gowns, washed cutting collars, cleaned and sterilized combs, brushes, clipper blades and scissors.

◆ Having them all at hand at the work station and ready for use.

◆ Ensuring that the client is comfortable and in a position where you can work safely.

Sharps box for the safe disposal of used razor blades

ACTIVITY

Every salon has its own way of doing things. Note down your salon's code of practice in respect of:

(a) meeting and greeting clients

(b) gowning

(c) maintaining tools and equipment

(d) disposal of sharps

(e) hygiene and preventing the spread of infection or infestation

(f) expected standards of service.

Disposal of waste and sharps

Barbers use many chemicals, such as shampoos, conditioners and styling products, and they are not necessarily potential hazards. In fact, much of the other chemical waste created is simply rinsed down the sink with lots of warm water.

However, sharp items such as disposable razor blades do need handling with extreme care. Used blades must be disposed of carefully to prevent any injury or cross-infection to others. After use, they should be put in the sharps box.

ACTIVITY

Find out about and note down: your salon's policy and that of your local authority for the safe disposal of sharp items.

ALWAYS REMEMBER

Disposal of sharp items

Used razor blades and similar items should be placed into a safe container ('sharps box'). This type of salon waste should be kept away from general salon waste. as special disposal arrangements are required and may be provided by the local authority.

Fixed blade razors Different local authorities have environmental health bylaws that apply to their region alone. Some authorities do not permit the use of fixed blade razors in barber shops / salons, as they are considered to be a potential risk to public health.

Remove product build-up

You cannot cut hair well if it is loaded with hairspray or it has product build-up. If the client uses a lot of finishing products on his hair, you will need to make sure that this is clean before you start. You should be able to comb the hair freely during sectioning so that you can achieve the correct holding angles and cutting angles without tangles or binding.

See **CHAPTER 15** Health and Safety for more information.

Preparation checklist

✓	Make sure that the styling section and chair is clean, safe, and ready to receive clients.
✓	Make sure that the seat is lowered, providing easier access for the clients whether they are young, old or with physical constraints.
✓	Make sure that the client is well protected with a clean fresh gown and a close-fitting cutting collar.
✓	Find out what the client wants. Men can often be more difficult during consultation as they are often reluctant to use a technical term that they are not sure about or express themselves clearly to people they don't know (see the section on communication below).
✓	Style books/files provide many male looks to help the diagnostic process.
✓	Make sure you consider the reasons and the purpose for the style. Hairstyles required for professional purposes have more restrictions on freedom and expressions than fashionable, trendy looks or more general wear.
✓	Assess the styling limitations – hair and skin problems or physical features.
✓	Avoid technical jargon or style names; if you do use them, always clarify in simple terms what you mean to avoid confusion; this will help to educate your clients for the future.
✓	Do not carry out a style if you think that it is wrong. If there are reasons why you think it will be unsuitable, it is best to tell the client straightaway.
✓	Always give some advice on how to maintain their hairstyle; men often need products to help them achieve similar effects themselves. Make sure you show them how they can use and apply any new product at home to maintain their own hair/skin condition or styling effect.
✓	Give them an idea of how long it will last and remember to re-book their next appointment before they leave. Alternatively, if they prefer just to pop in on the off chance, tell them when they should expect to revisit.

Preparation for cutting

Consultation

Clear communication Good communication is the key to successful barbering. During the consultation with the client, avoid jargon or technical terms as this might lead to misunderstandings and unexpected results.

Use visual aids Some men are not good at expressing what they want; they need your help in finding the right solutions that suit them and their needs. Try to find different ways of explaining a style or use pictures to show what you mean. Often a pictorial illustration will express far more than just a cutting style or technique. It creates an overall finished impression too.

Gain confirmation throughout the service Make a habit of summarizing the main points as you go through the consultation by asking:

◆ 'So you would like to keep the overall length at the back although you don't mind it shorter around the ears?'

◆ 'Do you want me to reduce the thickness in the sides so that it isn't so wide?'

◆ 'You want to keep the parting where it is because you find it doesn't lie very well elsewhere?'

It is also worth remembering that other people have a huge impact on the styles that we choose and that includes your male clients. Some men have no real opinions about their self-image and this makes it difficult to work out what they mean.

Intelligent communication Clear communication goes beyond speech, because as well as using words we show our interest, attitude and feelings through our body language. We express our innermost feelings via our posture, our eyes and mannerisms. Collectively, these can be *saying* something quite different from our verbal language. As an experienced barber, you will be well aware of the issues addressed here; you will have seen this daily in a busy salon.

TOP TIP

As many men have shorter hair, their nape hair growth patterns will have far more impact on what hairstyle you choose.

ALWAYS REMEMBER

Personal hygiene is especially important to hairdressers. You work in close proximity to the client so make sure that you shower each day, keep your hands and nails scrupulously clean and brush your teeth after eating.

ACTIVITY

Complete the table below to explain why each of these consultation factors are important and what might happen if these aspects are ignored.

Factors	Why is this important?	What might happen if this is ignored?
Client's requirements		
Hair growth patterns		
Hair texture		
Hair density		
Lifestyle		
Physical features		

Influencing factors affecting style choice

Male pattern baldness Your consultation will cover a wide variety of factors that influence what happens next and **male pattern baldness (MPB)** or the early signs of it should be high upon the list of things to look for.

MPB is a balding or thinning condition. The cause for it is due to high levels of the male hormone testosterone within the body. Many treatments have been developed with little or no long-term remedial effects. Hair transplants have been a possible option in the past, but this type of treatment is expensive and needs a lot of upkeep.

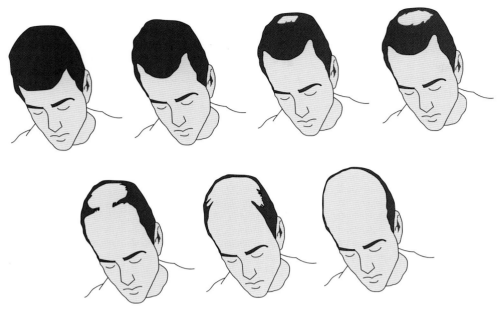

Consultation guide for MPB type 1

Consultation guide for MPB type 2

Depending on the stage of the MPB, you need to find out how your client feels about it. If the hair loss is relatively slow, there is no need to rush immediately for the clippers and a grade 2. There could be some considerable time before the condition requires a focused attention and so you need to provide advice and reassurance with a range of styling alternatives.

If, however, the MPB is at an advanced state then it is obviously going to impact on what styles are achievable. For example, if there is a significant general thinning or hair loss on top (MPB type 1) then your styling options are far more limited than if MPB is only apparent in the recession area around the forehead (MPB type 2).

If your client wears a hairpiece, you must account for this in your styling. Obviously, there has to be some **blending** between the natural, remaining hair and the added hair. Be careful not to leave the hair either too long or too short around the blending area. If there is any imbalance in lengths between the two, it is definitely going to show.

It is also important to consider facial shapes and physical features when choosing a style.

> ### ALWAYS REMEMBER
>
> Many men elect to have very short hair when faced with thinning or bald areas on the scalp. Provide them with other alternatives if you can; let them see the benefits of styling their hair in different ways.

Facial shapes

Facial shape	How best to work with it
Square and oblong facial shapes	Square and oblong are typically masculine and provide a perfect base for traditional classic well-groomed looks on shorter hair. These facial shapes have less impact on longer men's hairstyles.
Round faces	If shorter, classic styles are required, then a round face is improved by the introduction of angular or linear perimeters. Conversely, if the hair is to be worn longer the roundness of the face will be reduced, as more will be covered.

(Continued)

Facial shape	How best to work with it
Square angular features, jaw, forehead, etc.	Generally accepted as a feature of masculinity, they do not really pose any limitations for classic type work. They work well with longer hair too. Squarer, angular features are softened with beards and moustaches.
Flatter head at the back	Are improved by contoured graduation; this creates shaping and tapering that is missing from having a flatter occipital bone Sometimes the head is both flat and wide and this can make the problem harder to deal with. Wider, flatter heads are less noticeable with longer hair; if this is not possible then explain what the effect will look like if taken very short.

Physical features

Physical feature	How best to work with it
Prominent nose	Hair taken back away from the face accentuates this feature, whilst hair around the face and forehead tends to diminish the feature.
Protruding ears	Are better left covered rather than exposed. Sufficient hair should be left to cover and extend beyond their length if at all possible.
Narrow foreheads	Are disguised by softer fringes and side partings, whereas they are made more obvious by hair taken back away from the face.
Bushy, thick eyebrows	If the eyebrows are a different colour to the natural hair, the feature will be even more prominent. The hairstyle will be improved with some form of light trimming and grooming.
Large faces	Are framed by classic, short-cropped or sculpted hair.
Shapes of glasses	Generally, people will have had assistance in the selection and suitability of their frames. Therefore they should not work against the hairstyle that you want to create as they should already suit the shape and size of the face.
Long side-burns	Are made more prominent with shorter hairstyles, make sure that this is acceptable to your client first as he might be rather attached to his facial hair feature.
Beards	Make sure that the client's beard is still going to be balanced to the amount of hair on top of his head in the finished effect. If an imbalance is going to occur, mention it first and give him the option of taking the beard shorter to compensate.

For more on this see CHAPTER 4 Creative Cutting, page 85.

Ingrowing hairs are an uncomfortable aspect of closely cut wavy or curly hair.

Examination of hair and scalp

While you are looking at the client's hair and scalp, be particularly aware of the texture of the hair. If it is coarse and tightly curled, you will need stronger combs to stretch the hair out from the head before cutting, and firmer movements will need to be applied. The density of the hair is important too: if it is thick, then styles with varied hair lengths are possible. Conversely, sparse hair, particularly if it is fine, requires a great deal of attention and expertise. If finely textured hair has to cover sparse areas of the head, it will have to be longer than hair of coarser texture. The amount, type and growth patterns of hair are all-important too. Younger men may have distinctly higher forehead hairlines than women of similar age.

Hair growth patterns

Hair growth pattern	How best to work with it
Thinning hair/baldness	Younger men can be quite sensitive about thinning hair. Be tactful and try to find solutions that are realistic and sympathetic to the problem. When dealing with male pattern baldness most men will tend to opt for shorter 'close cropped' or clippered hairstyles rather than long.
Double crown	A double crown is particularly problematic when it is cut short. It sticks upwards, and will not lie down until it grows longer. This problem can be lessened if the hair is left longer to over-fall the opposing growth movement and sometimes benefits for a little thinning if the hair is thick.
Nape whorls	Nape whorls can occur on either or both sides of the nape; the movement caused by this growth pattern forces the hair to flatten and move towards the centre. This growth pattern is not necessarily a problem for short clippered lengths or longer over-falling hair, it affects shorter layered shapes that need to keep a perimeter baseline.
Cowlick	A cowlick appears at the hairline at the front of the head. It makes cutting a straight fringe difficult, particularly on fine hair, because the hair often forms a natural parting. This strong movement can often be improved by moving the parting over so that the weight over-falls the growth pattern. Sometimes a fringe can be achieved by leaving the layers longer so that they weigh down the hair.

(Continued)

Hair growth pattern	How best to work with it
Widow's peak	The widow's peak is a hair growth pattern that appears at the centre of the front hairline. The outline shape protrudes downwards in a 'v' shape and the hair grows upwards and forwards, forming a strong peak. It is often better to cut the hair into styles that are dressed back from the face, as any light fringes will separate around this area.

ALWAYS REMEMBER

Always look for contra-indications for styling and shaping; these could relate to infections, infestations, poor hair/skin condition, difficult hair growth patterns, face shape, and physical features.

ACTIVITY

The table below contains a list of terms used within cutting.

(a) Give an explanation of each term.

(b) Find examples or draw a sketch of the effects.

Cutting term	What does this term mean?	Examples of the effect
Texturizing		
Fading		
Reverse graduation		
Disconnection		

Finding out what the client wants

Finding out what the client wants is fundamental to achieving a satisfactory result. You need to consider factors such as practicality, suitability and the client's ability to cope with his hair. The final effects are influenced by the following:

◆ amount of hair

◆ distribution of the hair over the scalp

◆ texture of the hair

◆ condition of the hair and scalp

◆ tendency of the hair, i.e. the amount of wave or curl.

ALWAYS REMEMBER

How often do you find people using the wrong expression or term to explain what they want? Always make a point of correcting misused terms; it will show that you:

◆ listen to the client and you are hearing what he says

◆ are a highly skilled and knowledgeable stylist

◆ have pride and professional interest in your work.

Look out for	Why is it a concern?
Hair density	Scalps with densely populated hair can always be reduced, thinned or controlled in some way. Whereas thinner hair or male pattern baldness create a range of limitations that you will need to both express and contend with.
Hair tendency	Curly hair has more styling limitations than straighter hair. Wavy hair is always easier to direct or position than straight hair. Point these factors out before you start.
Hair texture	Fine hair is always difficult to handle, whereas coarser hair, when straight, will often appear spiky or blunt. Conversely, coarse, wavy hair can often appear dry regardless of natural condition. Each hair texture type creates a different problem.

Although men can wear longer hair as well as short, a whole range of modern contemporary styling effects has developed since the basic and traditional short back and sides. The application of hair products will often 'dress up' an otherwise professional- or classic-looking hairstyle, turning it into something with a more distinctive 'fashion look' for social and special occasions.

Now and again, a men's named style becomes fashionable. Some of these names, such as 'crew cut', the 'mullet', the 'wedge' or a 'mohawk', have passed into the general vocabulary. Always make sure that you know what your client means if he uses a name to describe a style; remember, it may be completely different from your idea of that style.

> " Remember, clients come from all age groups, races, cultures and all have characters. Thus, to ensure continuation of employability, the barber must be versatile and up-to-date with all male hair fashions, skills and knowledge. Be aware of influences and changes in male hairstyle fashion. Male hair fashion usually changes much more slowly than ladies hair fashion.
>
> *Kenton Yates*

Cutting tools and equipment

Good care and regular maintenance of your tools form an essential part of hairdressing and barbering. This covers:

◆ scissors

◆ thinning scissors

◆ combs, neck brushes and sectioning clips.

More information on scissors, etc. can be found in CHAPTER 4 Creative Cutting, page 88.

As clippers and razors are mainly associated with men's barbering techniques the relevant information is below.

Clippers The electric clippers cut hair by oscillation: the side-to-side movement of an upper metal blade passing over a lower rigid or fixed one. On each pass of the upper blade, the hair caught between the teeth of the lower blade is cut and falls away.

Regular cleaning and lubrication will prolong the useful life of the blades and keep the cutting edges sharp. Without this care the constant friction of one blade passing over another will affect their ability to work properly, i.e. electric clippers generate quite a lot of heat and, if they have not been maintained, their ability to cut cleanly and efficiently deteriorates over time. New blades are relatively expensive, as they can often cost half the price of a new pair of clippers. If the clipper blades are unable to cut keenly you will not be able to trim, shape and style neck or facial hair shapes accurately.

You should always take care not to drop them, as this can easily cause damage to the cutting teeth or even break. Any missing areas of teeth along the blades will be extremely dangerous and could easily cut the client if they were used. So, when they are not in use, hang them up out of the way or replace them back in the charger unit.

Clipper blades should always be checked for alignment before each time they are used. The fixed lower blade is adjustable and this allows for small adjustments to be made backwards, forwards or even side to side.

Loosening the small retaining screws underneath allows the blades to be adjusted. This also provides access to the upper blade, for removal, cleaning out the fragments of hair and essential oiling/lubrication.

When the blades are replaced, the retaining screws must be retightened properly, if this is not done, the vibration will dislodge the alignment, and this could easily take a chunk out of your client's hair, or worse, even cut him.

Well-maintained clippers will cut wet or dry hair with equal ease, although many stylists prefer to remove the hair first, and then wash the hair to remove any small fragments and make any final checks.

Razors The open or 'cut-throat' style razor used in shaving is made from a single steel blade, which is hinged and closed into a protective handle. The modern counterpart for this has disposable blades which can be removed and put into a sharps box after use.

The razor used for hair styling is called a shaper; it too has disposable blades, which are fitted into a hinged sheath that provides a handy, safe styling tool. Razor cutting is carried out on wet hair and with sharp blades. This is because of the way in which razors are used and the angle at which they cut through the hair; razoring should never be done on dry hair as this it will pull and tear the hair; causing it to split, even if the blades are new.

TOP TIP

Always take care when using any sharp items of equipment. Your safety and the safety of the client are in your hands; take care not to be distracted while you are working.

Types of razor

Type of razor	Description
Open (cut-throat) razor	This razor has a fixed/rigid blade that folds into its handle for safety. The blade is kept keen by regular stropping and honing, and must be sterilized on each use between clients.
Safety razor	This razor simulates the shape and feel of the open razor with disposable blades, which makes it more hygienic as blades can be replaced for each client.
Shaper	This is a popular razor with disposable blades that is used for cutting and styling hair but cannot be used for shaving.

ACTIVITY

Clipper maintenance

Note: This activity should be carried out with the electric clippers removed from the power supply or rechargeable clippers 'run down'.

Knowing how to remove, replace, realign and maintain a clipper's blades is an essential part of hairdressing. Get your supervisor to show you how the lower blade-retaining screws are undone and removed. This will give you access to both cutting blades and the area below the armature (the vibrating arm that works them) for cleaning purposes.

When you have dismantled the blades, check for signs of corrosion. If a rusted area exists, it will look like blackened areas around the blade edges. If blades are allowed to get to this stage, they should be replaced by new ones, as their ability to cut cleanly without friction has been greatly reduced.

With the blades stripped down, you can now use clipper oil to lubricate the two blades, wiping any excess away. Now you can replace the blades (the right way up) and partially re-tighten the retaining screws. Finally, readjust the alignment of the blades and tighten the screws. Check the alignment once more: the lower cutting blade should extend around 3 mm further than the upper cutting blade with the clippers adjusted fully forwards to the shortest cutting length.

Hand them back to your supervisor so that your maintenance can be checked.

Cutting tools and techniques

Cutting tools	Techniques that can be achieved	Explanation of technique
Scissors (straight or flat parallel blades)	Club cutting Blunt cutting	The most popular way cutting hair straight across, parallel to the index and middle finger. The blunt, straight sections of cut hair that it produces are ideal for precise lines. The different angles that the hair is held at will produce square, graduating and reverse graduating layer patterns.
	Freehand cutting	Cutting without holding with the fingers is known as freehand cutting. A technique that allows for cutting with natural fall and without tension. Its main uses are at the perimeter edges around ears, fringes and trimming awkward growth patterns.
	Pointing (point cutting)	Pointing is a texturizing technique that reduces bulk from the ends (2–3 cm) of the hair in order to create softer, more shattered, textured edges. It uses the point ends of the scissors and is more successful on straighter hair than wavy, and does not add any value to curly hairstyles at all.
	Deep chipping	Another texturizing technique that reduces fine sections of hair from much deeper, closer to the root (1–3 cm from the scalp). It will add texture but is better for creating and adding lift in medium to thicker hair types.
	Brick cutting	A texturizing technique that is a combination of pointing and chipping to gain benefits of both forms of cutting techniques. The cutting action would resemble the position of bricks in a wall.
	Disconnection	Disconnection creates an obvious demarcation area between two different levels. It is a deliberate style or styling feature that draws attention to the different levels and can be done on all one-length hair or layered effects. An example of disconnection is regularly used on short-layered styles that have longer hair over-falling the shorter perimeters.
'Japanese-style' scissors (hollow ground edges)	All of the above plus slicing or slider cutting (a technique only suitable for extra sharp scissors or a razor)	Slicing is a texturizing technique where the hair is held at an angle away from the head and cut downwards. Then a single blade of the scissors or a razor cut through the hair in one continuous and downwards motion, this reduces the weight from within the hair forming a tapered effect.
Thinning scissors	Thinning/texturizing	Thinning scissors will remove uniform bulk from any point between the root area and ends. However, they have more creative uses when they are used to 'feather' the perimeter edges of hairstyles (which is often more difficult with straight-bladed scissors).

Cutting tools	Techniques that can be achieved	Explanation of technique
Razors	Tapering	Tapering produces a similar effect to that produced by thinning scissors. Razors are often a better choice, as when used correctly they will produce a non-uniform effect.
	Slicing/slider cutting	See 'Japanese style' scissors.
Electric clippers	Clippering with grade attachments	Clipper grades (the attachments that provide uniform cutting lengths) are made in a range of sizes for different purposes, and are numbered accordingly. They will provide closely cut uniform layering or, if differing grades are used, they can provide graduation on hair that is too short to hold between the fingers.
	Fading	Fading is a way of blending short hair at the nape or edges of a hairstyle down or 'out' to the skin. It is achieved by using the clippers with the blade 'backed off' creating a very short, tapered effect with a smooth blended effect without any lines.
	Clipper over comb and scissor over comb	Both techniques are a popular way of layering very short hair into styles that cannot be held between the fingers. The hair is held and supported by a comb and the free edges protruding through are removed.

Combining the cutting techniques As covered throughout this book, there are a variety of cutting techniques: club cutting, freehand cutting, pointing, deep chipping, brick cutting, slicing or slider cutting, thinning/texturizing and tapering. None of these techniques are used in isolation; a precision or artistic cutter will use a combination of these cutting disciplines to personalize a finished effect. As you become more experienced in your work, you will probably find that you would need to use at least 30–50 per cent of these in each haircut in order to achieve satisfactory fashionable results.

Creatively restyle men's hair

Cutting rules

Keep the hair damp throughout the cut Rough-dry the hair so that you work with damp hair throughout the cut. The client should not be sitting with saturated hair; it is uncomfortable for them, as wet hair soon feels cold and it will drip onto the gown and their clothes.

More importantly, working with damp hair enables the natural tendencies and directions to be seen as you progress in the haircut.

Outline shapes Many short, layered cuts are graduated at the sides and into the nape sometimes by clipper over comb or, when left slightly longer, by scissor over comb techniques. On shorter hairstyles, the neck and hairlines become the main focal perimeters of the hairstyle. These require careful attention, as it is very easy to infringe into the hairline and remove hair that is needed for the outline shape.

Where possible, always use the natural hairlines as the limit for the hairstyle. This produces a smoother effect on the eye and produces styles that look balanced and right. If you ignore the natural hairlines and cut above them, you will find that the hair below will grow back very quickly and produce a stubbly effect within a few days. Or, if done on dark hair, it will produce a 'shadowed' effect within 24 hours.

However, natural necklines often lack consistency; the growth is often uneven, intermittent or sparse. Therefore, the outline shapes for these men wearing shorter hair need to be defined. The more natural the nape-line hair growth, the softer, and less severe, will be the look.

TOP TIP

Hair growth patterns

The movement and direction of hair can be particularly problematic on shorter hairstyles. Make sure that you take this into consideration during your consultation with your client.

> In men's hairdressing the length males wear their hair can vary significantly – as will the cutting and dressing skills required to create and maintain these looks. Male medium- and long-length hair looks require techniques more associated with ladies/unisex hairdressing, but with one main requirement – the hairstyle look should still appear masculine.
>
> *Kenyon Yates*

ALWAYS REMEMBER

In order to create clean lines around the ears on shorter hairstyles, you will need to hold the tops of the ears down so you can see what you are doing and ensure that you are working safely.

Round neckline

Tapered/faded neckline

Square neckline

The shaping of front hair into a fringe can produce a variety of facial frames and the focal point it creates changes the overall effect dramatically. In many men, the front hairline recedes and this is often a sign of MPB. This influences the choice and positioning of perimeter fringe shapes. Always give this some thought before cutting the hair.

In men, the side hairlines, side burns, or *sideboards*, bridge the hairstyle and beard shape. These need to fit, and care must be taken in shaping them. Lining the hair above the ears and along the sides of the nape is usually carried out with the scissor points, or carefully angled inverted clippers.

Common cutting problems

Necklines All of the classic perimeter neckline shapes already mentioned can be ruined through lack of attention. The detail of the outline of very short hair can easily be spoiled by careless layering or clippering. Every millimetre counts. You need to make sure that your outlining with the clippers is even and smooth throughout. If you do make a mistake and find that you have encroached on the outline shape, you would be better off re-cutting the outline slightly shorter to eliminate the fault.

Blending from clipper lengths to scissor length Another common problem on short hair occurs at the blend area between different clippered grades, or between the clippered and hand-held cut lengths. If a careful blending has not been made it will show cutting marks at the point where the two areas combine.

There are two ways of tackling this problem:

1   If the hair is still too long on the hand-cut/held side, you can re-fade the two zones together by scissor over comb methods. Be careful not to undercut the longer lengths as this will mean that you will have to re-cut all the clippered area.

2   If the lengths of hair between the two areas are slightly uneven it will definitely show unless you correct it. In any area where clippers fade out to club cut lengths you can re-surface the hair by using thinning scissors over comb just on the very tips of the hair. A light blending of thinned hair produces an optical illusion that cheats the eye by softening the two hard cut edges and the final effect appears correct.

Ears Nature does not guarantee symmetry, and this is true of faces and ears too. One may be larger, they may be irregular in shape or at different heights; you need to make sure that you have considered this before you start. Unevenness on long hair does not matter but, when it is on short hair, the imperfections will be obvious.

TOP TIP

Hairlines

The higher the cuts made into the hairline, the harsher and starker the look becomes.

TOP TIP

Always check the clipper blade alignment before using the clippers.

ALWAYS REMEMBER

Removing cutting marks

The thinning over comb technique can be used as a corrective method on most clipper-cut or scissor-cut lengths to 'join' areas of differing grades/lengths or to remove cutting marks on fine, medium and coarse hair textures.

TOP TIP

Men's short hairstyles often benefit from washing again after cutting. This removes all the shorter clippings and makes them more comfortable.

Tapering and thinning encourage the hair to curl at the ends, while club cutting increases density and reduces that tendency.

Feathering and texturizing can produce extra lift and bounce.

If your client has a build-up of wax, gel or moulding crème on their hair you must insist that the hair is washed to remove it before you attempt the haircut.

You need to find out how your client feels about his facial features. Sometimes these natural imperfections are not a concern; they are merely a characteristic of the client's personality. You should also remember to check whether your client wears glasses or a hearing aid; take all of these factors into your assessment.

Finally, you and your client will be able to agree exactly what look is required, and you will then have a basis on which to decide how the work is to be carried out.

Hair type

If your client's hair is very curly, do remember that it will coil back after stretching and cutting. Similarly, wavy hair, when cut too close to the wave crest, can be awkward to style as it tends to spring out from the head. Very fine straight hair will easily show cutting marks or can disclose unwanted lines from clippering if you take too large sections. Make sure that the sections you take are accurately divided and sectioned.

> " The artistic and technical ability to create traditional or classic barbered looks is deceptive – they are highly skilful. There is little room for error, so using a male practice head to perfect these skills is an excellent learning resource. All hairdressing techniques are best learned slowly and in bite-size chunks. Attempt and practise the skills in slow motion. You will learn quicker if you start slowly and build up your technique speed.
>
> *Kenyon Yates*

Final points to remember

1 If the hair is dirty, it must be washed before you cut it. Wet hair is a necessity for blow-drying and finishing, but not necessarily a convenient arrangement for a quick trim before work or during lunch.

2 Clean, dry hair should not be cut with a razor because of the discomfort to your client, due to the tearing and dragging action of the razor on the hair.

3 Accurate sectioning and **graduation** produces fine layering. This is partly determined by how much hair there is to cut. Longer lengths can be sectioned with the comb and taken between the fingers, while short lengths are best tackled either by clipper over comb or scissor over comb techniques. A section (that cannot be held between the fingers) is lifted with the comb and a guideline is created by cutting straight across. Subsequent lifting with the comb to the guideline length produces the next section to be cut.

4 Clippers must be used to tidy the necklines on short styles, graduating from the natural line out from the head. How far up the head and how short the cut needs to be is determined by the style and shape agreed with your client. If longer lengths are required higher in the back hair, then the clippers need to graduate away from the head sharply.

5   Cross-checking is an essential part of cutting. It is your way of including a quality control. As you progress through the cut, you obviously need to change your stance, holding position and holding angle. These factors can lead you to go wrong. Typically, problems might be that the back section does not blend with the sides properly, the top does not blend with the sides, or the fringe does not fit with the top. Whatever the potential problem is, the easiest way to compensate for this is to cross-check to make sure that the cut works well in different planes (vertically and horizontally).

Finishing products for men

Product	Application	Purpose	Suitability
Dry wax	Applied in small amounts by the finger tips into pre-dried hair.	A moderately firm hold, providing a non-'wet look', or greasy finish. Ideal for men who really don't like the look of product on the hair, but need the benefits of the control it provides.	Suited to short to medium length hair; the effects need to be created carefully and slowly by adding more as needed. It is very easy to add too much and overload the hair, particularly on finer hair types.
Defining clay Wella: Texture Touch Reworkable Clay	Work a small amount of product in your finger tips and sculpt into dry hair. Shape and rework as desired.	A strong defining clay that allows styles to be reshaped and recreated.	Ideal for textured looks and short hair; styles can be used to define key areas.
Hair varnish/high gloss gel	Applied in small amounts by the finger tips into pre-dried hair. Care needs to be taken in applying the product evenly, throughout the hair.	A high-gloss look with a greasy texture. The styles created are moisture repelling. Ideal for men who do like product effects on their hair.	Suits short hair with long-lasting, low maintenance looks, suitable for sports etc. Again the effect needs to be created slowly; it is easy to overload the hair and these types of product do produce a build-up upon the hair.
Hair gel Wella: Pearl Styler Styling Gel	Distribute one or two pumps into your palms and work into the hair.	A styler that lifts, texturizes and tousles hair into many different styles with a pearl shine, finish and hold.	Can be used on dry hair. For extra lift in short hair, work into wet hair and blow dry.

(Continued)

Product	Application	Purpose	Suitability
Styling glaze	Applied first to the hands and rubbed into wet or pre-dried hair all over. The hair is then styled and allowed to dry and fix into shape.	A wet look effect with firm to strong hold. Suitable for controlled or groomed looks with a mild, wet look effect.	Again – like gel – you can't overload the hair as the look is based on 100% coverage. The styles created are suited to short hair and are more resistant to moisture than gel but create less sculpted or high-hair effects than gel.
Hair paste Wella: Bold Move Matte Styling Paste	Work a small amount of product in your fingertips into dry hair to design your desired shape.	A matt styling paste to create casually textured styles.	Suitable on dry hair to add definition.
Defining crème Wella: Rugged Fix Matte Moulding Cream	Work a small amount of product in your fingertips into dry hair.	A moulding cream to construct a rugged texture with a strong definition.	Ideal for short hair for an edgy, matt finish.
Hairspray Wella: Stay Essential Finishing Spray	Applied to pre-dried hair by directional spraying from 30 cm away.	Provides mild, moderate and firm hold, can be used as a final fixative or as a styling product when scrunched in.	Easy to apply providing a touchable support effect on any hair length.

ALWAYS REMEMBER

Always use products sparingly. Most hair preparations are concentrated and can easily cause a product build-up as well as being an unnecessary waste.

Always follow the manufacturer's instructions and guidance for use when using any styling or finishing products.

STEP-BY-STEP: MEN'S CONTEMPORARY CUT – THE VARSITY SIDE PART

1 After greeting your client, conduct a full consultation to determine the style and length of cut that is desired.

2 Using a clipper with a grade 5/6 attachment, cut the back and sides from the nape to the occipital bone line.

3 Club cut the top sections using a uniform layer of 3–4 inches.

4 Blend the sides into the top section using scissor over comb and point cut techniques, thereby reducing bulk.

5 Graduate the back and sides using clippers set between grades 1 and 5.

6 Neaten the edges using the edge of your clippers (or small clippers).

7 Using a cut throat razor create a precision shape to the neckline.

8 Using clippers remove any unwanted neck hair.

9 The finished Varsity side part.

CourseMate video: Varsity side part haircut

STEP-BY-STEP: MEN'S CONTEMPORARY SHORT CUT

1 After greeting your client, conduct a full consultation to determine the style and length of cut that is desired.

2 Place a parallel parting from centre forehead to nape to place guide.

3 Cut using point cutting techniques to create texture.

4 Cut radial sections from crown, pivoting using previous guide set. Texture as required.

5 Work through either side of the front sections, following guide first placed to ensure hair cut is balanced and precise.

6 Cross-check your cut visually and check for balance.

7 Texturize throughout to remove weight and bulk as required.

8 Dry the hair and use texturizing cutting techniques to personalize the style.

9 The finished contemporary style is a classic.

CourseMate video: Men's contemporary short haircut

STEP-BY-STEP: MEN'S HAIR PATTERNING

1 After greeting your client, conduct a full consultation to determine the style and pattern desired.

2 Clipper cut to a grade 2 to achieve a short back and sides.

3 Use the edge of the clipper blade to create the requested design, taking great care with the accuracy of the pattern.

4 Fade the edges using a grade 1 attachment or clipper over comb to enhance the pattern and achieve shading.

5 The pattern looked at close-up shows the high level of precision required.

6 The finished look is unique.

CourseMate video: Men's hair patterning

Provide aftercare advice

Home and aftercare checklist

✓	Talk through the style as you work; that way the client sees how you handle different aspects of the look.
✓	Show and recommend the products/equipment that you use so that the client gets the right things to enable them to get the same effects.
✓	Tell the client how long the style can be expected to last and when they need to return for reshaping.
✓	Demonstrate the techniques that you use so they can achieve that salon hair look too.

See **CHAPTER 4** Creative Cutting for detailed aftercare advice, on page 104.

> Barbering services are mainly maintenance based and clients will often just require a regular repeated service – usually a haircut. It does however give the barber an opportunity to promote extra services such as head hair and facial hair restyles, patterns, shaving, face massaging, colouring, perming, etc. All these extra skills will help towards better career prospects.
>
> *Kenyon Yates*

REVISION QUESTIONS

Q1. Accuracy is achieved by _____ and cutting the hair at the correct angle.

Fill in the blank

Q2. A razor should be used on dry hair.

True or false?

Q3. Select the texturizing techniques from the following list.

Multi-selection

Club cutting	☐ 1
Graduation	☐ 2
Slice cutting	☐ 3
Fading	☐ 4
Point cutting	☐ 5
Chipping	☐ 6

Q4. Symmetrical styles produce outline shapes that are equally balanced.

True or false?

Q5. Which of the following is not a cutting term?

Multi-choice

Cross-checking	○ a
Thinning	○ b
Free hand	○ c
Free style	○ d

Q6. Disconnection is a term defining a continuous outline shape or layer patterning.

True or false?

Q7. Which of the following hair growth patterns does not affect the way that hair lays after it is cut?

Multi-selection

Nape whorl	☐ 1
Double crown	☐ 2
Widow's peak	☐ 3
Low hairline	☐ 4
Cowlick	☐ 5
High hairline	☐ 6

Q8. A graduated layer shape that tapers out onto the neck is referred to as _____.   Fill in the blank

Q9. Which of the following cuts would easily describe a disconnection?   Multi-choice

Graduation in a long hairstyle   ○   a

Reverse graduation in a long hairstyle   ○   b

A fringe in a shoulder-length bob style   ○   c

Texturizing in a short cropped style   ○   d

Q10. 'Personalizing' is the term that refers to any technique that is used to complete a style, tailoring it to the client's specific needs.   True or false?

6 Beards and Moustaches

LEARNING OBJECTIVES

◆ Be able to work safely when cutting beards and moustaches.

◆ Be able to create a range of facial hair shapes and styles.

◆ Be able to communicate professionally and provide aftercare advice.

KEY TERMS

avant-garde
cleanse

clipper over comb
ingrowing hairs

Design and create a range of facial hair shapes

INTRODUCTION

There is nothing new in the shaping and styling of men's facial hair. Many of the current designs seen etched over the scalps or around the faces of men have their roots in ancient tribal cultures. Ancient civilizations used this as a form of body art to distinguish one unique culture from another.

This type of adornment, unlike jewelry or fine textiles, has been seen as achievable by many people because its effects are short-lived and are easily changed or modified to create something new. Beards and moustaches have always been a feature of men's barbering and this chapter explores the different designs that can be created, ranging from the traditional, professional to the progressive and avant-garde.

Beards and moustaches

PRACTICAL SKILLS

Learn how to work with care, attention and safety at all times

Identify features that will influence the ways that you cut your client's beard and/or moustache

Learn how to work with different thicknesses of hair

Learn how to achieve a variety of creative effects and styles

Learn how to work with, or correct, simple faults and common problems when cutting

Provide aftercare advice to your clients

UNDERPINNING KNOWLEDGE

Know how to work safely during cutting

Know the common problems that can affect the way that you cut beards and moustaches

Know how the client's physical aspects and features can affect styling options

Know a variety of different facial hair shapes and how to achieve them

Know how the client's physical aspects and features can affect styling options

Know how to communicate effectively and professionally

Maintain effective and safe methods of working when cutting facial hair

Facial hair barbering

You will no doubt be familiar already with the men's barbering techniques of freehand, scissor over comb and **clipper over comb**. These techniques underpin nearly all of the skills needed to carry out this work; in addition you need care, control and creativity. This chapter covers:

◆ preparation and maintenance

◆ client consultation

◆ identifying the influencing factors that affect style choice

◆ cutting techniques and accuracy

◆ good customer care.

Note: There are many references to other chapters. For example, the tools used in styling facial hair are the same as in other barbering and cutting techniques. Therefore, you will find more information about tool care, maintenance and safety elsewhere in this book.

INDUSTRY ROLE MODEL

MUCKTARU KARGBO (MK) MK Hair Studio and Academy – Proprietor

"MK has a unique approach to cutting-edge barbering, which he describes as 'The Art and Science of Hair with Flair'. MK started barbering as a trainee, when he was 16. With some seriously hard work MK became manager two years later, with 14 staff working under him. MK was in charge of it all, from training sessions, shows, photo shoots and shift systems, etc. He stayed for three years before breaking out on his own to become a salon owner in 2006. MK has won a UK award every year from 2005 to 2010. He is making his second step-by-step educational DVD called 'Get 2 the Point', endorsed by Habia and travels the world delivering courses and demonstrations.

Preparing the client

Your duty and responsibilities towards health and safety are reinforced throughout all hair-related services and each service has specific procedures that you should follow. This section addresses all of the health and safety issues that you need to consider when styling facial hair.

Gowning the client Always use freshly clean, laundered protective equipment:

◆ Fasten a gown at the back, or secure the cutting square with a clip, ensuring that the covering is close fitting around the neck and protects the client from any clippings or spillages.

◆ Place a towel around the front of the client so that the free edges fasten at the back.

◆ Tuck a strip of neck wool (or neck tissue) into the top edge of the towel to stop hair fragments from falling inside the client's clothes.

◆ Cover the client's eyes with a cotton wool pad to prevent snippings and clippings from entering his eyes (after consultation and just before starting).

Positioning the client

Facial hair cutting requires the client to tilt his head back so that you can work safely and carefully at an angle that suits you. The barbering chair is specially designed for this, with its inbuilt headrest and reclining ability.

Ensure that the client is sitting in the chair with his feet squarely on the foot rests. This posture stops him from twisting or 'hunching up' and allows the client to sit safely for long periods of time without discomfort, injury or fatigue; it also gives you the access and freedom to do your work properly and safely.

Ask the client if he is comfortable and if you need to adjust the working height or angle, you can do so now.

Working position and posture

Barbering involves a lot of standing and you should always adopt a comfortable but safe work position.

Your work position The client's cutting position and height from the floor have a direct effect on your posture too. You must be able to work in a position where you do not have to bend 'doubled up' to do your work. Cutting involves many arm and hand movements and you need to be able to get your hands and fingers into positions where you can cut the hair unencumbered, without bad posture.

1 You should adjust the seated client's chair height to a position where you can work upright without having to over-reach on the top sections of his head.

2 You should clear trolleys or equipment out of the way, so that you get good all-round access (360°) around the client.

3 Your equipment should be close enough at hand so that you can reach it safely without putting you or the client at risk. The items should be clean, sterile and ready for use (this also demonstrates your organization skills and professionalism).

Your personal hygiene

Personal hygiene is vitally important for anyone working in personal services. Your personal hygiene (or lack of it) will be immediately noticeable to every client you work with. Daily showering and regular brushing of teeth are essential for you and your colleagues for the sake of your clients, as body odour and bad breath are unpleasant in any situation. Other strong smells are offensive too; the smell of nicotine from smoking is very off-putting to the client, particularly if he is a non-smoker.

Working efficiently, safely and effectively

Working efficiently and maximizing your time is essential, so you must make the most of the salon's resources and take care in the way that you handle the equipment and products that you use. Always treat the salon's materials in the same way that you would look after your own equipment; aim to minimize waste, being careful about how much product you use.

Safe disposal of sharps 'Sharps' is the term used to describe any blades used in safety razors or shapers. When these items have been used, and need to be replaced, they must be disposed of safely.

Preventing infection

A warm, humid salon can offer a perfect home for disease-carrying bacteria. If they can find food in the form of dust and dirt, they may reproduce rapidly. It is important to keep the salon clean, dry and well aired at all times, and work areas kept free from clutter or waste items. This includes clothing, work areas, tools and all equipment. Some salons use sterilizing devices as a means of providing hygienically safe work implements.

ACTIVITY

Avoiding cross-infection

1 Why is it important to avoid cross-infection/cross-infestation?

2 What things should you do to minimize the risk of cross-infection?

3 How does your work area affect hygiene and cross-infection?

ACTIVITY

What is your salon policy for the use of razors and the disposal of sharp items?

For more on sterilization and cleaning see CHAPTER 4 Creative Cutting, page 82.

ALWAYS REMEMBER

Always make sure that the clippers are cleaned before they are used. Any hair caught between the blades will limit their ability to work, and is unhygienic for the client.

ACTIVITY

Fill in the table below with the missing information.

Tools and equipment	What is it used for?	How is it used?
Barbicide™		
UV cabinet		
Autoclave		

Preparation for cutting

Prepare the client's facial hair

Many men with longer beards never comb them out because they do not want to lose the shape they naturally take on, or they simply cannot see the benefits of doing so. However, regular grooming keeps them free from debris and reduces the chance of infections or ingrowing hair. You should always make a point of giving clients advice on this or at least tell them how they can manage their own facial hair between visits.

Moustaches and beards can get matted as they get longer because the bristles tend to get curlier and lock together. These tangles have to be removed so that the longer hair is revealed; this allows you to style *all* of the hair and not just part of it.

Cleansing is important too; a beard with debris or grease cannot be styled until it has been cleaned. If you are shampooing the hair as part of another service, then the beard can be done at the same time or, alternatively, you can ask the client to wash his face and beard in the front washbasin, or **cleanse** with facial wipes.

Consultation

Check the client's requirements   Hairdressing and barbering require us to deliver a continuing, repeated service that satisfies our clients by finding out what they want and carrying out those instructions. It is therefore important to gain confirmation of what we are doing throughout the service.

- ◆ It ensures that we only do what the clients want.

- ◆ It makes clients more confident about our work.

- ◆ It removes any ambiguity or confusion.

- ◆ It gives us more confidence.

- ◆ It involves the client and helps to develop a professional relationship.

Effective communication is important in any hair service. Consultation does not just take place before a service; it is a continual process of reconfirming your intentions with those of the client. Therefore, during your discussions, you must find out what the client wants and weigh this against any limiting factors.

You need to understand your client's requirements and be able to discuss and seek agreement with him throughout the service.

> Discuss client's hair and lifestyle; don't just hear his request. There may be a better option for him and it gives you time to show off your knowledge and gain respect for the industry. When creating a design, explain to the client what you are going to do and draw pictures if possible. The main thing is that the style you create must complement the main haircut, whether it is a short back and sides, uniform layer mohawk, etc. Keep it within the main frame of the style; this will ensure you do not compromise it.

ALWAYS REMEMBER

Disposal of sharp items

Used razor blades and similar items are placed into a safe container (sharps box). When the container is full it can be discarded. This type of salon waste should be kept away from general salon waste as special disposal arrangements are required and may be provided by your local authority. (See Chapter 5 page 112 for an illustration of a sharps box.)

Using visual aids   Visual aids, such as men's styling magazines, Internet downloads and trade publications, are an excellent source for different effects. Pictures are a universal language and this is a good way to demonstrate ideas and themes. As a barber, you are used to people bringing pictures of their ideas, and you have probably spent considerable time trying to explain why some styles do not work. Most clients have no self-visualization; they need the reassurance of others to help them make their decisions. Use your skills to find the most suitable images for your client and, more importantly, try to explain why your selections will work for them, taking into account their physical features and personal image.

ACTIVITY

Develop a portfolio of effects

Finding examples of outline beard or moustache shapes is very difficult unless you have a pre-prepared selection of options. It is helpful to collect examples of different beard and moustache shapes so that you can use them as a visual aid with your clients.

When you have collected a suitable range, you can display them in a ring binder for use in the shop.

Identify factors that influence the service

Be sure to listen to your client's requests. Many mistakes can be avoided if you achieve a clear understanding of what the client is asking for.

The facial style that you choose with your client should take into account each of the following points:

◆   face and head shape/size

◆   facial physical features (including any scars or blemishes that need disguising)

◆   hair quality, abundance, growth and distribution

◆   age, lifestyle, and suitability.

Aspects to consider

Facial features

Facial feature	How best to work with it
Square and oblong facial shapes	Square and oblong are typically masculine and provide a perfect base for traditional classic, well-groomed looks. The angular features of the face can be featured with closer, shorter beards or moustaches and would probably benefit from fewer curves and more angular, linear effects.
Round face	The effects of a round face can be lessened, or increased, it depends what the client wants. If the plan is to lessen the effects, then choose beard designs that lengthen the jaw line and incorporate lines and angles rather than curves. If the round features suit the personality and image of the client then work with it by cutting uniform-length shapes.

Facial feature	How best to work with it
Square angular features, jaw, forehead etc.	Again, these are traditionally accepted as a feature of masculinity. These can be handled in a similar way to that of the square and oblong features already mentioned. Soften squarer, more angular features with beards and moustaches.
Small face	Balance smaller faces with facial hair designs that do not overpower the overall effect. Keep your designs close cut and uniform in length.
Wider head	This feature is more prominent with full–long beards. Create beard designs that are closer cut at the sides and extend to more length at the chin.
Scars, marks and blemishes	If the client has any scars or blemishes, this may be the reason for growing a beard or moustache in the first place. You need to ask if there are any features that the client wants to disguise. *Always make a point of looking for these during your consultation.*
Facial piercing	Facial piercings around the mouth and ears need consideration during your consultation. It is unlikely that the client will want to remove them/it so you have to be very careful in combing, detangling, and cutting, anywhere near to the area(s) of the piercing(s).

You need to find out how your client feels about his facial features. Sometimes these natural imperfections are not a concern; they are merely a characteristic of the client's personality. Do not forget to check whether your client wears glasses or a hearing aid; all of these factors must be taken into account.

Finally, you and your client will be able to agree exactly what look is required, and you will then have a basis on which to decide how the work is carried out.

Some men have a heavy, daily growth of facial hair and they find that they need to shave every day. This heavy growth can be obvious for a range of reasons and sometimes because of a combination of all of the following:

◆ the growth appears heavy because of the contrast against the skin due to natural colour

◆ the hair seems to grow particularly fast

◆ the density of hair distributed on the face is particularly thick.

Initially, men who are most likely to choose to grow beards and/or moustaches will have ticked two or more of the above. However, they are not the only ones who choose to do this, as many others with a poorer growth or definition will grow facial hair for other reasons.

Head shape and size The facial hair should not outweigh the proportions of hair on the head. This may be fine for people with plenty of hair, as this would allow them to grow their beards to a longer length. However, many men with male pattern baldness also

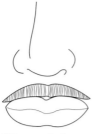

Thin moustache

Thick narrow moustache

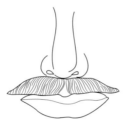

Wide thick moustache

The required cutting tools are also covered in **CHAPTER 4** Creative Cutting. However, the table on page 143 provides a quick reference guide to clipper grades/attachments and their effects when used on facial hair.

like to grow a beard or moustache too. In these situations, closer-cut effects seem to suit the wearer better than thick bushy cuts.

Mouth and width of upper lip to base of nose   The size and width of the mouth forms the basis for any moustache. The distance between the upper lip and the base of the nose creates a sort of canvas for the moustache. If the distance between the two areas is quite deep, it will provide more outline shape options for the wearer rather than if it were narrow.

Similarly, the width of the face at the cheeks will also determine the most suitable effect. Someone with a wide face will be able to wear a fuller moustache, whereas someone with a narrow face could be swamped by this much hair.

Bone structure and facial contours   You should take particular care for clients who have a well-defined bone structure, i.e. cheekbones, jaws and facial contour. If they have a particularly linear aspect to their facial features then it would be wiser to retain that similar effect with the overall shapes and outlines. (That is unless they wanted to disguise themselves or have physical features they want to cover up.)

Conversely, the client who has a rounder, fuller face can benefit, aesthetically, from a shape that defines the face with a more structured effect. Remember that these people can wear beards with fuller effects than those with narrower facial features.

Width of chin and depth of jaw line   Facial hair growth forms a frame for the physical features of the face and it is the width of the chin, and the depth down to the bottom of the jaw, that becomes the focal point of any facial hair shaping. The outlines of the shapes created here are more noticeable than others. Historically, beards were left relatively full; this meant that there was very little upkeep for the wearer, apart from keeping the beard from getting too bushy. Latterly, the fashion for wearing more chiseled effects has meant that not only thickness but also an outline shape has to be maintained too.

Tools and equipment

"When using electrical equipment such as clippers, trimmers, hairdryers etc., visually check every time you use them. Having them checked annually prevents hazardous injuries. There is nothing worse than having client complaints about something to do with your tools, which can easily be prevented. The most common problems are abrasions, whaling of the skin, cuts and pitches. Clippers and trimmers that are not correctly lined up generally cause these and happen because of negligence, bad storage in the salon and when in transit.

MK

ALWAYS REMEMBER

Always take care when using any sharp items of equipment. Your safety and the safety of the client are in your hands; take care not to be distracted while you are working.

Clippers

Standard clipper grades/attachments	
Clipper attachment size/ no. for attachment	**Length of cut hair**
Grade 1 = 3 mm (⅛ inch)	Very close to skin, almost as close as shaving Very short, on darker hair it will only leave a stubbly shadowing effect.
Grade 2 = 6 mm (¼ inch)	Close cut, will see some skin on finer hair types but short enough for the hair to appear straight even if it is naturally curly.
Grade 3 = 9 mm (⅜ inch)	Popular length grade for shorter, groomed effects. Typically cuts to that of short scissor over comb lengths.
Grade 4 =13 mm(½ inch)	Popular cutting length, which has a similar effect to longer, scissor over comb type effects.
Grades 6–8 =16 mm–25 mm (½–1 inch)	Popular longer length used for beard shaping.

Note: There are no set standard sizes for clipper attachment combs/grades; you will need to adapt the hair length required by your client in light of the make and model clippers that you have/or your salon provides.

Cleaning the equipment

	Tools	**Method of cleaning/ sterilization**
	Neck brush	Wash in hot soapy water and place in UV cabinet for 10 minutes.
	Sectioning clips	Wash in hot soapy water and immerse in Barbicide™ jar for 30 minutes.
	Cutting comb	Wash in hot soapy water and immerse in Barbicide™ jar for 30 minutes.

(Continued)

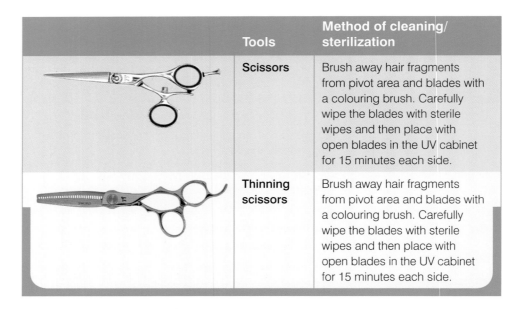

	Tools	Method of cleaning/ sterilization
	Scissors	Brush away hair fragments from pivot area and blades with a colouring brush. Carefully wipe the blades with sterile wipes and then place with open blades in the UV cabinet for 15 minutes each side.
	Thinning scissors	Brush away hair fragments from pivot area and blades with a colouring brush. Carefully wipe the blades with sterile wipes and then place with open blades in the UV cabinet for 15 minutes each side.

Create a range of facial hair shapes

Designing an outline shape

When a new look and outline shape are required, you will find it easier to draw the outlines first. This enables you to get an idea of balance and proportion in the mirror and it also enables the client to see a rough idea of where the new outlines will be in proportion to his face.

On fairer, lighter hair, you can use an eyebrow pencil to mark the outlines and on darker hair you can use a white pencil. (Do not forget to remove any leftover marking when you complete the look.)

Facial haircutting techniques

See CHAPTER 4 Creative Cutting, page 89 for more information.

Many of the techniques used to cut men's facial hair are common to barbering practices:

◆ scissor cutting

◆ freehand cutting

◆ scissor over comb

◆ clipper over comb

◆ controlling the cut

◆ cross-checking the cut.

Every client is different; they all have differing needs, features and requirements. Therefore, they all need individual care and attention. Your consultation and analysis will need to reflect this but, normally, the process of trimming and shaping beards or moustaches will be the same:

◆ you remove the bulk from the interior of the feature first, then

◆ you tidy and shape the outline to finish the effect.

Facial hair is bristle; it is stiffer than hair on the head and this is due to the frequency that it is cut in relation to hairstyles. This makes the hair coarser and creates its own problems, as it is more difficult to cut by the scissor over comb method. This leads stylists and barbers to choose clipper over comb, as the mechanical advantage makes the job far easier. However, as you need to use one hand to steady and position the comb, you can only have one other holding the relatively heavy clippers. This technique is more complex than using clippers held with two hands and clipper grade attachments.

Clipper over comb The clipper over comb technique should start by combing and lifting away the ends, then skimming over these to remove wispier bits first. The benefit of this will allow you to:

◆ see if any areas show less density in growth than others

◆ make sure that you do not reduce these areas resulting in skin showing through in a patchy effect.

Bristles are strong and when they are cut into smaller fragments they can fly around, over you and the client. This can be dangerous as bristles can stick in any areas of unprotected skin. Worse, they can enter into the eye. To prevent this from happening to the client, it would be safer get him to close his eyes while you trim.

> Clipper design techniques all reside within a full anchor beard shape. Perfecting all symmetrical beards is the best way to improve on pattern work. When cutting a design, regardless of whether you are doing a pictorial or geometric style, it's always best to have a story element to it (beginning, middle and end). This way the eye can have something to follow and decide whether it's attractive or not. Searching on the Internet is also a great way to pick up ideas.
>
> *MK*

When the interior of the facial hair shape has been cut, you can then concentrate on defining the shape by creating the outside perimeter line. Hair growth can often be uneven across the head as well as the face. Even if the client is a regular visitor to the salon, you will need to check for balance throughout the shaping, to make sure that the growth does not occur thicker and deeper on one side than on the other side.

Although comfort is always a major concern, for beard trimming it may be easier to start your outlining with the clippers, centrally up the neck to the point below the chin to start the profile shape. Doing this means you can define the exact position where you stop and you will find that you can then work on either side of the client to create an even symmetrical finish. After this, you can complete the shape behind or over the jaw and finally form the cheek area, down to the desired top profile of the beard.

On the other hand, most moustaches are trimmed at, or above, the upper lip by scissors; this is easier to handle and stops the vibration of the clippers tickling the client and causing him to pull back. The clippers give a clean, finished profile shape that can define the upper perimeter line.

ALWAYS REMEMBER

Safety

Always take care when using sharp implements on the client's skin. Concentrate on the job in hand and ignore other distractions.

However, if you do cut the client:

1   Put on a pair of disposable vinyl gloves.
2   Use a medical sterilized wipe to clean the area and remove any hair or bristle.
3   Apply pressure to the cut to stem the flow of blood.
4   When the wound has stopped bleeding, you can finish the service and give the client a clean, dry tissue for any minor cut.

 Subscribe to as many hair publications as possible. In order to keep up with trends you need to observe the best and emulate what you can. Attending trade shows will also help you know what the latest and best tool options are. Your client can then have the best service possible!

MK

Removing unwanted hair outside the desired style line

Finishing and tidying is the last part of the service; you need to clean-up any hair or stubble that lies outside the desired style line. After you have agreed with the client that the overall shape and design is OK, you take the electric clippers and turn them so that the fixed blade works away from the design line. You may find that you need to stretch the skin around the neck and face to get a cleaner, straighter cut close to the skin. Then, working back from the design line, draw the clippers away over the neck/cheek or jaw to create a clean finish. Take a neck brush and carefully brush away any fragments. Finally, recut any areas that show signs of uncut bristle and brush away any fragments.

STEP-BY-STEP: MAINTENANCE–BASED TRIM

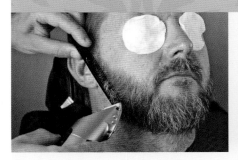

1 Remove excess hair using the clippers-over-comb technique. Note the pads to protect the eyes from clippings.

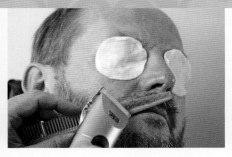

2 Remove excess hair from the moustache using the clippers-over-comb technique.

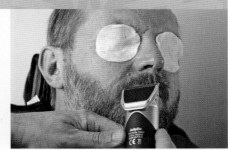

3 Use the clippers carefully to produce a tidy edge above the top lip.

4 Be aware of critical influencing factors such as moles, scars, infected spots, etc.

5 Mark out and create an edge using the clippers.

6 Remove unwanted hair up to the edge.

> " Always take pictures of your work; this enables you to see your own development as time goes by, making you your own critic. When you look back at the pictures, observe them in detail – taking note of all that is wrong or what you feel could be done better. Decide which techniques were used or should have been used and practise them. You can ask friends to have the cuts with those techniques or give discounts to your clients for a given period to increase your productivity and skills quickly.
>
> *MK*

Provide aftercare advice

No service is complete unless the client leaves in the knowledge that he can achieve the same result at home. If he cannot achieve a similar effect, he is unlikely to return. You can help make this happen and the real sign of client satisfaction is the rebooking of his next visit before he leaves the salon.

You can advise your client by telling him:

◆ How long the effect will last and when he needs to come back.

◆ Which products and equipment you have used and how they might benefit the client at home.

◆ How changes in hairstyle may affect the overall look.

How long will it last?

Hair grows at an average 12.5 mm per month so a shorter, closer styled beard or moustache will have grown out within a month whereas longer facial designs will last longer. Remember, if the effect incorporates a moustache too then that will need trimming anyway, as it will quickly grow over the upper lip.

Products and skin care

Explain to the client how he can manage the effect himself. You need to provide advice on cleansing – what to use, in relation to their hair and skin types, how often to use it and what products would not suit him and therefore should be avoided (and your reasons why).

Exfoliation Exfoliation is beneficial to the client; it removes dead skin cells from the epidermis and stimulates blood circulation, which will generally improve the skin's condition. There are many different products now available for men and these can be bought as grains that are mixed with water and applied as a paste or, alternatively, a wide range of ready-to-use products with a variety of bases such as fruit acids or herbal with essential oils.

ALWAYS REMEMBER

Avoid ingrowing hairs

Ingrowing hairs are an uncomfortable aspect of closely cut wavy or curly hair. When hair is razored or clippered close to the skin, say at the neckline, the action of the cutting implement tends to pull the hair slightly within the follicle. After the hair is cut, the hairshaft tends to retract back into the follicle giving a very close, smooth finish to the touch. Unfortunately, if the area is not kept thoroughly clean the open area of the follicle can become partially blocked with dead skin cells or dirt. This will change the growing direction of the hair within the follicle, often allowing it to find a new position within the skin. Bearing in mind that the average rate of hair growth is about 12.5 mm per month, if this is allowed to continue, an ingrowing hair will result.

The physical effects of this will:

◆ start to create a raised area often accompanied with a mild infected spot

◆ cause irritation beneath the skin and begin to itch.

If the ingrowing hair is not exposed, or removed, a longer-term, more serious condition/infection will result.

While you are finishing up, describe to your client the techniques used and any challenges you had to overcome, such as growth patterns, differences in density, along with what you needed to do to overcome them.

At the end of the service say 'Thank-you sir', don't wait till they get to the till. Politeness goes a long way in this world and even further when done with a smile.

MK

TOP TIP

Whatever the length or effect created, you need to tell your client from the outset how long he can expect it to last, so that he does not have any unrealistic expectations.

REVISION QUESTIONS

Q1. Barbicide™ is a form of chemical _____. Fill in the blank

Q2. UV radiation is suitable for hygienic preparation of scissors. True or false?

Q3. Which of the following is suitable for immersing into Barbicide™? Multi-selection

Scissors ☐ 1

Combs ☐ 2

Thinning scissors ☐ 3

Clipper blades ☐ 4

Plastic brushes ☐ 5

Open razor ☐ 6

Q4. Grade 1 is shorter than grade 2. True or false?

Q5. You are going to use all of the following clipper grades in a gent's clippered graduation. Which clipper grade would you start with? Multi-selection

1 ○ 1

2 ○ 2

3 ○ 3

4 ○ 4

Q6. Facial hair is stiffer than scalp hair. True or false?

Q7. Which of the following factors would influence the choice of shape of a beard? Multi-selection

Facial features ☐ 1

Hair length ☐ 2

Parting position ☐ 3

Facial hair density ☐ 4

Height ☐ 5

Weight ☐ 6

Q8. Quality combs are strong but _____. Fill in the blank

Q9. Which of the following will cause ingrowing hairs? Multi-choice

Leaving a beard too long ○ a

Repeated close cutting with a razor ○ b

Fading out to a hairline ○ c

Blunt clipper blades ○ d

Q10. Alopecia does not affect facial hair. True or false?

7 Shaving

LEARNING OBJECTIVES

◆ Be able to identify the client's wants and needs.

◆ Be able to prepare the skin for shaving.

◆ Understand the health and safety factors related to shaving.

◆ Be able to shave hair and apply finishing products.

◆ Be able to communicate professionally and provide aftercare advice.

KEY TERMS

astringent
backhand technique
carborundum

exfoliation
forehand technique
honing

moisturising balm
stropping

Provide shaving services

INTRODUCTION

The popularity of contemporary facial hair shapes has helped to promote the need for careful and skilled shaving. The intricate shapes that are popular and the needs of busy professionals have helped to drive forward a service that is luxuriant and soothing.

Shaving

PRACTICAL SKILLS

Learn how to work with care, attention and safety at all times

Look for things that will influence the ways that you carry out shaving

Learn how to work with different densities of hair growth

Learn how to work with common problems when shaving

Provide aftercare advice to your clients

UNDERPINNING KNOWLEDGE

Know how to work safely during shaving

Know how to prepare the client's skin for shaving

Know how the client's physical aspects and features can affect shaving

Know how to take suitable remedial action to resolve any problems arising during the shaving service

Know how to communicate effectively and professionally

Maintain effective and safe methods of working when shaving

The barber's shave

The everyday necessity of shaving for men can be tedious whereas the barber's shave is considered to be a luxury, but it is still a valuable service. This chapter covers:

◆ tool preparation and maintenance

◆ client preparation

◆ client consultation

INDUSTRY ROLE MODEL

KENYON YATES International Barbering Tutor, Assessor and Demonstrator

I served my three-year apprenticeship, gained a first class City & Guilds Award in Men's Hairdressing and, unlike many of my barbering colleagues, adapted to the cutting and styling of male medium and long-length hair. My boss retired, I bought the business and for the next 40 or so years progressed through competition work, to being an international jury person, local branch NHF [National Hairdressers' Federation] president, teacher and assessor at three FE [Further Education] Colleges, etc.

- the factors that affect the service
- shaving techniques, problems and accuracy
- good customer care.

Tool preparation and maintenance

The tools that you use must always be sterilized and ready for use; the razor should be kept nearby, but well out of the way of clients and children. If you use razors with disposable blades, make sure that any used blades are disposed of properly in the sharps box. Alternatively, if you use a fixed-blade razor it must be hygienic and sharp – ready for use.

Your main concern during shaving is damage to the skin. This is avoided by careful, shaving techniques and the use of sterile and sharp tools.

> Ensure your barbershop environment is safe, hygienic, warm, tidy, welcoming and industrious. Old and out-of-date reading material can give a negative image. Inappropriate magazines should not be on show. While attending to clients you must always be aware of colleagues near you or children coming up behind you. Keep sharp tools out of children's reach and never leave electrical equipment cables trailing. As a matter of course you should never use profane language or make uncomplimentary comments about other colleagues or clients.
>
> *Kenyon Yates*

Shaving tools and equipment

Shaving requires a range of equipment.

Clippers Clippers can be used as a pre-shaving procedure for men who have longer facial hair. Dense, heavy growth needs to be removed before shaving, so that only short, stubbly growth remains. This is particularly beneficial, as it not only saves time during the shaving process; it also enables you to see the client's skin and his facial hair growth patterns more clearly.

ALWAYS REMEMBER

Bristles are strong and, when they are cut into smaller fragments, they can fly over you and the client. This can be dangerous as bristles are stiff and sharp, and can pierce areas of unprotected skin or, even worse, can enter into the eye. To prevent this happening to the client, it would be safer to get him to close his eyes while you trim or, alternatively, place damp cotton wool pads or a damp warm flannel over the eyes to protect them from the cut fragments.

Open blade razor There are two types of razor that are used for shaving: the safety razor, which has a disposable blade, and the fixed-blade razor that has a blade kept sharp by honing and stropping.

TOP TIP

Safety first

Open cut-throat razors are extremely sharp and must always be handled with care and respect.

ALWAYS REMEMBER

Facial haircutting requires the client to tilt his head back so that you can work at an angle safely and carefully. The barber's chair is specifically designed for this, with its built-in headrest and reclining ability to ensure the client's comfort and safety.

TOP TIP

A new blade needs to be used on every client to prevent cross-infection.

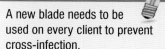

For more information on clippers, and clipper maintenance, or razors, see CHAPTER 5 Creative Barbering.

ALWAYS REMEMBER

Local bylaws

Different local authorities have different local bylaws. Some areas do not allow the use of fixed-blade open razors and only permit the use of disposable bladed types.

Find out what the policy is in your area.

If you use a safety razor, you must always replace the disposable blade with a new blade for every client. This hygienic practice prevents the risk of cross-infection and eliminates the need for any sharpening processes.

The fixed-blade razor can have two types of blade:

◆ **The hollow ground** – the more popular, lighter and durable type is made out of tempered steel. It has a narrower, waist-like profile that is more sensitive during use, which discloses resistance to cutting when in need of sharpening.

◆ **The solid blade** – which is made of a softer metal that 'dulls' more easily and therefore needs sharpening more frequently.

Choose your blade with care

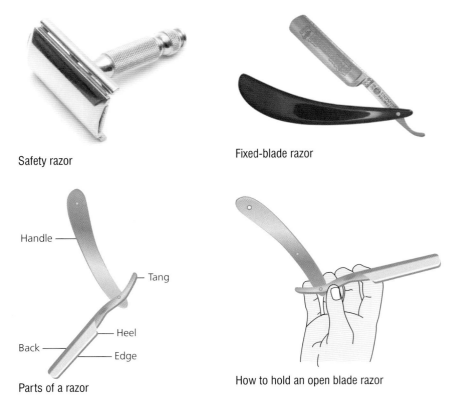

Safety razor

Fixed-blade razor

Handle

Tang

Heel

Back

Edge

Parts of a razor

How to hold an open blade razor

Honing (setting) fixed blade razors

If you looked at the cutting edge of a razor through a powerful magnifying glass, you would see that it has very fine teeth like a saw. During routine use, this cutting edge is 'dulled' and must be re-sharpened by the barber; this is done by **honing** and **stropping**.

There are three different types of hone: natural, synthetic and combination.

A natural hone is a rectangular block of stone and the most popular of these originate from Belgium and Germany.

A synthetic hone is made from a material called **carborundum**. It produces a sharp edge to a razor more quickly than that of natural stone. Care should be taken not to over-hone, as this will damage the edge of the blade.

The third type of hone is double sided and is a combination of the natural on one side and synthetic on the other. So if a razor has a badly damaged edge then the synthetic side of the hone is used to quickly bring back its keenness. Alternatively, if a razor only needs a slight re-sharpening than the natural stone side is used.

ALWAYS REMEMBER

A fixed-blade razor must always be sterilized between clients and therefore, for handiness, a barber may own a couple, so that one can be cleaned and prepared, always ready for use.

Honing technique

A fixed-blade razor is sharpened by a technique of drawing the edges over the lubricated surface on the stone in a particular way:

◆ Wipe over the surface of the hone to remove any oil or debris.

◆ Drop a little lubricating oil onto the surface of the hone.

◆ With the edge of the blade pointing left, start at the upper right area of the hone with *the heel* of the blade.

◆ Draw down diagonally towards the lower left area of the hone, keeping the blade flat and parallel to the abrasive so that it sharpens the edge up the blade towards the point.

◆ Turn the blade over and slide it upwards towards the upper left area of the hone.

◆ Then repeat the diagonal movement again from upper left to the lower right.

◆ The complete movement is like a figure of '8' on its side.

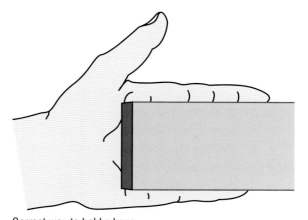

Correct way to hold a hone

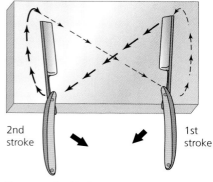

2nd stroke   1st stroke

Proper method to hone a razor

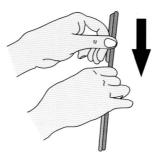

Test the sharpness of the razor on moistened thumbnail

Make sure you get the correct angle

Testing the razor's edge

During honing, the edge of the razor's blade needs to be tested to prevent over-honing. This is done by lightly drawing the razor over a moistened thumbnail to feel for any signs of roughness or snagging.

◆ A sharp razor will dig into the nail with a smooth, continual draw.

◆ A blunt razor will glide over the surface of the nail without digging in.

◆ Any nicks or damaged parts of the blade will feel rough and uneven.

Stropping

A fixed-blade razor is stropped to produce what is known as a *'whetted' edge*. Stropping cleans (but does not sterilize) a razor's edge and re-aligns the microscopic serrations or teeth, to provide an extremely sharp blade. There are two types of strop:

1 **Hanging strop** – this is the most popular type of strop which is flexible and has a canvas and leather side.

2 **Solid (French or German) strop** – these strops are used for solid-blade razors and are made from a combination of materials.

Solid strop

Hanging strop

A strop gives you a sharp edge to work with.

ALWAYS REMEMBER

Never try to strop a razor with the cutting edge facing towards the direction of sharpening. This will damage the razor's edge and also the surface of the strop.

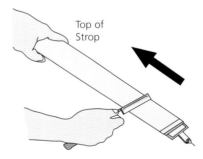

Top of Strop

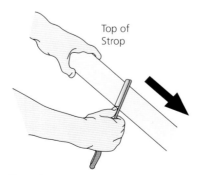

Top of Strop

Stropping a razor

ACTIVITY

Carry out some research to answer the following questions.

1 Why is it important to avoid cross-infection/cross-infestation?
2 What is meant by setting a razor's blade?
3 When would you do a sponge shave?
4 How would you dispose of the waste items produced during shaving?
5 How do you test a razor's edge?

Stropping a hollow ground razor

1 Hang the strop at one end by its hook.

2 Then, holding the other end, pull it out horizontally so that the strop is flat.

3 Position the razor near the hook end with the heel of the blade pointing towards you and holding it by its metal shank between the thumb and forefinger.

4 Draw the razor down the strop towards you; then as it reaches three-quarters of the way down, carefully turn the blade over.

5 Repeat the stroke in the opposite direction.

6 Repeat this for about ten strokes and the blade should be ready for use.

Correct stropping will extend the useable life of the razor and reduce the risk of damaging the client's skin. You should always sterilize the razor after stropping and honing to prevent the risk of cross-infection to the client.

Stropping a solid razor The movement or technique is similar to that used for the hanging strop, except the strop is placed on a flat, solid surface, rather than hung from a hook.

ACTIVITY

This activity relates to the variety of products used during shaving. Complete the table with the missing information.

Product	What does it do?	When is it used?
An astringent		
An exfoliating crème		
A moisturizing balm		
A solid hone		
A hanging strop		

Prepare the hair and skin for shaving

> Prepare for the service by ensuring a tidy sanitized work area and your good personal presentation. Observe the client's head and facial hairstyles and look for any factors that may be hidden by the beard, for example warts, pimples and scars. Consult fully with the client before reclining him.
>
> *Kenyon Yates*

Client preparation

After making sure that your hands and nails are scrupulously clean, check that your client is comfortable and correctly positioned, seated in the barber's chair. Find out what the client wants and agree your course of action. Check to see that the client's beard is clean, free from debris and short enough to work on. (If the beard is long, the excess length can be removed by clippering first.) After looking for contra-indications to shaving, as well as checking the beard's hair growth directions, you can make your adjustments to the position of the headrest to suit the client's neck and head. Then place a towel across the client's chest covering the gown and tuck the edges into the neckline.

> When carrying out the client consultation for shaving, acknowledge the client will probably know his facial hair better than you! Carry out the consultation before reclining the client and, always ask if the client has any history of back or neck problems – as he may be reclined for a lengthy period of time.
>
> *Kenyon Yates*

> Always safely recline the client to attend to his facial hair. Keep your posture straight to avoid injury problems and outstretch your arms rather than 'hovering' too closely. Keep the client comfortable by constantly brushing away clippings. Remember to take a step back occasionally to take in the whole effect and use a hand mirror to show your client the progress (while he is reclined) to confirm his requirements.
>
> *Kenyon Yates*

Contra-indications to shaving

There are several situations when shaving is not suitable for the client. Look for signs to see if the service *can* be provided and *ask* the client if he has any reasons he knows of that would *not permit* the service to be conducted.

TOP TIP

If you do need to remove excess beard length first, place damp cotton pads over the eye sockets to stop bristle clippings from entering the eyes.

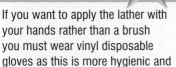

ALWAYS REMEMBER

If you want to apply the lather with your hands rather than a brush you must wear vinyl disposable gloves as this is more hygienic and prevents cross-infection.

TOP TIP

If, during the process of shaving the lather dries out, then re-apply it as otherwise you may cause damage to the client's skin.

Possible contra-indications can include:

◆ infections or infestations that you see during your examination

◆ cuts or abrasions within the facial hair areas

◆ uneven skin due to acne, eczema or other conditions

◆ skin sensitivity.

Lathering

Before applying the lather, the skin and beard need to be prepared; this is done by applying hot towels to the face. Hot towels can be prepared by pre-soaking them in a hot towel cabinet or basin of hot water; after wringing out the excess water, place around the facial area (but not covering the nose) to:

◆ soften the bristly hair

◆ open up the follicles and

◆ prepare the skin for lathering.

Always make sure that the towels are not dripping wet and they are not too hot for the client. Damp cotton pads should be placed on the eyelids under the hot towel to protect the eyes.

Hot lather is applied to the beard area using a brush and bowl, it needs to be done quickly and care taken not to cover the mouth, nose or go anywhere near the client's eyes. Begin lathering by placing the brush on the point of the chin and making small circulatory movements; extend the area of lather to cover all of the beard area. Keep the brush and lather hot by dipping it in hot water: a hot lather produces a better, closer shave.

ALWAYS REMEMBER

If you have any doubts about symptoms and contra-indications, always ask a more experienced member of staff for their assistance. You may be putting the client at health risk and the salon at risk from legal action if you do not follow this process properly. If you do offer shaving services and have covered the above contra-indications, make sure that you keep a record of your consultation and the responses made by the client after the service, for future reference.

Shaving

There are two main movements to shaving: the forehand and backhand techniques.

First time over

The first time shave is always in the same direction as the client's natural hair growth. If you are right-handed stand on the right-hand side of the client (left-handed stand on left) and begin at the outer cheek nearest to you.

Tension the skin with your free hand so that it is taut, keeping your hand free from lather so that you do not slip and allow the skin to relax. (This would be dangerous, as it would allow the skin to cut easily.) If the skin is taut, the razor will be able to glide easily over the skin and cut the hair cleanly and smoothly.

Holding the razor loosely with your thumb on the blade:

1 Begin the shave on the nearest sideburn area to you.

2 Move the razor in steady sweeping movement starting at the top and moving downwards.

3 Follow the sequence shown in the diagram.

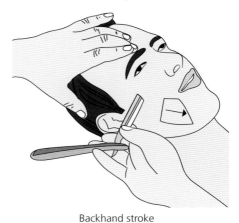

Backhand stroke

Forehand stroke

Shaving techniques

Second time over

The second time shave is done against the natural growth direction; this ensures that the shave is as close as possible. This will be the last shave unless the client has a particularly coarse, stubborn growth, which will then require a sponge shave.

Re-lather the face and start the shave at the neck area:

1 Work upwards towards the chin, then upwards again on each side of the face.

2 Clean the face thoroughly with a damp, warm towel then gently pat dry, without rubbing.

Shaving procedure

Finishing

1 Make sure the face is completely dry.

2 Then apply an astringent to close the pores or a cool towel if your client has sensitive skin, and then a **moisturizing balm** to cool and rehydrate the skin.

3 Sit the client back up to an upright position and get him to check the result.

ALWAYS REMEMBER

Hot and cool towels

Hot towels should not be used on a client with sensitive skin as this will irritate the skin further and prevent you from carrying out the service.

Cool towels are used to soothe the face after shaving, and to close the pores to finish the service.

Cool towels should not be used on a client who is going to have a facial massage as this will close the pores prematurely and prevent the client from gaining the full benefit of the massage service.

> When working alongside other barbers, consider yourselves a team. Support each other for the common good and profitability of the business. One instance of poor standards from an individual can lose income for everyone. **All** the team must give a client a reason to return!
>
> *Kenyon Yates*

TOP TIP

Remove any lather from the razor between each stroke so you can see clearly, in order to prevent damaging the client's skin.

Sponge shaving

If the client has particularly heavy, coarse or thick growth, it may be necessary to do a sponge shave after the second time over.

◆ Dip a sterilized sponge into hot water and wipe it over the skin directly before shaving.

◆ This opens the hair follicle and lifts the hair/bristle enabling the closest of shaves.

◆ Take particular care as you can damage the skin or cause ingrowing hairs.

◆ Do not do this type of shave unless absolutely necessary.

Shaving problems

Problem	Possible cause	Remedy	
Facial cuts	1 Poor skin tensioning 2 Blunt razor	1 Administer first aid – continue after bleeding stops (wear PPE). 2 Re-hone the blade and test for sharpness; then shave again with the correct tension	
Skin rashes	1 Blade is dull and is pulling the hair before cutting 2 Shaving too close 3 Towels too hot 4 Dragging the razor	1 Re-hone the blade and test for sharpness. 2 Let the skin rest. 3 Check heat and allow to cool first. 4 Incorrect cutting angle – change.	
Ingrowing hairs	1 Shaving curly hair too close 2 Blocked pores/follicle	1 Avoid cutting that close. 2 Face not deep cleansed enough – provide advice for exfoliation.	
Uneven skin	1 Acne, moles	1 Take extra care in uneven skin areas.	
Folliculitis	1 Inflamed follicles 2 Ingrowing hairs	1 and 2 Refer client to pharmacist for remedial treatment. Do not carry out shaving.	
Patchy shaving results	1 Blunt dull razor	1 Re-hone the blade and test for sharpness, then re-shave with an even tension.	

Disposal of waste materials

There will always be some waste materials at the end of a shave and sharps must be disposed of in line with local authority bylaws and regulations. All used razor blades need to be disposed of in a sharps bin, which must be clearly labelled.

Any unused lather should be washed away down the basin; it cannot be used again on another client as this may cause cross-infection.

Any used towels need to be laundered and should be removed from the work area or temporarily placed into a covered towel bin prior to washing.

Provide aftercare advice

Recommended re-shaving intervals

Hair growth on men differs greatly, some men have less dense facial hair that is fine in texture, and others have heavy, coarse growth that seems to show as a shadow within hours. Therefore, the advice and recommendations that you give to clients is individual to their needs. Some men do not mind showing a small amount of regrowth, whereas others wish to be clean-shaven all the time.

A lot depends on what your client does professionally; someone who has to meet people as part of his job needs to be seen as a smart professional. Stubble or several days' growth may be acceptable in certain circles, but will not work in others. Your advice should reflect their individual lifestyle and personal needs.

As the client will have been reclined for a while, after returning him back to an upright position you should attend to his hairstyle before he leaves your barbershop.

Kenyon Yates

Aftercare advice

You should provide advice to clients on suitable shaving equipment and skin care and cleansing – they need to know what products they should be using, in relation to their facial hair and skin types. You must advise them on how often to use the items and, particularly, the products that will not suit them and the reasons why they should be avoided.

Skin care for men is now a very popular and growing business area. It is as important to men as it is for women; the only difference is the range of products available. Men who shave regularly will already know that blunt razors and close shaving are a contributing factor in the causes of minor skin infections, blocked pores and follicles and the start of ingrowing hairs. When this occurs, a spot forms on the surface of the skin and the bacteria causes a small infection. Like ingrowing hairs it is uncomfortable and itchy and can easily be avoided with the correct advice.

Exfoliation is beneficial to the client; it removes dead skin cells from the epidermis and stimulates blood circulation, which will generally improve the skin's condition. There are

many different exfoliants now available for men and these can be bought as grains that are mixed with water and applied as a paste or, alternatively, a wide range of ready-to-use products with a variety of bases such as fruit acids or herbal with essential oils.

ACTIVITY

Shaving problems

Review this chapter then complete the table by filling in the missing information.

Problem	Possible cause	Remedy
	1 Poor skin tensioning 2 Blunt razor	1 Administer first aid – continue after bleeding stops 2 Re-hone the blade and test for sharpness; then shave again with the correct tension
Skin rashes		1 Re-hone the blade and test for sharpness; 2 Let the skin rest 3 Check heat and allow to cool first 4 Incorrect cutting angle – change
	1 Shaving curly hair too close 2 Blocked pores/follicle	
Uneven skin		
Folliculitis	1 Inflamed follicles 2 Ingrowing hairs	
		1 Re-hone the blade and test for sharpness, then re-shave with an even tension

REVISION QUESTIONS

Q1. An _____ produces sterilization by heat. Fill in the blank

Q2. Honing is also known as 'setting' a blade. True or false?

Q3. A hollow ground blade has the following features: Multi-selection

It is not as sharp as a solid blade razor. ☐ 1

It is more sensitive during use. ☐ 2

It is less sensitive during use. ☐ 3

It is narrower than a solid blade razor.   ☐   4

It is heavier than a solid blade razor.   ☐   5

It needs sharpening more often.   ☐   6

Q4.   Natural hones originate from Germany and Belgium.   True or false?

Q5.   Which of the following is not a contra-indication to shaving?   Multi-choice

Infections   ○   a

Infestations   ○   b

Uneven skin   ○   c

Alopecia   ○   d

Q6.   A solid strop is suitable for solid razors.   True or false?

Q7.   Which of the following are the main contributors to skin rashes?   Multi-selection

Dull razor blade   ☐   1

Sharp razor blade   ☐   2

Shaving too closely   ☐   3

Towels too hot   ☐   4

Towels too cold   ☐   5

Poor skin tensioning   ☐   6

Q8.   A _____ shave is suitable for particularly heavy, coarse thick growth.   Fill in the blank

Q9.   A 'first time over' shave: Multi-choice

Should be done in the same direction as the client's natural growth.   ○   a

Should be done in the opposite direction to the client's natural growth.   ○   b

Should be done very quickly.   ○   c

Should be done with a blunt razor.   ○   d

Q10. Alopecia does not affect facial hair.   True or false?

8 Creatively Style and Dress Hair

LEARNING OBJECTIVES

◆ Be able to identify what clients want.

◆ Be able to advise clients about suitable styling options.

◆ Be able to work safely during setting, styling and dressing.

◆ Understand the factors that can affect clients' styling options.

◆ Be able to style and dress hair creatively.

◆ Be able to communicate professionally and provide aftercare advice.

KEY TERMS

added hair

adverse hair or scalp condition

back-brushing

croquignole winding

non-conventional styling equipment

temporary bonds

UNIT TOPIC

Creatively style and dress hair

INTRODUCTION

Creative styling builds upon the skills that you learned at Level 2 and uses them with new techniques to create all sorts of fashion, day-wear, evening-wear and bridal effects. Most stylists will have favourite styles and lengths of hair that they like to work with.

Level 3 takes you out of your comfort zone helping you to:

◆ investigate the aspects of styling that you would like to be able to do

◆ further develop your skills to achieve those creative effects.

Creative styling and dressing

PRACTICAL SKILLS

Find out what clients want by questioning and visual aids

Learn to work safely and carefully when setting, styling and dressing hair

Look for things that will influence the ways that you do your client's hair

Learn the techniques that will enable you to achieve creative effects for your clients

Learn to use a variety of tools and equipment to create different effects

Learn how to work with different lengths and amounts of hair in order to optimize the final effects

Provide aftercare advice to your clients

UNDERPINNING KNOWLEDGE

Know how to communicate effectively and professionally

Understand the factors that can affect hairstyling options

Recognize the limiting factors that can affect styling options

Know how to prepare the hair for a variety of styling services

Know how to communicate effectively and professionally

Maintain effective and safe methods of working when styling hair

Styling and dressing hair are always carried out at the work station and therefore the main health and safety concerns should relate to the client's comfort, positioning and protection, as well as your posture, accessibility and care.

INDUSTRY ROLE MODEL

PATRICK CAMERON, Patrick Cameron Training School

"I didn't start my training until I was in my early twenties. I met an amazing lady called Lyndsay Loveridge, who today is still my mentor and biggest inspiration. I worked in the same salon in Plymouth, New Zealand for seven years before setting off to London.

When I arrived in London I found that there were not many hairdressers who dressed long hair and I set out to find out why this was. The answer seemed to be that some hairdressers were afraid of this aspect of hairdressing. I then decided to make it my goal to teach this art in the simplest way I knew — step-by-step.

Preparing the tools and equipment for setting and dressing hair

Make sure that you have prepared the area. Get everything that you need together beforehand, including the equipment that you need, as well as the products. You should have your trolley prepared with all the materials you will need. Styling materials should have been previously prepared by thorough washing/sterilizing and combs, brushes, sectioning clips, etc. should be all cleaned, sterilized and made ready for use.

As much of the essential information surrounding client preparation, health and safety, and tool care and maintenance are common throughout Level 3, you will find that these aspects are comprehensively covered in CHAPTER 15 on health and safety.

ALWAYS REMEMBER

Make sure that you always:

- ✔ Follow your salon's policy in respect to standards and public image
- ✔ Prepare yourself in a professional manner ready for work
- ✔ Wear the minimum of jewellery that can dangle or tangle in the client's hair
- ✔ Wear appropriate footwear
- ✔ Be aware of your personal hygiene
- ✔ Minimize the risk of cross-infection to your colleagues and clients
- ✔ Work responsibly and respectfully within the salon's service timescales

TOP TIP

Lateral thinking

Non-conventional styling equipment can include items such as highlighting foils, rags or pre-formed plastic containers (e.g. plastic cups, flower pots, cones, etc.) cut to shape. You can think creatively about applying materials in different ways; that is what creative thinking is.

Styling materials and tools

Equipment	Effects achieved	Precautions	Cleaning
Highlight foil (concertina folded)	Produces zig-zag set effects that can have uniform or tapering effects on wet or dry hair.	If you use a hood dryer to help set the hair, remember that foil retains heat, so it could burn the client.	It is unlikely that you would keep used foils, these can be disposed of in a sealed bin.
Rags	Creates twisted 4 spiralled-curl effects on wet hair.	No obvious problems as rags tend to be kinder on hair than any other setting equipment.	Machine (hot) wash (if they are to be retained).
Chopsticks	Hair wound in a figure of eight around two chopsticks produces a softer zig-zag effect than foils.	Be careful not to buckle the ends of the hair, it will produce *fish-hooks* (like perming can). Use end papers or foil.	Wash with hot, soapy water and scrub clean, then dry and put in UV cabinet and finally back in trolleys/trays.
Straws	Single, long straws can be bent in half to create two legs; on which the hair can be wound in a figure of eight (similar to chopsticks).	Be careful not to buckle the ends of the hair, it will produce *fish-hooks* (like perming can). Use end papers or foil.	It is unlikely that you would keep used straws, these can be disposed of in a sealed bin.
Preformed plastic items (e.g. conical vase shape)	Plastic formers can be used to create the bases for all sorts of avant-garde effects. Wet hair is wrapped or styled onto bases and fixed in place.	The initial fixing of the former to the head is important to the durability of the effect. Ensure it can be securely fastened without causing discomfort to the client.	Make sure that any item is washed thoroughly before it is used. It is unlikely that you would keep used formers after use; dispose of carefully.
Bendy foam-covered rollers	These can be used to create spiral curls on wet or dry longer hair, or used more conventionally to produce loose, set results like rollers.	Damaged rollers should be thrown away.	Wash with hot, soapy water and scrub clean, then dry and put in UV cabinet and finally back in trolleys/trays in sized order ready for future use.

(Continued)

Equipment	Effects achieved	Precautions	Cleaning
Velcro rollers	On dry hair they produce a softer (although not as durable) curl effect as wet-setting rollers.	Self-clinging rollers tend to lock onto finer hair types. Be careful when removing as they will not only pull the hair but damage it too.	Wash with hot, soapy water and scrub clean, then dry and put in UV cabinet and finally back in trolleys/trays in sized order ready for future use.
Heated rollers	On dry hair they produce a soft but longer-lasting, more durable effect than Velcro rollers.	Electrical item: remember to check the condition of the lead before use. These are very hot when they are first put into the client's hair. Use cotton neck-wool as an insulating base between the bottom of the curler and the client's scalp.	Wash with hot, soapy water and scrub clean, then dry and put in UV cabinet and finally back onto their appropriate heating stems ready for future use.
Heated tongs/Rik-Rak stylers	Provide long-lasting waves or 5 spiralled curls on dry hair.	Electrical item: remember to check the condition of the lead before use. Very hot when in use, be careful not to burn the client or yourself.	Spray with cleaner and dry thoroughly after use. Remove hairstyling product build-up from the curling surfaces.
Heated straighteners	Will produce super-flat effects on dry hair that is unruly, or curly hair. Can also be used to create spiral curls on longer hair.	Electrical item: remember to check the condition of the lead before use. Very hot when in use (170–220°C) be careful not to burn the client or yourself.	Spray with cleaner and dry thoroughly after use. Remove hairstyling product build-up from the styling surfaces.
Grips and hair pins	Hair ups, partial hair up/back effects.	Grips and pins are metal with sharp points; care should be used when putting them in and when removing them to avoid tugging or snatching the hair.	Wash and dry and put into trolleys or trays.
Pin clips	On wet hair they provide a narrow curl stem that can be positioned either flat against the skin or standing away.	Pin clips tend to have strong durable springs. These can pull the hair if not careful.	Wash with hot, soapy water and scrub clean, then dry and put back in trolleys/ trays.

Brushes

	Uses	Description	Technique	Cleaning and maintenance
Denman classic styling brush	General brushing, detangling hair before: shampooing, styling and blow-drying straight hair of any length	A cushioned, flat brush, with parallel rows of removable bristles. Available in small (five rows), medium (seven rows) and large (nine rows) rows of bristles.	Blow-drying is achieved by placing the leading edge of the bristles against the mesh of hair then turning the brush to engage the hair across all of the width. The brush is used from roots to points, with the dryer blowing across the cushioned surface.	Denman brushes can be dismantled by removing the rubber-cushioned head from the brush handle and the bristle rows can be removed and washed in hot soapy water, then dried and replaced. Rows of bristles can be replaced if damaged or overheated. Brushes can be placed in a UV cabinet for 15 minutes to complete the cleaning/ sterilization process.

	Uses	Description	Technique	Cleaning and maintenance
Vented brush	For blow-drying straight, short and mid-length hair	A parallel, flat brush with a double row of rigid plastic bristles (short and long) affixed to a brush head that is not solid allowing air to pass between the bristles.	Blow-drying is achieved by placing the leading edge of the bristles against the mesh of hair, then turning the brush to engage the hair across all of the width. The brush is used from roots to points with the drier blowing across the cushioned surface.	Vented brushes can be cleaned by raking out any loose or tangled hair from the bristles, then washing in hot, soapy water. The brush is then dried before use. Brushes can be placed in a UV cabinet for 15 minutes to complete the cleaning/ sterilization process.
Bristle (curved head) brush	General brushing, detangling hair and pre-dressing hair	A brush with a wide, curved head usually with a cushioned head with finer, closer teeth than a Denman. Sometimes the bristles are natural in composition, but generally are plastic.	Flatter, curved bristle brushes are not generally used for blow-drying, unlike their radial counterparts. They are more associated with general brushing and brushing out.	Bristle brushes can be cleaned by raking out any loose or tangled hair from the bristles, then washing in hot, soapy water. The brush is then dried before use. Brushes can be placed in a UV cabinet for 15 minutes to complete the cleaning/ sterilization process.
Radial brushes	Blow-drying with volume, lift, wave and curl on shorter- or longer-length hair	Radial brushes are completely round in section and come in a wide variety of sizes. The bristles are usually made of plastic, although pure bristle brushes are still available. The inner body of radial brushes is often made of metal, allowing the brush to heat up; improving the drying speeds of the underneath hair within a section.	Blow-drying volume and movement is achieved by placing meshes of damp hair around the brush and drying the hair in position from both sides. When dry, the curl or movement can be affixed or set into position by applying a cool shot to increase the durability of the set.	Radial brushes can be cleaned by raking out any loose or tangled hair from the bristles, then washing in hot, soapy water. The brush is then dried before use. Brushes can be placed in a UV cabinet for 15 minutes to complete the cleaning/ sterilization process.
Diffuser (although not a brush it is a piece of equipment used to style the hair)	Scrunch-drying and finger-drying hair to optimize the natural or permed movement within the hair	A diffuser is an attachment for a blow-dryer that suppresses the blast of hot air and turns it into a multidirectional diffused heat.	The hair is styled with the fingers by either pressing into the cupped diffuser to dry or by working through the hair with the fingers until the hair is dried with the required amount of texture or definition.	Diffusers are cleaned by spraying with an antibacterial spray then wiped with paper towels.

The classic dressing out brush has a cushioned, pure bristle head. This brush is an essential part of the dressing out collection as it is designed to fit into the hand, often with a shaped handle that fits the fingers, and provides good control for applying back-brushing. The bristled end of the brush can be 'spoon' shaped or narrow and parallel in section. This allows the user to take sections between their fingers and apply back-brushing across the whole section in one movement without missing areas. This is particularly useful because back-brushing is softer and less tangling to the hair than back-combing and, whichever technique is used, an even amount needs to be applied throughout the hairstyle to create volume and structure for the dress-out.

ACTIVITY

Ionic styling equipment

The term *ionic* appears many times with modern styling tools and equipment. It describes new types of hairdryer, heated styling tools and brushes and combs.

Q. What are the claims of ionic equipment technology and in what ways do they benefit the hair?

Tip. You could find more information about these products on the Internet.

ACTIVITY

Styling materials

For each of the items listed below, write down how these items could be used for styling hair and what sort of effects it will produce.

Styling item	What effects will it produce?	How could the item be used?
Bendy foam-covered rollers		
Highlighting foils		
Rik-Rak waving irons		
Rags		
Chopsticks		
Straws		

Hairdryers

The hairdryer is one of the most commonly used items of equipment in the salon. There is a huge range of models available with a variety of power outputs, speeds and heat settings. The latest ionic dryers can even reduce the 'flyaway' effect that is produced by static electricity when hair is heated.

A well-designed professional hairdryer needs to have a number of special features and you should bear this in mind if you buy one. A good professional hairdryer should:

◆ have at least two speeds and two heat settings

◆ have different-shaped nozzles to channel the heat onto the brush or comb

◆ have a lead long enough not to tangle around the chair or client

◆ be powerful enough to dry damp hair quickly (1300 w–1500 w)

◆ have a 'cool shot' button – to enable hot, dried hair to be quickly cooled and fixed (set) into shape around a brush

◆ be not too long so that it is balanced in the hand and can be held away from the client's hair during drying

◆ be light enough so that it can be manipulated easily and used for long periods without tiredness

◆ be quiet enough so that it allows natural conversation with the client.

Styling products

There is an ever-growing range of styling and finishing products available to the profession. This has been a major growth area, with careful market research and product development ensuring that each one is designed to do a specific job. The result is a huge range of products, supported by successful advertising and brand awareness. If you look closely at each manufacturer's range, you will find that the brand leaders all have similar and competing products, which can increase confusion for the client wanting to purchase products for use at home. At this stage, the only factor that can help someone to make a choice between one product and another is a reputable and recognizable name. This is where a good stylist can support their clients by informing them of the right products to use; when working in a salon stylists are expected to retail products that will enhance the salon's income and their own commission.

Styling products contain plasticizers and fixatives to hold and support the hair in its shape. Apart from hold, they often have other agents and additives within the products that can resist moisture, provide protective sunscreens, add shine and lustre or add definition and shape.

Setting lotions Setting products such as mousses protect the hair from excessive heat. They increase the time that the hair remains in its new shape and the volume and/or movement created, while being exposed to the blast from the dryer's nozzle. They can be in a variety of different strengths for differing hair types and holds.

Finishing products These are products that enhance the hair by giving hold, adding shine or gloss, and improving handling and control by removing static, fluffiness or frizziness from the hair. Certain finishing products like waxes will define the movement in hair, giving texture or spikiness, that could not be achieved any other way.

Heat protection Many products provide protection from heat styling. Regular use of straightening irons can damage the hair so there are a number of products that can be applied to eliminate any long-term effects. Other products provide protection from harsh UVA in sunlight in a variety of 'leave-in' treatments that can be used at any time. They are put on before exposure to harsh sunlight and can be removed afterwards by washing.

ALWAYS REMEMBER

Always use products sparingly and economically within the salon and avoid overloading the hair during styling.

Make a point of explaining the benefits of using professional products to your clients so that they can achieve similar effects themselves at home.

TOP TIP

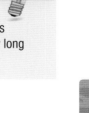

Moisture is the enemy of any finished set. It weakens the hold and therefore how long the overall effect will last.

This is particularly useful for clients who have coloured hair, as the bleaching effects of sunlight will quickly remove colour. Other products have the ability to resist or remove the effects of minerals on the hair, such as chlorine from swimming pools. This is particularly useful, as blonde hair that is regularly subjected to chlorine tends to look green.

Pros and cons for products

Properties → Product ↓	Positives	Negatives	Advice for clients
Mousse (styling product)	Good for general styling use; can be used in blow-drying or as a setting agent	Can make hair stiff and rigid if incorrectly applied. None, but mousses between manufacturers vary a great deal.	Explain that not all products are the same. Professional products are very different from those available in high street shops.
Setting lotion (styling product)	Good for setting, easy to apply and good distribution throughout the hair	Limited uses: not really suitable for any brush-dried styling, tends to 'bind' too easily.	If the client needs to get support from setting rather than blow-drying this could be a better option than mousse.
Styling gel/glaze (styling product)	Produces a firm hold that will provide long-lasting textured effects	Takes a while to dry so any 'high-hair' effects need to be held in place until dry. Absorbs moisture from the air readily, so tends to collapse quickly.	Match the client's needs with the product. If the client has an outdoor lifestyle the product will not hold height or volume in humid conditions.
Dressing cream (finishing product)	Good for applying texture to all hair types as a finishing product. Easy to apply, lasts well	Be careful in the application of finishing products. If you apply too much it can make the hair look greasy.	Match the client's needs with the product. Very easy to use and essential for quick and convenient hairstyling.

	Properties → Product ↓	Positives	Negatives	Advice for clients
	Serum (setting/blow-drying finishing product)	Good for achieving a smooth, controlled finish on straighter hair.	Be careful in the application of finishing products. Over-application will make the hair look limp and greasy.	Match the client's needs with the product. Great for improving the handling of dry, unruly hair. Adds lustre and shine.
	Wax (setting/blow-drying finishing product)	Good for applying texture into all hair types as a finishing product. Easy to apply, lasts well and repels moisture.	Be careful in the application of finishing products. Over-application will make the hair look wet and greasy.	Match the client's needs with the product. Choose the right wet or dry type wax to suit the client and tell them why you recommend your choice rather than another product.
	Hairspray (setting/blow-drying finishing product)	There are hairsprays for every type of hold and for producing all sorts of finishes. Still a great favourite with easy application.	Ensure that you use the right product for the right finish. Check the client does not have any respiratory problems before use.	Match the client's needs with the product. Recommendation is essential for the client to maximize the benefits of hairsprays.
	Heat protection	An essential product for straightening irons, tongs or crimpers. Protects and helps to set the hair all in one.	None. Essential recommendation for anyone using very hot styling equipment on a regular basis.	Use on dry hair prior to applying heat treatments.

ACTIVITY

Equipment and its uses

From the following list of equipment, indicate which ones are suitable for the lengths of hair indicated in the three columns.

	Short hair	Medium length	Long hair
Denman brush			
Vented brush			
Rags			
Chopsticks			
Heated straighteners			
Heated rollers			

For those that you have marked as suitable now describe what sorts of effects they will achieve.

ACTIVITY

Styling and finishing products

List the products that your salon uses and for each one complete the information in the table below.

Product name	What hair type is it for?	What benefits are there to the hair?	How is it used?

HAIR SCIENCE

The physical changes that take place during setting

Hair is both flexible and elastic. As alpha keratin, hair, in its natural state, is curled or waved, it is bent, under tension, into curved shapes. The hair is stretched on the outer side of the curve and compressed on the inner side. If it is dried and fixed in this new position, the beta keratin or stretched hair curl will be retained. This happens because when hair is set the hydrogen bonds and salt bonds between the keratin chains of the hair are broken. The linking system is moved into a new temporary position. (The stronger di-sulphide links remain unbroken.)

Hair, however, is hygroscopic – it is able to absorb and retain moisture. It does so by capillary action: water spreads through minute spaces in the hair structure, like ink spreading in blotting paper. Wet hair expands and contracts more than dry hair does because water acts as a lubricant and allows the link structure to be repositioned more easily. Therefore, the amount of moisture in hair affects the curl's durability. As the hair picks up moisture, the rearranged keratin chains loosen or relax back into their original shape and position. This is why the humidity – the moisture content of air – determines how long the curled shape is retained, *so moisture causes a set to drop*.

The condition and the porosity of hair affect its elasticity. If the cuticle is damaged, or open, the hair will retain little moisture, because of normal evaporation. The hair will therefore have poor elasticity. If too much tension is applied when curling hair of this type, it may become limp, overstretched and lacking in spring. Very dry hair is likely to break.

Creatively style and dress hair

Consultation

Consultation is the most essential part of every hairdressing process; you will always need to find out a variety of information before you start.

If you have already reviewed this information or covered it previously within your training, here is a quick checklist to cover the applicable aspects for setting and dressing hair.

For more information about consultation, see CHAPTER 1 Consultation and advice.

Consultation checklist before setting hair

✓	What is the texture, type, tendency and condition of the hair like?
✓	How much hair is there and how long is the hair?
✓	Are there any limiting factors or adverse hair or scalp conditions?
✓	What type of effect does the client want?
✓	Does the hair type/texture/tendency/amount and length support the client's ideas?
✓	What tools and equipment do you need?
✓	What lifestyle limitations are there?
✓	How much time will it take?
✓	How much will it cost?
✓	Do you need to use any products to achieve the effect?
✓	Will the client benefit from using these products too?
✓	Agree and confirm desired effects with the client prior to starting the service.

The principles of heat styling

Both setting and blow-drying are methods of forming wet or damp hair into a new shape and then fixing it with the aid of heat to create a finished look. These methods of styling

See the sections on pages 167–169 on styling materials and tools.

and dressing hair are temporary and the looks can be either classic or fashionable. You can make hair straighter, curlier, fuller, flatter or wavier.

The styling involves placing and positioning wet hair in to selected positions, and fixing the movement into it, while it is dried into shape.

Before removing rollers or curl formers from the hair, allow them to cool down in position after taking out of the dryer. The cooling allows the hair set to fix into position. In blow-drying this principle is assisted by a cool shot, this helps to fix the curl in after using round or curved brushes.

> It's important to learn how to use your left and right hands together. So having greater dexterity with techniques such as blow-drying, roller setting with different winding techniques, finger waving and pin curling, etc. is very helpful. All great hairdressers have this.
>
> *Patrick Cameron*

ACTIVITY

Using electrical equipment

Safety issues must be paramount when you use electrical equipment. List at least six safety precautions that should be considered when handling electrical equipment in the salon.

1
2
3
4
5
6

Checklist

Before styling:

✓	Prepare the client by carefully gowning them and making sure that they are comfortable.
✓	Find out what the client wants, the type of occasion/event.
✓	Look for factors that will influence your styling options, i.e. physical features, style suitability, growth patterns or density, for example.
✓	Use visual aids to agree on the effect.
✓	Get your materials, products and equipment ready.
✓	Check the condition of the electrical tools.
✓	Pre-wash, condition and detangle (if working on wet hair).

During styling:

✓	Check the heat from the dryer: is the client comfortable?
✓	Check the style's construction: is the shape and proportions correct?
✓	Do you need additional movement from heated styling equipment?
✓	Does the hairstyle need addition: added hair, ornamentation, grips, pins, etc.?
✓	Is the finished hairstyle OK; if not, what aspect needs to be changed?

Effects

Different effects can be produced by different techniques:

◆ **Increasing volume** – adding height, width and fullness, by lifting and positioning 'on base' when rollering or curling.

◆ **Decreasing volume** – producing a close, smooth, contained or flat style by pin curl stem direction, or by dragged or angled rollering 'off base'.

◆ **Movement** – variation of line waves and curls, produced by using different sized rollers, pin curls or finger waving.

Relaxed hair effects are achieved by wrapping hair or by using large rollers. Different techniques are used for hair of different lengths:

◆ Longer hair (below the shoulders) requires large rollers, or alternating large and small rollers, depending on the amount of movement required.

◆ Shorter hair (above the shoulders) requires smaller rollers to achieve movement for full or sleek effects.

◆ Hair of one length is ideal for smooth, bob effects.

◆ Hair of layered lengths is ideal for full, bouncy, curly effects achieved by, say, barrel or clockspring curls.

Different techniques can also be used to improve the appearance of hair of different textures:

◆ Fine, lifeless hair can be given increased body and movement. Lank hair can be given increased volume and movement.

◆ Coarse thick hair requires firmer control.

◆ Very curly hair can be made smoother and its direction changed.

Setting techniques

Curls and curling Curls are series of shapes or movements in the hair. They may occur naturally, or be created by hairdressing – this could be chemically by perming or physically by setting. Curls add 'bounce' or lift to the hair, and determine the direction in which the hair lies.

With styled curls, each has a root, a stem, a body and a point. The curl base – the foundation shape produced between parted sections of hair – may be oblong, square, triangular, etc. The shape depends on the size of the curl, the stem direction and the curl type. Different curl types produce different movements.

You can choose the shape, size and direction of the individual curls: your choice will affect how satisfying the finished effect is and how long it lasts. The type of curl you

On and off base diagrams

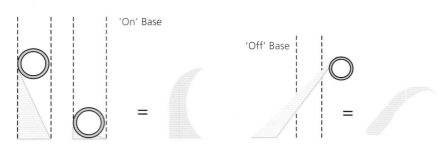

choose depends on the style you are aiming for – a high, lifted movement needs a raised curl stem; a low, smooth shape needs a flat curl. You may need to use a combination of curl types and curling methods to achieve the desired style – for example, you might lift the hair on top of the head using large rollers but keep the sides flatter using pin curls.

Winding root to point Root to point winding is normally carried out on longer-length hair to produce spiralling-type curls that can be formed on bendy foam-covered wavers, chopsticks, spiral wavers as a setting method or, alternatively, with heated tongs or spiral tongs. These tend to produce curl formations that have a uniform curl diameter throughout the length.

Winding point to root On the other hand, conventional winding (croquignole winding), which is used in normal setting, produces a curl formation that has a larger curl size at the root that tapers to tighter near the points.

 Always remember, fashion is like a circle and trends always come back around – so understanding different styling techniques is very important.

Patrick Cameron

Rollering hair There are various sizes and shapes of roller. In using rollers or non-conventional formers, you need to decide on the size and shape, how you will curl the hair on to them and the position in which you will attach them to the base.

◆ Small rollers produce tight curls, giving hair more movement. Large rollers produce loose curls, making hair wavy as opposed to curly.

◆ Rollers pinned on or above their bases so that the roots are upright, produce more volume than rollers placed below their bases.

◆ The direction of the hair wound on the roller will affect the final style.

ALWAYS REMEMBER

Common rollering problems

✔ Rollers not secured properly on base, either dragged or flattened, will not produce lift and volume in the final style.

✔ Too large a hair section will produce uneven curl.

✔ Too small a hair section will produce increased movement or curl in the final effect.

✔ Longer hair requires larger rollers unless tighter effects are wanted.

✔ Poorly positioned hair over-falling the sides of the roller will have reduced/impaired movement in the final effect.

✔ Incorrectly/poorly wound hair around the roller will create 'fish-hook' ends.

✔ Twisted hair around the roller will distort the final movement of the style.

TOP TIP

Evenly tensioned curls produce even movements. Twisted curl stems or ones where the tension *sags* produce movement that is difficult to style.

Pin curling

Pin curling is the technique of winding hair into a series of curls or flat waves, which are pinned in place with pin clips while drying. The two most common types of curl produced in this way are the barrel curl and the clockspring.

◆ The **barrel curl** has an open centre and produces an even effect. When formed, each loop is the same size as the previous one. It produces an even wave shape and may be used for *reverse curling*, which forms waves in modern hairstyles. In this, one row of pin curls lies in one direction, the next in the opposite direction. When dry and dressed, this produces a wave shape. When used in just the perimeter outline of a short hairstyle they can control the shape and stop the ends (that could otherwise be set on rollers) from buckling.

◆ The **clockspring curl** has a flat shape with a closed centre and produces a tight, springy effect. When formed, each loop is slightly smaller than the previous one. It produces an uneven wave shape throughout its length. It can be suitable for hair that is difficult to hold in place.

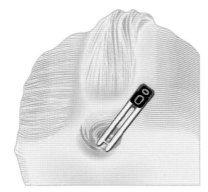
A clockspring curl

Barrel curls

ALWAYS REMEMBER

Common pin curl faults

✔ Tangled hair is difficult to control. Comb well before starting.

✔ If the base is too large curling will be difficult.

✔ If you hold the curl stem in one direction but place it in another the curl will be misshaped.

✔ If you do not turn your hand far enough it will be difficult to form concentric loops.

Step-by-step: pin curling

1 Neatly section the hair and comb through any setting lotion – the size of the section will relate to the degree of movement achieved.

2 Hold the hair in the direction it will lie after drying.

3 Hold the hair at the midpoint in one hand using the thumb and first finger, with the thumb uppermost. Using the thumb and first finger of your other hand and thumb underneath, hold the hair a little way down from the hair points.

4 Turn the second hand to form the first curl loop. The hand should turn right round at the wrist.

5 On completion of the first loop, transfer the hair to the finger and thumb of the other hand.

6 Form a series of loops until the curl base is reached. The last loop is formed by turning the curl body into the curl base. The rounded curl body should fit into the curl base.

7 Secure the curl with a clip without disturbing the curl formed in the process.

ALWAYS REMEMBER

✔ Never place grips or pins in your mouth – this is unhygienic and could cross-infect.

✔ Never place a tailcomb in your pocket – you could injure yourself and pierce your body.

✔ Never work on wet slippery floors and always clear up loose clippings of hair.

✔ Never use any items of personal equipment that have not been cleaned, or sterilized.

Curl body directions

A flat curl may turn either clockwise or anti-clockwise. The clockwise curl has a body that moves around to the right and the anticlockwise a body that moves around to the left. Reverse curls are rows of alternating clockwise and anticlockwise pin curls; these will produce a finish that has continuous 'S' waves, similar to the effect of finger waves throughout the style. Stand-up pin curls produce a waved movement the same as barrel curls but with raised crests and deep troughs

STEP-BY-STEP: FINGER WAVING

Finger waving is a technique of moulding wet hair into 'S'-shaped movements using the hands, fingers and a comb. It is sometimes called *water waving* or water setting. This technique is used as part of an overall finished style.

To form the wave:

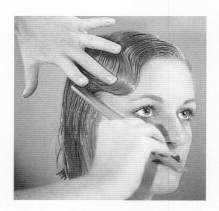

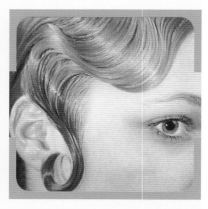

1 Use one finger of one hand to control the hair and to determine the position of the wave. Comb the hair into the first part of the crest, and continue along the head.

2 Place the second finger immediately below the crest formed, and comb the hair in the opposite direction.

3 Form the second crest similarly, to complete the final wave shape.

ALWAYS REMEMBER

Finger waves

- Finger waving is most successful on medium or fine hair that is about 10 cm long. Coarse or lank hair can be more difficult.
- Setting lotion, gel, mousse or emulsion will be needed to hold the waves.
- Keep your forearm level with or slightly higher than the wrist, to control the hair and your hand during waving. Hold the comb upright and do not use too much pressure when combing, to avoid tearing the scalp.
- Keep the waves the same size and depth. About 3 cm (the tips of two fingers) between crests is usually best.
- Pinching or forcing the crests will distort the waves. Correct control and angling will produce the best waves.
- Positioning is important. Comb the hair to make it lie evenly, and return it to this position after each wave movement is complete.
- Keep the hair wet (but not dripping) during waving. If you find that it is drying out, dampen it while you work and apply more setting lotion if necessary.
- Dry the completed shape under a hood dryer, preventing movement or slip.
- Dressing out should not disturb the waves. The hair is not normally brushed.

TOP TIP

Clean the gauze filters out on the back of the hairdryer regularly. This will prolong the dryer's life and enable it to work more efficiently too.

Blow-drying hair

When the client's hair is ready for styling the excess water must be removed so that:

- only damp and not saturated sections are worked upon
- the benefits of any styling products are not lost or diluted by excess water

- the client's hair is not unduly or excessively overheated
- the client remains comfortable and dry throughout the process rather than getting wet and cold.

TOP TIP

Remember that hairdryers only blow out what they suck in from the other end.

Always make sure that the filter is attached to the back of the dryer, as this will prevent your client's hair from being sucked in. This is not only embarrassing for you, it is also dangerous and unpleasant for the client.

TOP TIP

Alpha and beta keratin

The keratin bonds of un-stretched hair, hair in its natural state, are in alpha keratin state and the keratin bonds of stretched hair are in beta keratin state. This is the basis of cohesive or temporary setting.

Drying the hair – roots to points As you learnt in Level 2, blow-drying the hair from root to point ensures that the cuticle lays flat therefore reducing flyaway frizz and smoothing the overall result.

ALWAYS REMEMBER

If too much heat is applied to the hair, it will sustain permanent, un-repairable damage. Excessive heat will damage the cuticle, remove essential moisture from the hair and make hair very difficult to manage.

Practice always makes perfect. I still have mannequin heads at home and I practise the styles I am going to present to an audience. I also time myself, which helps me when I am putting a show together.

Patrick Cameron

Dressing hair

Dressing is the process of achieving a finish on previously set or styled hair. Dressing uses brushing and combing techniques, and dressing aids such as hairspray, to keep the hair in place. If you have constructed the set carefully and accurately, only the minimum of dressing will be required. The initial part of any dressing out requires a partial removal of the roller or styling marks so that the final dressing can begin. The hair is then finished with back-brushing and smoothing or the use of heated styling equipment such as tongs, straighteners, crimpers, etc.

TOP TIP

The temporary bonds (beta keratin state) within the hair are only reformed when the hair is completely dry. However, they will revert to their former (alpha keratin state) if moisture or water is introduced to the hair.

Back-brushing Back-brushing is a technique used to give more height and volume to hair. By brushing backwards from the points to the roots, you roughen the cuticle of the hair. Hairs will now tangle slightly and bind together to hold a fuller shape. The amount of hair back-brushed determines the fullness of the finished style.

Tapered hair, with shorter lengths distributed throughout, is more easily pushed back by brushing. Most textures of hair can be back-brushed because it adds bulk; the technique is especially useful with fine hair.

ALWAYS REMEMBER

Always let the hair cool before removing all the rollers. Hot hair may seem dry when you first check it, but when it cools it may actually be damp.

Step-by-step: back-brushing

1. Hold a section of hair out from the head; for maximum lift, hold the section straight out from the head and apply the back-brushing close to the roots.

2. Place the brush on the top of the held section at an angle slightly dipping in to the held section of hair.

3. Now, with a slight turn outwards with the wrist, turn and push down a small amount of hair towards the scalp.

4 Repeat this in a few adjacent sections of hair.

5 Smooth out the longer lengths in the direction required, covering the tangled back-brushed hair beneath.

Note: The more the hair is back-brushed the greater the volume and support will be.

Back-combing This technique is similar to back-brushing; however, in this situation a comb is used rather than a brush to turn back the shorter hairs within a section to provide greater support and volume. Back-combing is applied deeper towards the scalp than back-brushing and therefore provides a stronger result.

TOP TIP

Back-combing is applied to the underside of the hair section. Do not let the comb penetrate too deeply otherwise the final dressing and smoothing out will remove the support you have put in.

ACTIVITY

You will need to gather information from a variety of sources for this activity. You should collect together photographs, digital images and magazine clippings about different creative styling, blow-drying, setting and dressing techniques.

Make sure that you include styles for long hair as well as short; for weddings, special occasions and casual wear.

Then describe:

1 How the different styles were achieved.

2 Why each is suitable for its purpose.

3 The equipment (with examples) that were used to create the effects.

4 The products (with examples) used to help hold or define the effects.

Added hair Hairpieces, wefts and extensions have been around for many years and are a popular option for providing temporary, alternative styling effects. They are a relatively quick and cost-effective way of achieving:

◆ more volume, for people who have finer or lank hair, or

◆ additional length for people that want their hair to appear longer.

There are many types of hairpiece or extension available. People can choose between:

◆ added hair pieces that match their own hair – therefore disguising the fact that hair has been added in the first place

◆ added hair pieces that contrast against their hair to quickly achieve highlighted or partial block colour effects – a way of achieving colour for those people who may otherwise not be able to have colour

◆ added hair pieces that provide strong fashion statements as an extension of their personality and personal image.

ACTIVITY

1 What difference would there be from setting with rollers on base as opposed to off base?

2 What type of effect do you get from clockspring pin curls?

3 What type of effect do you get from barrel curls?

4 Why do you need to brush the hair first when combing out?

5 What is the difference between back-combing and back-brushing?

The base or fixing of the hair must be hidden well and should be secure. Your client must feel confident that the hair will stay in place for the duration it is worn. You can make sure that the fixing is secure by interlocking two hair grips into the hair (back-combed if necessary) and placing the comb or clip behind them.

Most hair pieces/extensions are artificial. They are made from a form of acrylic or nylon, which is more difficult to manage than real hair itself. Make sure that you tell your client this during the styling, as they cannot be styled after they have been placed. Show them how to remove them and any hair ornaments without putting any tension on the rest of the hair which may cause damage.

Acrylic pieces can only be cleaned in cold water using specialist products suitable for extensions. Advise clients that they should avoid using hairspray or other styling products on them, as this tends to bind them – making them difficult to separate or manage.

For more information about hairpieces, wigs, and so on, visit the website for Banbury Postiche www.banburypostiche.co.uk.

Not all hairpieces are artificial; interest is growing and there are more natural hair-type pieces available. These are more expensive, but if they are looked after properly they do pay back in the longer term, as they can be coloured, styled, permed, washed and dried just like normal hair.

Use the styling mirror to check your progress As you work, keep using the mirror to check the shape that you are creating. If you find that the outer contour is misshaped or lacking volume, do not be afraid to go back to resection and back-brush/comb again or even take down the area of hair that lacks shape and support.

Heated styling equipment

Electrical accessories health and safety checklist

✓	Never get too close to the client's head with hot styling equipment.
✓	Never leave the styling equipment on one area of hair for more than a few moments.
✓	Always replace the styling tools into their holder at the work station when not in use.
✓	Always check the filters on the back of hand dryers to make sure that they are not blocked (this would cause the dryer to overheat and possibly ignite).
✓	Look out for trailing flexes across the floor or around the back of styling chairs.
✓	Let tools cool down before putting them back into storage.
✓	Always check for deterioration in flexes or equipment damage.
✓	Never use damaged equipment under any circumstances.

ALWAYS REMEMBER

Heated equipment – advice for the client

Heated styling equipment such as straightening irons and tongs work at very high temperatures. When you are ready to use them, ask your clients to keep their head still as any sudden movement or twisting could draw the hot surfaces closer to the scalp or even burn them.

Straightening irons, heated brushes and tongs Electric curling tongs, heated brushes and straightening irons are popular ways of applying finish to a hairstyle. They are particularly useful in situations where:

◆ setting or blow-drying will not achieve the desired look

◆ the hair is not in a suitable condition to be dried into shape.

Professional heated tongs (and many hair straighteners) usually have a thermostatic temperature control, this is particularly useful so that you can *dial up* the heat setting required to achieve the desired effect for a particular hair type. This eliminates the chance of damage to the hair caused by excess heat.

Straightening irons, and particularly ceramic straightening irons, have been a very popular way of calming unruly hair. They work by electrically heating two parallel plates so that the hair can be run between them in one movement from roots to ends, smoothing out the unwanted wave or frizz in the process.

Ceramic straighteners have been particularly successful as they heat up in just a few moments and have a higher operating temperature than metal irons (220°C). This very high temperature is damaging to hair. However, because they can transfer heat quickly and smoothly to the hair they are very effective in creating smoother finishes with reduced adverse effects. Because of their temperature you must check them before you introduce them to the hair so that you do not permanently damage or break the client's hair.

When straightening is needed to complement the look on longer hair, it is often better to straighten each section as the blow-dry proceeds. If you start underneath, each section is completely finished before you move on up the head. The hair will stay flatter from the outset and each section is totally dry, stopping the hair from reverting to its previous state (i.e. reverting to alpha keratin).

The use of crimping irons tends to go through phases of popularity at least once every decade or so. They too have parallel fixed plates but these are wavy and produce flat 'S' waves on longer hair. They are a great styling accessory for competition and stage work, as crimped effects are visually striking and very unusual. In staged hairdressing shows, models with crimped hair will often accompany the look with strong fashion colours.

Unlike tongs and straightening irons, crimpers are not turned, twisted or drawn through the hair.

1   Each mesh of hair is started near the head and works down to the points of the hair.

2   The meshes should be no wider than the crimping irons and are crimped across the width of the plates.

3   After a few moments of heating each section of the mesh, the crimpers are moved to the last wave crest created and pressed again.

4   This is repeated down the lengths of the hair until all of the hair is crimped.

5   The final look is not combed out or brushed, but allowed to fall in waved sections.

Crimping is not advisable on shorter, layered hair unless a frizzy, fluffy look is wanted. The most successful results are on longer, one-length hair.

Provide aftercare advice

A good professional service is supported through sound advice and recommendation. The client appreciates what you do and this can be seen by the repeat business. However, you need to tell them how to achieve and maintain the same effects at home.

The checklist below outlines a summary of the things that you should cover.

Home and aftercare checklist

✓	Talk through the style as you work; that way the client sees how you handle different aspects of the look.
✓	Show and recommend the products/equipment that you use so that the client gets the right things to enable them to get the same effects.
✓	Tell them what products or styling practices they need to avoid and why.
✓	Explain how routine styling with heated equipment, such as tongs or straighteners, can have detrimental effects.
✓	Demonstrate the techniques that you use so they can achieve that salon hair look too.
✓	If you have put the client's hair up, or added a hair piece or extensions, give them advice on how to take the style down/remove the hair pieces.

REVISION QUESTIONS

Q1. Self-cling rollers are commonly known as _____ rollers.   Fill in the blank

Q2. Humidity in the atmosphere will help to retain set hairstyles.   True or false?

Q3. Which of the following are pin curling techniques?   Multi-selection

Rik-Rak winding   ☐   1

Off-base rollering   ☐   2

Barrel curls   ☐   3

Clockspring curls   ☐   4

Tonged curls   ☐   5

Chopstick winding   ☐   6

Q4. The keratin bonds of stretched hair are said to be in the beta state.   True or false?

Q5. Which chemical bonds within the hair are not affected during setting?   Multi-choice

Hydrogen bonds   ○   a

Di-sulphide bonds   ○   b

Salt bonds   ○   c

Oxygen bonds   ○   d

Q6. Heated rollers are a quick way of setting wet hair in to style.   True or false?

Q7. Hair set with rollers 'on base' produces which of the following results and effects?   Multi-selection

Increased body at the roots   ☐   1

No body at the roots   ☐   2

No movement at the ends   ☐   3

Straighter effects   ☐   4

Wavy effects   ☐   5

Same as blow-dried effects   ☐   6

Q8. Rik-Rak stylers produce a _____ waved effect.   Fill in the blank

Q9. Which item of equipment would smooth and flatten frizzy, unruly hair best?   Multi-choice

Curling tongs   ○   a

Ceramic straighteners   ○   b

Crimping irons   ○   c

Blow-dryer   ○   d

Q10. Hair should be brushed out with a radial brush before dressing.   True or false?

9 Creatively Dress Long Hair

LEARNING OBJECTIVES

◆ Be able to identify what clients want.

◆ Be able to advise clients about suitable, long hair styling options.

◆ Be able to work safely during setting, styling and dressing long hair.

◆ Understand the factors that can affect the clients' styling options

◆ Be able to style and dress long hair creatively.

◆ Be able to communicate professionally and provide aftercare advice.

KEY TERMS

cornrowing

knots

plaiting

pleat

roll

traction alopecia

twists

weaving

UNIT TOPIC

Creatively style and dress long hair

INTRODUCTION

Long hair can be daunting, particularly if you are not used to working with it on a regular basis. So it is important to maintain your skills. 'Hair ups' need not be too difficult, especially if you keep the basics firmly in your mind, so:

◆ have a clear idea of what you are trying to achieve

◆ build enough structure and support into the look to ensure that the finished effect is comfortable and durable.

It is also important that you:

◆ choose a style suitable for the occasion

◆ assess whether a particular look or effect is going to suit the client

◆ give the client enough visual information to help her get an idea of what can be achieved.

Creatively style and dress long hair

PRACTICAL SKILLS

Find out what clients want by questioning and visual aids

Learn to work safely and carefully when setting, styling and dressing long hair

Look for things that will influence the ways that you do your client's hair

Learn the techniques that will enable you to achieve creative long hair effects

Learn to use a variety of tools and equipment to create different long hair effects

Learn how to work with different densities of hair in order to optimize the final effects

Provide aftercare advice to your clients

UNDERPINNING KNOWLEDGE

Know how to prepare the hair for a variety of long hair dressings

Understand the factors that can affect hairstyling options

Recognize the limiting factors that can affect styling options

Know how to communicate effectively and professionally

TOP TIP

Long hair is easier to work with the day after it is washed because clean, well-conditioned, just-washed hair is often too 'silky and slippery' to work with when it has been dried.

Therefore, if you book a long 'hair-up' get your client to wash their hair the night before.

Maintain effective and safe methods of working when dressing long hair

Long hair dressings are always carried out on dry hair at the styling units; because of this, the main health and safety issues relate to work positioning and client comfort.

INDUSTRY ROLE MODEL

PATRICK CAMERON   Patrick Cameron Training School

" For me, education is the key to success. I now spend my time working with hairdressers, showcasing and demonstrating the techniques I have developed. Through demonstrations, I try to create long hair looks that are stunning but are so simple that even novice hairdressers can feel confident enough to try them for themselves.

As much of the essential information surrounding client preparation, health and safety, and tool care and maintenance are common throughout Level 3, you will find that these aspects are comprehensively covered in CHAPTER 15 on health and safety.

TOP TIP

Always make sure that your choice of non-conventional hair accessories are safe and hygienic to use. These can be anything from flowers and foliage to feathers, crystals and sequins.

Preparing the tools and equipment for setting and dressing hair

Make sure that you have prepared the area first. Get everything that you need together beforehand and this includes the equipment that you need, as well as the products. You should have your trolley prepared with all the materials you will need. Styling materials should have been previously prepared by thorough washing/sterilizing, and combs, brushes, sectioning clips, etc. should be all cleaned, sterilized and made ready for use.

Service timings Different services require different amounts of time; you need to be clear in your own mind how long these services take. For example, a French pleat is going to take less time than a full bridal look. This means that the cost implications in relation to materials used, cost of labour and profit margin required by the salon will all vary considerably. You need to be aware of how much your salon charges for each type of service so that you can inform clients about the range of options available to them.

ALWAYS REMEMBER

Make sure that you always:

✓ Follow your salon's policy with respect to standards and public image

✓ Prepare yourself and your materials in a professional manner ready for work

✓ Wear the minimum of jewellery that can dangle or tangle in the client's hair

✓ Be aware of your personal hygiene

✓ Minimize the risk of cross-infection to your colleagues and clients

✓ Work responsibly and respectfully within the salon's service timescales

Styling materials and tools

Equipment	Used for	Benefits	Precautions	Cleaning
Heated rollers	Adding more wave/curl to dry hair prior to dressing out	Makes hair easier to style, improves handling, gripping and pinning into place.	Electrical item: remember to check the condition of the lead before use. These are very hot when they are first put into the client's hair. Use cotton neck-wool as an insulating base between the bottom of the curler and the client's scalp.	Wash with hot soapy water and scrub clean, then dry and put back on to their appropriate heating stems ready for future use.
Heated tongs	Dry curling only. Used for adding more movement	Provides quick, added support for hair that lacks movement prior to dressing the hair. Or, alternatively, a quick, simple option for styling tendrils of hair as a way of finishing a 'hair up'.	Electrical item: remember to check the condition of the lead before use. Very hot when in use, be careful not to burn the client or yourself.	Spray with cleaner and dry thoroughly after use. Remove hairstyling product build-up from the curling surfaces. NOTE: Look out for product build-up on heated styling equipment. Clean before use to avoid 'grabbing'.

(Continued)

Equipment	Used for	Benefits	Precautions	Cleaning
Heated straightening irons	For use on dry hair; makes hair straighter or adds curls/movement	Provides a failsafe way of improving the look and smoothness of the hair before dressing. Provides quick, added support for hair that lacks movement prior to dressing the hair. Or, alternatively, a quick, simple option for styling tendrils of hair as a way of finishing a 'hair up'.	Electrical item: remember to check the condition of the lead before use. Very hot when in use so be careful not to burn the client or yourself.	Spray with cleaner and dry thoroughly after use. Remove hairstyling product build-up from the curling surfaces. NOTE: Look out for product build-up on heated styling equipment. Clean before use to avoid 'grabbing'.
Tail or pin comb	Sectioning hair into workable sizes dependent on the setting, plaiting or twisting technique used.	They will provide tension when combing through sections and help to manage the hair.	The point ends of tail combs can be metal or plastic, but both types are sharp. Be careful in sectioning or combing that you don't scratch the client's scalp.	Wash in hot, soapy water then keep in Barbicide™ until needed. When needed for setting they should be rinsed and dried first.
Straight combs	Dressing out hair; enabling the hair to be back-combed and smoothed	They will provide tension when combing through sections and help to manage the hair.	Be careful during combing or back-combing that you don't scratch the client's scalp.	Wash in hot, soapy water then kept in Barbicide™ until needed. When needed for setting they should be rinsed and dried first.
Flat, bristle brushes	Flat, bristle brushes when used on wet hair can work setting agents through, distributing the product more evenly	Used during dressing to remove roller/setting marks, smooth or shape the hair or introduce back-brushing into the hair.	Be careful not to brush too vigorously as it can make hair static or be painful to the client.	Wash in hot, soapy water and scrub clean to remove hair and particles. They should be dried thoroughly and put in a UV cabinet for sterilization.
Ornamental grips and hair pins	Fixing hair into position as part of the finished hairstyle	Provides simple ways of accessorizing hair to create, decorative or themed effects.	Most metal ornaments have sharp edges; use care when putting them in and when removing them to avoid tugging and snatching the hair.	Hand wash carefully in warm soapy water. Can be put into a UV cabinet for storage purposes.
Pin clips	Wet or dry setting hair	Provides a narrow curl stem that can be positioned either flat against the skin or standing away.	Pin clips tend to have fairly strong durable springs. These can pull the hair if not careful.	Washed with hot soapy water and scrubbed clean, then dried and put back in trolleys/trays.
Hair grips	Fixing hair into position as part of the finished hairstyle	Choose hair grips or similar with crimped leg for better grip. Hair ups, partial hair up/back effects.	Grips and pins are metal with sharp points; use care when putting them in and when removing them to avoid tugging and snatching the hair.	Wash and dry and put into trolleys or trays.
Hair 'doughnut' rings	Used for providing a base for chignons	Great for adding more bulk to hair that lacks density or volume. Makes **knots** and chignons an easy option for anyone with long enough hair.	Take care when removing as hair grips tend to bond on very well.	Washed in hot, soapy water. Put into washing machine.

Heated rollers

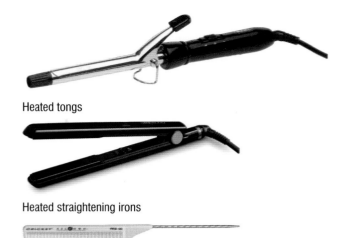

Heated tongs

Heated straightening irons

Tail or pin comb

ACTIVITY

Styling equipment

For each of the items listed below; write down how these items can be used for styling hair, what sort of effects each one will produce and how each item should be hygienically maintained in the appropriate column.

	How could the item be used?	What effects will it produce?	How should this item be hygienically maintained?
Heated rollers			
Heated tongs			
Heated straightening irons			
Crimping irons			

Added hair and hair pieces

Added hair type	Advantages	Disadvantages
Synthetic artificial wefts with clip-on or comb attachments 	Very easy and relatively quick to apply. Colours don't fade; they are pre-coloured in a variety of single or multi-toned effects Can be shaped and cut easily (not with best hairdressing scissors) to suit or achieve the desired effect. Bulks up fine hair, or hair that lacks volume. Can provide a quick cost-effective solution to people who want longer hair. Can be removed at home with care, providing the right advice has been given.	Needs careful handling, tends to get matted very easily. Needs to be brushed regularly once applied, to avoid tangling. Avoid heat styling as all synthetic extensions are very susceptible to becoming misshapen or damaged when excess heat is applied. Can cause traction alopecia if incorrectly applied.

(Continued)

Added hair type	Advantages	Disadvantages
Natural hair pieces with stitched bases	Lasts for a long time with care and attention. Can be shampooed, conditioned, styled and shaped with heated equipment just like natural hair. Provides a cost-effective solution for repeated long term use. Fairly easy to apply for the client as well as the professional. Colours can be changed just like natural hair (i.e. decoloured or lightened, toned and coloured). Can be shaped and cut easily to suit or achieve the desired effect. Bulks up fine hair or hair that lacks volume.	Need some care and attention with handling; needs to be shampooed, conditioned and detangled now and again. Needs to be gripped into place to secure the added hair base against the rest of the hair. Can cause traction alopecia if incorrectly applied.
Synthetic – artificial full-head (wigs)	A wide range of price options, providing choice to suit many budgets. Lots of choices in natural, fashion and multi-toned effects. Pre-styled with little maintenance, can be shampooed and reshaped back into style with little effort (not with best hairdressing scissors). Easy option for those who want to look different or would like 'instant' hair length change. Good-quality synthetic wigs are long lasting, can take daily wear with care for anything up to one year. (Ideal for clients who have lost hair through ill health, radiology or chemotherapy).	Need some care and attention with handling; need to be shampooed, conditioned and detangled now and again.
Natural full-head hair pieces (wigs)	Can be pre-made or made to order; provides options for any length depending upon needs. Beautifully crafted to suit the wearer. Can be shampooed, conditioned, shaped and styled just like normal hair. Easy option for those who want to look different or would like 'instant' hair-length change. Capable of daily wear (with care) for several years. (Ideal for clients who have lost hair through ill health, radiology or chemotherapy).	Expensive

ALWAYS REMEMBER

Traction alopecia

The tension on the hair is increased with the use of added hair; this can exert exceptional pressure on the hair follicle, particularly on scraped-back hairstyles or on free-hanging effects.

In extreme cases hair loss may be caused by this continued pulling action; areas of hair become thin and even baldness may be result!

For more information on added hair/hair pieces see CHAPTER 8 Creatively style and dress hair, pages 182–183. To review information on brushes see CHAPTER 8 Creatively style and dress hair, pages 168–169.

Brushes The variety of brushes available for combing out and hairdressing has been comprehensively covered elsewhere in this book.

ACTIVITY

Styling materials

For each of the items listed below; write down how these items could be used for dressing hair and how it should be maintained in the appropriate column.

	How could the item be used?	How should it be maintained?
Styling item:		
Artificial hair wefts		
Natural hair piece		
Full head hair piece (wig)		
Hair 'doughnut' ring		

TOP TIP

Always follow the manufacturer's instructions for using dressing and finishing products within the salon.

ALWAYS REMEMBER

Control is the secret when handling long hair, when you have hold of the hair while brushing do not let go or lose your grip. You need to keep an even tension upon the hair until it is fixed/pinned into position.

Styling products

There are many different products available for dressing and adding hair. As a rule of thumb, you should always try to use the minimum amount of product while you are dressing the hair as it is very easy to overload the hair. This will:

◆ make the hair harder to manage as it starts to lock together while you work with it

◆ make the hair look greasy, which you will not really be able to much about, apart from starting again

◆ look unprofessional – in providing a very average, amateur result

◆ look unnatural – which most people want to avoid.

Most hairdressers who use lots of product during the dressing out phase do so because they:

◆ do not have control of the hair they are trying to manage

◆ are taking sections that are too large

◆ have not applied sufficient back-brushing or interlocked grips

◆ lack the confidence to create the result in the first place, or

◆ have had very little experience in this type of work.

Specific styling products for hair ups There are some products that are extremely useful in helping you to create the right sorts of results and they fall into the following ranges:

Product	Do use . . .	Avoid . . .	
Hairspray	Finer, atomized, dry hairsprays. They are particularly useful during the dressing out as a way of smoothing unwanted frizziness. Hairsprays that repel/resist moisture.	Slow drying, hand-pump-type hairsprays as these tend to 'clog' up or spray globules onto the hairstyle, making it impossible to get a satisfactory finish on the hair. Applying too much hairspray as a finish, it's too easy to spoil the effect at the end when you can do little about it.	
Heat protection	Heat protection sprays on the hair if you are going to use heated styling equipment and particularly, straighteners, as these operate at very high temperatures.	Using these as well as other finishing products. It is easy to overload the hair and spoil the result.	
Serums	A little initially, as you work and brush the hair when you start. Or as a finishing product by applying a small amount on any free hanging tendrils, but not on any part of the fixed hairstyle.	Applying too much at any point during the dressing as it will make the hair VERY oily very quickly.	
Wax	Some dry wax as a finishing product to bind 'spiky' or sculpted areas.	Using them at any other point of the dressing.	

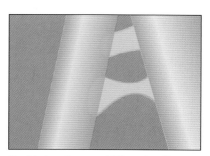

Flexible bonding capabilities of styling products on and between hair fibers

ALWAYS REMEMBER

Always use products sparingly and economically within the salon and avoid overloading the hair during styling. Make a point of explaining the benefits of using professional products to your clients so that they can achieve similar effects themselves at home.

How styling products bond the hair together Styling products contain plasticizers and fixatives to hold and support the hair in its shape. There are often other agents and additives within the products that can retain or repel moisture, provide protective sunscreens, add shine and lustre or add definition and shape.

TOP TIP

Moisture is the enemy of any finished hairstyle; it weakens the hold and therefore how long the overall effect will last. This can be improved with some moisture-repelling products such as hairspray. Make sure that you point this out to your clients, so that they can make the most of their hairstyle.

Creatively dress long hair

Consultation

The points that you need to find out about during consultation will alter with the type of service that you are going to carry out.

What are you trying to achieve?

You need clear ideas about what you are trying to achieve so that you do not waste both your time and the client's. Hair ups can be fun, as they tend to be non-routine work and individual in the design. However, you must make sure that you keep to time as over-running is unprofessional and unproductive.

If the work is for a forthcoming wedding, the bride will often try to get a package price deal that includes a series of trials too. From the very first consultation (or before that, if possible) you should ask the bride-to-be to start collecting images of the looks that she would feel happy with. Remember your creation is part of her total look, so ask her to bring in pictures of the dress and the theme of the wedding.

> To achieve a successful result, performance criteria must be learned, as this will help you to get the right result your client is hoping for. The information in this section will help you to succeed.
>
> *Patrick Cameron*

Assess the style's suitability The suitability of the style is the first aspect that you should consider. In most cases, hair up is a special situation. It is not a quick, casual throw up that the client does to get her hair out of the way. Clients come to the salon for special occasion hairstyles that they cannot achieve themselves. The issue is how the client will know if she is going to like her hair up, if she seldom has it styled that way. People who do not normally wear their hair up may have underlying reasons for this, such as:

◆ their hair is too thick

◆ they do not like the shape of their ears

◆ it makes their nose appear bigger

See **CHAPTER 8** Creatively style and dress hair, for more information about:

◆ hair styling on the hair structure

◆ the effects of humidity on hair and hairstyles

◆ products in general.

See **CHAPTER 1** Consultation and advice, for general information on dealing with:

◆ different hair textures

◆ different hair types

◆ different densities of hair

◆ contra-indications to styling

◆ client communications

◆ recording client information.

Other information that specifically relates to dressing long hair can be found here.

◆ they prefer their hair to have volume so they do not like it scraped back

◆ their hair is not really long enough in the first place.

Do not cut corners! Make sure you look at the style from all different angles.

Take your time. You should not hurry the hairstyle as this will have a detrimental effect on the overall finished effect. Check and re-check – it is worth it in the end.

When you have selected a suitable style, and you have shown the client examples of how this would look, you will need to help with her self-visualization. You need to try to rearrange the hair loosely, so that she can get an idea of the balance, proportions and weight distribution.

Simple, quick placement can eliminate problems or unwanted effects right from the start such as:

◆ added height

◆ extra width

◆ revealed physical features – forehead, ears or jaw lines.

If you can convey to your client what it should look like, and she likes what she sees, you are halfway there. It will also save lots of time later.

General rules for style suitability

Physical features	Facial shapes					Protruding ears	Prominent nose	Short neck
	Oval ◯	Round ◯	Heart ♥	Triangular ▽	Square ☐			
Hairstyle								
Vertical **roll**/pleat	✓	With height to compensate	✓	With height to compensate	With height and width to compensate	Volume at the sides to cover	Volume at the sides	Needs to be sleek
Barrel spring curls	✓	✓	✓	✓	✓	Volume at the sides to cover (not triangular)	Volume at the sides	Needs to be sleek
Low knot or chignon	✓	With height	✓	✗	With height	Volume at sides except triangular	✗	✗
High knot	✓	✓	✓	✓	✓	Volume at sides except triangular	✗	Needs to be sleek
Plaits	✓	With height	✓	✓	✗	Volume at sides except triangular	✓	Needs to be sleek
Twists	✓	With height	✓	✓	✓ Use designs with curves and not straight lines	Volume at sides except triangular	✓	Needs to be sleek

Building enough structure and support into the look

The mistake that some stylists make with 'hair ups' involves not making sure that there is enough secure support from the very beginning.

This occurs because they have not:

◆ put enough back-combing into the hair to start and the client is not used to that type of technique

◆ created a comfortable, secure base from which to position and fix the hair into style

◆ maintained an even tension on the hair during dressing, allowing lengths or ends to fall out during the dressing.

The style needs support; it cannot be durable without it. It needs to be secure as well as creative in its effect. It can only be secure if you use back-combing, grips or bands.

Reassure your client first; then do not be afraid to back-comb the hair. It may look as if the whole thing is getting too big, but do not forget you can take out as much as you like within the dressing. Back-combing provides you with a solid base that you can grip to without the fear of the grips dropping out.

Maintain your grip and an even tension on the hair as you work: not too tight or rough, but firm enough to gain a grip on all the hair that you are trying to fold, pin or fix into place.

As you become more experienced in handling long hair, you will find that you will not need to use much spray in the styling stage but only later in the finishing off.

The other main tool for giving structure and support is grips. Kirby grips have one leg with a serrated profile; this helps them to interlock together to stay in the hair much better or to comfortably hold large amounts of hair together without pulling or tugging.

> For more information on back-brushing, back-combing and other dressing techniques see **CHAPTER 8** Creatively style and dress hair, pages 181–182.

TOP TIP

Back-combing is applied to the underside of the hair section. Do not let the comb penetrate too deeply otherwise the final dressing and smoothing out will remove the support you have put in.

ALWAYS REMEMBER

Do not put grips in your mouth, it is an unhygienic practice and could cross-infect your client.

ACTIVITY

Styling and finishing products

Note the products that your salon uses and, for each one, complete the information in the table below.

Product name	How is it used?	What hair type is it for	Benefits to the hair

It is very important to understand creative and classic styling for long hair because this is an area of hairdressing which will continually come up throughout your career, and learning these skills will set you on the road to success.

Patrick Cameron

Interlocking grips

Consultation checklist before dressing hair

✓	What is the purpose for the dressing/hair up?
✓	What is the texture, type, tendency and condition of the hair like?
✓	How much hair is there? How long is the hair? Is there enough to achieve the result?
✓	Are there any limiting factors or adverse hair or scalp conditions?
✓	What type of effect does the client want?
✓	Does the hair type/texture/tendency/amount and length support the client's ideas?
✓	What tools and equipment do you need?
✓	How much time will it take?
✓	How much will it cost?
✓	Do you need to use any products to achieve the effect?
✓	Will the client benefit from using these products too?
✓	Agree and confirm desired effects with the client prior to starting the service.

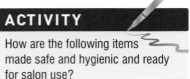

ACTIVITY

How are the following items made safe and hygienic and ready for salon use?

1 Heated curling tongs
2 Crimpers
3 Ceramic straighteners
4 Hot brushes
5 Heated rollers
6 Hairdryers

ALWAYS REMEMBER

Because longer hair takes more time to dry, it is easy to apply too much heat during the drying. If you have set the hair prior to adding hair or dressing the hair, make sure that you do not burn the client's hair or head.

ACTIVITY

It is important to know which sort of hairstyle will suit which facial shape. For this activity, simply tick to identify which type of hairstyle is suitable for the type of facial shape.

Hair Style	Facial shapes				
	Oval ◯	Round ◯	Heart ♥	Triangular ▽	Square ☐
Vertical roll/pleat					
Barrel curls					
Low knot or chignon					
High knot					
Plaits					
Twists					

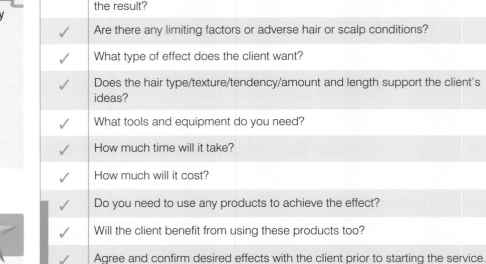

CourseMate video: Vertical roll (French pleat) with ornamentation haircut

Vertical roll (French pleat)

The vertical roll is a formal classic dressing that suits many special occasions. The hair can be enhanced further by the additions of accessories or fresh flowers. As we saw earlier, back-combing is an essential aspect for creating a solid foundation. This should be your starting point for the step-by-step procedure.

STEP-BY-STEP: VERTICAL ROLL (FRENCH PLEAT) WITH ORNAMENTATION

1 After greeting your client, conduct a full consultation to determine the style and effect desired.

2 Place heated rollers in the direction in which the hair is to be dressed. This is called a directional set.

3 Taking each section and working throughout the head, back-comb the hair at the root area only.

All hair can be pulled back into this style, or some in the front area can be left out to be incorporated into the style later.

4 Smooth over the hair using a soft bristle dressing brush, gently sweeping the hair from one side back and slightly upwards.

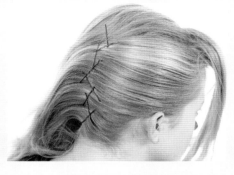

5 Secure the hair in the centre from the nape to the crown with interlocking grips. The final grip should be placed just below the crown, facing downwards.

6 If hair has been left out in the front area, this can now be personalized into the style. Smooth the hair back and secure with pins to create your finished style.

7 Fold the hair neatly over the grips, tucking hair under to form a roll. The end of a pin tail comb can be used to tuck in any stray hairs.

8 Secure the roll using pins, making sure that no grips or pins are visible. Use a dressing product and spray to smooth and hold the style.

9 The finished style is seen here with added ornamentation and finishing products applied.

Plaiting hair

Plaiting is a method of intertwining three or more strands of hair to create a variety of woven hairstyles. When this work is done for specific occasions, it is often accompanied by ornamentation: fresh flowers, glass or plastic beads, coloured silks and added hair are also popular.

The numerous options for plaited effects are determined by the following factors:

◆ number of plaits or **twists** used

◆ positioning of the plait or twist across the scalp or around the head

◆ the way in which the plaits are made (under or over)

◆ any ornamentation/decoration or added hair applied.

'Plaits' usually refers to a free-hanging stem(s) of hair that is left to show hair length. This length can be natural or can be extended by adding hair during the plaiting process; an example is the 'French' or 'fish tail' plait.

Step-by-step: Loose plaiting (three-stem (loose) plait) The three-stem plait is easily achieved and demonstrates the basic principle of plaiting hair.

1   Divide the hair to be plaited into three equal sections.

2   Hold the hair with both hands, using your fingers to separate the sections.

3   Starting from either the left or the right, place the outside section over the centre one. Repeat this from the other side.

4   Continue placing the outside sections of hair over the centre ones until you reach the ends of the stems.

5   Secure the free ends with ribbon thread or a covered band.

ALWAYS REMEMBER

The tension used in plaiting can exert exceptional pressure on the hair follicle and scalp-type plaits/cornrows create more vulnerability than free-hanging plaits. In extreme cases hair loss may be caused by this continued pulling action; areas of hair become thin and baldness may even result.

This condition is called 'traction alopecia' and is particularly obvious at the temples of younger girls with long hair who regularly wear their hair up for school, sport or dancing.

Step-by-step: Three-stem 'French' plaiting

1   Brush the hair to remove all tangles.

2   With the head tilted backwards, divide the foremost hair into three equal sections.

3   Starting from the left, or right, cross an outside stem over the centre stem, repeat this action with the opposite outer stem.

4   Section a fourth stem (smaller in thickness than the initial three stems) and incorporate this with the next outside stem you are going to cross.

5   Cross this thickened stem over the centre, and repeat this step with the opposite outer stem.

6   Continue this sequence of adding hair to the outer stem before crossing it over the centre.

7   When there is no more hair to be added, continue plaiting down to the ends and secure them.

"
Long hair skills are more important now than they have ever been. So don't become the hairdresser who doesn't feel confident in dressing long hair. Remember you are perfecting many skills as a hairdresser and long hair is the artistry of your craft, so enjoy it.

Patrick Cameron

Weaving

Hair **weaving** is a process of interlacing strands of hair to produce a wide variety of effects. A small area of woven hair can be very effective by itself, or used to highlight a particular part of a style. Hair weaving is also used to place and hold lengths of hair. At its simplest, hair weaving may be used to hold long hair back from the face. This may be done by taking strands of hair from each side, sweeping them over the hair lengths, and intertwining them at the back. More intricate is the *basket weave*, which uses a combination of plaiting, twisting and placing to form many shapes and patterns. It is important to wet or gel the hair before starting to weave. Weave tightly or loosely according to the effect you are aiming for. The hair may be woven as follows.

Step-by-step: Weaving

1   Use six meshes of hair, three in the left hand and three in the right.

2   Start with the furthest right-hand mesh. Pass this *over* the inner two meshes.

3   From the left, pass the outside mesh *under* the next two and *over* one.

4   Continue to the ends of the hair.

5   Tuck in the hair ends and secure them in position.

TOP TIP

Practise on colleagues or models and experiment with different woven shapes before you attempt to weave hair for clients.

ACTIVITY

You will need to gather information from a variety of sources: collect together photographs, digital images and magazine clippings about different types of creative dressings. Look for styles covering a range of themes and for different occasions. You should include styles for adding hair to short hair as well as long hair.

Then describe:

1   How the styles were achieved.

2   Why each is suitable for its purpose.

3   The equipment (with examples) that were used to create the effects.

4   The products (with examples) used to help hold or define the effects.

STEP-BY-STEP: CONTEMPORARY HAIR UP PARTY EFFECT 1

1 After greeting your client, conduct a full consultation to determine the style and effect desired.

2 Create a curved section from ear to ear across the crown of the head, secure with a band and section out the way with a clip.

3 Part another parallel section across the occipital bone, secure and section away.

4 Secure nape hair into a centre pony using a band.

The three pony tails must be in line vertically.

5 Tie a piece of weaving thread around the base of each pony tail, leaving a length of thread hanging which is longer than the client's hair.

6 Loosely plait each pony entwining the weaving thread, leaving 2–3 inches of hair out at the end, and secure with a braiding band.

7 Holding the end of the thread, gently push the plait down towards the base of the pony and secure.

Repeat this process on the remaining two plaits.

8 Mould and sculpt the plaits into the desired shape and secure with pins, back-comb the ends of the hair and spray to create texture and style.

9 The finished style is ready for the party.

CourseMate video: Contemporary hair up party effect 1

STEP-BY-STEP: CONTEMPORARY HAIR UP PARTY EFFECT 2

1 After greeting your client, conduct a full consultation to determine the style and effect desired.

2 Create a triangular parting from nape to crown and secure with a band.

3 Create a parting design from the front recession and section in triangular sections. Leave two fringe sections free.

4 Starting at the front, begin flat twisting the hair and secure around the base of the centre banded pony.

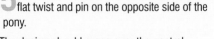

5 When working on the side and back sections, flat twist and pin on the opposite side of the pony.

The design should cross over the central scraped-up triangular panel.

6 Repeat on the opposite side. The design should now cross over the previously twisted sections creating a criss-cross effect.

7 Visually check for close-up detailing.

8 Once all the twisted sections have been firmly secured, twist and pin the remaining hair leaving 2–3 inch ends sticking out to create texture.

9 The finished look. Personalize the style using finishing products.

STEP-BY-STEP: CLASSIC BARREL CURLS (BRIDAL) WITH ORNAMENTATION

1 After greeting your client, conduct a full consultation to determine the style and effect desired.

2 Create a curved diagonal parting from ear to ear and crossing through the crown. Secure back hair out the way and divide the top section into five.

3 Gently brush each section back and secure in place with a braiding band.

4 Lightly spray each section with a holding spray and set either on a heated roller or curl with a heated barrel tong. Then place 3 or 4 heated rollers into the back section of hair, rolling just the ends.

5 If using rollers, allow these to cool before removing one at a time. If using barrel tongs, allow the hair to cool before styling.

6 Knot each of the five sections by wrapping the hair over and pulling the ends through (facing forwards). Pin the ends in place using Kirby grips.

Create loose knots shaped like barrel curls.

7 Take the end of each strand, sculpt and shape into a barrel curl and pin in place.

8 Remove the heated rollers from the back, smooth and shape into the desired style.

9 This elegant style is perfect for your client's wedding day.

CourseMate video: Classic barrel curls (bridal) with ornamentation

STEP-BY-STEP: ASYMMETRIC CHIGNON

1 After greeting your client, conduct a full consultation to determine the style and effect desired.

2 Create a side parting stemming from the crown, then section the fringe out of the way.

3 Smooth all the remaining hair into a lower pony behind the ear and secure in the nape area.

4 Comb the fringe section forwards and spray with hairspray then sweep back in a curve and pin around the base of the pony.

5 Divide the hair into five sections and lightly back comb or brush to create a cushioned bed to place pins. Then smooth the sections and fold them in to secure the base.

A bun ring can be used if required.

6 The classic chignon is smart and stylish.

CourseMate video: Asymmetric chignon

STEP-BY-STEP: WOVEN EFFECT (BASKET WEAVE)

1 After greeting your client, conduct a full consultation to determine the style and effect desired.

2 Create a parting from the crown to the fringe.

3 Create four even meshes woven from the top parted section of hair.

4 Take a front panel the same width from the fringe area and secure it underneath the lifted woven meshes.

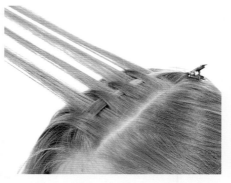

5 Place the top woven sections back down and lift alternating weave sections from in-between.

Repeat the process following the triangled section.

6 Secure the weave into the finished style.

STEP-BY-STEP: CORNROWING

1 After greeting your client, conduct a full consultation to determine the style and effect desired from cornrowing.

2 Plan your design, image shows pattern to be created.

3 Section the next parting to plait a straight cornrow and repeat the pattern. Divide the hair into three strands and carry out an underhand individual braid stitch.

4 Bring the left strand under the middle strand, but between it and the right strand, so that it becomes the middle strand.

5 Now, bring the current right strand under the current middle strand.

6 Once again, bring the current left strand under the current middle strand.

Continue this process taking up sections of loose hair from underneath as you move along the track.

7 Section the hair ready for the final crossed over plait.

8 Following the parted sections plait the cornrow crossing from one side to the other to create a chain effect.

9 The finished cornrow style requires real dexterity.

CourseMate video: Cornrowing

Provide aftercare advice

A good professional service is supported through sound advice and recommendation. The clients appreciate what you do and you can tell that by the way they keep coming back on a regular basis. However, unless you tell them how to achieve and maintain the same effects at home, it is unlikely that anyone else will get to see the endeavours of your work.

Before the clients leave the salon, you need to give advice on what they should be doing at home themselves and the products and bad practice that they should be avoiding.

In your summing up you should at least cover the following.

Aspect to cover	What to say
The effects of humidity on the hairstyle	Tell the client what will happen if they have a bath before going out. If they are in a steamy environment for a long period, the style could drop. Showering might be a better option.
How traction alopecia can occur	Advise the client to avoid putting their own hair up with too much tension on their hair. Explain the signs of traction alopecia and long-term effects.
Products for home use	Explain the benefits of buying the products that you use with the salon, as opposed to buying from the supermarket without professional recommendation or knowledge.
How to take down the style – remove the grips and pins	Explain to the client how they can remove the pins without tension or pulling, tell them where to start and what to avoid.
What sorts of materials and ornaments they should avoid	Tell them about the benefits of using covered bands and fabric covered 'bungees'. Tell them what sorts of ornaments would be suitable and how they can be positioned/fixed without excess weight or damage to the hair.

(Continued)

Aspect to cover	What to say
What sorts of tools and equipment they should avoid	Explain how the use of heated styling equipment can damage hair if temperature settings are too high. Tell them about the benefits of heat-protecting sprays.
Hair condition	Explain how the hair can be maintained in good condition, what products to use, how they should brush their hair to avoid tangling or damage.

"

Aftercare advice is very important to aid the client in taking down or maintaining the long hair up. I find a little aftercare advice really helps the client to brush down their style and control any damage which might have been done to the hair through back-combing, etc.

Patrick Cameron

REVISION QUESTIONS

Q1. A _____ pleat is a type of vertical roll.   Fill in the blank

Q2. Humidity in the atmosphere will make a set drop.   True or false?

Q3. Which of the following dressings are traditionally long 'hair up' styles?   Multi-selection

Plaits   ☐   1

Knots   ☐   2

Twists   ☐   3

Rolls   ☐   4

Braids   ☐   5

Pleats   ☐   6

Q4. Un-stretched hair is said to be in the alpha state.   True or false?

Q5. What condition is created by excessive tension being made on the root area of the hair?   Multi-choice

Alopecia areata   ○   a

Male pattern baldness   ○   b

Traction alopecia   ○   c

Alopecia totalis   ○   d

Q6. Heated rollers are a quick way of setting dry hair in to style.   True or false?

Q7. Hair styled with tongs produces which of the following results and effects? Multi-selection

Increased body at the roots ☐ 1

No body at the roots ☐ 2

Movement at the ends ☐ 3

Straighter effects ☐ 4

Wavy effects ☐ 5

Same as blow dried effects ☐ 6

Q8. On long hair, a chignon is a form of _____. Fill in the blank

Q9. Which item of equipment would provide spiralled curls on long hair? Multi-choice

Curling tongs ○ a

Ceramic straighteners ○ b

Crimping irons ○ c

Blow-dryer ○ d

Q10. 'Hair ups' are easier to perform on hair that has just been washed, conditioned and dried off. True or false?

10 Hair Extensions

LEARNING OBJECTIVES

◆ Be able to work safely during hair extension services.

◆ Be able to plan and prepare for hair extension services.

◆ Be able to create a variety of effects by attaching hair extensions.

◆ Know the advantages and disadvantages of using different hair extension systems.

◆ Be able to identify a range of factors that can limit or stop a planned service.

◆ Be able to cut and style hair extensions.

◆ Be able to maintain and remove hair extensions.

◆ Be able to communicate professionally and provide aftercare advice.

KEY TERMS

blending

block colouring

contra-indications

hair extension

mixing mats

multi-mixing

polymer resin adhesive stick

real hair extensions

reconstructive conditioners

silicone pads

synthetic fibre extensions

traction alopecia

INTRODUCTION

There are many different types of hair extension services and clients now have many options available to them. The variety of choice in terms of quality, quantity, colour and cost can make extensions available to all clients. People can choose to have temporary or long-lasting effects, on a budget that suits their pocket.

Hair extensions

PRACTICAL SKILLS

Find out what clients want by questioning and visual aids

Look for things that will influence the way that you carry out an extensions service

Learn how to carry out tests on the hair and how the results will affect the service

Make recommendations and give advice to clients during consultation

Learn how to do a variety of hair extension services and systems

Learn how to cut and style different extensions to create the desired effect

Provide aftercare advice to your clients

UNDERPINNING KNOWLEDGE

Know how to communicate effectively and professionally

Understand the factors that can affect your service options

Understand different extension systems and their suitability in different situations

Know the problems associated with extensions and how they affect the intended service

Know how to work with different types of hair extensions

Know how to style extensions after they have been applied

Know how to communicate effectively and professionally

Maintain effective and safe methods of working when adding hair extensions

Preparation for extension services is quite different from that of other salon services, although the aspects that affect your safety and that of the client are similar to any salon procedure.

INDUSTRY ROLE MODEL

ROSS MILLER Managing Director/Artistic Director at Renella

"I have worked with extensions from the very start of my career. At first all my extension work was secondary to my hairdressing salon work, as I gained experience in them and using them on clients. I then started training in my salon, teaching others hairdressing and then onto giving training on extension work. Ten years later I started training the clients of a manufacturer of extensions. This gave me a very good insight into the world of extensions. It's fun and hard work too, but today I still apply for clients, I still teach and train and I still enjoy it!

The factors that affect your client in relation to comfort are particularly important as most extension services involve a lot of sitting in one position for long periods. Because of the time spent standing, you are also at risk from injury from poor posture.

You therefore need to take care with your personal health and safety and make sure that you do not cross-infect your client from poor hygiene.

You must also take precautions in the way that you work as:

◆ it is easy to spill bonding chemicals on to the client

◆ you may be using hot equipment that could easily burn you or your client

◆ you will need to keep an even tension upon the client's hair while you work, without putting too much pressure upon their hair or scalp.

> Be professional at all times, especially in these additional services. The reason is simple – extensions cost a lot of money and take more work, so people will need you more and expect more, so being professional is paramount.
>
> *Ross Miller*

Potential risks with different hair extension methods

Hot bonded extension systems The tools used with this system reach very high temperatures and can cause serious burns if used incorrectly or accidentally touched. The heated tools have a high operating temperature ranging from 120°C–250°C.

When applying the system ensure that the client's gown is securely fastened; this will protect clients from accidental spillage of the hot polymer resins dispensed from the extension tools. Do not bring the hot applicators that dispense heated resins directly to the client's head; use scalp protectors or scalp shields when working closely to the client's head.

In addition, while you work you should ensure that the heated extension tools, such as dispensers, applicators and heat clamps, are placed safely on a flat work surface, capable of withstanding the high operating temperatures.

Cold bonded extension systems These systems can cause allergic reactions. You should always conduct a skin sensitivity test before applying this system to a client.

Braided/plaited extensions The braiding extension system can cause injury by putting too much tension onto small areas of the scalp. These tension areas are at the very least painful and at worst can cause traction alopecia. Areas of tension are created when hair is pulled excessively at the scalp, which puts a strain on the hair root. At these points, infection can occur as the hair follicles are opened by the excessive tension, which means that bacteria can enter and multiply. It can cause the client severe discomfort and medical treatment may be required.

Sewn-in hair extensions When working with a sewn-in system, take particular care as the needles are very sharp and could inadvertently pierce a client's skin. Where possible use a curved needle as this reduces the risk of injury.

Sewing with thread can put excessive tension on the scalp area causing tension spots; so, again, traction alopecia is also a risk with this system.

General health and safety considerations

◆ Keep your nails trimmed, hands clean and any hand jewellery to a minimum; there is the risk of cross-infecting the client by passing on infections with dirty hands and nails.

◆ Nails and jewellery can easily be caught in the extension hair, which is uncomfortable and painful. If jewellery catches in the extension hair, it may result in having to cut it out.

◆ Wear comfortable clothes that are not too tight as working with hair extensions involves a lot of standing for long periods.

◆ Always wear comfortable shoes that will enable you to stand for long periods.

◆ Consider your posture and work position. Avoid bending over or down to the client – use a hydraulic or gas lift chair to adjust the client to the correct working height. This will lessen your chances of any skeletal injury, fatigue or repetitive strain.

◆ The work area must be kept scrupulously clean. Work stations, hairdressing trolleys, any work surfaces and the surrounding floor area must be spotless.

◆ Extension hair can be a hazard if it falls to the floor and is not swept up; you or others could slip on it.

◆ When applying the hair extension service on a client ensure that gowns or capes are always worn. A gown can protect them from hair clippings, accidental spillage of hot polymer resins, cold fusion products or removal solutions. All of these products can cause burns or an allergic reaction.

◆ Any extension hair that has fallen on the floor *must not be applied* into a client's natural hair.

◆ Always ensure tools are sterilized and cleaned when they have been in use *before* using them on another client.

◆ If you use a razor for restyling the hair after applying the extensions, make sure that you dispose of used blades in a sharps container.

◆ Ensure that the working environment is adequately ventilated so that any vapours or odours from hair extension materials do not make you, your client or others feel ill.

◆ Clear up any spillages of chemicals or materials immediately. Wear appropriate PPE (personal protective equipment: apron and disposable non-latex gloves) if you are handling chemicals or hazardous substances.

> Practise thoroughly on model extension clients and look at the results – what works and what doesn't? Learn through these results and you will go a long way. Make sure you keep abreast of fashion and what's new in extensions. Most importantly, know the health and safety aspects of the role.
>
> *Ross Miller*

ACTIVITY

There are different ways of bonding extensions to natural hair. You can complete this activity by filling in the missing information in the table below. Keep the information for future use within your portfolio.

Extension system	What precautions should you take?
Hot bonded extensions	
Cold fusion extensions	
Plaited extensions	
Sewn-in extensions	

Products and equipment for hair extensions

There are many different products and equipment used to perform the hair extension service and depend upon which application technique you choose to work with.

Below is a list of products and equipment needed when working with several extension systems.

Hair extension connector tools

1 **Professional heat clamp** Heat clamps are designed to attach and seal fibre extensions. (They will not attach real hair extensions.) This tool has two heated tips that melt fibre hair at a specific temperature. It heats from 140°C to 220°C. When the fibre has melted it creates a hard heat-sealed bond that holds a fibre extension in place at the root area or seals the end of a fibre extension when the stylist has created a braid or dreadlocks (see below).

2 **Professional bonding applicator** This tool allows for pre-shaped polymer resin adhesive sticks to be inserted into the tool. It heats to a temperature of 180°C. The resin is dispensed from a nozzle on the applicator, enabling the stylist to deposit the resin onto real hair or fibre hair and then attach an extension in place by creating a polymer resin bond.

3 **Heated pre-bonded extension applicator** This tool has one or two heated tips that reach a temperature of 100°–140°C. It melts pre-bonded extensions, which are extension strands with a wax, protein or keratin polymer resin already applied to the end of a piece of extension hair. Pre-bonded applicators melt the resin on to the natural hair, creating a bond that attaches the extension in place. Pre-bonded applicators will attach pre-bonded fibre hair and pre-bonded real hair.

4 **Needle and thread** A curved mattress needle can be used to sew weaves or wefts on to hair. The weave is sewn on to a scalp plait or corn braid. The thread used is designed to sew weaves onto hair. It is a silk thread that can be purchased in a variety of colours.

HEALTH & SAFETY

◆ Perform an allergy alert or skin test before using cold fusion liquids as some of these products can cause allergic reactions.

◆ Read the manufacturer's instructions contained with these tools and products before using them.

TOP TIP

Bonding applicator tools must be placed on a flat stable work surface when in operation as the heated resin that drips from this tool will deliver serious burns if accidentally spilled on skin.

A bonding applicator

A heated extension applicator

Hair extension connector products

1 **Liquid, cold fusion adhesives** These products vary greatly from gum to rubber- or latex-based liquids. They are applied from bottles that have brushes attached to the lids, similar to the application of a nail polish. Cold fusion liquids will attach fibre hair and real hair in place by painting the liquids onto the real hair and extension hair, then placing the two hair types together. These products are not heated.

2 **Polymer resin adhesive sticks** This is a resin (*glue*) stick inserted into a bonding applicator that is melted and then dispensed from the applicator. They are designed to create extension bonds on individual fibre or real extension hair.

3 **Acrylic fibre (artificial) hair** This extension hair is made from acrylic and designed to be used for extension hairstyles. The fibre is manufactured in a variety of deniers (thickness or fineness). The deniers are designed to look and feel as similar to natural hair as possible. Fibre hair is manufactured in specific colours that, once produced, cannot be altered. The fibre comes in a variety of structures – straight, soft wave, deep wave, braids and dreadlocks, and crinkled structures that mimic straightened African hair textures.

◆ Fibre structurally changes using heat from a hot hairdryer or heated rollers.

◆ Fibre hair will be damaged or melted when excessive heat is applied to it.

◆ Fibre is bought in a variety of lengths, and in any colour ranging from natural colours to neon fantasy colours.

◆ Fibre is bought as bales, packets or is sewn into weaves/wefts.

4 **Real-human extension hair**

◆ **Asian hair** Asian real hair comes mainly from China or India. This hair is cleansed in a caustic soda solution to remove bacteria or infestation. It is then rinsed and dried, bleached and coloured to match the manufacturer's colour shade charts. When wavy or curly structures are required, the Asian hair is permed. After the chemical treatments have been carried out, the cuticle layer of the Asian hair is damaged and must be treated as chemically damaged hair when applied onto natural hair. Asian hair is coarse in structure and very strong, which is why it is used for extensions, wigs, toupées and hair pieces. Asian hair comes in a variety of lengths from 10–24 inches. It is normally packaged in 4 oz packets or sewn into weaves/wefts. The weaves/wefts are sold in 4 oz packets.

◆ **European hair** This hair comes mainly from former Soviet states such as Russia. It is coloured to match extension colour shade charts. If a wave or curl structure is required it is permed. European hair is finer than Asian hair and should be treated as chemically processed when applied onto natural hair. It comes in a variety of natural colours and fantasy colours plus several lengths from 10–24 inches. It comes in 1 oz packets and is sometimes sewn into weaves/wefts in 1 oz packets. European virgin hair has had no chemical treatment. It comes in only natural colours and structures. This hair comes in lengths from 10–24 inches. It is very rare and very expensive. It comes in 1 oz packets and is sometimes sewn into weaves/wefts in 1 oz packets.

5 **Pre-bonded extensions** These are strands of fibre or real hair with wax, protein or keratin resin applied at the root end of an extension by the manufacturers. These are applied using a heated pre-bonded tool. These extensions come in a

TOP TIP

Real hair prepared for hair extensions is root point correct. It is important that the roots and tips are aligned together or tangling of this loose hair will occur.

TOP TIP

Real hair that is not prepared on weaves or wefts comes loose. This loose hair is tied in 1 oz or 4 oz bundles called *bulk hair*.

range of structures and a variety of colours and lengths. Pre-bonded extensions come in packets of 5, 10, 20 or 25 strands.

6 **Silicone pads** These are small pieces of heat-resistant silicone sheet measuring 2 cm by 4 cm. They are used during the application of bonded extensions. The silicone pads protect stylists' fingers from the intense heat of resin bonds and are used to roll resin into bonds at the root area of the natural hair attaching extensions in place.

7 **Scalp protectors or scalp shields** These are circular plastic discs 4 cm in diameter. They are placed at the root area of an extension. They are secured on to the natural hair before attaching an extension in place. Scalp protectors have a small hole in the centre where natural hair feeds through. This strand of natural hair will have the extension attached. Scalp protectors prevent loose strands of natural hair becoming trapped in an extension bond and keep the natural hair sections clean and neat.

8 **Soft bristle brush** This is a brush that has a padded face and soft bristles which protrude from the padded base. Soft bristle brushes should be used when blending, mixing and brushing extension hair.

9 **Mixing mats** are designed to hold real hair root point correctly and they assist in the blending of real hair colours. They are used to hold real hair safely whilst applying extensions. They are two square mats, 15 × 15 cm. Each mat has small wire teeth or pins protruding from one side. These teeth are slightly bent. They are bent backwards (away from you) on the bottom mat and forwards (towards you) on the top mat. The two mats are then placed together so that the teeth can interlock, trapping the real hair in between. The real hair is then drawn out of the mats in small pieces the size of an extension. This hair then has a resin bond dispensed on the root end of the real hair ready to be attached to the natural hair as an extension.

10 **Removal tool** This tool is similar to a small pair of pliers. It crushes and breaks extension seals and resin bonds during the removal of an extension.

11 **Removal solutions** These are acetone, alcohol, oil- or spirit-based solutions that are recommended by the extension product companies to break down the resins, acrylics, rubber, latex and wax bonds which attach extensions in place.

12 **Resin drip tray** A metal dish or silicone mat is placed underneath the nozzle of a bonding applicator to catch dripping resin.

Client aftercare products

◆ **Clarifying shampoo** Shampoo designed to remove sebum, oil, wax, styling products and pollutants from the extension hair and natural hair. This shampoo must be oil- and silicone-free.

◆ **Light conditioner** A conditioner specifically designed for fibre hair and natural hair, produced by manufacturers and designed to coat the fibre hair, forming a surface barrier and protecting the fibre from any heat or friction damage caused by brushing and styling. It is a light conditioner for the natural hair.

◆ **Daily maintenance spray** A spray mist that is applied to the mid-lengths and ends of the fibre. It untangles the fibre and protects the fibre from heated appliances used for styling. Produced by manufacturers, the daily maintenance

spray must be used before brushing fibre and before any heated styling tools are used to style the fibre.

◆ **Reconstructive conditioner** A conditioner that works within the hair shaft designed to strengthen and rebuild hair. Reconstructive conditioners are manufactured by extension product companies specifically for real hair extensions.

◆ **pH-balanced rinse** An acid-balanced rinse that has the same pH as hair and skin (4.5–5.5). This product is diluted in water: 1 part rinse to 10 parts water. It is used after shampooing and conditioning and applied through the mid-lengths and ends of the hair. The acid-balanced rinse is designed to close the cuticle layers of the real hair, therefore reducing tangles and matting that can occur when the real hair is wet. A pH-balanced rinse is manufactured by extension product companies specifically for real hair.

In addition to the specialist equipment list, you would also be using a variety of general hairdressing tools and equipment such as: combs, brushes, scissors and a flat, stable work surface on which to place heated bonded tools and a hairdressing trolley.

TOP TIP

Do not place fibre hair in a mixing mat, as it will tangle if drawn through the interlocking teeth.

ACTIVITY

Use the table below to explain what each of the pieces of equipment does.

Item of equipment	What is it for?	What does it do?
Heat clamp		
Bonding applicator		
Heated pre-bonded extension applicator		

TOP TIP

Do not use your best pair of hairdressing scissors, as extension hair will blunt scissors.

Plan and prepare to add hair extensions

Consultation

A pre-service, thorough hair extension consultation is essential; it is one of the most important parts of the hair extension service, as there are many issues to cover and explain. A consultation appointment should be booked for a minimum of 30 minutes although you may need longer.

During the analysis, you will be checking the client's hair condition to see that the natural hair is strong and healthy enough to hold an attachment of extension hair in place for a three-month period.

You will be looking at the first 5–10 cm of the client's natural hair, as this area is where an extension will be secured. The analysis should establish whether the client's hair is normal, fine, coarse, dry or greasy. If a client has a greasy hair condition that requires shampooing on a daily basis, then a hair extension service is *not* suitable. The sebum will either break down the bonds securing the extension in place or make the natural hair too slippery and the attached extensions will slip down the hair shaft and fall out.

TOP TIP

You have to be satisfied that a client's fine hair will be strong enough to hold a secured extension in place for up to three months.

ALWAYS REMEMBER

The porosity of the client's natural hair has to be assessed throughout the length of the hair, to check whether the client's hair is normal, dry, chemically treated or lightened.

Consultation considerations

◆ What hair extension hairstyle is to be achieved – length, thickness, volume, body, colour, decoration or texture?

◆ Establish what style is to be achieved before starting this service as this will determine which hair extension application technique can be used and whether the client's natural hair length, condition and porosity are suitable for the required hairstyle.

◆ A client's natural hair should be a minimum of 8 cm (3 inches) long to apply a textured extension hairstyle.

◆ A client's natural hair should be a minimum of 10 cm (4 inches) long to apply hair additions that create decorative looks, e.g. crystal strands, highlights or flashes of colour.

◆ A client's natural hair should be a minimum of 13 cm (5 inches) long to create volume and thickening extension hairstyles.

◆ A client's natural hair should be a minimum of 15 cm (6 inches) long to create a natural-looking lengthened extension hairstyle. When applying a weave or weft extension to African hair types, the client's natural hair must be long enough to enable the stylist to create a continuous and firm tight corn braid or scalp plait.

You need to work out whether the client's natural hair is strong enough to withstand extension hair attached at the root area, and whether the mid-lengths and ends of the client's hair will withstand the friction and wear and tear of added extension hair for up to three months. If the client's natural hair is delicate or broken then it will not be strong enough to wear a hair extension hairstyle. (See hair tests for hair extension services, page 220.)

> ❝ The most important part of any extension service is the communication that you have with the client. You need to be able to ask and, importantly, listen to what is said. Extensions can be great but I also consider them to be extra work for the client and if this is not communicated well, or you haven't asked and listened, they can also become a problem for both client and stylist. Always record your communication and get your client to sign the record card. Give them a copy to take away as this will help resolve potential issues in the future.
>
> *Ross Miller*

Contra-indications for hair extensions

The most important part of any extension service is the communication that you have with the client. You need to be able to ask and, importantly, listen to what is said. The following are the **contra-indications** (issues that adversely affect a service being performed) to hair extensions; these factors must be taken into account during a hair extension consultation.

Do not provide hair extensions when your client:

◆ is suffering hair loss or is having medication or treatment for hair loss

◆ shows signs of alopecia

◆ shows signs of thin or thinning hair

ALWAYS REMEMBER

Hair extension hairstyles can take from 30 minutes to 10 hours to apply. During the consultation, clients should be informed about the length of time needed to apply their hairstyle.

TOP TIP

The price of the extension hair normally increases as the length of the hair increases.

◆ has any breakage through the first 5–10 cm of their hair at the root area

◆ has weakened or damaged hair

◆ is taking medication or having treatment for cancer

◆ is pregnant or during the first six months after giving birth

◆ shows signs of psoriasis or eczema on the scalp

◆ has skin allergies or excessive skin sensitivities

◆ has excessively oily hair and scalp.

As part of the consultation and before any extension service, you will need to identify:

◆ strength and any weakened hair/hair damage

◆ ability to stretch and return to original length.

You will also need to test the client's skin for sensitivity or allergies to the attachment and removal chemicals that you may be using.

Hair tests for extension services

Tests	What does this test do?
Pull test	A pull test will identify whether the hair is capable of retaining the root within the hair follicle. If you pull a single hair and it comes away from the scalp it could indicate that the hair is either in a catagenic stage (see hair growth, page 21) or is unable to sustain any additional weight. Either way, this is a contra-indication to hair extension services.
Elasticity test	By taking a hair between the fingers and stretching it you can assess the amount of spring it has. Hair without spring has no elasticity and lacks sufficient strength or structure to enable extension services to take place.
Skin test	The skin test is used to assess the reaction of the skin to chemicals or chemical products. To find out whether a client's skin reacts to the chemicals in the removal agent, a skin test should be carried out at least 24 hours prior to the chemical process.

See **CHAPTER 1** Consultation and advice, pages 22–23, for more information on contra-indications and hair tests.

Choosing extension hair

After establishing that the client's hair condition, porosity, and length are suitable for the requested hair extension hairstyle, you must then choose what extension hair is suitable for the client.

Real human hair, either Asian or European hair (real hair), or synthetic acrylic fibre hair (fibre) can be applied. The choice of extension hair to be used to create the extension hairstyle depends on the client's budget, suitability of the client's natural hair or the choice of application tools and attachment techniques.

You need to find out the length of extension hair needed for the required style to be achieved. Real and acrylic artificial hair are produced in several structures: straight, soft wave, deep wave, curly or spiral. Acrylic fibre is also available as pre-made dreadlocks, braids and crimped hair.

During the consultation, establish the correct structure suitable for the client's extension hairstyle and the colours needed.

Using a **colour ring** or colour shade chart, select the base colour, major tone and minor tone required (see colour selecting, mixing and blending section, page 223).

During the consultation, the stylist should recommend the correct aftercare products and home maintenance suitable for the type of extension.

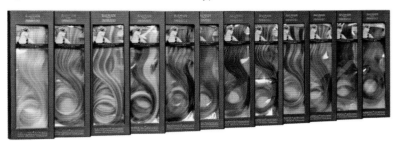

ACTIVITY

Copy out the table below and explain the purpose of the following items in the blank column.

Product	What is it for?
Liquid adhesive	
Polymer resin sticks	
Silicone pads	
Scalp shields	
Mixing mats	
Removal tool	
Resin drip tray	

Maintenance appointments

Each hair extension hairstyle needs regular maintenance appointments at the salon. These appointments are every two, four or six weeks depending on the hairstyle, application technique and extension hair used to create the style. During the consultation, the client should be told about the maintenance appointments required for their hairstyle.

ALWAYS REMEMBER

Hair growth

The rate of hair growth is an important factor in assessing how long extensions will last, as they grow away from the root area at the same rate as the natural hair; this is on average 1.25 cm per month. If the hair extension is allowed to grow too far away from the scalp, it will become very difficult to manage and should be professionally removed. However, all hair extension hairstyles should be removed from the client's natural hair after three months' wear. All removal should be conducted in the salon by a professional trained stylist, as the removal of extensions, if poorly performed, will break and damage the client's natural hair.

ALWAYS REMEMBER

Natural loss of hair

Each of us naturally loses 80–100 hairs per day. While wearing extensions this hair cannot fall out and becomes trapped at the root area above the extension attachment. If the extension strand is left in the hair for more than three months, the trapped hair will begin to mat. When the matting occurs it becomes impossible to remove an extension hairstyle without damaging the natural hair.

How long will it take?

To get a rough idea of how long a hair extension service will take, you could time how long it takes to apply one extension then multiply that by the total amount that you will use to create the whole effect.

On average, a trained extension stylist should take one minute to apply one extension. A lengthened hairstyle needs approximately 150–250 single extensions. So, the time of each appointment will be relevant to the number of extensions to be applied.

Consultation: Points to remember

- Analyse the client's natural hair condition, porosity, strength, length and required style.
- Select the type of extension hair, the style, length, colour and structure.
- Advise the client about the correct aftercare products, maintenance appointments and removal procedures.
- Work out the length of time needed to provide the service.
- Provide a quote for the total price, including application, maintenance appointments, removal service and aftercare products.
- Book the correct time for each appointment and ensure that you have the trained support staff available to complete the service.
- Order in advance the correct extension hair materials, length and colour for the client.

> To become a true extensionist, I normally recommend at least one year's experience handling hair prior to becoming one. This really gives confidence to you and employers alike. Ideally you want to have at least gone through manufacturer training, but I would strongly recommend that you also go through a vocational qualification on extensions. Doing this will give you the best possible career in extensions and employability.
>
> *Ross Miller*

Finalizing the consultation and completing the records Make sure that you quote the full price of the extension service:

- Recommend the products that will be required to maintain the look.
- Explain that a deposit needs to be taken in order to book the service.
- Remember the time that needs to be allowed for the service.
- Tell the client how long the extensions will last before they need to be removed.

Finally, book an appointment date and time, and make sure that the record is completed, covering all the following aspects.

The client record will:

- assist the salon reception when booking the appointment time, maintenance appointments and the removal appointment

◆ help in the ordering of materials, i.e. the extension hair required and the correct aftercare products

◆ provide evidence of the responses given to you by the client at the time that the consultation was carried out.

ACTIVITY

Aftercare advice and maintenance is a very important part of the total extension service. Explain the benefits for each of the products listed in the table below.

Aftercare product	What benefits does it provide
Clarifying shampoo	
Light conditioner	
Daily maintenance spray	
Reconstructive conditioner	
pH balanced rinse	

Selecting and blending extensions and colours together

Creating extension hairstyles requires learning new skills over and above the stylist's existing hairdressing knowledge. Selecting, mixing, and blending fibre hair and real hair is an additional skill that requires a new understanding of selecting colours. When applying a natural-looking extension hairstyle it is important to create extension hair that is the same colour as the natural hair. If there is seen to be even a fractional difference in colour the extension hairstyle will look false. Therefore, a vital new skill to learn for this service involves selecting, mixing, and blending colours.

Selecting extension hair colours to match to the natural hair colour Take a colour ring or shade chart. Place the colour ring swatches of hair colour against the client's natural hair. Select the extension hair colour that is the nearest colour to the client's own hair colour. This is the base colour or first colour. Use the colour ring to identify the second colour that is the nearest colour to the client's own hair colour. This is the major tone or second colour. Use the colour ring and look closely for the third colour that matches the hue (glint) of the client's natural hair colour. This is the minor tone or third colour.

You have now selected the base colour, major tone and minor tone of the natural hair. Blending the extension colours in their correct proportions to match natural hair is a very visual technique. Below is a formula to follow that will assist you in calculating the correct proportions of colour to mix in order to make an exact colour match.

Colour formula for mixing extensions

◆ Adding 25% (25 g) of a second colour to the base colour will lighten or darken the base colour.

◆ Adding 12.5% (12.5 g) of a second colour to the base colour will give a strong tone.

◆ Adding 6% (6 g) of a second colour to the base colour will give a hue or glint of colour.

◆ Adding equal amounts (25 g + 25 g) of two different extension hair colours together will change the base colour.

TOP TIP

Fibre extension hair is packaged in 100 g bales.

Real extension hair is packed in 4 oz bulk or weaves/wefts.

ALWAYS REMEMBER

It is very rare to find extension hair that matches natural hair exactly. Often a second or third colour will have to be selected and blended into the base colour to get an exact match of natural colour. A natural head of hair is not one colour but made up of several colours. Extension hair must mimic this natural phenomenon.

Blending extensions together Remove the selected base colour of extension hair from its packaging, then divide 25%, 12.5%, or 6.25% of the second colour and third colour from the packing (using the formula on page 223). After selecting the quantities of extension hair, place the colours together. Ensure the ends of the extension hair are together. When blending colours together, hold the fibre hair in the centre. Lightly spray fibre with a daily conditioning spray. Use a soft bristle brush to blend the fibre hair colours together. Start brushing from the ends of the fibre, working towards the centre.

When blending real hair colours together, always place real hair in a mixing mat with the root ends together.

Place the first colour or base colour onto the mixing mat. Then place the second colour and third colour on top of the base colour. Place the lid of the mixing mat on top of the selected colours. As you draw the real hair out of the mixing mat during application, the colours will mix together.

Multi-mixing Multi-mixing is the process of mixing a number of colours together until the fibres are totally blended.

1 Decide on the colour or colours to be achieved.

2 Select the appropriate fibres to be mixed. Carefully remove the required amounts of fibre from the packs.

3 Take the base colour – the greatest amount – and place the fibres in the palm of your hand, holding them near one end. Place any secondary colours – the lesser amounts – on top. Close your fingers and hold the fibres tightly in your hand.

4 With your thumb, fan out the fibre along your first finger.

5 Now, using a bristle brush, brush the fibre downwards. This will tend to mix the fibres.

6 Hold the opposite end of the fibres in your palm and repeat this process. This will mix the fibres further.

7 Continue brushing and changing ends until the colours blend together and the final colour is uniform.

Block colour Block colouring gives a more defined colour or highlighting effect.

1 Take the base colour – the greater amount – and hold it centrally in your palm.

2 Lay the secondary colour – the lesser amount – on top, keeping the ends together.

3 Starting with your hands about 20 cm apart, bring your hands together. Divide the fibre into two equal amounts and separate your hands. Slight mixing will have occurred.

4 Again, place the fibre in one hand on top of the fibre in the other. The fibre is now all in one hand, partially mixed.

5 Bring your hands together and again divide the fibres into two. Further mixing will have occurred, but the fibres will still be in blocks of colour.

6 Repeat until the colour is sufficiently mixed.

7 Gently brush the fibre to remove any tangles, but do not mix the fibres any further.

Attach hair extensions

Adding hair extensions

There are different methods of attaching extensions to natural hair. The method chosen by a stylist and client will vary depending on whether the stylist is attaching fibre or real-hair extensions.

The following section contains a number of step-by-step guides to the process of adding hair extensions. They include a step-by-step guide to attaching synthetic hair, a step-by-step guide to attaching real hair and a pre-bonded method, which can be used to attach either synthetic or real hair. There are also descriptions of two extension hairstyles and diagrams that will demonstrate how to undertake the planning and placement of extensions to create hairstyles from synthetic fibre or real hair.

Attaching synthetic (artificial) fibre extensions

One stylist and one assistant are required to work on this service.

1   Prepare the fibre to be used.

2   Prepare the heat-sealing device – clean it, select the temperature setting and plug into the power point.

3   Prepare other tools and materials: combs, clips, brushes, scissors, styling sprays and bonding solution (if required).

4   For hair extensions over the whole head, section the head into five areas.

5   Leave a 7 mm section of natural hair out around the hairline.

6   Start at the nape area. Take a band of hair above the hairline; secure the remainder out of the way with clips.

7   Starting in the centre of this band, take a section of hair 5 mm by 5 mm and again clip the remainder out of the way.

8   Divide the section of natural hair into two.

9   Your assistant now takes a similar amount of fibre and lays it centrally in between, forming a cross.

10   Cross the two pieces of natural hair over, right over left and hold the hair apart while your assistant crosses left over right.

11   Cross again, right over left.

12   Your assistant leaves the top weft of fibre out and subdivides the bottom weft into two, pulling these apart so that you can cross over between them.

13   You both continue crossing until you have 12 mm of braided 'hair'.

14   With your assistant holding the top of the plait between thumb and index finger, wrap the weft of fibre left out (step 12) around the braid.

15   With the heat sealer, close the tips over the bound braid approximately 20 mm down the braid. To close the tips, gently press them on to the fibre for 2 seconds. Lift the top tip and give a half turn, then close your tips again. Remove the heat after 2 seconds.

16   Pinch and roll the heated area between your fingers. Ensure that you have a smooth, round seal.

17   Repeat steps 1 through 16.

18 When the nape row is complete, continue up the head row by row.

19 After all the extensions have been applied, cut and dress the hair into the desired style.

Attaching processed hair extensions

1 Prepare the high-frequency equipment according to the manufacturer's instructions.

2 Prepare the other tools – the plastic strand shield, brushes, combs, clips and whatever else you will need.

3 For hair extensions over the whole head, section the head into five working areas.

4 Leave a 7 mm section of hair out around the hairline.

5 Start at the nape area. Take a band of hair above the hairline. Secure the remainder out of the way with clips.

6 Starting in the centre of this band, take a section of hair 5 mm by 5 mm and again clip the remainder out of the way.

7 Slide on the plastic protection shield and push it near to the scalp area.

8 Place the polymer-bonded end of the extension into the centre of the hair section, approximately 12 mm from the scalp and forming a V-shaped wedge of hair around the bond. An even distribution of natural hair should surround the bonded end of the hair extension, to prevent uneven tension or breakage.

9 Place the grooved tip of the high-frequency device below the hair section.

10 Wait for the polymer to bubble before rolling it smoothly between your index finger and thumb. (Bubbling will occur in just a few seconds.)

11 Check that the bottom end of the bond is adequately sealed.

12 Continue the process, repeating steps 1–11 until the complete row is finished.

13 Continue working up the back of the head until the section is complete.

14 After all the extensions have been applied, cut and dress the hair into the desired style.

STEP-BY-STEP: ATTACHING WEAVE HAIR EXTENSIONS

1 After greeting your client, conduct a full consultation to determine the style and effect desired.

Plan the track placement, according to the length and density of your client's own hair.

2 Part the hair horizontally in the nape area to create a base for your first track, make sure you leave up to one inch free around the hair line.

3 Create two corn rows starting from the outside and meeting in the middle; secure with a braiding band.

4 Measure the weft of weave to the exact length of your track and cut.

5 Using a weaving needle; sew the weft onto the track using a stocking stitch technique.

6 Repeat steps 2–5 creating and sewing another track just below the occipital bone and then again from temple to temple.

Ensure the parted sections curve with the shape of the head.

7 Parted sections.

8 corn row.

9 Analyse the client's hair to decide if a fourth row is required above the temple.

10 Prior to cutting/blending.

11 As with any hair extension technique the look is not complete unless the hair has been cut into shape.

Use texturing techniques to blend the extensions with the client's natural hair.

12 The finished extended look.

CourseMate video: Attaching weave hair extensions

Attaching pre-bonded hair extensions

Natural hair extensions bought from the supplier arrive prepared by the manufacturer. They are available in a variety of strand colours, sizes and types. The finer strands are for use around hairlines and partings, thicker strands are for use in other areas. You can also choose between straight, wavy and curly hair types. The wefts of hair are 'gummed' together with a polymer resin, so no colour blending is required. They are ready to attach to the hair.

STEP-BY-STEP: ATTACHING PRE-BONDED HAIR EXTENSIONS

1 Section the client's hair following the contour of the head.

2 Take up some of the client's hair in a triangular shape around the diameter of the extension. Place the protector disc around the section of hair at the roots.

3 Slide the protector disc down slightly and evenly so that the hair hangs down loosely, without any strain. Secure the hair protector with a clip. Hold the strand of the client's hair between the thumb and forefinger and place the extension right under it. Hold the extension and the client's own hair at the right distance 1–1.5 cm from the scalp. This will ensure the extensions are comfortable to wear.

4 Place the bond in the connector. Keep it still; do not slide it up and down. Melting the bond is the most critical part in applying extensions. Wait until the bond begins to bubble and becomes soft. Tap the connector twice. This has to be a short quick movement otherwise the bond will stay on the cold part of the connector.

5 Roll the bond with your fingers. Take care that the hand that is rolling the bond rests on the client's head (palm turned towards the scalp) and that you roll downwards away from the scalp.

6 Seal/heat again if there still are white spots on the bond. Always seal down, never up, as you can stretch the bond and reduce the bond's effective holding power.

Although the procedures for natural hair extensions may seem similar to those for synthetic hair extensions, in reality the processes are quite different. The polymer resin is activated by equipment that emits ultra-high-frequency sound waves: once activated it moulds around the section of hair and creates a strong permanent bond.

Natural hair extensions can be styled by blow-drying, tonging, or using heated rollers. It is possible to use semi-permanent or temporary colours, but perming and colouring are not recommended.

Planning and placement

Many extension hairstyles can be created with this service using fibre hair and real hair. Once the principles of the planning and placement of the extensions are understood, you can create virtually any style.

Sectioning of client's natural hair The diagram below shows the sectioning of the client's natural hair and the areas where extensions can be placed. Hair extensions are applied in the interior of the natural hairstyle. They are placed 1–2 cm behind the hairline, parting, and crown area. The natural hair is divided into six areas.

Area 1 is the nape section. Take a section using a tail comb from the top of the ear to the top of the ear across the occipital bone.

Area 2 is at the back section of the head above the occipital bone up to the crown. Take a section using a tail comb from the top of the ear to the top of the ear over the top of the head and through the crown area.

Area 3 is at the right-hand side or temple area section of the head from the ear to the top of the recession area. Take a section from the top of the recession at the front hairline straight back to the crown area.

Area 4 is from the top of the right-hand recession to the client's parting and reaching back to the crown area.

Area 5 is the left-hand side or temple area section of the head from the ear to the top of the recession area. Take a section from the top of the recession at the front hairline straight back to the crown area.

Area 6 is from the top of the left-hand recession to the client's parting and reaching back to the crown area.

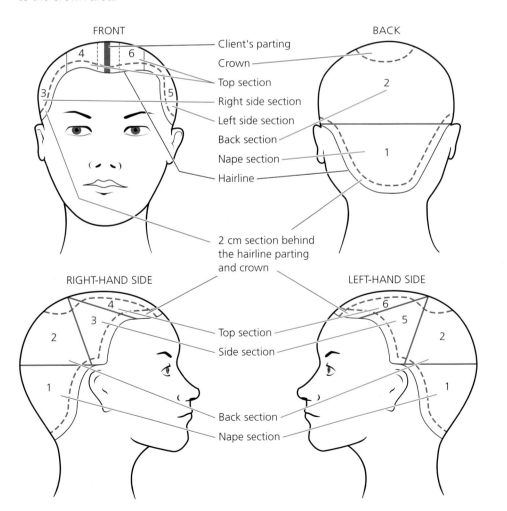

Sectioning of client's natural hair

Extensions are placed in the interior of a hairstyle always placed 1–2 cm away from the hairline and placed 1–2 cm away from the parting and crown.

Planning and placement of extensions
This diagram shows a natural-looking lengthened hairstyle using fibre. The extensions are placed 2 cm behind the hairline. The natural hair is divided into six sections. A heat clamp is used creating individual extensions, held in place with fibre heat seals.

Area 1 has two rows of extensions applied. They are placed 2 cm away from the hairline. A 2 cm section of natural hair is left extension free across the occipital bone. This section is left extension free. If extensions are applied on this bone, they will protrude and create a distorted shape.

Area 2 has three rows of extensions applied right up to the crown, placing the extensions 2 cm away from the crown area.

Planning and placement of extensions

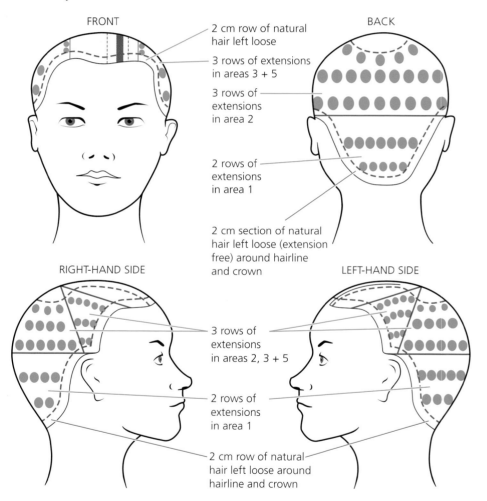

FRONT

2 cm row of natural hair left loose

3 rows of extensions in areas 3 + 5

3 rows of extensions in area 2

2 rows of extensions in area 1

BACK

2 cm section of natural hair left loose (extension free) around hairline and crown

RIGHT-HAND SIDE

3 rows of extensions in areas 2, 3 + 5

2 rows of extensions in area 1

2 cm row of natural hair left loose around hairline and crown

LEFT-HAND SIDE

Areas 3 and 5 have three rows of hair extensions applied. A 1 cm section of hair is left extension free in areas 3 and 5 above the second row. This area is the widest point of the head and is often left loose as extensions applied here will protrude and distort the hairstyle shape.

There are 100 extensions in this hairstyle. It takes three hours to apply these extensions, using 150 g of fibre hair, which is one-and-a-half packets. The colours used are 50 g of pale blonde, 50 g of cool blonde/ash blonde, 25 g of light brown and 25 g of gold.

The fibre is blow-dried using a warm hairdryer. The style is cut using scissors, point cutting the perimeter lines and slide cutting to create some layers at the front of the hairstyle. This hairstyle has to be completely removed after three months' wear.

Textured extension hairstyle The diagram shows the planning and placement of extensions to create a large, curly, textured extension hairstyle. This is a lengthened spiral curled look. The extensions are placed 1 cm behind the hairline. A heat clamp is used creating individual extensions held in place with fibre hot seals. Two stylists were required to work on this hairstyle.

The natural hair is sectioned into six sections.

See attaching synthetic fibre extensions, pages 225–226.

Application of extensions

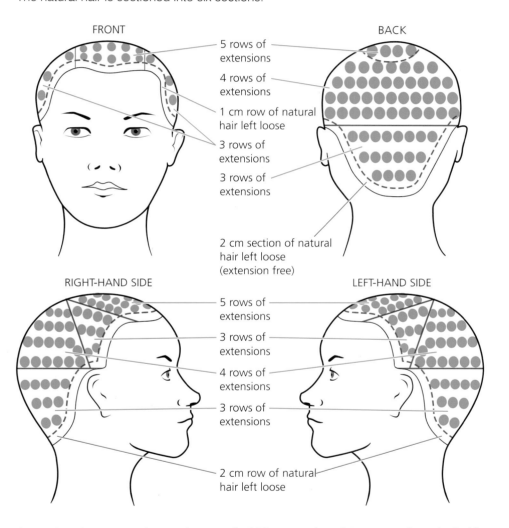

FRONT

BACK

5 rows of extensions

4 rows of extensions

1 cm row of natural hair left loose

3 rows of extensions

3 rows of extensions

2 cm section of natural hair left loose (extension free)

RIGHT-HAND SIDE

LEFT-HAND SIDE

5 rows of extensions

3 rows of extensions

4 rows of extensions

3 rows of extensions

2 cm row of natural hair left loose

Area 1 has three rows of extensions applied. They are placed 1 cm away from the hairline. A 2 cm section of natural hair is left extension free across the occipital bone. This section is often left extension free. If extensions are applied to this area, they will stick out, and create a distorted shape.

Area 2 has four rows of extensions applied right up to the top of the head.

Areas 3 and 5 have three rows of hair extensions applied. A 1 cm section of hair is left extension free in areas 3 and 5 above the second row. This area is the widest point of the head and often left loose as extensions applied here will protrude and distort the hairstyle shape.

Areas 4 and 6 have five rows of extensions applied covering the top area, right back and over the crown.

There are 250 extensions in this hairstyle. It takes four hours to apply these extensions, using 300 g of fibre hair, which is in three packets. The colours used are 50 g of white, 50 g of light blond, 100 g of cool/ash blonde, 50 g of light brown, 25 g of ginger and 25 g of gold.

The fibre is curled on heated bendy rollers. The style is cut using scissors. The curled fibre is blunt cut and shaped visually. Gel is used to reduce fizzy ends and hold the curls in shape. The fibre has to be re-curled every 4–6 weeks as the curl will drop. This hairstyle has to be completely removed after three months' wear.

ACTIVITY

Knowing when or if an extension service can be carried out is critical to the success of the service and the ongoing repeated business. Answer the following questions:

1   What is traction alopecia?
2   What things should you do when conducting a consultation for an extension service?
3   What things would be contra-indications to an extension service?
4   What preparations should you make?
5   How would you go about selecting the colour(s) of fibre for a client?
6   How do you mix the fibres together?
7   What aftercare advice should you give to clients?

ALWAYS REMEMBER

If the extension hair is wet or damp while cutting, when it dries it will shrink into a distorted finished result. Fibre hair has no elasticity and therefore needs no tension while cutting. Real hair is chemically damaged and will therefore stretch when wet. If cut while wet it will dry in uneven lines.

Cut and finish hair with extensions

Cutting extension hairstyles

Hairdressing scissors, razors, clippers, and thinning scissors can all be used on hair extensions. All extension hairstyle cutting is carried out on dry extension hair, as creating the styles are *visual* haircuts using soft blending techniques and not *technical* blunt haircuts. The principles of hairstyle balance and proportion must be retained in creating a finished result.

Use a razor or clipper on very straight extension hair as these tools will create soft lines in the hairstyle. If scissors are used on straight fibre it will exaggerate blunt or straight lines that are cut into them which can make the finished hairstyle look very false.

Use hairdressing scissors or clippers on wavy or curly extension hair as these tools will give a softer line in the wavy or curly hair without thinning the ends of the extension hair to such a degree that the ends will frizz, mat or tangle.

Cutting techniques for extension hairstyles

1   **Blunt or club cutting** This is a technique that cuts a heavy straight line and is created using scissors.

2   **Soft tapering** This technique cuts a soft, thin layer into the mid-lengths and ends of the extension hair and is created using a razor or thinning scissors.

3 **Spiral tapering** This technique cuts uneven lengths through the mid-lengths and ends of extension hair and is achieved by using a razor or scissors.

4 **Layering** This is a technique used to connect short hair into long hair by sliding a razor or scissors down the hair length creating a soft lightly layered profile line.

5 **Surface graduated layering** This is a technique used to layer extension hair without removing weight and bulk from the extension hair and is created using scissors or clippers.

6 **Skim or surface clippering** This is a technique used to break-up surface layers of extension hair or reduce length and bulk at speed and is created using electric clippers.

7 **Point cutting** This is a technique used to cut an even perimeter line into extension hair without cutting a false-looking straight line, created using scissors or thinning scissors.

Styling hair extensions

Only use styling tools that can deliver a medium heat temperature to the fibre. Anything hotter may damage or change the handling properties of the hair fibre. Avoid heat styling equipment such as straighteners, tongs, etc. as they tend to be far too hot for extensions. Even if they do not distort the fibre, they will definitely make future handling a problem for you or the client.

You will be incorporating the natural hair in with the extension hair during styling. Do not try to separate the extensions away from the natural hair, as this will make blow-drying and setting very difficult. It will also give variable structure to the finished hairstyle, making the result look *unrealistic*.

Wavy or curly fibre extension hair will drop after a couple of weeks. The curl or wave will fall out as the fibre is softened from the heat of the client's body temperature. The fibre will need re-curling regularly throughout the hairstyle's life span of three months.

For more information relating to controlling the accuracy of the cut see CHAPTER 4 Creative cutting techniques.

TOP TIP

If wavy or curly extension hair is thinned or tapered too much it will tangle and frizz at the ends.

ALWAYS REMEMBER

Hot hairdressing tools will melt and irreparably damage fibre hair. Do not use hot hairdryers, heated curling tongs, hot brushes or straightening irons on fibre hair.

ALWAYS REMEMBER

For the best styling results on extension hair, always style the extension hair in a dry or virtually dry state. This will ensure you achieve a stronger set or shape that will hold in place for a longer time. Setting or drying extension hair from wet will take a very long time to dry and will result in a weak result lasting only a couple of hours.

ALWAYS REMEMBER

Before re-curling fibre hair, smooth the fibre out using a soft bristle brush and warm hairdryer and then re-curl fibre from a virtually straight structure. This will ensure that you do not double-curl the fibre. Double-curling will give a frizzy, unattractive result. Double-curling is achieved when you attempt to curl fibre on top of an existing curl or wavy structure.

Styling products for use on extension hair

◆ Mousse

◆ Gel

◆ Setting lotions

◆ Hairspray

◆ Pomade – use sparingly

Apply styling products to the mid-lengths and ends of the extension hair. Avoid the root area as some styling products contain oil, wax, silicone and alcohol ingredients, which can break the extension bonds down, making the extension fall out, or making the natural hair very slippery so the extensions slip out down the hair shaft.

Maintain and remove hair extensions

Different hair extension types have different techniques of removal. You need to check with the manufacturer for more details for their particular hair extension services.

The following information provides general guidelines for the removal of extensions for:

◆ cold fusion systems

◆ hot bonded systems

◆ cold fusion (using adhesive tape).

Cold fusion systems

1   Section the hair so that you can work on the extension weft without getting removal product on the rest of the hair.

2   Place a cotton wool pad beneath the weft to protect the scalp.

3   Apply the removal chemicals directly from the applicator onto the weft.

4   Wipe the weft with the pad to evenly apply the chemical.

5   Leave to penetrate through the extension and weaken the bond.

6   Heat the weft near to the root area to activate the removal product.

7   When the bond has disintegrated, remove the weft by carefully peeling downwards and away from the hair.

Hot bonded systems

All hot bonded systems have similar removal techniques, the only differences being in products and removal tools.

1   Section the hair so that you can work on the extension without altering the rest of the hair extensions.

2   Using the removal tool, crush the bond throughout the length of the seal.

3   Place a cotton wool pad beneath the bond and apply the removal liquid evenly across the seal.

ALWAYS REMEMBER

Hair extension hairstyles must be dried immediately at the root and bonded area. Do not brush or comb extensions when they are still wet as the bonds can break down and fall out. You should also remember not to apply excess tension on the root area by tying tight ponytails.

HEALTH & SAFETY

Some removal solutions can irritate the skin or dissolve nail polish and nail extensions – ensure you wear disposable (non-latex) gloves.

TOP TIP

Put on your disposable (non-latex) gloves on before handling any removal chemicals.

4 Leave for a minute or two, so that the chemicals can break down the adhesion to the hair.

5 Re-crush the bond throughout the seal.

6 Then remove the extension by pulling it gently along the hair.

There may be some residue of resin left in the client's natural hair. Brushing the natural hair, and shampooing and conditioning it, will easily remove this.

Other systems

Cold fusion (attached by adhesive tape) Self-adhesive extensions use special chemical sprays or solutions to dissolve the bond attaching them to the hair. When this is applied/sprayed onto the self-cling tape, it quickly reduces the adhesion and therefore releases the weft from the hair. Look at the removed weft to see if it is worth keeping or not.

The self-cling strips can be removed from the weft in the same way that the extension is removed from the hair and, depending on the quality of the weft, it can be washed, re-conditioned and retained for future use or otherwise discarded. If the weft is to be re-used, new double-sided self-adhesive tapes will need to be applied after the wefts have been dried.

1 Section the hair so that you can work on the extension without getting removal product on the rest of the hair.

2 Pour a small amount of removal solution onto a cotton wool pad and evenly apply along the length of the adhesive tape.

3 Allow the chemicals to penetrate for about 30 seconds.

4 Hold the end of the extension weft and pull it down and away from the natural hair.

Sewn-in extensions

On straight hair These can be easily removed by the careful cutting of the stitches that hold the weft in place.

On plaited hair Which appears as free-hanging braids, simply cut above the hard bond at the ends of the extension and unravel the plaits using the end of a tail comb. After removing, comb the hair through to detangle the lengths, then shampoo and condition to prepare the hair for the next service.

Provide aftercare advice

Home-care products

There are two types of extension hair that can be used to create hair extension hairstyles:

◆ synthetic, acrylic fibre extension hair

◆ real-human extension hair that is either Asian or European hair.

The two hair types need different home-care products. The following client advice and home-care procedures should be given to clients who wear extension hairstyles.

Acrylic, synthetic fibre extension hair As described on pages 217–218, the fibre extensions require the use of a clarifying shampoo, a light conditioner and a daily maintenance spray; a soft bristle brush will also be needed.

Real extension hair The home-care products required for real hair extensions:

1 **A clarifying shampoo** This is a shampoo that is designed to remove sebum, oil, wax, styling products and pollutants from the extension hair and natural hair. This shampoo must be oil-and silicone-free.

2 **A reconstructive conditioner** This is a conditioner that works within the hair shaft, designed to strengthen and rebuild hair. Reconstructive conditioners are manufactured by extension product companies specifically for real hair. Real hair has gone through several chemical processes before application, making it porous and chemically damaged. Reconstructive conditioners should be used to assist in maintaining the strength, shine and manageability. Light conditioners are not recommended for real hair as these products would build up on the real hair, making it dull, lifeless and heavy.

3 **A pH-balanced rinse** This is an acid-balanced rinse that has the same pH as hair and skin (4.5–5.5) and is diluted in water 1 part pH rinse to 10 parts water. This product is used after shampooing and conditioning, and is applied through the mid-lengths and ends of the hair. The acid-balanced rinse will close the cuticle layers of the real hair, so reducing the tangles and matting that can occur when the real hair is wet. This rinse can be used as a daily product contained in a water spray and applied to damp-down real hair before restyling. A pH-balanced rinse is manufactured by extension product companies specifically for real hair.

4 **A soft bristle brush** Soft bristle brushes should be used when brushing real-hair extension hairstyles.

Home-care advice for extensions

Advice for clients:

1 Shampoo at least twice a week to reduce natural oil build-up at the root area.

2 Before shampooing, brush gently with a soft bristle brush. Begin brushing at the ends of the hair in downward strokes until you reach the root area. Then brush from the root through to the tips.

3 Using your fingers, ensure all the extensions are separated at the root area.

TOP TIP

Do not use brushes that have balls on the end of the bristle as this can rip and tear fibre and damage the bonds holding the extensions in place.

4   While shampooing real hair, the head should be in an upright position. Standing in the shower is an ideal position. The hair and water should flow in a vertical downwards direction. The water temperature should be warm.

5   Use a clarifying shampoo recommended by your stylist specifically for your extension hair. Using your fingertips, stroke the shampoo gently into the hair from the roots to the tips. Do not massage or rub the extensions when wet.

6   Apply a recommended conditioner to the mid-lengths and ends of the extension hair. Then rinse thoroughly.

7   After shampooing and conditioning, wrap a towel around the hair and pat gently to remove excess water. Wrap hair in a towel and leave for 20 minutes for the towel to absorb all the moisture from the extensions and natural hair.

8   Separate the extensions at the root area with your fingers before drying.

9   Dry the extensions with a *warm* hairdryer, using a diffuser if required.

10 After drying the root area, brush gently from the ends towards the root area in a downwards direction. Always hold the hair extensions at the roots while brushing to avoid placing unnecessary tension on the bonds and root area.

11 Long or lengthened hair extensions must be plaited and secured with a covered band on the ends before going to bed at night. Do not go to bed with damp or wet extensions as matting and tangling will occur.

ALWAYS REMEMBER

All home-care products and tools must be recommended to the client by the stylist who has created the extension hairstyle. Products used on extensions that are not recommended by a stylist could damage or make extensions fall out.

TOP TIP

Do not allow natural oils to build up at the scalp area as this can break down the bonds and then the extensions slip out.

Styling extension hairstyles do's and don'ts

✓	Always explain the benefits of using professional brushes and styling products.
✓	Explain to the client the ways that they should be handling their own hair extensions at home.
✓	Always tell the client what types of products to use and those that should be avoided and why.
✓	Explain to the client the things that they should look out for in respect to hair damage, weakened hair and traction alopecia.
✓	Use hairdryers and heated rollers on the fibre hair. Heat will straighten or curl fibre extensions. Once heated and cooled down, the fibre will retain this texture until more heat is applied to re-curl or straighten. Water will not alter the movement of fibre extensions. Do not use curling tongs, curling brushes, straighteners or crimpers directly on the fibre.
✓	Use hairdryers and heated rollers on real hair. Electric curling tongs, straighteners and hot brushes may be used on Asian or European hair extensions.
✓	Always avoid the bonded areas with heated styling tools, as direct heat will soften the bonds.
✓	Do not back-comb the hair extensions, as this will cause irreparable tangling.
✓	Do not use styling products that contain oils, wax, silicone or excessive alcohol as these will break down the extension bonds.

Typical problems with artificial/synthetic hair extensions

Problem	Advice to give
Hair extensions tend to tangle or get knotted easily	Suggest suitable products and tools that will help to make grooming and brushing easier Explain that the longer lengths should always be combed from points to roots. Long hair and particularly synthetic hair needs to be disentangled by working and freeing up from the ends back through the lengths towards the bonded area of the weft with a wide tooth comb rather than a brush.
Hair lengths tend to get matted when they are shampooed	Suggest conditioning products that are designed to work on synthetic/artificial hair. Sometimes the conditioners for acrylic hair are best applied before the hair is wetted; this reduces the locking and matting result caused by the action of rubbing during shampooing Suggest that the hair is shampooed in a different way along the lengths by a smoothing action, rather than rubbing at the scalp.
Longer lengths get matted or knotted during sleep/ overnight	People who are more restless during sleep may find that the rubbing action of the hair on the pillow makes the extensions lock together overnight. Suggest tying the hair in a ponytail with a soft fabric ribbon. This reduces the movement and the effects caused by chafing on the pillow. Alternatively, if the hair is very long, you can advise wrapping the hair in a silken scarf. This holds the lengths together and stops any chance of knotting.
Hair gets in the way at work or in sport	Clients who are used to shorter hair will not be familiar with the problems associated with long hair. Explain the benefits of wearing hair up as opposed to down during work/sport as this may have a more professional or beneficial effect.
Limited styling options after extensions have been applied	When artificial extensions have been applied to the hair, the client's options for future styling/ finishing in other ways are limited. Synthetic hair does not respond well to hot styling. The hair cannot be moulded or shaped in similar ways to that of natural hair. *Remember to advise the client* that any excessive heat will cause their extensions to distort, mat together or melt. If this occurs, the only course of action is to cut the damaged lengths off or to remove them altogether.

"

There are a number of career options in hair extensions. Like myself you can become a trainer for a manufacturer and/or the industry authority. There are some fantastic salons that specialize in extensions and they have a demand for extensionists. There are options to specialize in long hair, like Patrick Cameron (Chapters 8 and 9) and create amazing looks, or Angelo Seminara, who creates the most mind blowing avant-garde styles; all of which can stem from extension work.

Ross Miller

REVISION QUESTIONS

Q1. The condition of the client's hair near the _____ is particularly important when considering a hair extension service.

Fill in the blank

Q2. Clients who have to wash their hair on a daily basis are suited to hair extension services.

True or false?

Q3. Which of the following are contra-indications for hair extensions?

Multi-selection

Psoriasis ☐ 1

Dandruff ☐ 2

Alopecia ☐ 3

Excessive oily scalp ☐ 4

Infection ☐ 5

Gel, mousse or hair wax ☐ 6

Q4. Artificial fibre and synthetic fibre are the same thing.

True or false?

Q5. A full head of extensions would on average amount to:

Multi-choice

75–125 single extensions ○ a

150–250 single extensions ○ b

300–500 single extensions ○ c

500–1000 single extensions ○ d

Q6. The cost of real hair extensions is proportional to their length.

True or false?

Q7. Which of the following are not connector products for hair extensions?

Multi-selection

Heated pre-bonded extension applicator ☐ 1

Cold fusion adhesive ☐ 2

Polymer resin adhesive sticks ☐ 3

Mixing mats ☐ 4

Acetone ☐ 5

Spirit-based solutions ☐ 6

Q8. In making colour choices the main or first colour is known as _____ colour.

Fill in the blank

Q9. Which of the following is not recommended as home-care maintenance for real hair extensions?

Multi-choice

Washing the hair with a clarifying shampoo ○ a

Refreshing the hair with a daily maintenance spray ○ b

Applying conditioners to the root area ○ c

Drying the hair with a warm dryer ○ d

Q10. You should always hold the hair extensions at the roots while brushing.

True or false?

11 Colouring Hair

LEARNING OBJECTIVES

◆ Be able to work safely during colouring procedures at all times.

◆ Be able to make preparations for colouring and lightening services.

◆ Be able to colour and lighten hair creatively.

◆ Know how to resolve basic colouring problems.

◆ Be able to identify a range of factors that can limit or stop a planned service.

◆ Be able to communicate professionally and provide aftercare advice.

KEY TERMS

accelerator

activator

allergy alert test

alkalis

artificial colour

canities

colour coordination

colour restorer

eumelanin

melanin

metallic salts

oxidation

para dyes

pheomelanin

pre-pigmentation

pre-soften

progressive dyes

quasi-permanent

synthetic colour

toning

trichosiderin

UNIT TOPICS

Colour and lighten hair

Creatively colour and lighten hair

INTRODUCTION

Colouring is arguably the most exciting and often the most difficult aspect of modern fashionable hairdressing. The increasing demands and expectations of clients have made colouring and, in particular, the creation of special colour effects the 'must-have' of hairdressing.

Our clients are more informed than ever about hair styling, hair condition and the benefits of using certain products. Colouring is an essential part of their image; they know that techniques and products are developing all the time, so the possibilities are boundless.

Colouring hair

PRACTICAL SKILLS

Find out what clients want by questioning and visual aids

Look for things that will influence the ways that you colour or lighten your client's hair

Learn how to carry out tests on the hair and how the results will affect the service

Learn how to carry out a variety of colouring and lightening techniques

Make recommendations to clients about colouring and lightening services

Provide aftercare advice to your clients

Keep client's records up-to-date

UNDERPINNING KNOWLEDGE

Know your salon's ranges of products and services

Understand the factors that can affect your colouring options

Understand how different colouring products work and their suitability in different situations

Know the current fashions for colouring and lightening hair

Know how to correct colouring problems

Know the principles of colour and how they affect colouring and lightening services

Know how to communicate effectively and professionally

The principles of colour and colouring

What is colour?

We see colour because of light, however light waves are not in themselves coloured. Our perception or *recognition* of colour arises in the human eye and brain. A particular light ray (of electromagnetic energy) defines each hue or pure colour.

INDUSTRY ROLE MODEL

DAVID MORGAN   Partner at Xposure, Penkridge, Staffordshire

"I've been hairdressing for 26 years, since the age of 14. I started as a Saturday boy, after I'd watched a programme called *In at the Deep End*, which was about a celebrity learning to cut hair in a month. I was so inspired at how the industry appeared; I decided to get a job to learn the skills myself. I'm now a salon owner and member of the Habia skills team 26 years later. I have travelled all over the world, demonstrating and teaching, as well as being a Wella master colour expert.

The human eye sees visible light, falling between wavelengths of 400 and 700 nanometres (nm). We are not able to see (perceive) other forms of light rays, such as infrared or ultraviolet light, without the use of specialist equipment.

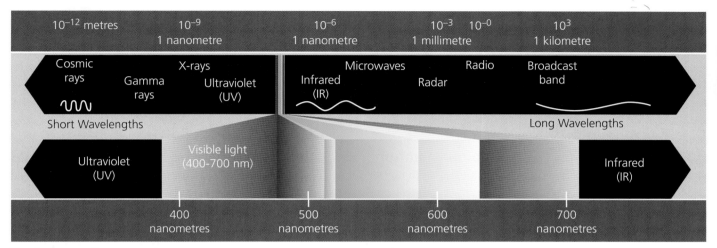

The electromagnetic wave spectrum

Daylight (white light) is made up of numerous waves or impulses, each having different dimensions or wavelengths. When separated, any single wavelength will produce a specific colour impression to the human eye. What we actually see as colour is known as its *colour effect*. When light rays illuminate an object, the object absorbs certain waves and reflects others; this determines the colour effect.

The coloured light in the visible spectrum ranges from violet to red. We can see this process by passing sunlight (white light) through a prism. Upon entering the prism, white light refracts (is bent), causing light waves of different lengths to be revealed (red having the longest wave length and violet having the shortest) into the visible spectrum. This splitting of white light creates what we see as seven different colours: red, orange, yellow, green, blue, indigo and violet. (Remember: **R**ichard **O**f **Y**ork **G**ave **B**attle **I**n **V**ain – for the colours of the rainbow.)

White light refracted through a prism into colours of the rainbow

A white object reflects most of the white light that falls upon it; a black object absorbs most of the light falling on it. A red object reflects the red light and absorbs everything else.

White reflects light; black absorbs it

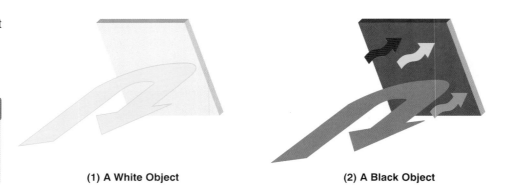

(1) A White Object

(2) A Black Object

Colour addition If we take the three primary colours of red, yellow and blue, pairs of these give the secondary colours, so red and blue mixed together creates violet, yellow and blue creates green and yellow and red creates orange. White and black can be added to vary the tone of the colour.

Adding further colours together, secondary colours can be added to primary colours to give further variations of tertiary (third level) colours.

Mixing primary and secondary colours

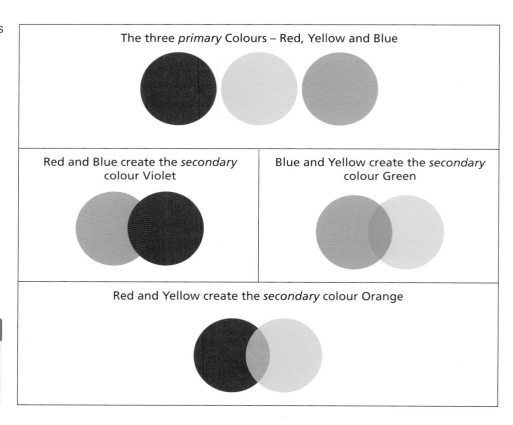

The three *primary* Colours – Red, Yellow and Blue

Red and Blue create the *secondary* colour Violet

Blue and Yellow create the *secondary* colour Green

Red and Yellow create the *secondary* colour Orange

The colour wheel

When you know about the basic principles of colour, and the relationship that colours have to one another, you can appreciate which colours contrast, and which work harmoniously, in your personal colour choices.

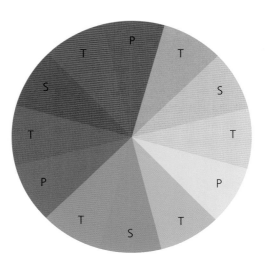

The primary, secondary and tertiary colour wheel:
P = primary
S = secondary
T = tertiary

A colour wheel provides a fast visual guide to how colours interact. The colours on opposite sides of the wheel contrast against one another, whereas colours next to each other harmonize.

Contrasting colours Contrasting colours are known as complementary colours and they are positioned on the opposite sides of the wheel. When they are included in the same colour theming, they will clash (and cancel/neutralize each other) adding impact to the resulting colour effect.

Note: This neutralizing or cancelling-out effect is used in advanced colouring to balance out unwanted tones within the hair. For example, if a client's hair is looking *green*, a colourist will use *red* tones to balance out and cancel the unwanted effect. Similarly, if a client has had a lightening service and their hair is looking *yellow*, a *violet* toner will neutralize the unwanted tones.

TOP TIP

The colour directly opposite on the colour wheel, when used together, will neutralize its effect.

Depth and tones

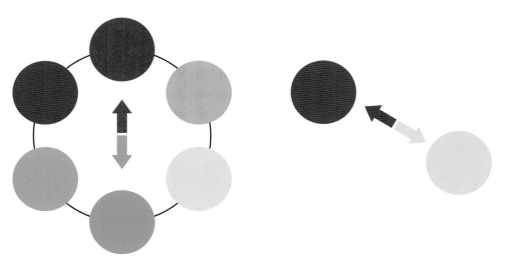

Dominant and recessive colours (warm tones and cool tones) When choosing colours for hair it is important to remember that some have more impact than others. Warm tones, such as, red, yellow and orange, stand out more and are therefore regarded as dominant or advancing colours In contrast, cool tones, such as shades of blue and green, fade into the background and are commonly known as recessive colours.

The choice and use of colour is often an instinctive, intuitive skill. So the same advancing colours that will make hair look vibrant and bold may sometimes appear inappropriate and too dramatic! Conversely, recessive colours in hair tend to *sink* backwards and have far less visual impact.

The effects of light and lighting upon hair Hair colour depends chiefly on the pigments in the hair, which absorb some of the light and reflect the rest. The colour that we see is also affected by the light in which it is seen and (to a lesser extent) by the colours of clothes worn with it.

◆ **White light** from halogen bulbs and full daylight will show the hair's natural colour.

◆ **Yellow light** emitted from standard electric light bulbs adds warmth to hair colour, but neutralizes blue ash or ashen effects.

◆ **Blue or green light** from fluorescent tubes and *long-life* energy saving bulbs, reduces the warmth of red/gold tones in hair.

Natural hair colour

When looking at the natural colour of hair, what we are really seeing are microscopic pigments within the hair's cortex. This naturally occurring colouration is created when nutrients in the blood supply are converted to form the protein, keratin. This pigmentation is then added into the newly formed keratin at the germinal matrix.

The natural hair pigments are collectively called **melanin** and different quantities of these pigments, which varies between individuals, gives us all the colour of our hair.

Facts about colour pigments and their effects

Pigment	Facts	Effects	
White hair (grey hair) has lost all natural pigments and therefore appears colourless.			
Eumelanin	◆ Produces cool tones ◆ Brown or black in colour	◆ Dark hair/base – has high levels of **eumelanin**, little **pheomelanin**	
Pheomelanin	◆ Produces warm tones ◆ Yellow or red in colour	◆ Light hair/base – has high levels of pheomelanin, little eumelanin	
Trichosiderin	◆ Is very rare ◆ Produces warm tones ◆ Golden red or red in colour	◆ Red, *Celtic* hair – has high levels of **trichosiderin** and pheomelanin	

TOP TIP

Pheomelanin pigments are larger than eumelanin and are harder to remove from hair during lightening

The hair colour you actually see is affected by the amount and proportion of the pigments present. But remember that the type/amount of light or lighting also affects how it is seen.

With age, or after periods of stress, the production of natural pigments may be reduced. The hairs already on the head will not be affected, but the new ones will. As hairs fall out and are replaced, the proportion that have the original pigmentation reduces and the hair's overall colour changes. It may become lighter. If no pigment is produced at all, then the new hairs will be white/grey.

The proportion of white hairs among the naturally coloured ones causes the hair to appear grey. Grey hair or greyness (**canities**) is often referred to as a percentage; for example, '50% grey' means that half of the hairs on the head are white and the rest are pigmented. It is not uncommon for young people to show some grey hairs – this does not necessarily mean that they will go grey, or completely white, at an early age.

Types of synthetic or artificial hair colour

Natural hair colour is made up from melanin but these pigments, or the appearance of these pigments, can be changed or modified. Hair colour can be changed by the *addition* of artificial pigments, i.e. colour application, or the reduction, i.e. the removal of artificial or natural pigments, through lightening techniques.

A colour board

Different effects and the colouring products linked with them

Effect that can be achieved	Colouring product
You can **add** *artificial colour* pigments to hair on a temporary basis	◆ Temporary colour
You can **add artificial** colour pigments to hair that last for several washes	◆ Semi-permanent colour
You can **add artificial** colour pigments to hair so that they fade over time or stay there permanently	◆ Quasi-permanent colour ◆ Permanent colour
You can **remove natural** colour pigments permanently	◆ Lightener ◆ High-lift colour
You can **remove artificial** colour pigments permanently	◆ Colour strippers

ALWAYS REMEMBER

Hair colour and the under 16s

All chemicals that are used within hair and beauty products are controlled under European laws (EU Directives). Some chemicals found in permanent and quasi-colours, lighteners, permanent waving, relaxers and chemical straighteners, are restricted and no longer permitted for young people under 16 years of age. Any hair preparation that falls within this group is now prohibited and you will see that these products are clearly marked: 'This product is not intended for use on persons under the age of 16.'

Temporary colour: uses, applications and facts

Temporary colours are available in the form of lotions, creams, mousses, gels, lacquers, sprays, hair mascara, crayons, paints and glitter dust. If the hair is in good condition, they will not penetrate the hair cuticle, nor do they directly affect the natural hair colour: they simply remain on the hair until washed off, so are ideal for a client who has not had colour before. They can be used to produce subtle colour effects, or bright, bold options, without any adverse effects on the quality and condition of the hair.

However, if they are used on badly damaged or very porous hair, the temporary colour may quickly be absorbed into the cortex, producing unwanted, long-lasting, uneven patchy results.

Features and benefits of temporary colour

Features	Benefits
Have large pigments/molecules and sit on the surface of the hair.	Easy to remove as can be washed away during the next shampoo.
Come in a variety of types as mousses, setting lotions, gels, creams, colour sprays and colour shampoos.	Easy to apply either during the shampoo process or alternatively as a styling or finishing product.
Come in a variety of shades and colours.	Can be used as a fashion statement or alternatively to enhance natural tones by either adding depth to faded hair or neutralizing unwanted tones from hair.

Facts about temporary colour

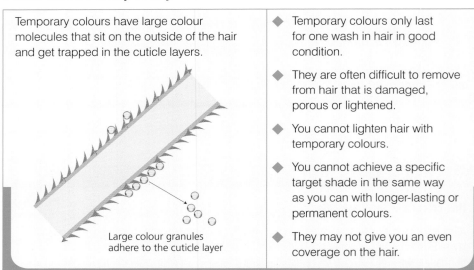

Temporary colours have large colour molecules that sit on the outside of the hair and get trapped in the cuticle layers.

Large colour granules adhere to the cuticle layer

◆ Temporary colours only last for one wash in hair in good condition.

◆ They are often difficult to remove from hair that is damaged, porous or lightened.

◆ You cannot lighten hair with temporary colours.

◆ You cannot achieve a specific target shade in the same way as you can with longer-lasting or permanent colours.

◆ They may not give you an even coverage on the hair.

ALWAYS REMEMBER

Always follow the manufacturer's instructions when using any colouring products.

ALWAYS REMEMBER

Compound henna is incompatible with hairdressing materials. It should not be confused with natural (vegetable) henna, which is compatible with other hairdressing services.

Semi-permanent colour: uses, applications and facts

Semi-permanent colours are made in a variety of forms – some are used as shampoos or conditioning rinses, which make them easier to apply, others are applied directly from the bottle onto dry hair.

Semi-permanent colours have pigments which deposit into the hair cuticle and outer cortex. This type of colour gradually fades each time the hair is shampooed. Some colours will last through six washes, others longer.

These colours are not intended to cover white/grey hair – they will only produce translucent effects, masking some of the grey.

Features and benefits of semi-permanent colours Semi-permanent colours are ideal for those people who want to try colour but who do not want the maintenance of permanent colour effects. They generally last up to six or eight shampoos and do not produce any regrowth; the hair loses the colour on each shampoo so the effect lessens each time.

Semi-permanents can also provide an ideal solution for livening up faded mid-lengths and ends for clients who have permanent colours; this is particularly useful if the hair is not really ready yet for further permanent processing.

Some semi-permanents will colour white/grey hair to some extent, although the penetration does not normally extend beyond the cuticle layer, so colour density is relatively poor. (White hair tends to have a very smooth cuticle so there are fewer spaces for the pigments to bond on to.) The colour range is varied, ranging from fashion effects to many of the shades you would expect to see in a standard shade chart. They are simple to use and require no developer and hence no mixing.

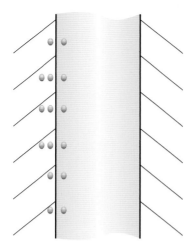

Temporary colour hair strand

Note: Some of the darker semi-permanent shades do contain Paraphenylenediamine (PPD) and this needs to be checked within the ingredients, as a skin test WILL be required.

Features and benefits of semi-permanent colour

Features	Benefits
Has large molecules that sit on the surface of the hair while other smaller ones penetrate deeper into the hair.	Good for introducing clients to colour without any long-term commitments. Fairly easy to remove as it is washed away in six or eight shampoos.
Comes in a variety of types such as mousses, liquids, gels, creams.	Easy to apply, normally requires no mixing, takes a short time to apply and leaves no regrowth.
Comes in a variety of colours as fashion effects or as standard shade chart references.	Can be used as a fashion statement or alternatively as a trial for a permanent colour effect.
Can be used in colour correction work. Adds tone to white/grey hair.	A simple and quick pre-filler and pre-pigmentation shade. Provides *some* masking/coverage for unwanted greys.
Provides an alternative to permanent colour.	Can provide a colour choice for those people who, because of sensitivity or allergy, may not be able to have permanent colours.

Facts about semi-permanent colours

Semi-permanent colours have large colour molecules (but smaller than temporary colour) that get trapped in the cuticle layer and penetrate just to the outer cortex.	◆ Semi-permanent colours only last for up to six or eight washes. ◆ They are often difficult to remove totally from hair that is extremely porous or lightened. ◆ You cannot lighten hair with semi-permanent colour. ◆ They will not cover white/grey hair. ◆ They cover with far better results than temporary colours. ◆ Some darker shades contain Paraphenylenediamine (PPD) check ingredients prior to application. ◆ Colours containing PPD require an allergy alert test 48 hours prior to application.

ALWAYS REMEMBER

Quasi-colour should be treated like a permanent colour or lightener; it does require an allergy alert test 48 hours beforehand.

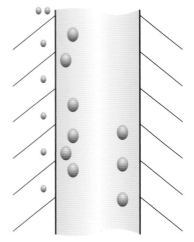

Quasi-permanent colour hair strand

Quasi-permanent colour (tone on tone colour): uses, applications and facts

Quasi-permanent colours have large ranges that provide plenty of choice. The effects are long lasting and they are mixed with lower strengths of hydrogen peroxide than permanents. However, because of this, they do produce a regrowth, and the colour effect fades over time. They have a larger colour molecule than permanent colours and are ideal for colour matching and refreshing worn, tired hair.

Features and benefits of quasi-permanent colour Quasi-colours make up the *bulk* of hair colour bought by people from shops, for use at home. They are not true permanent colours but do require mixing; they last for at least 12 washes and anything up to 24, but they do leave a regrowth. These types of colours do have a better ability to cover white/grey hair and this is the main reason why they are popular for home colouring.

Features	Benefits
They are processed with a developer and have molecules that swell inside the hair.	Easy solution for all-over colouring; last a long time, generally up to 12–24 shampoos.
Come as gels or creams.	Require mixing with developer and have been made easy to apply for home use.
Come in a variety of colours as fashion effects or as standard shade chart references.	Can be used for a fashion effect or as an alternative to more permanent-based colour.
Can be used in colour correction work.	Are an alternative to longer-lasting pre-pigmentation shades.
Add depth and tone to white/grey hair.	Provide up to 80% coverage for unwanted greys.
Provide a different alternative to permanent **para dyes**.	Tend to be used regularly and more often than salon-provided treatments.
Good conditioning properties, add shine and improve manageability.	Leave hair in good condition, manageable and with added shine.

Facts about quasi-permanent colour

Quasi-permanent colours have smaller colour molecules and are mixed with lower strength hydrogen peroxide. They enter the cortex of the hair and are oxidized during the processing. This makes them swell and then they are trapped inside the cortex.

◆ Quasi-permanent colours always require an allergy alert test 48 hours prior to application.

◆ They last for at least 12 washes, leaving a regrowth.

◆ They can only be removed by colour strippers, not by colour or lightener.

◆ They often provide the basis for colour correction work if wrongly used at home.

◆ They offer good coverage; colour white/grey with better saturation than semis.

◆ They have similar effects to permanent colours.

Permanent colour: uses, applications and facts

Permanent colours have the largest variety of shades and tones. They can cover white and natural-coloured hair to produce a range of natural, fashion and fantasy shades.

Hydrogen peroxide is mixed with permanent colour; this oxidizes the hair's natural pigments and joins the artificial pigments together. The hair will then retain the colour permanently in the cortex. Hair in poor condition, however, may not hold the colour and colouring could result in patchy areas and colour fading.

Facts about permanent colours

Permanent colours have small colour molecules that are mixed with hydrogen peroxide. They enter the cortex of the hair, and are oxidized during the processing. This makes them swell and then they are trapped inside the cortex.	◆ Permanent colours (para dyes) are the only colours that cover white/grey hair with 100% accuracy. ◆ Permanent colours always require an allergy alert test 48 hours prior to application. ◆ They can only be removed by colour strippers (not by colour or bleach). ◆ People can be allergic to Paraphenylenediamine (PPD) (this is contained in quasi-, permanent and some semi-permanent colours). ◆ They are resistant to fading and have to grow out – leaving a regrowth. ◆ Have the largest choice of colours for clients. ◆ Darker colours contain more PPD than lighter shades.

Permanent colours contain a chemical compound called paraphenylenediamine (PPD). This is a known allergen. For this reason, it is essential that you follow the manufacturer's instructions in carrying out an allergy alert test 48 hours before a permanent colouring service.

There are other chemicals within permanent colours and they have different properties and effects:

◆ Ammonia is alkaline and when it comes into contact with the hair it swells the hair shaft in preparation for the pigmentation. This provides a better penetration but will need pH balancing to return the hair to its normal state.

◆ Conditioning agents improve the hair during the colouring process, enabling it to be smoother and shinier as a result.

◆ Hydrogen peroxide oxidizes the natural pigments of the hair and this enables the artificial pigments to bond with them, creating a permanent change within the hair's cortex.

Vegetable-based colour

As well as being a popular source for conditioning agents, plant extracts have been used as dyeing compounds for thousands of years. These were the only sources of colour until chemists developed **synthetic colour** alternatives. Natural henna (Lawsonia) is still used widely today in many countries and it is used for dyeing skin as well as the hair. Natural plant-based dyes do not present any problems for hairdressing treatments; however,

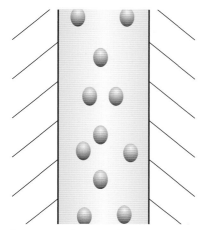
Permanent colour strand

these ingredients are sometimes added to other elements to form compounds, mixtures of vegetable extracts and mineral substances. One that is still available is compound henna – vegetable henna mixed with **metallic salts**. This penetrating dye is incompatible with professional products used in hairdressing salons and will react with professional salon colours and perming products.

Progressive (metallic) dyes

Progressive dyes work in a different way to other organic hair colours; the depth of colour builds up over several applications, so the effects can be subtle to start with. These are metallic dyes containing *lead acetate* which only affect the cuticle layers of the hair. However, because of the compounds within them, these types of dye are *incompatible* with chemical hairdressing services. Metallic elements can be found in some **colour restorers** such as 'Just for Men' and 'Grecian 2000'.

ALWAYS REMEMBER

Metallic and compound dyes are incompatible with hairdressing materials. Always carry out a test before using a lightener, colour or perm on any new clients whom you do not know, or on clients who have been recently having their hair done overseas. Lead acetate is toxic!

Depth and tone

When talking about colour, we often use the words *depth* and *tone*. Depth is used to describe how light or dark the colour is and tone is used to describe the colour or hue that we see such as brown, golden red, etc.

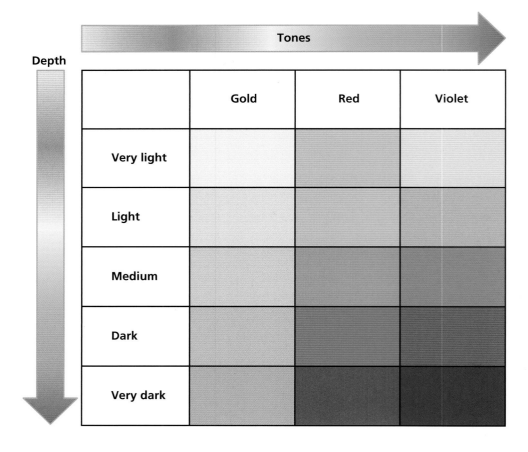

International Colour Chart System (ICC)

The International Colour Chart (ICC) is a table that provides a way of defining hair colours in logical, sequential way. In this table, the *depth* of colour is divided and numbered, with black (**1**) at the lower end and lightest blonde (**10**) at the upper end. The *tones* of colours (/1/3 or also shown as •1 •3) spread across the table from left to right, and when the depth is combined with tones it produces a wide variety of colours.

The ICC numbering system During colour consultation, a shade chart is used to:

1   First, identify the client's natural hair colour depth.

2   Then, find the client's natural colour tones.

3   Discuss the client's desired target colour(s).

4   Finally, work out how the effect can be achieved.

The ICC numbering system								
NO.	**Shade**	**Ash** **•1**	**Violet** **•2**	**Gold** **•3**	**Copper** **•4**		**Red** **•6**	**Metallic** **•7**
10	Extra light blonde							
9	Very light blonde							
8	Light blonde							
7	Mid blonde							
6	Dark blonde							
5	Brown				5.4			
4	Dark brown							
3	Darkest brown							
1	Black							

So if your client has **brown** hair (**5**) and you colour with a copper tone (**•4**), the result should be a rich copper brown (**5.4** shown in the ICC table). There are many options. Some other possible tones could be:

◆   to produce ash shades, add blue or •1

◆   to produce matt shades, add green •7

◆ to produce gold or copper shades, add yellow or orange •**3 or** •**4**

◆ to produce warm shades, add red •**6**

◆ to produce purple or violet shades, add mixtures of red and blue •**2**

Primary and secondary tones We have already established that the first number before the decimal point (or forward slash) refers to depth. If we break the numbering down we see that there are three parts to the number of a colour. The depth, then the primary and secondary tone.

Shade	Depth		Primary tone	Secondary tone
6.64	6	•	6	4

1 The **primary tone** indicates the main tonal colour that you see (about 70%).

2 The **secondary tone** indicates the additional pigmentation within the shade and, like the colour wheel, this extra 30% of colouration provides many extra colour permutations.

Sometimes colour manufacturers want to increase a shade's intensity and vibrancy. This is achieved by adding double the tone to a particular shade. This double tone effect will always denote a high intensity, chromatic effect.

Therefore, in the table (primary and secondary tones) we see that we have a colour of Base 6, or dark blonde. It has a primary (or majority tonal effect) of •**6** or red, it also has further colour properties as its secondary tone is •**4** or copper.

Double strength tones

Shade	Depth		Primary tone	Secondary tone
6.66	6	•	6	6

In this illustration we see that the shade has the same base and primary tones as the example above, i.e. **6•6** but in this scenario the secondary tone is the same as the primary i.e. **6.** These double strength hair colours mean that there is extra red in the colour, so it denotes this shade as extra bright red.

The effects of different strengths of hydrogen peroxide on hair

Hydrogen peroxide is an acidic solution used for developing both, quasi- and permanent colours. Its chemical make-up is hydrogen and oxygen and in a solution the elements link together to create molecules of H_2O_2, so each molecule contains two hydrogen atoms and two oxygen atoms. The only difference between the chemical compositions of hydrogen *per*-oxide and water is that hydrogen peroxide has two atoms of oxygen whereas water only has one.

ALWAYS REMEMBER

Never mix colour or put it into a bowl before you need to use it. Permanent colours will oxidize in the air; this expands the colour pigments and they will not be able to enter the hair shaft through the narrow cuticle layers, thus rendering the colour useless.

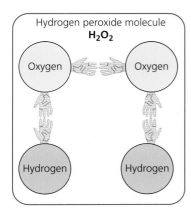
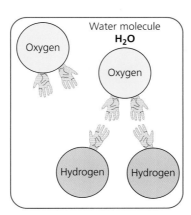

This is useful in hairdressing chemistry, as oxygen is the element that does all the work. It oxidizes natural and artificial colour pigments, getting them to change their properties. Therefore, when oxygen is *given up* during chemical processes, the hydrogen peroxide is reduced to create water as a by-product of the process. This simple chemical reaction minimizes any damage to the hair during the process. So under normal circumstances, this acidic solution (pH 3.7) is reduced to something harmless, (pH 7) but only providing that the manufacturer's instructions are followed.

Hydrogen peroxide strength	Effect upon the hair
6 vol. (1.9%)	◆ Will deposit colour and tone into the hair, adding depth, making it darker when using quasi-permanent colours or bleach toners.
9 vol. (2.7%)	◆ Will deposit colour and tone into the hair, adding depth, making it darker when using quasi-permanent colours.
15 vol. (4.7%)	◆ Will deposit colour and tone into the hair, adding depth, making it darker when using quasi-permanent colours. ◆ Will lighten 1 level when using quasi-permanent colours.
20 vol. (or 6%)	◆ Will deposit colour and tone into the hair, adding depth, making it darker when using permanent colours. ◆ Enables coverage of white/grey hair. ◆ Will lighten 2 levels above base 6 (on fine hair). ◆ Will lighten 1 level below base 4.
30 vol. (or 9%)	◆ Will lighten hair 3 levels above base 6. ◆ Will lighten hair 2 levels below base 4.
40 vol. (or 12%)	◆ Will lighten hair 4 levels above base 6 (with high-lift colour). ◆ Will lighten up to 7 levels of lift with lightening product.

Special note: When hydrogen peroxide lightens hair to any level it will remove the smaller, (eumelanin) brown/black pigments first. The larger, warmer pigments (pheomelanin) within the hair are more difficult to remove, so unwanted golden or even orange undercoat tones are often left within the hair.

ALWAYS REMEMBER

Mixing colour

Don't mix permanent colours together until you are ready to use them. Mix the colours carefully, making sure that you measure the amounts accurately. If the proportions are wrong, the final effect will be wrong.

Diluting hydrogen peroxide

30 vol. (or 9%)	Will lighten hair 3 shades above base 6
	Will lighten hair 2 levels below base 4

Because hydrogen peroxide has different strengths based upon the amounts of free oxygen within it, (e.g. 6% of free oxygen within any given amount or 12% of free oxygen within any given amount), it can be diluted with water in times when the salon has run out of stock. The following table shows you how the strengths that you have in stock can be diluted to make a reduced strength hydrogen peroxide.

The first column refers to the strength peroxide that you have, the second refers to the peroxide you want to create. The last three columns show you how many parts of the peroxide you need and how many parts of distilled water you need to add to it.

Diluting different strengths of hydrogen peroxide

Strength you have	Strength you want to create	Peroxide	Add	Water
40 vol. (i.e. 12%)	30 vol. (i.e. 9%)	3 parts	+	1 part
40 vol	20 vol. (i.e. 6%)	1	+	1
40 vol	10 vol. (i.e. 3%)	1	+	3
30 vol. (i.e. 9%)	20 vol. (i.e. 6%)	2	+	1
30 vol	10 vol. (i.e. 3%)	1	+	2
20 vol. (i.e. 6%)	10 vol. (i.e. 3%)	1	+	1

Lightening hair

Lighteners have alkaline chemicals within them that achieve lightened effects by dissolving the natural tones (pigments) within hair. Similar to para dyes, lightening products are mixed with hydrogen peroxide to activate the oxidizing process, and they are used in three main forms:

◆ High-lift colour – which is a non-lightener process for lightening hair partially or whole head.

◆ Powder lightener – which is used for highlighting and partial lightening techniques.

◆ Gel/oil lightener – which is suitable for on-scalp application.

The alkaline compound acts upon the hair by swelling and opening up the cuticle. This enables the peroxide at 6% or 9% to release oxygen and oxidize the natural pigments of melanin from within the cortex. This creates oxymelanin and is seen as it reduces the natural colour through the different degrees of lift.

The colour control during lightening is not the same as with colouring; often, when full-head lightening is done, the result is quite yellow (although the control of warm pigments during lightening is better with high-lift colour).

ALWAYS REMEMBER

The degree of lift required in emulsion lightener is controlled by the number of activators added into the mixing bowl with the oil. It is not boosted by stronger hydrogen peroxide levels. Always follow the manufacturer's instructions when using lightening.

Choice of lightening method/product

High-lift colour High-lift colour provides a non-lightener solution for lightening hair. These types of colour are gentler than lightener and can be mixed with 6% or 9% volume hydrogen peroxide.

The colour control or colour targeting is generally better than lightening too, as high-lift colour can deposit tones (e.g. ash, beige or warm tones) as the hair lightens. Moreover, because you are using a hair colour and not lightener, there is less moisture removed from the hair during the process, so a better hair condition can be guaranteed at the end.

High-lift colours are similar in composition to normal hair colours with one exception: they use an alkaline compound (e.g. resorcinol) which swells the hair shaft enabling a better penetration of the lightener into the polypeptide chains within the cortex.

Removing high-lift colour You can remove high-lift colour by emulsifying the colour. You do this by adding a little warm water to the colour and gently massaging all over. This mixes the colour with the water and helps to release the products from the hair enabling you to use a lighter action within the shampooing process. The hair can be conditioned as normal at the end or an antioxidant can be applied to help close the cuticle and lock in the colour results.

Oil/gel (emulsion) lightener Oil, or emulsion lightener is a slower acting lightening process that is made up from two compounds that are added together and then mixed with hydrogen peroxide:

◆ oil (or gel) lightener

◆ activators (also known as boosters or controllers).

This type of lightening product is formulated for use directly onto the roots of the hair, and is suitable for contact with the scalp. It is kinder and gentler during the lightening process, is mixed with 20 vol. hydrogen peroxide for root, mid-length and ends application. The lift through the *undercoat* shades is aided and controlled by the addition of activators. These boost the power of the lightener while maintaining relatively low hydrogen peroxide strength.

Emulsion lighteners also contain additives, which control the resultant colour as the lightening product lightens the hair. As mentioned earlier, these tend to make hair yellow, so they have matt emulsifiers which neutralize unwanted yellow tones while the lifting process takes place. Heat may be used during the development of the process, but the process must be monitored closely (particularly if a colour accelerator or steamer is used to aid development) as these types of lightener can often be more 'slippery' and mobile, and might drip.

Removing emulsion lightener Make sure that when the lightening product is removed you rinse without massaging, using only tepid or warm water. The client's scalp has been subjected to chemicals and could be sensitive; the cooling action of the rinsing will stop the lightening process and make the client more comfortable. After the emulsion has been rinsed away, the hair can be shampooed with a mild colour shampoo and conditioned with an anti-oxidizing conditioner.

Powder lightener

Powder lightening product can be mixed with 6%, 9% or 12% hydrogen peroxide, depending upon the level of lift required. Powder lighteners are fast acting and are used

ALWAYS REMEMBER

Avoid inhalation of powder lightener

Be very careful when you dispense powder lightener into a bowl. The particles are very small and tend to 'dust' into the air very easily. This is a hazardous chemical compound which can cause respiratory conditions. You must avoid contact through inhalation; wherever possible only use 'dust-free' powder lightener.

TOP TIP

The degree of lift required in emulsion lightener is controlled by the number of activators added into the mixing bowl with the oil. It is not boosted by stronger hydrogen peroxide levels. Always follow the manufacturer's instructions when using lightening products.

for a variety of highlighting techniques. When they are mixed in the bowl the consistency is that of a thick, 'porridge-type' paste. The stiffness of the consistency prevents spillages and enables the lightener to work like a poultice. As the process continues the lightener/peroxide mix will expand. This action is speeded up more if accelerated by heat, so a careful eye should be kept on the development timings.

Removing powder lightener The removal of powder lightener is similar to that of emulsion lightener. Make sure that when the lightening product is removed from the hair you rinse it without massaging, using only tepid or warm water. The removal of powder lightener can be far more problematic than that of emulsion lightener and this has more to do with the colouring technique that has been used. If different-coloured highlights have been done with Easi-Meche™, foil, wraps, etc. these need to be removed carefully and individually as one colour may affect another. Although lightening product has not been directly applied to the client's scalp, it still might be sensitive from the colouring technique. Therefore, the cooling action of the rinsing will stop the lightening process and make the client more comfortable. Afterwards the hair can be shampooed with a mild colour shampoo and conditioned with an anti-oxidizing treatment.

HEALTH & SAFETY

Powder lightener can cause breathing problems. Use with care and do not inhale.

Lightening hair – quick reference table

Lightening service required	What you need to check for	Technique/application	Lightener type
Whole head (on virgin hair)	**Test results:** ◆ (Skin tests, etc.) **Natural hair depth:** ◆ Lighteners will lift five levels quite happily on hair with brown/ash pigments. However, strong red content will be difficult to remove. ◆ Hair beyond base 5 will not lift safely beyond base 9. Suggest other colouring options.	Lightening product mixture must be applied to mid-lengths and ends first. A plastic cap should envelop the contents and can be developed with gentle heat until ready. When the lightened hair has lightened two to three levels of lift, the root application can be applied.	Only emulsion lightener or cream is suggested for application to the scalp. These are used with 6% hydrogen peroxide and sachet controllers to handle levels of lift.
	Hair length: ◆ Lengths up to 10 cm lighten evenly, provided manufacturer's instructions are followed. ◆ Lengths over 15 cm are not recommended, as evenness of colour will be difficult to guarantee. **Hair texture:** ◆ Finer hair needs extra care and lower hydrogen peroxide strengths, i.e. 6%. ◆ Medium and coarser hair present fewer technical problems. **Hair condition:** ◆ Only consider hair in good condition for lightening. (Lightening removes moisture content during the process, hair that is porous or containing low moisture levels has insufficient durability for lightening.)	◆ Always follow manufacturer's instructions.	

Lightening service required	What you need to check for	Technique/application	Lightener type
Full head (on previously coloured)	◆ Not recommended		
Root application (pre-lightened ends)	**Existing client:** ◆ Yes, check previous records and current hair condition and carry out service. ◆ No, new client: go through all the checks in the **full-head application** table and find out the previous treatment history.	Roots only without overlapping previous lightened ends. ◆ Always follow manufacturer's instructions.	Only emulsion lightener or cream is suggested for application to the scalp.
Highlights (fine even meshes on virgin hair)	**Test results:** ◆ (Skin tests, etc.)	◆ Plastic self-grip meshes (e.g. Easi-Meche™, L'Oréal) ◆ Foil meshes ◆ Colour wraps	High-lift powder lightener with suitable hydrogen peroxide developer at 6%, 9% or for highest lift 12% (providing no product is allowed to make contact with the skin/scalp).
Note – The success of highlights on coloured hair is very poor in comparison. This work is often undertaken in salons, but ends seldom lighten effectively, while the roots lighten very quickly. (Colour should be removed with a synthetic colour remover or decolour).	**Natural hair depth:** ◆ Lightening products will lift five levels quite happily on hair with brown/ash pigments. However, strong red content will be difficult to remove and require stronger developer and/or additional heat. ◆ Hair beyond base 5 will not lift safely beyond base 9. Suggest other lightening technique. **Hair length:** ◆ Hair length will have an impact on evenness of colour. However a small tolerance is acceptable and 'visually' indistinguishable on longer hair lengths. **Hair texture:** ◆ Finer hair needs extra care and lower hydrogen peroxide strengths, i.e. 6%. ◆ Medium and coarser hair present fewer technical problems but generally take longer. **Hair condition:** ◆ Only consider hair in good condition for lightening. (Lightening removes moisture content during the process, hair that is porous or containing low moisture levels has insufficient durability for lightening).		

Cap highlights Highlight caps are a safe, simple and popular choice for cost-effective, single-colour or lightened highlights on short layered hair; even a single colour application can achieve an attractive, multi-toned effect.

Woven highlights Woven highlights in foil, Easi-Meche™ or wraps are the preferred technique for multi-toning hair. The visual effects are unlimited and new exciting colour combinations and techniques are happening each year.

Toners and lightener toning Permanent (and most quasi-permanents) colours are unsuitable for depositing tones on pre-lightened hair. If they were used, the effects could damage, if not destroy the hair. Permanent colours and some quasi colours contain alkaline compounds and if you look back at the properties of pH upon hair, you will remember that **alkalis** swell the hair. If this were to happen after the hair has already been weakened from lightening, the hair is likely to break off.

Full head lightening seldom provides satisfactory results. There are, usually, some unwanted (golden or yellow) tones to be neutralized and this is achieved by using specially formulated lightening product toners. Lightener toners are ranges of pastel shades that can be used to balance out and neutralize unwanted tones.

Toners and their effects

Toning is the process of adding (depositing) colour to previously lightened hair. A variety of pastel shades, such as silver, beige and rose, are used to produce subtle effects. Different types of toners are available; read the instructions provided by their manufacturers to find out what is possible.

Toners are often used after lightening as a form of colour correction to counteract unwanted yellow or gold tones. (See the Colour Wheel, pages 244–245.)

Level of depth	Tone required	Tonal effect	Pre-lighten to
10	Silver, platinum, mauve/violet	Cool	Very pale yellow
9	Ashen blondes, light beige blonde	Cool	Pale yellow
8	Beige blonde	Cool	Yellow
8	Sandy blonde	Warm	
7	Golden blonde	Warm	
7	Chestnut, copper gold	Warm	Orange
6	Red copper	Warm	
5	Mahogany	Cool	Red/orange
4	Burgundy, plum	Cool	

Safe methods of working for colouring and lightening services

Safety and preparation

Chapter 15 on health and safety covers much of the general aspects that you need to know, but each technical procedure has specific points that relate to that area of hairdressing alone. Hair colouring is particularly problematic as it involves the application of a variety of potentially harmful chemicals. Therefore, the care that you take in handling products and preparing yourself and the client is absolutely critical to safe and successful colouring.

Records

These should be found and put ready at the beginning of the day. The appointment book identifies all the expected clients, so all their treatment history – dates of visits, who provided the services, previous chemical services, records of any tests and any additional comments – can all be collated long before the client arrives.

Similarly, when clients have been into the salon, any results of tests or notes following treatments must be updated as soon as possible.

Tests Always refer to your test results before carrying out any colouring service. Check with the client that they have not had any positive reaction to the chemicals from the patch test and refer to your strand test for the agreed target shade. Finally, recheck the porosity and elasticity of the hair and update the client's record accordingly.

Materials After the records have been found, it is advisable to get all tubes, cans or bottles of colour put aside and ready, along with the client's record information. Doing this early has several benefits. It can also save valuable time later when you need to mix them, particularly if you are running on a tight schedule.

Gowning Always make sure that the client and the client's clothes are adequately protected before any process is started. Most salons have special 'colour-proof' gowns for colouring and lightening processes. These gowns are resistant to staining and are made from finely woven synthetic fabrics that will stop colour spillages from getting through onto the client's skin or clothes. When you gown the client, make sure that the free edges are closed and fastened together. On top of this and around the shoulders you can place a colouring towel and over this a plastic cape. This needs to be fastened but loose enough for the client to be comfortable throughout the service.

Using barrier cream Barrier cream is less popular now than in the past, however it can be used as a physical barrier to prevent staining around the client's face/hairline. It is also particularly useful if the client has any *general* sensitivity to products on their skin. (Allergy Alert Tests must always be done 48 hours before oxidation colours are applied.) Remember it is not an excuse for poor slapdash application, allowing you to extend the colour application beyond the root area to the skin. However, it will help in areas where the colour seeps off the hair onto the skin.

Apply barrier cream to the skin with a finger or cotton wool close to the hairline, taking care not to get it onto the hair, as this could stop the colour from taking evenly.

It can be removed later after you have shampooed the colour from the hair and before any other services are conducted.

ALWAYS REMEMBER

Always gown your client properly so that they are protected from spillages of chemicals and always wear the PPE provided by the salon every time you apply colour.

TOP TIP

Always carry out the necessary tests before you start, they could have a critical impact on the end results.

ALWAYS REMEMBER

After any treatment or tests have been carried out you should always update the client's records immediately. These tests are critical to the client's well-being and the salon's good name. You should record the:

- date
- the test carried out
- development times
- results and any comments from the client
- recommended home-care or follow-up advice given.

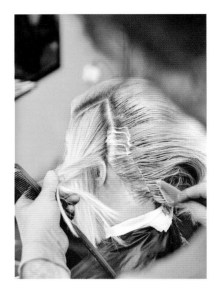

Seating position The chair back should be protected with a plastic cover. If this is not available, a colouring towel could be folded lengthwise and secured with sectioning clips at either end and the client should sit comfortably, in an upright position, with their back flat against the cushioned chair pad.

Trolley You should have your colouring trolley prepared and at hand with the materials you will need. Foils for highlighting should have been previously prepared to the right lengths and combs, brushes, sectioning clips, etc. should be all cleaned and sterilized and ready for use.

Protecting yourself Your personal hygiene and safety are also important. The care you take in preparing for work should be carried through in everything you do and this is made even more important when you are about to handle hazardous chemicals. Put on a clean colouring apron and fasten the ties in a bow. Then take a pair of disposable (non-latex) gloves and put them on ready for the application.

Use your time effectively; each salon allocates different times for different services. A retouch may only take 20 minutes to apply on shorter hair, whereas a long hair set of full-head woven highlights could be booked for an hour.

Preparation dos and don'ts

Dos	Don'ts
✓ When a client's hair is developing under a Climazone, Rollerball or any other colour accelerator, do check at intervals during the processing to see that they are comfortable and that the equipment is not too hot.	✗ Never handle electrical equipment with wet hands, always dry them first.
✓ Do check the equipment controls so that the timing and temperature settings are correct.	✗ Never leave colour spillages until later. Mop them up straight away while you still have your protective gloves on.
✓ Do check the manufacturer's instructions before you mix any products. They will give you the recommended amounts and quantities to mix together.	✗ Don't mix up too much product at one time, it is wasteful and expensive. If you need more you can always mix when you need it.
✓ Always do a skin test on the client before any colouring process.	✗ Never mix products up before they are needed. Colour products have a set development time and oxidization will start if they are exposed or mixed too soon.
✓ Do put screw tops and lids back on colouring products immediately. Their effectiveness will be impaired if they are exposed to the air for any longer than needed.	✗ Never attempt to do any colouring procedure without wearing the correct PPE.
✓ Do make a note of low levels of stock as product is removed from storage.	✗ Don't work in a cluttered environment. Always make sure that the work area is prepared properly and ready for use.
✓ Do make good use of your time. Always prepare your work area and the materials you will need before the client arrives; this saves valuable time later.	✗ Don't forget to complete the client details/records after doing the service. Make sure that all aspects – dates, times, changes in materials, etc. – are recorded accurately.

Consultation

Your choice of colour is crucial: take time to make it carefully. A hurried choice may give disastrous results. It would be incorrect to say that one form of consultation is more important than another, but colouring is a complex area of work and if you do not make the right assessment before you start, the outcome could be disastrous.

A number of questions need to be answered before the final choice of colour is made.

ACTIVITY

Visual aids are essential for showing clients some of the different themes, ideas and impressions that are available. The diversity of colouring effects available for your client is only limited by your imagination. Build a creative portfolio of effects so that you can use it with your clients. You can find sources for your ideas from the Internet, style magazines, colour swatches or almost anywhere.

What does the client require? Clients look to colour as a solution for many issues.

Many clients requesting a permanent colour are seeking to disguise their greying hair. A client who wants something to tone a few grey hairs may be successfully assisted with temporary, semi-permanent or longer-lasting colourants.

> "Listening to and understanding your client is the most important part of what we do; if we don't listen, it can make or break a colour or style.
>
> *David Morgan*

Most colour work undertaken within the salon has been stimulated by fashion and calls for partial colouring techniques. The multi-toning permutations present many different options, carried out by an ever-growing range of techniques and applications. With the decline in full-head colouring (probably much to do with the variety and choice of home colouring products) these partial, varietal colouring options remain a 'professionals only' option for the client.

What other factors are relevant? During your consultation with your client, you will need to consider the following points:

1 their age and lifestyle
2 their job or occupation
3 their fashion and dress sense, and the colours they prefer to wear
4 their natural hair colour and their skin colour
5 the hair's texture, condition and porosity
6 the colourant you could use
7 the techniques you would employ
8 the time and cost involved.

HEALTH & SAFETY

Before using permanent or any long-lasting colourants, you must always perform an allergy alert test. If there is any reaction, you cannot carry out the colouring. To do so could result in an allergic reaction and a possible personal injury claim. Allow 48 hours, or follow the manufacturer's guidelines, when carrying out an allergy alert test.

When you have taken these points into consideration, you should be able to determine which hair colour shade, colourant and process to recommend to your client.

Outdoor natural light against indoor artificial light Lighting should always be considered when selecting colours with the client. Strong white halogen light may give a true likeness to actual depth and tone. However, people seldom stand for long periods of time within these spotlights. Make sure that you point this out during consultation as most clients (and many hairdressers) don't look closely enough at hair images in style and fashion magazines. Good photography (and later digital editing) will always deceive the non-professional eye; always look closely at the effects within images to see if colours are:

◆ **Realistic** – does the image result from clever colouring or a lighting effect?

◆ **Achievable** – does the positioning of the colours that you see reflect something that can be adapted into a technique?

◆ **Attractive** – would the overall colour of the hair still look as good against the client's skin tones?

Take your visual aids or magazines into natural daylight so that a more realistic colour comparison can be made. This will help dispel any misunderstanding later, particularly if you are matching up a shade from an image created in a studio under specialist lighting for a fashion magazine.

Don't be afraid to recommend ideas you have. Don't forget you are the professional – a client doesn't have the knowledge that you have.

David Morgan

HEALTH & SAFETY

Preparing yourself and the client

◆ Always refer to the results of tests first.

◆ Always gown and prepare your clients properly so that they are protected from spillages of chemicals.

◆ Prepare your work area so that you have everything at hand.

◆ Always wear the PPE (i.e. the disposable (non-latex) gloves and aprons) provided by the salon every time you apply colour.

◆ Apply barrier creams as or where necessary.

◆ Always follow the manufacturer's instructions; never deviate from the tried and tested formulae.

◆ Make sure that your work position is clear and that your posture is correct.

◆ Make sure that the client is comfortable throughout as they will be sitting for some considerable time.

ACTIVITY

Colour coordination

Colour coordination is an important part of creating a total colour effect. Most clients prefer to have colours that suit them and your ability to find colour solutions for them is therefore an important part of your job.

Create a colour portfolio of themes that cover both harmonizing and contrasting effects. You can do this by collecting pictures from magazines and from the Internet that show different hair colours worn in different ways.

Now answer these questions:

1 Which ones work well and why?
2 Which ones didn't work well together and why?
3 What aspects of colour create harmonizing effects?
4 What aspects of colour create contrasting effects?

> "
> Remember, if you combine more than four to five colours they could merge together giving a flat look to the hair. This is the same as when we were children painting, mixing the colours in the water made it look like mud.
>
> *David Morgan*

Colour tests

Do not forget that the following tests are designed to help you identify contra-indications and to protect your client:

◆ **allergy alert test** – to check to see if the client is allergic to colour

◆ **porosity test** – to assess the smoothness or roughness of the cuticle

◆ **elasticity test** – to determine the condition of the hair

◆ **incompatibility test** – to see if metallic chemicals are present

◆ **colour test/strand test** – to check the process of colouring.

Colour consultation considerations

First of all	Things to consider
Does the client know of any reasons that would affect your choice of service?	Ask the client about their hair to find out if there are any known reasons why the service cannot continue – are there any contra-indications?
What colour would be best to suit their needs?	Should you be using temporary, semi-permanent or permanent colouring?
How can the desired effect be best achieved?	Does the colour need to be applied to the roots first, the mid-lengths and ends, or can it be applied all over? Would the effect benefit more from partial colouring such as highlights or lowlights?
How long will it last?	Will the colour fade off or does it have to grow out?
How much will it cost?	Is this affordable and something that can be kept up in the future?
How will it affect the hair?	Will the long-term effects be what the client expects?
Is the hair suitable for colouring?	Have you tested the hair and skin beforehand to see if there are any contra-indications or hair condition issues that will affect the result?
Now consider	
What are the client's expectations?	How will the colour enhance the style and natural colour of the hair? What are the benefits for them?
What are the results of your tests?	Examine the hair: does it present any limitations for what you intend to do?
What is the hair condition like?	Are there any factors that will change the way in which colouring will work on the hair? What previous information is available?
What do the client's records say?	Does this information influence the choice and colour process?
How will you show the effect to the client?	Have you got any illustrations of the finished effect? Does the colour chart give a clearer picture of the shade the hair will go?
How long will the process take?	Is there enough time to complete the effect? Has anything changed as a result of the consultation? Would this service now need to be rebooked or do you have the time to complete it?

What contra-indications are you looking for?

Following on from your initial considerations, you should now be looking for any contra-indications.

Colour contra-indications

Contra-indications – look for the following	How could you find out more?	How else would you know?
Skin sensitivity	Ask the client if they have ever had a reaction to hair or skin products in the past.	Allergy alert test
Allergic reaction	Ask the client if they have ever had a reaction to hair or skin products in the past.	Allergy alert test
Skin disorder	Ask the client if they know about any current skin disorders.	Examine the scalp to see if there are any physical signs of skin abrasions, discoloration, swellings, infestation or infections.

Contra-indications – look for the following	How could you find out more?	How else would you know?
Incompatible products	If you see the results of any previous colour ask what type it was, how it was done.	Look for discoloration or unnatural colour effects on the hair. Test for incompatibles.
Medical reasons	Ask the client if there are any current medical reasons why colouring cannot be performed.	Examine the hair; look for signs of healthy active growth. If there are signs of weakened, damaged, broken or missing hair, ask for more information. Test for elasticity and porosity.
Damaged hair	Ask the client if there are any current known reasons why the hair is in its current state/condition.	Examine the hair; look for signs of healthy active growth. If there are signs of weakened, damaged, broken or missing hair, ask for more information. Test for elasticity and porosity.

What type of colour should I use?

Type, PPE and timings	Preparation	Suitability	Effects
Temporary colour PPE – wear gloves and apron. Whole-head application done at workstation takes 5 minutes.	No mixing required, colour applied straight from the can, bottle, etc. as coloured mousses, setting lotions, hair mascara.	No skin test required. Most hair types (including coloured and permed) although it can be more difficult to remove from bleached hair. Colour control – poor, shade guide targeting can only be used as an approximation.	The colour only lasts until the next wash. Subtle toning on grey hair. Hair condition may be improved. Surface colour without chemical penetration. Does not lift natural colour, only deposits.
Semi-permanent colour PPE – wear gloves and apron. Whole-head application done at workstation or at basin takes 5 minutes. Left on up to 15 minutes	No mixing required, although transference to an applicator may be necessary.	Skin test may be required. Most hair types (including coloured and permed) often used as a colour refresher between permanent colour treatments. Can cover small amounts of greying hair. Colour control – poor, shade guide targeting can only be used as an approximation.	Lasts up to six shampoos. Colour fades/diffuses after each wash. Does not lift natural colour, only deposits. No regrowth, natural colour unaffected.
Quasi-permanent (longer-lasting) colour PPE – wear gloves and apron. Whole-head brush application done at workstation, takes up to 25 minutes. Alternatively, using an applicator bottle can save time and takes up to 15 minutes. Left on up to 40 minutes	Mixed with developer or activators. These can be in liquid or crystal form. Measurement and mixing must be accurate.	Skin test required. Most hair types (including coloured and permed) often used as a colour refresher between permanent colour treatments. Will cover up to 30% grey hair. Colour control – good, will achieve shade guide targeting.	Lasts up to 12 shampoos. Colour fades a small amount after subsequent shampoos. Does affect natural colour, bonds with natural pigments. Can produce a regrowth.
Permanent colour (para dye) PPE – wear gloves and apron. Regrowth brush application done at work-station, takes up to 25 minutes. Left on up to 40 minutes. Whole-head colouring will depend on length and order of application.	Mixed with hydrogen peroxide at 10, 20, 30 or 40 volumes (3%, 6%, 9% or 12%). Measurement and mixing must be accurate.	Skin test required. All natural hair types and most coloured and permed hair (providing hair not too porous or damaged). Will cover all grey. Can lift up to two shades – high-lift colour will lift three or four shades.	Permanent colour or para dyes are made in a wide variety of shades and tone. Long-lasting and grows out. Will change natural hair pigments.

TOP TIP

If you are in doubt about the timing of colouring always follow the manufacturer's instructions.

Note: All timings are approximated. Partial colouring techniques – highlights, slices, dip ends, etc. – may take longer depending on operator experience, the amount of colour applied and the technique used.

Measuring flasks and mixing bowls Accurate measurement of hydrogen peroxide at any strength is essential. The amount used in relation to colour is a critical factor to a successful outcome and different types of colour are formulated to be used with particular developers. For example, a L'Oréal DiaLight should be mixed with DiaActivator™ developer. If you use a different developer, the consistency will be wrong and this will make the application difficult. All gel and cream colours, when mixed, will be stiff enough not to run or drip when either on the brush or on the hair. Using unmatched, alternative developers will do the opposite and could be a potential hazard for the client.

Note: When you measure developer into a measuring flask, you must make sure that your eye-line is at the same level as the liquid in the flask so that the measurement is accurate.

When you mix developer with colour from tubes, you will notice that all tubes have markings on the side showing the ¼, ½ and ¾ points. These enable you to squeeze from the bottom of the tube up to these points, knowing that your measurement will be accurate.

If you are mixing two or more shades of colour together, always mix these well in a non-metallic bowl first before adding any developer. This allows the different pigments to be evenly distributed throughout the colour and also throughout the hair when it is applied!

Skin and hair tests

Allergy alert test

The allergy alert test is used to assess the skin's sensitivity or reaction of the skin to chemicals or chemical products. In the salon it is mainly used before colouring. Some people are allergic to external contact of chemicals such as paraphenylenediamine (PPD) found in permanent colour.

To find out whether a client's skin reacts to chemicals in permanent colours, carry out the following test 48 hours (or the manufacturer's recommended time) prior to the chemical process.

Allergy alert test procedure

1.	Select and apply a small amount of a **dark, natural shade,** within the range to be used, on a cotton bud. (Follow the manufacturer's explicit details).
2.	Clean a small area of skin with warm water about 8 mm square, behind the ear.
3.	Apply a little of the colour* directly to the skin.
4.	Do not cover the area and ask your client not to wash the area or touch. Ask them to report any discomfort or irritation that occurs over the next 48 hours. Arrange to see your client at the end of this time so that you can check for signs of reaction.
5.	Record the details on the client's record file (they may be useful for future reference).
6.	Make an appointment for the future service.

*Not mixed with any hydrogen peroxide

7. If there is a positive response – a contra-indication, a skin reaction such as inflammation, soreness, swelling, irritation or discomfort – do not carry out the intended service.

Note: *Never* ignore the result of a skin test. If a skin test shows a reaction and you carry on anyway, there may be a more serious reaction that could affect the whole body.

8. If the result is negative, i.e. no result, the service may proceed as planned.

Hair tests

Incompatibility test

Incompatibility test	When is it done	How is it done
This will indicate if any **metallic salts** or other mineral compounds are present within the hair.	Prior to colouring, highlighting and perming treatments	Place a small sample of hair in a mixture of 20 parts hydrogen peroxide (6%) and 1 part ammonium-based compound from perm solution. If the mixture bubbles, heats up or discolours do not carry out the service.

TOP TIP

Lightened hair has very little wet strength and often locks together making detangling very difficult. The hair strength improves as the hair is dried and therefore there is less chance of damaging the hair.

Elasticity test

Elasticity test	When is it done	How is it done
This determines the condition of the hair by seeing how much the hair will stretch and return to its original length. Overstretched hair will not return to the same length and indicates weakness and damage.	Before chemical treatment(s) and services. (Ideal for hair that has poor elasticity, e.g. from lightening or colouring).	Take a dry strand of hair between your fingers, holding it at the root and the end. Gently pull the hair between the two points to see if the hair will stretch and return to its original length. (If the hair breaks easily it may indicate that the cortex is damaged and will be unable to sustain any further chemical treatment). **Note:** *Do not conduct this test on **wet hair** as the hydrogen (and salt bonds) are already broken. This will give you a false result as the hair is bound to stretch.* **If the hair is lightened, it would probably snap and break off altogether.**

Porosity test

Porosity test	When is it done	How is it done
This test also indicates the hair's current condition by assessing its ability to absorb or resist moisture from liquids. (Hair in good condition has a tightly packed cuticle layer which will resist the ingress of products.) Hair that is porous holds onto moisture; this is evident when you try to blow-dry it as the hair takes a long time to dry.	Before chemical services. If the cuticle is torn or damaged, the absorption of moisture and therefore hydrogen peroxide is quicker, so the processing time will be shorter. Over-porous hair will quickly take in colour but will not necessarily be able to hold colour, as the cuticle is damaged and allows the newly introduced pigments to wash away.	Take a single strand of dry hair between your fingers and thumbs, holding it out away from the head. Now run your finger and thumb, backwards, down the hair from point to root. If it feels roughened, as opposed to coarse, it is likely that the hair is porous. **Note:** *As cuticle layers lie with their free edges towards the point ends, you are far more likely to be able to feel the tiny ridges of any lifted cuticles.*

Strand test

Strand test	When is it done	How is it done
Most colouring products just require the full development time recommended by the manufacturer – check their instructions. (However, some hair conditions take on the colour faster than others do. A strand test will check the colour development and see if it needs to come off earlier).	A strand test or hair strand colour test is used to assess the resultant colour on a strand or section of hair after colour has been processed and developed. A strand test is also useful prior to lightening natural pigments from hair or prior to removing synthetic pigments (i.e. decolour or colour reducer) to see how the hair will respond.	1 Rub a strand of hair lightly with the back of a comb to remove the surplus colour. 2 Check whether the colour remaining is evenly distributed throughout the hair's length. If it is even, remove the rest of the colour. If it is uneven, allow processing to continue, if necessary applying more colour. If any of the hair on the head is not being treated, you can compare the evenness of colour in the coloured hair with that in the uncoloured hair.

Recording the results Make sure that you record the details of any test that you conduct. Update the client's record card or computer file immediately after you have done the test. Do not leave it until later, you might forget. These records are essential information that will be needed again and help to show that a competent service has been provided at that time. This would be vitally important if there were a problem at some later stage, particularly if it involved any legal action taken against the salon.

Colour selection

Colour selection, i.e. the process you go through in choosing the right target shade for your client's hair and the correct mixture of products to achieve that target shade, is based upon:

◆ Their personal choice (initially).

◆ The condition and quality of your client's hair (i.e. if it has already undergone processes such as highlights).

If the hair has been regularly coloured before and there is a clear regrowth, with ends that have faded, you may only need to do a straightforward regrowth application with the same colour. Then, later in the development process, the residual colour can be diluted and taken through to the rest to refresh the total effect. Therefore, in this instance a regrowth that takes 20 minutes to apply can be left for 30 minutes' development (depending on manufacturer/colour type). Then, in the last 15 minutes, it can be taken through to the ends, until it is ready to be removed.

However, if your client's hair has been coloured before, you also need to remember that it will not be possible to make the hair lighter by colouring. Permanent colour does not reduce permanent (synthetic) pigments in the hair. (If this is required, you will have to use a colour remover first.)

If you need or want to counteract and neutralize unwanted tones in the hair, you will need to apply the principles of the colour wheel. If the client wants to reduce or 'calm down' unwanted red tones then you will be choosing a colour slightly darker in depth but which has the matt tones capable of neutralizing that effect. Conversely, if your aim is to eliminate ashen 'green' tones (e.g. the discoloration resulting from swimming in chlorinated swimming pools) then you will be introducing warmer tones to the hair. So in this situation a 'greeny'-looking base 6 blonde will be improved by a shade depth 6 but with a tone warmth .03 (for more information see the section on depth and tone earlier in this chapter).

If you had to reduce a tonal effect that was too yellow, say on a head that had been lightened, then (although the principle of toning lightened hair is slightly different) you would still be applying the principles of the colour wheel. Therefore, a violet-based ash colour should be used to neutralize the unwanted tones.

If your client has never had any colour on their hair before (*virgin hair*) then colour targeting is easy. Your client will be able to choose practically any shade on the permanent shade chart, providing it is at the same depth or darker than their own. (It is possible to lighten a shade or two with colour in certain situations.)

If your client has grey or greying (i.e. white) hair then you will have to decide and agree on what reduction of grey is necessary. If the client wants to cover all the grey, then this is only achievable by using or adding base shades to the target colour, i.e. a natural shade or a natural shade plus the target shade.

Note: *Each colour manufacturer produces a range of natural shades that are guaranteed to cover resistant, white/grey hair. Check your shade charts to identify which ones you can use.*

Assessing the amount of grey The amount of base added to the target shade is directly proportional to the amount of grey. Grey hair is referred to as a percentage of the whole head; therefore, a client who has about a quarter of their hair that is grey is referred to as 25% grey. Similarly, a client whose hair is one-tenth grey is 10% grey.

Existing hair condition The hair's existing condition is a major contributing factor to the way it will respond when it is coloured. Porous hair will absorb the colour at a different rate. The porosity of hair is never even along the hair length, let alone across the hairs throughout the head. This is because the porosity of the hair relates to areas of damaged cuticle. Areas of high porosity occur at sites along the hair shaft where cuticle is torn or missing. At these points, moisture or chemicals can easily enter the inner hair without cuticle layer resistance.

This changes the rate of absorption, which ultimately affects the final evenness of the colour and the hair's ability to retain colour in subsequent washing, etc. pH balancing helps to even out the hair's porosity and return chemically treated hair back to a natural pH 5.5.

During processing the only other factors that affect the achievement of an even and expected final colour result (assuming your selection is correct) are:

◆ timing

◆ temperature.

Pre-softening resistant grey hair before colouring White hair can be very resistant to colouring, especially lighter shades ranging from bases 8 up to 10. Sometimes it is necessary to prepare the hair, prior to colouring, by lifting the cuticle slightly, so that the target shade can work better.

If your client is 100% grey, you may have problems getting an even result. You can **pre-soften** the client's hair by using 20 vol. 6% hydrogen peroxide, by applying it directly to the resistant areas. The easiest way to do this would be at a backwash where you could apply the peroxide directly to the areas on dry hair with a colouring brush. (Alternatively, if

your salon uses a *thinner* form of developer rather than crème, you may be able to apply it to the hair via a water spray.)

Leave the peroxide on the hair for 5 minutes then remove any excess with a towel (remember to put the towel immediately into a basin of water or the washing machine).

Finally, move the client back to the workstation and dry the client's hair with a hairdryer; then you are ready to apply the permanent/quasi-permanent colour as planned.

Timing the colour development

Colour saturation is proportional to the length of time that the colour is on, and under-processed hair cannot achieve the same saturation as hair that has had full development. So, the longer that the colour is left on, the more density (saturation) it has.

The effects of temperature during colouring Temperature is also a contributing factor to colour development. The warmer the salon environment the quicker the colour processing will be. We know that when colour is introduced to heat, it 'takes' even more quickly; this can be localized to the client or relate to the whole salon. For instance, the salon temperature may be cool but the colour can be speeded up, by putting the client under a Climazon™ or Rollerball™.

Nevertheless, remember, the human body produces heat too. In fact, up to 30% of body heat is vented through the top of your head. This heating effect has a critical impact on the development of colour and makes it even more precarious when lightening.

Therefore, with this extra heat around the scalp area, you can see there are potential problems in controlling the colour and aspects of the client's safety. To help control the process you must make sure that, when the colour is applied to the root area, the hair is lifted away from the scalp so that the air is able to circulate and ventilate the scalp evenly. This ensures that there are no 'hot spots' anywhere that might take more quickly or become a safety hazard to the client.

HEALTH & SAFETY

- When the scalp becomes warm during the processing, the skin attempts to regulate the overheating by producing sweat. At this point, when the skin is moistened by sweat, the colouring products become more mobile and will spread more easily onto and even into the skin. This is extremely dangerous and will cause scalp burns as chemicals enter the skin through the hair follicles.
- Initially, the client will not be able to distinguish between the heat from the processing and the burning sensation from the chemicals. By the time that they do, the longer-term damage is done. Chemical burns continue to act upon the skin long after the colour has been removed. You *must* avoid this happening.
- However, if this does occur, the client must have medical attention immediately.

Pre-colour and post-colour treatments

The unevenness of porosity is also a major factor in the achievement of good colouring results. If the hair has different porosity levels throughout its length then the resulting colour will also be uneven because the hair will absorb the colour at different rates. In areas along the hair where the porosity is higher, then the colour will take more quickly. Conversely, in areas where the cuticle is intact, then the colour will develop at the normal rate.

This unevenness can be counteracted by preparing the hair with a pre-colour treatment. The treatment is applied to dry hair before colouring throughout the lengths to even out the colour absorption rate and enable the hair to achieve an even colour result.

A post-colour treatment can also be used in order to seal areas of damaged or missing cuticle, helping the colour to reduce fading by locking in the synthetic para dye pigments. These treatments will also neutralize the action of hydrogen peroxide upon the hair by:

◆ removing excess oxygen as an anti-oxidant (anti-oxidizing agent) another factor in reducing colour fade

◆ closing the cuticle layer

◆ balancing the pH of the hair by returning it to pH 5.5.

Colour coordination: harmonizing and contrasting effects

Understanding how colours harmonize or contrast in a hairstyle is fundamental to advanced colour selection. In the illustrations that follow, the same colours have been used in a variety of different ways. Note how the effects convey very different images. In each of the illustrations, the same two colour combinations have been used. But the way in which they have been used differs in each case. (Each of the possible colour variables is denoted in *italic* type.)

1   (Far left) Illustration depicts the balanced *light* and *dark* effect that is created when *vertical*, yet *uniform* sections of *contrasting* colour are *evenly* applied to a plain gold background.

2   (Centre left) This is the predominately *darkened* result when *more vertical* and *varying* sections of *contrasting* colour are *randomly* applied to a plain gold background.

3   (Centre right) This predominantly *lighter* effect is produced when *fewer vertical* and *varying* sections of *contrasting* colour are *randomly* applied to a plain gold background.

4   (Far right) This *darkened* effect results from *more vertical* yet *uniform* sections of *contrasting* colour being *evenly* applied to a plain gold background.

Before looking at the next four illustrations showing examples of different weaving and partial colour techniques, think about the visual permutations and variables that can occur. (See table below).

	Effect	Sections	Direction	Amount	Positioning	Intensity
Colouring effects	Light or dark	Uniform or varying	Vertical or diagonal or horizontal	Less or more	Even or random	Harmonizing or contrasting

In this series of illustrations, new variables have been introduced; they have a major impact on the total effect too. Here a *third*, yet *harmonizing* colour is added. This new addition harmonizes with its companion colours in two ways:

◆ It provides a mid-tone, somewhere halfway between the light gold and the dark brown.

◆ It has a natural, tonal 'fit', i.e. it harmonizes with the two other colours.

The purpose of the illustration is to show that the same background is totally changed with just the addition of one other shade. Stark contrasts of colour are softened by the addition of a harmonizing tone. The reverse happens when the added tone does not harmonize with its background. There are two other aspects to now consider:

1 Added tone does not harmonize.

2 Uneven background colour.

The next illustration introduces these criteria.

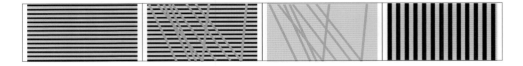

◆ We can now see that in the first illustration (far left) the image has changed direction. The colours are exactly the same but the resultant effect is very different.

◆ This is compounded in the next image where the third colour is reintroduced again, but now appears multi-directional.

◆ The next depicts the removal of contrast colour, just leaving random directional movement.

◆ Finally, there is the added dimension of uneven colour background.

Collectively, these illustrations show a variety of colour correction situations such as:

◆ colour banding, produced by horizontal, contrasting colours

◆ highlights placed on colour banded background

◆ poorly executed colour positioning on to an even background

◆ strong, contrasting highlights placed on uneven and faded background hair.

This is aesthetics – there are no rules here. However, there are a number of points that have a bearing on taste, good design, artistic appreciation and therefore your colour planning.

Taste is subjective; whatever you find appealing may not be acceptable to others. In colour consultation you must get a feel of what the client finds attractive.

Good design is again subjective; the images displayed within the text bear no relationship to hair colouring. On the other hand, the planning of styles, particularly new fashions, must originate from somewhere. Our clients use our artistic knowledge and skills to get the look they want.

Wherever your inspiration comes from, remember that it has to be applied in a commercial context (unless it is purely for promotional purposes when the sky is the limit). Bearing this information in mind here are the variables that you need to consider before colouring hair.

Effect	Monochrome (single colour)	Dual-toned	Multi-toned	Light/dark	Subtle	Strong
Sections	uniform	varied	narrow	wide	narrow	blocked
Direction	vertical	horizontal	angled		vertical	
Amount	singles	less than 20%	20–40%	40–60%	less than 20%	over 60%
Positioning	evenly placed	randomly placed	over other colours	with other colours	with other colours	below other colours
Intensity	harmonized	contrasting	vibrant	muted		

Colour selection checklist

✓	What is the client's target shade?
✓	What is the percentage of white/grey hair, does it need to be pre-softened?
✓	What is the difference in depth between the natural hair and the target colour?
✓	If the target shade is lighter than the natural shade, is it achievable by colouring alone?
✓	What colouring products would be needed to achieve the effect?
✓	What developer will be needed to achieve the effect?
✓	If the hair appears porous or porous in areas, do you need to do a test cutting first?
✓	If the hair has a small percentage of white/grey, what amount of base shade will you need to add to the target shade to stabilize the effect?

TOP TIP

You do not have to recolour hair as a full head every time. You can refresh the colour when you do a retouch service.

Colour applications

Refreshing the colour of the lengths and ends

When you are re-colouring the roots on longer hair you will often find that the ends of the colour need refreshing too. This does not mean that a full head colour is necessary: the refreshing can be done during the application.

Appearance	1st step	2nd step	3rd step	4th step
When the colour looks the same at the ends as the target shade	Apply to regrowth	Allow to develop and then 15 minutes before full development time.	Add 15–20cc of tepid water to the mixture in the bowl then apply this to the lengths and ends.	Leave for a further 5–10 minutes, to complete the development process.
When the tonal quality has faded but the colour is still the same depth	Apply to regrowth	Allow to develop then 25 minutes before full development time.	Add 15–20cc of tepid water to the mixture in the bowl then apply this to the lengths and ends.	Leave for a further 15–20 minutes, to complete the development process.
When both the tonal quality and depth has faded on the ends	Apply to regrowth	Add 15–20cc of tepid water to the mixture in the bowl then apply this to the lengths and ends immediately.	Leave all the colour on for full development 35–40 minutes.	

Regrowth checklist

Consultation	◆ Find out what needs to be done.
	◆ Are there any modifications needed to do the regrowth?
Prepare the client	◆ Make the usual protective preparations.
	◆ Brush the hair through to remove the tangles.
	◆ Apply a barrier cream to the hairlines (if necessary).
Prepare the materials	◆ Make sure that you have everything you need at hand.
	◆ Put on your disposable gloves and apron.
	◆ Mix the products correctly.
Method/technique	◆ Divide the hair into four equal quadrants.
	◆ Start the application with a brush to the roots at the top of the head, working down and along each quadrant.
	◆ Pick up a horizontal section of hair within a back quadrant and with the tail of the brush.
	◆ Apply the colour/lightener to the regrowth evenly.
	◆ Repeat down the back of the head and through the sides.
Development	◆ Monitor the colour development throughout the processing.
	◆ Apply heat if needed to speed up the development process.
	◆ Check with the client throughout to ensure their comfort.
Removal	◆ When processing is complete, take the client to the basin and rinse the hair with tepid/warm water, gently massage to emulsify the colour.
	◆ Rinse thoroughly until the colour is removed.
	◆ Shampoo and condition with an anti-oxidizing agent.

STEP-BY-STEP: REGROWTH APPLICATION

1 Gown the client before the service begins.

2 Prepare the client's hair by brushing through to remove any tangles. Then divide the hair into four quadrants and section each one out of the way.

3 Start by applying the colour directly to the regrowth without overlapping onto previously coloured hair.

4 Use your brush to apply product right up to the roots.

5 After applying the colour to the 'hot cross bun' area you can start to take sections within the quarters, horizontally.

6 Work down through each section and move the hair that has been coloured over, and out of the way.

7 Using your brush applicator, work methodically through each section in turn.

8 With all of the sections completed, check where you have applied the colour and the hair lines to make sure that everywhere has been covered. Then allow the colour to develop for the product manufacturer's recommended time.

9 The final effect is an evened-out tone.

Full-head colour checklist

Consultation	◆ Find out what needs to be done. Are any changes needed for the application, colour(s) or lightener?
Prepare the client	◆ Make the usual protective preparations. ◆ Brush the hair through to remove the tangles. ◆ Apply a barrier cream to the hairlines.
Prepare the materials	◆ Make sure that you have everything you need to hand. ◆ Put on your disposable gloves and apron. ◆ Mix the products correctly.
Method/technique	◆ Divide the hair into four equal sections. ◆ Start the application with a brush to the mid-lengths and ends at the top of the head, working down and along each quadrant. ◆ Pick up a thin 5 mm horizontal section of hair within a back quadrant and with the tail of the brush. ◆ Apply the colour/lightener to the mid-length and ends evenly. ◆ Repeat down the back of the head and through the sides.
1st part of development	◆ Allow the mid-length and ends to develop sufficiently first. ◆ Monitor the colour development throughout the processing. ◆ Apply heat if needed to speed up the development process. ◆ Check with the client throughout to ensure their comfort.
2nd part of development	◆ Pick up each of the horizontal sections of hair with the tail of the brush. ◆ Apply the colour/lightener to the root area evenly. ◆ Repeat down the back of the head and through the sides. ◆ Monitor the colour development throughout the processing. ◆ Apply heat if needed to speed up the development process. ◆ Check with the client throughout to ensure their comfort.
Removal	◆ When processing is complete, take the client to the basin and rinse the hair with tepid/warm water, then apply gentle massage to emulsify the colour. ◆ Rinse thoroughly until the colour is removed. ◆ Shampoo and condition with an anti-oxidizing agent.

STEP-BY-STEP: FULL HEAD APPLICATION

1 Gown the client before the service begins.

2 Prepare the client's hair by brushing through to remove any tangles. Then divide the hair into four quadrants and section each one out of the way.

3 A full head colour on longer hair must be applied to the mid-lengths and ends first because these area will need the longest time to develop.

Mix your colour according to the product manufacturer's instructions. Start applying colour to the lower back sections first. Apply colour to the mid-lengths and draw down to the ends.

4 Once you have applied colour to the back sections, repeat the process on the hair at the sides of the head. When the mid-lengths and ends are complete, leave the colour to develop for the product manufacturer's recommended time. You can use an accelerator to speed up the process.

5 After the mid-lengths and ends have developed, apply the root application. Allow the product to process following the manufacturer's guidelines.

6 The final effect is a rich tone with full coverage.

CourseMate video: Full head lightening and toning

STEP-BY-STEP: FULL HEAD LIGHTENING AND TONING

1 After greeting your client, conduct a full consultation to establish the level of lift desired plus your choice of product and peroxide strength.

2 Apply the lightening product to the mid-length hair and ends.

3 Work methodically throughout the head.

4 Visually check the development until the desired tone has been reached (orange to dark yellow).

5 Once the desired tone has been reached, apply the lightening product to the root area.

6 Continue root application ensuring speed and even coverage.

7 A plastic cap can be used to help speed up development. Do not exceed the manufacturer's development time.

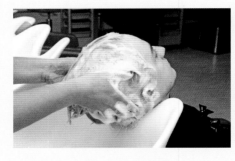

8 Once the target tone is reached, remove the product by shampooing. Do not use conditioner.

9 Assess the tone of the hair and select a suitable ash toner to neutralize any unwanted yellow tones.

10 Apply the ash toner and visually check for development. Do not exceed the manufacturer's development times.

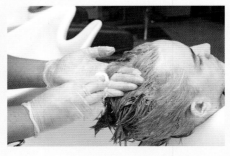

11 Once development is complete, remove the product over the basin, wash thoroughly and condition.

12 The finished result is an eye-catching full head of lightened hair.

Woven highlights checklist

Consultation	◆ Find out what effect your client is trying to achieve. ◆ How much lightened and/or coloured hair in relation to natural colour is expected? What percentage of each is needed – 5%, 10%, 25%? ◆ How will you explain the effect to the client? ◆ Do you have any visual aids to help? ◆ Explain everything that you are going to do.
Prepare the client	◆ Make the usual protective preparations. ◆ Brush the hair to examine the growth patterns and to remove any tangles, look for areas where highlights would be conspicuous or unsightly. ◆ Look for natural part/parting areas; confirm how the hair is to be worn.
Prepare the materials	◆ Make sure that you have everything you need at hand, including foils cut to the required length. ◆ Put on your disposable gloves and apron. ◆ Mix the products correctly.
Method/technique	◆ Divide the hair into four equal sections. ◆ Start at the back of the head at the nape. ◆ Divide the remaining hair and section and secure it out of the way. ◆ Pick up a horizontal section of hair and, with your pin-tail comb, weave out of the section a mesh of fine amounts of hair. ◆ Place underneath the mesh a foil long enough to protrude beyond the hair length. ◆ Apply the colour/lightener to the mesh evenly. ◆ Fold in half and half again (fold the edges too, if required). ◆ Continue onto next section with the alternating colour(s) or lightener. ◆ Repeat up the back of the head and through the sides.
Development	◆ Monitor the colour development throughout the processing. ◆ Apply heat if needed to speed up the development process. ◆ Check with the client throughout to ensure their comfort.
Removal	◆ When processing is complete each foil must be removed individually by rinsing thoroughly until all the product is removed (this ensures that the colours do not run and bleed together). ◆ Shampoo and condition with an anti-oxidizing agent.

ALWAYS REMEMBER

If you use dry heat to accelerate the development of permanent colours, don't let the heat dry out the products as this will stop the colour from developing further.

STEP-BY-STEP: WOVEN HIGHLIGHTS

1 After greeting your client, conduct a full consultation to establish the levels of lift required plus the depth and tone of tint to be applied.

2 Prepare the hair by sectioning it ready for colour placement foils or meche packets. Plan where the colour is to be placed to complement the finished style.

3 Take fine sections of hair and use a pin tail comb to weave close to the scalp. Use a traditional weaving technique or swirl the pin tail end across the scalp.

4 Place the foil or meche underneath the woven section.

5 Keeping the foil or meche as close to the roots as possible, paste on the product.

6 Seal the corners by gently folding, taking care not to dislodge the packet or foil.

7 Work methodically through all the sections, alternating colour where required.

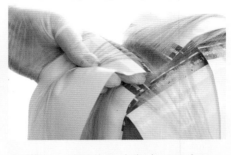

8 When working through the front packets, angle them forwards over the client's forehead to ensure closer root coverage.

9 Follow the manufacturer's instructions for development time, visually checking any lightening products applied.

10 Once target shades are achieved, remove the foils one at a time over the basin.

11 Emulsify, shampoo and condition the hair thoroughly. Rinse well.

12 The final result is a pleasing blend of multi-tonal high- and low-lights.

Colouring problems

Colouring and lightening are complex chemical processes and even senior, experienced stylists can make mistakes. The following list provides the main reasons for faults that can occur during processing.

Poor consultation

◆ Not finding out what is really needed in the first place

◆ Lack of experience in dealing with the technical issues

◆ Not understanding what you see

◆ Not understanding what the client sees or wants.

Poor execution of the service

◆ Incorrect choice(s) of colour(s)

◆ Incorrect choice of hydrogen peroxide strengths

◆ Incorrect or careless application

◆ Unrealistic timescale allowed for the service (rushed service)

◆ Unnecessary handling or disturbance of meches packets, etc. during processing

◆ Poor removal of colour products (colours mix or merge causing discoloration).

Processing/development

◆ Insufficient time allowed for product development (taking off too soon)

◆ Poor heat/air ventilation around all areas that need processing (hot spots)

◆ Too much heat (causing over-development, hair damage or breakage)

◆ Too much time (causing over-processing, hair damage or breakage).

So what went wrong?

Sometimes things do go wrong, and you need to think quickly about how you will resolve the issues. There are all sorts of colouring problems and the table on page 284 addresses some of the more common faults.

Problem or fault	Possible reasons why	Corrective actions	
Colour patchy or uneven	Insufficient coverage by colour Poor application Poor mixing of chemicals Sectioning too large Overlapping, causing colour build-up Under-processing (colour was not given full development)	Spot colour the patchy areas	
Colour too light	Incorrect colour selection Peroxide strength too high causing lightening Peroxide strength too low Under-processed Hair in poor condition	Choose a darker shade Check strengths and re-colour Check strengths Re-colour Apply restructurants	
Colour fades quickly	Effects of sun or swimming Harsh treatment: over-drying, ceramic straighteners, etc. Hair in poor condition Under-processing	Recondition before next application	
Colour too dark	Incorrect colour selection Over-processing Hair in poor condition Metallic salts present	Process correctly Senior assistance required	
Colour too red Root glare	Peroxide strength too high revealing undertone colour Hair not lightened enough Under-processing	Apply matt/green tones	
Discoloration	Hair in poor condition Undiluted colour repeatedly combed through Incompatibles present	Use colour wheel to correct unwanted tones Senior assistance required	
White hair not covered	Resistance to peroxide/colour Lack of base shade within the mixed colours	Pre-soften Re-colour with correct amount of base and tones	
Hair resistant to colouring	Cuticle too tightly packed Under-processed Incorrect colour selection Poor mixing/application	Pre-soften Re-colour Senior assistance required Senior assistance required	
Scalp irritation	Chemicals not removed from hair properly after processing Peroxide strength too high	Wash hair again and condition with antioxidants	
Skin reaction or burn	Poor quality materials causing abrasions to the scalp Client allergic to chemicals	Senior assistance required Refer to doctor/hospital	
Breakage	Lightening/highlighting hair that has previously had lightener on it before	Use restructurant on remaining hair to strengthen the weakened hair	
Too yellow	Under-lightened Base too dark Wrong toner used Wrong lightener	Re-lighten Try stronger lightener Use violet toner Use other than oil lightener	

Problem or fault	Possible reasons why	Corrective actions
Too red	Under-lightened	Re-lighten
	Too much alkali	Use different lightener
	Wrong toner used	Use green matt or olive
Roots not coloured	Under-lightened	Re-lighten
	Under-timed	Re-lighten
	Toner too dilute	Reapply
	Unclean or coated	Clean and reapply
Hair breakage	Over-processed	Recondition remaining hair
	Incompatibles present	Re-test to make sure
	Harsh treatment	Provide hair care advice
	Sleeping in rollers	Provide hair care advice
	Tied-back long hair	Provide hair care advice
Discoloration	Under-processed	Colour match or develop further
	Excessive exposure	Recondition and keep covered
	Home treatment	Provide hair care advice
Green tones	Re-colouring previously lightened hair without pre-pigmentation	Test hair
	Incompatibles	Use warm or red shades
	Blue used on yellow	Use violet
	Too-blue ash used	
Too orange	Under-processed	Apply blue ash
	Pigment lacking	Add blue
Too yellow	Under-processed	Add violet and/or lighten further
Hair tangled	Over-lightened	Use antioxidants/treatment reconditioners
	Raised cuticle/over-porous	
Colour not taking	Over-porous	Recondition and pre-fill hair
	Chemicals masking the hair	Use deep cleanser to remove build-up
	Colour bounce/red pigment grabbing at roots	Recolour with brown ash pigments (not deposited into hair correctly)
Colour build-up	Over-porous	Recondition and pre-pigment
	Excessive over-application	Consider colour removal

Provide aftercare advice

Clients need your help to look after their hair at home; you need to give them the right advice so that they can make the most of their new colour and style between visits.

You should tell them what sorts of products they could use that would make their colour last and reduce the risk of fading. Also, make a point of telling them what they should avoid: some products will reduce the effects of your colouring, causing it to fade prematurely or lose its intensity or vibrancy.

You should also explain the benefits to clients of maintaining their hair in good condition, as hair in good condition is easier to manage, it looks better and is noticeable to everyone else as well.

Home and aftercare checklist

✓	Talk through the colour effect as you work; tell the client how they can maintain their hair by optimizing both the look and the condition of their hair.
✓	Explain how lifestyle factors can affect their hair colour and ways that they can combat these factors.
✓	Show and recommend the products/equipment that you use so that the client gets the right things to enable them to get the same effects.
✓	Tell the client how long the effect can be expected to last and when they need to return for re-colouring.
✓	Warn them about the products that will have a detrimental effect on the colour by telling them what these products would do.
✓	Warn the client about how incorrect handling or styling may reduce the length of time that the new effect will last.

Explain how routine styling tools can have detrimental effects

Maintaining the hair in good condition is the single, most important factor for making the most of new colouring effects. You know how difficult it is to make dry, damaged hair look good and your clients can recognize the difference between good and poor condition; they want hair with lustre, shine, flexibility and strength.

You need to warn your clients of the pitfalls of incorrect use of hot styling equipment. Ask them if they use heated straighteners or tongs on a daily basis and then tell them about the benefits of using heat protection sprays or treatments.

Show and recommend the products/equipment that you use

As you talk to your client about the ways in which you have styled the hair, make a point of talking through the products that you have used as well. You know from experience the products that you would use on their hair to achieve different results and you know the ones that you would avoid. So tell the client too, because they have not had the benefit of your training and they do not know.

When you use a particular product, you could hand it to them for a closer look. This way they get to see, smell and feel the product too and, subconsciously, this has a very powerful effect. You are involving the client in what you are doing by giving them a greater experience of the service. They will be able to see a direct link between the effects that you are achieving on their hair, with the added benefits of buying those particular products that will help them to recreate a similar effect.

Demonstrate the techniques that you use

Clients want to be able to recreate the effects that you achieve in the salon and this is your chance to show them how to do it. Clients have not had the benefit of your training; so show the client how to do the routine things of applying conditioner, combing and detangling, blow-drying and styling. We know what happens if these are not done properly, but you can give them a few tips on how they can achieve similar results themselves and how long they can expect it to last.

REVISION QUESTIONS

Q1. An _____ alert test will identify a client's sensitivity to colour products.
Fill in the blank

Q2. A quasi-permanent colour lasts longer than a semi-permanent colour.
True or false?

Q3. Which of the following products are likely to be an incompatible?
Multi-selection

Permanent colour containing PPD ☐ 1

Retail permanent colour containing PPD ☐ 2

Vegetable henna ☐ 3

Compound henna ☐ 4

Single-step applications for covering grey, e.g. 'Just for Men' ☐ 5

Single-step toners for application to lightened hair ☐ 6

Q4. Lighteners and high-lift colours are the same.
True or false?

Q5. Which of the following tests do not apply to colouring services?
Multi-choice

Allergy alert test ○ a

Incompatibility test ○ b

Porosity test ○ c

Curl test ○ d

Q6. Permanent colours alter the pigmentation of hair within the cortex.
True or false?

Q7. Which of the following colouring products do not require the addition of hydrogen peroxide as a developer?
Multi-selection

Powder lightener ☐ 1

Semi-permanent colour ☐ 2

Quasi-permanent colour ☐ 3

Temporary colour ☐ 4

Vegetable henna ☐ 5

High-lift colour ☐ 6

Q8. Green tones within hair are neutralized by adding _____ tones.
Fill in the blank

Q9. Hair lightened from natural base 7 should be capable of maximum lift to:
Multi-choice

White ○ a

Pale yellow ○ b

Yellow ○ c

Yellow/orange ○ d

Q10. Lightened hair that appears too yellow can be neutralized by adding mauve.
True or false?

12 Colour Correction

LEARNING OBJECTIVES

◆ Be able to work safely during colouring procedures at all times.

◆ Be able to make preparations for colour correction services.

◆ Be able to carry out a variety of colour correction services.

◆ Know how to resolve a variety of colouring problems.

◆ Be able to identify a range of factors that can limit or stop a planned service.

◆ Be able to communicate professionally and provide aftercare advice.

KEY TERMS

banding
colour reducer
colour remover
colour stripper
colouring back

decolouring
discoloured hair
gradated colour
incompatible
interleaves

pre-pigmentation
spot colour
T–section highlights
tinting back

INTRODUCTION

Colour correction services are becoming more frequent within the salon but will never simply be routine. Your experiences with colour correction will continue to build your confidence; the more that you are involved with the service the better prepared you will be in undertaking this essential work. The most important part of a colour correction service is the consultation to find out exactly what has taken place and over what length of time. Planning a course of corrective action is essential; unless you understand what you are seeing, you cannot provide the right course of action.

Colour correction

PRACTICAL SKILLS

Find out what clients want by questioning and examination

Look for things that will influence the ways that you do any colour correction processes

Learn how to carry out tests on the hair and how the results will affect the service

Learn how to do a variety of colour correction procedures

Make recommendations to clients about the ways in which colour correction can be done

Provide aftercare advice to your clients

UNDERPINNING KNOWLEDGE

Know your salon's ranges of products and services

Understand the factors that can affect your colour correction plan of action

Understand how different colour correction products work and their suitability in different situations

Know the ways in which you can correct colouring problems

Know how to correct hair colouring problems

Know the principles of colour and how they affect colour correction services

Maintain effective and safe methods of working when colour correcting hair

ALWAYS REMEMBER

There will be times when what you see before you does not make sense. In other words, there is a discrepancy between what you see and what your client claims has taken place. Clearly, some vital information is missing. This could be because your client is embarrassed about things they have done but do not wish to admit to. If so, do not make an issue of the situation. You will get a lot more help if you do not apportion blame or look for a scapegoat.

Preparation and safety

The safety and preparation procedures needed for both you and your client for colour correction work is the same as for colouring hair.

See CHAPTER 11 Colouring Hair, pages 261–262, for more on safety and preparation procedures.

Records It is unlikely that you will have a record for your client covering the same colour correction work twice, as each colour correction process is unique and dependent on the problems that are apparent at the time of the visit. However, the maintenance of client records is very important and you should make sure that you provide a full detailed account of any work undertaken.

ALWAYS REMEMBER

Client records are essential for:

◆ monitoring the effectiveness of any colour correction work undertaken

◆ keeping things going smoothly and for maintaining services even if key staff are away

◆ a back-up if things go wrong and they are needed as evidence in any legal action taken against the salon.

Tests Undertake your tests to find the contra-indications that confirm your visual and verbal information. In most cases the hair will be in a poorer condition than that of unprocessed (virgin) hair, so testing for elasticity, porosity or signs of damage are vital to the success of the operation. It is very unlikely that a skin test will be required as you will be correcting a previous colour problem and your salon records should show when this test was previously carried out.

ALWAYS REMEMBER

If the client is new and not from your salon you will need to undertake a skin test, regardless of whether you can see that the hair is coloured. You must protect the salon and yourself legally in the event of something going wrong.

Materials Sometimes the products or materials needed for colour correction work will be out of stock. Check your materials to see that you have everything that you need. Where items are missing make sure that you bring this to your manager's attention so that they can be ordered and ready for use.

Gowning Always make sure that the client and the client's clothes are adequately protected before any process is started. Most salons have special 'colour-proof' gowns for colouring and lightening processes. These gowns are resistant to staining and are made from finely woven synthetic fabrics that will stop colour spillages from getting through onto the client's skin or clothes. When you gown the client, make sure that the free edges are closed and fastened together. On top of this and around the shoulders you can place a colouring towel and over this a plastic cape. This needs to be fastened but loose enough for the client to be comfortable throughout the service. Remember that this may be a couple of hours or so.

Using barrier cream Barrier cream can be used as a physical barrier to prevent staining around the client's face/hairline. It is also particularly useful if the client has any general sensitivity to chemical based products.

Apply barrier cream to the skin with a finger or cotton wool close to the hairline, taking care not to get it onto the hair as this could stop the colour from taking evenly.

It can be removed later after you have shampooed the colour from the hair and before any other services are conducted.

Seating position The chair back should be protected with a plastic cover. If this is not available, a colouring towel can be folded lengthwise and secured with sectioning clips at either end. The client should be sitting comfortably, in an upright position, with their back flat against the cushioned chair pad.

Trolley You should have your colouring trolley prepared and at hand with all the materials you will need. Foils for highlighting should have been previously prepared to the right lengths and combs, brushes, sectioning clips, etc. should be all cleaned and sterilized and ready for use.

ALWAYS REMEMBER

Salon cleanliness is of paramount importance: the work area should be clean and free from clutter or waste items. Any used materials should be disposed of and not left out on the side, which is unprofessional and presents a health hazard to others through cross-infection.

Protecting yourself Your personal hygiene and safety are also important. The care you take in preparing for work should be carried through in everything you do and this is made even more important when you are about to handle hazardous chemicals. Put on a clean colouring apron and fasten the ties in a bow. Then take a pair of disposable (non-latex) gloves and put them on ready for the application.

What if you think you have dermatitis?

Dermatitis is an occupational health hazard for hairdressers. It is avoided by wearing non-latex disposable gloves for all processes or services that involve contact with chemicals. If you think you are suffering from dermatitis, then you should visit your doctor for advice and treatment. If you believe it has been caused or made worse by your work as a hairdresser, then you should mention this to your doctor and you must also tell your employer as employers are required by law to report a case of work-related dermatitis among their staff.

Colour correction preparation checklist

✓	Find the client's treatment record/or prepare a new one.
✓	Make sure that the styling section and chair is clean, safe and ready to receive clients.
✓	Make sure that the client is well protected with a clean fresh gown, a colouring towel and cape.
✓	Make sure you comb or brush the hair thoroughly before you start, to remove tangles, see if all product can be removed and check for cuts or abrasions on the scalp.
✓	Carry out your consultation to find out what needs to be done.
✓	Look for contra-indications.
✓	Conduct your tests based on the contra-indications found.
✓	Record the responses to your questions (and update the treatment history).
✓	**Note:** If you need to do a skin test then no further action can be taken until the results of the tests are available.
✓	If work has to be deferred, is a deposit required?
✓	Do any specialist materials need to be ordered such as colour strippers, treatments, etc.?
✓	Make sure that you have your tools and products prepared and close at hand.
✓	Make sure that you have checked the equipment that you need to use and it is ready and safe.

Find out what is wrong

Colour correction considerations

Colour correction is proving to be the largest growing area of the hair colouring market and much of this is because:

◆ more people are having colouring services

◆ more people are experimenting with home colouring products

◆ clients are expecting more and their sometimes unrealistic expectations can push the boundaries of safe, guaranteed practice to the limits.

Clients might want:

◆ a modification of what they have had before

◆ something totally new

◆ their darker roots lightened to match the over-processed, lightened ends (which are the result of one set after another of highlighting services).

These increasing salon colouring services inevitably result in colour correction and it is therefore an essential part of Level 3 work.

Highlights/lowlights are a very attractive option for professional colouring. However, they can create some serious long-term problems. Highlights are almost *addictive*; clients can feel compelled by the colour's benefits to keep coming back for endless top-ups.

Highlights are a great way to:

◆ cleverly and subtly cover grey hair

◆ introduce clients to colour

◆ personalize colour to individuals

◆ advertise your colouring skills

◆ enhance the features of a hairstyle.

However, hair regrowth can create a problem. As the client's hair grows, they become more conscious of the demarcation line between the coloured hair and the natural roots. When the hair has grown 2.5 cm a natural parting will show 5 cm of regrowth, so a further colouring application would now be necessary. In most cases it would not help just to highlight/lowlight the roots, particularly if several colours have been introduced to the hair. Even if the hair was only lightened, general shampooing, drying and weathering will have modified the colour on the ends, to the extent where newly introduced colour at the root will have little effect. If you redo the service, this reintroduces colour to hair that has already been processed, even if it is to be taken through to the ends for the last few minutes. Now you have hair that will continue to change in porosity and condition, creating a more difficult task for the future, that is, until you and your client sort out the problem with colour correction.

 Keep things simple, on many occasions (not all, but many) less is more.

David Morgan

Consultation for colour correction

The hardest and most crucial part of colour correction work is establishing what has happened and how you are going to correct it.

You must look at the current state of the hair and from this, and before you do any tests, ask the client questions to get a true picture of what has happened.

Example questions that need to be answered and recorded

Question to ask	Reason for asking
1 How long has your hair had this problem?	This will tell you if you have a worsening problem that could be linked to: 1 gradual deterioration from product fading over time or 2 swift product failure where there has been a sharp degradation/deterioration
2 What products do you normally use on your hair?	This will tell you if you have a product usage problem that could be linked to: 1 something that has been bought by the client and incorrectly/inappropriately used. 2 poor or lack of home care advice.
3 Have you tried to correct the problem yourself?	This will tell you if something has been bought and inappropriately used.
4 How do you style your hair?	This will indicate whether the handling, styling or equipment has played any part in the problem.
5 What sort of style do you want?	This will give you an idea of how much the client is prepared to have cut and give you an idea of how much of the problem is going to be removed by cutting alone.
6 Would you be prepared to have a different style?	If they haven't indicated that they want a different look, it at least addresses the point in very clear terms.
7 What sort of colour effect are you trying to achieve?	This is the most important question that you need to ask. You will need to show examples and get a clear idea of what they want their hair to look like.

At this point (although you haven't conducted any tests) you should already have enough basic information to know whether the request is:

A Achievable

B Desirable

C Realistic

What other factors are relevant?

Later in your consultation you should consider the following points:	
1 Age and lifestyle	Do these factors have any bearing on your decision? If so why?
2 Job or role	Will the result complement what they do?
3 Fashion, look or image	Will the colour suit the client when it is finished?
4 Natural hair colour and skin tones	
5 Hair's texture, condition and porosity	Is the colour effect going to work?
6 Colour and technique you choose	Are your choices correct for the desired effect? Is what they want realistically achievable? Can the hair condition stand the client's wishes/expectations?
7 Time and costs involved	Be thorough and accurate, work out the time and costs. Is it still OK?

When you have considered these points, you should be able to determine which hair colours and processes you need to use.

When correcting colour, try to break it down into component parts. Rarely will a universal approach to colour correction give you your desired result.

David Morgan

Colour tests

There are a number of tests to think about, before you carry out any colour correction:

TOP TIP

Don't rush the consultation; you need to have a clear idea of what is needed. If you are unsure ask for a second opinion.

◆ **Strand/test cutting** – carried out before attempting to colour the full head

◆ **Allergy alert test** – to assess the client's sensitivity to an allergic reaction from the colour.

◆ **Porosity test** – to assess the smoothness or roughness of the cuticle.

◆ **Elasticity test** – to determine the hair's state or condition.

◆ **Incompatibility test** – if metallic chemicals are present.

◆ **Development test/strand test** – to check the process of colouring.

Record the details of your consultation Every aspect of the work you do must be accurately recorded and this includes the responses to the questions that you ask during consultation.

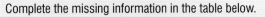

ACTIVITY

Complete the missing information in the table below.

Test	When is this test done?	Why is this test done?	How is this test done?
Strand test			
Allergy Alert Test			
Porosity test			
Elasticity test			
Incompatibility test			
Colour development test			

Consultation checklist

✓	Look at the client's treatment history.
✓	Identify any contra-indications by asking questions and looking through the hair for visible signs.
✓	Find out what the client wants – use visual aids to establish requirements.
✓	Assess suitability and achievability of the desired look.
✓	Look for colouring problems or pre-service requirements.
✓	Conduct any tests – check the results of hair tests against planned services.
✓	Avoid confusing technical jargon; always clarify in simple terms what you mean.
✓	Agree on a suitable course of action and give an idea of service timings and costs.
✓	Summarize the main points of the consultation and get ready to start.
✓	While the service is being done, give the client some advice on how to maintain their hair at home and recommend those products that are suitable to use and those that are not.
✓	Give them an idea of how long the effect will last (and prompt them to re-book their next appointment before they leave).

Principles of colour within colour correction

You can review your colour theory by looking again at **CHAPTER 11** Colouring hair, pages 240–287.

Plan and agree a course of action to correct colour

When you have completed the consultation you will know what you need to do, but now you have to explain this to the client.

Many people (and some hairdressers too) consider colour correction to be a simple task, similar to 'repainting the walls with a different colour'. You need to make sure that this is not the case and that any work that you undertake is both accurate and looked after for the future.

Which colours do I use for colour correction?

Colour correction work is complex and it is essential to have the right products to do the job. **Colour strippers** are an essential part of the work, but as they relate to lightening, they are covered elsewhere in this chapter (see pages 298–299).

Remember that, when doing colour correction, you will always be working on hair that is more porous than natural virgin hair, and this will have some bearing on what you should be using.

For pre-pigmentation   You will have better results if you use colours that naturally have larger colour molecules, as that way they stay in the hair longer. Therefore, for the most successful results use semi-permanent colours: they are kinder on the hair and more readily absorbed into porous hair. Another benefit of using semi-permanents is that they can be diluted with water, this allows you to *water down* stronger golden shades to create, paler, yellower tones. This is particularly useful in colour correction with highlights, as you can apply the **pre-pigmentation** first, and then go back over it with the target shade, without having to wash and dry first.

For target colours   During consultation, you choose a target shade matched up against your client's natural base, in doing this you should always use quasi-permanents and not permanents. Quasis have larger colour molecules than permanent colour, so they will stay in the hair longer. This reduces the risk of colour fading problems and bear in mind that they are always mixed with lower strengths of hydrogen peroxide. Typically, 6 vol./1.9% or 10 vol./3% hydrogen peroxide.

Colour perfection?

Sometimes a perfect solution is unachievable; if this is the case, you need to make this clear to your client. If you feel that you can get close to the required result, tell them, explain the reasons why perfection is not an option today. Make sure that you give advice on how the colour can be optimized in the interim and how they can look after it themselves. You also need to tell them about the things that they need to avoid: some products or styling routines will work against your plan so tell them about this from the outset.

Finally, before starting, do not forget to give your client a full run-down on what the costs will be and how long it all will take. Any senior stylist knows that correction work is difficult to quantify in terms of timings, but consider it thoroughly as you do not want to end up trying to rush it.

Pre-pigmentation

In the majority of colour correction situations, you will find that the client has lightened hair. If they want to change this back to their natural, darker tones, you will have to carry out pre-pigmentation before you can apply the target shade. This is the crucial stage that, if left out, will lead to shades of *green*.

If natural depth, say base 5, is applied to pre-lightened hair, with 3%–10 vol. hydrogen peroxide, it will look a *khaki* green colour when it is dry. This happens because the hair is missing the essential larger, warm pigments (pheomelanin). Therefore, any hair colouring process that involves **colouring back** (or **tinting back**) from previously being lightened requires a pre-pigmentation, using a semi-permanent golden tone.

Colouring back: reintroducing colour into lightened hair

When recolouring (colouring back) lightened hair back to its original or a darker colour, you need to consider the condition of the hair – how porous it might be and, in particular, whether there is sufficient colour pigment left in the hair for the hair to retain new pigment. Think back to how permanent colours work within the hair. Synthetic para dyes, for example, work within the cortex of the hair, bonding into the hair's structure where the natural pigments are. Conversely, lightening products simply remove all natural pigment from the hair and weaken the hair's internal structure.

Therefore, if the internal structure of the hair has been impaired then there will be less structure for reintroduced pigments to bond to. However, this problem can be resolved by pre-pigmentation. During lightening most of the natural pigments are removed. When these pigments are dissolved they create colourless, air-filled spaces a bit like the inside of a natural sponge. Some of these spaces would have contained pheomelanin, the larger, naturally occurring golden pigments. Others would have contained the smaller, darker, eumelanin. These gaps are not uniform in shape and, when synthetic pigments are put back into the hair, not all will fit in the spaces they occupy. Some will come away during washing, leaving a very uneven and unattractive result. So, to avoid green hair and to ensure that recolouring is successful, it is usual to pre-pigment (colour-fill) the hair. This is done by applying red or warm shades that have smaller molecules to colour fill all the spaces within the hair shaft, before the final shade is applied.

Reintroducing pre-pigmentation backgrounds for highlights on long hair

Highlights have been the most popular form of professional colouring for over 30 years. Therefore, you are going to have more colour correction work relating to highlights, rather than from any other form of colouring. However, this makes the correction work far more complex, as there are so many different variables to consider.

Take a typical client who has had their hair highlighted for many years. You will see unevenness in the overall colour effect from the root area, down towards the points. The upper root area will be darker than the ends, as the ends will have had many sets of highlights and be very light. When this effect occurs, the client will at some point want to return to something more natural that shows highlights against her natural base. But it is not possible to colour back the hair after pre-pigmentation and then consider highlights from scratch.

The way to do it is to pre-pigment the hair as you work through the highlights. As meches are taken for highlighting, the lightener is applied to the necessary areas of the hair, and placed into packets or foils. As you work up through the head, you will have the areas intended to be backgrounds left out and hanging down.

These sections of hair can be pre-pigmented as you work. You need to mix up the correct semi-permanent golden tone with some water and pour the colour solution into a water spray. Then, as you finish each packet, you lightly spray the background hair with the pre-pig spray, and move on to the next row of packets/foils. When all of the packets/foils have been applied, you will also have prepared all the hair in-between. You can then choose a matching base shade from your quasi-colour range, mixed with 6 vol./1.9% and apply it directly over the slightly damp pre-pigged hair and leave it to develop until the highlights are ready.

This interleave method of highlighting is a popular way of solving a complex problem. See pages 302–303 on longer highlighted hair within this chapter for more information.

Decolouring: colour strippers

Synthetic (para dyes) permanent hair colour can only be removed with a decolour treatment. All of the main colour manufacturers produce at least one. The **decolouring** process is also known as colour stripping and the products involved in this work are quite different from lightening products.

Colour removers If a client's hair has been previously permanently coloured and the client wishes for the colour to be removed or to be changed, in order to go to a lighter shade, then only a **colour remover** (decolour) may be used. The chemical formulation of a colour remover is very different from powder and emulsion lightener, hence, the colour remover is specially developed to seek out and remove (dissolve) only the synthetic molecules within the hair.

ALWAYS REMEMBER

The colour wheel

Unwanted yellow tones are lessened by violet tones, green tones are lessened by introducing red (warmth) and orange tones are lessened by blue.

Colour reducers Another form of colour stripper is the colour reducer. These work in a different way from the colour remover in that they still seek out the artificial molecules within the hair but in finding them, they reduce their size so that they can be washed from the hair.

Process preparation

Always make sure that you protect your client before undertaking any permanent colouring operation. The removal of hair colour becomes more problematic in proportion to the length of hair: the longer the hair, the more difficult the task and the easier it is for spillages to occur. Clients waiting for colours to process often have to wait for long periods. Make sure that they are regularly attended to and not just for checking development; check also for clothing protection, placement of towels and capes, and especially for poorly secured hair and product drips!

ACTIVITY

Hydrogen peroxide strengths

Complete the table below by finding as many reasons as you can for the uses of the following hydrogen peroxide strengths.

Hydrogen peroxide strength	For what purposes would this strength be used?
3%	
6%	
9%	
12%	

Reapplying to natural depth

The removal of synthetic hair colour is usually only a part of a larger technical operation. Generally, it will be the client's wishes to return to her natural colour. Clients who merely want to change their hair colour to a different, darker, new shade would not normally have to go through the process of colour reduction first.

The reapplication of base shade follows the removal of the synthetic tones. However, hair that has undergone the reduction process does now respond to the application of new colour in a different way. This new state of the hair is called double processed hair. This degree of treatment will make the hair respond to reintroduced colour more readily. Therefore the porosity levels are increased, which means that the hair becomes absorbent.

In these situations, there are certain rules that should be observed:

◆ Do a colour test on the hair first mixed only with water.

◆ Try using a liquid-based quasi-permanent colour (e.g. Diacolour) as these are easier to apply evenly and more quickly.

◆ Choose a colour one shade lighter than the expected target shade to avoid *colour grab*.

◆ Check development every few minutes (scrape colour off the hair and rub between fingers to see the colour development clearly).

Removal of permanent, synthetic hair dyes

Checklist	Special attention
◆ Client expectation/ target colour	What is the purpose for removing the permanent hair dye, is it to: 1 Recolour back to natural depth and tone? 2 Recolour to a lighter shade? 3 Remove unwanted/discoloured tonal effects from the hair?
◆ Treatment history	If you have no previous history available for the client and you want to undertake the technical operation, you must undertake an Allergy Alert Test and take a test cutting for incompatible chemicals and ability to achieve target shades before conducting any chemical process.
◆ Condition	What are the existing hair condition attributes (elasticity, porosity and strength)? Will these limit the effectiveness of the treatment?
◆ Hair length	The longer the hair, the more difficulty will be encountered in stabilizing the evenness of the lightening.
◆ Natural hair colour	Do you know what the natural hair colour is? 1 If the natural hair colour is darker than the resultant, permanent hair colour, can the target colour be achieved without removing previous colour? 2 If the natural hair colour is lighter than the resultant permanent colour, how many levels of lift are required and is this feasible? (Reducing synthetic hair dye above four levels of lift is not recommended.)
◆ Uneven, permanent hair colour	Where worn lengths have produced an uneven colour effect the darker bands/areas must be lightened to match the other lighter areas first.
◆ Work method	Always follow the manufacturer's instructions when mixing and applying the product. Start on the darkest areas first, then on to lighter areas.* Often the consistency of colour reducers makes it more difficult to work with on pre-coloured hair, so make sure that the product is applied evenly.
◆ Development	Slow development is easier to control, so develop without thermal acceleration. Ensure adequate air circulation do not paste hair flat; lift sections and separate the hair to assist an even lifting process.
◆ Removal of product	Always follow the manufacturer's instructions and remove the product with suitable shampoo and tepid water. After conditioning, dry the hair before further processing.

*It is normally not necessary to apply to mid-lengths and ends first, as worn hair colour tends to lighten on the ends anyway.

ACTIVITY

This activity covers some of the common colouring problems encountered every day in the salon. You are asked to help someone else with a consultation and you encounter the following problems.

Fill in the table with the missing information for what you think has occurred and how it can be rectified.

Colour problem	What is the likely reason for this?	How can it be resolved?
Colour not taking on white hair		
Hair appears green or khaki in certain lights		
Hair colour has faded a lot at the ends but has stayed on target nearer the root area		
Hair colour should be a natural base tone throughout, but appears redder near the roots		

Banded hair colour

Horizontally banded hair colour will occur from one of the following situations:

◆ poor home product application

◆ highlights/lowlights product seepage/bleeding during processing

◆ excessive heat/'hot spots' occurring during development

◆ uneven porosity of the hair (possibly from **masking**, where conditioning treatments have bonded to the surface of the hair unevenly).

Horizontally banded hair colour usually appears on long hair at, or near, one length. If the hair is to be restyled or layered it is worth cutting and drying first to see if the problem still exists.

If decolouring is still necessary in these areas, it would be advisable to apply the product by the following methods:

◆ Use a thicker consistency: this will allow you to spot colour, applying freehand to the central areas of 'patchiness'.

◆ Apply with the natural fall of the hair so that any natural light shadowing can easily be seen.

◆ Later, during development, extend the product to the edges of the gradated patches.

◆ Remove occasionally with warm water, then dry smoothly to see when the development is complete.

◆ Reapply to those areas requiring further development.

◆ Finally remove at the backwash with suitable shampoo and condition the hair with an anti-oxidizing treatment.

Gradated colour

Gradated colour or gradation of colour is where the hue changes with the levels of saturation. For example, black and white mixed together create grey and, depending how much white or black is added, the colour either deepens or lightens, moving from shade to shade seamlessly.

This merging of colour occurs in hair naturally after colouring and it is difficult to remedy. It occurs partly by colour fade, i.e. the natural wear and tear and weathering. However, this subtle colour fading is far easier to tackle than the gradated banding that occurs after incorrect colouring. Imagine what happens when a dark colour, say base 5, is applied to natural hair base 7. As the hair grows, the regrowth becomes more obvious. Initially, this appears as a solid line of demarcation. However, hair does not all grow at the same rate, and this becomes more apparent the further away from the roots the hair grows. The inconsistent growth blurs the edge of demarcation creating a gradated effect. In this scenario, the recolouring is simple: just retouch with base 5. However, if the hair were to be colour stripped, then the complexity would be increased because of the uneven edge of colour to be removed.

Discoloured highlights/lowlights and partial applications

Historically, we used the term 'discoloured hair' as a description for hair that has resulted from the presence of incompatible chemicals, i.e. metallic salts. As cases of this happening are few and far between, the term is now more usually applied in a different context. Partial colouring techniques are extremely popular and much of the colour correction work undertaken within the salon is a result of previous highlighting and/or lowlighting services. Problems will tend to occur in the following instances:

◆ on longer hair over a period of time following several, subsequent treatments

◆ on hair that involves two or more newly introduced colours on an uneven background colour

◆ poorly executed application or removal of product/s

◆ over-porous hair

◆ hair subjected to excessive sunlight, ultraviolet (UV) or chlorinated swimming pools.

Longer highlighted hair

Longer hair that has had several highlighting and/or lowlighting services is the most frequent reason for discoloured hair. A typical example might be a regular client with shoulder length hair of base 6 (dark blonde). The client has her first colouring service, which involves two colours:

1 a lightener with 9%

2 a beige mid-tone to harmonize the final effect.

After 6–8 weeks, with regular washing and blow-drying the mid-tones fade off. Some pigments are washed away and some are absorbed into the more porous lightened hair, discolouring the original highlights. At the 12-week stage the client returns for T–section highlights (recolouring of the roots at the parting and sides). What do you do? You are unable to do a reapplication of the mid-tone just at the roots as the ends have faded off and similarly you may not be able to locate the initial lightened highlights. Most stylists will in this situation end up recolouring the hair with both colours again just within the T section but roots to ends.

The client is happy once again. However, a longer-term problem will start to develop. Much of the surface hair is now double processed (it has had two applications of the same chemicals). Double processed hair is more porous and this impaired condition leaves the client's hair in a state less able to retain colour pigments. So again a fade-off occurs. This problem is compounded by the fact that the ends will now be predominantly lighter. The new roots will be darker and the hair in between will be lighter than the roots but not as light as the ends.

This colour state has now created an uneven background colour: a gradation from dark at the roots to light at the ends. Before any further partial colouring takes place this must be taken into consideration. At the point, where this client wishes to once again have the same colours applied, a colour correction of the discoloured, background hair will have to be undertaken. **Interleaves** are done at the same time as the highlighting service. When the meshes of coloured hair are secured in their packets/foils, the remainder hair is coloured back to the natural shade. (However, this may make the problem far worse. If this newly introduced colour bleeds into the lightened hair, all of the hair will be discoloured and the problem will be compounded further.)

Newly introduced colours

A similar problem occurs when newly introduced colours are applied to an uneven background. If the unevenness of the background colour is not corrected during the service, the impact of the newly applied colour will be greatly devalued.

The hardest choice to make in this type of colour correction is whether to colour back the uneven background colour first, or to attempt the correction at the same time as the application of the new colours. If the hair needs pre-pigmentation you would be better advised to complete the task in two separate phases:

1   pre-pigment and then colour back the hair to natural base, then

2   carry out the partial colouring technique.

However, remember that hair that has been coloured back will be more porous and less able to sustain further colouring operations. This is particularly problematic if you intend to lighten portions of the hair (it just won't work satisfactorily). If you intend to lighten as well as reintroduce colour to the background, it is better to conduct the process in a single phase; i.e. to recolour back the remaining hair that is not part of the partial colouring effect.

Over-porous hair

The condition of the hair dictates what should and can be done. Never ignore the obvious. If you believe that a treatment cannot be undertaken or is unachievable, you must tell the client so. You might also get another opinion from a colleague but, if they agree, you must refuse the service.

Removing artificial colour

Make sure that when the colouring product is removed from the hair you rinse it without massaging, using only tepid or warm water. For example, the removal of a lightener can

TOP TIP

Colour stability

The stability of a synthetic colour within the hair is directly affected by the condition of the hair. The better the condition the more able the hair to retain the pigments. Conversely, the poorer the condition the less able the hair to retain the pigments.

be far more problematic than that of colour reducer (colour stripper) and this has more to do with the colouring technique that has been used. If different coloured highlights have been done with Easy Meche, foil, wraps, etc. these need to be removed carefully and individually as one colour may affect another.

Although the client's scalp has not been subjected to chemicals, it still might be sensitive from the colouring technique; again, the cooling action of the rinsing will stop the lightening process and make the client more comfortable. Afterwards the hair can be shampooed with a mild colour shampoo and conditioned with a pH-balancing antioxidizing treatment.

Poor removal of colour products

As partial colour techniques become more involved the complexity of processes and procedures increases too. This is all too apparent in poorly removed colour products. When two or more colours are applied in the same operation, particular care must be taken in how they are removed. For example, consider a client who has lightened highlights with vibrant copper lowlights on a reintroduced background of base 6.

In this example, three separate colour combinations exist:

1 highlight

2 lowlight

3 background.

Imagine what would happen in the following scenarios:

A Highlight packets are removed and shampooed at the same time as the lowlight packets.

B Highlight packets are removed and rinsed with the background colour.

C All packets are removed and collectively shampooed at the same time.

Potentially, the effects created by clever techniques and artistic application can be totally wiped out if, during removal, the individual colour combinations are not removed one by one.

ACTIVITY

Complete the table below to indicate which and what amounts of pigments are present within each scenario.

Natural hair colour	What colour pigments are present within the hair?
Black hair	
Brown 'chestnut' hair	
'Celtic' red hair	
Blonde hair	
White hair	

STEP-BY-STEP: COLOUR CORRECTION (BANDED HAIR COLOUR)

1 After greeting your client, conduct a full consultation to establish the areas which require colour correction.

2 Section the hair and clip the sections into place.

3 Mix the required colour correction products together and carefully apply to the banded areas.

Work quickly and methodically throughout the head of hair until all areas have been treated.

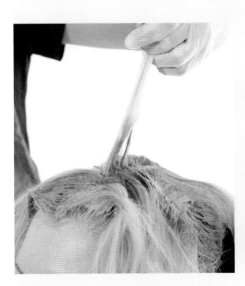

4 Visually check the development until the desired target tone is reached.

Follow the manufacturer's instructions so as not to exceed the stated development time.

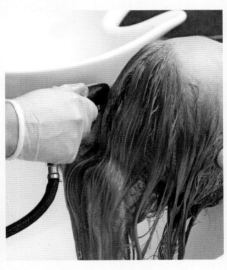

5 Rinse, emulsify and shampoo the hair to remove all product thoroughly. Apply a toner if required.

6 The finished result restores the hair's colour.

CourseMate video: Colour correction (banded hair colour)

STEP-BY-STEP: REMOVAL OF ARTIFICIAL HAIR COLOUR: DECOLOURING

1 After greeting your client, conduct a full consultation to assess the level of lift required and the ultimate target depth and tone.

Ask yourself: is a gentle cleanse or a deep cleanse required?

2 Mix the colour removal product as required following the manufacturer's instructions.

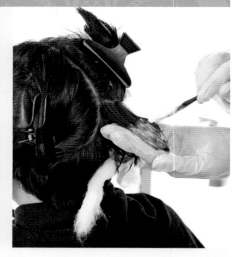

3 Using a tint brush or sponge, apply product directly to dry unwashed hair, starting with the darkest areas first.

Do not let the product touch the scalp.

4 Massage in and closely watch the product develop until the desired shade has been reached.

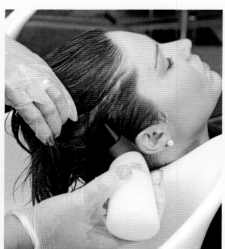

5 Rinse well and apply an optimizing treatment.

6 The final decoloured result is now ready to be recoloured.

CourseMate video: Colour removal

STEP-BY-STEP: HIGHLIGHT COLOUR CORRECTION

Due to the imbalance of light and dark hair, these highlights have lost all definition. This is a common problem which occurs daily in hairdressing salons. When highlighting fine hair, it is all too easy to overdo it. This technique will correct the light–dark balance problem while redefining the definition with two harmonizing colours.

1 Several sets of highlights have created over-lightened ends while the roots remain dark.

2 Carefully select highlights from the previously lightened areas. Position the foil carefully below the individual sections of hair. Only paste the lightener on at the roots as the ends are already light enough.

3 Once the foils around the parting are in place, the ends can be coloured back in one easy step.

4 Paste the product onto the hair with your brush, supporting the hair with your gloved hand.

5 When the colour has been applied to the lengths of the hair, a strip of foil can be placed so that it keeps the colour away from the skin while it develops.

6 The final effect.

Colour correction

A colour correction process can be applied to any length of hair. In this scenario, the hair has been lightened and the client wants to return to her natural colour. However, the hair has too little yellow pigment present and the hair has to be pre-pigmented with red/gold pigment before natural depth can be achieved or the hair will go green. Although the client wishes to return to her natural depth, experience shows that many clients find that type of dramatic change too much of a contrast. Here is a new, exciting alternative to recolouring. Flat, single dark colours only look good if the hair has lots of shine. Unfortunately, shine does not last very long so the best course of action here would be to introduce a variety of tones into the hair.

STEP-BY-STEP: REINTRODUCING COLOUR INTO PREVIOUSLY LIGHTENED HAIR

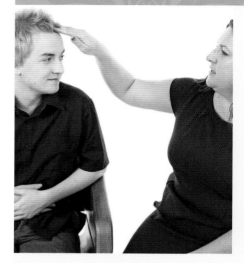

1 After greeting your client, conduct a full consultation to assess the depth, tone and target shade desired.

Is a porosity leveller required?

2 Select pre-pigment product and tone according to the depth and tone of your selected target shade.

3 Apply the chosen pre-pigment to the target areas.

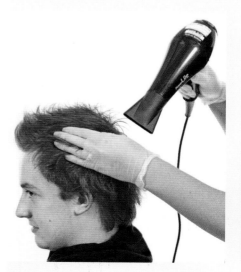

4 Dry the chosen pre-pigment product into the hair.

5 Apply the chosen shade of tint to the re-growth area.

6 Apply to mid-length hair and ends. Follow the manufacturer's instructions for development.

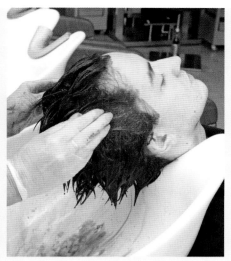

7 Carry out a strand colour test to check for even development.

8 Emulsify, rinse and remove the colour product, then shampoo and condition.

9 The result restores the hair's colour depth.

Provide aftercare advice

The corrective service does not end with the salon processing; the condition of the client's hair will be impaired and the client must help to maintain the colour's ability to last and the condition during routine daily handling. Therefore you should tell clients what sorts of products that they could use that would make their colour last and reduce the risk of fading. Also make a point of explaining what they should avoid, as some products will reduce the effects of your colouring – causing it to fade prematurely or lose its intensity or vibrancy.

Explain the benefits of maintaining their hair in good condition: it is easier to manage, looks better and is noticeable to everyone else as well.

> Always use professional aftercare products and ensure your clients understand how important aftercare is.
>
> *David Morgan*

Lifestyle

Normal everyday wear and tear is a major reason for colouring problems that require colour correction. Lifestyle has the biggest impact upon hair condition and client expectation; this can include busy work schedules, social lives and the search for the sun.

For example: you might have recommended a shampoo/conditioner to a client, who used it, but then went back to using her own particular favourite. The truth of the matter is that what you recommended actually did its job. The client has forgotten the original problem and now focuses on a new one; so the client might think that the product has stopped working but this is often because their needs have changed. For instance they might:

1   Go on an exercise binge and swim at the local pool every day.

2   Spend two weeks in the Caribbean.

CourseMate video: Reintroducing depth into previously lightened hair.

ALWAYS REMEMBER

You must provide the correct advice so that your client can make the most out of their colour correction. Recommending the right products and ways in which they should style and manage their hair is essential to a long-lasting effect.

3 Spill paint on their hair during redecorations at home.

4 Try a new home colour because a friend recommended it.

Home and aftercare checklist

✓	Talk through the colour effect as you work; explain to clients how they can maintain their hair by optimizing both the look and the condition of their hair.
✓	Explain how lifestyle factors can affect their hair colour and ways that they can combat these factors.
✓	Show and recommend the products/equipment that you use so that the client gets the right things to enable them to get the same effects.
✓	Tell the client how long the effect can be expected to last and when they need to return for recolouring.
✓	Warn them about the products that will have a detrimental effect on the colour by telling them what these products would do.
✓	Warn the client about how incorrect handling or styling may reduce the length of time that the new effect will last.

REVISION QUESTIONS

Q1. A _____ /test cutting is carried out before colour to assess what colour can be achieved.   Fill in the blank

Q2. A lightener lasts longer than the effects of a high lift colour.   True or false?

Q3. Which of the following are definite contra-indications to a colour correction service?   Multi-selection

Sensitive skin   ☐   1

Damaged hair   ☐   2

Poor elasticity   ☐   3

Lightened ends   ☐   4

Banded hair colour   ☐   5

Gradated colour   ☐   6

Q4. Decolouring requires pre-pigmentation.   True or false?

Q5. Which of the following tests is carried out during colour/lightening services?   Multi-choice

Allergy Allert Test   ○   a

Incompatibility test   ○   b

Porosity test   ○   c

Development   ○   d

Q6. Lighteners alter the pigmentation of hair within the cuticle.   True or false?

Q7. Which of the following affect the development of colouring?   Multi-selection

Heat   ☐   1

Poor application   ☐   2

Time   ☐   3

Sectioning   ☐   4

Combing   ☐   5

Peroxide strength   ☐   6

Q8. Yellow tones within hair are neutralized by adding _____ tones.   Fill in the blank

Q9. Which of the following could produce a green effect if not done correctly?   Multi-choice

Lightening hair   ○   a

Colouring back   ○   b

Decolouring   ○   c

Root retouch   ○   d

Q10. Lightened hair that appears too gold can be neutralized by adding mauve.   True or false?

13 Perming Hair

LEARNING OBJECTIVES

◆ Be able to work safely during perming procedures at all times.

◆ Be able to make preparations for perming services.

◆ Be able to create a variety of permed effects.

◆ Be able to identify a range of factors that can limit or stop a planned service.

◆ Know how to resolve basic perming problems.

◆ Be able to communicate professionally and provide aftercare advice.

KEY TERMS

acids	incompatibility	pull burns
alkalis	piggyback (double) wind	root perm
depilatory	polypeptide chains	spiral wind
disulphide bonds	post-perm treatments	straighteners
double winding	pre-perm test	straightening
fish-hooks	pre-perming treatments	weave winding

UNIT TOPICS

Perm and neutralize hair using basic techniques

Create a variety of permed effects

INTRODUCTION

Perming and colouring are arguably the most complex aspects of contemporary hairdressing. Perming is a very problematic technical procedure and the combination of client expectations, reality and ability to maintain their hair carefully and competently between salon visits is crucial.

Perming has not been a popular choice for several decades, but at last the interest is growing again. Curls are everywhere: in music, films and TV. If you cannot perm, you are likely to miss a big opportunity.

Perming hair

PRACTICAL SKILLS

Find out what clients want by questioning and visual aids

Look for things that will influence the ways that you perm your client's hair

Learn how to carry out tests on the hair and how the results will affect the service

Learn how to do a variety of perming techniques to create a range of different effects

Make recommendations to clients about looking after their perm

Provide aftercare advice to your clients

UNDERPINNING KNOWLEDGE

Know your salon's ranges of products and services

Understand the factors that can affect your perming options

Understand how different perming and neutralizing products work and their suitability in different situations

Know the current fashions for permed hair

Know how to recognize and correct perming problems

Know the key principles of perming and how perms affect the chemical and physical properties of different types of hair

Effective and safe methods of working when perming hair

Note: *If you have recently completed your Level 2 you will already be familiar with the underpinning principles and knowledge for perming, neutralizing and straightening hair.*

If you want to refresh your memory for this topic continue from this point onwards; alternatively, look at Chapter 15 on health and safety pages 364–389.

INDUSTRY ROLE MODEL

ANNE MILLER Creative Director at Renella

"I was influenced in my career choice by my mother, who had been a hairdresser. I started my apprenticeship at the age of 15 and became Scottish junior champion at the age of 19. I started Renella with others at the age of 24 and it quickly expanded from three team members to 40.

Lifelong learning continued for me, including Level 3, trainer assessor and internal verifier awards. I have also been awarded several training, individual and business awards. My favourite experience has been working with Habia as a skills team member and being on stage at Salon International. I now share my job with others. It includes training and assessing at Level 2 and Level 3 and developing our own in-house transition programme for graduates, senior stylists and trainers.

How perms work

Polypeptide chains and disulphide cross links The polypeptide chains have three different cross-linkages or bonds that join them to other polypeptide chains.

Two of them: The salt bonds and hydrogen bonds are temporary and are broken down when the hair is wet.

One other: The disulphide bond is permanent and is only changed during chemical services such as perming and relaxing.

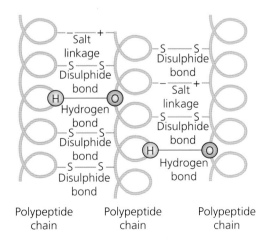

Polypeptide chain   Polypeptide chain   Polypeptide chain

Changing the keratin Keratin in hair contains cross-links between polypeptide chains and it is the disulphide bonds or bridges that give hair its shape. Each disulphide bridge is a chemical bond linking two sulphur atoms, between two polypeptide chains lying alongside each other. During perming (and straightening), some of these links or bridges are chemically broken, making the hair softer and more pliable, allowing it to be

moved into a new position of wave or curl. Only 10–30% (depending on lotion strength) of the disulphide bridges are broken during the action of perming. If too many are broken, the hair will be damaged beyond repair. You need to keep a check on the progress of the perm to ensure that:

◆ the desired curl shape or strength is achieved

◆ the chemical action is stopped at the right time.

You do this by rinsing away the perm lotion and neutralizing the hair.

During neutralizing, pairs of broken links are joined up again at different sites along the hair. The newly formed cross-links (**disulphide bonds**) hold the permed hair firmly into its new shape.

Changing the position of the bonds
The hair is first wound with tension onto some kind of former, such as a curler or rod. This is the *moulding* stage. Then you apply perm lotion to the hair, which makes it swell. The lotion flows under the cuticle and into the cortex. Here it reacts with the keratin, breaking some of the disulphide bonds between the polypeptide chains. This *softening* stage allows the tensioned hair to take up the shape of the former or rod: you then rinse away the perm lotion and neutralize the hair. This fixing stage permanently rearranges the *disulphide bonds* into the new shape.

This process can also be described in chemical terms. The softening part that breaks some of the cross-links is a process of *reduction*. The disulphide bridges are split by the addition of hydrogen from the alkaline perm lotion. (The chemical in the perm lotion that supplies the hydrogen is called a 'reducing agent'.)

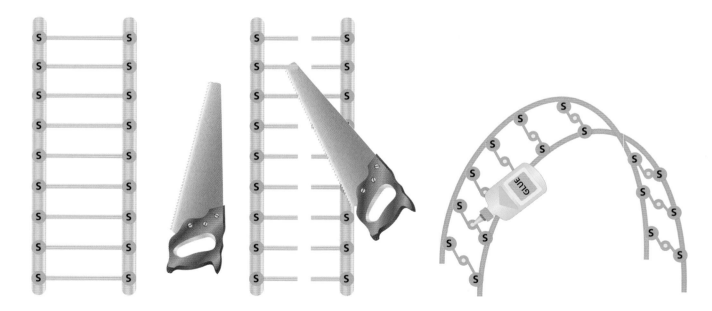

The illustration above tries to explain it in another way:

1 The ladder represents the hair and disulphide bonds as rungs holding the ladder together.

2 When perm lotion is applied, the rungs/disulphide bonds are broken.

3 During perming, the hair is bent around a curler. The disulphide bonds are then re-formed in the neutralizing process, into their new locations; permanently waving the hair.

The final part of the process, the *fixing* stage, recreates bonds at different positions between polypeptide chains. It occurs by an oxidation reaction. New disulphide bridges form and these rearranged links hold the hair in a different shape. The hydrogen reacts with the oxygen in the neutralizer, forming water. (The chemical in the neutralizer that supplies the oxygen is called an 'oxidizing agent' or 'oxidizer'.)

> **TOP TIP**
>
> Always bear in mind how your perm will affect the client's hair structure.

Acidity and alkalinity: the pH scale The pH scale measures acidity or alkalinity. It ranges from pH 1 to pH 14. **Acids** have pH values from 1–6, whereas **alkalis** have values ranging from pH 8–14. Substances that are neither acid nor alkaline, i.e. neutral, have a pH value of 7. The higher the pH number, the more alkaline the substance; the lower the pH number, the more acid the substance.

The normal pH of the skin's surface is 5–6, referred to as the skin's acid mantle. This acidity is due in part to the sebum, the natural oil produced by the skin. The levels of pH can be measured using pH universal indicators; Litmus papers will indicate whether something is acid, alkaline or neutral.

If hairs are placed in alkaline solution, they swell and the cuticle lifts. In slightly acid solutions the hair contracts and the cuticle is smooth. However, stronger acidic or alkaline solutions will impair the hair's structure, causing it to break down. This has both beneficial and negative effects in modern-day applications, as we all like to have well-groomed, easy to manage hair. For example, in the case of excess hair on arms and legs, an alkaline (high-pH) compound can be applied as a crème to act as a **depilatory**.

However, perming can be hazardous when combined with other chemical treatments such as colouring, and especially highlights. If you apply a perm for a client today, you will be selecting from normal, resistant, sensitized or coloured formulations and these form the basis of all permutations. However, how do you perm highlighted hair safely? The truth is that you cannot and that is because of the multiple porosities of individual hairs on the same head. A perm cannot differentiate between normal, coloured and lightened hair, it will process them all equally and in the same way. Therefore, some hair will perm well, some will be in poor condition, and other lightened hair may break off all together.

Admittedly, you can prepare the hair by applying a pre-treatment to even out the porosity; this can reduce the amount of hair to sustain damage. But there will still be some impairment to the overall condition and this will be due to the alkaline solution, which swells the hair beyond a point where it can maintain its polypeptide structures and causes it to disintegrate.

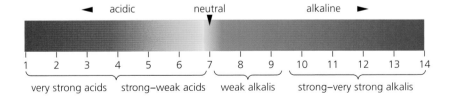

The pH scale

Protecting the client and yourself

Gowning the client It is important to protect the client from the substances used in perming. Give them a chemical-proof gown to wear and secure into place a clean, fresh towel around their shoulders. On top of this you should fix a plastic cape, ensuring that it is comfortable around the neck. Perms are generally applied to hair by the post-damping method (where the lotion is applied and easily absorbed into previously wound sections of hair) and the solutions tend to be very watery and mobile. This, potentially, could be a hazard to the client unless you take adequate precautions. The majority of perm lotions are alkaline. If they drip or soak into textiles they will be held against the skin like a poultice. This is potentially very dangerous as it could cause irritation, swelling and even cause burns! This is also a problem if it does go onto the client's clothes as it may discolour the fabrics. So you must make sure that you protect your client well so that this never happens.

Barrier cream   After gowning, you can now apply the barrier cream to the hairline.

Protecting yourself   Your salon must provide all the personal protective equipment (PPE) that you may need in routine daily practices. Perming involves the handling and application of chemicals you must protect yourself from their harmful effects. Always read the manufacturer's instructions and follow the methods of practice that they specify. You are obliged to wear and use the PPE provided for you: non-latex disposable gloves, a waterproof apron and barrier cream.

The Control of Substances Hazardous to Health Regulations (COSHH) 2003 lay out the potential risks from hairdressing chemicals. You need to make yourself aware of the information provided by the manufacturers about their handling, storage and safe disposal. Perming solutions should be stored in an upright position, in a cool, dry place, away from strong sunlight, and in a metal lockable cupboard to comply with chemical storage requirements. When they are used they should be applied in a well-ventilated area and if there is any waste (materials that cannot be saved and used another time) it should be disposed of by flushing down the basin with plenty of cold water.

Working effectively

Experienced hairdressers work in a way that makes the most out of the time available: they optimize the way in which they work. You need to learn the ways in which you can optimize your time and effort so that your work is both productive and effective. There are several ways that you can do this.

Minimize waste   Get into the habit of eliminating waste. All the resources that you use cost money and affect profitability; you can be more effective by maximizing your time and efforts while minimizing the cost of carrying out your work.

Maximize your time   Always make good use of your time. For example you can:

◆ Prepare the salon by organizing work areas ready to receive the clients.

◆ Prepare the materials, look out for stock shortages and record them.

◆ Prepare the equipment, organize the cleaning and washing of combs, rods flexible foam and rigid curlers.

◆ Prepare client records, and get things ready for when they arrive.

◆ Prepare the trolleys; get the right materials organized and ready.

ALWAYS REMEMBER

For clients – barrier cream provides a resistant layer to the skin, helping to prevent the action of chemicals upon their skin.

For you – wearing non-latex disposable gloves provides a barrier from chemicals having any contact with your skin.

ALWAYS REMEMBER

Disposable gloves

Your salon will provide non-latex disposable gloves for your personal safety when you are handling any chemicals in the workplace. Make sure that others use them too when they carry out any duties involving the handling of chemicals.

TOP TIP

Remove and dispose of waste items as soon as possible; do not leave used cotton wool, plastic caps, etc. around at the basins. Put them into a covered bin.

Wash, dry and replace perm rods/equipment in the trays as soon as possible. Check perm rubbers for chemical damage and replace worn, split or broken ones.

ACTIVITY

Complete the table to help you to review the preparations that you should make before carrying out any perm service.

Essential aspect	Why should you do this?
Wear PPE	
Prepare the work area	
Check the equipment	
Look for contra-indications	
Record test details	
Prepare the client	

> " When learning about perming, you will find all the information you need to know for best practice is in the manufacturer's instructions and training manuals. These are researched and developed over many years by experts. Trainers and experienced peers can help interpret them for you. Remember, manufacturers upgrade their products from time to time and training manuals are continually revised. So be sure you are always using the latest version.
>
> *Anne Miller*

Types of perm

For the client a perm is a major step – they will have to live with the result for several months. They may not be familiar with the range of perming solutions available so you will need to explain what the differences are and what is involved in each.

◆ **Cold wave solutions** – such as L'Oréal's *Dulcia* or *Synchrone* – are mild alkaline perms. These types of perm are widely available and simple to use with applications for all hair types and most conditions. These solutions tend to have a pH at around 9.5 so they are a fairly strong alkali that will swell the hair and affect around 20% of the disulphide bridges. They reduce the natural moisture levels within the hair and are therefore better on normal to greasy hair types. They are particularly good for achieving strong, pronounced movement and curl, and therefore create lasting effects that can withstand the high maintenance of regular blow-drying, setting, etc.

◆ **Acid wave solutions** – such as Zotos's *Acclaim* – provide alternatives for perming when the hair is sensitized and needs to retain higher moisture levels or requires softer, gentler movement. They have lower pH values at around 6–7 and are therefore much gentler in the way that they work. They are suited to drier, more porous hair types. Acid perms are two-part solutions and require the components to be mixed together just before application so that the perm is self-activated. Any residual lotion left over after application will not last and must be discarded.

◆ **Exothermic perms** – such as Zotos's *Warm and Gentle* – tend to be similar to acid waves in their chemical composition and therefore can have similar benefits. The only difference is that these perms need heat to be activated and will self-generate this when the two chemical parts (reagents) are added together, without the need for accelerators or hood dryers.

Prepare for perming

HEALTH & SAFETY

Contra-indications for perming The following list indicates situations when perming should not be undertaken:

◆ When the hair is particularly porous (possibly over-lightened), with varying levels of porosity throughout the lengths (poorly coloured or lightened).

◆ When the scalp has abrasions or sensitive areas.

◆ When the hair is weakened, broken or damaged.

◆ When the hair does not have any ability to stretch and return to the same length.

◆ When incompatible chemicals (lead acetate) are used on the hair (*Just for Men, Grecian Formula,* compound henna, etc.).

◆ When there is any evidence of physical or chemical changes on the hair or scalp and the client is unable to provide you with a full, satisfactory account of what actions have been taken.

◆ Any evidence of scalp disease or disorder.

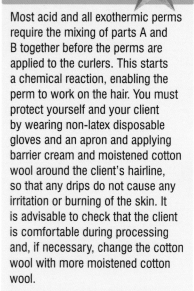

ALWAYS REMEMBER

Most acid and all exothermic perms require the mixing of parts A and B together before the perms are applied to the curlers. This starts a chemical reaction, enabling the perm to work on the hair. You must protect yourself and your client by wearing non-latex disposable gloves and an apron and applying barrier cream and moistened cotton wool around the client's hairline, so that any drips do not cause any irritation or burning of the skin. It is advisable to check that the client is comfortable during processing and, if necessary, change the cotton wool with more moistened cotton wool.

Note: If a perm requires parts A and B to be mixed together, you will have to dispose of any residual, unused lotion after the perm. It will not keep.

ACTIVITY

Consultation is an essential part of any hairdressing service and perming has its own particular aspects and considerations. Complete the table below to identify how you would find out about the following information.

Contra-indications	How would you identify this?
Skin sensitivity	
Allergic reaction	
Skin disorder	
Incompatible products	
Damaged hair	

Consultation

Find out your client's requirements – what they expect from a perm – and determine whether this is the best solution for them, bearing in mind the added maintenance, care and attention needed to achieve the desired effect.

◆ Consider the style and cut, together with your client's age and lifestyle.

◆ Examine the hair and scalp closely. If there are signs of inflammation, disease or cut or grazed skin, do not carry out a perm. If there is excessive grease or a coating of chemicals or lacquer you will need to remove these by washing with a pre-perm shampoo first. Previously treated hair will need special consideration.

◆ Analyse the hair texture, condition and porosity.

◆ Carry out the necessary tests to select the correct perm lotion.

◆ Always read manufacturer's instructions carefully.

◆ Determine the types of curl needed to achieve the chosen style.

◆ If this is a regular client, refer to the records for details of previous work done on their hair.

◆ Advise your client of the time and costs involved. Summarize what has been decided, to be sure there aren't any misunderstandings.

◆ Minimize combing and brushing, to avoid scratching the scalp before the perm.

◆ Update the client's treatment records for future reference.

Clients new to perming, especially those with long hair, often have expectations that may not be achievable, so organizing a consultation the week before their appointment to ascertain suitability is a good idea. If perming is not suitable this saves appointment space being wasted.

Anne Miller

Match the perm to the needs of the hair   It is important to make sure you choose the most suitable perm lotion, the correct processing time and the right type of curl for the chosen style. Consider the following factors:

1   **Hair texture** For hair of medium texture, use normal strength perm lotion. Fine hair curls more easily and requires weaker lotion; coarser hair can often be more difficult to wave and may require a stronger lotion for resistant hair. (Although this is not true for oriental hair types.)

2   **Hair porosity** The porosity of the hair determines how quickly the perm lotion is absorbed. Porous hair in poor condition is likely to process more quickly than hair with a resistant, smooth cuticle. (See the section on pre-perming treatments on page 325 in this chapter.)

3   **Previous treatment history** 'Virgin' hair – hair that has not previously been treated with chemicals – could be more resistant to perming than hair that has been treated. It would therefore require a stronger lotion and possibly a longer processing time.

4   **Length and density of hair** Longer, heavier hair requires a tighter curl than shorter hair because the hair's weight will cause it to stretch. Short, fine hair may become too tightly curled if given the normal processing time.

5   **Style** Does the style you have chosen require firm curls, or soft loose waves? Do you simply wish to add body and bounce?

6   **Size of rod, curler or other former** Larger rods produce larger curls or waves; smaller rods produce tighter curls. Longer hair generally requires larger rods. If you use very small rods in fine, easy-to-perm hair, the hair may frizz; if you use rods/curlers that are too large, you may not add enough curl.

7   **Incompatibility** Perm lotions and other chemicals used on the hair may react with chemicals that have already been used – for example, in home-use products. Hair that looks dull may have been treated with such chemicals. Ask your client what products have been used at home and test for incompatibility.

TOP TIP

Always record the details of the consultation/service for future reference.

"Clients do not always use the same words as professionals; curly may mean corkscrew to one person and waves to another. Style books are invaluable for showing clients different sizes of curls and perms that are dried naturally, set or spiral curled.

Anne Miller

ALWAYS REMEMBER

Some medical conditions affect the way that hair responds. For example, clients with thyroid problems may find that perms do not seem to take properly or last. Also, clients who have been taking health supplements, such as cod liver oil, over long periods will notice that they affect the way that the perm takes in the hair. (When cod liver oil supplements are taken, increased levels of moisture are deposited into the hair that ultimately overload the hair resulting in limp curls.)

What should you find out before you start?

First of all	Things to consider
Does the client know of any reasons that would affect your choice of service?	Ask the client about their hair to find out if there are any known reasons why the service cannot continue – are there any contra-indications?
What type of perm would be best to suit their needs?	Should you be using a cold wave, acid or exothermic?
How can the desired effect be best achieved?	What type of wind should you use, conventional, brick directional? What size curlers should you use, what curl or movement is required? How long does it need to develop?
How long will it last?	Is perming suitable for the hair type, condition and texture?
How much will it cost?	Is perming a cost-effective solution for the client?
How long will the process take?	Is there enough time to complete the effect? Has anything changed as a result of the consultation? Would this service now need to be rebooked or do you have the time to complete it still?

First of all	Things to consider
How will it affect the hair?	Will the long-term effects be what the client expects?
Is the hair suitable for perming?	Have you tested the hair and skin beforehand to see if there are any contra-indications or hair condition issues that will affect the result?
Now consider:	
What are the client's expectations?	How will the perm enhance or support the style and the hair? What are the benefits for them?
What are the results of your tests?	Examine the hair: does it present any limitations for what you intend to do?
What is the hair condition like?	Are there any factors that will change the way in which perming will work on the hair? What previous information is available?
What do the client's records say?	Does this information influence the choice and perm process?
How will you show the effect to the client?	Have you got any illustrations of the finished effect?

What contra-indications are you looking for?

Contra-indications	How could you find out?	How else would you know?
Skin sensitivity	Ask the client if they have ever had a reaction to hair or skin products in the past.	Patch test/allergy alert test.
Allergic reaction	Ask the client if they have ever had a reaction to hair or skin products in the past.	Patch test/allergy alert test.
Skin disorder	Ask the client if they know about any current skin disorders.	Examine the scalp to see if there are any physical signs of skin abrasions, discoloration, swellings, infestation or infections.
Incompatible products	If you see the results of any previous colour ask what type it was, how was it done?	Look for discoloration or unnatural colour effects on the hair. Test for incompatibles.
Medical reasons	Ask the client if they know of any current medical reasons that they have that could affect a perming service.	Examine the hair; look for signs of healthy active growth. If there are signs of weakened, damaged, broken or missing hair, ask for more information. Test for elasticity and porosity.
Damaged hair	Ask the client if there are any current known reasons why the hair is in its current state/condition.	Examine the hair; look for signs of healthy active growth. If there are signs of weakened, damaged, broken or missing hair, ask for more information. Test for elasticity and porosity.

Hair tests

Elasticity test This tests the tensile strength of the hair. Hair in good condition has the ability to stretch and return to its original length, whereas hair in poor or damaged condition will stretch and will not return to original length.

To do an elasticity test: Take a single strand of dry hair and hold at either end. Now gently stretch and let the hair return to its original length.

Porosity test

The purpose of this test is to find out how well protected the inner cortex is by the cuticle layers. Porous hair has a damaged cuticle layer and readily absorbs moisture; this presents a problem when drying, as this hair takes longer to dry and often lacks an ability to hold a style well.

To do a porosity test: Take a single strand of dry hair and, between your fingertips, slide it from the root through to the points; you can feel how rough or smooth the cuticle is. Rougher hair (as opposed to coarse hair) is likely to be more porous and will therefore process more quickly.

Incompatibility test

Professional hairdressing products are based upon organic chemistry formulations. These are incompatible with inorganic chemistry compounds and will cause damage to the client's hair. A typical example can be found in **progressive dyes** such as Grecian Formula™. These **colour restorers** will deposit metallic compounds onto the surface of the hair (lead acetate) and this will react with perms and colours, causing hair damage or breakage.

To test for incompatibility**:** Protect your hands by wearing non-latex disposable gloves. Place a small cutting of hair in a mixture of 20 parts hydrogen peroxide and 1 part ammonium hydroxide. Watch for signs of bubbling, heating or discoloration: these indicate that the hair already contains incompatible chemicals. The hair should not be permed, nor should it be coloured or lightened. Perming treatment might discolour or break the hair, and might burn the skin.

Pre-perm test curl

If you are unsure about how your client's hair will react under processing you could conduct a pre-perm test curl. Sometimes this can be done on the head and in other situations you will need to cut a sample of hair for testing.

To do a pre-perm test**:** Wind, process and neutralize one or more small sections of hair. The results will be a guide to the optimum rod size, the processing time and the strength of lotion to be used. Remember, however, that the hair will not all be of the same porosity.

Development test curl

This test is always carried out after the hair has been dampened with perm solution and during the processing time. It will determine the stage of curl development so that the processing does not continue beyond the optimum curl.

To do a development test curl: Unwind a curler during processing to see how the curl is developing. When you have achieved the optimal 'S' shape you want, stop the perm process by rinsing and neutralizing the hair.

Note: If the salon is very hot, or cold, this will affect the progress of the perm: heat will accelerate it, cold will slow it down.

Record the results

Make sure that you record the details of any test that you conduct. Update the client's record card immediately after you have done the test. Do not leave it until later, you might forget. These records are essential information that will be needed again and help to show that a competent service has been provided at that time. You should record the:

- ◆ date

- ◆ test carried out

◆ development times

◆ results of tests, positive or negative

◆ recommendations for home care or follow-up advice.

This would be vitally important if there was a problem at some later stage, particularly if it involved any legal action taken against the salon.

> A Level 3 qualification in perming will certainly interest a potential employer, so keep your skill active as pre-employment tests are common practice. An important feature of a skill or pre-employment test would be attention to detail. For example, make sure you quote prices before you start the perm, check the labels on perm and neutralizer bottles, fill in all details on record cards and advise clients on suitable products to maintain their new perm.
>
> *Anne Miller*

TOP TIP

Record the client's responses to your questions and the comments about how the results of any tests affected their hair and skin.

ALWAYS REMEMBER

Temperature has a major impact on perming. This could be from the environment, i.e. general salon temperature, or from added heat from a hood dryer. In either case remember that processing times will be reduced considerably.

Pre-perm and post-perm treatments

Matching the correct perm lotion to hair type is an essential part of the hair analysis. However, many perming solutions come in only a coloured, normal or resistant formula and this alone will not cater for all hair conditions. Dry, porous hair will absorb perming solutions more readily; therefore, special attention needs to be given in these situations. **Pre-perming treatments** are a way to combat these conditioning issues. Porous hair that is suitable for perming will have an uneven porosity throughout the lengths. Hair that is nearer the root will have a different porosity level to that at mid-length hair or that of the ends. Therefore, the hair's porosity levels will need to be evened out, i.e. balanced, before the perm lotion is applied. This enables the hair to absorb perm lotion at the same rate, evening out the development process and ensuring that the perm does not over-process in certain areas. A pre-perming treatment is applied to damp hair before winding, and then combed through to the ends. Any excess is removed and the hair is wound as normal.

Post-perm treatments are used after perming and neutralizing. They are also necessary to rebalance the hair's pH value back to that of 5.5. Post-perm treatments do this by removing any traces of residual oxygen from the neutralizing process.

Sensitized hair

Colouring and highlights are very popular, so it is unlikely that you will find many female clients without some sort of previous processing on their hair. This can create a variety of problems.

Any previous chemical processing on the client's hair is going to be a critical factor in:

◆ deciding whether perming is a suitable service

◆ selecting products that will do the job satisfactorily without causing any further damage.

Previous processing is not necessarily a contra-indication, but it is a major issue for retaining or re-establishing a good condition. Dry hair loses at least one of the key indicators of good conditioned hair, and that is shine. (Remember it may lose others as well, such as elasticity or porosity.)

The hair's natural 'background' moisture is important for the following reasons:

◆ it provides shine and lustre on the hair

◆ it stops the hair looking dry or porous

◆ it helps to provide flexibility and therefore the hair's elasticity.

Moisture is directly linked with the key indicators of good condition and, without it, one or more of those features are diminished.

Sensitized hair lacks this natural moisture and is therefore more susceptible to over-processing during perming. During processing, the curl/movement develops more quickly (as with any porous hair) and is generally a weaker/limper result than on unprocessed hair. This does not mean that it is resistant to perming; on the contrary, it soaks up the lotion far more readily.

If during your consultation you find these signs on your client's hair; then you should protect the hair with a pre-perm treatment to assist in evening out the porosity and then select a solution that is developed for the type of hair you are working with. For example, if you are considering a perm on highlighted hair you must use a lotion that is developed for hair with varying degrees of porosity on the *same* head, such as an acid-balanced perm like ZOTOS *Acclaim*. A general cold wave lotion will *fry* the hair and you will have a ball of frizz to contend with!

Likewise, if you are perming coloured hair then, again, you need to match the requirements of the hair to the product you want to use.

Hair can also become sensitized from the misuse of heated styling equipment. Excess heat will raise the hair's cuticle and make it act in a similar way to that of chemically treated hair.

ACTIVITY

Hair tests

Complete this activity by filling in the missing information in the spaces provided.

Type of test	What is the purpose of the test?	How is the test carried out?
Elasticity		
Porosity		
Incompatibility		
Development test curl		
Pre-perm test curl		

Perming techniques: quick reference guide

Perm technique	Final effect	Ideal length	Lotion type	Equipment
Root	Lift and body at root area only	Layered hair or graduated hair 100–150 mm long	Acid or alkaline, often used as thick cream or paste	Conventional rods, often used with non-porous end papers
Directional	Lift and body with definite forced movement	Layered hair or graduated hair 100–150 mm long	Acid or alkaline	Conventional or oval rods
Weaving	Textured soft and stronger movement at ends	Layered hair or graduated hair over 75 mm long	Acid or alkaline	Conventional or oval rods
Piggyback (double) wind	Textured curl with varying curl diameters	Layered hair or graduated hair over 75 mm long	Acid or alkaline	Conventional rods
Stack wind	No root lift but strong end movement/curl	Graduated hair 150 mm down to 70 mm	Acid or alkaline	Conventional rods
Zigzag	Strong geometric, angular movement	One length or long layered hair 250 mm	Alkaline	Perming chopsticks/u-stick rods
Spiral	Vertical cascade curls with uniform diameter	Long layered hair or one length over 250 mm	Alkaline	Spiral rods or foam covered flexible wavers

Create a variety of permed effects

Perming is a straightforward procedure – you just need to be organized. Once you have consulted your client and made the necessary tests, you are ready to start.

Checklist: The basics

✓	Prepare your materials: trolley, curlers, lotions, etc.
✓	Protect your client and clothes as necessary with a gown, towel and cape.
✓	Refer to records and your consultation to identify any special requirements.
✓	Shampoo the hair to remove product build up, grease, etc.
✓	Towel-dry the hair (excess water would dilute the perm lotion, but if the hair is too dry the perm lotion won't spread thoroughly through the hair).
✓	If the client requires a pre-perm treatment, apply it now. Make sure you have read the instructions carefully.
✓	If you are pre-damping (if the styling requirements dictate this) as opposed to post-damping, apply your lotion evenly and carefully (see page 330)

TOP TIP

Wear disposable non-latex gloves from the beginning. It is inconvenient to have to put them on later.

ACTIVITY

Perming techniques

Complete this activity by filling in the missing information.

Technique	How is this technique done?	What effects does it achieve?
Spiral wind		
Double wind		
Weave wind		
Root perm		

Perming tools and equipment

◆ The pin-tail comb is useful for directing small pieces of hair onto the curler. The pin-tail comb is narrower than a plastic tail comb so you can guide the wound hair around the wound section to make sure that all the hair has an even tension.

◆ Fibre end papers or wraps are specially made for winding perms. Very few hairdressers would consider winding without them as they ensure control of the hair when it is wound. Fold them neatly over the hair points (never bundle them). The wrap overlaps the hair points and prevents **fish-hooks**. For smaller or shorter sections of hair, half an end wrap is sufficient – a full one would cause unevenness. Other types of tissue may absorb the perm lotion and interfere with processing, and these are best avoided.

Many kinds of rod/curler are suitable for perm winding. Plastic and PVC foam are commonly used. The manufacturer uses different colours to indicate size. The greater the diameter or the fatter the curler, the bigger the wave or curl produced. The smallest curlers are used for short nape hair or for producing tight curls. Most curlers are of smaller diameter at the centre: this enables the thinner, gathered hair points to fill the concave part evenly and neatly as the hair is wound, widening out to the shoulder of the curler as you wind closer to the head.

Basic sectioning

The first part of the process is to divide the hair into sections that will be easy to manage and wind. Done properly, sectioning makes the rest of the process simpler and quicker. But if it is not done well you will have to resection the hair during the perm and this may spoil the overall result.

Cold perm sectioning

1 Following shampooing and towel-drying, comb the hair to remove any tangles.

2 Make sure you have the tools you will need, including a curler to check the section size.

3 Now divide the hair into six sections, as follows, using clips to secure the hair as you work:

◆ divide the hair at or near the crown into a horizontal mesh, no wider than a perm rod, secure the top hair out of the way

◆ divide the back hair into a vertical, yet parallel section down into the nape, secure in sectioning clip(s)

◆ divide the front hair approximately above both mid-eyebrows to create a central parallel section meeting the first division at the crown; secure in sectioning clip(s)

◆ divide the sides into two equally wide sections, one continuing down into the nape and the other terminating above the ears

◆ divide the opposite side likewise to give two equal width sections.

Basic wind

The basic wind is a classic technique for producing an even amount of movement from points to roots all over the head. The basic principles of winding with a firm and even tension (but without stretching the hair) underpins all the other alternative techniques (see also Alternative winding techniques, page 332).

Basic method

1   Divide off a section of hair of a length and thickness to match the curler being used.

2   Comb the hair firmly, directly away from the head. Keep the hair together, so that it doesn't slip.

3   Place the hair points at the centre of the curler. Make sure the hair isn't bunched at one side and loose at the other, or twisted.

4   Hold the hair directly away from the head. If you let the hair slope downwards, the curler won't sit centrally on the base section: hair will overlap and the curler will rest on the skin.

5   Before winding, make sure the curler is at an angle suited to the part of the head against which it will rest when wound.

6   Hold the hair points with the finger and thumb of one hand. The thumb should be uppermost.

7   Direct the hair points round and under the curler. Turn your wrist to achieve this. The aim is to lock the points under the curler and against the main body of hair. If they do not lock, they may become buckled or fish-hooked. Do not turn the thumb too far round or the hair will be pushed away from the curler and will not lock the points.

8   After making the first turn of the curler, pass it to the other hand to make the next turn. The hands need to be in complete control: uncontrolled movement or rocking from side to side may cause the ends to slip, the hair to bunch, or the firmness to slacken.

9   After two or three turns, the points will be securely locked. Wind the curler down to the head in a steady and even pressure, keeping the curler horizontal. (If it slips, wobbles or bunches the curl result will be uneven.)

10 At the end, the curler should be in the centre of the section. If it is not it will need to be rewound.

11 Secure the curler. Do not let the rubber band pinch or press the hair as it may cause damage or cause 'pull burns'.

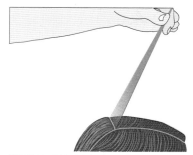
Winding: taking a hair section.

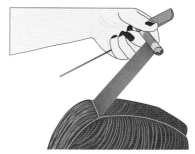
Winding the section onto the curler

Winding: securing the curler

Winding: depth of section

Winding: width of section

Alkaline and acid perming solutions

The systems used to perm hair rely on alkaline- or acid-based solutions. Their benefits and suitability for use are listed here.

Alkaline

◆ Effective on strong, coarse, resistant hair, which is difficult to wave.

◆ Alkaline lotion, up to pH 9, is suitable for different hair textures.

◆ The lotion swells the hair, lifts the cuticle and penetrates to the cortex.

◆ Less winding tension is required.

◆ Suitable for all winding techniques.

◆ The test curl forms a stronger, sharper shape.

◆ The hair must be neutralized.

◆ The higher the pH and the stronger the lotions, the more potential there is for damage.

◆ No additional heat is required.

Acid

◆ Suitable for fine, delicate or porous, and previously chemically processed hair.

◆ Shrinks hair and smoothes the cuticle.

◆ Some require additional heat to be applied: climazone, rollerball, accelerator or infrared dryer.

◆ Make sure that the reagents are activated by mixing the solutions correctly: check with the manufacturer's instructions.

◆ The test curl forms a softer, looser shape – a crisp, snappy test curl could result in over-processing.

◆ Generally needs longer processing times than alkaline perms.

◆ Pre-damp or post-damp – more often post-damp.

Processing and development
Perm lotion may be applied by:

◆ **Pre-damping method** – this is more useful when winding long hair and lotion penetration from the outside of the curler to the inner centre is difficult. *A perm carried out this way must always be done wearing non-latex disposable gloves so can be more difficult or slower to do.*

◆ **Post-damping method** – which is generally more convenient and the time taken in winding does not affect the overall processing time.

Applying the perm lotion
Most modern perming systems come in individually packed perm lotions, ready for application. Others may need to be dispensed from a litre-size bottle to a bowl, before applying to the wound head using cotton wool, a sponge, or a brush.

◆ Underlying hair is often more resistant to perming (e.g. at the nape of the neck), so you could apply lotion to those areas first.

ALWAYS REMEMBER

Pull burns

This happens when perming curlers/rods are wound in and fastened too tightly, causing the follicle to be pulled open and perm solution collecting in the rim – causing a burn.

TOP TIP

Always read the instructions carefully before applying products.

- Keep lotion away from the scalp. Apply it to the hair section, about 12 mm from the roots.

- If post-damping, apply a small amount of the perm lotion to each rod; do not oversaturate as the lotion will flood onto the scalp and will drip on to the client, possibly causing either irritation or burning on the scalp or skin.

- It is better to apply the lotion again once the first application has started to absorb into the hair.

- Do not overload the applicator, and apply the lotion gently. You will be less likely to splash your client.

- If you do splash the skin, quickly rinse the lotion away with water.

Processing time Processing begins as soon as the perm lotion is in contact with the hair. The time needed for processing is critical. Processing time is affected by the hair texture and condition, the salon temperature and whether heat is applied, the size and number of curlers used and the type of winding used.

The perm needs to be checked during the development so that over-processing is avoided. The optimum processing ensures that the curl is maximized whilst there is no detrimental effect to the hair condition.

The figure opposite shows two intersecting lines; both have a time element, but each one has a different resultant effect.

The green line is an increase in curl development over time, whereas the red line is a decrease in hair condition/damage over time. Ideally, where the two lines cross, it will denote the optimum perm processing. At this point a curl development check will show a good 'S' movement without loss of essential hair moisture and subsequent impaired condition.

Hair texture and condition Fine hair processes more quickly than coarse hair and dry hair more quickly than greasy hair. Hair that has been processed previously will perm faster than virgin hair.

Temperature A warm salon cuts down processing time; in a cold salon it will take longer. Even a draught will affect the time required. Usually the heat from the head itself is enough to activate perming systems. Wrap your client's head with a plastic cap to keep in the heat. Don't wrap the hair in towels: these will absorb the lotion, slow down the processing or create an uneven, patchy result.

Some perm lotions require additional heat from computerized accelerators, roller balls or dryers. Do not apply heat unless the manufacturer's instructions tell you to – you might damage both the hair and the scalp. In addition, do not apply heat unless the hair is wrapped; the heat could evaporate the lotion or speed up the processing too much.

Curlers Processing will be quicker with many small sections on small curlers than with large sections on large curlers. (The large sections will also give looser results.)

Winding The type of winding used, and the tension applied, can also affect processing time. Hair wound firmly processes faster than hair wound slackly – in fact, if the winding is too slack it will not process at all. Hair wound too tightly may break close to the scalp. The optimum is a firm winding without tension.

HEALTH & SAFETY

Do not pack curlers with dry cotton wool. This absorbs the perm lotion; it also keeps it in direct contact with the skin, causing irritation.

Remember to use a barrier cream across the hairline. Don't let barrier cream get on the hair, however, as it will prevent the lotion from penetrating into the hair.

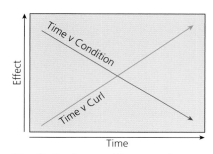

The effects of perm while processing

Development test curl This involves testing the curl during processing. As processing time is so critical, you need to use a timer. You also need to check the perm at intervals to see how it is progressing. If you used the pre-damping technique, check the first and last curlers that you wound. If you applied the lotion after winding, check curlers from the front, sides, crown and nape.

Checking standard or round curlers, bendy curlers

◆ Unwind the hair from a curl former. Is the 'S' shape produced correct for the size of curler used?

◆ If the curl is too loose, rewind the hair and allow more processing time. (However, if the test curl is too loose because the curler was too large, extra processing time will damage the hair and won't make the curl tighter.)

If the curl is correct, stop the processing by rinsing.

Checking other perming systems If you have used another perming system, such as U sticks, where a bend is not so easy to see, you will need to use your judgement by slightly unravelling the former from the hair to assess the degree of movement achieved. With alternative systems, more accents are put upon development timing rather than curl checks (This is also true for straightening processes where no curl is evident.)

HEALTH & SAFETY

Take care not to splash your client's face while rinsing. Even diluted perm lotion can irritate the skin. If perm lotion enters the client's eye, flush out immediately with cold running water. Ensure the water drains downwards away from the face. Seek help from a qualified first aider.

Alternative/creative winding techniques

Basic sectioning and winding techniques create lateral waves and curl that are in the horizontal plane to the head: this is not the only curl placement option.

Spiral (vertical) curl movement

Selecting the size and position of the curls The spiral curl is dependent on the length of the client's hair. If the hair is less than 10 cm long it will be difficult, and perhaps impossible, to form spiral shapes of any size. Hair longer than 10 cm will permit reasonable spiral formations: longer hair will enable fuller, thicker, and longer curls to be shaped.

The position that these spiral curls take and the overall effect they produce must be discussed with your clients: you must ensure that they understand what is being done. Because the handling and maintenance of this type of perm is quite different from traditional methods, it is important that they know how to maintain the effects before they leave the salon.

Spiral curls may be formed all over the head, length permitting, or they may be formed and positioned to make a cascade in the nape. Alternatively, bunches of spiral curls may be positioned asymmetrically. The degree of the spiral curl shape and the effect finally produced is for you and your client to determine jointly at the outset.

Starting the wind The spiral wind can be started at the root end of the hair or from the hair points. If you use a curl former that is of the same thickness overall, the curl you produce will be even throughout. If you use a former that tapers or is concave, the results will be uneven.

For the resultant curl to be even, springy and smooth, your winding must be firm without undue tension, wrapped cleanly over the former and secured without indenting or marking the wound hair.

If you apply uneven tension to your winding, the spiral formation will be inconsistent – the loops and turns will not follow on and there may be gaps in the shape.

Securing the wind When you secure the hair formers, be careful not to cut into the wound hair as this could breakage. Follow the recommendations of the makers of the curl former that you are using. You must also ensure that the formers are secured firmly: if they are loose, the hair may drop or unwind.

Monitoring the perm process Once you have completed your winding and secured the formers, you must monitor the perm process. If the perm lotion is applied to the hair before winding – a technique called pre-damping – the winding must be carried out without delay. Alternatively, the lotion can be applied after the winding is complete – post-damping. Perm processing is always timed from the moment the lotion is applied.

You will need to check the development of the perm process. You can achieve this by taking a test curl. By gently unwinding the hair partway, you can check the development of the 'S' shape. If the shape is loose then further development may be required. If the shape is well formed, begin neutralizing straightaway: the perm is said to have 'taken'. As well as monitoring the timing carefully you must check the following:

◆ Ensure your client is comfortable.

◆ Keep excess lotion off the skin to avoid skin irritation.

◆ Remove damp cotton wool (used to protect the skin) when it has absorbed the lotion or scalp 'burns' will occur.

◆ Continually reassure the client so that she never feels she has been forgotten.

◆ Use a timer which makes an audible noise when the time has elapsed.

Typical problems These are some typical problems when producing spiral curls:

The root end is straight	You can avoid this by securing the former firmly at the root end of the hair. The helical loops of hair will be too loose if they are not firmly wound and in close contact with the former.
Hair flicks out	Ensure that the hair is not twisted when you form the spiral curl. After each turn, the hair should be repositioned. If you allow the hair to twist, an irregular spiral will be formed: this could cause the hair to stick out from the head. It is difficult to remedy this afterwards.
The spiral curl is too loose	Provided that the hair condition permits it, the hair may be reprocessed. You must take special care if you do this as the hair will be far more receptive to the perm lotion and could easily become over-processed.

Perming hair of different lengths

Short lengths of hair (less than 10 cm) are not suitable for permanent spiral curls because it is impossible to form the helical shape on the curl former.

ALWAYS REMEMBER

When the perm has taken, you must stop the perming process quickly. This is an important stage. If you are supervising a junior hairdresser, be sure to double-check with her that she is monitoring whether the perm has taken and is ready to neutralize the hair.

HEALTH & SAFETY

Take care not to splash your client's face while rinsing. Even dilute perm lotion can irritate the skin.

Medium hair lengths (10–15 cm) do allow spiral formations. These are likely to be short and narrow.

Longer lengths of hair (15 cm and longer) are the most suitable. Here there is sufficient hair to produce a variety of full, long, springy shapes. Greater lengths allow the hair to be placed onto the former more easily and a wider variety of curl formers may be used.

Direction and degree of movement

The direction of perm movement is determined by the angle that you wind and position the curlers or rods. If a forward direction of the fringe area is required, the wound curlers must be positioned accordingly.

The degree of perm movement is the 'tightness' or 'looseness' of the wave or curl. This is determined by the:

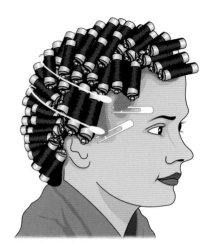

Directional winding

- size of curlers, rods or formers used
- time for which the hair is processed
- amount of tension used
- hair texture and its condition
- perm lotion strength
- type of winding used.

Directional winding

The hair is wound in the direction in which it is to be finally worn. This technique is suitable for enhancing well-cut shapes. The hair can be wound in any direction required and the technique is ideal for shorter hairstyles.

Staggered or brick winding

Staggered winding or brick winding

The wound curlers are placed in a pattern resembling brickwork. By staggering the partings of the curlers, you avoid obvious gaps in the hair. It is suitable for short hairstyles.

Weave winding

Weave winding

In **weave winding** the normal size section is divided into two and then the hair is woven. A large curler is used to wind the upper subsection and a smaller one is used for the lower subsection. This produces two different curl sizes, giving volume without tight curls. Alternatively, one subsection is wound and the other left unwound. With short hair, this produces spiky effects.

Double winding

Double winding consists of winding a section of hair halfway down on a large curler, then placing a smaller curler underneath, and winding both curlers down to the head. This produces a varied curl effect.

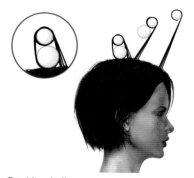

Double winding

Piggyback winding

This is winding using a small and a large curler. The normal size section is wound from the middle onto a large curler, down to the head. The ends are then wound from the points onto a smaller curler, which is placed on top of the large curler. This produces softly waved roots and curly points. Alternatively, this technique can be used to produce root movement only by not winding the point ends.

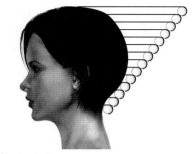

Piggyback winding

Stack winding

This is used where fullness of long hair is required, with little curl movement on top – it is ideal for bobbed hair lengths. The sections are wound close to the head in the lower parts; the upper sections are part wound only at the points. This allows the curlers to stack one upon another.

Stack winding

Other types of perm which give volume support

Root perms A **root perm** creates movement at the lower root end of the hair. The hair is wound at the root ends only: the point ends are left out and not processed. This allows the hair to produce fullness and volume. Reperming must be kept strictly to the regrown root ends.

Body perms The root and middle hair lengths can be processed to give added body to the hair.

Roller perms, semi- or demi-perms These involve the application of a weaker form of perm lotion, which lasts for 6 to 8 weeks. Reprocessing can take place through the hair lengths after this time has elapsed. These are not intended to be permanent, but to produce body fullness.

Other types of perming equipment

Foam rollers and formers

1 Take a small rectangular section of hair.

2 Secure the hair points in an end paper.

3 Wind the hair around the foam roller.

4 Secure the roller in position by bending over the ends.

5 Repeat steps 1–4 to complete the entire head.

Chopsticks

1 Take a small square section of hair and protect it with one or more end papers.

2 Place the hair section through the loop and hold it securely.

3 Separate the chopstick legs and wind in a figure of eight.

4 Secure the end paper on to the chopsticks using a rubber band.

5 Repeat steps 1–4 to complete the entire head.

Chopsticks

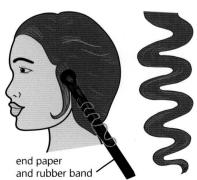

end paper
and rubber band

Position the chopsticks on the head   **The expected curl using chopsticks**

U-stick rods

1 Take a small square section and pull it through the middle of the u-stick.

2 Wind the hair in a figure-of-eight movement around the u-stick.

3 Protect the ends with one or more end papers.

4 Secure the end papers on the u-stick with a rubber band.

5 Repeat steps 1–4 to complete the whole head.

U-stick rods

rubber band over and papers

Position the u-stick
rod on the head

The expected curl
using u-stick rods

Straightening hair

Straightening or **relaxing** processes have always, in one form or another, been applied to hair because people with very tightly curled hair have wanted less curly or smoother looks. Most early relaxing processes were physically based and temporary in their effects, but today's chemical **straighteners** can produce effective and permanent results.

Preparation

In addition to carrying out the normal preparation of your client (see pages 317–318 at the beginning of this chapter) and covering their hair, and ensuring you have all the tools/materials required, you should double-check the points below for this service.

Preparation: Points to consider

✓	The client's needs
✓	Your client's hair type (curly or wavy) and hair texture (fine, medium or coarse)
✓	Whether your client's hair is 'virgin' (chemically untreated) hair; if so, it may be more resistant to relaxing
✓	The condition of the hair if it has previously been chemically treated, e.g. coloured
✓	The hair and scalp for signs of poor condition, sensitivity or disease
✓	Contra-indications are present, then refer to your seniors so a decision can be made
✓	With your client, exactly what is to be done, about how long it will take and what it will cost
✓	That the client is comfortable and that they remain so throughout the service

---

HAIR SCIENCE

Straightening hair: Two-step process

The chemistry of hair relaxing with a thioglycolate derivative is a two-step process, similar to permanent waving. The disulphide bridges in the cysteine links between the keratin chains of the hair are reduced (broken) by the action of the ammonium thioglycolate in the relaxing cream/gel/lotion. This softens the hair, which can then be moulded into its new relaxed shape. This is followed by neutralization, which is an oxidation process (a reaction with oxygen). Cysteine groups pair up again to form cystines, and the disulphide bridges reform in new positions. (See the section on neutralizing on pages 338–341 in this chapter.)

Test the hair Always carry out tests on your client's hair to ensure that it is in a suitable state for relaxing, particularly when dryness, brittleness or breakage of the hair are evident. The following tests are recommended (see pages 323–324 on testing hair):

◆ Test cutting, to check the likely result of the intended process.

◆ Elasticity check, to determine the hair condition.

◆ Porosity check, to determine the rate of absorption.

◆ Testing a strand, to check on process development.

◆ Incompatibility test, to detect the presence of metallic compounds.

Factors affecting product choice and application Thorough product knowledge is essential. You must study the manufacturer's instructions for use before your client arrives or before you attempt to apply the product. (This also applies to your tools and equipment.) You should only decide on the most suitable strength of chemical product after doing the following:

◆ Consultation with your client and making sure you know exactly what your client requires.

◆ Checking to determine whether your client is taking any prescribed medication and if they have any allergies.

◆ Examining the hair and scalp condition.

◆ Finding out the results of the relevant tests.

◆ Checking with a salon senior or specialist (proceed only after agreement is reached).

◆ Ensuring products are in stock to avoid disappointing your client.

◆ Deciding whether the hair is fine, medium, coarse, thick, thin, porous or resistant (coarse hair requires the longest processing time and fine hair the shortest; grease or heavy chemical build-up on hair can block the relaxer product; hair that has been previously lightened, permed, straightened or relaxed can be very receptive and may process very fast).

You can begin the straightening process once you have considered the following factors:

◆ Whether the hair is in a suitable condition for processing (for instance, a rough cuticle could indicate uneven porosity, which would be likely to affect the result).

◆ The salon temperature – a hot salon could speed processing, a cold one could delay it.

◆ The hairstyle required after the hair has been straightened, taking into account your client's head and face shape and hair growth patterns. If the client's hair is to change from very curly to very straight, they may need guidance from you about managing it afterwards and about home maintenance products.

A straightening method

The following method uses ammonium thioglycolate, i.e. perm lotion, to straighten hair, but this should not be used in place of the manufacturer's instructions.

◆ Section the hair into four: centrally, from forehead to nape, and laterally, from ear to ear.

◆ Apply the basing product.

ALWAYS REMEMBER

Most of the methods of curling hair can be used to relax hair. 'Straightening' is the term given to describe curl and wave reduction. As with curling, straightening hair may be temporary or permanent.

◆ Subdivide the nape sections into smaller ones.

◆ Apply the cream, gel or lotion, avoiding the skin. Do not go closer than 12 mm from the scalp.

◆ Comb the hair gently. Use a comb with widely spaced teeth. Some manufacturers advise you to wait until the hair has softened before combing.

◆ Do not continually comb the hair when it is soft. Treat it gently at this stage – it can easily break. Leave the hair as straight as the client requires.

◆ Processing time depends on the product and the hair. Softly curled hair relaxes quickly. Tighter curled hair takes longer. It is safest to monitor continuously throughout the process. Do not exceed the manufacturer's recommended time for processing.

◆ When processing is complete, you may apply neutralizer to rebalance the pH values of the hair and return it to a stable condition.

◆ After final rinsing, condition with antioxidant treatment and style as planned.

Neutralizing

Introduction   The successful outcome of a perm depends on the correct processing and the way the hair is rebalanced during the action of neutralizing.

In this section we will look at:

◆ Principles of neutralizing perms

◆ How neutralizing works

◆ Choosing a neutralizer

◆ Neutralizing techniques

◆ What to do after perming

Rebalancing the hair   Neutralizing is the process of fixing the curl or movement into the hair, while returning the hair back to a balanced chemical state. An industry term, 'neutralizing' could be misleading. In chemistry, a 'neutral' chemical condition is neither acidic nor alkaline (pH 7.0). But during the hairdressing treatment of 'neutralizing', the previously processed hair is returned to the skin's healthy, slightly acidic natural state of pH 4.5–5.5. Rebalancing the pH value of the hair is essential for maintaining hair in good condition; if the hair is not rebalanced the hair will be dry, porous and the perm will be very difficult to manage afterwards.

How neutralizing works   As described earlier, perm lotion acts on the keratin in the hair. The strongest bonds between the polypeptides are the disulphide bonds. Perm lotion breaks some of these, allowing the keratin to take up a new shape. This is how new curls can form.

Neutralizing makes new disulphide bonds. If you did not neutralize the hair it would be weak and likely to break, and the new curls would soon fall out. Neutralizing is an oxidation process – a process that uses oxidizing agents such as hydrogen peroxide H_2O_2 to release oxygen to reform the disulphide bonds.

Chemical reaction taking place H_2O_2 (hydrogen peroxide) loses an oxygen atom, which is used in the chemical process to reform the disulphide bonds in new locations on the polypeptide chains, and is subsequently reduced to H_2O (natural water).

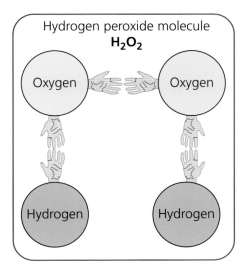
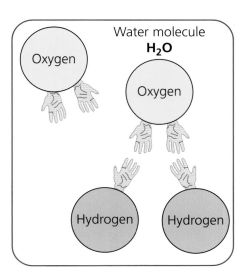

Choosing a neutralizer Manufacturers of perm lotions usually produce matching neutralizers. These are designed to work together. Always use the neutralizer that matches the perm lotion you have used. As most perms are individually packed, you will find a perm lotion and its matched neutralizer in the box.

A neutralizer may be supplied as an emulsion cream, foam or liquid. Always follow the manufacturer's instructions. Some can be applied directly from the container, others are applied with a sponge or a brush.

Neutralizing technique Neutralizing follows directly on from perming. Imagine that you have shampooed, dried and wound the hair. The hair is now perming, and you are timing the perm carefully and making tests to check whether it is complete. You will also be reassuring the client that they have not been forgotten. As soon as the perm is finished, you need to be ready to stop the process immediately.

Preparation

1   Prepare the materials you will need.

2   Make sure there is a back basin free. (This makes it easier for you to keep chemicals away from the client's eyes.)

First rinsing

1   As soon as the perm is complete, move your client immediately to the backwash. Make sure they are comfortable.

2   Carefully remove the cap. The hair is in a soft and weak stage at this point; do not put unnecessary tension on it. Leave the curlers in place.

3   Run the water. You need an even supply of warm water. The water must be neither hot nor cold as this will be uncomfortable for the client. Hot water will also irritate the scalp and could burn. Check the pressure and temperature against the back of your hand. Remember that your client's head may be sensitive after the perming process.

4   Rinse the hair thoroughly with the warm water. This may take about 5 minutes or longer if the hair is long. It is this rinsing that stops the perm process – until you rinse away the lotion, the hair will still be processing. Direct the water away from the eyes and the face. Make sure you rinse *all* the hair, including the nape curlers. If a curler slips out, gently wind the hair back onto it immediately.

Applying neutralizer

1   Make sure your client is in a comfortable sitting position.

2   Blot the hair thoroughly using a towel (you may need more than one). It may help if you pack the curlers with cotton wool.

3   When no surplus water remains, apply the neutralizer. Follow the manufacturer's instructions. These may tell you to pour the neutralizer through the hair, or apply it with a brush or sponge, or use the spiked applicator bottle. Some foam neutralizers need to be pushed briskly into the hair. Make sure that neutralizer comes into contact with all of the hair on the curlers.

4   When all the hair has been covered, time the process according to the instructions. The usual time is 5 to 10 minutes. You may wrap the hair in a towel or leave it open to the air – follow the manufacturer's instructions.

5   Carefully remove the curlers. Do not pull or stretch the hair. It may still be soft, especially towards the ends, and you do not want to disturb the curl formation.

6   Apply the neutralizer to the hair again, covering all the hair. Arrange the hair so that the neutralizer does not run over the face. Leave for the time recommended, perhaps another 5 to 10 minutes.

Second rinsing

1 Run the water, again checking temperature and pressure.

2 Rinse the hair thoroughly to remove the neutralizer.

3 You can now treat the hair with an after-perm (anti-oxidant) or conditioner. Use the one recommended by the manufacturer of the perm and neutralizer, to be sure that the chemicals are compatible.

Perm aids or conditioner and balanced conditioners (anti-oxidants) help neutralize the effect of the chemical process by helping to restore the pH balance of the hair to pH 4.4–5.5 and smooth down the hair cuticle improving the hair's look, feel, comb-ability, and handling.

Step-by-step: Neutralizing

1 First rinse; rinse each curler to remove perming lotion.

2 Towel dry; blot dry each curler to remove excess moisture.

3 First application of neutralizer; carefully apply neutralizer to each curler – leave for 5 minutes.

4 Remove curlers; carefully unwind curlers and remove end papers.

5 Second application of neutralizer; apply more neutralizer to ends of hair and leave for 3–5 minutes before final rinsing with warm water.

At the end of the neutralizing process, you will have returned the hair to a normal, stable state.

◆ The reduction and oxidation processes will have been completed.

◆ The hair will now be slightly weaker – fewer bonds will have formed than were broken by the perm.

◆ Record any hair or perm faults on the client's record card. Correct faults as appropriate.

◆ **Under neutralizing** – not leaving neutralizer on for long enough – results in slack curls or waves.

◆ **Over-oxidizing** – leaving the neutralizer on too long or using oxidants that are too strong – results in weak hair and poor curl.

The hair should be ready for shaping, blow-drying or setting.

After the perm

Check the results of perming.

◆ Has the scalp been irritated by the perm lotion?

◆ Is the hair in good condition?

◆ Is the curl even?

Dry the hair into style.

◆ Depending on the effect you want, you may now use finger-drying, hood drying or blow-drying.

◆ Treat the hair gently as the hair may take a few washes to settle in. If you handle it too firmly the perm may relax again.

ALWAYS REMEMBER

When applying a conditioner, apply to the palms of the hands first and gently work the conditioner through the hair. Do not massage the scalp, pull or comb the hair as it may soften the newly formed curl.

Advise the client on how to manage the perm at home.

◆ The hair should not be shampooed for a day or two.

◆ The manufacturer of the perm lotion may have supplied information to be passed to the client.

◆ Discuss general hair care with your client.

Clean all tools thoroughly so that they are ready for the next client.

Complete the client's record card.

◆ Note details of the type of perm, the strength of the lotion, the processing time, the curler sizes and the winding technique.

◆ Record any problems you have had. This information will be useful if the hair is permed again.

Perming problems and solutions

	Possible cause	Immediate action	Future action
Hair/scalp damage Breakage	Too much tension or bands on curlers too tight. Hair over-processed – chemicals far too strong	Apply restructurant or deep-action conditioner to remainder of hair.	Use less tension. Review choice of lotion, timing, etc.
Pull burn	Perm lotion allowed to enter follicle. Tension on hair excessive. Poor rinsing of surplus perm lotion	Apply soothing moisturizer to affected area. If condition serious, refer to doctor.	Use less tension. Take smaller meshes.
Sore hairline, skin irritation	Chemicals in contact with skin. Poor scalp ventilation	Consult regarding allergies, then apply soothing moisturizer to affected area. If condition serious refer to doctor.	Curlers to rest on hair not skin. Keep lotion away from scalp. Renew cotton wool after damping.
Straight frizz	Lotion too strong for hair. Excessive winding tension. Hair over-processed	Cut ends to reduce frizz. Apply restructurant or penetrating conditioner.	Ensure appropriate lotion is used in future. Wind with less tension. Time carefully.
Perm result/effect Too curly	Curlers too small. Lotion too strong	If hair condition allows, reduce curl amount by relaxing.	Ensure appropriate curlers and lotion are used.
No result	Lotion too weak or not enough used. Curlers too large. Poor neutralizing. Hair under-processed	If hair condition allows, re-perm hair with suitable lotion.*	Use appropriate lotion and rods. Process perm and neutralizer in line with manufacturer's instructions.
Fish-hooks	Hair points not wrapped properly. No end papers	Remove ends by cutting.	Check points of hair are wrapped correctly. Use end papers.
Perm weakens	Poor neutralizing. Hair stretched excessively while drying	If hair condition allows, re-perm hair.*	Check method and timing of neutralizer. Do not overstretch while drying hair.
Good result when wet, poor when dry	Hair stretched while drying. Ineffective neutralizing. Over-processed	If hair condition good, reperm.* Apply conditioning agents to moisturize hair.	Check method and timing of neutralizer. Avoid stretching while drying.

	Possible cause	Immediate action	Future action
Uneven curl	Uneven winding technique. Uneven tension. Uneven lotion application. Ineffective neutralizing	If hair condition allows, re-perm affected areas.*	Check wound curlers before applying perm lotion or neutralizer.
Straight pieces	Lotion not applied evenly. Rods too large	If hair condition allows, re-perm affected area.**	Ensure even lotion application.

* Do not re-perm the hair unless its condition is suitable. For example, you should not re-perm if the hair is over-processed. Conditioning treatments and/or cutting may help. Discuss the problem with a senior or your trainer.

** Before attempting to correct this fault, make sure that the hair is not over-processed. Dampen the hair to see how much perm there is.

Not all hairdressers use perming, so focusing on perming could be a career opportunity. For example you could become:

◆ an in-salon specialist

◆ a technician for a manufacturing company where you would teach other hairdressers how to use their products

◆ a trainer–assessor either in a salon or in a further education establishment.

◆ Focus on your dream – do whatever it takes to achieve it and don't be side-tracked!

Anne Miller

Provide aftercare advice

Clients need your help to look after their hair at home. You must give them the right advice so that they can make the most of their new perm and style between visits.

You should advise them on what sorts of products that they could use that would benefit the condition and manageability of their hair. You should also make a point of telling them what they should avoid, as some products will work against the new perm.

Explain the benefits of maintaining their hair in good condition: hair in good condition is easier to manage; it lasts longer, looks better and is noticeable to everyone else as well.

Home and aftercare checklist

ALWAYS REMEMBER

Remember to explain to clients how a new perm needs particular care in the way that it is handled and that stretching during styling will weaken the result and could even cause the perm to fail prematurely.

✓	Talk through the permed or straightened effect with your client as you work; tell them how they can maintain their hair by recreating the look and maintaining the condition of their hair.
✓	Explain how delicate the hair is after perming and how excessive stretching during styling will reduce the life of the perm.
✓	Explain how lifestyle factors can affect their hair colour and ways that they can combat these factors.
✓	Show and recommend the products/equipment that you use so that the client gets the right things to enable them to get the same effects.
✓	Tell the client how long the effect can be expected to last and when they need to return for re-perming or straightening.
✓	Warn the client about the products that will have a detrimental effect on their hair by telling them what these products would do.
✓	Warn the client about how incorrect handling or heated styling may reduce the length of time that the new effect will last.

REVISION QUESTIONS

Q1. A development test _____ will identify when optimum movement is achieved.   Fill in the blank

Q2. Cold wave perms are usually post-damped.   True or false?

Q3. Which of the following factors are likely to be affected by perming?   Multi-selection

Elasticity   ☐   1

Natural colour   ☐   2

Thickness   ☐   3

Texture   ☐   4

Porosity   ☐   5

Abundance   ☐   6

Q4. All perm solutions are alkaline.   True or false?

Q5. Which of the following chemical bonds are permanently rearranged during perming?   Multi-choice

Salt bonds   ○   a

Hydrogen bonds   ○   b

Disulphide bonds   ○   c

Oxygen bonds   ○   d

Q6. Time and temperature have a direct impact upon perm development.   True or false?

Q7. Which of the following tests are *not* applicable to perming?   Multi-selection

Strand test   ☐   1

Incompatibility test   ☐   2

Peroxide test   ☐   3

Porosity test   ☐   4

Elasticity test   ☐   5

Skin or allergy alert test   ☐   6

Q8. The rearrangement of chemical bonds take place within the _____.   Fill in the blank

Q9. The chemical compound responsible for modifying the hair's structure during perming is?   Multi-choice

Hydrogen peroxide   ○   a

Ammonium hydroxide   ○   b

Ammonium thioglycolate   ○   c

Sodium perborate   ○   d

Q10. Smaller perming rods produce tighter curl effects.   True or false?

Notes

14 Develop Your Creativity

LEARNING OBJECTIVES

◆ Be able to research ideas in relation to specific themes.

◆ Be able to plan and create a mood board.

◆ Present your ideas and themes to others.

◆ Be able to re-create your ideas as finished hair effects.

KEY TERMS

artistic interpretation	emphasis/accent	mood boards
asymmetrical	geometric	movement
avant-garde	harmonizing	proportion
balance	harmony	storyboard
contrasting	maquette	symmetrical

UNIT TOPIC

Develop and enhance your creative skills

INTRODUCTION

As hair stylists for men or women, we have to develop a number of different skills, but all focus on the needs of the client.

We need to have:

◆ an appreciation of shape, dimension, image, colour and textures

◆ an understanding of balance or imbalance

◆ the skills to express our creativity by moulding, shaping and forming our client's hair

◆ the ability to explain our ideas and interpretations to others through mood boards and/or storyboards or other visual media.

Hair styling involves skills that are aesthetic, artistic, sometimes scientific but always practical. The hair stylist who can bring these skills together and can communicate effectively with their clients will always be in demand.

This chapter will help you to develop your own creativity for use with your clients, in competitions, photo shoots and public demonstrations.

Develop your creativity

PRACTICAL SKILLS

Learn the basic principles of design

Learn how to create images based on themes

Learn how to research source information and ideas

Create visual representations of ideas, like mood or storyboards

Learn how to present ideas to others

UNDERPINNING KNOWLEDGE

Understand how shape and form affect the aspects of design

Understand how colours and textures can be used in design

Understand the importance of research in creativity and design

Know the basic principles of design

Know how to communicate effectively and professionally

Plan and design a range of images

Hair creativity

'Hair creativity' is a term associated with **artistic interpretation**, **geometric** understanding and design.

This chapter aims to capture the essential skills of artistic perception, practical ability and sound knowledge, helping you to create your own template for success.

Where do I get my creative ideas?

Inspiration can come from almost anywhere, at any time. Films, TV, magazines, videos, the Internet, even from someone you see on the street – anything, anywhere can trigger

INDUSTRY ROLE MODEL

MALACHY WEATHERALL Manager/Creative Director, Academy Salon, Brunswick Street, Belfast

Being a hairdresser for the last 16 years, I've been very lucky to be involved with many aspects of our wonderful industry. The highlights include: being a finalist in the British Hairdressing Awards; hairstylist for Belfast Fashion Week; working with Habia, doing stage presentations for the past 12 years; working my way up to creative director and managing the step-by-step photo collections, which arrive every six months.

the creative process. One of the biggest sources for inspiration from popular culture is music, because it touches everyone in different ways and inspires many different feelings. Some people find inspiration in nature, such as the sea, and the shapes, colours, patterns and textures of plants, animals, as well as minerals, are also a great source of visual ideas. At times, you may find yourself looking to the past for inspiration. A hairstyle from an earlier era might inspire you to reinvent it in a way that works for today.

Modern inspiration in fashion often starts on the streets and in the clubs. Hair design usually follows the fashion trends and helps to complete the total look.

Once inspired, you will need to decide which tools, such as scissors, razors, clippers, straighteners, tongs, you will need. Then you will have to think about which techniques you will use to create the effects.

It is always a good idea when working out a design to first practise on your modelling block. As you develop or practise a technique, there is always the chance that your original concept will turn into something entirely different. There are no failures if the experience is a lesson learned. If you are open to change, the creative process will be exciting and satisfying.

As a creative stylist, you will need to develop a visual understanding of which hairstyles work best on different face shapes and body types. It takes time and experience to train your eye to recognize the best design elements.

How is an image created?

Different people understand 'image' in different ways. It can be defined as 'representation, likeness, semblance, form, appearance, configuration and structure'.

You need to have the opportunity to experiment and *play* with creative effects in order to improve. You have to learn by experience which means that sometimes you will get things wrong. So if you need to practise, get a modelling block out. You may have many more disasters than successes, but your block is not going to complain or be upset.

Over time, your practice will pay off and you will find techniques that work and things that do not. The difference between you and a very experienced stylist is that they can visualize the final effect of what they are trying to create and then work out the process to achieve that final effect. You need to understand the basic elements of design, and how they can be put together to create a finished image.

Make a maquette The term maquette (sounds like 'tennis racquet' starting with an 'm') is from the French word, meaning to make a scale model. This is an ideal way of seeing how a hair accessory or hair adornment, such as a headdress or tiara, might work at the planning stage. Rather than going to the expense of buying expensive feathers, jewels, metals, etc., why not try to make a scaled-down model? This will help you to:

◆ accessorize a smaller area and get an idea of the final object

◆ assess the balance, shape, weight and proportions of the item

◆ see how well different colours or themes work

◆ work out the costs of making one, or several, full-size replicas for your event.

Scale modelling your ideas is almost like trialling your hair plans on a training block and may save you a lot of wasted energy, time and money.

Pictures and images

For an example of how you and the client might see different things, see the section on using visual aids in consultation in CHAPTER 1 Consultation and advice, pp. 35–36.

A picture can paint a thousand words, and this happens every day within the salon. Good photography is an immensely important visual aid and another form of language that hairdressers understand very well. Different people get different messages from the same picture. Your trained eye sees things quite differently from your client.

> When dealing with our clients during the consultation process it is important to be honest about what can be achieved on that particular appointment and how long it will take to achieve their goal (colour).
>
> *Malachy Weatherall*

This is the fundamental part of consultation and probably the hardest thing to do. However, we use all the aspects of accessories, make-up and clothes to enhance the total look that stimulates the audience.

ACTIVITY

Building a portfolio of ideas and themes

Use the information covered within this chapter to create your own portfolio of work. How you choose to display this information depends upon your skill and artistic abilities. Whether you want to collect examples of work you find, or sketches from ideas, or pictures of your own work, it is all up to you.

The basics of good hair design

To begin to understand the creative process involved in hairstyling, it is essential to learn the five basic elements of three-dimensional design. These elements are:

- ◆ line
- ◆ form
- ◆ space
- ◆ texture
- ◆ colour.

Line

Line defines form and space. The presence of one nearly always means that the other two are involved. Lines create the shape, design, and movement of a hairstyle. The eye naturally follows the lines in a design. They can be straight or curved. There are four basic types of lines:

Horizontal lines Create width in hair design. They extend in the same direction and create different levels, or elevations, as well as baselines for frames (see photos).

Vertical lines Create length, height or depth in hair design. They make a hairstyle appear longer and narrower as the eye follows the lines up and down (see photos).

Diagonal lines Are positioned between horizontal and vertical lines. They are used to emphasize and accentuate, or minimize and diminish facial features. They do this by diverting attention from one area or detail and moving the eye-line to another. Therefore, diagonal lines can also be used to create interest in hair design (see photos).

Curved lines Curved lines which move in a circular or semi-circular direction, soften a design. They can be large or small, a full circle or just part of a circle. Curved lines may move in a clockwise or counter-clockwise direction. They can be placed horizontally, vertically or diagonally. Curved lines repeating in opposite directions create a wave (see photos).

Designing with lines The type of line, direction or combination you choose defines a hairstyle.

Single lines An example of this is the one-length hairstyle. These hairstyles produce classic effects and are best for clients requiring the lowest maintenance when styling their hair.

Parallel lines Repeating lines in a hairstyle. They can be straight or curved. The repetition of lines creates more interest in the design. A finger wave is an example of a style using a series of curved, parallel lines.

Contrasting lines Horizontal and vertical lines that meet at a 90-degree angle. These lines create a dramatic hard edge. Contrasting lines in a design are usually for confident clients who are able to carry off a strong look.

Transitional lines Usually curved lines that are used to blend and soften horizontal or vertical lines.

Convergent or divergent lines Lines with a definite momentum, with forward or backward movement.

ACTIVITY

Build a portfolio of models

Start to build a model portfolio for your salon. You can approach suitable clients directly, finding out whether they would be able to take part or be interested in competitions, promotional work or photographic sessions. Remember to keep all details on file for future reference and don't forget to get a model release/disclaimer signed if you want to use their hair in any published media, i.e. newspapers, magazines and the Internet.

Shape and form

Form is the mass or general outline of a hairstyle. It is three-dimensional and has length, width and depth. Form or shape creates mass, which can also be called volume. The two-dimensional silhouette is usually the part of the overall design that a client will respond to first. This can be in the negative, where the outline produces a dark silhouette against a light background, or the reverse – a positive, as shown opposite. Generally, simple forms are best to use and are more pleasing to the eye. The hair form should be in proportion to the shape of the head and face, the length and width of the neck, and the shoulder line.

Space

Space is the area surrounding the form or the area the hairstyle occupies. We are more aware of the (positive) form than the (negative) spaces. In hair design, with every movement the relationship of the form and space changes. A hairstylist must keep every angle in mind – not only of the shapes being created but of the spaces surrounding the shapes as well. The space may contain curls, curves, waves, straight hair or any combination.

Textural content

Textural content refers to wave patterns that must be taken into consideration when designing a style for your client. All hair has a natural wave pattern – straight, wavy, curly or extremely curly. For example, straight hair reflects light better than other wave patterns, and we see that as shine. It is also worth mentioning that straight hair produces the most shine when it falls as a single sheet and is cut to a single length. Wavy hair can be combed into waves that create horizontal lines. Curly hair and extremely curly hair are not able to reflect much light and can sometimes be coarse to the touch. Curly hair creates a larger form than straight or wavy hair does.

> **TOP TIP**
>
> Texture in design concepts has a different meaning to texture in reference to the thickness of individual hairs.

Creating texture with styling tools Texture can be created temporarily with the use of heat and/or wet styling techniques. Hair straighteners or hot rollers can be used to create a wave or curl. Curly hair can be straightened with a blow-dryer or straightening irons.

Crimpers can be used to create interesting and unusual wave patterns like zig-zags. Hair can also be wet-set with rollers or pin curls to create curls and waves. Finger waves, braids and plaits are another way of creating temporary textured pattern changes.

Creating texture with chemicals Wave pattern changes can be permanent through the chemical services of perming and relaxing. They last until the new growth of hair is long enough to alter the design. Curly hair can be made straighter with relaxers and straight hair can be curled with permanent waves.

Tips for designing with wave patterns

◆ When using many wave pattern combinations together, you create a look that is very busy. This is fine for the client who wants to achieve a multi-textured look, but may be less appropriate for more, classic, professional effects.

◆ Smooth wave patterns accent the face and are particularly useful when you want to narrow a rounder head shape.

◆ Curly wave patterns take attention away from the face and can be used to soften square or rectangular features.

Hair colour

Hair colour plays an important role in hair design, both visually and psychologically. It can be used to make all or part of the design appear larger or smaller. Hair colour can help define texture and line, and it can tie design elements together.

Warm colours/tones	Cool colours/tones	
Reds (if orange based or tomato red)	Reds (if blue or violet based)	
Oranges	Blues	
Yellows (if golden based)	Pinks or beige	
Browns	Greys (ashen-based colours)	
	Greens	

Dimension with colour Light colours and warm colours are advancing and demand to be noticed, or create the illusion of volume. Dark and cool colours recede or move backwards, towards the head, creating the illusion of less volume. The illusion of dimension, or depth, is created when colours that are lighter alternate with those that are darker.

You should avoid mixing warm and cool colours within the same hair effect, as they are discordant with what the eye *expects to see* and accepts as normal. However, if you want to create, strong contrasts and you have the model/client with the confidence to wear it, you can produce some very striking effects!

Lines and linear effects with colours Because the eye is always drawn naturally to the lightest colour, you can use a light colour to draw a line in a hairstyle in the direction you want the eye to travel. A single line of colour, or a series of repeated lines of colour, can create a bold, dramatic accent.

To give another explanation of the linear effects of light lines within in hairstyle, the most popular form of colouring over the past 40 years is highlighting. So if you now think of which colour options for highlighting have been the most popular, then lightened highlights always come 'top'.

Colour selection The choice of colour for clients is another important aspect to consider. There are two ways of providing colour:

◆ **harmonizing**

◆ **contrasting** colour.

Most clients want a hair colour with harmonizing tones as these are colours that are compatible and complimentary with the skin tone of the client. For example, if a client has a gold tone to her skin, warm hair colours are more flattering than cool hair colours. Similarly, if a client has an olive skin tone, then cooler colours will be more suitable. For a more conservative or natural look when using two or more colours, choose colours with similar tones within two levels of each other.

Some clients will always want to follow fashion, whether it suits them or not. Contrasting effects occur when one colour is placed against another and the result does not compliment the client, or the colours create stark effects. It does not mean that they should never be used, but it does require a bit more time during consultation, so that these effects can be explained and you can find out if the client has the confidence to wear them.

When using high contrast colours in most salon situations, you should use one colour sparingly. A strong contrast can create an attention-grabbing look and should only be used on clients who have the confidence to wear a *bold* look.

As a hairdresser for over 16 years, I've learnt that the most important thing you can do for your client is to listen intently. Give them exactly what they want and then add advice (but only after listening).

Malachy Weatherall

Style design

In day-to-day salon work, hair stylists seldom have the opportunity to let their creativity have a free rein with their clients' hair. So having the opportunity to do something really creative is quite special. The opportunities tend to come around in certain situations:

1 Your client wants a total restyle and a new colour effect.

2 Through hair demonstrations or promotional activities.

3 When you take part in hair competitions.

What aspects of a hairstyle create a great image?

Have you ever studied great hair images closely? What is it about those images that gives them impact or appeal? Now, what is it about 'every day' commercial work that makes it seem routine? You might say that it is make-up or clothes – or the lighting – or say that it's something to do with the client. All these factors contribute to the final effect, but we haven't yet been looking at the hair.

The hair design elements

There are five elements of hair design:

◆ **proportion**

◆ **balance**

◆ **movement**

◆ **emphasis/accent**

◆ **harmony**.

The more you understand about these design elements, the more confident you will feel about creating styles that please your clients and the people who comment about those styles to your clients.

Proportion Proportion is the comparative relationship of one thing to another. So a person with a very small chin and a very wide forehead might be said to have a head shape that is not in proportion. A well-chosen hairstyle could create the illusion of better proportion for such a client.

Considering your client's body proportions is an essential part of consultation. So the design of a hairstyle must take into account the client's body shape and size. Challenges in body proportion become more obvious if the hair form is too small or too large. When choosing a style for a woman with large hips or broad shoulders, for instance, you would normally create a style with more volume. But the same large hairstyle would appear out of proportion on a petite woman. A general guide for classic proportion is that the hair should not be wider than the centre of the shoulders, regardless of the body structure.

Balance You can establish balance with equal or appropriate proportions to create symmetry. In hairstyling, it can be the proportion of height to width. But balance can be symmetrical or asymmetrical. To measure symmetry, divide the face into four equal parts. The lines cross at the central axis, the reference point for judging the balance of the hair design. You can then decide if the hairstyle looks pleasing to the eye and is in correct balance.

Symmetrical balance occurs when an imaginary line is drawn through the centre of the face and the two resulting halves form a mirror image of one another. Both sides of the hairstyle are the same distance from the centre, the same length and have the same volume when viewed from the front.

Balance or imbalance in art The term symmetry, or symmetrical, refers to things being even and asymmetry, or asymmetrical, refers to things being uneven. Now consider the following:

◆ Does balance relate to symmetry?

◆ Does imbalance relate to asymmetry?

If balance and symmetry convey harmony, does imbalance and asymmetry portray discord?

The vast majority of commercial work undertaken in salons leads us, as hairdressers, to ensure that both sides of the haircut are of an even length, that weight is proportionally distributed and that degrees of curl or straightness are maintained throughout the hairstyle.

Most clients require us to produce a finished effect that may have originated from a picture, which is then modified to suit the client and becomes the perfect example of symmetry. This may be a nice style, but does it still have the same impact as the original? What often happens with the best fashion images is that we automatically convert a high impact, dynamic, **asymmetrical** image into a recessive, passive **symmetrical** style.

The best images, which are often asymmetrical in appearance, contain the visual excellence in artistic or aesthetic balance.

The following illustrations try to explain this key principle or concept in art.

In the first diagram we have a 'see-saw' with equal masses on each of the opposing ends. As these masses are of equal weight, we arrive at equilibrium: an apparent balance which is symmetrical and harmonized.

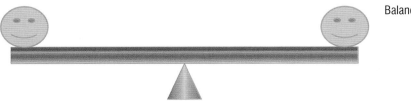

Balance-Symmetry

In the second diagram we have a 'see-saw' which has a weight on one end and another heavier weight counterbalanced across the pivotal point. We now arrive at another form of equilibrium or artistic balance that is asymmetrical. However, this image has far more impact on the eye because the image is *advancing*.

Imbalance-Asymmetry

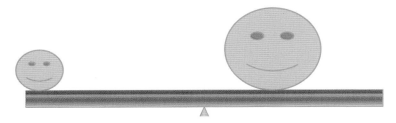

This key principle of artistic balance can be used in many ways in your hairdressing too. Not only in the ways of apportioning balance or volume but in colours as well. When you put slices of colour into hair, never put the same amount on either side of the parting, use different amounts on either side. Second, if you do use slices as a colour effect, *always* use odd numbers either side instead of even ones. So try 1 and 3 or 3 and 5 but not 2 and 4 or 4 and 6.

Movement The direction that the hairs take, individually and collectively, affects the overall style. The position and line of the hair gives direction to the style. The variation of this line produces direction within the style. The more varied the line direction, the more movement will be seen, showing as texture, wave or curl.

A fluid or flowing line gives a softer effect, whereas broken lines of movement create a harder visual impact. The more breaks within the style continuity, the greater the contrasts produced.

As far as movement is concerned within hair design, texture also plays a part in the visual effect of the hair. Texture is the term given to the way an object feels: rough or smooth, fine or coarse. In hairdressing, we can *see* the textural effects and we can also *feel* the textural aspects.

Emphasis/accent Emphasis creates focus and, in a design, it is emphasis that draws the eye first, before it continues to the rest of the design. A hairstyle may be well balanced, with movement and harmony, and yet still be boring.

Emphasis or accent within a hairstyle can be created by the following:

◆ wave patterns

◆ colour

◆ change in form/shape

◆ ornamentation.

Choose an area of the head or face that you want to emphasize. Keep the design simple so that it is easy for the eye to follow from the point of emphasis through to the rest of the style. You can have multiple points of emphasis as long as you do not use too many and as long as they are decreasing in size and importance. Remember, *less is often more*.

Harmony Harmony is the creation of unity in a design and is the most important of the art principles. Harmony holds all the elements of the design together. When a hairstyle is harmonious, it has the following elements:

◆ an overall form of interesting shapes

◆ coordinated colours and textures

◆ balance and movement that together strengthen the design.

A harmonious design is never too busy, and it is in proportion to the client's facial and body structure. A successful harmonious design includes an area of emphasis from which the eyes move to the rest of the style.

The principles of design may be used in modern hairstyling and make-up to guide you as you decide how to accentuate a client's best features and to minimize features that do not add to the person's appearance. Every hairstyle you create for every client should be properly proportioned to body type and correctly balanced to the person's head and facial features. The hairstyle should attractively frame the client's face.

An artistic and suitable hairstyle will take into account physical characteristics such as the following:

◆ the shape of the head, including the front view (face shape), profile and back view

◆ the features (perfect as well as imperfect features)

◆ body posture.

TOP TIP

Harmony in a hairstyle refers to a form of balance.

> Remember that all of the 'celeb' hairdressers started in the salon, training and picking up tips and tricks from the stylists within the salon they worked at. Getting involved with your salon's creative team is a great way to develop your skills and will help you build a great clientele. Learn how to separate your time from clients and your creative work, so you equally give 100% to both.

Malachy Weatherall

Produce a range of creative images

Demonstrations, photo shoots and competitions

For more information about internal promotions, see **CHAPTER 17** Promotional Activities.

When you understand the aspects and elements of good design, you will want to apply them and show off your skills in the best possible ways. Public demonstration, hair competition and photo shoots are the *normal avenues* for this creative outlet.

Public demonstrations Hairdressing demonstrations form a very important part in training and salon promotion. Public demonstration provides the opportunity to generate new sales through increasing the numbers of clients within the salon and helping to promote the products sold and used within the salon.

> " As a manager of a successful salon, I have a checklist of things that I am looking for in a person I might employ. Firstly – how they are turned out. Are they smartly dressed, do they have their own individual style and are they approachable? Secondly – are they ambitious and enthusiastic about our industry? What motivates them? Thirdly – do they have any experience working in our industry and where?
>
> *Malachy Weatherall*

Photography and photo sessions

The power of a good photograph is undeniable. It instantly says more about your work and the image you want to project than any *advertorial* will. However, while fun, photo-sessions are not easy. They can be time-consuming, expensive – and sometimes disappointing if not properly coordinated.

Think about themes Define your look, questions to consider:

- ◆ Will the look be classic, fashionable, **avant-garde** or themed?
- ◆ Will the finished effect be the result of a service, say a colouring or cutting, or created by specific products?
- ◆ Will the look have more impact in black and white or colour?
- ◆ What clothes and accessories are best suited to the look?
- ◆ What image or effect are you trying to create – natural, classic, dramatic or romantic?
- ◆ Have you created a mood board to provide a visual representation of your intended effects?

Once you have decided on the look you want you can start to create your photographic team.

Your model Picking a suitable model can be a tricky task. A common mistake is to choose a pretty girl with unsuitable hair, or vice-versa; ideally, she should have a combination of both. Remember that a conventionally pretty face isn't always photogenic, so study each prospective model carefully.

◆ Look for regular features and bright, clear eyes.

Avoid prominent chins and noses, over-full lips or dark circles under the eyes. The skin should be clear (even the most skilful of make-up artists will not be able to disguise very obvious blemishes), and she should have a long, slim, unlined neck and a good profile to give the photographer maximum scope.

◆ Make sure that the hair suits the type of work you plan to do.

All models have limitations on what they will let you do in relation to cut, colour or perm. So your choice needs to be the right length, shade, texture and style. The model must also have the right features to fit your look – a sweet face is no good if you want an 'aggressive', moody or 'edgy' image.

The photographer If you can get the help of a professional, or someone who is learning to be a professional, choose someone who specializes in hair, beauty or fashion photography. All photographers, even if they are students will have a portfolio, so have a look at the effects that they like to create to check their ideas are in tune with yours.

Make-up Good make-up is vital for a successful shoot, whereas bad make-up will ruin your work. If you know someone in this area of work, or a keen amateur, use them, but do not expect the impossible – a make-up artist, however good, cannot completely change a model's face.

Clothes and accessories The clothes that you dress your model in are going to have a dramatic impact on the overall success of the final images. Do not forget, if you choose current fashions they will get out of date very quickly. If you choose historical themes or fantasy ideas they can take away the impact of what you are trying to focus upon, or miss your target audience completely.

What type of clothes work best? Obviously, this depends on the image you want to achieve and whether you are working with a professional stylist. If you do not want your shots to date too quickly then use neutral fashions. Necklines should be simple and jewellery effective. However, if in doubt, leave it out.

Have a plan Think about the designs and put together a 'mood board' by cutting out images you like from magazines. Once you have decided on the styles, work out how you are going to achieve them. By creating a mood board you will be able to convey a visual representation to everyone else involved.

Draw up a list of the equipment and products you will need, and check them off when packing your session tool kit. The general rule is to take everything – and then add anything else that might come in handy!

On the day Have a clear idea of the looks you want to create but have in mind several alternatives as back-up. Pay attention to detail and make sure you see the digital stills, perhaps as output to a laptop's screen before changing and moving on to the next look. Picking up on faults is not always easy. Those to look out for are gaps in the style; stray hairs on clothes/face; rumpled clothes; pins showing or too much product/make-up. The larger the image is shown on the screen, the more obvious the defects will be.

TOP TIP

Photographic students are always keen to get involved in interesting work, and fashion photography students do not get many options during their training to put their skills into action!

Be decisive and do not settle for second best. If you are not completely happy with an image, say so nicely and make the necessary changes. Keep the backgrounds or backdrops simple, so that they do not distract from the hair. White backdrops are good for any hairstyles and are a *classic* look for fashion effects. If you do want to use colour, keep to lighter colours or pastel shades so that they do not detract from the purpose of the shot.

Hairdressing competitions

Entering hairdressing competitions can be good fun and a great way of showing your creativity too. It is, however, very challenging and requires a lot of personal discipline, dedication and thorough practice in order to achieve the right look that will catch the eye of the judges. Competitions vary enormously between internal college events, to regional and national heats, and these vary greatly in the way that entrants take part.

The **L'Oréal Colour trophy** is a national competition. It is initially short-listed at a regional level by photographic entry. Entries are sent out to participating salons or colleges early in the New Year and the closing date for final entries is in March. Then, after a preliminary judging, selected entrants are invited to take part at the regional finals, where they have to demonstrate their work '*live*' in front of a large audience, and against the clock. This high-energy show takes place in seven regions across the UK and has student ticket options. Winners from each of the regions are then invited to take part in the grand final in a London hotel in late spring.

The National Hairdressers' Federation (NHF) holds competitions at regional levels and is very popular in supporting students as well as the experienced professionals. Their competitions allow all-comers to participate and finalists from individual regions are then invited to take part at national level.

At the top of British hairdressing, the **British Hairdressing Awards** attract the very best stylists from the top salons. These again are shortlisted by photographic entry. In this competition, entrants enter a variety of categories ranging from Regional, Avant-Garde, Artistic Team, London and British Hairdresser.

If you visit **Salon International** each year in autumn at **ExCel** (London Exhibition Centre), you can see the nominees' work as they take a prominent gallery position within the main hall.

Many hairdressing organizations, colleges and major manufacturers run or sponsor competitions. If this is something that you would like to do, find out who is organizing competitions in your area and send off for a competition brief.

Good practice – tips for taking part in competitions

◆ Monitor the trade press for news about when and where competitions are taking place.

◆ Go along to competitions and watch what happens. See what type of work is successful in competitions and keep an eye on emerging trends and fashions.

◆ Ask trainers and tutors for advice. Listen to people who have entered or know about competitions.

◆ Read the rules carefully and know exactly what is required.

◆ Take time to find exactly the right model: one with the right type of hair, the right age and with looks that fit into the competition rules. A beautiful girl with good deportment helps considerably, but if her hairline is not up to scratch she may put you out of the competition.

◆ Understand that competition work is very different from salon work, Colouring in particular can often be a lot stronger on a competition floor than the salon floor.

Regular competitors stress the importance of preparation

◆ Check and prepare your equipment.

◆ Take time to find the right model, particularly if you are trying to express a specific image or theme.

◆ Product knowledge and application are imperative; never attempt to style a model's hair without testing the products effects on her hair beforehand.

◆ Practise, practise and practise!

Regular entry to competition keeps you up-to-date; you get a feel for the emerging fashions and trends. The motivation gained by attending competitions is infectious and can then be passed on to younger members of staff. Competitions give you the opportunity to see what other salons are doing. There is always something to be learned by watching other salon teams and stylists work.

Competition day! You have prepared your model and you have practised the look for hours. When the day of the competition arrives keep calm; remember that everyone, including the 'great names', suffer from nerves at this time. It is not just the stylists but the models too.

Stick to the rules The style you do must conform to the competition rules. For example, if a day style is required, do not go over the top with elaborate 'hair up' or hair ornaments. If it is free style, a wider choice is allowed.

Once you and your fellow competitors have finished your models, you will be asked to leave the floor so the judges can take over. These people are usually qualified hairdressers, hair and beauty journalists and, occasionally, previous winners, and they will choose the most competently designed and dressed head of hair. Depending on the type of competition and the marking criteria, the judges will award points covering all aspects of style ranging from technical detail, shape, movement, use of colour and artistic adaptation.

Creating a mood board for competitions or photo shoots

In order to take part in any of these events, you will need to create a visual plan of what you are trying to produce. This will help you to develop your ideas and offer a visual representation that you can share with others.

Purpose of a mood board A mood board is an alternative way of communicating ideas or methods of work to a target audience. It provides a way of setting the scene and prepares a storyline, or narrative to share concepts, and express moods or feelings behind image.

TOP TIP

Sharing your mood board and its themes and details with others is important, as the feedback might lead you to modify initial thoughts and therefore influence the final project.

What is a mood board? A mood board is a visual representation of your ideas and thought processes. The simplest form of a mood board would be a large poster, probably A2 in format/size, and could be made up as a collage: a pastiche of different media including images, text, objects, textiles and accessories.

 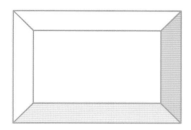

Two-dimentional and three-dimensional mood boards

Other mood boards could take on the form of a *mini-installation*, for example a designed set or stage on which you could add objects and other media to create a three-dimensional representation of your ideas.

Developing a theme The hardest part of the planning process for a creating a mood board is developing the theme. This is where you need to make decisions about your project and its purpose.

You will need to consider the following from the outset.

TOP TIP

A mood board is a collection of ideas, translated through a variety of different media. When you find a textile with an interesting design or texture, save it as you could use it at some other time.

Factors influencing your decisions	Aspects to consider
Where can you find sources of information?	◆ Internet – searches, YouTube, Flickr ◆ Library ◆ Magazines ◆ Films ◆ TV ◆ Shows and demonstrations ◆ Other leading stylists' work ◆ Photographs
What genre of hairdressing will your theme address?	◆ Avant-garde ◆ Current fashion ◆ Theatrical ◆ Historical ◆ Fantasy ◆ Film or famous people ◆ Futuristic
What is the purpose of the hairstyle?	◆ School prom ◆ Public demonstration ◆ Competition ◆ Photo shoot

After collecting the variety of objects and elements that you will use to create your mood board, you can then start building the image. At this stage, you might find that your original ideas have changed quite considerably; don't worry as this always happens and the more time that you build-in to the creation stage, the more likely you are to change your ideas.

Finally, when you have pieced together all the textures, colours, information and feelings, you should try out your ideas by doing a small evaluation to a small audience of people. If you have stimulated the right feelings, you will find that your work will create lots of conversation and 'prompt' several questions. If all this happens, you know that you are on the right track.

To find out how you can carry out evaluation, see CHAPTER 17 Promotional Activities.

Evaluate your results against the design plan objectives

Evaluation is an essential part of any planning process; without knowing how effective you have been, how will you know what to change for future improvements?

REVISION QUESTIONS

Q1. Which of the following is not part of the five basic elements of design? — Multi-choice

Line	○ 1
Form	○ 2
Texture	○ 3
Tendency	○ 4

Q2. A _____ line creates an illusion of height or depth in hair design. — Fill in the blank

Q3. Which of the following words could describe an image? — Multi-selection

Configuration	☐ 1
Vertical	☐ 2
Diagonal	☐ 3
Three-dimensional	☐ 4
Appearance	☐ 5
Contrast	☐ 6

Q4. A curved line within a hairstyle will create waves or movement. — True or false?

Q5. A contrasting line is said to impact against another line within a hairstyle when it meets at what angle? — Multi-choice

30°	○ a
60°	○ b
90°	○ c
180°	○ d

Q6. The L'Oréal _____- trophy is a national competition that is initially short listed at a regional level by _____ entry. · · · · · · · · · · · · · · · · · Fill in the blanks

Q7. Which of the following are *not* aspects of style design? · · · · · · · · · · · Multi-selection

Shape ☐ 1

Form ☐ 2

Growth ☐ 3

Lines ☐ 4

Angles ☐ 5

Lifestyle ☐ 6

Q8. You can publish images of models without their permission. · · · · · · · · · True or false?

Q9. The eye is naturally drawn to the _____ tone(s) within a hair style. · · · Multi-choice

lightest ○ a

darkest ○ b

cooler ○ c

warmer ○ d

Q10. A dramatic look produces a _____ effect within a hairstyle. · · · · · · Fill in the blank

PART THREE

Supporting management

Chapter 15
Health and Safety

Chapter 16
Personal Effectiveness

Chapter 17
Promotional Activities

As a Level 3 practitioner, you set an example to others in the profession. It is therefore vital that you are fully in the know when it comes to health and safety, Habia standards, salon management – and that you keep abreast of industry developments on a continuous basis.

In Part Three we take a look at the legislation that governs practice in our industry and what we can do to ensure all aspects of salon operations run smoothly and efficiently. Finally, we present useful advice on marketing your services and organizing special promotional events, which will raise your profile within your local community and across the profession too.

15 Health and Safety

LEARNING OBJECTIVES

◆ Be able to identify any potential hazards within your salon.

◆ Be able to assess and reduce risks to health and safety in your salon.

◆ Know how to monitor health and safety within the salon.

◆ Know who you should report health and safety issues to, in situations that are beyond your normal work remit.

◆ Know your salon's policies and legal obligations.

KEY TERMS

autoclave
control
databank
HASAWA

hazards
monitoring
occupational dermatitis
risk

risk assessment
tone-on-tone colour

UNIT TOPICS

Industry regulations

Employer and employee responsibilities

INTRODUCTION

Your role at Level 3 requires that you actively take part in the day-to-day monitoring of health and safety for everyone within the salon environment. Your responsibilities extend beyond a duty of care that you must show at work, you must now also:

◆ keep up-to-date with health and safety developments

◆ make sure that the health and safety training needs of less experienced staff are identified and communicated to management

◆ ensure that health and safety records are kept up-to-date and made available for those who need to access them.

This chapter looks at the current health and safety legislation and how you will be following it in the workplace. The process of monitoring workplace health and safety is the responsibility of all senior staff in a salon, not just that of the manager or proprietor. This responsibility extends beyond the salon's staff to all people entering the business, e.g. clients, suppliers, contract cleaners, etc.

At Level 2 you had a duty of care to your employer for taking responsibility for your own actions in the ways that you worked. Now, at Level 3, you must broaden your role by monitoring workplace health and safety for everyone.

Health and safety

PRACTICAL SKILLS

Identify the hazards within the salon

Minimize the risks to health and safety within the salon

Learn how to monitor the health and safety of others in the salon

Learn how to carry out a risk assessment within the salon

Follow written instructions and procedures

Learn how to record accidents and incidents

UNDERPINNING KNOWLEDGE

Know the health and safety legislation relevant to your salon's needs

Know where to find current health and safety guidance and information

Understand the implications of health and safety regulations for the employer and employees

Know the salon's policy in relation to emergency evacuation and first aid

Know how to report accidents and incidents

Know the occupational health hazards associated with hairdressing and barbering

Check that health and safety instructions are followed

What is required by the law?

The basis of British health and safety law is the **Health and Safety at Work Act 1974 (HASAWA)**. This Act sets out the general duties which employers have towards employees and members of the public, and employees have to themselves and to each other.

INDUSTRY ROLE MODEL

WENDY NIXON   Senior Business Development Manager, Habia

" I am an Environmental Health Practitioner, specializing in health and safety enforcement. I have developed health and safety initiatives, including HSE training materials, the Bad Hand Day Campaign and the Health and Safety Award Scheme for the hair and beauty sector, to raise standards and recognize excellence within the industry.

My current role at Habia is to develop health and safety standards and guidance for the sector. I am involved in government and industry initiatives at both national and European level and produce guidance and training for both enforcement officers and industry professionals.

ACTIVITY

For this activity you will create a list of resources for health and safety (H&S) information that you can use now and in the future. When completed the table below should cover a broad range of areas affected by health and safety regulations. You need to find out the missing information and then keep this reference table safe for future use. The first one is completed for you.

Area of H&S	Applicable regulations	Website address	Associated leaflets	Other sources of information
Handling chemicals	COSHH	www.coshhessentials.org.uk	*Control of substances hazardous to health. The Control of Substances Hazardous to Health Regulations 2002 (as amended)*. HSE Books 2005 ISBN 0 7176 2981 3	HSE: www.hse.gov.uk HSE Advisory Team: tel. 0300 0031747

These duties are qualified in the Act by the principle of '*so far as is reasonably practicable*'. In other words, an employer does not have to take measures to avoid or reduce the **risk** if they are technically impossible or if the time, trouble, or cost of the measures would be grossly disproportionate to the risk.

What the law requires here is good management, and common sense, and this is what employers do anyway. They look for potential risks within the workplace and then take sensible measures to eliminate them. Your role is to assist your employer in fulfilling these duties.

The Management of Health and Safety at Work Regulations 1999 (the Management Regulations) generally make more explicit what employers are required to do to manage health and safety under the Health and Safety at Work Act. Like the Act, they apply to every work activity.

The main requirement on employers is to carry out a **risk assessment**. Employers with five or more employees need to record the significant findings of the risk assessment and this could be part of your role. Risk assessment should be straightforward in a work environment such as a typical salon or barber's shop (see risk assessment, page 373).

Besides carrying out a risk assessment, employers also need to:

◆ make arrangements for implementing the health and safety measures identified as necessary by the risk assessment

◆ appoint competent people (this could be you) to help them to implement the arrangements

◆ set up emergency procedures

ALWAYS REMEMBER

Health and safety laws are continually reviewed and updated. Make sure you are aware of the latest information and look at the health and safety posters within your salon.

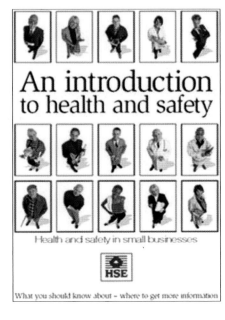

An introduction to health and safety

Health and safety in small businesses

HSE

What you should know about – where to get more information

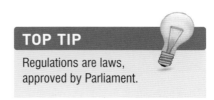

TOP TIP

Regulations are laws, approved by Parliament.

◆ provide clear information and training to employees

◆ work together with other employers sharing the same workplace.

How regulations apply

Some regulations apply across all work locations, such as the **Manual Handling Regulations**, which apply wherever things are moved by hand or bodily force, and the **Display Screen Equipment Regulations**, which apply wherever visual displays/ computer screens are used.

Besides the Health and Safety at Work Act itself, the following apply across any place of work in the UK:

1 **Management of Health and Safety at Work Regulations 1999**: require employers to carry out risk assessments, make arrangements to implement necessary measures, appoint competent people and arrange for appropriate information and training.

2 **Workplace (Health, Safety and Welfare) Regulations 1992**: cover a wide range of basic health, safety and welfare issues such as ventilation, heating, lighting, workstations, seating and welfare facilities.

3 **Health and Safety (Display Screen Equipment) Regulations 1992**: set out requirements for work with visual display units (VDUs/computers).

4 **Personal Protective Equipment at Work Regulations 1992**: require employers to provide appropriate protective clothing and equipment for their employees.

5 **Provision and Use of Work Equipment Regulations 1998**: require that equipment provided for use at work, including machinery, is safe.

6 **Manual Handling Operations Regulations 1992**: cover the moving of objects by hand or bodily force.

7 **Health and Safety (First Aid) Regulations 1981**: cover requirements for first aid.

8 **The Health and Safety Information for Employees Regulations 1989**: require employers to display a poster telling employees what they need to know about health and safety.

9 **Reporting of Injuries, Diseases, and Dangerous Occurrences Regulations 1995 (RIDDOR)**: require employers to notify certain occupational injuries, diseases and dangerous events.

10 **Noise at Work Regulations 1989**: require employers to take action to protect employees from hearing damage.

11 **Electricity at Work Regulations 1989**: require people in control of electrical systems to ensure they are safe to use and maintained in a safe condition.

12 **Control of Substances Hazardous to Health Regulations 2002/3 (COSHH)**: require employers to assess the risks from hazardous substances and take appropriate precautions.

HEALTH & SAFETY

Potential areas of risk to people

◆ Manual handling
◆ Electricity and electrical equipment (clippers, etc.)
◆ Scissors and razors
◆ Dermatitis
◆ Chemicals
◆ Slips and trips

ALWAYS REMEMBER

Website help from HSE Direct and COSHH Essentials

These offer easy online steps to control health risks from chemicals.

COSHH Essentials has been developed to help firms comply with the COSHH regulations. The COSHH Essentials website is easy to use and available free as part of HSE Direct – a databank of all health and safety legislation and HSE's priced guidance.

For more information visit: **www.hse.gov.uk/coshh/essentials/**

Employees' responsibilities

The word responsible is used many times and in many ways at work. You play an important part in spotting potential **hazards** and preventing accidents, so helping your salon and colleagues to avoid problems.

In addition to your general responsibilities as an employee, with your experience you also have a duty to help monitor the health and safety of others with whom you work.

Communication This is the first and most vital aspect of your role. You may be responsible for making the team aware of the health and safety aspects through the following:

◆ Information displayed around the workplace, e.g. posters, leaflets, employer liability insurance, etc.

◆ Telling staff things that your employer wishes them to know, e.g. new developments, courses or health and safety training.

◆ Looking for potential situations and telling others when unsafe working practices occur, e.g. clients or staff not wearing the provided PPE.

◆ Identifying hazards and taking prompt action yourself, e.g. sweeping up clippings on the salon floor, removing equipment with loose plugs or unravelling coiled leads, clearing up spilt chemicals quickly and safely.

Record keeping You may be required to complete and maintain the following records:

◆ Risk assessments made in the workplace by you or your employer, e.g. COSHH assessments, manual handling, contact dermatitis.

◆ Accidents that have occurred and the actions that were taken, e.g. maintaining the accident book.

◆ Tests for equipment (made by relevant, competent people), e.g. PAT tests (portable appliance tests see: **www.hse.gov.uk/electricity/faq-portable-appliance-testing.htm**), general checks.

Health and safety training

◆ Identifying when team members have particular health and safety training needs and reporting that back to management, e.g. COSHH, manual handling, fire safety, emergency evacuation, PPE.

◆ Updating your own knowledge by using relevant sources of information such as websites, e.g. **www.hse.gov.uk**, **www.hse.gov.uk/coshh/essentials/**

Employer's responsibilities

Management of Health and Safety at Work Regulations 1999 The main regulation requires the employer to appoint competent personnel to conduct risk assessments for the health and safety of all staff employed or otherwise, and other visitors to the business premises. Staff must be adequately trained to take appropriate action, eliminate or minimize any risks. Other regulations cover the requirement to set up procedures for emergencies, reviewing the risk assessment processes. In salons where five or more people are employed, it is necessary to set up a system for recording the findings of risk assessment; this could be written on paper or on computer.

The main requirements for management of health and safety are as follows:

◆ identification of any potential hazards

◆ assessing the risks which could arise from these hazards

◆ identifying who is at risk

◆ eliminating or minimizing the risks

◆ training staff to identify and **control** risks

◆ regular reviewing of the assessment processes.

Young workers at risk There is also a requirement to carry out a risk assessment for young people. Any staff member who is under school-leaving age must have a personalized risk assessment kept on file. This would be applicable for those on work experience or Saturday staff.

Risks for young people when starting work may arise because of their lack of experience or maturity and not having the confidence to ask for or knowing where they can get help.

Before a young person is employed, the employer's health and safety risk assessment must consider these specific factors:

◆ fitting-out and layout of the workplace and the particular areas where they will work

◆ any physical, biological and chemical agents they will be exposed to, for how long and to what extent

◆ what types of work equipment will be used and how this will be handled

◆ how the work and processes involved are organized

◆ level of health and safety training given to young people

◆ risks from the identified agents, and the work processes.

COSHH Risk Assessment

Staff member responsible: Natasha Smith   Date: 1st September 2015   Review Dates: 10th January 2016

Hazard	What is the risk?	Who is at risk?	Degree of risk high/med/low	Action to be taken to reduce/control risk
Aerosols (List aerosols used in your salon)	These can contain flammable gases and irritant chemicals. There is a risk of fire, explosion and intoxication.	Everyone in the salon, but in particular the user of the aerosol and the client.	Low	Look for aerosols with non-flammable gases if possible. Do not expose to temperatures above 50°C. Do not pierce or burn containers. Do not inhale.
Permanent wave neutralizer (List products used in your salon)	Irritant to the skin and eyes. Moderately toxic if swallowed or inhaled.	Stylists, juniors, trainees and clients.	Medium	Store in a cool place. Reseal after use. Do not use on damaged or sensitive skin. Avoid breathing in. Never place in an unlabelled container.

HSIP2a

ACTIVITY

Conducting a risk assessment

Procedure for conducting a risk assessment:

1 Take a walk around the salon looking for any hazards, i.e. anything with the potential to cause harm. By using a similar form to the example shown above, it will help to organize the necessary information.

2 Decide what the risks are and who could be harmed by those hazards. Broken tiles at the backwash may only affect the staff, whereas loose carpet in reception would affect all visitors and staff.

3 For each of the listed hazards decide what level of risk exists, e.g. low, medium or high. Then, looking at each entry, ask yourself if the risk can be eliminated or reduced.

4 Write down the findings of your risk assessment (salons with less than five employees do not have to record these findings). However, you may get a visit from Environmental Health and written records will prove that the assessments have been made.

5 Review your risk assessments at regular intervals. The introduction of new equipment, different product ranges or chemical processes will potentially create new hazards. Setting a review date within the assessment process will ensure that your salon is kept up to date in the future.

Workplace (Health, Safety and Welfare) Regulations 1992 These regulations superseded the Offices, Shops, and Railway Premises Act 1963 (OSRPA) and cover the following workplace key points:

◆ maintenance of the workplace and the equipment in it

◆ ventilation, temperature and lighting

◆ cleanliness

◆ sanitary and washing facilities

◆ drinking water supply

◆ rest, eating and changing facilities

◆ storage of clothing

◆ glazed windows, screens, mirrors

◆ traffic routes (work thoroughfares)

◆ work space.

Personal Protective Equipment at Work Regulations (PPE) 1992

The PPE Regulations 1992 require employers to make an assessment of the processes and activities carried out at work and to identify where and when special items of clothing should be worn. In hairdressing environments, the potential hazards and dangers revolve around the task of providing hairdressing services – that is, hairdressing treatments and associated products. (Many requirements under this Act will have been met in complying with COSHH regulations.)

Potentially hazardous substances used by hairdressers include:

◆ solutions such as hydrogen peroxide in varying strengths

◆ caustic alkaline such as perming solutions of varying strengths

◆ flammable liquids such as hairsprays, which are often in pressurized containers

◆ vapours from chemical products

◆ colouring products – para dyes

◆ shampoos and conditioners (see dermatitis on page 389).

All these items require correct handling and safe usage procedures and for several of them this includes the wearing of suitable items of protective equipment.

Control of Substances Hazardous to Health Regulations 1999 (COSHH)

The purpose of COSHH Regulations is to make sure that people are working in the safest possible environment and conditions. A substance is considered hazardous if it can cause harm to the body.

It only presents a risk if it is:

◆ in contact with the skin or eyes

◆ absorbed through the skin or via the eyes (either directly or from contact with contaminated surfaces or clothing)

◆ inhaled, i.e. breathing in substances in the atmosphere

◆ ingested via contaminated food or fingers

◆ injected or introduced to the body via cuts and abrasions.

> When undertaking your COSHH assessments in the salon, don't forget to include the general cleaning or laundry products. Also ensure that you record the precautionary measures to reduce risks to the client as well as the staff, e.g. skin sensitivity tests. Your written procedures should note that you undertake skin sensitivity tests under the heading 'What action should I take?' on your assessment forms.
>
> *Wendy Nixon*

The **Cosmetic, Toiletry & Perfumery Association** (CTPA) represents all types of companies involved in the UK cosmetics industry. Hair products must comply with stringent safety regulations. The regulations detail how ingredients such as permanent and tone-on-tone colours can be used. Manufacturers have to list on the label all the ingredients that are used in their products.

Therefore, employers make a risk assessment for the products they use:

◆ What products are used?

◆ What is the potential of a product for causing harm?

◆ What is the chance of exposure?

◆ How much are people exposed to it, for how long and how often?

◆ Can the exposure be prevented and, if not, how is it adequately controlled?

Useful website: www.ctpa.org.uk/

Wherever safer products are available, they should be used; where not, the exposure should be controlled. Exposure can be controlled by:

◆ providing good ventilation

◆ using the product only in recommended concentrations

◆ clearing up spillages or splashes immediately

◆ resealing containers immediately after use

◆ providing safe storage

◆ using personal protective equipment.

Dispose of salon waste carefully

HEALTH & SAFETY

Further guidance for health and safety from Habia The Hair and Beauty Industry Authority (Habia) produces and continually updates a comprehensive information guide for hairdressers.

The Habia Health and Safety for Hairdressers and Barbers Habia has a separate, designated website for salon health and safety. From this website you can order a variety of materials and ready-to-use systems for hairdressers and barbers.

Visit: www.habia.org/healthandsafety/ for more information.

Habia's Health and Safety pack for hairdressers includes the following:

◆ Official forms and notices for you to copy and use in your salon.

◆ How to write and produce a health and safety policy with a blank version for you to personalize.

◆ An accident book adhering to the new guidelines.

◆ Health and Safety Risk Assessments, Fire Risk Assessments and COSHH Risk Assessments in example formats for guidance and as blank forms for you to complete.

◆ Large print and a clear uncluttered layout, which make the information easy to read.

◆ Fully up-to-date as of the date of purchase.

◆ Fire exit signs.

ACTIVITY

Health and safety

1 Read the instructions on the label of each chemical or product in your salon.

2 Further information for each individual type of product can be found in the Guide to Health and Safety in the Salon booklet, or by using the manufacturers' data sheet.

3 If the product could cause harm, list it on the risk assessment form (see example risk assessment form on page 375) together with the risk and who is at risk.

4 Using the information provided on the label, decide on the level of risk. (From your previous experience with the routine chemicals that you use at work, such as lightener, colour or perming solutions, you will be able to decide the degree of risk.)

5 Decide how you could minimize and control the risk. Suggest a possible replacement for a high risk product with a lower risk product. If this is not possible, try to decide how you go on to control risk, remembering that personal protective equipment (PPE) is only to be used as a last resort.

6 In some cases your manager may need to replace certain products. See if you can find information, or source better personal protective equipment. This can be documented as an ongoing action plan, making sure that all relevant members of staff understand the actions that have been taken.

7 Discuss the completed risk assessment with staff and make sure that they are fully trained to use all products safely.

8 Remember to review your COSHH assessment on a regular basis and do not forget to add in any new products that have been introduced to the salon and keep all data sheets filed safely for future reference.

Electricity at Work Regulations 1989 The Electricity at Work Regulations 1989 address the installation, maintenance and use of electrical equipment and systems in the workplace. Equipment must be checked by a qualified person on a yearly basis and if any maintenance is required this should be carried out by a qualified electrician.

An electrical testing record should be kept for each piece of equipment and should clearly show:

◆ electrician's/contractor's name, address, contact details

- ◆ itemized list of salon electrical equipment, along with serial number (for individual identification)
- ◆ date of inspection
- ◆ date of purchase/disposal.

Health and Safety (First Aid) Regulations 1981 The Health and Safety (First Aid) Regulations 1981 require employers to provide equipment and facilities that are adequate and appropriate in the circumstances for administering first aid to their employees. Remember that any first aid materials used from the kit must be replaced as soon as possible. All accidents and emergency aid given within the salon must be documented in the accident book.

What should a first aid box in the workplace contain? There is no mandatory list of contents for first aid boxes and HSE does not 'approve' or endorse particular products. Deciding what to include should be based on an employer's assessment of first aid needs.

As a guide, where work activities involve low hazards, a minimum stock of first aid items might be:

- ◆ a leaflet giving general guidance on first aid, e.g. HSE's leaflet: Basic advice on first aid at work
- ◆ 20 individually wrapped sterile plasters (assorted sizes), appropriate to the type of work (you can provide hypoallergenic plasters, if necessary)
- ◆ two sterile eye pads
- ◆ four individually wrapped triangular bandages, preferably sterile
- ◆ six safety pins
- ◆ two large, individually wrapped, sterile, unmedicated wound dressings
- ◆ six medium-sized, individually wrapped, sterile, unmedicated wound dressings
- ◆ a pair of non-latex disposable gloves.

See **www.hse.gov.uk** for more information on first aid.

How often should the contents of first aid boxes be replaced?
Although there is no specified review timetable, many items, particularly sterile ones, are marked with expiry dates. They should be replaced by the dates given and expired items disposed of safely. In cases where sterile items have no dates, it would be advisable to check with the manufacturers to find out how long they can be kept. For non-sterile items without dates, it is a matter of judgement, based on whether they are fit for purpose.

Recording accidents and illness All accidents must be recorded in the accident book. The recording system should always be kept readily available for use and inspection. When you are recording accidents, you will need to document the following details:

- ◆ date, time and place of incident or treatment
- ◆ name and job of injured or ill person

◆ details of the injury/ill person and the treatment given

◆ what happened to the person immediately afterwards (e.g. went home, hospital)

◆ name and signature of the person providing the treatment and entry.

HEALTH & SAFETY

General guidance on first aid

The following basic information is available in leaflet form from HSE:

Basic rules for first aiders

REMEMBER: YOU SHOULD NOT ATTEMPT TO GIVE ANYTHING MORE THAN BASIC FIRST AID!

When giving first aid it is vital that you assess the situation and that you:

◆ Take care not to become a casualty yourself while administering first aid (use protective clothing and equipment where necessary).

◆ Send for help where necessary.

◆ Follow this advice from the HSE.

What to do in an emergency

Check whether the casualty is conscious. If the casualty is unconscious or semi-conscious:

◆ Check the mouth for any obstruction.

◆ *Open the airway* by tilting the head back and lifting the chin using the tips of two fingers.

◆ If the casualty has stopped breathing and you are competent to give artificial ventilation, do so. Otherwise send for help without delay.

Unconsciousness

In most workplaces, expert help should be available fairly quickly, but if you have an unconscious casualty it is vital that his or her airway is kept clear. If you cannot keep the airway open as described above, you may need to turn the casualty into the recovery position. The priority is an open airway.

Reporting of Injuries, Diseases and Dangerous Occurrences Regulations 1995 (RIDDOR)
Under these regulations, there are certain situations, and diseases, that if sustained at work are notifiable by law (see Reportable injuries table on page 381). Therefore, if any employees suffer a personal injury at work which results in:

◆ death

◆ major injury

◆ an incapacity to work for more than seven calendar days

you must complete the online report available from the HSE website.

Certain industrial diseases are reportable using the online form and these include occupational dermatitis (a condition connected with the hairdressing/barbering industry). Up to 70% of hairdressers suffer from work-related skin damage such as dermatitis at some point during their career; but most cases are absolutely preventable by wearing protective non-latex disposable gloves (see page 389).

Other accidents (and 'near-misses') that occur within the salon must also be recorded. Entries must be kept up-to-date within the accident record book.

Reportable major injuries

◆ Fracture, other than to fingers, thumbs and toes

◆ Amputation

◆ Dislocation of the shoulder, hip, knee or spine

◆ Loss of sight (temporary or permanent)

◆ Chemical or hot metal burn to the eye or any penetrating injury to the eye

◆ Injury resulting from an electric shock or electrical burn leading to unconsciousness, or requiring resuscitation or admittance to hospital for more than 24 hours

◆ Any other injury leading to hypothermia, heat-induced illness or unconsciousness, or requiring resuscitation, or requiring admittance to hospital for more than 24 hours

◆ Unconsciousness caused by asphyxia or exposure to a harmful substance or biological agent

◆ Acute illness requiring medical treatment, or loss of consciousness arising from absorption of any substance by inhalation, ingestion or through the skin

◆ Acute illness requiring medical treatment where there is reason to believe that this resulted from exposure to a biological agent or its toxins or infected material

Health and Safety (Information for Employees) Regulations 1989

The regulations require the employer to make available to all employees, notices, posters and leaflets in either the approved format or those actually published by the Health and Safety Executive (HSE).

The *Health and Safety Law* leaflet is available in packs of 50 from HSE Books Box 1999, Sudbury, Suffolk CO16 6FS; tel. 01787 881165. Other useful HSE publications include:

◆ *Essentials of Health and Safety at Work* (ISBN 0 7176 0716)

◆ *Writing your Health and Safety Policy Statement* (ISBN 0 7176 0425)

◆ *Successful Health and Safety Management* (ISBN 0 7176 0425)

◆ *A Guide to RIDDOR* (ISBN 0 7176 0432 2)

◆ *Step by Step Guide to COSHH Assessment* (ISBN 0 11886379 7) or online information at **www.hse.gov.uk/coshh/essentials/**

◆ *First Aid at Work* (ISBN 0 7176 0426 8).

> " Hairdressing receptionists are often required to answer the phone and make appointments at the same time, trying to balance the phone on their shoulder while writing or typing. This can cause musculoskeletal problems, such as neck and back pain, but can be solved easily by providing a headset for the phone. This allows free movement of both hands and encourages a good posture.
>
> *Wendy Dixon*

Fire Precautions Act 1971 Employers must carry out a fire safety risk assessment and keep it up-to-date. This can be carried out either as part of an overall risk assessment or as a separate exercise.

Based on the findings of the assessment, employers need to ensure that adequate and appropriate fire safety measures are in place to minimize the risk of injury or loss of life in the event of a fire. To help prevent fire in the workplace, a risk assessment should identify what could cause a fire to start, i.e. sources of ignition (heat or sparks) and substances that burn, and the people who may be at risk. It is essential to do the following:

◆ Carry out a fire safety risk assessment.

◆ Keep sources of ignition and flammable substances apart.

◆ Avoid accidental fires, e.g. make sure heaters cannot be knocked over.

◆ Ensure good housekeeping at all times, e.g. avoid build-up of rubbish that could burn.

◆ Install smoke alarms and fire alarms or bells. Do not smoke or allow smoking.

◆ Have the correct fire-fighting equipment for putting a fire out quickly.

◆ Keep fire exits and escape routes clearly marked and unobstructed at all times.

◆ Ensure your workers receive appropriate training on procedures they need to follow, including fire drills.

◆ Review and update your risk assessment regularly.

Health and Safety (Display Screen Equipment) Regulations 1992

These regulations cover the use of computers and similar equipment in the workplace. Although not generally a high risk, prolonged use can lead to upper limb disorders (ULDs), or eye strain. As more hairdressing salons now use computers it is becoming a consideration for employees.

See APPENDIX 2 for more information.

It is the employer's duty to assess display screen equipment and reduce the risks that are discovered. They will need to plan the scheduling of work so that there are regular breaks or changes in activity and provide information training for the equipment users. Computer users will also be entitled to eyesight tests that are paid for by the employer.

Manual Handling Operations Regulations 1992 These regulations apply in all occupations where manual lifting occurs. It is the employer's responsibility to provide employees with training on the correct methods of lifting. The regulations require employers to carry out a risk assessment of the work processes and activities that involve lifting. The risk assessment should address detailed aspects of:

◆ any risk of injury

◆ the manual movement that is involved in the task

◆ the physical constraints that the loads incur

◆ the work environmental constraints that are incurred

◆ the worker's individual capabilities

◆ steps and/or remedial action to take in order to minimize the risk.

Provision and Use of Work Equipment Regulations (PUWER) 1998 These regulations refer to the regular maintenance and monitoring of work equipment. Any equipment, new or secondhand, must be suitable for the purpose that it is intended. In addition to this they require that anyone using this equipment must be adequately trained.

HEALTH & SAFETY

PUWER

◆ Is all salon equipment checked regularly?

◆ Are maintenance logs kept for items of equipment?

◆ Has introduced secondhand equipment been checked?

◆ Have all the staff been trained to use the equipment?

Make sure that risks are controlled safely and effectively

Acting responsibly

You share a responsibility with your work colleagues for the safety of all the people within the salon (clients, visitors and staff) so you need to be aware of the types of hazards that could exist. Knowing *what* to do and *who* to approach in different situations is particularly important.

A risky situation can develop at any time and you may be the only one who sees it happen, so your swift action may make all the difference. You have to decide whether:

◆ you are able to deal with the situation yourself

◆ you need to inform someone else.

You may be keen to help out but you need to know what it is that you are dealing with.

For example, if you see that the floor is wet from a spillage:

◆ What liquid has been spilt on the floor?

◆ How do you know if this is something that you can deal with?

◆ What action do you need to take about it?

Is it a simple spillage like water or something more hazardous like a chemical? If it is water, then you know that this is something that you could deal with yourself, if it is a chemical; it has to be handled in a certain way with the right protective equipment. The action you now take depends on the hazard, you will be either clearing the spillage yourself, or blocking the area, as a temporary measure, say with a chair, so that others are aware of the hazard, while you speak to the manager.

A sharps box is for the safe disposal of razor blades and other sharp objects.

Reducing risks to health and safety

People may be exposed to different types of hazardous substances at work. These can include chemicals that people make or work with directly, and also dust, fumes and bacteria, which can be present in the workplace. Exposure can occur through breathing them in, contact with the skin, splashing them into the eyes or swallowing them. If exposure is not prevented or properly controlled, it can cause serious illness including cancer, asthma and dermatitis, and sometimes even death.

Health and safety law requires employers to control the risk of workers or trainees to the exposure of hazards in the workplace. This is done through the process of risk assessment. Risk assessment can cover all sorts of areas, but in hairdressing there are known risks that apply in our industry, that provide specific areas for scrutiny.

> "
> When writing or reviewing your risk assessments always look for shortcuts staff may take and assess the impact on health and safety, e.g. plugging in electrical appliances before removing protective gloves, following a chemical treatment. Remember that the exterior of the glove is wet and so you are putting yourself at risk when handling electrical equipment.
>
> *Wendy Dixon*

Hazard and risk

Almost anything may be a hazard, but it may or may not become a risk. For example: a trailing electric cable from a piece of equipment is a hazard; if it is trailing across a passageway it presents a high risk of someone tripping over it, but if it is along a wall out of the way, the risk is much less.

Poisonous or flammable chemicals are hazards and may present a high risk. However, if they are kept in a properly designed secure store and handled only by properly trained and equipped people, the risk is much less than if they are left about for anyone to use.

A failed light bulb is a hazard. If it is just one of many in a room, it presents very little risk, but if it is the only light on a stairwell, it is a very high risk. Changing the bulb may be a high risk, if it is up high or if the power has been left on, or a low risk if it is in a table lamp that has been unplugged.

A box of heavy materials is a hazard. It presents a higher risk to someone who lifts it incorrectly than to someone who uses the correct manual handling techniques.

Sweep up hair clippings promptly.

ACTIVITY

Hazard and risk

For this activity, think of examples of hazards that could exist within your workplace. For each one, specify the risk involved and the measures that should be taken to control the risk.

An example is shown to help you start.

Identify the hazard	What is the risk?	How can this be controlled?
Hair clippings on the floor	Slipping	Make sure that they are swept up and put in the bin

Potential hazards in the workplace

Working with equipment

What could be a potential hazard?	What is the risk?	What action should I take?	Relevant legislation other than management of health and safety at work regulations 1999
Razor blades, scissors	Injury – cuts and abrasions	1 Inform staff of the correct ways to handle sharps 2 Ensure that a sharps box is available for safe disposal 3 Check with local authority for bylaws affecting your premises	Health and Safety (First Aid) Regulations 1981
Loose plugs or coiled leads on blow-dryers, tongs, straighteners	Electric shock	1 Inform staff on how to check for faults before plugging into the mains 2 Show staff how to put equipment away safely 3 Regular maintenance of equipment test records	Electricity at Work Regulations 1989
Faulty or worn equipment	Electric shock Accidental injuries	1 Visually check equipment regularly 2 Report findings to management 3 Stop others from using equipment until replaced or mended	Workplace (Health Safety and Welfare) Regulations 1992 The Provision and Use of Work Equipment Regulations 1992 Electricity at Work Regulations 1989

Handling chemicals

What could be a potential hazard?	What is the risk?	What action should I take?	Relevant legislation other than management of health and safety at work regulations 1999
Handling hydrogen peroxide Inhalation of powder lighteners Chemical spillage	Burns Contact dermatitis Asthma or respiratory conditions Injury Slipping	Check manufacturer's instructions Inform/show staff the correct ways to handle chemical products. Check storage arrangements. Monitor staff for safe handling.	Control of Substances Hazardous to Health 1999 RIDDOR 1995 (re. occupational dermatitis) Personal Protective Equipment at Work Regulations 1992

Process for undertaking COSHH risk assessment The table on page 386 is the guidance from HSE for undertaking COSHH risk assessment. If your employer designates you to undertake this assessment, you should follow these steps.

Step 1	Assess the risks	Think about the risks to health from hazardous substances used in or created by your workplace activities.
Step 2	Decide what precautions are needed	Do not carry out work that could expose others to hazardous substances without first considering the risks and the necessary precautions, and what else you need to do to comply with COSHH.
Step 3	Prevent or adequately control exposure	Prevent staff from being exposed to hazardous substances. Where preventing exposure is not reasonably practicable, then you must adequately control it (e.g. shampooing – wear non-latex disposable gloves).
Step 4	Ensure that control measures are used and maintained	Ensure that control measures are maintained properly (e.g. aprons, gloves, plastic capes ready for use, etc.).
Step 5	Monitor the exposure	Monitor the exposure of employees to hazardous substances if necessary (e.g. adequate ventilation in dispensary areas where powder lightener is mixed).
Step 6	Carry out appropriate health surveillance	Carry out appropriate health surveillance where your assessment has shown this is necessary or where COSHH sets specific requirements. (Look for unsafe practices, e.g. shampooing without non-latex disposable gloves).
Step 7	Prepare plans and procedures to deal with accidents, incidents and emergencies	Prepare plans and procedures to deal with accidents, incidents and emergencies involving hazardous substances, where necessary.
Step 8	Ensure employees are properly informed, trained and supervised	You should provide the staff with suitable and sufficient information, instruction and training.

Working environment

What could be a potential hazard?	What is the risk?	What action should I take?	Relevant legislation other than management of health and safety at work regulations 1999
Hair clippings on the floor	Slipping	1 Inform staff of the risks of leaving floors un-swept 2 Monitor staff in the work environment	Workplace (Health Safety and Welfare) Regulations 1992
Obstructions	Slips or trips Injury	1 Inform staff of the risks of leaving obstructions in work areas 2 Check staff knowledge for safe handling methods 3 Identify training needs 4 Inform management of staff training needs	Manual Handling Operations Regulations 1992 Workplace (Health Safety and Welfare) Regulations 1992
Poor hygiene, cleaning, and maintenance of salon resources	Cross-infection Cross-infestation	1 Inform staff of the risks of poor salon/shop maintenance 2 Show staff methods of disinfecting and sterilizing surfaces, tools, equipment 3 Monitor staff in the work environment	Workplace (Health Safety and Welfare) Regulations 1992 Personal Protective Equipment at Work Regulations 1992
Poor personal hygiene	Cross-infection Cross-infestation	1 Inform staff of the risks of poor personal hygiene 2 Monitor staff in the work environment	
Computers	Eye strain, Upper limb disorders (ULDs)	1 Inform staff of the risks of using computers for long periods of time 2 Monitor staff ensure that they take regular breaks	Health and Safety (Display Screen Equipment) Regulations 1992

Controlling risks in the workplace

As a salon professional there are many ways you can help to control risks within the workplace.

General salon hygiene

Personal hygiene

1 Hands should be washed before every client. Poor personal hygiene could lead to cross-infection for the client so excellent personal hygiene must be a priority for all staff.

2 Any skin problems, cuts or boils must be covered with a waterproof, impervious dressing. Dermatitis (see page 387) or similar condition on the hands or arms should be covered using non-latex disposable gloves changed for each customer.

3 Clean, washable overalls/uniforms must be worn and must be changed if contaminated or soiled by any blood or other body fluid.

Disinfection and sterilization procedures

1 All equipment once used should be disinfected or sterilized (according to contamination) before reuse and should be divided into plastic items – combs, brushes, etc. – and metal items – scissors, razors, etc.

2 Combs, brushes, rollers and curlers need to be washed in hot soapy water. Scissors and razors should be carefully wiped with alcohol or spirit. Any items contaminated by blood or body fluid must be handled with care to ensure that the contaminated area does not come into contact with any broken skin.

3 These should then be placed in a solution of hypochlorite until required.

4 The metal equipment should be sterilized, ideally in an **autoclave**.

> " I have seen too many jars of disinfection fluid which have turned black and are full of floating debris. This not only gives a poor visual message to the clients but it will not be effective. Always wash equipment to remove visual contamination prior to disinfection or sterilization. Disinfection fluids must be changed daily in accordance with the manufacturer's instructions. If you use an autoclave, run a test strip through the system each morning.
>
> *Wendy Dixon*

An autoclave is used to sterilize metal equipment

Autoclave The autoclave is a very effective method for the sterilization of metal or glass items. It works by superheating water under pressure in the same way that a pressure cooker does; heating the water to 120°C. This very high temperature kills all organisms within 20 minutes. Autoclaves may be automatic and go through a cycle with appropriate holding times, otherwise the times and temperatures shown in the table must be achieved.

Temperature	Holding times
121°C	15 minutes
126°C	10 minutes
134°C	3 minutes

An alternative to an autoclave is to boil all equipment for 30 minutes in a boiler specially designed for instruments.

The third alternative method for dealing with metal instruments is to rinse away visible contamination. Care must be taken to avoid touching any contamination, especially with areas of broken skin. The instruments must then be placed in a disinfectant solution of glutaraldehyde for at least 30 minutes. It is advisable to use disposable razors.

Disinfectants The two most commonly used disinfectants are hypochlorite and glutaraldehyde. These do not sterilize, i.e. kill all known germs, but they do reduce the number to the extent that there is very little danger of infection.

Hypochlorite (sold commercially as Milton or bleach) is corrosive to metals and therefore can only be used for wiping down work surfaces, chairs, etc. and for soaking combs, brushes, etc. The solution should be made up freshly each day with a dilution of one part Milton to ten parts water or one part good quality bleach to one-hundred parts water.

Glutaraldehyde (sold commercially as Cidex™ or Totacide™) can be used for wiping down all surfaces including metals and for soaking metal implements. The solution can be made up weekly according to manufacturer's instructions.

Ultraviolet radiation Ultraviolet (UV) radiation provides an alternative sterilizing option. Items should be cleaned to remove grease and dirt and placed in wall- or worktop-mounted cabinets fitted with UV-emitting light bulbs and exposed to the radiation for at least 15 minutes. If your scissors or combs are sterilized in a UV cabinet, remember to turn them over to make sure both sides have been done.

Lifting and handling large objects

The incorrect handling of large and/or heavy items can result in low back pain/problems, repetitive strain injuries, and strain disorders. Think about the situations that can occur in a salon environment:

◆ moving stock into storage

◆ unpacking heavy or awkward items

◆ lifting equipment and moving salon furniture

◆ working heights – chairs, trolleys driers, etc.

The regulations require employers to:

◆ *avoid* the need for hazardous manual handling, so far as is reasonably practicable

◆ *assess* the risk of injury from any hazardous manual handling that can't be avoided

◆ *reduce* the risk of injury from hazardous manual handling, so far as is reasonably practicable.

See APPENDIX 2 for more details about manual handling techniques.

Contact dermatitis

Up to 70% of hairdressers suffer from skin damage and many trainee hairdressers have to give up hairdressing because of this common occupational health hazard. But you can avoid this condition by making sure that you always use non-latex disposable gloves. Always dry your hands thoroughly and moisturize them to keep them healthy.

> Always dry your hands thoroughly using a soft clean and dry towel. Remember to dry between the fingers and around the thumb and wrist. Hairdressers often dry their hands on damp towels previously used on clients. This is not effective and can lead to skin damage and dermatitis. Your hands are your tools, so take a few minutes to protect your skin and ultimately your career.
>
> *Wendy Dixon*

Five steps to prevent dermatitis

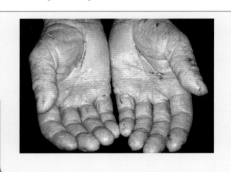

Step 1 Wear non-latex disposable gloves when rinsing, shampooing, colouring, lightening, etc.

Step 2 Dry your hands thoroughly with a soft cotton or paper towel

Step 3 Moisturize after washing your hands, as well as at the start and end of each day. It is easy to miss fingertips, finger webs and wrists

Step 4. Change gloves between clients. Make sure you do not contaminate your hands when you take them off

Step 5. Check skin regularly for early signs of dermatitis

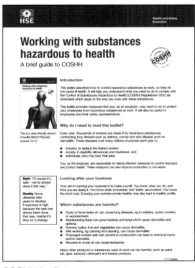

COSHH leaflet

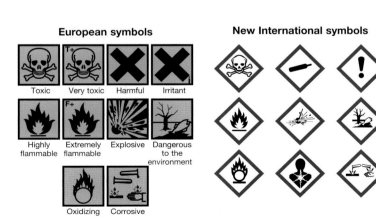

Symbols showing types of hazardous substances

Handling chemicals Many of the hairdressing services involve some contact with chemicals and salons have safe ways of dealing with this. Employers have a legal obligation to control the exposure of hazardous substances in the workplace and you

are protected by the **C**ontrol **O**f **S**ubstances **H**azardous to **H**ealth Regulations (2003) (COSHH). Most hairdressing chemicals are safe and only propose a risk to health at the point when they are handled and used.

COSHH precautions Risk assessments will indicate the types of chemicals available within the salon and the safe ways in which they may be handled or used. These chemical products will vary from cleaning items, such as washing materials, bleach and polish, to the more typical salon-specific items such as colours, lighteners, hydrogen peroxide and general styling materials.

Make sure that staff are aware of the assessments; they will indicate the level of risk that each of the chemical products presents to you, i.e. a hazard rating and details on how they can be handled safely with the PPE provided by your employer.

Keep floors clean and thoroughfares clear The most common cause of injuries at work is the slip or trip. Resulting falls can be serious and a busy salon means many people and the more clients there are, the more hair clippings there will be. Loose clippings left on the salon floor present a hazard to staff and clients alike.

See **APPENDIX 2**, Manual handling section.

Both wet and dry hair clippings are easily slipped on; make sure that working areas are swept regularly and do not wait for stylists to finish: get rid of clippings before they build up. Clear them away from areas where people are working or walking and then brush them into a dustpan and put them into the waste bin.

Keep thoroughfares clear of obstructions; move any hazards away from busy traffic areas.

Working with computers Using a computer for long periods can give rise to back problems, repetitive strain injury or other musculoskeletal disorders. These health problems may become serious if no action is taken. They can be caused by poor design of workstations (and associated equipment such as chairs), insufficient space, lack of training or not taking breaks from display screen work.

Work with a screen does not cause eye damage, but many users experience temporary eye strain or stress. This can lead to reduced work efficiency or taking time off work.

Working with electricity Electricity can kill. Although deaths from electric shocks are very rare in hairdressing salons, even a non-fatal shock can cause severe and permanent injury. An electric shock from faulty or damaged electrical equipment may lead to a fall (e.g. down a stairwell).

Those using electricity may not be the only ones at risk. Poor electrical installations and faulty electrical appliances can lead to fires that can also result in death or injury to others.

Get into the habit of looking for loose cables and plugs on tongs, straighteners and hairdryers *before* plugging them in for use. If you think that a piece of electrical equipment is faulty or damaged, tell your supervisor immediately. Make sure that no one else tries to use it.

> Get actively involved in daily checks, which quickly become part of your routine work. Encourage staff to get involved by discussing health and safety issues in staff meetings, report back on findings or promote discussion and recognize good practice.
>
> *Wendy Dixon*

REVISION QUESTIONS

Q1. A _____ is something with potential to cause harm.   Fill in the blank

Q2. Risk assessment is a process of evaluation to ensure safe working practices.   True or false?

Q3. Which of the following are hazards? Select all that apply: Multi-selection

Shampooing with disposable gloves on   ☐   1

Hair clippings on the floor   ☐   2

Wet or slippery floors   ☐   3

Shampooing products   ☐   4

Styling mirrors   ☐   5

Trailing flexes from electrical equipment   ☐   6

Q4. A standard first aid kit should contain sterilizing tablets.   True or false?

Q5. Which of the following regulations relate to the safe handling of chemicals?   Multi-choice

PPE   ○   a

RIDDOR   ○   b

COSHH   ○   c

OSRPA   ○   d

Q6. All salons must have a written health and safety policy.   True or false?

Q7. Which of the following records must a salon keep up-to-date by law: Multi-selection

Stock records   ☐   1

Accident book   ☐   2

Appointment book   ☐   3

Electrical equipment annual test records   ☐   4

Health and safety at work checklist   ☐   5

Fire drill records   ☐   6

Q8. Safety regulations require employers to provide adequate equipment and facilities in case of a _____ occurring.   Fill in the blank

Q9. What sign/poster is required by law to be displayed on the premises?   Multi-choice

Latest hairstyles   ○   a

Health and safety information   ○   b

Employer's name, address and contact details   ○   c

Forthcoming events   ○   d

Q10. Contact dermatitis is an occupational health hazard.   True or false?

16 Personal Effectiveness

LEARNING OBJECTIVES

◆ Be able to work efficiently and effectively.

◆ Be able to monitor salon resources.

◆ Be able to maintain salon recording systems.

◆ Understand ways for developing yourself in your job role.

◆ Know your own productivity targets and how you can achieve them.

◆ Be able to work as a team member with your fellow staff.

KEY TERMS

appraisal

consumer protection

effectiveness

induction

job description

productivity

resources

UNIT TOPICS

Develop and maintain your effectiveness at work

Contribute to the financial effectiveness of the business

INTRODUCTION

It is essential to be effective in our work; the way in which we use salon resources and maintain productivity has a direct impact on the profitability of the business. This chapter focuses upon the following:

◆ the salon's resources

◆ developing yourself in your work

◆ being productive.

The resources of a business are its key assets. When they are managed well, the business should be productive and can make a profit; when they are managed poorly, or business processes break down, then the business loses money and fails.

This chapter describes how you can contribute to the financial effectiveness of the business. It focuses upon the use of the salon's resources – human, stock, tools and equipment – as well as making the best use of your time while you are at work. If you can optimize these resources, you will be helping your own self-improvement and the achievement of your personal targets. The people employed within a business – its human resources – are its most important asset. This chapter looks at the legalities of employment and performance monitoring, then at stock management.

Personal effectiveness

PRACTICAL SKILLS

Find out how to monitor the salon's resources

Look for ways to improve how resources are used

Learn how to develop yourself within your role

Make recommendations to management about how things can be improved

Work with your manager to evaluate your effectiveness

Work towards your own productivity targets

UNDERPINNING KNOWLEDGE

Know your salon's resources and where efficiencies can be made

Understand how the use of resources can impact the business effectiveness

Know how to find sources to develop yourself within your role

Know your own strengths and weaknesses

Understand why you need to work towards targets

Contribute to the effective use and monitoring of resources

Human resources

Hairdressing and barbering is a labour-intensive service industry. It relies solely on the profits generated from the sales of services/treatments and products provided by stylists to their clients. The people working in a hairdressing salon are therefore the most essential part of the business and it involves many people. The receptionist meets and greets the clients, handles bookings over the telephone, operates the till and assists in the selling of retail products. The stylists tend to their clients, providing them with services and treatments. The

INDUSTRY ROLE MODEL

ROSS MILLER Managing Director/Artistic Director, Renella

"In 1990 I started in the family business, Renella, which was first established in 1969 by my parents. Our focus has always been on training and we have gained a number of awards and recognition for this. We believe strongly in qualifications and insist on trainees gaining qualifications in our salon. I train along with two others in the salon.

In 1997, I started training for Great Lengths (a hair extension company), then with Habia from 2005, which has been amazing.

I have learned so much from this and I still love it. I now also train for Wella on all their retail products for their many salons across the UK.

Since 2011, we have been a certified centre and want to expand this over the next two years. We are adding more courses for our own staff and for external hairdressers.

juniors are trained to create business for the future of the salon. The manager delegates the tasks, collects information and makes decisions in order for the business to function. The cleaners ensure that all areas are fit for their purpose and hygienically safe. They are all part of one team and work together collectively so that the business can function effectively and make a profit. Human resources are a salon's most important resource.

One of the key skills is flexibility, the ability to adapt and not ever to say 'NO'. Be like this as much as possible and no matter where you work, you will improve your career no end. The other key skill is a positive attitude, get this right and you will always succeed.

Ross Miller

TOP TIP

Find out what your salon's policies are in relation to the use of resources.

Induction When people start a new job there are many things that they need to learn about before they can join the rest of the team; this happens during induction. During the induction process, new members of staff will learn about health and safety policies, the expectations of the employer and the details of their duties.

New staff are given many important documents, including their job description. This document describes the remit of the job, the duties involved and the responsibilities that the job holder will have within their work. New staff will also be given a contract of employment, health and safety information and grievance and disciplinary procedures. Collectively, this information sets out what they should be doing, when they will be doing it and under what conditions it will all take place.

Job description: Stylist

Location:	Based at salon as advertised
Main purpose of job:	To ensure customer care is provided at all times
	To maintain a good standard of technical and client care, ensuring that up-to-date methods and techniques are used following the salon training practices and procedures
Responsible to:	Salon Manager
Requirements:	To maintain the company's standards in respect of hairdressing/beauty services
	To ensure that all clients receive service of the best possible quality
	To advise clients on services and treatments
	To advise clients on products and aftercare
	To achieve designated performance targets
	To participate in self-development or to assist with the development of others
	To maintain company policy in respect of:
	• personal standards of health/hygiene
	• personal standards of appearance/conduct
	• operating safely while at work
	• public promotion
	• corporate image as laid out in employee handbook
	To carry out client consultation in accordance with company policy
	To maintain company security practices and procedures
	To assist your manager in the provision of salon services
	To undertake additional tasks and duties required by your manager from time to time

TOP TIP

Arbitration and Conciliation Advisory Service (ACAS)

Helpline 08457 47 47 47

www.acas.org.uk

Each member of staff will be given their own personal targets; these are the efficiency, productivity and/or development targets that will be monitored and reviewed over time. These targets are set and mutually agreed by the staff member and their supervisor, so that each member of staff knows what is expected of them.

Working conditions

Contracts of employment (ACAS) Key points:

◆ A contract of employment is an agreement between an employer and employee and is the basis of the employment relationship.

◆ Most employment contracts do not need to be in writing to be legally valid, but it is better if they are.

◆ A contract 'starts' as soon as an offer of employment is accepted. Starting work proves that you accept the terms and conditions offered by the employer.

◆ Most employees are legally entitled to a Written Statement of the main terms and conditions of employment within two calendar months of starting work. This should include details of things like pay, holidays and working hours.

◆ An existing contract of employment can be varied only with the agreement of both parties.

You might assume that a contract of employment consists of only those things that are set out in writing between an employer and an employee and many of the main issues, such as pay and holidays, are usually agreed in writing. But contracts are also made up of terms that have not been spelt out. This might be because they are:

◆ **too obvious to mention**: for example, you would not expect a contract to say that 'an employee will not steal from an employer'; it is implicit

◆ **necessary to make the contract work**: for example, if you are employed as a senior stylist it is assumed that you hold that level of qualification

◆ **custom and practice**: some terms of a contract can become established over time.

TOP TIP

The Working Time Regulations (1998) govern the hours anyone can work. They determine the maximum weekly working time, pattern of work and holidays, plus the daily and weekly rest periods. They also cover the health and working hours of night workers. There are a small number of exceptions: certain regulations *may be excluded or modified by a collective or workforce agreement,* and certain categories of worker are excluded.

In general, the Working Time Regulations provide the following rights:

◆ Limit to an average 48 hours a week on the hours a worker can be required to work, although individuals may choose to work longer.

◆ 5.6 weeks' paid leave a year.

◆ 11 consecutive hours' rest in any 24-hour period.

◆ 20-minute rest break if the working day is longer than six hours.

◆ One day off each week.

◆ There are special regulations for young workers, which restrict their working hours to 8 hours per day and 40 hours per week. They are entitled to two days off each week.

◆ Limit the normal working hours of night workers to an average 8 hours in any 24-hour period, and an entitlement for night workers to receive regular health assessments.

Rest breaks during the working day Employees are entitled to regular breaks in the working day.

◆ Workers aged 18 or over should be offered a minimum 20-minute break for every shift lasting more than 6 hours. (The break cannot be taken at the start or the end of a work shift.) It is the employer's duty to ensure that their workers can take their breaks.

◆ Young workers, aged 16 and 17, should take at least 30 minutes' break if they work more than 4.5 hours.

Holiday entitlement Under the Working Time Regulations 1998 most workers are entitled to paid holidays or annual leave.

A full-time worker is entitled to holiday from their first day of employment and the entitlement is 5.6 weeks (or 28 days) annually. Pay is based on a normal week's pay. Part-time workers are entitled to the same holidays as full-time workers but this is calculated on a pro rata basis.

Public and bank holidays When the Christmas and New Year public holidays fall at a weekend, other weekdays are declared public holidays. Paid time off does not legally have to be given for public holidays, and if it is it can be included in the employee's minimum leave entitlement.

Part-time workers have the same entitlement to leave as full-time workers, so if full-time workers are given paid leave for bank holidays, part-time workers should also be granted payment on a pro rata basis.

Employment of young workers between the ages of 13 and school-leaving age

Part-time work The youngest age a child can work part-time is 13 years, except children involved in employment areas such as:

◆ television

◆ theatre

◆ modelling.

Children working in these areas will need a performance licence.

Restrictions on child employment There are several restrictions on when and where children are allowed to work. School-aged children are not entitled to the National Minimum Wage.

Children **are not allowed** to work:

◆ without an employment permit issued by the education department of the local council, if this is required by local bylaws

◆ in places like a factory or industrial site

◆ during school hours

◆ before 7 am or after 7 pm

◆ for more than one hour before school (unless local bylaws allow it)

◆ for more than 4 hours without taking a break of at least 1 hour

◆ in most jobs in pubs and betting shops and those prohibited in local bylaws

◆ in any work that may be harmful to their health, well-being or education

◆ without having a two-week break from any work during the school holidays in each calendar year.

Local council rules for child employment permits Most local councils say that businesses intending to employ school-aged children must apply for a child employment permit before they can be employed. If a child is working without a child employment permit, it is possible that the employer will not be insured against accidents involving the child.

School leavers starting full-time work After a child has reached the minimum school-leaving age, they can work up to a maximum of 40 hours a week. In England the school-leaving age is 17 (since 2013) and it will be 18 years in 2015.

Staff training

Good technical skills and sound product knowledge are essential for success. The employer makes the investment in staff training as this will repay the business through:

◆ increased productivity

◆ improved services and treatments

◆ better financial performance.

We have to always see ourselves as professionals. I believe we are in one of the most professional industries there is. If you look at all the top salons, they all consider professionalism in everything they do. Act this way, be this way and you will succeed.

Ross Miller

On the other hand, a lack of staff training will affect all of these aspects negatively. If the staff try to carry out technical services that they are not trained or able to do, the results will show through:

◆ dissatisfied clients

◆ lack of confidence and poor self-esteem

◆ loss of business through lost clients.

Good working relationships are also vitally important to the overall success of the business. Every member of staff needs to feel part of the team as each one plays an important role in ensuring that success. On the other hand, poor working relationships create an unpleasant working environment for the staff and the clients; one way of controlling this is through good effective communication.

Good communication

Effective communication takes place when information is passed from one person to another with a clear, unambiguous message. The style of communication could be oral or written.

Oral communication Oral or verbal communication occurs when you speak to clients over the telephone or face to face, during consultation or when carrying out the service. As a more senior member of staff, it also occurs when you instruct junior staff, handle complaints, and deal with external bodies such as trade suppliers.

Oral communication is a good technique for providing information quickly. When you communicate verbally with others remember:

◆ that the tone of your voice and your body language will also be taken into account by the listener

◆ that the place in which you provide the information may affect the listener's response

◆ that clear information will not be misunderstood.

Informing clients about all levels of your business is a key component to the cogs of the business. Not only will it bring them to your door, it will allow a much wider spread of knowledge about your salon and what you are about. Never think it doesn't matter, because it always does.

Ross Miller

Written communication Written communication may not be as quick at providing information, but the results of it are much longer lasting. In many situations, a detailed or documented record is far more useful than '*I thought you said I could . . .*'

Different forms of written salon communication include:

◆ formal instructions to staff

◆ reports prepared for management

◆ formal details or instructions to external bodies

◆ details of treatments, tests and responses from clients

◆ records of complaint.

There are times when information has to be recorded; for example: client records, taking messages and stock procedures.

Client records can be manual or computerized but will contain similar information:

◆ client name, title, address and contact information

◆ previous service, treatment, tests and product information

◆ date, costs and timings of previous visits

◆ stylist/operator details and any other additional memos

◆ detailed responses from clients during consultation for potentially more problematic services, e.g. hair extensions, hair patterning and design.

Messages and notes An effective message or note will be clearly written and should include the following:

◆ for whom it is intended

◆ who took the message

◆ the date and time

◆ purpose

◆ clear details or instructions.

Body language

For more information on positive and negative body language see CHAPTER 1 Consultation and Advice, pages 9–10.

Code of conduct at work

Be polite and courteous with colleagues and clients at all times.

Specifically:

◆ Never talk down to staff members – treat them how you would wish to be dealt with yourself.

◆ Never get cross with staff members in front of clients – if you need to speak sternly, do it away from the salon floor. (If the matter is serious, you may need a witness.)

◆ Never ever ridicule or make fun of a member of staff in front of clients or other staff. You should also never discuss clients with other clients or staff members.

◆ If you do have a contentious situation or personal issues with a colleague, do not let your professionalism fail. Settle grievances as soon as possible and move on.

Dealing with client complaints A client has every right to expect to receive the services that were agreed and paid for; if an unexpected result occurs, the client will be right to complain. However, dealing with a dissatisfied client is not easy and the situation should therefore be approached with consideration and care.

If a client comes to you with a complaint, you should first move your client away to a quieter area of the salon. You then need to:

◆ Find out exactly what the problem is

◆ Assess the validity of the complaint

◆ Mutually agree on a suitable course of action (you may need the help of a senior manager)

◆ Carry out/organize any corrective work

◆ Record the occurrence and the remedial action that was taken.

Consumer rights and legislation

Equality Act 2010

The Equality Act 2010 legally protects people from discrimination in the workplace and wider society. It replaced previous anti-discrimination laws with a single Act, making the law easier to understand and strengthening protection in some situations. It sets out the different ways in which it is unlawful to treat someone and provides the UK with a new discrimination law which protects individuals from unfair treatment and promotes a fair and more equal society.

The Equality Act has replaced the Equal Pay Act 1970, the Sex Discrimination Act 1975, the Race Relations Act 1976, the Disability Discrimination Act 1995 and legal protection for people on grounds of religion or belief, sexual orientation and age. It requires equal treatment in access to employment as well as private and public services, regardless of the protected characteristics of age, disability, gender reassignment, marriage and civil partnership, race, religion or belief, sex and sexual orientation. In the case of gender, there are special protections for pregnant women.

Data Protection Act (1998)

Any organization that records information about staff or clients, whether on a card index system or a computer, must comply with the Data Protection Act. The law requires people to keep safe and secure any information that is held on file about their customers. In salons, this means that information about clients must be kept confidential and handled with the utmost professional care. All salon staff have a responsibility to maintain this confidentiality at all times, even after working hours. You are not at liberty to discuss any details of your clients with anyone.

Data Protection Act 1998

The Data Protection Act (DPA) applies to any business that uses computers or paper-based systems for storing personal information about its clients and staff.

It places obligations on the person holding the information (data controller) to deal with it properly.

It gives the person that the information concerns (data subject) rights regarding the data held about them.

The duties of the data controller

There are eight principles put in place by the DPA to make sure that data is handled correctly. By law, the data controller must keep to these principles. The principles say that the data must be:

1 fairly and lawfully processed
2 processed for limited purposes
3 adequate, relevant and not excessive
4 accurate
5 not kept for longer than is necessary
6 processed in line with your rights
7 secure
8 not transferred to other countries without adequate protection.

For more information see http://www.ico.gov.uk/

Summary of the Data Protection Act

Sale of Goods Act (1979)

Under the Sale of Goods Act, when a salon sells something to a customer it has an agreement or contract with them.

A customer has legal rights if the goods they purchased do not conform to contract (are faulty). The Act says that to conform to contract goods should:

1   **Match their description** – by law everything that is said about the product must not be misleading – whether this is said by a sales assistant, or written on the packaging, in-store, on advertising materials or in a catalogue.

2   **Be of satisfactory quality** – this quality of goods includes:

  ◆ appearance and finish

  ◆ freedom from minor defects (such as marks or holes)

  ◆ safe to use

  ◆ in good working order

  ◆ durability.

3   **Be fit for purpose** – if there is disagreement with the customer about a particular purpose, it should make this clear, perhaps on the sales receipt, to protect the business against future claims.

Trades Descriptions Act 1968/1972

Products must not be falsely or misleadingly described in relation to their quality, fitness, price or purpose, by advertisements, orally, displays or descriptions. Since 1972 it has also been a requirement to label a product clearly, so that the buyer can see where the product was made.

Briefly, a retailer cannot:

  ◆ mislead consumers by making false statements about products

  ◆ offer sale products at half price unless they have been offered at the actual price for a reasonable period.

Consumer Protection Act (1987)

The **Consumer Protection** Act follows European laws to protect the buyer in the following areas:

  ◆ Product liability – a customer may claim compensation for a product that doesn't reach general standards of safety.

  ◆ General safety requirements – it is a criminal offence to sell goods that are unsafe; traders that breach this conduct may face fines or even imprisonment.

  ◆ Misleading prices – misleading consumers with wrongly displayed prices is also an offence.

The Act is designed to help safeguard the consumer from products that do not reach reasonable levels of safety. Your salon will take adequate precautions in procuring, using and supplying reputable products and maintaining them so that they remain in good condition.

Resale Prices Act (1964 and 1976)

The manufacturers can supply a recommended price (MRRP or manufacturer's recommended retail price), but the seller is not obliged to sell at the recommended price.

Stock and stock control

Stock is a valuable resource of the business and, at any one time, the business may have large amounts of money tied up in stock. It is essential that stock is used appropriately, that wastage is kept at an absolute minimum and that it is kept secure while on the premises.

Stock levels can only be maintained if accurate records are kept of how much stock the business has. Stock records should be able to accurately show:

◆ minimum holding levels of each product line that the business needs

◆ current levels of each of those product lines

◆ products/items that need to be reordered.

Good stock-keeping practice

It is important to monitor the usage of products against the minimum holding levels, so that stock needs are anticipated and reordered before the business runs out. Good practice involves placing orders when stock levels run low.

Check deliveries when they arrive against the delivery note. The delivery note is not necessarily a copy of what has been ordered; it is a list of what has been dispatched by the supplier. It can be different if:

◆ products are out-of-stock at the suppliers

◆ products have been substituted (as ordered items are out of stock)

◆ products are missing.

In cases where the delivery does not tally with the stock delivered, the stock should always be checked to see that it is not damaged or that it has not deteriorated.

Stock control

The stock control systems that your salon uses will provide management with up-to-date information. These systems will deal with:

◆ reordering stock

◆ movements of stock

◆ usage of stock

◆ shortages of stock

◆ safety and security of stock.

Consumables may be used during salon services or sold to clients for home use. The salon must keep sufficient stock and will purchase products in varying quantities, for short-term or long-term availability.

To ensure that the products remain usable or saleable, the stock controller must monitor them and will need to be aware of:

◆ shelf life – how long a product will last

◆ handling – how it will be moved, stored, etc.

◆ losses – missing items, theft, out-of-date, etc.

◆ damages – products lost in transit or during handling.

Stock held in store is a valuable asset to the company. The stock controller is responsible for its safe storage between delivery and use or sale.

ACTIVITY

Dealing with resources

This activity looks at the problems associated with shortages, surpluses and breakdowns of salon resources. Complete the table below with the missing information.

Resource issue	What are the possible reasons for this problem?	What could happen if this is not addressed?	Suggest possible ways to resolve the issue.
Stock shortages			
Stock excess			
Stylist overbooked			
Stylist absent due to illness			
Stylist running late			
Equipment breakdowns			
Backlog of clients waiting for a basin			

Stock rotation

Many items sold in shops have a short or limited shelf life. This quality control ensures that the product is sold safely, and in its optimal condition. If the sell-by date has passed, the item must be removed from the shelves by law.

In a salon, however, the products that are sold or used have much longer lives as their ingredients do not deteriorate very quickly. However, salons still rotate their stock as a matter of good customer service. So when new stock is placed on shelves for sale or use, ensure that old products are brought forward so that they may be sold or used first.

Stocktaking

Items for use such as tools, small pieces of equipment and potentially hazardous chemicals should be kept in a locked store, the size of which will depend on the salon and its needs. Individual items are accounted for by stocktaking.

Stocktaking at regular intervals provides management with up-to-date information of stock movement. Without regular stocktaking, individual items and product lines could run out, creating a situation in which services and treatments normally offered were not available. This would mean loss of profit to the salon, both at the time and later through a damaged reputation. Every business requires accurate, reliable accounting systems which:

◆ organize their products into different categories, e.g. backwash, styling, retail

- ◆ monitor the usage of the products
- ◆ identify shortages
- ◆ report damages or defects when products are delivered
- ◆ update records.

These guidelines provide the basis for a simple yet effective stock management system.

Product coding

Many salons now use computer technology to produce management information. Stock control is one of the facilities available in software systems for salon management. Salons turning over large quantities of stock find it helpful to devise coding systems for the products they use and sell.

The product's manufacturer, its category, name and size can all be stored as a single alpha numerical code – the product code. These codes can streamline the processes of stock control, monitoring, pricing and tax calculation.

Products received into storage are individually itemized and allocated the relevant product code. All the information about a product is then held on the computer. The computer continually recalculates the stock levels, providing management with automated stock control information and printouts for use in manual stocktaking checks. This coded system is one form of point-of-sale (POS) management; another system uses barcoding. The principle here is exactly the same, but the product information is converted into a series of stripes printed on labels or directly onto the product. The barcodes can be read directly by the computer via a scanning barcode reader, which recognizes the product and makes the necessary stock control adjustments.

Ordering stock

Products are purchased either directly from the manufacturer or via a wholesaler, on a credit- or cash-based agreement. Credit account terms are arranged with the supplier, usually on a monthly payment system.

ALWAYS REMEMBER

If your salon uses tubes of permanent colour, encourage staff to replace part-tubes in original packaging with a clear indication of how much is left.

Stock records and recording stock movements

Placing a stock order A salon's order may be placed with a company representative, who completes a purchase order on the salon's behalf. The purchase order is a paper system documenting all the manufacturer's product listings and categories. This is returned to the company so that the order can be processed and dispatched.

Taking delivery When the stock order arrives at the salon it will be accompanied by a delivery note that will list the items so far dispatched and any that are to follow, such as items temporarily out of stock. The delivery note must be checked against the contents of the consignment and discrepancies or damages in transit identified before countersigning the order and confirming the delivery. Any discrepancies between the documents should be referred to the management for later adjustment. The incoming stock should be moved immediately from reception to a secure location away from the working area of the salon. At a convenient time, the salon stock systems can be updated and stock put into storage.

After a short while, the supplier will send an invoice, i.e. a request for payment. Details of the invoice must be checked against the delivery note and the stock actually received.

Choice of stock supplier

Wholesalers carry stock from a wide range of manufacturers, providing the salon owner with a choice of products and differing prices to suit various budgets. When orders are placed through a manufacturer's representative, the salon is restricted to buying the products available from that manufacturer.

ACTIVITY

A manual stock recording system

This activity will provide you with a working system for operating a small but effective manual stock record system.

Keep copies of the stock control sheets that you create over an 8 to 12-week period so that you can collect a reasonable amount of data.

The example shown in the figure below provides a simple format for a paper-based stock recording system. We can see that in the first column there is a range of product types: each family of products – shampoos, conditioners, etc. – is grouped together. In the next column the product's unit size is identified.

The next two columns are each repeated several times. These contain space to enter the date, product minimum holding levels, amount in stock and quantities for order. Stock is then ordered when the amounts fall below the minimum holding levels. When repeated over several columns it is easy to identify faster moving products and trends or patterns.

Stock master		Date 12/Sept/14			Date			Date		
Salon retail products	Size	Minimum holding level	In stock	Order	Minimum holding level	In stock	Order	Minimum holding level	In stock	Order
Shampoos										
Moisturizing	250 ml	4	3	1 box						
Enriching	250 ml	4	4	0						
Revitalizing	250 ml	3	7	0						
Oil control										
Conditioners										
Protein	250 ml	4	2	1 box						
Colour care	250 ml	4	2	1 box						
Moisturizing	400 ml	2	4	0						
Frequent use	400 ml	2	4	0						

You may be able to visit a nearby wholesale cash-and-carry warehouse. Such warehouses provide an alternative service to the salon, holding stocks ranging from consumable product lines to sundry items such as towels, gowns and hair ornaments, and even coffee and washing powders.

Wholesalers such as this provide the salon with a one-stop shopping facility.

For more on lifting stock see CHAPTER 15 Health and Safety, Manual Handling Regulations 1992, page 382.

Stock handling

Most products used by salons are packaged and many are chemicals. Movements of stock into or within the salon may involve lifting, stacking, dispensing, displaying or pricing, all of which are subject to stringent legislation.

The Health and Safety at Work Act 1974 relates to all workplace health and safety, although the Act has specific requirements for the employer. Employees have a duty under the law not to endanger their own health or safety, or that of other people who may be affected by their actions.

All cosmetic products come under strict legislation (Cosmetic Products Regulations 1989) and a specific guide to health and safety in the salon relating to the control of substances hazardous to health (the COSHH regulations) has been written by the Cosmetic, Toiletry and Perfumery Association (CTPA) with the cooperation of the Hairdressing and Beauty Suppliers' Association (HBSA).

The responsibilities of the employer can be found in CHAPTER 15 Health and Safety, page 372.

This guide assesses substances potentially hazardous to health and provides information to employers about exercising adequate controls. Apart from basic rules for hairdressers relating to hair product and salon safety, substances are categorized as 'potential' or 'unlikely' hazards. Each type of product identified is specified by:

◆ name – including ingredients and a general description

◆ health hazard – inhalation, ingestion, absorption, contact or injection

◆ precautions – during work activity, storage and disposal or spillage

◆ first aid – in relation to eyes, skin or ingestion

◆ fire risk – if applicable.

Security

Stock in storage is a valuable asset to the company. Thieves are often opportunist, not always planning their activities. They will seize opportunities as they arise: such as money left around, products on display, unlocked doors and open windows. You should take all necessary precautions to maintain a secure working environment.

TOP TIP

If you are unsure about the contents of your salon's products, contact your supplier for relevant COSHH information.

Avoiding waste and damage

Regular checks on goods through careful stock control will assist in minimizing shortages, but shortages can still occur if items are carelessly wasted, such as preparing a colour using a whole tube of colour when half a tube would have been enough. Applications should be carefully measured. Manufacturer's recommendations can be found on all products.

Utilities Staff should be given clear guidelines about the efficient use of resources, as any form of waste increases the costs for the salon and decreases the profits. For example, taps should not be left on between shampoos. Hood-dryers should not be left running after the client has finished. Personal calls should not be made from the salon telephones.

Tools and equipment Regular checks on tools and equipment will help to minimize problems. These may still occur, however, if items are misused. Using tools for purposes other than those intended could be negligent, if not dangerous. Staff must know how to use and maintain tools and equipment correctly and should be given relevant health and safety training.

It is important to monitor tools and equipment to:

◆ maintain the correct numbers of working items that are needed to provide an uninterrupted, smooth salon operation should any damages or broken items occur

◆ identify which items need replacing or updating to maintain the levels of service.

Space Effective use of space should also be monitored. Turnover can be measured against the square metre to gauge the productivity of a given area; for example, retail sales.

Time and time management

Time is money

Time is a resource that, although not tangible, is crucial to the financial **effectiveness** of the business. It affects issues such as pricing structure and staff training. As the financial income of the salon is largely based on client service, the price structure will reflect the length of time a service takes.

For example, a cut and blow-dry may have a time allowance of 45 minutes, while a highlighting service may have an allowance of two hours and will therefore be correspondingly more expensive, irrespective of other resources that have been used, such as light and heat, laundered items, equipment and products.

Ineffective use of time comes from not doing the right job at the right time. Wasted lost time cannot be made back up in the normal allocation of work. So having to spend extra time getting back to where we were before, we expend more efforts and usually more money in the process. This ineffective work method is inefficient and is a burden on other team members.

Be organized . . . Plan . . . Become an efficient person and this will always see you through in all that you do.

Ross Miller

ACTIVITY

Work survey

Ask your fellow work colleagues if they mind taking part in a work survey.

Over a period of six observations, see how long it takes them to complete the following services:

◆ Cut and blow-dry (short hair including shampoo/conditioning, etc.).
◆ Cut and blow-dry (shoulder length hair including shampoo/conditioning, etc.)
◆ Blow-dry (short hair including shampoo/conditioning, etc.).
◆ Blow-dry (shoulder length hair including shampoo/conditioning, etc.).

Now make your evaluation of the work study by answering these questions:

1 What was the average time taken for each service?
2 What was the longest and shortest time taken in each case?

In answering questions 1 and 2 you can now draw your conclusions by answering these final questions:

a. Have you any suggestions how the results of (Q2) can be improved?
b. What did the stylists feel about the time available for each instance?
c. How does this fit in with the salon's expected timescales?

Time management

We might like to think that as hairdressers we are good managers of time; our working life requires us to work and keep to time. We may never keep clients waiting, not even five minutes, but it is also important to manage time effectively so that we do not negatively impact on the work of others – getting in their way or interrupting their work flow.

Get organized If you do not take control and organize your time at work effectively, you will never have time for anything. You need to be well organized.

Prioritizing things to do Tasks need to be prioritized. For example, it may be your job to check the stock levels on a weekly basis; so make sure that this is done at the right time. Do not waste time dealing with non-urgent tasks. Make a list of things to do and then put them into order of priority.

Lists are very useful time management tools, but they only work if you stick to them rigidly. Find a system that works for you and a way of keeping your list to hand so that you can work with it, add to it and finally cross things off when completed.

Write things down If you do not write down what you have to do you may forget some of your tasks, only remembering at the last minute or too late. Build list writing into a daily routine and set aside time to review the items on the list on a regular basis. Attend to the important issues as soon as possible. People who are really in control of their time plan their activities, remembering that social and leisure time are just as important as their working life.

Don't put things off Do not get into a habit of pushing a task that you do not want to do over into another day. It will not go away. So get organized and tackle the things that you do not like doing; as well as those you enjoy.

ACTIVITY

Create a list of things to do

Writing things down is a very easy and effective way of remembering what has to be done and is a major contribution to effective time management.

Create your own things to do list – what needs to be done and when; for example, things to do today, things to do this week or this month. Update the list as you go along and cross things off when you have completed them so you also benefit from seeing what you have achieved.

Meet productivity and development targets

Fierce competition is the driver for improvement. In order to keep ahead of the competition, it is essential that you strive to improve your own skills and learn new things – changing what you do now to meet the expectations of the clients tomorrow.

Your employer already has a plan of how they will take the business forward or keep abreast of the competition. But you might also be able to see gaps in the range of services that your salon delivers to its clients, or think of ways in which there is room for improvement; if you do have ideas, report them to your line manager.

Ideally, you should be able to analyse your own performance in the salon and identify areas where there are weaknesses; which provides you with room to improve.

Maintaining productivity

Low productivity is the result of poor service, lack of training and ineffective use of materials and time. The guaranteed outcome of this is easy to forecast. It also signifies a failure of management. If people are not given the time, skills and materials initially, they cannot achieve their targets.

Good productivity is the result of achievement, so it is easy to see that in order to achieve there must be clear objectives. Targets should be clearly understood and attainable. Virtually all salons work on the basis that stylists earn a basic salary with a commission incentive scheme. Commission is payable as a bonus on top of wages when individual or group targets have been achieved.

Targets

We all need targets. We all like the benefits that come with achievement: credit and praise, higher self-esteem, increased confidence, an ability to please others and, last but not least, rewards. All these are positive outcomes for doing what is expected from us. Collectively, well-defined targets forge a unity within a working team that act as an incentive for everyone involved.

In order for people to respond to the challenge, the targets should be realistic, achievable and tailored to the individual. Unrealistic targets, which from the outset are over-optimistic, unrealistic and unachievable, will have a very negative impact. The result will be a reduction in the bond of the working team and demotivation in the struggling individual.

Targets are not just about selling; they can also be relevant to personal learning.

ACTIVITY

Strengths and weaknesses chart

This activity provides a way for you to study and assess your strengths and weaknesses at work and to provide a course of action for the future.

Personal skill	My strengths	My weaknesses	Action to take
Dealing with clients			
Dealing with complaints			
Communicating with work colleagues			
Organizing work for others			
Helping others in their work			
Sorting out problems			

SMART productivity

For a hairdressing business to be successful, the salon owner has to take an overall view of productivity. This requires a continuous analysis of personal performance which can only be measured against a target figure. First, the salon owner has to set an overall salon target. This figure can then be divided between departments of stylists, technical and retail. Each person in each department is then given a personal target which they understand and agree with. The personal target can be worked out as follows:

Target = service price × number of clients

For example, if stylist Kerry charges £40 for a cut and blow-dry and can take eight clients a day, her daily takings for this service would be £320 and her weekly takings would be £1,600 (based upon a five-day week). This may be adjusted to allow for different daily performances. We cannot ensure consistent bookings, although through analysis we can establish high and low points. In addition to the styling takings we would also expect some retail sales, so the overall personal target would include this. Target setting should follow the SMART principle and should be:

◆ **S**pecific – clearly defined.

◆ **M**easurable – quantifiable in some way.

◆ **A**greed – between both parties.

◆ **R**ealistic – able to be achieved.

◆ **T**imed – for the duration of a fixed-time period.

Targets may be confidential between the manager and employee, in which case salon procedures relating to confidentiality must be observed. Personal reviews or appraisals provide an opportunity for management to establish an employee's performance level, to compare it against their target and to discuss ways of improving productivity. Most hairdressing businesses work on an incentive payment scheme. This can have a major impact on the overall salary rates. The salon owner needs to establish fixed costs and variable costs, and the wage percentage needs to be established in order for the necessary profit margins to be maintained. From the overall wage portion of income, an individual target is set for each stylist.

See **CHAPTER 17** Promotional Activities, for more information.

Recognizing achievement This is really about developing people's confidence and creating an environment of encouragement, but it has to be translated in the right way. Individuals should be allowed to make mistakes as well as given praise and recognition. There should be both financial targets and learning targets. Training is a major incentive – with relevant rewards.

Evaluating results The benefits must motivate individuals and prove worthwhile. Typical amounts spent vary from 3 to 5% of turnover on training budgets, to 30% salary equivalent on total packages. Measuring the return is notoriously difficult since incentives can have a ripple effect. However, health care benefits (check-ups, dental treatments, eye tests, counselling) are also a real benefit.

Working together

Always remember that your work colleagues also need your help to meet their targets. Sharing the workload is working as a team and can be achieved by:

◆ providing support

◆ anticipating the needs of others

◆ maintaining harmony

◆ communicating effectively.

In some salons, you might see some staff busy attending to their clients while others hang about around reception, flicking through magazines or disappearing off to the staffroom for a coffee. Teamwork is about making an active contribution, seeking to assist others even if only by passing up rollers. It is good for staff morale and presents a good image to the clients.

Anticipating the needs of others follows on from providing support. Clean and prepare the work areas ready for use, locate and prepare products as and when they are required. This will help the smooth operation of the salon. Cooperate with your colleagues. Make a positive contribution to your team by assisting them to provide a well-managed and coordinated quality service. Be self-motivated and keep yourself busy. Do not wait to be asked to do things.

Maintain harmony and try to minimize possible conflicts. Most good working relationships develop easily. However, others may need to be worked at. Whatever your personal feelings about your fellow workers, the clients must never sense a bad atmosphere within the salon caused by a friction between staff. You will spend a lot of time in the company of people you work with, although you will not always like everyone you meet. At work, in order to maintain teamwork, a mutual respect for others is more important than close friendships. So remember to treat others with respect, and be sensitive and responsive to other's feelings. Show concern and care for others.

Personal development

Managers use *performance* appraisal or *progress reviews* to evaluate the effectiveness of the work team. An appraisal is a system whereby you and your manager, in an interview situation, review and evaluate your personal contribution and/or progress over a predetermined period, as measured against expected targets or standards (see below).

A similar process would take place at suitable points within a personal programme of training in order to review progress and training effectiveness, measured against specific training objectives.

Measuring effectiveness In order to measure progress towards overall work contributions, as well as training targets, it is important to have clearly stated expectations of the performance required. These standards should show:

◆ the tasks that need to be performed

◆ training activities that will take place

◆ expected achievement levels

◆ when things will be assessed

◆ planned review of progress towards the agreed targets.

In normal, ongoing work situations, performance appraisal will be based on the following factors:

◆ results achieved against targets and job requirements

◆ additional accomplishments and contribution

◆ contribution made by the individual compared with those of other staff members.

" Contribute to financial success: you are key to any financial success, so don't ever think that you can't make a difference – you can! From the smallest amount of money made, it has an impact on the success of a salon. Even if you are a non-earner in a salon, you still make an impact, so always be aware of this and see how you can improve the salon as a whole, it is in your interest as well.

Ross Miller

For more information about job descriptions see page 395.

The job requirements are outlined in the employee's job description and the standards expected from the job holder will often include behaviour and appearance. If these have been stated from the outset, the job holder will know what is expected of them.

ACTIVITY

Identify strengths and weaknesses

Use this self-check-system before your appraisal as a way of monitoring your own performance.

Area of work	I am good at this	I am not so good at this	Supervisor's comments
Consultation			
Communication			
Customer care			
Retailing			
Cutting			
Colouring			
Perming			
Styling and dressing			
Long hair ups			
Extensions			
Health and safety			
Barbering			
Shaving			
Creative development			

The appraisal process

At the beginning of the appraisal period, the manager and employee discuss jointly, develop and mutually agree the objectives and performance measures for that period.

An *action plan* will then be drafted, outlining the expected outcomes. During the appraisal period, if there are any significant changes in factors such as objectives or performance measures, these will be discussed between the manager and employee and any amendments will be added to the action plan.

At the end of the appraisal period, the results are discussed by the employee and the manager, and both sign the appraisal. A copy is prepared for the employee and the original is kept on file.

An appraisal of performance will contain the following information:

◆ employee's name

◆ appraisal period

◆ appraiser's name and title

◆ performance objectives

◆ job title

◆ work location

◆ results achieved

◆ identified areas of strength and weakness

◆ ongoing action plan

◆ overall performance grading (optional).

ALWAYS REMEMBER

Be positive in the way that you receive feedback on performance, even if it is negative.

Dealing with negative feedback It is always hard to take criticism. However, negative feedback can be used positively, if it is seen as an opportunity to improve.

It is important that you try to remain positive and look upon it as part of the 'bigger picture'. Consider negative feedback as a learning experience that you use, like a tool, to move onwards and upwards.

> Never pass up an opportunity, always be eager to learn. There are so many chances and things to be done in this industry. Try to be around like-minded people. Always strive to be the best you can and better than everyone else around you. Enter as many competitions as you can, push your limits of creativity, and always ask questions; if you do this you will progress.
>
> *Ross Miller*

Self-appraisal In order for you to manage yourself within the job role, you need to identify the areas where you meet the expectations of your job and also the areas where there is room for improvement. Measuring your own strengths and weaknesses against laid-down performance criteria (as found in the Level 3 standards) is one way of monitoring your own progress. Use these to help you to identify where: further training is required, further practice is needed and competence can be achieved.

REVISION QUESTIONS

Q1. A worker aged 18 or over cannot be forced to work for more than _____ hours a week (on average). Fill in the blank

Q2. A key feature of good customer service is being customer-focused. True or false?

Q3. Which of the following would be considered an indication of poor communication? Multi-selection

Avoiding eye-to-eye contact	☐ 1
Smiling	☐ 2
Standing over the client and talking to them through the styling mirror	☐ 3
Talking with your hand covering your mouth	☐ 4
Being polite	☐ 5
Being courteous	☐ 6

Q4. Employees are entitled to regular breaks in the working day. True or false?

Q5. Which of these laws has specific relevance to holding people's private information on computer? Multi-choice

Equality Act (2010)	○ a
Disability Discrimination Act (2005)	○ b
Data Protection Act (1998)	○ c
The Prices Act (1974)	○ d

Q6. Stock rotation means turning products around on the shelves so they don't get too dusty. True or false?

Q7. Which of the following are advisable when dealing with a client's complaint? Multi-selection

Find out what the problem is.	☐ 1
Keep them waiting so that they cool off.	☐ 2
Assess the validity of the complaint.	☐ 3
Let them have a rant then you can retaliate.	☐ 4
Move them to a quieter part of the salon.	☐ 5
Let someone else deal with it.	☐ 6

Q8. A _____ note should accompany ordered stock when it arrives at the salon. Fill in the blank

Q9. Which legislation protects the clients from defective purchases? Multi-choice

Data Protection Act (1998)	○ a
The Sale of Goods Act (1979)	○ b
Disability Discrimination Act (2005)	○ c
The Prices Act (1974)	○ d

Q10. A barcode is the same as having a personal ID card. True or false?

17 Promotional Activities

LEARNING OBJECTIVES

◆ Be able to assist in the planning of promotional activities.

◆ Know the current legislation affecting the implementation of internal and external events.

◆ Be able to promote and sell salon services, treatments and products.

◆ Know your salon's policies and legal obligations.

◆ Be able to evaluate the effectiveness of promotional activities.

◆ Be able to communicate professionally and provide aftercare advice.

KEY TERMS

demonstration   in-salon campaign   SMART

UNIT TOPIC

Contribute to the planning and implementation of promotional activities

INTRODUCTION

Promotional activities are designed to stimulate customer interest, provide information and generate sales for the business. However, the work involved in the planning and execution is not just aimed at the clients. The whole team benefits, as everyone is motivated by doing something new. It promotes a sense of renewal and provides a break from the normal routines, which everyone finds exciting.

Promotional activities

PRACTICAL SKILLS

Learn how to plan internal and external promotions

Learn the aspects and features of good promotion and display

Make recommendations or provide alternatives to management based upon your plans

Learn how to sell, based upon clients' needs

Obtain feedback from clients by written and oral questioning techniques

Learn how to record feedback information for management purposes

UNDERPINNING KNOWLEDGE

Know your salon's ranges of products and services

Know the aspects and techniques of professional selling

Know how to plan and organize information for evaluation purposes

Understand the different ways that you can gain client feedback

Know where to find sources for current legislation affecting proposed events

Understand how to recognize strengths and weaknesses within your team

Know how to communicate effectively and professionally

Contribute to the planning and preparation of promotional activities

INDUSTRY ROLE MODEL

CARL MITCHELL Managing Director, Bonce Salons, West Midlands

I began my career within the high street retail sector, and soon became determined to own and develop my own brand. I started to work with my wife, Donna Mitchell (Creative Director) and we have taken her freelance hair salon, operating out of her parents' garage, into a flourishing salon environment. I do all of the marketing for our two salons and in 2012 Bonce Salons were announced as winners of the Marketing Award at the British Hairdressing Business Awards.

INDUSTRY ROLE MODEL

SAM GROCUTT Managing Director, Essence PR

" I established Essence PR in 2004, following ten years in the industry. A specialist hair and beauty agency, we work with professional, high street, niche and mass market brands. Our expertise spans both the industry press and consumer media and we are heavily involved in all aspects of the PR arena from product launches, artist liaison, media relations, marketing, branding, social media and press office management. I'm also heavily involved in the industry bodies such as the HABB, Fellowship for British Hairdressing, and Women in PR (WPR). My career highlight would have to be working with the talented Richard Ward and his team during the Royal Wedding. A once-in-a-lifetime opportunity and so much fun!

Initial preparation

There are all sorts of promotional events that you can get involved with, which may be internal or external. The bigger the ideas, the more complex it becomes. The table below helps you to think about the types of promotions that you can do and some of the issues that need to be addressed.

Promotional planning model

What type of event do you want to put on? What is the purpose of the event?	
Is it an internal event? For example:	Is it an external event? For example:
1 Window displays	1 Demonstration
2 Internal product promotion	2 Hair show
3 Internal display	3 Seminar
4 Demonstration	
What is the budget?	
What resources will you need?	
When will it take place?	
Who will be involved?	
What preparation or training is needed for participating staff?	
What advance notification, publicity or advertising is needed?	
How will you get the message across to the target group?	
What advantages will there be by running the event?	
How will you evaluate the event?	

You will need to cover the answers to each question before you can put any proposal to the management.

Finding a comprehensive answer to all of those questions may take you a lot of time, particularly if you have not had any experience in planning an event before. Nevertheless, unless you can be convincing with your proposition to management, your plans are unlikely to be taken forward.

The first things to think about are: what, why, where, who and how (you might also consider when). If you want to put forward an idea for a promotional activity, you need to be clear in your own mind why the event is useful at all.

ACTIVITY

Preparing your promotional ideas

Useful suggestions for finding ways of stimulating business are always going to interest management. So what ideas do you have? Try to focus your plans by filling in the table below. This will help you to provide a comprehensive and compelling case.

What?	Why?	Where?	Who?	How?

> For a simple but effective marketing plan consider breaking your plan into months and pencil in the key dates of the year – Valentine, Christmas, Summer, Mother's Day, Father's Day, etc. – then think in advance what you can do to promote these special occasions. Can you speak to your suppliers for ideas/help? Think of something that you can promote each month that will have a knock-on effect to your services as a whole, but don't discount, complement with another service instead. Consider the different mediums to promote, i.e. advertising, text campaigns, online, in-salon promotions, PR, retail offers, direct marketing, press releases and reader offers.
>
> *Sam Grocutt*

With that exercise completed, you can now consider the following before making your pitch.

Working to a budget Working out the cost of a promotion is the first thing that management need. It is pointless putting together a big flashy **demonstration** if the cost of putting the event on will far outweigh the income that it will generate. It is one thing working out what an event will cost to put on, but it is quite another to estimate how much income it will generate.

Your financial planning has to be accurate, although a tolerance of +5 or −5% should take care of unforeseen costs. Any overspend beyond this, regardless of the success of the promotional event, will be considered a dismal failure. Similarly, any dramatic under-spend shows that you cannot do sums, or proper planning, so it will be considered a failure too.

Getting the sums right means that you have planned down to the last photocopied leaflet or product used within the event. If you involve external suppliers, make sure that you obtain written quotes to back-up your plans. If they change the rules at some later time at least you have given management something to bargain with.

If several people are involved with the financial side then you will need regular reports from them to check on progress.

> First of all make sure you invest in marketing and treat it as your salon's sales person. The more you keep marketing active, the more you will see sales and clients. Some salons make the mistake of only trusting friend recommendations in order for the salon clientele to grow; while this is effective and cheap, it's certainly not quick.
>
> *Carl Mitchell*

Decide on the type of promotion

There are many possibilities and following are a few examples:

◆ Themed window display for general salon promotion

◆ In-salon display to promote a new range of products

◆ Introductory offers – e.g. offering a free retail conditioner for permed hair with each perm treatment

◆ Evening seminars – a targeted promotional event to a segment of the salon's clients

◆ External demonstration 'hair show': either as an internal evening seminar, or as an external event at a venue suitable for the target group.

> Never miss trying to sell to your current clients. They are visiting you week in week out so look to promote certain services. A loyalty card or scheme is good at keeping your clientele regular but instead of giving them money off services they already use introduce them to different services or retail. To me, it seems pointless from a business outlook to reward them for something they are already prepared to pay for.
>
> *Carl Mitchell*

> Social media is huge nowadays but you need to use it in the right way; after all, you are selling your company. This tool is free, effective and will help grow business. Make sure you take time to update the company website and keep it fresh. Remember that too much information is not clear enough communication.
>
> *Carl Mitchell*

Setting your objectives

Your objectives should fulfil the overall aims of the business:

◆ Generating customer interest and loyalty

◆ Increasing the salon's sales and therefore its profit

◆ Motivating salon staff

◆ Improving the salon's professional image.

If you can address these topics, you will be testing your plans against the main aims of the business. Remember that there will be many ways in which those objectives can be met.

Examples:

◆ You could create an **in-salon campaign** to promote a new product range.

◆ Run a seminar to a small group of clients (perhaps who are not yet colour clients) on the features and benefits of a new colour range.

◆ Devise a client loyalty scheme, where clients who return sooner can get a discount.

Be SMART When you set objectives you should follow the **SMART** principle. Your objectives should be:

◆ **S**pecific – clearly defined stating what it is that you want to achieve.

◆ **M**easurable – have some form of evaluation that can measure the success.

◆ **A**chievable – do you have the resources to attain the objectives.

◆ **R**ealistic – relate to something that is achievable.

◆ **T**imed – set within a fixed timeframe without over-running.

Most importantly, focus upon your target market or group: is your plan something that they *would or could* benefit from?

If you can honestly say yes to this question, you have a starting point.

> Make sure when you market your business you know what your USP (unique selling point) is. This is what makes you stand out from your competitors. If you want to find out what it is simply ask your clients. Every stylist in the salon should have a USP.
>
> *Carl Mitchell*

Know who your customers are

Different promotions or campaigns are going to be suitable for different client (or potential client) groups. For example, you would not plan an in-salon promotion for a new treatment for male pattern baldness if the majority of the salon's clients are female. You need to look closely at your target group and focus upon their needs and social patterns.

If you were planning a catwalk *showcase* of new looks, say for example, at a local nightclub, what time would be preferable to maximize the audience?

ACTIVITY

Features and benefits

With your colleagues, select six of the services provided by your salon and jointly consider what the features and benefits to the client are in each case. Make a record of the collective opinions and keep this within your portfolio for future reference.

ALWAYS REMEMBER

When there is a salon promotion, everyone needs to be informed about the campaign. Make sure before the promotion is rolled out that there are sufficient staff meetings to make everyone aware of what is going on.

When you have defined the target group and the age range that it relates to, you need to think about the visual content of your programme.

◆ Is it going to stimulate your audience?

◆ Is it going to promote feedback from your audience?

◆ How will you respond to the feedback from your audience?

◆ Is it going to enhance the salon's *good* name?

> " Always understand who your clients are and know who to target. If you carry out a marketing campaign, make it strategic and focused on the type of clients you wish to reach. If you market to everyone, this can be very costly, no matter how good your literature is.
>
> *Carl Mitchell*

Pitch your idea

The term 'pitch' is *marketing speech* that describes a meeting where an individual, or group, presents their ideas to potential customers. An advertising agency *pitches* its ideas to prospective clients in the hope that their detailed outline plan is *bought* by the client. If the agency is successful, it means that they win the contract for a set length of time.

You are doing something similar. It may not be for the same scale of campaign or involve anything like the same sorts of budget. However, you are putting forward your idea.

Typically, any pitch for support will need some presentation medium:

◆ PowerPoint or Prezi presentation – the most popular way of presenting ideas to individuals, or with a projector, to a group of people.

◆ Flip chart – a large-format drawing board, providing a clear way for writing down ideas or concepts.

◆ Handouts – A4 paper-based plans that each meeting delegate can have and take away after the presentation.

Ideally, if you want to make the most of your presentation to management, you would use all of these media. It can be daunting to stand up and present ideas to others but as your confidence develops you might find it similar to talking to your clients.

Checklist for making your presentation

✓	Have you created an outline plan?
✓	Have you timed your presentation?
✓	Are your objectives clear and understandable?
✓	Have you budgeted correctly?
✓	Have you provided handouts for management to take away?

Produce a detailed plan

The previous topics have given you the basic, starting points:

◆ Initial planning and preparation

◆ Working within a budget

◆ Deciding on the type of promotion

◆ Setting your objectives (i.e. aligning them to the business aims)

◆ Knowing who your customers are

◆ Presenting your ideas to management.

This is the starting point. Now you have to produce a detailed plan that will still be acceptable to management.

ACTIVITY

Detailed planning

This activity will help you to produce a detailed plan for your selected promotional event. Any materials that are produced, such as designs, photographs, storyboards, checklists, copies of invoices and quotations, should all be retained for future reference.

First of all, go back to the beginning, to the section covering initial preparation, then:

1 For each aspect covered on the promotional planning model, produce a new sheet of paper and put them all into a ring binder or presenter.

2 Label each sheet of paper with a title from the model.

3 For each sheet, mark out the page with a table with the following headings.

Planning element: .... (What is it?)

How does it fit into the plan?	Who needs to be involved?	When does it need to be done by?	What does it cost?	Done ✓

4 Complete each section with as much information that you have at this stage.

5 Add into each of these sections any pictures, magazine cuttings, drawings, etc. to illustrate your ideas as a storyboard.

6 Then create your master list (a composite document) that will go at the front of the binder; it will be a table or list that covers all the pages and components of your plan listed in chronological (timed) order.

Event master list

Item	Date	Event element	Running total £	Done ✓	Page

The size of the list of items will depend on the scale of the event, but by doing this you will have a detailed binder that you can use to show to management as a progress report and finally presentation folder that has documented evidence of your planning abilities.

Other things to consider

Order stock and promotional materials You might need to have leaflets printed or place an advertisement to inform potential clients. You may need additional materials: for instance, for a perm promotion you will need to order additional perms. Most product houses will provide advice and merchandising support.

Inform staff Ensure that all staff are fully aware of the promotion. This may require refresher training, for example, for their product knowledge, new skills or treatments. You cannot ask people to take part in an event if they do not have the skills to carry it out.

Your presentation to the staff needs to make them motivated and enthusiastic, they need to feel part of a team event, and they too need to have the same positive outlook so that you all pull together in a united front. Only when they have the correct skills and advanced knowledge, will they then be able to sell the concept on to clients.

Advertise the event Select the best medium of communication. You may inform clients verbally. You may try a mailshot, distribute leaflets, display posters or notices, or advertise in local papers or selected magazines. In any written literature, remember to highlight the features and benefits of the service.

Evaluate and record the results To make full use of the results, you need to keep a permanent record of the facts. This record can then be referred to later. Gradually you will identify strengths and weaknesses and be able to repeat or amend the promotion, depending upon the results.

Promotional outline: ....

Date .... Leaflets mailshot (No.) ....

.... Posters (No.) ....

.... Advertisement (type/place): ....

Leaflets returned: Week ending: .... No.: ....

     Week ending: .... No.: ....

     Week ending: .... No.: ....

     Week ending: .... No.: ....

New clients recorded:

Revenue increases: Cutting: .... %

     Styling: .... %

     Perm: .... %

     Sales: .... %

     Beauty: .... %

Comments: ....

Implement promotional activities

" Are you using your supplier effectively? It is a benefit for all parties to encourage footfall into the salon, so it's always worth speaking regularly with your sales person about any forthcoming opportunities – maybe they have a new product launching and there are some samples available. Using their point-of-sale material – posters, leaflets, shelf talkers, etc. – can all help to bring focus to a product/service. Talk to your supplier and ask how they can help.

Sam Grocutt

Hairdressing demonstration

Hairdressing demonstrations are a very compelling visual representation for sales and promotion. As a display, a demonstration shows the skills of the people involved and produces finished styles that others can see *before their eyes.* A hair show is a *formal* presentation that can be expensive, but do not forget that, *informally*, every member of staff is continually demonstrating their skill in the salon every day. However, there is also a place for the formal demonstration. To organize a demonstration, you again need to keep to a structure in your planning.

Select the venue

Is the demonstration going to be carried out in the salon, or at a local hall or hotel?

◆ Will the audience be able to see the demonstration area?

◆ Is the lighting and sound provided?

◆ Can you make use of a raised platform or stage?

◆ Will you need to take some heavy equipment? (Such as large portable mirrors or hydraulic chairs.)

◆ Are shampooing and model dressing areas/facilities available?

◆ Do you need to arrange transport to the venue for your models and equipment?

◆ Is the venue easily accessible to your audience by public transport?

◆ What are the parking arrangements?

◆ Will you offer refreshments?

The requirements and restrictions of any external locations are more complicated than running an event on the premises. You will also have to consider the following aspects:

◆ **Accessibility** – how will your audience get in, where do they park?

◆ **Health and safety** – who is responsible for public and staff safety?

◆ **First aid/emergencies** – what facilities are provided for emergencies?

◆ **Insurance** – has the venue got a public liability policy?

◆ **Risk assessment** – commercial venues have assessments they can show you. However, if your activity presents a different use you need to do an assessment too.

◆ **Legislation** – are there any local or national laws affecting your proposal?

◆ **Contracts** – are there any contractual obligations with external contacts that tie you into timescales, cancellation clauses, hidden costs or penalties?

Plan the demonstration

Now consider the following: models, show content, methods of presentation and how you expect the audience to participate. Depending on the scale of the demonstration, you may need to build in opportunities for the audience to participate and ask questions. This is not possible when demonstrating to a very large number of people, but good communication skills are always essential in creating a rapport with your audience and you should always be prepared to answer their questions. It is good, professional customer service.

Expect the unexpected To maintain control of the event, your preparation needs to include contingency plans in the event of things going wrong. For example, what would you do in the following situations?

◆ Your models did not arrive on time.

◆ Your guest artist was delayed.

◆ There is a power failure.

◆ You have to evacuate the building.

◆ You do not achieve your aims for the event.

Give thought to each component of the event and be mentally prepared for any eventuality. You will then feel confident and your demonstration will be a success.

Plan the timings Once you have planned the event, it is advisable to make a checklist with a time schedule. To do this, start with the event time and work backwards, for example:

6.00 pm	The event. Guest artist will demonstrate the latest long hair fashion.
5.30 pm	Facility open to the audience
5.00 pm	Final check: platform, demonstration area, lighting, microphone/sound. Offer refreshments to models and platform artist
4.30 pm	Check platform tools and equipment. Prepare tray: this should contain everything that the artists will use during their demonstration. Check that all sprays work. (For a cutting demonstration it is a good idea to have a first aid kit handy – even the best hairdressers cut their fingers occasionally.)
3.30 pm	Model to be made up
3.00 pm	Arrival of make-up artist
2.00 pm	Arrival of model and guest artist
12.00 noon	Organizer to arrive at venue. Arrange seating, erect display material. Check preparation area

Alongside this schedule, make a checklist of each item needed for the event.

ALWAYS REMEMBER

Storyboarding

Be professional. Always create a storyboard of the event processes and actions, starting with the listings of equipment and resources, people, contact information and other external factors, right through to the expected running pattern of the event.

During the planning of the promotion, always remember to update your storyboard in line with any changes.

For more information on storyboards/mood boards see CHAPTER 14 Develop your creativity, pages 362–363.

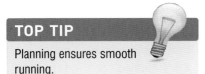

TOP TIP

Planning ensures smooth running.

You will find information about dealing with these selling opportunities in CHAPTER 2 Promote services and products, pages 44–57.

Getting your message across The style of communication that you use will depend upon the promotional event. For example, if you are running an in-salon promotion then you will probably be dealing with your potential customers on a one-to-one basis.

However, if you are planning an external event, you will now be dealing with a one-to-many situation. The style of communication in this scenario is completely different.

Presentation and demonstration skills

Regardless of how interesting the content of your presentation is; you need to remember that your audience will only have an attention span of 20 minutes at best. So you need to schedule the content of your event such that it re-awakes your audience with something interesting and different. People are really impressed by visual magic. The elements of surprise, wonder and vibrant colour are psychologically very entertaining. Therefore, showcases of hair work teamed with the current season's clothes are always a winner as they form an entertainment show. Similarly, makeovers are really popular and visually stimulating.

Entertain your audience The most impressive makeover done to a large TV audience was initially, by Vidal Sassoon (for *The Generation Game* in 1973) and then copied by Lee Stafford in a live show 30 years later. The response to the original 1970s' TV programme was huge. It made an already successful, professionally respected hairdressing company into a national household name, and Vidal Sassoon (who died in 2012) into a legend.

The makeover took just 60 seconds! Vidal Sassoon is seen talking with the game show host; the host asks to see a demonstration of his expertise. So a tall, beautiful, blonde model with beyond-the-shoulder, one-length hair is brought onto stage. She kneels on the stage facing the studio audience. V.S. asks her to put her head forwards and then carefully brushes all of her hair into a single ponytail, about a hand's width away from a position just above the forehead. He then asks the model to reach up and hold the pony above his hand and then takes a pair of scissors from his suit top pocket and cuts the whole pony off, just above his hand, and below hers.

She stands up and shakes her head; her hair then falls into a beautifully layered, graduated transformation and she holds up the pony to the audience. The whole thing has taken less than a minute but visual impact was stunning and it produced the first live makeover. Lee did the same show live but with a young child, not quite the same impact as with an adult but still impressive.

The point behind this story: it's not about great hairdressing it's about showmanship. It is an example of magic and you too can create the same long-lasting impressions *if* you can do something similar.

When you are demonstrating or presenting remember the checklist below.

Presenting: points to remember

✓	Be clear about what it is that you want to sell.
✓	Break down your presentation into a sequence of logical parts or steps.
✓	Keep the presentation at a steady pace; don't rush, take your time.

✓	Encourage your audience to participate in some way; questions, volunteers, etc.
✓	Keep lengthy technical processes to 20 minutes (or have other things going on too).
✓	Be professional, confident and assertive (*no shrinking violets*).
✓	Provide answers to audience questions in a way that everyone will understand.
✓	If there are too many questions, say that you will make yourself available afterwards so that the amount of feedback doesn't de-rail your programme.
✓	If there isn't a flow of questions, then you can *fish* for a few by asking the obvious ones, like 'Does anyone here tonight find that their hair lacks volume?'

ACTIVITY

Handling problems

Even the best planners cannot plan for every eventuality. You can try to build contingencies into your planning, but there will always be the unexpected. What you can do is work together with the team during the planning stage to brainstorm the sorts of things that could go wrong.

If you create a list from the brainstorming event, work together with a flip chart to find solutions to the ones that you have found. Try to cover a range of issues; here are a few to start:

1 How do you go about releasing a member of the team who is locked in the toilet and cannot get out?

2 You are halfway through the hair show and the lighting rig fuses and the stage lighting goes out.

3 You have one member of the audience who is becoming difficult, and keeps asking questions and is stopping others from getting their points across.

4 Someone lets off the fire extinguisher in the reception area.

5 Progress through the programme is slow and the team is starting to run over time.

Advertising, PR and the press

Advertising

Advertising is always a useful way of promoting your salon's services. It is important to define the purpose of the advertisement so that you can choose the most appropriate form of advertising:

◆ Do you want to attract new customers?

◆ Do you want to advertise a new service?

◆ Do you want to increase the salon's profile?

◆ Do you want to increase retail sales?

◆ Do you want to maintain loyalty with regular customers?

◆ Do you want to draw customers' attention to other services on offer?

ACTIVITY

Local research

Make a list of all the venues that can accommodate a seated audience of over 100 people in your local area.

What additional facilities do they have to offer?

Venue/size	Accessibility/parking	Facilities available

There are many different forms of advertising, including:

- ◆ magazines
- ◆ newspapers
- ◆ directories
- ◆ leaflets/posters
- ◆ local radio
- ◆ calendars.

Advertising can be very expensive. Therefore, the form of advertising you choose will also depend on the budget available. If you decide to advertise your business through printed media, consider carefully the content of the message you wish to convey. As this is a very special and important form of promotion, you do need to think about whether the content as well as the origination is handled by the professionals.

Local newspapers and business directories provide comprehensive advertising services, but it can be hard for them to get to know the needs of your particular business. Here is a simplified overview of the different processes involved:

- ◆ **Commercial artist** – responsible for creating graphical illustrations through image manipulation and typography.
- ◆ **PR consultant/agency** – responsible for content and media management.
- ◆ **Reprographics** – the intermediaries, between the originators and printer, who create the film or digital output.
- ◆ **Printer** – produces the finished material.

Public relations

Public relations (PR) is an effective tool with which to promote the business or product. It targets the media best suited to the company image or product profile, bringing the finished package to the eye of the consumer and thus increasing your business potential. Salons can handle their own PR or employ the services of a PR consultancy.

ALWAYS REMEMBER

If you make contact with any representative of the local press, remember to get their direct email address for swift access to the right pair of eyes!

> With PR you need to think about what it is you want to achieve. Ultimately everyone wants to get more success for their salon, which will hopefully transfer to increased footfall and sales. Positive PR is the best method of creating awareness; it may not give you immediate sales results and should not be mistaken for marketing. However, unless you have a clear view on why you're building that profile, and how this will benefit your business, then PR may not be a worthwhile exercise for you.

Sam Grocutt

DIY PR With the right contacts and more importantly the time, it is possible to promote the business effectively as a team. However, be aware that dealing with the press is not necessarily a simple case of a phone call and then a press release. It is a question of knowing whom to contact, working on them and then how to get your message across. Email is a particularly useful medium for putting information onto the right desk and in front of the right eyes. However, it is also a question of degree: how much promotion do you want? Is it for one specific project or is it ongoing? If the latter, then be prepared for PR to take up a lot of your time.

If you're considering hiring a PR agency consider the following:
1. Ask a journalist for recommendations – they talk to PRs every day and know who is proactive and helpful.
2. Look at the PR's editorial accomplishments to gauge their experience.
3. Does the PR talk to you or **at** you – communication and understanding of your needs is vital to ensure a successful PR campaign.

The press

Trade and consumer press are completely different so it is therefore essential that they are each approached in the appropriate manner.

Trade press Aimed at other businesses within the same industry, the trade press is interested in news items within the trade (e.g. new salons, trends, techniques, etc., plus charity events and product launches). It is also very *warm* to launching new photographic collections, or showcasing salon interiors and conducting business profiles. Trade journalists tend to be easier to deal with and more accessible; they are already sympathetic to your salon's products and services and require your salon's help to fill their pages.

Consumer press Aimed at the general public, consumer magazines reach a great number of people who may never have heard of you or your product, but are about to do so through effective PR. Public relations is about editorial endorsement. This is quite different from advertising. It means the journalist is giving your salon magazine space without expecting any payment for it. Such editorial endorsement can be much harder to secure, but the benefits can be huge. The consumer press is not interested in new staff appointments, but it does want to know about innovative techniques, upcoming seasonal trends, latest product advancements and new salons. More and more consumer magazines are coming around to the idea that the consumer is interested in their hair, and, through increased hairdressing standards and higher stylist profiles, are becoming increasingly confident in our industry. Magazines now have hair supplements banded to their issues and these have to be filled.

Hair salons and product companies with effective PR are the ones who help to fill these supplements and hair features, supplying press releases to the appropriate journalists outlining new techniques and products, photographic material, seasonal trends and quotes.

> If you're looking to build your profile locally, invite the key press contacts into the salon for a complimentary treatment and get to know their needs and interests. Or, work with a local charity to see if there are any opportunities you can link in with to get your name out there.
>
> *Sam Grocutt*

Participate in the evaluation of promotional activities

After any event or promotional activity, you will need to work out how effective it was. It may have been fun, it may have been a change to the routines, but unless it has achieved the aims, it will not have been worth doing.

You can measure success in a number of ways, and to think of achievement in simple sales terms is only part of the story. People do things for all sorts of reasons; such as a corporate day out, or a charity dress-down day. Not every benefit can be measured in monetary terms, people can benefit too – providing that was part of the plan.

If we look at the purpose of evaluation, we see that it is an organized process:

◆ **Collecting** customer/client feedback.

◆ **Collating** the feedback into organized, measurable, indicative information.

◆ **Analysing** the information to see where goals have been met, or where further improvements can be made.

The purpose of your evaluation

The purpose of your evaluation is to provide management with enough information to help them make informed decisions about services, treatments and products in a report-style format that presents the facts without long-winded details. Obviously if you are asked to provide extra details, then that would be your opportunity to expand on your findings.

Your evaluation should be able to show ways that you can:

◆ Learn more about your customers – such as their needs, their purchasing decisions, their likes and dislikes.

◆ Improve the services to your customers – identifying areas for staff development, changes to existing services or products.

◆ Determine/gauge customer satisfaction – what works well, or otherwise.

◆ Measure the success or impact of your promotion.

Evaluating your promotional response

How will you know if the event was a success? Do you ask people if they enjoyed the show or if they have noticed the new product promotion in reception? They might say yes, but is that good feedback? If it has not necessarily turned into a sale, how useful is it? Unless you find some way of measuring the impact and outcome, then the event will have little or no meaning at all.

A comprehensive evaluation is the only way to measure the effectiveness of the investment in time, resources and money to put the event on.

The methods that you use to evaluate the success of the event will depend on the type of event that you arrange. For example, it is relatively straightforward to distribute questionnaires in a small seminar scenario, but how do you get a town hall audience to fill in questionnaires before they dash home?

You need to channel the responses in ways that suit the needs of the event. The box below provides a list of the sorts of things that you can use.

TOP TIP

Evaluation systems

- A written questionnaire (simplest standard evaluation system)
- Telephone follow-up (pre-prepared standard questionnaire)
- Pre-paid business reply service (mailshot to attendees or left on seats in the event)
- Focus group feedback (invite a sample of the audience to attend a meeting)
- Sales reports, sales summaries
- Increase of clients within the salon database

Collect, collate and analyse the information

If you have collected the feedback as a multiple choice, or Likert scale, it is easier to record in a computerized way (by means of spreadsheet or tabular document). For those with the expertise and inclination, a database provides the easiest way to look for patterns in responses. Large marketing companies gauge customer feelings or experiences by using this sort of business tool, but this may not be appropriate for you.

You need to use a system that will show you clear comparisons and feedback opinion. This way, your data can be managed easily to provide charts and reports to show trends, likes and dislikes. However, if this is not available, you can record the information in a Word™ document.

When compiling your data, be honest with the response. If you find that the results are not what you expected, do not be too concerned as any negative results or general indicators are all indicative of people's feelings. This will help management to tailor their offerings in future promotional events, so that they are closer to meeting the needs of your customers in the future.

TOP TIP

If you are familiar with databases, spreadsheets and word-processing software; you can really give your final presentation impact, by importing your reports and charts into PowerPoint. If you do have the skills to utilize computers but do not have access to Microsoft Office a very comprehensive suite of freely available software can be found at **www.openoffice.org.**

Presenting an evaluation report

When you have completed your analysis, you can present your findings to management. Remember, your detail needs to be clear and concise; you should present information 'as is' without providing speculative comments. The whole idea is to provide the information to others so that they can draw *their own* conclusions from the data. The report should provide facts that are positive; the facts should raise issues which can then be discussed.

See CHAPTER 16 for more information about people's rights and consumer legislation.

Be prepared to provide answers to questions. Where you do know the answer, be clear and concise; where you have not got an answer, say you do not know, or you haven't been able to evaluate that/those response(s).

Finally, make notes of what has been said and, if there are any questions that still need to be answered; you can re-evaluate your data to find your response.

REVISION QUESTIONS

Q1.	Selling opportunities will occur when the features and _____ of products are explained to the client.	Fill in the blank
Q2.	A budget for a promotion can normally be ignored.	True or false?
Q3.	Which of the following are not aspects of setting objectives?	Multi-selection
	Specialize	☐ 1
	Spontaneous	☐ 2
	Specific	☐ 3
	Metered	☐ 4
	Measurable	☐ 5
	Achievable	☐ 6
Q4.	Objectives should be set within a timeframe.	True or false?
Q5.	When pitching your ideas to management, which of the following should you *never* do?	Multi-choice
	Have an outline plan	○ a
	Time your presentation	○ b
	Have your objectives clearly defined	○ c
	Make a guess at the costings	○ d
Q6.	An external event should be stimulating for the audience.	True or false?
Q7.	Which of the following are essential to an effective external presentation?	Multi-selection
	Being clear about what you want to sell	☐ 1
	Skipping over difficult questions	☐ 2
	Ignoring problems encountered by the team	☐ 3
	Rest breaks	☐ 4
	Encouraging the audience to participate	☐ 5
	Being professional and confident	☐ 6
Q8.	A smooth event should always _____ to time.	Fill in the blank

Q9. If, during an external demonstration there isn't a good flow of questions, what should you do?   Multi-choice

Forget it and finish early   ○   1

Drag out the presentation to fill the time   ○   2

Prompt for questions by cautious 'fishing'   ○   3

Just say 'Have you lot lost your tongues?'   ○   4

Q10. The success of an external event is evenly 'weighted' across the whole participating team?   True or false?

Appendix 1: People's rights and consumer legislation

Equal opportunities

The Equality and Human Rights Commission (EHRC) has the statutory duty to:

◆ work towards the elimination of discrimination

◆ promote equality of opportunity between men and women (and in relation to persons undergoing gender reassignment)

◆ keep the relevant legislation under review.

The legislation within the remit of the EHRC is wide-ranging; however, the main considerations are:

◆ age equality

◆ disability equality

◆ gender equality

◆ race equality

For more information visit http://www.equalityhumanrights.com/

Disability Discrimination Act 2005 (DDA 2005)

The Act makes it unlawful to discriminate against disabled persons in connection with employment, the provision of goods, facilities and services or the disposal or management of premises; to make provision about the employment of disabled persons; and to establish a National Disability Council.

Data Protection Act (1998)

Your clients have the following rights which can be enforced through any county court:

◆ **Right of subject access** – This is the right to find out what information about them is held on computer and in some paper records.

◆ **Correcting inaccurate data** – They have the right to have inaccurate personal data rectified, blocked, erased or destroyed. If your client believes that they have suffered damage or distress as a result of the processing of inaccurate data they can ask the court to award compensation.

◆ **Preventing junk mail** (from salons that market to their customer base) – Your client has the right to request in writing that a data controller does not use your personal data for direct marketing by post (sometimes known as junk mail), by telephone or by fax.

Diversion

For more information on consumer rights in relation to the Data Protection Act http://www.legislation.gov.uk/ukpga/1998/29/contents

The Data Protection Act (DPA) applies to any business that uses computers or paper-based systems for storing personal information about its clients and staff. It places obligations on the person holding the information (data controller) to deal with it properly. It gives the person that the information concerns (data subject) rights regarding the data held about them.

The duties of the data controller

There are eight principles put in place by the DPA to make sure that data are handled correctly. By law, the data controller must keep to these principles. The principles say that the data must be:

1 fairly and lawfully processed

2 processed for limited purposes

3 adequate, relevant and not excessive

4 accurate

5 not kept for longer than is necessary

6 processed in line with your rights

7 secure

8 not transferred to other countries without adequate protection.

The Sale of Goods Act (1979) and Sale and Supply of Goods Act (1994)

The Sale of Goods Act 1979 and the later Sale and Supply of Goods Act (1994) are the main legal instruments helping buyers to obtain redress when their purchases go wrong. It is in the interest of anyone who sells goods or services to understand the implications of these Acts and the responsibilities they have under them. Essentially, these Acts state that what you sell must fit its description, be fit for its purpose and be of satisfactory quality. If not, you – as the supplier – are obliged to sort out the problem.

Briefly these Acts require the vendor:

◆ To make sure that goods *conform to contract*. This means that they must be as you describe them, e.g. highlight shampoo stops your highlights from fading.

◆ The goods must also be of *satisfactory quality*, meaning they should be safe, work properly, and have no defects.

◆ You must also ensure the goods are *fit for purpose*. This means they should be capable of doing what they're meant for. For example, in the case of a brush it shouldn't fall apart when it is first used.

For more information visit http://www.legislation.gov.uk/ukpga/1994/35/contents

The Consumer Protection Act (1987)

This Act follows European laws to protect the buyer in the following areas:

◆ product liability – a customer may claim compensation for a product that doesn't reach general standards of safety

◆ general safety requirements – it is a criminal offence to sell goods that are unsafe; traders that breach this conduct may face fines or even imprisonment

◆ misleading prices – misleading consumers with wrongly displayed prices is also an offence.

The Act is designed to help safeguard the consumer from products that do not reach reasonable levels of safety. Your salon will take adequate precautions in procuring, using and supplying reputable products and maintaining them so that they remain in good condition.

The Prices Act (1974)

The price of products has to be displayed in order to prevent a false impression to the buyer.

The Trades Descriptions Act (1968 and 1972)

Products must not be falsely or misleadingly described in relation to their quality, fitness, price or purpose, by advertisements, orally, displays or descriptions. And since 1972 it has also been a requirement to label a product clearly, so that the buyer can see where the product was made.

Briefly, a retailer cannot:

◆ mislead consumers by making false statements about products

◆ offer sale products at half price unless they have been offered at the actual price for a reasonable length of time.

The Resale Prices Act (1964 and 1976)

The manufacturers can supply a recommended price (MRRP or manufacturer's recommended retail price), but the seller is not obliged to sell at the recommended price.

Appendix 2: Answers to revision questions

Chapter 1

Consultation and advice questions pp. 42–43

Q1	Catagen	**Q6**	False
Q2	False	**Q7**	2, 5
Q3	1, 2, 4	**Q8**	Germinal
Q4	False	**Q9**	c
Q5	d	**Q10**	True

Chapter 2

Promote services and products questions p. 56

Q1	Features	**Q6**	False
Q2	True	**Q7**	1, 2, 6
Q3	2, 5	**Q8**	Listening
Q4	True	**Q9**	b
Q5	b	**Q10**	True

Chapter 3

Customer service questions p. 73

Q1	Closed	**Q6**	False
Q2	True	**Q7**	1, 2
Q3	1, 2	**Q8**	Suggestion
Q4	True	**Q9**	d
Q5	c	**Q10**	True

Chapter 4

Creative cutting questions pp. 105–106

Q1	Holding	**Q6**	False
Q2	False	**Q7**	4, 6
Q3	3, 5, 6	**Q8**	Base
Q4	True	**Q9**	3
Q5	d	**Q10**	True

Chapter 5

Creative barbering questions pp. 132–133

Q1	Holding	**Q6**	False
Q2	False	**Q7**	4, 6
Q3	3, 5, 6	**Q8**	Fading
Q4	True	**Q9**	3
Q5	d	**Q10**	True

Chapter 6

Beards and moustaches questions p. 149

Q1	Sterilizer	**Q6**	True
Q2	True	**Q7**	1, 2, 4
Q3	2, 5	**Q8**	Flexible
Q4	True	**Q9**	2
Q5	4	**Q10**	False

Chapter 7

Shaving questions pp. 162–163

Q1	Autoclave	**Q6**	True
Q2	True	**Q7**	1, 3, 4
Q3	2, 4	**Q8**	Sponge
Q4	True	**Q9**	1
Q5	5	**Q10**	False

Chapter 8

Creatively style and dress hair questions p. 185

Q1	Velcro	**Q6**	False
Q2	False	**Q7**	1, 5
Q3	3, 4	**Q8**	Spiral
Q4	True	**Q9**	b
Q5	b	**Q10**	False

Chapter 9

Creatively dress long hair questions pp. 208–209

Q1	French	**Q6**	True
Q2	True	**Q7**	3, 5
Q3	2, 4, 6	**Q8**	Knot
Q4	True	**Q9**	a
Q5	c	**Q10**	False

Chapter 10

Hair extensions questions p. 239

Q1	Root	**Q6**	True
Q2	False	**Q7**	5, 6
Q3	3, 4, 5	**Q8**	First
Q4	True	**Q9**	c
Q5	b	**Q10**	True

Chapter 11

Colouring hair questions p. 287

Q1	Allergy	**Q6**	True
Q2	True	**Q7**	2, 4, 5
Q3	4, 5	**Q8**	Red/gold
Q4	False	**Q9**	b
Q5	d	**Q10**	True

Chapter 12

Colour correction questions pp. 310–311

Q1	Strand	**Q6**	False
Q2	False	**Q7**	1, 3, 6
Q3	1, 2, 3	**Q8**	Violet
Q4	False	**Q9**	b
Q5	d	**Q10**	False

Chapter 13

Perming hair questions p. 344

Q1	Curl	**Q3**	1, 4, 5
Q2	True	**Q4**	False

Q5	c	**Q8**	Cortex
Q6	True	**Q9**	c
Q7	1, 3	**Q10**	True

Chapter 14

Develop your creativity questions pp. 364–365

Q1	a	**Q6**	Colour, Photographic
Q2	Vertical	**Q7**	c, f
Q3	a, e	**Q8**	False
Q4	True	**Q9**	a
Q5	c	**Q10**	contrasting

Chapter 15

Health and safety questions p. 391

Q1	Hazard	**Q6**	False
Q2	True	**Q7**	2, 4
Q3	2, 3, 6	**Q8**	Fire
Q4	False	**Q9**	b
Q5	c	**Q10**	True

Chapter 16

Personal effectiveness questions p. 415

Q1	48	**Q6**	False
Q2	True	**Q7**	1, 3, 5
Q3	1, 3, 4	**Q8**	Delivery
Q4	True	**Q9**	b
Q5	c	**Q10**	False

Chapter 17

Promotional activities questions pp. 434–435

Q1	Benefits	**Q6**	True
Q2	False	**Q7**	1, 5, 6
Q3	1, 2, 4	**Q8**	Run
Q4	True	**Q9**	3
Q5	d	**Q10**	True

Appendix 3: Useful addresses and websites

Business

Arbitration, Conciliation and Advisory Service (ACAS)

Acas National (Head Office)
Euston Tower, 286 Euston Road, London NW1 3JJ
Tel: 08457 38 37 36
www.acas.org.uk

Call the Helpline on 08457 47 47 47.
Monday–Friday, 8 am–8 pm and Saturday, 9 am–1 pm

Hairdressing Employers Association (HEA)
10 Coldbath Square, London EC1R 5HL
Tel: 020 7833 0633

Training & Education

Hairdressing and Beauty Industry Authority (Habia)
Oxford House, Sixth Avenue, Sky Business Park, Robin
Hood Airport, Doncaster DN9 3GG
Tel: 08452 306080
Fax: 01302 774949
www.habia.org

Association of Colleges (AOC)
2–5 Stedham Place, London WC1A 1HU
Tel: 020 7034 9900
Fax: 020 7034 9950

City and Guilds (C&G)
1 Giltspur Street, London EC1A 9DD
Tel: 020 7294 2800
www.city-and-guilds.co.uk

Department for Education and Skills
www.education.gov.uk/

Further Education and 6th Form Colleges in the UK
FIND FE (a website listing all FE colleges in England, Wales,
Scotland and N. Ireland)
http://findfe.com/

Salonstudies – an on-line knowledge base for hair education
http://salonstudies.com

ITEC
2nd floor, Chiswick Gate, 598–608 Chiswick High Road,
London W4 5RT
Tel: 020 8994 4141
www.itecworld.co.uk/

The Institute of Trichologists
107 Trinity Road, Upper Tooting, London SW17 7SQ
Tel: 0845 604 4657
www.trichologists.org.uk

Vocational Training Charitable Trust (VTCT)
Prysmian House, Dew Lane, Eastleigh,
Hampshire SO50 9PX
Tel: 02380 684 500
Fax: 02380 651493
www.vtct.org.uk

World Federation of Hairdressing and Beauty Schools
PO Box 367, Coulsdon, Surrey CR5 2TP
Tel: 01737 551355

Publications/Fashion Forecasting

Black Beauty and Hair
Culvert House, Culvert Road, London SW11
Tel: 020 7720 2108
www.blackbeautyandhair.com

Creative Head
21 The Timberyard, Drysdale Street, London N1 6ND
Tel: 020 7324 7540
Fax: 020 7739 7789
www.creativeheadmag.com

Hairdressers Journal International (HJ)
Quadrant House, The Quadrant, Sutton, Surrey SM2 5AS
Tel: 020 8652 3500
www.hji.co.uk

Runway Magazine
www.runwaybeauty.com/

Trade Associations

British Association of Beauty Therapy and Cosmetology Limited (BABTAC)
Ambrose House, Meteor Court, Barnett Way, Barnwood,
Gloucester GL4 3GG
Tel: 01452 623110
Fax: 01452 611599
www.babtac.com

Cosmetic, Toiletry and Perfumery Association (CTPA)
Josaron House, 5–7 John Princes Street, London W1G 0JN
Tel: 020 7491 8891
www.ctpa.org.uk
www.thefactsabout.co.uk

Fellowship for British Hairdressing
Bloxham Mill, Barford Road, Bloxham, Banbury, Oxfordshire
OX15 4FF
Tel: 01295 724579
www.fellowshiphair.com/

Freelance Hair and Beauty Federation
FHBF Head Office, The Business Centre, Kimpton Road,
Luton, Bedfordshire LU2 0LB
www.fhbf.org.uk

The Hairdressing and Beauty Suppliers Association
Greenleaf House, 128 Darkes Lane, Potters Bar,
Hertfordshire EN6 1AE
Tel: 01707 649499
www.thehbsa.uk.com

The Hairdressing Council (HC)
30 Sydenham Road, Croydon, Surrey CR0 2EFT
Tel: 020 8771 6205
www.haircouncil.org.uk

**Health and Beauty Employers Federation
(part of the Federation of Holistic Therapists)**
18 Shakespeare Business Centre, Hathaway Close,
Eastleigh, Hampshire SO50 4SR
Tel: 023 8062 4350
www.fht.org.uk

Incorporated Guild of Hairdressers, Wigmakers and Perfumers
Langdale Road, Barnsley, South Yorkshire S71 1AQ
Tel: 01226 786 555
Fax: 01226 731 814

National Hairdressers' Federation (NHF)
One Abbey Court, Fraser Road, Priory Business Park,
Bedford MK44 3WH
www.the-nhf.org
Tel: 01234 831965 or 0845 345 6500

Legal and Regulatory

Equality and Human Rights Commission (EHRC)
Equality Advisory Support Service
Tel: 0808 800 0082
www.equalityhumanrights.com

Health and Safety Executive
Publications:
PO Box 1999, Sudbury, Suffolk CO10 6FS
HSE Infoline:
Tel: 0845 345 0055
www.hse.gov.uk

Union of Shop, Distributive and Allied Workers (USDAW)
188 Wilmslow Road, Fallowfield, Manchester M14 6LJ
Tel: 0161 224 2804 / 249 2400

Glossary

accelerator A machine that produces radiant heat (infrared radiation); can speed up chemical hair processes such as colouring or conditioning.

acid A substance that gives hydrogen ions in water and produces a solution with a pH below 7.

activator A chemical used in bleaches or some perm lotions to start or boost its action.

added hair A general term that covers the addition of hair pieces, wefts and extensions.

adverse hair or scalp condition In hairdressing terms: a condition which indicates a contra-indication.

alkali A substance that gives hydroxide ions in water and produces a solution with a pH above 7.

alopecia Baldness.

alpha keratin Hair in its natural state.

anagen The stage of hair growth during which the hair is actively growing.

appraisal A process of reviewing work performance over a period of time.

arrector pili The muscles that raise the hair (in humans they are very feeble).

astringent A substance applied after shaving to close the pores.

asymmetrical Unevenly balanced, without an equal distribution of hair on either side.

artificial colour The term refers to any form of colour that is not a naturally occurring pigment; also called synthetic colour.

autoclave A device for sterilizing items in high temperature steam.

avant-garde A genre of fashion that is considered progressive or exaggerated.

back-brushing/back-combing Pushing hair back to bind or lift the hair using a comb or brush.

backhand technique A method of razoring where the cutting action occurs with a backhand technique.

banding An unwanted effect that appears as distinct bands of uneven colour.

benefits The ways in which the functions of products or services provide advantages.

beta keratin Hair in its moulded shape, i.e. curly hair that has been dried straight.

blending A technique for mixing different colours of hair extension fibres to create more naturally occurring effects, multi-toned effects, highlighted effects.

block colouring Colouring areas of hair in a way that is intended to enhance the cut style.

blunt cutting See **club cutting.**

brick cutting A way of point cutting into a held section of hair in different positions to create the scatter pattern like brick work.

canities Hair that is without pigment and therefore grey or white.

carborundum A hard, abrasive material that is used for smoothing, shaping and even sharpening other materials (e.g. metal instruments such as knives or blades).

catagen The stage of hair growth during which the hair stops growing, but the hair papilla is still active.

cleanse Remove dead skin cells, sebum and debris from the skin.

clipper over comb A technique of cutting hair with electric clippers, using the back of the comb as a guide, especially on very short hair and hairline profiles.

club cutting Cutting a hair section straight across, producing blunt ends.

colour stripper A colouring product that is specially formulated to remove synthetic/artificial colour from previously coloured hair.

colouring back A process of recolouring previously lightened hair (e.g. highlights) back to the hair's natural hair depth and tone.

colour restorer See **progressive dye (incompatibles).**

concave A concave perimeter slopes inwards.

confidential Private information, not for general use.

consumer protection The legislation protecting customers from unlawful sales practices and mishandling of personal information.

contra-indications Reasons why a proposed course of action or treatment should not be pursued because it may be inadvisable or harmful.

control The ways in which risks identified are eliminated or reduced to an acceptable level.

convex Sloping outwards.

corn row Fine plaits running continuously across the scalp.

croquignole winding Winding a curl from point to root.

customer feedback The information retrieved by a variety of methods, used as a mechanism for evaluating the customers' experiences.

Data Protection Act (1998) Legislation designed to protect the client's right to privacy and confidentiality.

databank A manual or computerized store of data or records.

decolouring Removing synthetic colour from hair.

demonstration A display and explanation of a physical instruction.

depilatory A hair-removing compound.

discolouration Unwanted colour produced by a chemical.

disconnection An area within a haircut where there is a distinct difference between two levels within the layering patterns or perimeter baselines.

double wind A variation of a weave wind – where the hair left out of the rod is wound on another rod.

dreadlocks The naturally occurring look of hair aided by twisting to create a matted effect.

effectiveness The quality of output achieved in a work setting.

effleurage A gentle stroking movement used in shampooing.

eumelanin Black and brown pigment in the skin and hair.

exfoliation The removal or shedding of a thin outer layer of skin from the epidermis.

fading A cutting term that refers to the blending of short layered hair, usually from a neck outline into the graduated shape.

features The aspects of a product or service that state its functions, i.e. what it does.

fish-hook A point of hair that has been bent back during rollering or winding.

folliculitis Inflammation of the hair follicles; may be caused by bacterial infection.

forehand technique A method of razoring where the cutting action occurs with a forehand technique.

fragilitas crinium Splitting of the hairs at their ends.

freehand cutting A method of cutting without holding between the fingers, or below or above a comb.

graduation A sloping variation from long hair to short, or from short to long, produced by cutting the hair ends at a particular angle.

hair extension Real or synthetic fibre added to existing hair.

HASAWA Health and Safety at Work Act 1974.

hazard Something with a potential to cause harm.

humid Something that contains a high amount of moisture or water vapour – being noticeably moist.

humidity The levels of moisture within the air.

hygroscopic Readily absorbing moisture from the atmosphere.

incompatible Causing a chemical reaction on mixing; as between a chemical being added to the hair and another chemical already on the hair.

incompatibility In hair science terms: A chemical process that is unable to co-exist with previously applied processes/treatments (e.g. containing metallic salts – lead acetate).

induction A process that takes place at the beginning of employment to introduce new employees to the employer's workplace policies and procedures (e.g. health and safety training, workplace rules, disciplinary and grievance procedures, etc.).

interleaves The meshes of hair left out between packets/foils during highlighting that are processed later within the service to provide e.g. natural base tones, contrasting colour etc.

inversed clipper cutting A technique of holding the clippers with the cutting edges inverted so that hair is removed on downward, rather than upward strokes.

job description An official written description of the responsibilities and requirements of a specific job (usually made available during the recruitment phase).

keratin The principal protein of hair, nails and skin.

knot The effect produced when long hair is wound, positioned and secured to take on a tied or knotted rope-like effect.

maquette (French) A small, scaled model of a planned sculpture or artwork. Applied to hairdressing, this would be an investigative trial piece to determine suitability and complexity.

melanin The pigments that give colour to skin and hair.

metallic salts Found in some colouring products containing lead acetate. See **progressive dye, colour restorers, incompatibility.**

moisturising balm Cooling, soothing and moisture replenishing lotion applied after shaving to counteract the abrasive effects of the process.

monilethrix Beaded hair.

monitoring Being responsible for checking incorrect practices.

non-conventional styling equipment Items that can be used to style hair other than rollers and pin clips such as rags, chopsticks, straws, Rik-Rak, etc.

occipital bone Bone forming the back of the head.

outliner See **T-liner.**

oxidation Reaction with oxygen, as in the neutralizing of a perm.

para dyes A term that refers to permanent colours containing paraphenylenediamine or PPD.

paraphenylenediamine See **para dyes.**

personalize A term which refers to a variety of cutting techniques applied to a style dependent on the client's specific needs.

petrissage A kneading massage movement of the skin that lifts and compresses underlying structures of the skin.

pheomelanin A natural hair pigment.

piggyback wind A technique of winding curlers/rods into hair to create a multi textured effect that has curls/movement of differing diameters

pleat A visual description of hair that is folded, e.g. a 'French pleat'.

point cutting A texturizing technique for using the point ends of the scissors to remove hair nearer the root area.

pointing See **point cutting.**

polymer resin adhesive stick A resin or glue stick inserted into the bonding applicator and melted, for use during hair extension services.

PPD Reference/abbreviation for paraphenylenediamine found in permanent and quasi-permanent hair colours or colours with an ability to cover grey/white hair.

pre-perm test A test where an appropriate curler and lotion are applied to the hair before the main service to determine how the hair will react under processing.

pre-pigmentation Applying a preliminary colouring of red to hair so that new colour will adhere.

Pre- and post-perm/colour treatments These are special products that can even out the porosity before perming or colouring.

pre-softening A process of softening resistant white hair with hydrogen peroxide.

productivity The levels of output achieved in a work context.

progressive dye A type of hair colour that builds-up a colour effect over several applications (colour restorers).

quasi-permanent A colour that is mixed with a developer for a longer-lasting effect.

real hair extensions Naturally occurring hair types derived from organic proteins found in humans and animals used for extending hair.

referral The situations where you need to re-direct people to other sources of treatment or service.

resources The variety of means available to a business that can be utilized or employed within any given task or project: time, money, people, etc.

risk The likelihood of the harm occurring.

risk assessment A careful examination of what could cause harm to people in the workplace.

roll A visual description of hair that is rolled to create a bulked, rounded shape. This can be aided by using a bun ring or similar styling aid.

root perm A technique for winding perming rods near the root area to produce lift without end movement.

scissor over comb A technique of cutting hair with scissors, using the back of the comb as a guide (usually when the hair is at a length that cannot be held between the fingers).

silicone pad A small piece of heat-resistant silicone sheet used during the application of bonded extensions.

skin test A test done prior to colouring to establish whether a client has a sensitivity or reaction to chemical products.

slicing A texturizing technique for cutting hair using the sharp blades of scissors (without open and closing) like using a razor/shaper.

SMART An acronym used for setting objectives i.e. **S**pecific, **M**easurable, **A**chievable, **R**ealistic and **T**imed.

spiral wind A perming technique of winding longer hair from root to point.

storyboard A way of pictorially and verbally collating ideas and concepts into a visual flow chart or schedule of events.

straightener An ammonium-based lotion similar to perms that can be used to remove wave in hair.

straightening Reducing the curl or wave in hair.

symmetrical Balanced by means of an even and equal distribution of hair on either side.

synthetic colour Another professional term that can be used instead of **artificial colour.**

synthetic fibre extensions A range of alternative, fibrous materials (nylon, acrylic, kerakalon etc.) used for extending hair.

T-liner / or outliner A type of clipper with a different blade type to standard clippers, enabling closer cut outlines around ears, necklines and facial hair shapes.

T-section highlights A partial highlighting technique around the hairline and along the parting only.

tapering Cutting a hair section to a tapered point (i.e. a point like that of a sharpened pencil).

Tapotement A percussive massage movement done with the fingertips.

tariff A displayed list of fixed charges.

telogen The period during which a hair ceases to grow before it is shed.

temporal Bones forming the lower sides of the head.

temporary bonds The hydrogen bonds within the hair that are modified and fix the style into shape.

texturizing A term which refers to a variety of cutting techniques.

tinting back See **colouring back.**

total look A term that is often used to describe a visual themed effect that incorporates hair, clothes, accessories and make-up.

tone-on-tone colour The common industry name for quasi-permanent colours

toning Adding colour to bleached hair.

traction alopecia An area of baldness resulting from the stress or pull applied to hair.

tramliner A specialist clipper with a tapered, narrow blade for detailing designs on hair.

trichologist An expert specializing in the treatment of diseases affecting the hair and scalp.

trichosiderin Rare naturally occurring, iron oxide red pigmentation, Celtic hair origins.

twist A technique of styling hair (or multiple stems of hair) by twisting together.

unique selling point (USP) In a sales/selling context, these are features, aspects or benefits about a product or service that make it stand out, or differ from, a competitor's products or services, therefore providing a competitive advantage or selling opportunity.

weave wind A perming technique for winding rods into the hair so that part of the mesh taken leaves hair out providing a multi-textured effect.

whorls Hair growth patterns.

Index

ACAS 396
accelerator 257
accidents 379–80
acid 20, 317
acne 27
activators 257
adhesive tape 235
advertising 425, 429–30
aftercare advice/service 104
    colour correction 309–10
    creative styling/dressing 184
    demonstrate techniques used 105, 286–7
    equipment used 104–5, 286
    exfoliation 148
    explain routine styling tools/detrimental
        effects 105, 286
    hair colouring 285–7
    hair extensions 235–6
    long hair 207–8
    products/skin care 104–5, 148, 286
    shaving 161–2
alkaline/acid perming solutions 330
    applying lotion 330–1
    checking other systems 332
    checking standard/round curlers/bendy
        curlers 332
    curlers 331
    development test curl 332
    hair texture/condition 331
    processing time 331
    processing/development 330
    temperature 331
    winding 331
alkalis 260, 317
allergy 23
    quasi-colour testing 249
Allergy Alert Test 22, 268–9
alopecia areata 28
alopecia totalis 29
alpha keratin 17
anagen 21
analytical skills 14–15
appraisals 413–14
arrector pili muscle 21
artificial colour 247
    decolouring step-by-step guide 306
    removal 303–4
artistic interpretation 348
Asian hair 216
asymmetric cut 102–3
autoclave 82, 387–8
avant garde 359

back-brushing 181
    step-by-step guide 181–2

back-combing 182
banded hair colour 301
    colour correction step-by-step
        guide 305
barbering
    blending from clipper lengths to scissor
        length 125
    client preferences 118–19
    consultation 113–14
    creative restyling 124–7
    cutting rules 124–5
    cutting tools/equipment 119–21
    ears 125–6
    effective/safe methods 110–13
    examination of hair/scalp 117
    facial shapes 115–16
    finishing products 127–8
    gowning the client 111
    hair growth patterns 117–18
    hair patterning 131
    hair type 126
    influencing factors 114–16
    necklines 125
    outline shapes 124–5
    physical features 116
    points to remember 126–7
    preparation for cutting 113–23
    products 127–8
    short cut 130
    step-by-step techniques 129–31
    varsity side part 129
barrel curls step-by-step guide 204
barrier cream 261
basic sectioning 328–9
basic wind 329
beards and moustaches
    aftercare advice 147–8
    client position 137
    client preparation 137
    consultation 139–40
    cutting techniques 144–6
    effective/safe working methods 136–8
    facial features 140–1, 142
    facial hair shapes 144–7
    factors affecting 140–2
    head shape/size 141–2
    maintenance-based trim 147
    personal hygiene 138
    personal working position/posture 137–8
    preparation for cutting 139–44
    preventing infection 138
    removing unwanted hair outside desired
        style line 146–7
    step-by-step guide 147
    tools/equipment 142–4

benefits 47
beta keratin 17
blending 115, 223, 224
block colouring 224
blow-drying 180–1
    roots to points 181
blunt cutting 122
body language 9
    during consultation 10
    gestures 9–10
body perm 335
body shape 33–4
brick cutting 97
British Hairdressing Awards 361
brushes 193
    bristle (curved head) 169
    Denman classic styling 168
    flat bristle 190
    neck 91
    radial 169
    soft bristle 217, 236
    vented 169

canities 247
cap highlights 260
carborundum 154
catagen 21
change implementation 69
    be positive/explain changes 71
    evaluating change 71–2
    identifying possible changes to services
        69–70
    presenting/sharing ideas 70–1
    sharing information 69
chemical sterilization 82
chemicals, handling 385, 389–90
chignon (asymmetric) step-by-step guide 205
child employment
    local council rules 398
    part-time 397
    restrictions 397–8
    school leavers in full-time work 398
chopsticks 167, 335
cicatrical alopecia 29
client preparation 79, 111–12
    barrier cream 261, 318
    chair adjustment 80
    checklist 113
    clean hair 80
    cover up 79–80
    gowning 111, 137, 261, 317–18
    materials 261
    positioning 80, 137–8
    protecting yourself 262
    removal of product build-up 113

seating position 262
tests 261
trolley 262
client records 41, 222–3, 261–2, 324–5
client/s
    advice on hair maintenance/management 38–9
    age group 34
    agree services with 40
    dealing with complaints 400
    deposits 41
    identify needs 8
    informing through recommendation 48–9
    losing 62–3
    make-up 34–5
    manageability 35
    preferences 118–19, 263
    questionnaires 65
    relevant factors 263–4
    surveys 66
clipper over comb 95, 136, 145
club cutting 94
code of conduct 400
cold fusion systems 234
cold perm sectioning 328–9
colour
    addition 244
    contrasting 245
    description 242–4
    dominant/recessive (warm/cool tones) 265–6
    effects of light/lighting on hair 264
    reducers 299
    remover 298
    restorers 252
    stability 303
    strippers 296, 298–9
    wheel 244–6
colour contra-indications
    allergic reaction 266
    damaged hair 267
    incompatible products 267
    medical reasons 267
    skin disorder 266
    skin sensitivity 266
colour coordination 265
colour correction
    aftercare advice 309–10
    banded hair colour 301
    barrier cream 291
    client record 290
    colour perfection 297
    colour tests 295
    considerations 292–3
    consultation 293–5
    decolouring/colour strippers 298–9
    dermatitis 292
    discoloured highlights/lowlights 302
    effective/safe methods 290–2
    gowning 291
    graded colour 302
    lifestyle 309–10
    longer highlighted hair 302–3
    materials 291
    newly introduced colours 303
    over-porous hair 303
    partial applications 302

plan/agree course of action 296–304
poor removal of colour products 304
pre-pigmentation 296–7
preparation 290–5
process preparation 299
protecting yourself 292
reapplying to natural depth 299
reintroducing colour into lightened hair 297–8
reintroducing pre-pigmentation backgrounds for highlights on long hair 298
removal of permanent/synthetic hair dyes 300
removing artificial colour 303–4
seating position 291
step-by-step guides 305–9
target colours 297
tests 291
trolley 291
colouring back 299
colouring hair
    aftercare advice 285–6
    applications 275–82
    checklists 275, 278, 281
    client requirements 263
    colour coordination 265, 273–5
    depth/tone 252–4
    full head application 279
    full head lightening/toning 280
    harmonizing/contrasting effects 273–5
    health and safety 264
    legislation for under 16s 247
    lightening 254–60
    measuring flasks/mixing bowls 268
    mixing colour 255
    outdoor natural light vs indoor artificial light 264
    pre-colour/post-colour treatments 272–3
    principles 242–52
    problems 283–5
    refreshing lengths/ends 275–6
    regrowth application 277
    regrowth checklist 276
    relevant factors 263–4
    safe methods 261–8
    skin/hair tests 265, 268–70
    step-by-step guides 277, 279, 280, 282
    timing colour development 272
    type of colour 267
    under-processed colour 272
    woven highlights 281, 282
colouring problems 283
    breakage 284
    colour build-up 285
    colour not taking 285
    colour too dark 284
    colour too light 284
    colour too red 284
    discoloration 285
    fading too quickly 284
    green tones 285
    hair breakage 285
    hair resistant to colouring 284
    hair tangled 285
    patchy/uneven 284
    poor consultation 283

poor execution of service 283
processing/development 283
root glare 284
roots not coloured 285
scalp irritation 284
skin reaction/burn 284
too orange 285
too red 285
too yellow 284, 285
white hair not covered 284
combs
    straight 190
    tail/pin 190
communication 53
    clear 113
    effective 7, 8–9
    good 399–400
    intelligent 114
    listening skills 8
    messages/notes 400
    non-verbal 53
    oral 399
    questioning styles 53
    reading skills 9
    speaking well 8
    verbal 53
    written 399–400
competition (in business) 67–8
    knowledge of business 68
    mystery shopper 68
competitions (in hairdressing) 361
    competition day 362
    good practice 361
    mood board 362–4
    organizations involved in 361
    preparation 362
    stick to the rules 362
compound dyes 252
computers 390
concave 92
confidential 41
consultation 7, 84
    advice on hair maintenance/management 38–41
    analytical skills 14–15
    body language 9
    checklist 41–2
    client requirements 8, 139
    colour correction 295–6
    communication 7, 8–9, 15, 113, 114
    creative styling/dressing 175
    current fashions 37
    eye contact 9
    face/head shape 32–4
    hair colouring 262–4, 266
    hair extensions 218–22
    influencing factors/features 30–6
    lifestyle, personality, age 34–5
    long hair 195, 198
    male 113–14
    misunderstandings 13–14
    perming 320–2
    physical contact 9
    posture, body position, gestures 9–10
    questioning techniques 10–13
    recommendations to clients 36–8
    referrals 37–8

consultation (continued)
  review 7
  senior stylist 7
  services/products 37
  signs of a professional 7–8
  visual aids 35–6, 113–14, 140
Consumer Protection Act (1987) 402, 438
consumer rights/legislation 401–2, 436–8
contact dermatitis 389
  preventing 389
contra-indications
  hair colouring 266–7
  hair extensions 219–20
  perming 323
  shaving 157–8
contrasting colour 355
Control of Substances Hazardous to Health
    Regulations (COSHH) (1999, 2002/3)
    372, 376, 385–6
convex 92
cornrowing step-by-step guide 206–7
Cosmetic, Toiletry & Perfumery Association
    (CTPA) 377
creative development
  basics of good hair design 350–8
  inspiration/ideas 348–9
  plan/design range of images 348–50
  produce range of creative images 359–64
creative styling/dressing
  aftercare advice 184
  back brushing 181–3
  blow-drying 180–1
  consultation 175
  curl body directions 179–80
  effective/safe methods of working 166
  effects 177
  finger waving 180
  hairdryers 170–1
  heated styling equipment 183–4
  pin curling 179
  preparation of tools/equipment 167–9
  principles of heat styling 175–6
  pros/cons for products 172–3
  setting techniques 177–8
  styling products 171–2
croquignole winding 178
curl body directions 179–80
curlers 331
curls/curling 177–8
customer feedback 61, 62–4
cutting
  accurate sectioning 91
  aftercare service 104–5
  checklist 91, 100
  combining techniques 99, 123
  common problems 125–6
  controlling the shape 92–3
  creative restyle 91–4
  cross-checking 93
  cutting with natural fall 92
  dealing with cutting problems 93–4
  dry 111
  effective/safe methods 78–84
  hair extensions 232–3
  points to remember 126–7
  preparation 84–91, 139
  problems 93–4

quality, quantity, distribution of hair 86–7
reason/purpose for hairstyle 86
rules 124–5
sources of information 99
step-by-step techniques 101–4
style suitability 87–8
styling limitations 84–6
styling techniques 94–9
tools 88–90, 122–3
wet 111
cutting guides 93
cutting lines 93
cutting lines/angles 93
cutting techniques (hair extensions)
  blunt/club 232
  layering 233
  point 233
  skim/surface clippering 233
  soft tapering 232
  spiral tapering 233
  surface graduated layering 233

damaged cuticle 30
dandruff 28
Data Protection Act (1998) 64, 401, 436–7
databank 373
decolouring 298
demi-perm 335
demonstration see hairdressing demonstration
depilatory 317
dermatitis 27
dermis 19–20
Disability Discrimination Act (DDA) (2005) 436
discoloration 24
discoloured hair 302
disconnection 98, 122
disinfectants 388
  glutaraldehyde 388
  hypochlorite 388
disinfection 387
disulphide bonds 316
disulphide cross links 315
double strength tone 254
double winding 327, 334

elasticity test 24, 220, 265, 269, 295, 323–4
Electricity at Work Regulations (1989) 372,
    378–9
employee's responsibilities 373
  communication 373
  health and safety training 374
  record keeping 373–4
employers' responsibilities 374
employment
  contracts 396
  holiday entitlement 397
  human resources 394–5
  induction 395
  job description 395
  part-time 397
  productivity 396
  public/bank holidays 397
  rest breaks 397
  staff training 398
  working conditions 396–7
  young workers (between 13 and school-
      leaving age) 397–8

epidermis 19
equal opportunities 436
Equality Act (2010) 401
Equality and Human Rights Commission
    (EHRC) 436
equipment see tools/equipment
eumelanin 246
European hair 216
ExCel (London Exhibition Centre) 361
exfoliation 148
extensions see hair extensions

face/head shape 32, 115–16, 141–2
  bone structure/facial contours 142
  ears, nose, mouth 32–3, 142
  eyes 33
  male 115–16
  neck/shoulders 33
  physical features 85–7, 116, 140–1
  width of chin/depth of jaw line 142
facial expression 31
facial hair barbering see beards and
    moustaches
fading 93, 123
fashion 37
features 47
feedback
  customer 61, 62–4
  negative 414
finger waving 180
finishing products
  defining clay 127
  defining crème 128
  dry wax 127
  hair gel 127
  hair paste 128
  hair varnish/high gloss gel 127
  hairspray 128
  styling glaze 128
Fire Precautions Act (1971) 382
first aid
  contents of box 379
  legislation 372, 379
  recording accidents/illness
      379–80
  replacement of contents 379
fixed blade razors 112
folliculitis 25
freehand cutting 95
French pleat (vertical roll) 198
  step-by-step guide 199
furunculosis 25

geometric 348
gestures
  folded arms 9
  holding hand in front of mouth while
      talking 10
  inspecting fingernails 10
  open palms 10
  scratching behind ear/bak of
      neck 10
  shifting from foot to foot 10
gradated colour 302
graduated layers 101–2
graduation 31, 126
grey hair 271–2

hair
    condition 331
    dry/frizzy 86–7
    drying 80
    length/density 321
    natural loss 221
    natural moisture levels 24
    porosity 303, 321
    straight 87
    texture 22, 321, 331
    tight/curly 87
    wavy 87
hair analysis
    causes of physical damage 18
    chemical damage 18
    chemical properties 17
    good condition 18, 86
    health/condition 17–18
    physical properties 17
    poor condition 18
    structure 15–17
    weathering 18
hair colour
    assessing amount of grey 271
    contrasting 355
    convergent/divergent lines 352
    dimensions 354
    effects of different strengths of hydrogen
        peroxide 254–6
    existing hair condition 271–2
    harmonizing 355
    lines/linear effects 354
    natural 246–7
    permanent 251
    pre-softening resistant grey hair before
        colouring 271–2
    progressive (metallic) dyes 252
    quasi-permanent (tone on tone) 250
    selection 355
    semi-permanent 248–9
    synthetic/artifical 247
    temporary 247–8
    transitional lines 352
    vegetable-based 251–2
hair design
    balance 356
    contrasting lines 352
    curved lines 351
    diagonal lines 351
    elements 356–8
    emphasis/accent 358
    evaluating results against plan objectives
        364
    harmony 358
    horizontal lines 350
    line 350
    movement 357
    opportunities 355
    parallel lines 352
    proportion 356
    shape/form 353
    single lines 352
    space 353
    symmetry 356–7
    textural content 353–4
    vertical lines 351
    wave patterns 354

hair extensions
    aftercare advice 235–6
    attaching pre-bonded 227–8
    attaching processed 226
    attaching synthetic (artificial) fibre 225–6
    attaching weave 226–7
    blending 224
    block colouring 224
    braided/plaited 213
    choosing 220–1
    cold bonded systems 213
    cold fusion systems 234
    colour formula for mixing 223
    colour matching 223
    consultation 218–19, 222–3
    contra-indications 219–20
    cutting techniques 232–3
    cutting when wet/damp 232
    do's and don'ts 237
    effective/safe methods of working 212–14
    health and safety 215
    home-care advice 236–7
    hot bonded systems 213, 234–5
    length of time 222
    maintain/remove 234–5
    maintenance appointments 221
    multi-mixing 224
    planning/placement 228–32, 230–1
    potential risks 213–14
    problems 238
    products 235–6
    products/equipment 215–18
    re-curling 233
    sectioning natural hair 229–30
    selecting/blending colours 223–4
    self-adhesive 235
    sewn-in 214, 235
    styling 232–4
    textured hairstyle 231–2
hair growth 21, 221
    cowlick 31, 117
    direction/distribution 87
    double crown 31, 117
    nape whorl 31, 117
    patterns 31
    stages 21–2
    thinning/balding 117
    widow's peak 31, 118
hair tests 22
    allergy alert 23, 265, 295
    colour 23, 265
    curl check/development test curl 23
    development strand 22, 295
    elasticity 24, 220, 265, 269, 295, 323–4
    hair extensions 220
    incompatibility 24, 265, 269, 295, 324
    porosity 24, 265, 269, 295, 324
    pull test 220
    strand test (hair) 265, 270, 295
    test curl 23, 324
    test cutting 23
hair up step-by-step guides 202, 203
hair/scalp diseases 25
    infectious 25–7
    non-infectious 27–30
hairdressing demonstration 420, 426
    entertain your audience 428

expect the unexpected 427
    get message across 428
    handling problems 429
    plan 427–8
    presentation points 428–9
    select venue 426–7
    timings 427
hairdryers 170–1
hairpieces 182–3
    natural full-head (wig) 192
    natural pieces with stitched bases 192
    synthetic artificial full-head (wig) 192
    synthetic artificial welts 191
harmonizing colour 355
hazards 373, 384
    COSHH assessment 385–6
    handling chemicals 385
    work environment 386
    working with equipment 385
    in workplace 385–6
head lice 26
health 18–19
health and safety
    basic rules 380
    electrical accessories 183
    emergency procedures 380
    employees' responsibilities 373–4
    employers' responsibilities 374–83
    general guidance 380
    hair colouring 264
    hair extensions 214, 215
    legislation 370–2
    perming 332
    regulations 372
    risk control 383–90
    unconsciousness 380
Health and Safety at Work Act (HASAWA)
    (1974) 370–2
Health and Safety (Display Screen Equipment)
    Regulations (1992) 372, 382
Health and Safety Executive (HSE) 381
Health and Safety (First Aid) Regulations
    (1981) 372, 379–80
Health and Safety (Information for Employees)
    Regulations (1989) 372, 381
heat styling 175–6
heated styling equipment
    electrical accessories health and safety
        checklist 183
    grabbing effect 184
    straightening irons, heated brushes, tongs
        183–4, 189
henna 248
herpes simplex 26
high-lift colour 257
highlights 260
    colour correction step-by-step guide 307
    reintroducing pre-pigmentation
        backgrounds on long hair 298
    T-section 302–3
honing 154
    double sided 154
    natural 154
    synthetic 154
    technique 155
    testing razor's edge 155
hot bonded systems 234–5

human resources 394–5
humid 82
hydrogen peroxide 254–5
    diluting 256
hygroscopic 17

illness 379–80
image creation 349
    competitions 359, 361–2
    photography/photo shoots 359–60
    producing 359–64
    public demonstrations 359
impetigo 25
in-salon campaign 422
in-salon promotion 49
    checklist 50
incompatibility 324
incompatibility test 24
incompatible chemicals 24
induction 395
infection, preventing 82, 138
information
    formal collection 65–7
    from client 64
    informal collection 64–5
    non-verbal 65
    questions 65–7
    sharing 69
    verbal during routine discussion 64–5
ingrowing hairs 148
interleaves 303
International Colour Chart (ICC) system
    253–4
inverted bob 103–4

job description 395

keratin 17, 315–16

legislation 72, 401–2
lifestyle 34–5, 309–10
lifting/handling large objects 388
lightening hair 256
    cap highlights 260
    effect of toners 260
    highlights 259
    method/product 257
    oil/gel (emulsion) lightener 257
    powder lightener 257–8
    reintroducing colour 299–300, 308–9
    removing powder lightener 258
    toners/lightening toning 260
    whole head (previously coloured
        hair) 259
    whole head (virgin hair) 258
    woven highlights 260
liquid/cold fusion adhesives 216
long hair
    aftercare advice 207–8
    consultation 195, 198
    effective/safe methods of working 188–9
    materials/tools 189–93
    plaiting 200–1
    preparation of tools/equipment 189
    reintroducing pre-pigmentation
        backgrounds for highlights 298
    service timings 189

step-by-step guides 202–7
structure/support for 197
style suitability 195–6
styling products 193–5
T-section highlights 302–3
vertical roll (French pleat) 198–9
wash hair the day before 188
weaving 201
L'Oréal Colour trophy 361
lying 10

male pattern baldness (MPB) 29, 114–15
Management of Health and Safety at Work
    Regulations (1999) 372, 374
Manual Handling Operations Regulations
    (1992) 372, 382
maquette 349
melanin 246
metallic dyes 252
misunderstandings
    avoiding 12–14
    have empathy 13–14
mixing mats 217
model portfolio 352
moisturizing balm 159
monilethrix 29
mood board
    creating 362–4
    description 363
    developing a theme 363–4
    purpose of 362
moustaches see beards and moustaches
multi mixing 224
mystery shopper 68

National Hairdressers' Federation (NHF) 361
neutralizing 338
    applying neutralizer 341
    chemical reaction 339
    choosing a product 339
    how it works 338
    over-oxidizing 341
    preparation 340
    rebalancing the hair 338
    rinsing 340, 341
    step-by-step 341
    technique 340
    under 341
Noises at Work Regulations (1989) 372
non-conventional styling equipment 167
non-verbal communication 53

occipital bones 32
occupational dermatitis 380
oil/gel (emulsion) lightener 257
oxidation 261

permanent colour 251
    facts 251
perming
    acid wave solutions 319
    after the perm 341–2
    aftercare advice 343
    alkaline/acid solutions 330–2
    alternative/creative winding techniques
        332–6
    basic wind 329

chemical changes 315–17
cold perm sectioning 328–9
cold wave solutions 319
consultation 320–2
contra-indications 323
effective/safe methods 314–19
effects 327–32
exothermic 320
hair porosity 321
hair texture 321
health and safety 332
incompatibility 322
length/density of hair 321
neutralizing 338–41
post-damping method 330
pre-damping method 330
preparations 320–6
previous treatment history 321
problems 333
size of rod/curler 322
straightening/relaxing 334–6
style 321
tools/equipment 328
types 319–20
perming problems
    fish-hooks 342
    good when wet/poor when dry 342
    hair/scalp damage/breakage 342
    perm weakens 342
    pull burn 342
    skin irritation 342
    sore hairline 342
    straight frizz 342
    straight pieces 343
    too curly 342
    uneven curl 343
perming techniques
    directional 327
    piggyback (double) wind 327
    root 327
    spiral 327
    stack wind 327
    weaving 327
    zigzag 327
personal effectiveness
    appraisal process 413–14
    code of conduct at work 400
    dealing with negative feedback 414
    employment of young workers 397–8
    evaluating results 411
    good communication 399–400
    human resources 394–6
    personal development 412–13
    productivity 409–11
    recognizing achievement 411
    self-appraisal 414
    staff training 398
    working conditions 396–7
    working together 411–12
personal hygiene 81, 138, 387
Personal Protective Equipment (PPE) at Work
    Regulations (1992) 372, 376
personality 34–5
pheomelanin 246
photography/photo sessions
    clothes/accessories 360
    on the day 360–1

make-up 360
model 360
photographer 360
planning 360
themes 359
piggyback winding 327, 335
pin curling 179
step-by-step guide 179
plaiting 200
three-stem (French) plait 200–1
three-stem (loose) plait 200
planogram 51
pleat 189
point cutting (pointing) 96
point-of-sale 51
polymer resin adhesive sticks 216
polypeptide chains 315
porosity test 24
post-perm treatments 325
posture 9–10
mirroring 10
work position 82, 83–4
powder lightener 257–8
removing 258
PPD (paraphenylenediamine) (para dyes) 22, 23
pre-bonded extensions 216–17
pre-bonded hair extensions guide 228
pre-perm test 324
pre-perm treatments 325
pre-pigmentation 249, 296–9
reintroducing background for highlights on long hair 298
pre-soften 271–2
press 431
consumer 431
trade 431
Prices Act (1974) 438
primary tone 254
product costs 41
productivity 396
maintaining 410
SMART 410–11
targets 409–14
working together 411–12
products 37, 148
clarifying shampoo 217, 236
connector products (hair extensions) 216–18
daily maintenance spray 217–18
light conditioner 217
pH-balanced rinse 218, 236
reconstructive conditioner 218, 236
removal solutions (hair extensions) 217
styling for hair extensions 234
see also styling products
progressive dyes 252
promotion
client commitment to using additional services/products 52–5
communication 53
features/benefits 47–8
in-salon 49
inform client through recommendation 48–9
materials 51
personal 46–7

product/service implementation 50
recognizing interest/buying signals 52
salon services/products 46–52
sensory event 51
website 52
what is available 47
window displays 51
promotional activities 418
advertise event 425
advertising 429–30
collect, collate, analyse information 433
decide on type of promotion 421
evaluate/record results 425, 432–4
implement 426–8
inform staff 425
initial preparation 419–21
know who are your customers 422–3
order stock/promotional materials 425
participate in evaluation 432–4
pitch your ideas 423
the press 431
produce detailed plan 424
public relations 430–1
set objectives 422
Provision and Use of Work Equipment Regulations (PUWER) (1998) 372, 383
psoriasis 27
public relations 430–1

quasi-colour allergy test 249
quasi-permanent colour 250
facts 250
features/benefits 250
questions 10, 65
choices 12
client surveys 66
closed 11–12, 53
feeling 12, 53
limitations/influencing factors 12
misunderstandings 12–13
open 11, 53
questionnaires 66
salon website 67
styles 65
suggestion boxes 66
telephone survey 67

razors 89, 112, 120, 123
honing (setting) fixed blade 154–5
open blade 153–4
open (cut-throat) 121, 153
safety 121
shaper 121
real hair extensions 215
rebalancing the hair 338
reconstructive conditioners 218
referrals 37
salon 38
specialist remedial 38
regulations 372
relaxing process 336
Reporting of Injuries, Diseases, and Dangerous Occurrences Regulations (RIDDOR) (1995) 372, 380
Resale Prices Act (1964/1976) 402, 438
resources 393
Rik-Rak stylers 168

risk
acting responsibly 383
assessment 371, 375
hazard and 384
potential hazards in workplace 385–6
reducing 383
in workplace 387–90
roller perm 335
rollering hair 178
rollers
bendy foam-covered 167, 335
heated 168, 189
velco 168
root perm 335

Sale of Goods Act (1979) 402, 437–8
Sale and Supply of Goods Act (1994) 437–8
Salon International 361
salon requirements 72
scabies 26
scalp diseases see hair/scalp diseases
scalp protectors/shields 217
scissor over comb 95, 124
sebaceous cyst 30
sebaceous gland 20
seborrhoea 28
secondary tone 254
semi-perm 335
semi-permanent colour 248
facts 249
features/benefits 249
senior stylist 7
sensitized hair 325–6
service
changing 60–1
collecting information 64–7
competitor activity 67–8
costs 41
evaluating change 71–2
feedback 62–4
gain confirmation throughout service 114
implementing change 69–71
improvement 60–8
timings 40–1
unique selling point 61–2
services 37
setting techniques
curls/curling 177–8
rollering hair 178
winding point to root 178
winding root to point 178
sharps disposal 112
shaving
aftercare advice 161–2
client preparation 157
contra-indications 157–8
disposal of waste materials 161
effective/safe working methods 152–6
honing (setting) fixed blade razors 154–5
lathering 158
local bylaws 153
recommended re-shaving intervals 161
sequence 158–9
sponge shaving 160
stropping 155–6
tools/equipment 153–4

shaving problems
  facial cuts 160
  folliculitis 160
  ingrowing hairs 160
  patchy results 160
  skin rashes 160
  uneven skin 160
silicone pads 217
skin
  arrector pili muscle 21
  dermis 19–20
  epidermis 19
  hair follicle 20
  sebaceous gland 20
  sweat glands 20
skin care 148
skin test 23
slicing 97
SMART 422
spiral (vertical) curl movement
    332–3
split ends 29
stack winding 335
staff training 398
sterilization 387
stock/stock control 403–4
  avoiding waste/damage 407–8
  choice of supplier 406
  good practice 403
  ordering 405
  product coding 405
  records/recording stock
      movements 405
  security 407
  space usage 408
  stock handling 406–7
  stock rotation 404
  stocktaking 404–5
straighteners 336
straightening hair 326
  method 337–8
  preparation 326–7
  product choice/
      application 337
  test hair 337
  two-step process 336
stropping 154, 155
  hanging strop 155
  hollow ground razor 156
  solid (French or German) 155
  solid razor 156
style suitability 87
  age 87
  balance 87–8
styling limitations 84
  moisture 171
  physical features 85–6
styling products 171, 193–4, 481
  dressing cream 172
  finishing products 171
  gel/glaze 172
  hairspray 173, 194
  heat protection 171–2, 173, 194
  mousse 172
  pros/cons 172–3
  serum 173, 194

setting lotion 171, 172
  wax 173, 194
  see also products
sweat glands 20
sycosis 25
synthetic-based colour 251–2

T-section highlights 302–3
tapering 89
target colour 299
telogen 21
temporal 32
temporary bonds 181
temporary colour 247–8
  facts about 248
  features/benefits 248
texturizing
  brick cutting 97
  point cutting/pointing 96
  slicing 97–8
thinning 95–6
time/time management 408–9
  don't put things off 409
  get organized 409
  prioritizing 409
  write things down 409
tinea capitis 27
tinting back 299
tone-on-tone colours 377
toners 260
tools/equipment 81, 119–20
  bendy foam-covered rollers 167
  bonding applicator 215
  brushes 91, 168–9, 190
  checking regularly 407
  chopsticks 167
  cleaning 143–4
  clipper attachments 90
  clippers 120, 143, 153
  combs 190
  cutting comb 89
  diffuser 169
  electric/rechargeable clippers 89, 123
  grips/hair pins 168, 190
  hair doughnut rings 190
  hairdryers 170–1
  heat clamp 215
  heated brushes/tongs 183–4
  heated rollers 168, 189
  heated straighteners 168, 183
  heated tongs/Rik-Rak stylers 168, 189, 190
  highlight foil 167
  maintaining 90
  measuring flasks 268
  mixing bowls 268
  needle and thread 215
  pin clips 168, 190
  plastic/rubber cutting tools 81
  pre-bonded extension applicator 215
  preformed plastic items 167
  razors 89, 112, 120–1, 123, 153–5
  removal tools (hair extensions) 217
  resin drip tray 217
  scissors 88–9, 90, 122
  sectioning clips 91
  straws 167

styling 167–8
  thinning scissors 89, 122
  velcro rollers 168
  water sprays 91
  working safely with 385
towels
  cool 159
  hot 159
traction alopecia 28, 191, 193, 213, 214
Trades Descriptions Act (1968/1972)
    402, 438
training 398
trichorrexis nodosa 30
trichosiderin 246
tricologist 38

u-stick rods 336
ultraviolet radiation 82, 388
unconsciousness 380
unique selling point (USP) 61–2
utilities 407

vegetable-based colour 251–2
verbal communication 53
vertical roll (French pleat) 198
  step-by-step guide 199
visual aids 84, 113–14
  colour charts 36
  computer-generated images 36
  pictures 35–6
volume
  decreasing 177
  increasing 177

warts 26
waste disposal 112
weave hair extensions guide 226–7
weave winding 334
weaving 201
websites 38, 52, 183, 373, 374, 377, 381, 401,
    436, 437, 438, 441–2
whorl 31
wigs 192
winding 331
  direction/degree of movement 334
  directional 334
  double 334
  perming hair of different lengths 333–4
  piggyback 335
  spiral (vertical) curl movement 332–3
  stack 335
  staggered/brick 334
  weave 334
winding root to point/point to
    root 178
window displays 51
work environment
  autoclave 387–8
  cleanliness of floors 390
  clear thoroughfares 390
  contact dermatitis 389
  controlling risk 387–90
  COSHH precautions 390
  disinfectants 388
  disinfection/sterilization
      procedures 387

handling chemicals 389–90
hazards 386
lifting/handling large objects 388
personal hygiene 387
ultraviolet radiation 388
working with computers 390
working with electricity 390

working efficiently, safely, effectively 83,
    138, 318
  maximize your time 318
  minimize waste 318
Working Time Regulations (1998) 396
Workplace (Health, Safety and Welfare)
    Regulations (1992) 372, 376

woven effect (basket weave) step-by-step
    guide 205–6
woven highlights 260
  checklist 281
  step-by-step guide 282

young workers at risk 374